Terpyridines

Terpyridines

Multifaceted Ligands for Diverse Applications

Ulrich S. Schubert, Andreas Winter, and George R. Newkome

WILEY-VCH

Authors

Prof. Dr. Ulrich S. Schubert
Friedrich Schiller University Jena
Humboldtstr. 10
Jena
07743 Germany

Dr. Andreas Winter
Friedrich Schiller University Jena
Humboldtstr. 10
Jena
07743 Germany

Prof. Dr. George R. Newkome
Florida Atlantic University
Jupiter Campus
5353 Parkside Drive
Jupiter
Florida
33458 United States

Cover Images: © Tomasz Zajda/Alamy Stock Photo, © Gyuszko-Photo/Shutterstock, © Inna Dodor/Shutterstock, © Black_Kira/Shutterstock, © Andreas Winter, Friedrich Schiller University Jena, Germany

Typesetting: Straive, Chennai, India
Printing and Binding: CPI Group (UK) Ltd, Croydon, CR0 4YY

Library of Congress Card No.: applied for

British Library Cataloguing-in-Publication Data A catalogue record for this book is available from the British Library.

Bibliographic information published by the Deutsche Nationalbibliothek
The Deutsche Nationalbibliothek lists this publication in the Deutsche Nationalbibliografie; detailed bibliographic data are available on the Internet at <http://dnb.d-nb.de>.

Print ISBN: 978-3-527-35452-8
ePDF ISBN: 978-3-527-84976-5
ePub ISBN: 978-3-527-84975-8
oBook ISBN: 978-3-527-84977-2

C9783527354528_031225

In memory of Ian Manners, who passed away on December 3, 2023.

In memory of Sir J. Fraser Stoddart, who passed away on December 30, 2024.

Contents

About the Authors

Ulrich S. Schubert was born in Tübingen (Germany) in 1969. He studied chemistry in Frankfurt and Bayreuth (both Germany) and at the Virginia Commonwealth University, Richmond (USA). He completed his PhD at the Universities of Bayreuth and South Florida. After a postdoctoral training with J.-M. Lehn at the University of Strasbourg (France), he moved to the Technical University of Munich (Germany) and obtained his Habilitation in 1999. From 1999 to 2000, he was a Professor at the University of Munich, and from 2000 to 2007, he served as a Full Professor at the Eindhoven University of Technology (the Netherlands). Since 2007, he is a Full Professor at the Friedrich Schiller University Jena (Germany). Prof. Schubert is the founder and director of the Jena Center for Soft Matter (JCSM) and the Center for Energy and Environmental Chemistry Jena (CEEC Jena). He received a VICI award of the Netherlands Organisation for Scientific Research (NWO) and is a fellow of the ACS Division of Polymer Chemistry (USA), the Royal Society of Chemistry (FRSC, UK), and the National Academy of Inventors (NAI, USA). Prof. Schubert is an elected member of the German National Academy of Science and Engineering (acatech) and external scientific member of the Max-Planck-Gesellschaft (MPI for Colloid & Interfaces, Golm). In 2023, he was awarded an ECR Advanced Grant and became the founding director of the new Helmholtz Institute HIPOLE Jena. He currently has an h-index of 120 (Google Scholar 140).

Andreas Winter was born in Herne (Germany) in 1972 and studied chemistry at the University of Dortmund (Germany) where he graduated in organic chemistry in 1999. In 2003, he received his PhD in chemistry at the University of Paderborn (Germany) for his work on applications of the Mannich reaction in the synthesis of pyridine derivatives under supervision of Prof. N. Risch, and continued as a postdoctoral researcher. In 2005, he joined the group of Prof. U. S. Schubert at the Eindhoven University of Technology (the Netherlands). In 2010, he moved with group from there to the Friedrich Schiller University Jena (Germany). His research is focused on the synthesis of metallo-supramolecular assemblies for optoelectronic and energy-conversion applications.

George R. Newkome was born in Akron, Ohio, and graduated from Kent State University with a PhD in chemistry. After a postdoctorate at Princeton, he moved to Louisiana State University and rose through the ranks to a Research Master Professor. In 1986, he moved to the University of South Florida as their Vice President for Research and Dean of the Graduate School. In 2000, he became the Vice President for Research, Dean of the Graduate School, and President of the Research Foundation at The University of Akron. In 2017, he moved back to Florida as a Research Professor in the Center for Molecular Biology and Biochemistry at Florida Atlantic University. He is the recipient of many honors including Distinguished Alumni Award, Kent State University; Docteur Honoris Causa, de l'Université de Bordeaux, France; Fellow, National Academy of Inventors, USA; Fellow, Royal Society of Chemistry, United Kingdom; Visionary Award, Smart Business Innovation in Business Awards, USA; and Fellow, Ohio Academy of Science, USA. He is one of the founders of cascade polymers (dendrimers), including their construction, functions, and supramolecular applications. He is the author of 550 publications, 11 books, holds 61 patents, and for 5 years he had his own TV show called "Science Adventures" on National Public Television.

Preface

Why are we presenting a third text book on terpyridine chemistry? Hasn't everything already been said beforehand in our previous books on this topic? The parent 2,2′:6′,2″-terpyridine ligand was discovered in the early 1930s rather incidentally and terpyridines have become one of the most prominent class of *N*-heterocyclic chelating ligands. The ease of synthesis and predictable coordination chemistry have been employed by numerous groups of authors to assemble complexes with metal ions. These complexes feature manifold properties, which have been utilized in diverse applications. The fundamental aspects of terpyridine chemistry were the main topic of the first book from 2006; whereas, the second book from 2011 already had a strong focus on potential applications: Polymer, surface and nanoparticle science, as well as catalysis and biomedicine. Nowadays, in the mid-2020s, research on terpyridine complexes is ongoing almost undamped, and research topics, which resemble contemporary challenges and needs, have moved into the focus of interest: Smart and responsive polymers, metallodrugs, energy interconversion, (photo)catalysis, energy storage, and so on. We have decided to address these subjects in a new book, which interconnects the historic and fundamental aspects of terpyridine chemistry with their cutting-edge applications.

We have focused on a number of topics, which, in our opinion, represent the most relevant ones in current research in the broad field of terpyridine chemistry. In each chapter, we attempted to cover the respective topic by highlighting the key examples to give a comprehensive overview on what has been done and to stimulate future research in this area. However, the sheer volume of available scientific literature called for a selection of examples to account space constraints. We kindly apologize in advance to all those authors whose valuable research papers have not been cited herein.

We, the authors, as always, would be most grateful to be made aware of any errors that may have crept into the manuscript in spite of extensive proofreading at the various stages of productions. We also acknowledge our coworkers and colleagues (i.e., Martin D. Hager, Michael Jäger, Stefan Zechel, and Uwe Köhn) for their assistance and fruitful discussion during the completion of this work.

Jena and Jupiter
November 2025

Ulrich S. Schubert
Andreas Winter
George R. Newkome

List of Abbreviations

A-2780	human ovarian carcinoma cell line
A-2780r	cisplatin-resistant variant of A-2780
A-375	human malignant melanoma cell line
A-549	human lung carcinoma cell line
A-549r	cisplatin-resistant variant of A549
AC	alternating current
ACA	alloxazine 7/8-carboxylic acid
acac	acetylacetonate
AFM	atomic-force microscopy
AI	artificial intelligence
AIE	aggregation-induced emission
AIQ	aggregation-induced quenching
Alq_3	*tris*(8-hydroxyquinoline)aluminum
AM	air mass
AN	acceptor number
aRFB	aqueous redox-flow battery
aRFID	active radio frequency identification
ASC	asymmetric supercapacitor
ATRP	atom-transfer radical polymerization
B. subtilis	*Bacillus subtilis*
BASF	Badische Anilin- und Sodafabrik
BCP	2,9-dimethyl-4,7-diphenyl-1,10-phenanthroline ("bathocuproine")
BD	black dye
BEL-7402	human hepatocellular carcinoma cell line
BHJ	bulk heterojunction
bmim	1-butyl-3-methylimidazolium
BODIPY	boron-dipyrromethene
bpy	2,2′-bipyridine
Br-tpy	4′-bromo-2,2′:6′,2″-terpyridine
BSA	bovine serum albumin
C	capacity or capacitance
CAL-51	human breast adenocarcinoma cell line

CAN	ceric ammonium nitrate or covalent adaptive networks
CBM	conduction-band minimum
CBD	chronic beryllium disease
CBP	4,4′-*bis*(9*H*-carbazol-9-yl)biphenyl
C^C^C	*tris*-cyclometalating tridentate ligand (coordination *via* three C-atoms)
CCDC	Cambridge Crystallographic Data Centre
C^C^N	*bis*-cyclometalating tridentate ligand (coordination *via* two C- and one N-atoms)
CD	circular dichroism
CDCA	3a,7a-dihydroxy-5b-cholic acid
cds	cadmium sulfate
CDSA	coordination-driven self-assembly
CE	coulombic efficiency, coloration efficiency or current efficiency
chdc	cyclohexyl-1,4-dicarboxylic acid
CID	collision-induced dissociation
CIE	Commission Internationale d′Eclairage
CIGS	copper indium gallium (di)selenide
Cl-tpy	4′-chloro-2,2′:6′,2″-terpyridine
CMC	carboxymethyl cellulose
CN	coordination number
C^N	mono-cyclometaling bidentate ligand (coordination *via* one C- and one N-atom)
C^N^C	*bis*-cyclometalating tridentate ligand (coordination *via* two C- and one N-atoms)
C^N^N	*mono*-cyclometalating tridentate ligand (coordination *via* one C- and two N-atoms)
CNT	carbon nanotube
cod	cycloocta-1,5-diene
COF	covalent-organic framework
CONASH	coordination nanosheet
CORM	CO-releasing molecule
COSY	correlated (NMR) spectroscopy
cryo-TEM	cryogenic transition-electron microscopy
CRP	controlled radical polymerization
CS	charge-separated
CSS	charge-separated state
CT	charge-transfer
CuCAAC	Cu^I-catalyzed alkyne-azide cycloaddition (reaction)
CV	cyclic voltammetry
CVP	chemical vapor deposition
Đ	dispersity
DA	donor-acceptor
DBN	1,5-diazabicyclo(4.3.0)non-5-ene
DCA	deoxycholic acid

dcbpy	2,2′-bipyridine-4,4′-dicarboxylic acid
dceb	4,4′-di(carboxyethyl)-2,2′-bipyridine
dcpp	2,6-*bis*(2-carboxypyridyl)pyridine
DCvC	dynamic covalent chemistry
ddpd	*N*,*N*′-dimethyl-*N*,*N*′-di(pyridin-2-yl)pyridine-2,6-diamine
DDQ	2,3-dichloro-5,6-dicyano-1,4-benzoquinone
DFT	density functional theory
DHP	1,4-dihydropyridine
DIB	dual-ion battery
DLP	digital light processing
DLS	dynamic light scattering
dmbpy	4,4′-dimethyl-2,2′-bipyridine
DMC	dimethyl carbonate
DMF	*N*,*N*-dimethylformamide
DMFC	direct methanol fuel cells
DMSO	dimethylsulfoxide
DN	donor number
DNA	deoxyribonucleic acid
DoCS	density of coordination sites
DOSY	diffusion ordered spectroscopy (NMR)
DPA	dipyrido[4,3-*b*;5,6-*b*]acridine
dpb	2,6-di(pyridin-2-yl)benzene
DPEPO	*bis*[2-(diphenylphosphino)phenyl] ether oxide
dpp	2,4-di(pyridin-2-yl)pyrazolate or 3,5-di(pyridin-2-yl)pyrazolate or 2,9-diphenyl-1,10-phenanthroline or 2,6-diphenylpyridine
dppep	*cis*-1,2-*bis*(diphenylphosphino)ethene
dppp	1,3-*bis*(diphenylphosphino)propane
DPV	differential pulse voltammetry
dppz	dipyrido[3,2-*a*: 2′,3′-*c*]phenazine
dpq	dipyrido-[3,2-*d*:2′,3′-*f*]-quinoxaline
dqp	2,6-di(quinoline-8-yl)pyridine
DSC	differential scanning calorimetry
DSIR	Department of Scientific and Industrial Research
DSSC	dye-sensitized solar cell
E. coli	*Escherichia coli*
EA	ethylammonium
EC	electrochromic or ethylene carbonate
Eca-109	human esophageal carcinoma cell line
ECD	electrochemical doping
ECE	pathway of two one-electron transfers with an in-between chemical reaction
ED	electrodynamical doping
EDLC	electric double-layer capacitor
EDOT	3,4-ethylenedioxythienyl
EDTA	ethylenediaminetetraacetic acid

EE	energy efficiency
EEC	pathway of two consecutive one-electron transfers and a subsequent chemical reaction
E_g	energy-band gap
EIL	electron-injection layer
EL	electroluminescence
emim	1-ethyl-3-methylimidazolium
en	1,2-ethylenediamine
EN	electronegativity
EPR	electron paramagnetic resonance (spectroscopy)
EQE	external quantum efficiency
ERC	electrochemical reduction of CO_2
ESI	electrospray ionization
ET	electron transfer or energy transfer
EthBr	ethidium bromide
ETL	electron-transport layer
ETM	electron-transport material
EU	European Union
EV	electric vehicle
F	Faraday constant (96,485 As mol^{-1})
FA	formamidinium
fac	facial
F_C	electrostatic force
Fc/Fc^+	ferrocene/ferrocenium cation
FDA	(United States) Food and Drug Administration
FF	fill factor
FIA	fluoride-ion affinity
FID	fluorescent intercalator displacement
FRP	free-radical polymerization
FTO	fluorine-doped tin oxide
fwhm	full width at half maximum
G′	storage modulus (rheology)
G**	loss modulus (rheology)
G4	G-quadruplex (DNA)
GalSL	galactosphingolipid
GlcSL	glucosphingolipid
GCE	glassy-carbon electrode
gMS^2	gradient tandem mass spectrometry
GM05757	normal human fibroblast cell line
GM07492	normal human fibroblast cell line
GO	graphene oxide
GS	ground state
GSL	glycosphingolipids
G0/G1	phases of the cell cycle
H-bonding	hydrogen bonding

H_0	Hammet acidity function
HBC	hexa-*peri*-hexabenzocoronene
HCT-116	human colon carcinoma cell line
HC71T-8	human colon carcinoma cell line
HDA	hetero-Diels-Alder
HEK-293T	human embryonic kidney cell line
HeLa	cervical cancer cell line from Henrietta Lacks
HepG2	human liver hepatocellular carcinoma cell line
HER	hydrogen-evolution reaction
HETTAP	heteroleptic terpyridine and phenanthroline (complexation)
HEV	hybrid electric vehicle
HIA	hydride-ion affinity
HL-7702	→ LO2
HOMO	highest occupied molecular orbital
HOPG	highly ordered pyrolytic graphite
HOR	hydrogen-oxidation reaction
HOS	human osteo-carcinoma cell line
HS	high-spin (state)
HSAB	hard and soft acids and bases
HTE	high-throughput experimentation
HTL	hole-transport layer
HTM	hole-transport material
I2M	bimolecular radical oxo-coupling
IC	internal conversion
IC_{50}	half maximal inhibitory concentration
ICT	intrachain charge-transfer or intramolecular charge-transfer
IL	intraligand
ILCT	intraligand charge-transfer
IoT	internet of thinks
IPCE	incident photon-to-current conversion efficiency
IQE	internal quantum efficiency
IR	infrared
ISC	intersystem crossing
isox	isoxazole
iTMC-LEEC	LEC based on an ionic transition-metal complex
ITC	isothermal titration calorimetry
ITO	indium-tin oxide
IUPAC	International Union of Pure and Applied Chemistry
J_{SC}	short circuit current
K_a	association constant
K-562	human leukemia cell line
L-asp	L-aspartic acid
L1210	murine leukemia cell line
L920	murine fibroblast cancer cell line
LBL	layer-by-layer

LC	ligand-centered or liquid crystalline
LCD	liquid-crystal display
LED	light-emitting diode
LEEC	light-emitting electrochemical cell
LIB	lithium-ion battery
LLCT	ligand-to-ligand charge-transfer
LMCT	ligand-to-metal charge transfer
LO2	(presumptive) human fetal hepatocyte cell line
LPB	lithium-polymer battery
LS	low-spin (state)
LSB	lithium-sulfur battery
LSM	laser-scanning microscopy
LT_{50}	50% of the initial luminance
LUMO	lowest unoccupied molecular orbital
MA	methylammonium
MALH	methylammonium lead halide
MALDI	matrix-assisted laser desorption/ionization
MC	metal-centered
MCF-7	human breast-cancer cells (MCF: Michigan Cancer Foundation)
mCP	*N*,*N*-dicarbazoyl-3,5-benzene
MCPBA	m-chloroperoxybenzoic acid
MDA-MB-231	human breast carcinoma cell line
MDS	molecular-dynamics simulation
Mebip	2,6-*bis*(1′-methylbenzimidazol-2-yl)pyridine
MeCN	acetonitrile
MEH-PPV	poly[2-methoxy-5-(2-ethylhexyloxy)-*p*-phenylene vinylene]
mer	meridonal
MGC-803	human gastric carcinoma cell line
MGE	main-group element
ML	machine learning
MLCT	metal-to-ligand charge transfer
MLLCT	metal-ligand-to-ligand charge transfer
MM	molecular modeling
MMA	methyl methacrylate
MMLCT	metal-metal-to-ligand charge-transfer
M_n	number-average molar mass
MnSOD	manganese superoxide dismutase
MOLT-4	human acute lymphoblastic leukemia cell line
MOF	metal-organic framework
MPEG	poly(ethylene oxide) monomethyl ether
MS	mass spectrometry
mtDNA	mitochondrial DNA
MV^{2+}	1,1′-dimethyl-4,4′-bipyridinium (“methyl viologen”)
NAD	nicotinamide adenine dinucleotide
naRFB	nonaqueous redox-flow battery

nbo	niobium-oxide type
NC	nanocrystal
NC-CoP	nitrogen-doped carbon-embedded cobalt phosphite
N^C^N	*mono*-cyclometalating tridentate ligand (coordination *via* one C- and two N-atoms)
NHC	*N*-heterocyclic carbene
NHE	normal hydrogen electrode
NiMH	nickel-metal hydride (battery)
NIR	near-infrared
NLO	nonlinear optics
NMP	nitroxide-mediated polymerization or *N*-methylpyrrolidone
NMR	nuclear magnetic resonance (spectroscopy)
N^N	bidentate ligand (coordination *via* two *N*-atoms)
N^N^N	tridentate ligand (coordination *via* three *N*-atoms)
NPC	nitrogen-doped porous carbon
NQR	nuclear quadrupole resonance
NREL	(United States) National Renewable Energy Laboratory
NRR	nitrogen-reduction reaction
OER	oxygen-evolution reaction
OFET	organic field-effect transistor
OLED	organic light-emitting diode
OPV	organic photovoltaic (cell)
OPV	oligo(phenylenevinylidene)
P. aeruginosa	*Pseudomonas aeruginosa*
P4VP	poly(4-vinylpyridine)
P3HT	poly(3-hexylthiophene)
PAni	polyaniline
pbpy	6-phenyl-2,2′-bipyridine
PBD	2-*tert*-butylphenyl-5-biphenyl-1,3,4-oxadiazole
PC-3	human prostatic adenocarcinoma cell line
PCBM	methyl [6,6]-phenyl-C_{61}-butyrate
$PC_{71}BM$	methyl [6,6]-phenyl-C_{71}-butyrate
PCE	power-conversion efficiency
PCET	proton-coupled electron-transfer
PDT	photodynamic therapy
PE	phosphorescence efficiency
PEC	photoelectrochemical
PEG	poly(ethylene glycol)
PEDOT	poly(3,4-ethylenedioxythiophene)
PeLED	perovskite light-emitting diode
PEMFC	proton-exchange membrane fuel cell
PET	photoinduced electron transfer
Ph-tpy	4′-phenyl-2,2′:6′,2″-terpyridine
phen	1,10-phenanthroline
PIB	potassium-ion battery

pic	4-picoline
PIL	poly(ionic liquid)
PIN	preferred IUPAC name
PiPOx	poly(2-isopropenyl-2-oxazoline)
pK_a	logarithmic form of the ground-state acidity constant
pK_a*	logarithmic form of the excited-state acidity constant
PL	photoluminescence
PLEEC	polymer-based light-emitting electrochemical cell
PLED	polymer-based light-emitting diode
pmbp	4,4′-diphenyl-6,6′-dimethyl-2,2′-bipyrimidine
PMMA	poly(methyl methacrylate)
PmPy	poly(pyridine-2,6-diyl)
POM	polyoxometalate
ppm	parts per million
PPV	polymer photovoltaic (cell) or poly(*p*-phenylene vinylidene)
ppy	2-phenylpyridinato
PPy	poly(pyridine-2,5-diyl) or polypyrrole
pRFB	polymer-based RFB
PS	polystyrene
PS-II	photosystem II
PSRFB	polysulfide redox-flow battery
PSS	poly(styrene 4-sulfonate)
PTAA	poly[*bis*(4-phenyl)(2,4,6-trimethylphenyl)amine]
PTE	periodic table of elements
PTHF	poly(tetrahydrofuran)
PUR	polyurethane
PV	photovoltaic
PVA	polyvinylalcohol
PVK	poly(*N*-vinylcarbazole)
py	pyridine
QM	quantum mechanical
qtpy	2,2′:6′,2″:6″,2‴-quaterpyridine
R	gas constant ($8.314\,J{\cdot}K^{-1}{\cdot}mol^{-1}$)
RAFT	reversible addition-fragmentation chain-transfer (polymerization)
rDA	retro-Diels–Alder
RFB	redox-flow battery
rGO	reduced graphene oxide
RISC	reverse intersystem crossing
RNA	ribonucleic acid
ROMP	ring-opening metathesis polymerization
ROS	reactive oxygen species
S. aureus	*Staphylococcus aureus*
S. marcescens	*Serratia marcescens*
salen	*N*,*N*′-*bis*(salicylidene)ethylenediamine

SAXS	small-angle X-ray scattering
SC	supercapacitor
SCC	supramolecular coordination complexes
SCE	saturated calomel electrode or solar-cell efficiency
SCO	spin-crossover
SEC	size-exclusion chromatography
SEM	scanning-electron microscopy
SET	single-electron transfer
SH-SYSY	human neuroblastoma cell line
SIB	sodium-ion battery
SiHa	human uterus carcinoma cell line
SK-MEL-2	human skin melanoma cells
SK-OV-3	human ovary adenocarcinoma cell line
SME	shape-memory effect
SMP	shape-memory polymer
S_NAr	nucleophilic aromatic substitution
SOC	spin-orbit coupling
SOD	superoxide dismutase
SPAN	sulfonated polyaniline
SPECT	single-photon emission tomography
spiro-MeOTAD	2,2′,7,7′-tetra*kis*[*N*,*N*′-di(4-methoxyphenyl)amino]-9,9′-spirbiuorene
$S_{RN}1$	radical nucleophilic aromatic substitution
STED	stimulated emission depletion
STM	scanning tunneling microscopy
SWCNT	single-walled carbon nanotube
SW-480	human colon adenocarcinoma cell line
SW-602	human colorectal cancer cell lines
T	temperature
TADF	thermally activated delayed fluorescence
TBA	tetrabutylammonium
TBP	*tert*-butylpyridine
TC	thiophene-2-carboxylate
TCA	thiophene-2-carboxylic acid
TCNQ	7,7,8,8-tetracyanoquinodimethane
TCS	transparent and conductive substrate
TCTA	*tris*(4-carbazoyl-9-ylphenyl)amine
TD	time-dependent
TEA	tetraethylammonium or triethylamine
TEM	transmission-electron microscopy
TEMPO	2,2,6,6-tetramethylpiperidinyl-1-oxyl
Tf	trifluoromethylsulfonyl
TFA	trifluoroacetic acid
T_g	glass-transition temperature
TGA	thermogravimetric analysis

THF	tetrahydrofuran
T_m	melting temperature
TM-AFM	tapping-mode atomic-force microscopy
ToF	time-of-flight
TOF	turnover frequency
TON	turnover number
TPA	triphenylamine
TPE	tetraphenylethene
tpy	2,2′:6′,2″-terpyridine
tripy	2-(1*H*-1,2,3-triazol-4-yl)pyridine
trt	trityl (*i.e.*, triphenylmethyl)
TTA	triplet-triplet annihilation or 2-thenoyltrifluoroacetone
TTF	tetrathiafulvalene
ttpy	4′-tolyl-2,2′:6′,2″-terpyridine
TWIM	travelling-wave ion-mobility
T-24	human urinary bladder cancer cell line
T-47D	human breast cancer cell line
UK	United Kingdom
UPS	ultraviolet photoelectron spectroscopy
US	United States
UV	ultraviolet
U2OS	human osteosarcoma cell line
VBM	valence-band minima
vbpy	4-vinyl-4′-methyl-2,2′-bipyridine
VE	voltage efficiency (η_V)
VERO	normal African green monkey's kidney cell line
vis	visible (light)
VRFB	vanadium redox-flow battery
V_{oc}	open circuit voltage
WNA	nucleophilic attack of water
WOC	water-oxidation catalyst
WOLED	white light-emitting diode
WOR	water-oxidation reaction
X-bonding	halogen bonding
XMCT	halogen-to-metal charge transfer
XMLCT	halide-metal-to-ligand charge-transfer
XPS	X-ray photoelectron spectroscopy
XRD	X-ray diffraction
Znq_2	zinc(II) *bis*(8-hydroxyquinoline)
β_{21}	cumulated stability constant
δ	chemical shift
Φ	bond angle
φ	torsion angle
μ	dipole moment
Φ_{PL}	photoluminescence quantum yield

τ	excited-state lifetime
Γ	surface coverage
1,4-bdc	1,4-benzenedicarboxylic acid (*i.e.*, terephthalic acid)
^{1}MLCT	singlet MLCT
2PA	two-photon absorption
^{3}MLCT	triplet MLCT
^{3}MC	triplet MC
6ctpy	2,2′:6′,2″-terpyridine-6-carboxylic acid

1

2,2′:6′,2″-Terpyridine – An Introduction, a Short Historic Overview, and Fundamental Aspects*

1.1 Introduction

Modern terpyridine chemistry brings together three worlds – coordination chemistry (from an inorganic chemist's perspective), organic chemistry, and supramolecular chemistry. The latter field was pioneered by Cram, Lehn, and Pedersen, who demonstrated the selectiveness of host–guest chemistry with the examples of crown ethers, spherands, cryptands, and related macrocyclic systems. The significance of this work was acknowledged when the *Nobel Prize in Chemistry* 1987 was awarded to these three researchers [1–3]. Since then, supramolecular chemistry has evolved tremendously and represents one of the most active fields of current research. Lehn described supramolecular chemistry as the "chemistry beyond the molecule," which is "based on organized entities of higher complexity that result from the association of two or more chemical species held together by intermolecular forces" [4]. Supramolecular chemistry relies on three fundamental processes, *e.g.*, self-recognition, self-assembly, and self-sorting, which enable one to prepare advanced structures on the basis of mainly non-covalent interactions (*e.g.*, van-der-Waals forces, hydrogen bonding, and ionic or coordinative interactions). As detailed elsewhere [5], such interactions are intrinsically weaker when compared to traditional covalent bonds and feature a certain degree of reversibility. Nature itself has set the benchmarks for functional supramolecular systems, and most biological processes, such as protein functions or enzyme catalysis, rely on sophisticated combinations of inter- and intramolecular non-covalent interactions [6]. The same also holds true for data storage and transfer in animate beings. In this respect, DNA is the most notable example, where self-recognition of complementary base pairs *via* hydrogen bonding (*H*-bonding) results in the self-assembly of the famous double-helical structure. Examples from nature have inspired researchers worldwide to assemble functional supramolecular systems for various purposes: molecular machines, *i.e.*, molecules, which might undergo controllable movements upon stimulus (*e.g.*, by energy), represent one notable application. Feringa, Sauvage,

* Parts of this chapter have previously been published elsewhere: A. Winter, G. R. Newkome, U. S. Schubert, The Chemistry of the s- and p-Block Elements with 2,2′:6′,2″-Terpyridine Ligands, *Inorg. Chem. Front.* **2024**, *11*, 342–399. The content is reproduced with kind permission; © 2024 The Royal Society of Chemistry.

Terpyridines: Multifaceted Ligands for Diverse Applications, First Edition.
Ulrich S. Schubert, Andreas Winter, and George R. Newkome.

and Stoddart received the *Nobel Prize in Chemistry* 2016 for their distinguished contributions to this field [7–9].

Metal-to-ligand coordination is widely used in modern supramolecular chemistry. Unlike van-der-Waals forces or hydrophobic-hydrophilic interactions, the interaction between metal centers and organic ligands is rather strong (40–120 kJ mol^{-1}) and highly directional [10]. These properties enable one to construct advanced metallo-supramolecular polymers [11–16], dendrimers [17–20], 2D/3D nanostructures [21–25] or coordination nanosheets (CONASHs) [26, 27] *via* cautious choice of metal ions and organic ligands. Among others, complexes derived from chelating *N*-heteroaromatic ligands, in particular based on pyridine derivatives [*i.e.*, 2,2′-bipyridine (bpy), 1,10-phenanthroline (phen), and 2,2′:6′,2″-terpyridine (tpy)], have found widespread use in metallo-supramolecular applications. As detailed in Section 1.2, the provable discovery of terpyridine dates back to the studies by Morgan and Burstall in the 1930s [28]. For the following decades, up to the 1980s, terpyridine chemistry remained a niche topic with only sporadic reports on the formation of complexes with transition metals or main-group elements being published [29]. This, however, changed when the remarkable photophysical [30–34], electrochemical [33, 35], electrochromic [36, 37], magnetic [38], pharmaceutical [39–42], and catalytic properties [25, 43, 44] of tpy complexes were discovered. These properties have been utilized by numerous research groups to assemble functional materials and devices for various fields of application [29]. The ongoing intense research is reflected by the enormous number of scientific publications per year (*i.e.*, journal articles and reviews), which deal with terpyridine or its complexes regarding synthesis, properties, and applications (March 2025: *ca.* 10.8 k hits in SciFinder^n for a substructure search; Figure 1.1). Moreover, the tpy substructure is mentioned in more than 3000 patents.

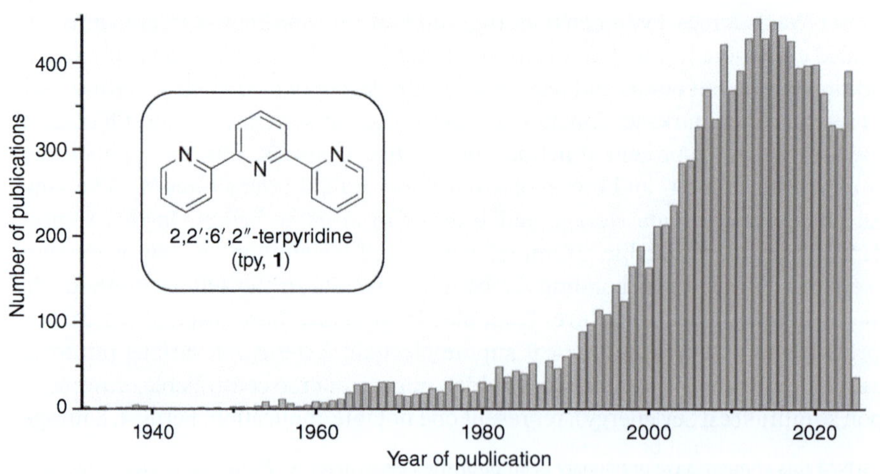

Figure 1.1 Histogram of the number of scientific publications containing the substructure of 2,2′:6′,2″-terpyridine (tpy, **1**); only journal articles and reviews were counted (SciFinder^n search performed on March 1st, 2025).

Before turning to the utilization of terpyridine and its complexes in the various applications, the reader is invited to step back and have a look at the ligand's early history. Subsequently, the fundamental aspects of terpyridine chemistry, such as structure and bonding as well as synthesis, will be discussed.

1.2 The Discovery of 2,2′:6′,2″-Terpyridine – A Story Often (But not Fully) Told

In 1925, the *Department of Scientific and Industrial Research* (DSIR) opened the *Chemical Research Laboratory* (CRL) at their campus in Teddington, United Kingdom (Figure 1.2a) [49]. There, research was focused on diverse topics, which were believed to meet the public interests (thus, filling the gap between academic research and industry). The laboratory, whose title was changed to the *National Chemical Laboratory* (NCL) in 1958, operated until 1965, when it was merged into the *National Physical Laboratory*. Morgan joined the CRL in 1925 and was appointed as the executive director only two years later; he remained in this position until his retirement in 1938 (Figure 1.2b). During this period, he established three main research schemes, which were devoted to organic reactions under high pressure, synthetic ion-exchange resins, and the exploration of low-temperature fractions of coal tar, as feedstock for organic fine chemicals [46, 49]. Against this background, Morgan and Burstall, his long-term co-worker whom he already knew as a student from his time at the University of Birmingham, performed the dehydrogenation

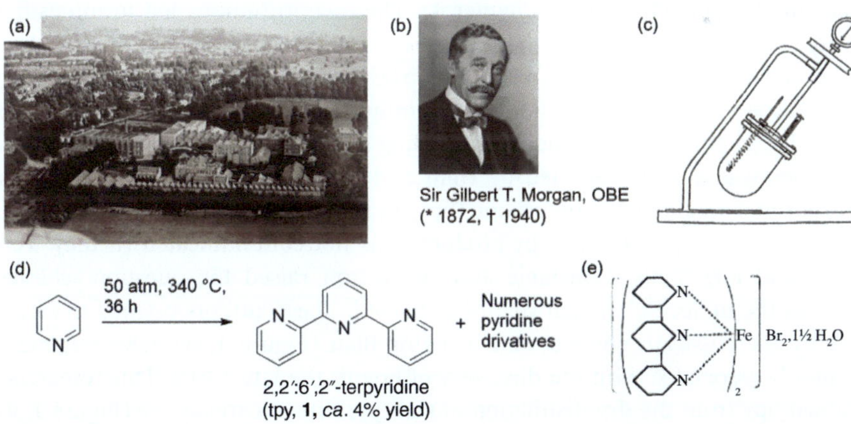

Figure 1.2 (a) Where it all began: Aerial photograph from 1919 showing the Teddington research campus, where the *Chemical Research Laboratory* was later also located [45] / Grace's Guide Ltd / CC BY-SA 4.0. (b) Photograph of Sir Gilbert T. Morgan [46] / with permission of Royal Society of Chemistry; regrettably, a photograph of Francis H. Burstall (1903–1955) could not be provided [47]. (c) Illustration of an in-house-built rotating autoclave, which was used for the high-pressure reactions [48] / with permission of John Wiley & Sons. (d) Schematic representation of the original tpy synthesis. Adapted from [28]. (e) Schematic representation of the proposed structure of $[Fe(tpy)_2]^{2+}$ [28] / with permission of Royal Society of Chemistry.

of pyridine using $FeCl_3$, as the oxidant, in an autoclave (*i.e.*, 50 atm at 340 °C; Figure 1.2c,d) [28]. Under these forcing reaction conditions, 2,2′:6′,2″-terpyridine (tpy, **1**) was formed, even though in very low yield, along with *ca.* 20 other pyridine derivatives, including various constitutional isomers of tpy (the in-total 48 constitutional tpy-like isomers are summarized in Appendix A [50, 51]). Notably, Morgan and Burstall also reported the first transition-metal complex, presumably purple-colored $[Fe(tpy)_2]^{2+}$ (Figure 1.2e) [28]. Thus, the capability of tpy to serve as a (presumed) tridentate ligand for metal-ion complexation was already recognized in this primary study. In subsequent years, these authors intensified their research on this topic and reported diverse complexes of **1** [52–54]. These results also found entrance into the inorganic chemistry textbook, which these authors published in the late 1930s [55]. However, after Morgan's retirement and the outbreak of World War II, research in Teddington on this topic was discontinued. In Morgan's perception, the discovery of tpy represented a notable, although not outstanding, result of his scientific work, which counted *ca.* 350 publications [48]. Burstall, who spent his whole working life at the CRL, continued working in NCL's radiochemical group until his death in 1956 (the chromatographic separation of actinoid-metal ions was his last research topic) [47]. Until 1960, other groups of researchers were already paying attention to important properties of tpy complexes in solution, such as their kinetic stability [56, 57] and the kinetics of ligand-exchange processes [58] as well as electrochemical [59, 60] and photophysical aspects [61]. Likewise, the scope of tpy complexation was extended toward other metal cations (*e.g.*, Cu^{I} [62] or In^{III} [63]). Interestingly, the first study also dealing with the biological activity of tpy was performed in that period of time [64], far before the emergence of tpy complexes, as potential metallodrugs (see Chapter 4). These contributions led to intensified research on tpy complexes, which then geared up by the mid-1960s (Figure 1.1).

Undoubtedly, Morgan and Burstall have to be credited for the verifiable discovery of tpy; accordingly, these authors are commonly regarded as the "fathers" of terpyridine chemistry. The oxidative coupling of pyridine under similar reaction conditions was already reported by Hein and Retter beforehand [65]. However, these authors obtained 2,2′-bipyridine (bpy) in reasonable yield (*ca.* 50%); whereas the nature of the presumptive by-products was not communicated (if they were analyzed at all). Thus, Constable and Housecroft raised the question whether 1932 was the molecule's actual year of birth [66]. These authors pointed to earlier work by the Austrian chemist and inventor Blau (Figure 1.3a), whose name is commonly associated with the discovery of bpy in the late 1880s. This researcher obtained bpy from the dry-distillation of Cu^{II} pyridine-2-carboxylate (Figure 1.3b) [68, 70]. Admittedly, Blau was not the first who performed the dry-distillation of picolinate salts [71], but he paid attention to a by-product, which he named "tetrapyridyl" [70]. This compound had a similar elemental constitution as bpy and revealed the characteristic purple coloration upon treatment with a Fe^{II} salt. Thus, one might speculate that Blau's "tetrapyridyl" was indeed tpy [66]. However, these experiments were performed when coordination chemistry was far from being understood – Werner established the ground-breaking theory of coordination chemistry only a few years later [72] – and further evidence for this assumption is

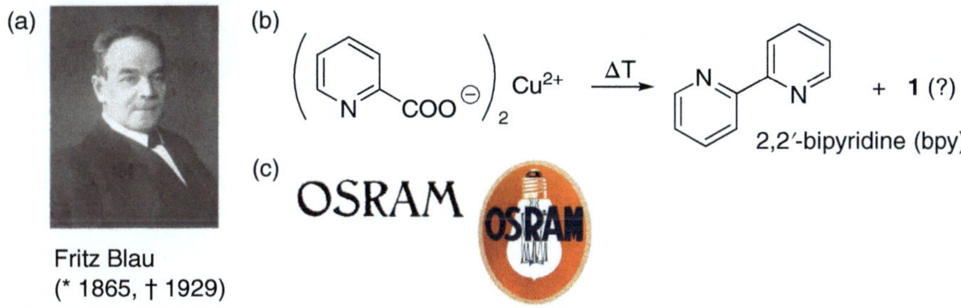

Figure 1.3 (a) Photographs of Fritz Blau. Reproduced with permission from [67] / with permission of Springer Nature. (b) Schematic representation of Blau's bpy synthesis, where tpy might have been formed as a by-product. Adapted from [68]. (c) Representation of OSRAM's original word logo and picture mark, which were registered in 1906 and 1919, respectively [69] / OSRAM SYLVANIA Inc.

not at hand. Blau concluded his scientific career after achieving his habilitation. His last scientific work, which dealt with bpy complexes, was published in 1899 [73]; his first patent on the "manufacture of electric glow lamps," was published in 1901 [74]. By that time, Blau was already involved in the advisory board of the *Elektrische Glühlampenfabrik "Watt"* and he eventually joined the *Auergesellschaft* in 1902. Later, he became the lead scientist of the *OSRAM GmbH* and was one of the inventors of the OSRAM™ light bulb [67]. The neology of the famous trade and company name *OSRAM*, as a frankenword of osmium and wolfram, is often ascribed to him (Figure 1.3c) [69]. As his focus of interest had shifted, studies on the coordination behavior of bpy or tpy were discontinued by this author. To the best of our knowledge, terpyridine was not mentioned further in the available literature until Morgan's and Burstall's experiments in Teddington (*vide supra*).

In the 21st century, science has become increasingly influenced by computers and artificial intelligence (AI), which play an important role in today's research. Computational chemistry (*i.e.*, calculations and simulations), automatization, and machine learning (ML) have brought significant progress that goes far beyond humans' learning and handling capabilities [75–80]. Looking back on those days where tpy chemistry started feels like looking at a world of difference, where scientific discovery was still often equalized with serendipity.

1.3 Terpyridine and Its Complexes – Nomenclature, Structure, and Bonding

Ambiguity in nomenclature represents one complicating factor regarding the identification of early contributions in the field of terpyridine chemistry. Burstall coined the generic term "polypyridyl" ligands for "bases, in which two or more pyridine rings are linked (but not fused) together" [54]. The various constitutional isomers of bpy and tpy are the simplest representatives of this ligand family. However, the term "polypyridyl" might already lead to the first misunderstanding. For a polymer chemist, a polypyridine is a rigid, π-conjugated

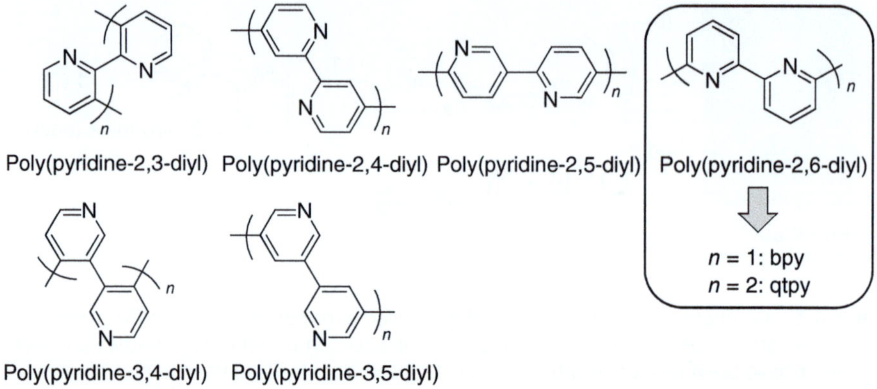

Figure 1.4 Schematic representation of the various regioregular "polypyridines."

chain of pyridine rings; one might distinguish between various regioregular isomers – poly(pyridine-2,3-diyl), poly(pyridine-2,4-diyl), poly(pyridine-2,5-diyl), poly(pyridine-2,6-diyl), poly(pyridine-3,4-diyl), and poly(pyridine-3,5-diyl) – as well as irregular ones with mixed substitution patterns (Figure 1.4). From this perspective, bpy, tpy, and 2,2′:6′,2″:6″,2‴-quaterpyridine (qtpy) represent the shortest oligomers of poly(pyridine-2,6-diyl) with two, three, or four pyridine rings, respectively. From these polymers, poly(pyridine-2,5-diyl) (PPy) and poly(pyridine-2,6-diyl) (PmPy) have been studied to a significant extent [81, 82]. However, the discussion of these conjugated polymers goes well beyond the scope of this Chapter; their utilization, as multidentate ligands, has – to the best of our knowledge – not yet been envisaged. Nonetheless, the generic term "polypyridyl" is commonly used throughout the scientific literature. On the contrary, Constable *et al.* prefer the more practical term "oligopyridine" in order to distinguish the short oligomers (up to the sexipyridine), which exhibit a defined coordination chemistry, from the "real" polypyridine polymers [83–85].

What about the nomenclature of the tpy ligand itself? Starting from the initial report from 1932, many names have been used for this molecule in a highly inconsistent fashion. The long list of alternative names includes tripyridyl, terpyridyl, tripyridine, 2,2′,2″-terpyridine, 2,6-dipyridin-2-yl-pyridine, *etc.* Similarly, various other abbreviations have been (and actually are still) used (*e.g.*, terpy, trpy, *etc.*). According to IUPAC recommendations from 2014 (IUPAC: International Union of Pure and Applied Chemistry) [86], one should prefer the systematic name $1^2,2^2$:$2^6,3^2$-terpyridine over 2,2′:6′,2″-terpyridine, which was listed as the preferred IUPAC name (PIN) in IUPAC's "Blue Book" of 1979 [87]. However, this alteration, which circumvents the use of multiple primers, is hardly recognized by the scientific community and the "traditional" notation is widely retained (and remains accepted by IUPAC). In order to be consistent with the majority of the available literature and following our habit also to do so, 2,2′:6′,2″-terpyridine, abbreviated as tpy, is uniformly used in this Book. By doing so, all the other "terpyridinic" constitutional isomers, which are depicted in Appendix A, are largely left out of consideration. From these, the 3,2′:6″,3″- and 4,2′:6′,4″- isomers have found

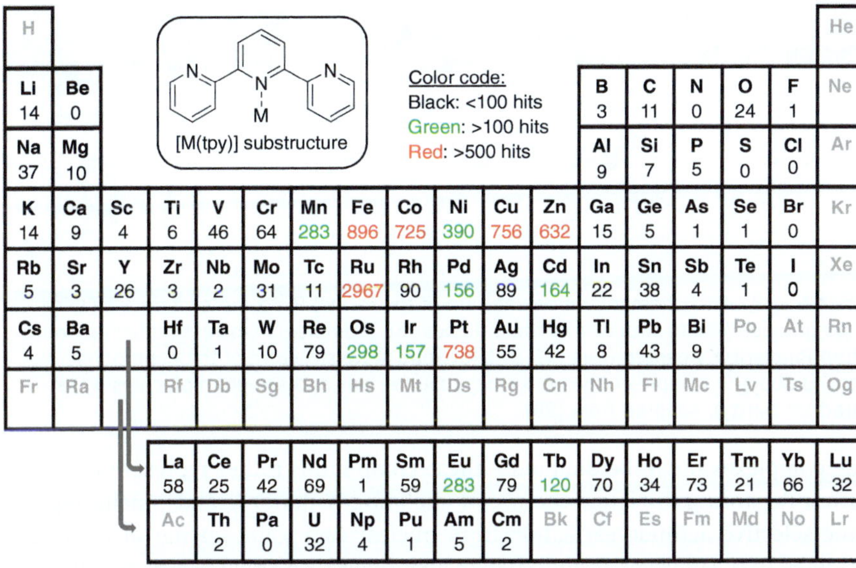

Figure 1.5 The periodic table of elements. For each element, the figure denotes the number of journal articles in which compounds with the "M(tpy)" substructure were mentioned. (SciFinder^{n} research performed on 1st December 2024).

application, as directing ligands, for the formation of coordination polymers based on transition-metal ions; in these assemblies, the central pyridine rings of the ligands typically do not participate in coordination [88–91]. Klein *et al.* identified in total of 21 isomers which could basically serve as tridentate ligands, *i.e.*, binding involves all three rings of the ligand framework (Appendix A). In this respect, one has to further consider the phenomenon of "rollover cyclometallation," which gives rise to a range of potential binding motifs: N^N^N, C^N^N, N^C^N, C^C^N, C^N^C, and C^C^C [51]. According to this notation, tpy represents a neutral N^N^N ligand, which uses its three *N*-atoms for coordination; whereas in the C^C^C motif, all three centers are formally carbanions. Using divalent group-10 metal ions, complexes with some of these binding motifs were recently reported [51]. The research, which deals with the coordination chemistry of tpy's constitutional isomers, is discussed in Chapter 2.

Starting from the very first tpy complex, *i.e.*, the $[Fe(tpy)_2]^{2+}$ *bis*-complex [28], a massive wave of complexes containing ions of main-group elements, transition metals as well as lanthanide or actinide metals has been reported over the last decades. The plethora of structures ranges from mononuclear complexes and dyads/triads to polynuclear assemblies. Figure 1.5 provides an overview of the number of scientific publications per element. Notably, for almost every element – excluding the noble gases and highly radioactive, mostly synthetic elements – examples for complexes (or compounds, to be more general) can be found.

At least in the case of the alkaline-metal, some of the early transition-metal (*e.g.*, Sc^{III}, Ti^{IV}, Zr^{IV}) and the actinide ions [*e.g.*, $U(O)_2{}^{VI}$, Am^{III}, Cm^{III}], a practical use

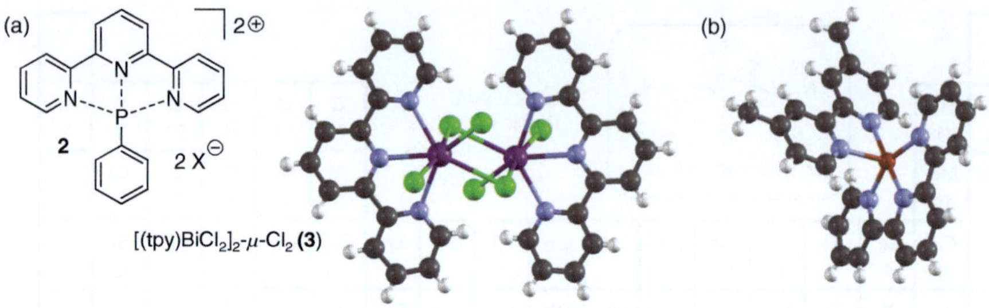

Figure 1.6 (a) Left: Schematic representation of the formal P^{III} complex **2** (X = TfO^- or PF_6^-); right: Solid-state structure of the dinuclear Bi^{III} complex **3** (solvent molecules were omitted for clarity). Adapted from [95, 97]. (b) Solid state structure of the pentagonal $[Cu(tpy)(dmbpy)]^{2+}$ complex (dmbpy: 4,4-bpy; counterions and solvent molecules were omitted for clarity). Adapted from [98].

of the complexes might be doubted – these compounds rather represent chemical oddities. Regarding the latter, the use of chelating *N*-donor ligands, including tpy, for the selective actinide extraction from nuclear fuel solutions has at least been considered [92]. However, ligands, which contained 1,3,5-triazine moieties, were by far more selective than tpy-based ones and have even been tested under "real" process conditions [93]. At least for some of the main-group elements, assemblies with notable properties have been reported. In this respect, the air-stable organo-catalyst **2**, which contains a formal P^{III} ion [94–96], as well as the Bi^{III}-containing solid-state emitters, such as **3**, are worth mentioning (Figure 1.6a) [97, 99]. Overall, the field of terpyridine chemistry with the main-group elements appears to be less explored and not even fully understood. Besides our recent review on this topic [29], the only systematic survey dates back to 1986 [85]. An overview of this often disregarded topic is presented in Chapter 3. On the other hand, the late transition-metal ions and some rare-earth-metal ions (*i.e.*, Eu^{III}) have attracted enormous interest. This holds particularly true for the ions with d^6 (*e.g.*, Fe^{II}, Ru^{II}, Os^{II}, Ir^{III}), d^8 (*e.g.*, Ni^{II} and Pt^{II}) or d^{10} electron configuration (*e.g.*, Zn^{II}).

According to the bond theories in coordination chemistry [100], which basically go back to Werner's visionary ideas [72], the coordination geometry of a complex will strongly depend on the nature of the employed metal ion. The (distorted) octahedral geometry represents the most common motif, which prevails *inter alia* for all d^6-configured ions. In the case of tpy, the coordination sphere of such ions can be saturated either by two tpy ligands or one tpy ligand and additional (ancillary) ligands. Accordingly, Morgan and Burstall already assumed a pseudo-octahedral geometry for their *bis*-complexes with Fe^{II} or Ru^{II} ions [53]. These authors further concluded that, for geometric constraints, the *mer*-isomer should be favored over the *cis-fac* and *trans-fac* ones (*mer*: meridional, *fac*: facial; Figure 1.7). Based on today's knowledge, tpy ligands are not able to form *fac*-isomers at all; these isomers are only accessible for more flexible ligands, such as the related 2,6-di(quinolin-8-yl)pyridine (dqp, see Section 1.6). Besides the octahedral geometry, other coordination environments are also possible: trigonal-bipyramidal (*e.g.*, with Cu^{II} centers; Figure 1.6b) [98] and square-planar complexes (*e.g.*, with Pd^{II}, Pt^{II}, or Au^{III} centers) with tpy ligands are known [39]. Moreover, rare-earth-metal ions can – due to their f-orbitals – reach

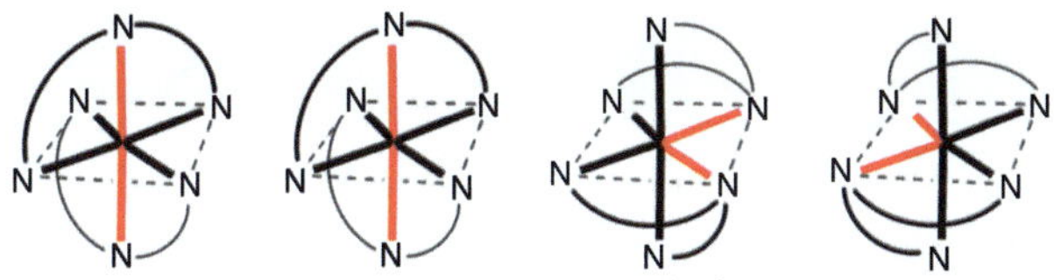

Figure 1.7 Schematic representation of the possible configurations, which might occur in *bis*-tridentate complexes of octahedral geometry [101] / with permission of Royal Society of Chemistry.

coordination numbers of up to nine and, thus, form complexes with three tridentate ligands, such as tpy [102, 103]. Such a behavior, *i.e.*, formation of complexes with similarly high coordination numbers, can also be observed for the heavy Pb^{II} and Bi^{III} ions (*e.g.*, complex $[Pb(tpy)_3]^{2+}$) [29]. The enormous structural diversity, in particular for all the heteroleptic complexes, reveals that the prediction of a certain coordination geometry is difficult. To address this issue, Moore *et al.* presented a tabulated overview of the coordination geometries of transition-metal ions, thereby taking the various oxidation states, coordination numbers, and geometries of the respective elements into account [104]. This qualitative analysis provides comprehensive guidance to appraise the complex geometry. Such *a priori* considerations can be supported by a cautious evaluation of the manifold of solid-state structures, which are accessible from the Cambridge Crystallographic Data Centre (CCDC) [105].

Which are the most recognized tpy complexes in view of practical use? From the unmanageable variety of structures, three examples are highlighted (Figure 1.8). Firstly, Ru^{II} complex **4**, Grätzel's "black dye," is a prominent sensitizer (*i.e.*, light absorber) in dye-sensitized solar cells (DSSCs) [106]. This topic, the utilization of tpy ligands in the context of photovoltaics, is discussed in Chapter 6. Secondly, tpy complexes, such as Meyer's Ru^{II} complex **5** [107], are utilized as catalysts in the context of artificial photosynthesis, *i.e.*, the (electrochemical) splitting of water [43]. This notable aspect of terpyridine chemistry is covered in Chapter 8. Thirdly, Jennette, Lippard *et al.* found that Pt^{II} complexes, such as **6**, were capable of intercalating into the DNA [108]; thus, complexes of this type are considered alternatives to the well-known metallodrug cisplatin (see Chapter 4) [109].

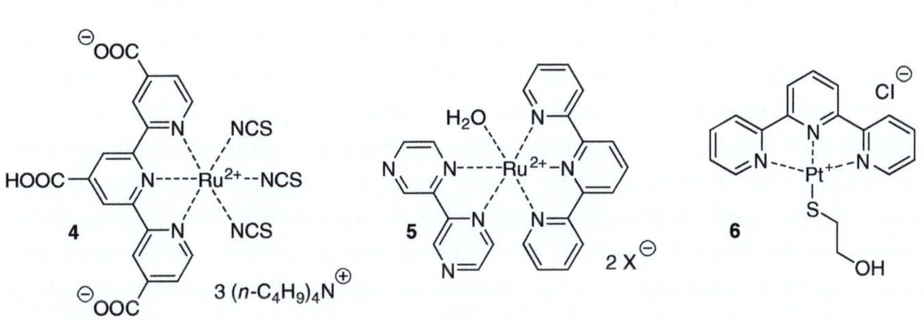

Figure 1.8 Schematic representation of Grätzel's "black dye" **4**. Meyer's water-oxidation catalyst **5** (X = Cl^- or PF_6^-), and Lippard's DNA intercalator **6**.

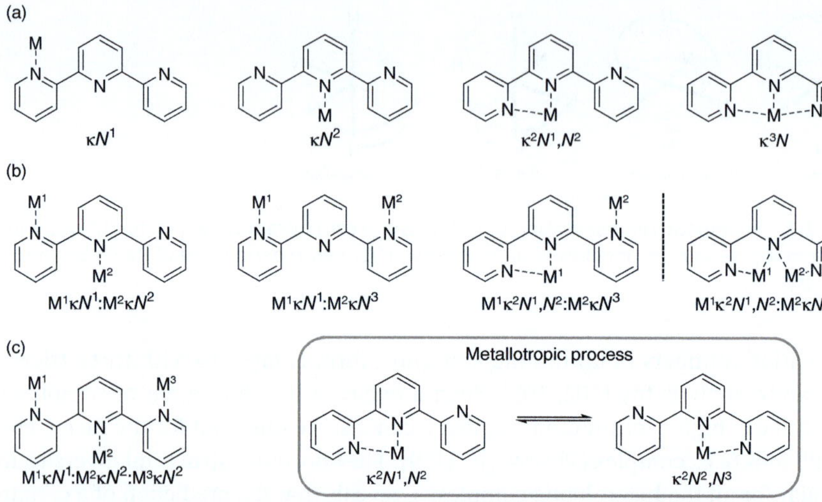

Figure 1.9 Schematic representation of the various coordination modes, which might occur in metal-terpyridine complexes; the respective "kappa notation," according to IUPAC's recommendations, is also given. (a) Mononuclear species, (b) binuclear species, and (c) trinuclear species. In the case of (b) and (c), the metal (*i.e.*, M^1, M^2, and M^3) might be the same or different. Adapted from [84].

Even though the previous considerations might suggest that tpy might always act as a tridentate ligand, the coordination behavior of this ligand is by far more complicated. In order to distinguish between the various possible binding modes, IUPAC recommends one to consequently follow the so-called "kappa notation" (Figure 1.9) [110]. Accordingly, tridentate binding of a tpy ligand to a metal center *via* its three *N*-atoms corresponds to the κ^3N mode.

Utilization of all three *N*-atoms might also give rise to other assemblies – binuclear as well as trinuclear ones (the respective notations are $M^1\kappa^2N^1,N^2{:}M^2\kappa N^3$ and $M^1\kappa N^1{:}M^2\kappa N^2{:}M^3\kappa N^3$). In addition to these, a range of nonclassical complexes has been reported, in which the tpy ligands are coordinated in hypodentate fashion (*i.e.*, not all *N*-atoms are used.) [83, 111]. Such complexes correspond to the mononuclear κN^1, κN^2, and κ^2N^1,N^2 as well as binuclear $M^1\kappa N^1{:}M^2\kappa N^2$ and $M^1\kappa N^1{:}M^2\kappa N^3$ modes. The chelate effect, which is addressed later in this section, can be overcome if the *mer*-configuration of a tpy ligand is disabled for geometric constraints or pre-saturation of the metal's coordination sphere. Bidentate coordination (*i.e.*, in the κ^2N^1,N^2 mode) has been found in some Re^I, Ru^{II}, and Ir^{III} complexes – representatively, complex $Re(tpy)(CO)_3Cl$ (**7**) is depicted in Figure 1.10 [112]. Such complexes often exhibit a metallotropic rearrangement, in which the κ^2N^1,N^2 and κ^2N^2,N^3 motifs are rapidly interconverted (as shown in Figure 1.9, the outer pyridine rings switch between an active and a passive role); it is commonly believed that species of higher coordination numbers are involved in this process [83]. Schatzschneider *et al.* reported one notable example of a complex in which the unreckoned monodentate κN^1-type coordination of tpy was observed (Figure 1.10) [113]. In this case, the azido ligand, as a strong σ- and π-donor, was

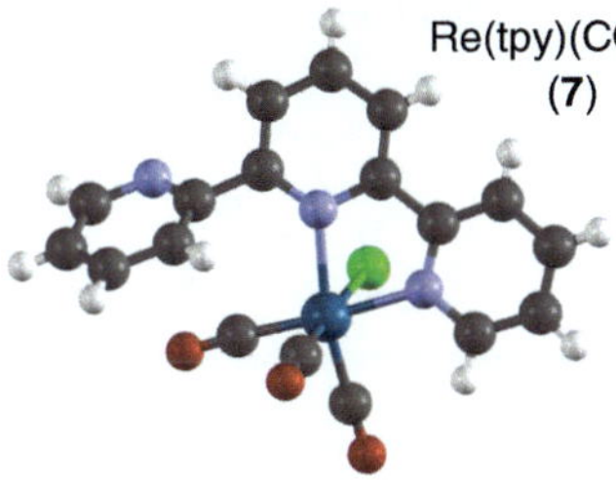

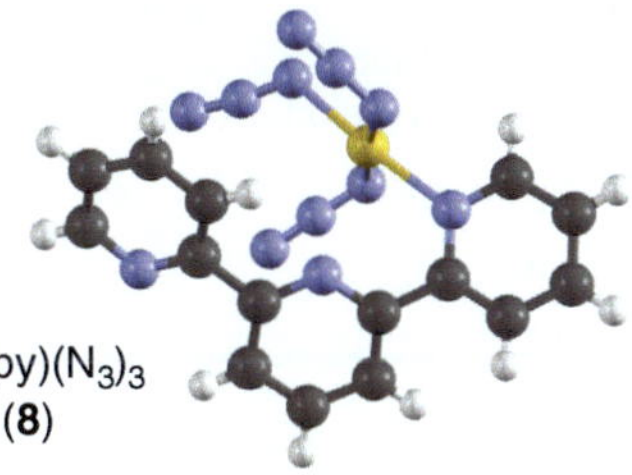

Figure 1.10 Solid-state structures of Re^{I} complex **7** and Au^{III} complex **8** (solvent molecules were omitted for clarity). Adapted from [112, 113].

even able to repel two initially coordinated pyridine rings in course of the synthesis [*i.e.*, square-planar Au(tpy)$(N_3)_3$ (**8**) was obtained instead of Au(tpy)(N_3)]. It has to be emphasized that the tridentate κ^3N mode is only possible for **1**; whereas bidentate coordination *via* the $M^1\kappa^2N^1,N^2$ mode is also possible for numerous of its constitutional isomers, which possess a bpy-like substructure (Appendix A) [114]. The coordination chemistry of those ligands (with mainly transition-metal ions) is comparatively less explored and will be discussed briefly in Chapter 2.

In some respect, tpy protonation can be compared to its complexation by metal ions. ^{1}H-NMR titration studies and single crystal X-ray diffraction (XRD) analysis revealed that the bidentate $M^1\kappa^2N^1,N^2$ mode was also favored in the singly protonated $[tpyH]^+$ state [115, 116]. Whereas, doubly protonated $[tpyH_2]^{2+}$ featured an unique fixation of the two protons, which both interacted with the *N*-atom of the central pyridine rings and each with an *N*-atom of one of the outer pyridine moieties [116, 117]. Such a binding, which formally corresponds to an $M^1\kappa^2N^1,N^2{:}M^2\kappa N^2,N^3$ mode, is, for steric constraints, hardly possible for binuclear metal-tpy complexes.

When talking about the coordination of tpy ligands to a metal center, one has to take the occurrence of steric repulsion into account. Steric repulsion between the ligand's protons in 3/3″-position on the one hand and 3′/5′-position on the other comes into play as soon as the molecule rotates into the *cis,cis*-conformation, which is the premise for coordination in the tridentate κ^3N mode (Figure 1.11a). Computational studies revealed that electronic repulsion of the inward-pointing *N*-atoms further destabilizes the all-*cis*-isomer. For the *trans,trans*-conformer, repulsive forces are minimized; thus, this isomer is highly favored in the absence of metal ions [117]. A comprehensive and differentiated analysis was provided by Batista, Hazari *et al.*, who computed the entire energy surface of tpy reflecting the rotation of its outer rings (Figure 1.12) [120]. According to these authors, the *cis,trans*-configuration is energetically disfavored by *ca.* 31.4 kJ mol^{-1} when compared to the all-*trans*-configuration, as the global energetic minimum. Notably, the conformational change to reach the thermodynamically least favorable all-*cis*-conformation is energetically even more uphill: The rotation of the second pyridine ring requires *ca.* 46.0 kJ mol^{-1}, thus *ca.* 14.6 kJ mol^{-1} more than for the rotation of the first pyridine ring. The predicted preference for the all-*trans*-configuration has been verified experimentally at the example of

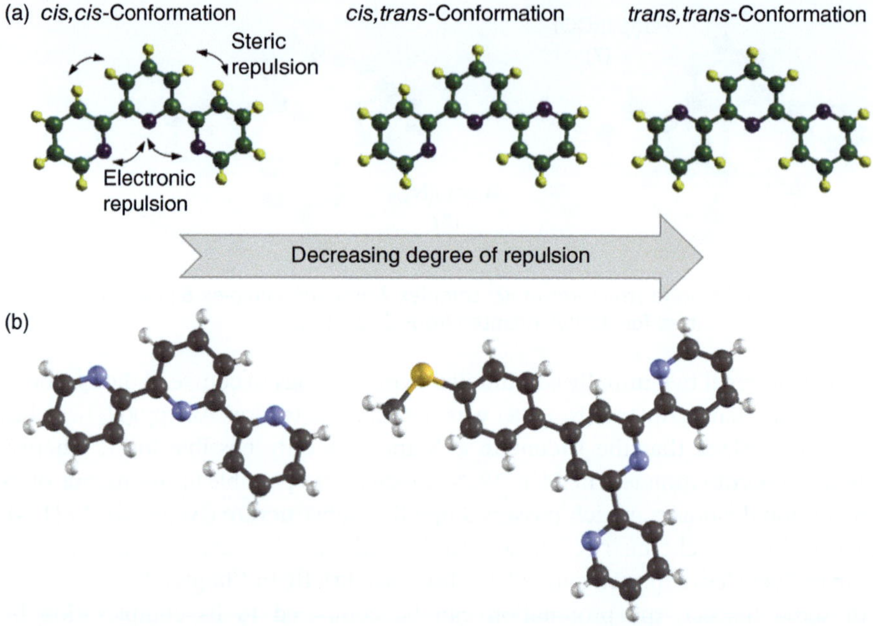

Figure 1.11 (a) Schematic representation of the energy-optimized structure of **1** in three different conformations – *cis,cis*, *cis,trans* and *trans,trans*. Adapted from [117]. (b) Solid-state structures of parent tpy (left) and its 4′-(4-methoxyphenyl) derivative (right). In both cases, solvent molecules were omitted for clarity. Adapted from [118, 119].

numerous single-crystal XRD analyses; two representative structures are depicted in Figure 1.11b [118, 119].

In transition-metal complexes, the length of the metal-to-ligand bond represents an important parameter regarding its nature. Basically, the bond length l can be derived from the solid-state structure; this, in turn, requires the growth of single crystals, which are suitable for XRD analysis. In the case of chelating ligands, such as tpy, two further values can be read from its solid-state structure: the bond and torsion angles (Φ and ϕ, respectively; Figure 1.13). The former, which is misleadingly also named bite angle, does not represent a ligand's fundamental property, since Φ depends on the bond length and, thus, on the properties of the coordinated metal (*e.g.*, ionic size). The torsion angle, on the other hand, is of importance regarding the binding mode: As a rule-of-thumb, a ϕ value of *ca.* 30° is a prerequisite for tpy to bind in its tridentate $\kappa^3 N$ mode (larger torsion angles will result in other binding modes; see Figure 1.9).

The bond strength, on the other hand, can be determined experimentally by titration experiments. As elaborated by Thordarson, nuclear magnetic resonance (NMR) as well as UV/vis absorption or emission spectroscopy might be utilized in this context [122]. Very recently, these established methods have been integrated into a robotic workflow for titrations, thus enabling the automated screening of complex formation in solution [123]. Furthermore, thermochemical analyses, using isothermal titration calorimetry (ITC), represent a versatile, yet demanding, method

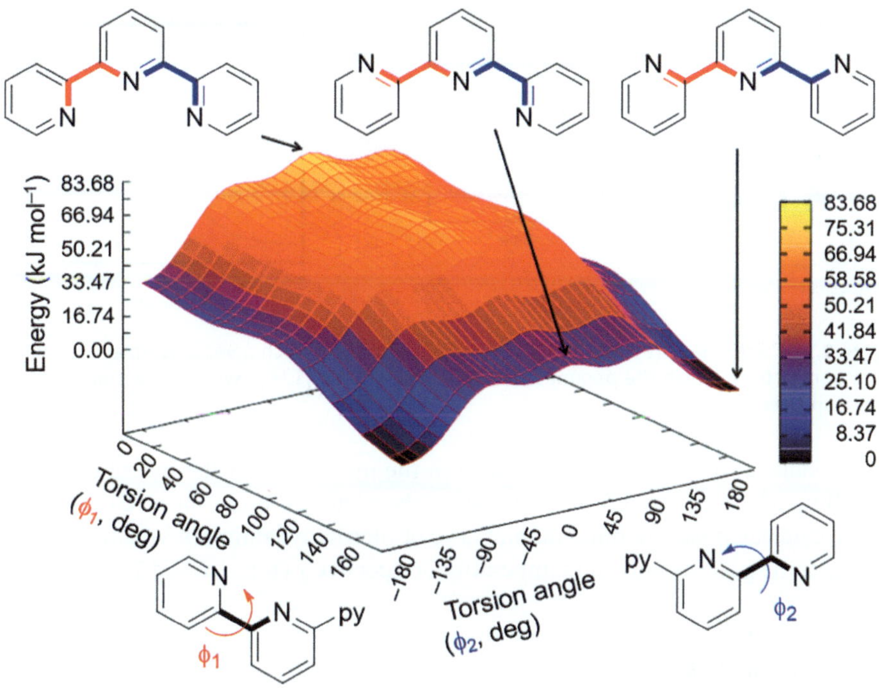

Figure 1.12 Energy-surface profile for the rotation of the outer pyridine rings of **1** in the gas phase. The all-*anti*-conformation, as the global energetic minimum, was set to 0 kJ mol^{-1}. The torsion angle between the central ring and one of the outer ones (ϕ_1) was varied from 0° to 180°; whereas, the torsion angle of the second outer ring (ϕ_2) was scanned from −180° to 180°. Thereby, the torsion angle of 0° corresponds to a *cis*-conformation; whereas, the *trans*-form is reached for the torsion angles of 180° and −180° [120] / with permission of Royal Society of Chemistry.

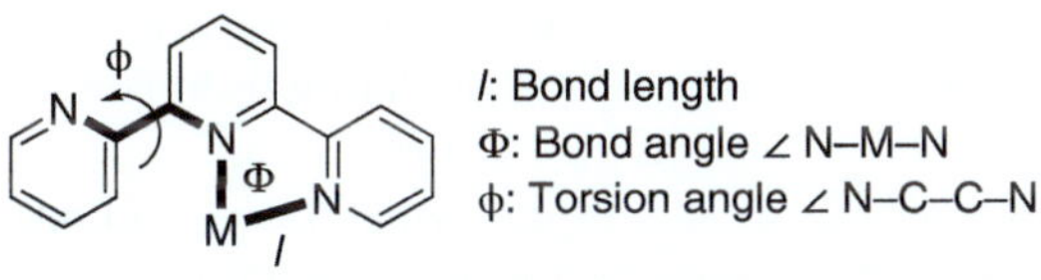

Figure 1.13 Schematic representation of a tpy complex showing the parameters, which can be deduced from the solid-state structure [121] / with permission of Elsevier.

to extract the association constant (K_a), stoichiometry (n) as well as the enthalpy (ΔH), entropy (ΔS), and Gibbs free energy (ΔG) from the fitted thermogram (Figure 1.14) [124]. One must be aware that cautious optimization of various experimental conditions is required in order to allow an ITC experiment to deliver reliable thermodynamic data [125, 126]. This will be the case if the ITC curve has a sigmoidal shape; the most accurate curve fitting will be possible if the value of Brandt's (or Wiseman's) constant (c) is in the range from 10 to 500. This constant is function of the stoichiometry, K_a and the receptor (*i.e.*, the tpy ligand) concentration (Eq. 1.1). Three values are directly accessible from the thermogram: K_a from the

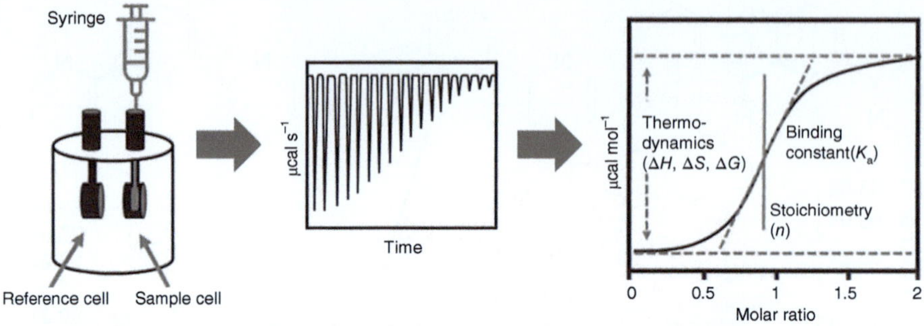

Figure 1.14 Schematic representation of an ITC workflow, which gives a sample thermogram, from which the parameters can be determined [124] / with permission of American Chemical Society.

curve's slope at the inflection point, n from the molar ratio at the inflection point as well as ΔH from the difference between the plateaus. From these, ΔG and ΔS can be calculated *via* the fundamental laws of thermodynamics, in which R and T represent the gas constant and temperature, respectively (Eqs. 1.2 and 1.3).

$$c = nK_a\,[\text{tpy}] \tag{1.1}$$

$$\Delta G = RT \ln K_a \tag{1.2}$$

$$\Delta G = \Delta H - T\Delta S \tag{1.3}$$

From a thermodynamic perspective, K_a gives a good indication of the stability of a complex [121]. However, in the case of tpy, the formation of *bis*-complexes proceeds *via* two distinct equilibria (Eq. 1.4a,b), which cannot always be deconvoluted. For convenience, the cumulated stability constant (β_{21}), which corresponds to K_a, is typically preferred over the individual constants, K_1 and K_2 (Eq. 1.5).

$$\text{M} + \text{tpy} \overset{K_1}{\rightleftarrows} [\text{M(tpy)}]$$

$$K_1 = \frac{[\text{M(tpy)}]}{[\text{M}]\,[\text{tpy}]} \tag{1.4a}$$

$$[\text{M(tpy)}] + \text{tpy} \overset{K_2}{\rightleftarrows} [\text{M(tpy)}_2]$$

$$K_1 = \frac{[\text{M(tpy)}_2]}{[\text{M(tpy)}]\,[\text{tpy}]} \tag{1.4b}$$

$$\beta_{21} = \lg K_1 + \lg K_2 \tag{1.5}$$

Beyond these fundamental aspects, which are based on the mass-action law, stability might be affected by further effects. In this respect, the so-called chelate effect has to be mentioned first. This effect is based on the formation of chelate rings (as in the case of tpy's tridentate κ^3N binding mode). In simple words, the cooperativity of binding results in an increased binding strength: The κ^3N-type binding of one tpy ligand to a metal center will, owing to entropic and kinetic reasons, always be stronger

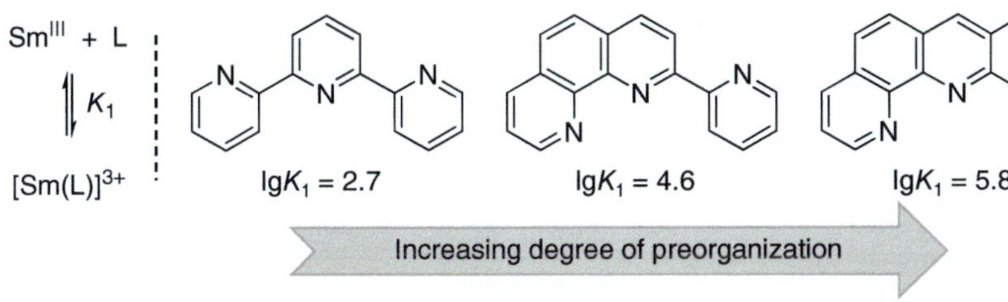

Figure 1.15 Schematic representation of parent tpy and its benzannelated derivatives. The respective lgK_1 values for the complexation of Sm^{III} ions are also listed. Adapted from [129].

than that of three individual pyridine (py) ligands [127]. Dobrawa and Würthner used the interaction of Zn^{II} with oligopyridyl ligands to elaborate the chelate effect. In the case of the simple Zn^{II}–py interaction, K_a values in the range of 10^3 M^{-1} were observed; whereas the related Zn^{II}–tpy provided significantly higher K_a values of $>10^8$ M^{-1} [128]. However, the chelate effect represents only one side of the story. As mentioned beforehand, chelation of **1** installs steric repulsion of adjacent *H*-atoms, which, in turn, has a destabilizing influence. Hancock showed that ligand preorganization might also have a tremendous effect [129]. Ring annulation not only cancels out the steric repulsion but also already offers the proper binding environment and unfavorable conformational changes, as for the parent ligands, become obsolete. Thus, phen is a stronger ligand than bpy, and the same trend holds true for **1** and its benz-annelated derivatives (Figure 1.15). Accordingly, the rigid DPA ligand provided the highest K_a values (DPA: dipyrido[4,3-*b*,5,6-*b*]acridine). This was shown by Thummel, Hancock *et al.* for a wide range of ions from s-, d-, and f-block metals [130–132].

The binding strength not only depends on the ligand's structure and denticity but also on the nature of the metal ion to be complexed. Firstly, the interaction of metal ions with pyridine-based ligands can be viewed in the context of Pearson's HSAB concept, which he established in 1963 (HSAB: hard and soft acids and bases) [133]. Even though key messages of this concept are often thwarted by other factors (particularly in organic chemistry [134]), it can be utilized to gauge the strength of metal-to-ligand interactions. It is well-known that alkali-metal ions, as hard Lewis acids, have a strong affinity toward oxygen atoms, as hard Lewis bases (*e.g.*, ether moieties). This behavior represents one cornerstone of supramolecular chemistry, *i.e.*, the size-selective complexation of group-1 ions by macrocyclic or cage-like polyether compounds (*i.e.*, crown ethers and cryptands, respectively). As mentioned beforehand, the 1987 *Nobel Prize in Chemistry* was awarded to Cram, Lehn, and Pedersen for their pioneering work in this field [1–3]. But what about pyridine? This molecule is ranked as an "intermediate" Lewis-basic ligand and, thus, better compatible with intermediate or soft Lewis bases, as the transition- and f-block-metal ions are (Figure 1.16) [133]. In complexes with these metal centers, σ-donation from the py-based ligand to the metal center *via* the *N*-atom's lone pair is accompanied by back bonding from the metal to the ligand's π* orbital. However, this effect is rather weak, at least in the case of the divalent

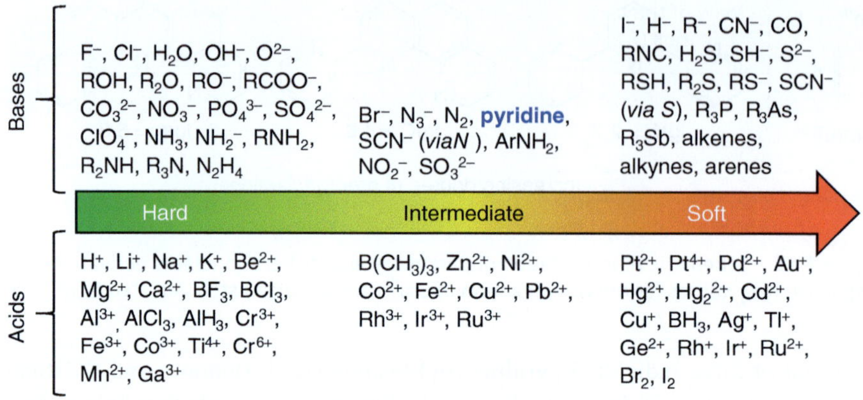

Figure 1.16 A compilation of important hard, intermediate, and soft Lewis-type bases and acids according to Pearson's HSAB classification (the overview makes no claim to be exhaustive). Adapted from [121].

transition-metal ions [135], and must even be neglected for the s-block-metal ions. Thus, the donor atom's electronegativity (EN) gives good evidence for its σ-donor capability. Compared to a sp^3-hybridized *O*-center (as present in crown ethers), pyridine's sp^2 *N*-atom is considerably less electronegative (*i.e.*, 3.44 *vs.* *ca.* 3.2). Thus, pyridine represents a weaker donor-type ligand and those complexes, which mainly rely on electrostatic forces, will be less stable (at least when compared to their crown-ether complexes) [121].

Secondly, the actual size of the complexes metal ion plays an important role. Tridentate coordination of tpy ligands will only be favored if the ionic radius matches with the binding pocket of the ligand in the all-*cis*-configuration. This, for example, explains why the Be^{II} ion is hardly used in the context of coordination chemistry [136]. The cation's remarkably small size of only 0.31 Å is hardly compatible with binding pockets offered by common chelating ligands, such as the oligopyridyl ones. Moreover, the Be^{II} cation represents one of the hardest Lewis acids (charge-to-radius ratio of *ca.* 6.45); for this reason, Be^{II} will strongly interact with hard Lewis acids (*e.g.*, *O*-donors), but unfavorably with weaker sp^2-hybridized *N*-donors (Figure 1.16). A thorough analysis of the matching/mismatching of metal ions with tpy's binding pocket was provided by Thummel, Hancock *et al.* [130–132]. These authors pointed out that the greater stabilization of complexes of large metal ions with DPA was due to the ligand's best-fit size, which is considerably larger than that of tpy (*i.e.*, 2.60 *vs.* 2.41 Å). Accordingly, the larger lanthanide(III) ions featured good stabilization in the case of DPA, whereas tpy was more compatible with the smaller Mg^{II} or Zn^{II} ions (Table 1.1).

Furthermore, Würthner *et al.* analyzed the thermodynamics of tpy complexation by the first-row transition-metal ions by ITC and NMR titration [139]; the trend of the thusly obtained K_a values matched well with the often-cited Irving-Williams series (*i.e.*, $Mn^{II} < Fe^{II} < Co^{II} < Ni^{II} < Cu^{II} > Zn^{II}$) [140]. Beforehand, similar trends in terms of kinetic and thermodynamic stability of metal complexes were also observed by Wilkins *et al.* [141, 142]. In this context, one has to be reminded not

Table 1.1 Overview of the binding constants ($\lg K_1$) of some divalent metal ions with organic chelating ligands (2×10^{-5} M in water, pH = 7) [130, 131].

	Mg^{2+}	Ca^{2+}	Sr^{2+}	Ba^{2+}	Zn^{2+}	Cd^{2+}	Gd^{3+}	Pb^{2+}
r^+ (Å)[a)]	0.72	1.00	1.18	1.35	0.74	0.95	0.94	1.19
$\lg K_1$(tpy)	0.74	0.85	−0.5	−0.6	7.7	6.4	2.6	6.0
$\lg K_1$(DPA)	0.7	3.68	2.20	1.5	7.9	8.1	5.7	8.3
$\Delta \lg K_1$[b)]	≈0	3.23	2.7	2.1	0.2	1.7	3.1	2.3

a) Shannon's ionic radii for the six-coordinate cations [137, 138].
b) $\Delta \lg K_1 = \lg K_1(\mathrm{DPA}) - \lg K_1(\mathrm{tpy})$; this value represents an estimate for the equilibrium constant of the ligand-exchange reaction $[\mathrm{M(tpy)}]^{2+} + \mathrm{DPA} \leftrightarrows [\mathrm{M(DPA)}]^{2+} + \mathrm{tpy}$.

to confuse the thermodynamic stability of complexes with the kinetic stability. The latter is expressed by the rate constant (often also named the exchange constant), which quantifies the rate at which a complex is formed, or ligands are exchanged. One must keep these aspects in mind when dealing with tpy complexes. On one hand, kinetic lability can be beneficial for materials which rely on the reversibility of bonding. This behavior is, *e.g.*, utilized in the context of self-healing or shape-memory polymers, *i.e.*, materials that are (at least partially) capable of restoring their initial properties, functions, or shapes if an appropriate stimulus is applied [15, 143]. Metallo-supramolecular polymers, which contain labile tpy *bis*-complexes, represent one possible approach to realize this (this topic is covered in Chapter 9). Moreover, reversibility of bonding can be used in metallo-supramolecular chemistry to assemble advanced 2D/3D structures *via* multicomponent self-sorting strategies (here, mainly Zn^{II} and Cd^{II} ions are used; see Chapter 10) [23]. However, kinetic lability can also be highly detrimental: Kinetically labile complexes will readily dissociate in solution and, thus, complex formation might be concentration-dependent. In such cases, purification by, *e.g.*, conventional adsorption or size-exclusion chromatography (SEC), is hardly possible. Moreover, the fast ligand exchange in solution might prevent one from preparing heteroleptic *bis*-complexes. Solution-stable complexes, which, *e.g.*, contain two different tpy ligands, can only be obtained reliably for a handful of metal ions: Ru^{II}, Os^{II}, Rh^{III}, Ir^{III}, Co^{III}, and Cr^{III} [144].

1.4 Fundamental Aspects of Terpyridine-Complex Synthesis and Characterization

1.4.1 Synthesis of Octahedral and Square-Planar Complexes

Bis-complexes of distorted octahedral geometry represent the most prevalent type of structure, for which homoleptic as well as heteroleptic complexes can be found. For the synthesis of homoleptic complexes of the general formula $[\mathrm{M(tpy\text{-}R)_2}]^{2+}$ (R denotes any substituent), the ligand is commonly combined with a specific

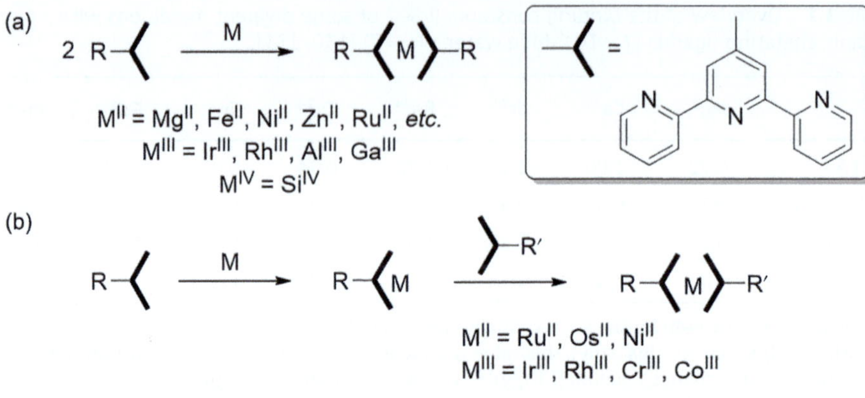

Figure 1.17 Schematic representation of the synthesis of homoleptic and heteroleptic M–tpy *bis*-complexes. (a) Homoleptic complex *via* one-step procedures and (b) heteroleptic complex *via* two-step procedures. Adapted from [144].

metal ion (*e.g.*, Mg^{II}, Zn^{II}, Cd^{II}, Co^{II}, Ni^{II}, Fe^{II}, *etc.*) in a metal-to-ligand ratio of 1:2 (Figure 1.17a). In many cases, coordination occurs at room temperature or in a refluxing solvent. However, forcing solvothermal reaction conditions might sometimes be required. In particular, the synthesis of Ru^{II} complexes is conveniently performed in sealed vials under microwave irradiation. To the best of our knowledge, the first protocol for a microwave-assisted synthesis of $[Ru(tpy)_2\}^{2+}$ was reported by Green and Mingos in 1991 [145]. Today, domestic microwave ovens have been replaced by laboratory ones, which allow for a better-controlled synthesis. Numerous examples for this application can be found in available literature [146]. As one representative example, the microwave-assisted reaction of $RuCl_3{\cdot}xH_2O$ with Cl-tpy in ethylene glycol proceeded within 5 min (Cl-tpy: 4′-chloro-2,2′:6′,2″-terpyridine; 90% isolated yield), whereas the conventional open-flask synthesis under reflux required at least 1 h [147]. The synthesis of homoleptic *bis*-complexes is typically followed by an anion-exchange step (*e.g.*, to PF_6^-, BF_4^-, BPh_4^-, *etc.*), in order to achieve solubility in polar-aprotic organic solvents (*e.g.*, acetone or MeCN). Crystallization by vapor diffusion is commonly applied to obtain single crystals suited for XRD analysis. Purification by liquid chromatography is only applicable in those cases where the complexes exhibit sufficient kinetic stability.

Heteroleptic *bis*-complexes can seldom be prepared *via* an one-step synthesis since the addition of a metal salt to an equimolar mixture of two different tpy ligands would afford a statistical mixture of both homoleptic complexes and the heteroleptic one [148]. This limitation can be overcome by applying a directed two-step procedure (Figure 1.17b). Owing to the special coordination chemistry of their M^{III}/M^{II} ions, the heavy group-8 metals (*i.e.*, Ru and Os) are particularly suited for this purpose. In many cases, the *mono*-complexes can be isolated and even purified [33]. In order to improve the synthesis of Ru^{II} complexes, $RuCl_3{\cdot}xH_2O$, as the traditional starting material, the use of precursor complexes, which allows for milder reaction conditions and improved reaction control, has attracted significant attention.

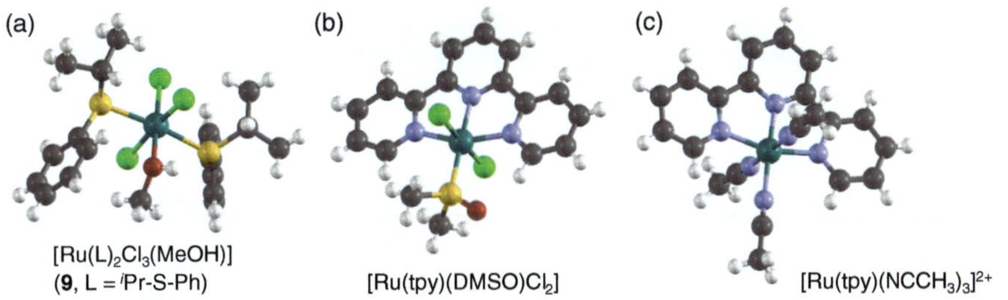

Figure 1.18 (a) Solid-state structures (excluding counterions and solvent molecules) of the Ru^{III} precursor **9**. Adapted from [149], (b) [Ru(tpy)(DMSO)Cl$_2$]. Adapted from [150] and (c) [Ru(tpy)(NCMe)$_3$]$^{2+}$. Adapted from [151].

In this respect, the molecular Ru^{III} complex **9** as well as the Ru^{II} *mono*-complexes [Ru(tpy)(DMSO)Cl$_2$] and [Ru(tpy)(NCCH$_3$)$_3$]$^{2+}$ are worth mentioning (Figure 1.18) [149–151].

Besides the widespread use of Ru^{II} and Os^{II} ions to form heteroleptic *bis*-complexes, assemblies with Rh^{III} and Ir^{III} centers have also been prepared [152, 153]. Regarding the complexation of first-row transition-metal ions, Fustin, Gohy *et al.* demonstrated that solution-stable heteroleptic complexes can also be obtained from Ni^{II} or Co^{III} ions (in the latter case, the *bis*-complex was locked by oxidizing the metal center from the +II to the +III state) [154]. Even though one might expect Cr^{III} complexes, as d^3-electron species, to be kinetically stable [155], the heteroleptic complexes were found to be stable under pH-neutral conditions, but ligand scrambling occurred in basic solution or in the presence of strong *O*-donor solvents [156]. The synthesis of heteroleptic Fe^{II}–tpy is demanding and has, to the best of our knowledge, not been realized *via* a two-step protocol. Fallahpour *et al.* employed the one-step procedure; in this case, the heteroleptic complex could be separated from the two homoleptic ones by column chromatography [157]. Whereas heteroleptic complexes with one tpy and a different type of tridentate ligand have been prepared. For example, Canton, Bauer *et al.* combined tpy ligands with the bulky *bis*-NHC ligand 2,6-*bis*[3-(2,6-diisopropylphenyl)imidazol-2-yldiene]pyridine (NHC: *N*-heterocyclic carbene) [158, 159]. And what about heteroleptic Zn^{II} and Cd^{II} complexes? The *bis*-complexes of these metal ions with tpy ligands are kinetically labile, feature relatively low K_a values, and have, thus, not been considered suitable for the assembly of heteroleptic complexes. Recently, Chan *et al.* presented an elegant approach, in which stabilization of such complexes was achieved by supporting non-covalent interactions within the compounds [160–162]. These authors prepared the substituted tpy ligands **10** (Figure 1.19a), which – owing to steric constraints – could not form homoleptic complexes with, *e.g.*, Zn^{II} or Cd^{II} ions. However, the heteroleptic complex [Zn(tpy)(**10a**)]$^{2+}$ exhibited a remarkable kinetic stability, which was due to strong $\pi\pi$-stacking interactions between the π-electron systems of the anthracene moieties and the sandwiched tpy ligand (Figure 1.19b) [160]. The ability to form solution-stable, heteroleptic complexes was utilized further to assemble metallo-supramolecular polymers: Diblock as

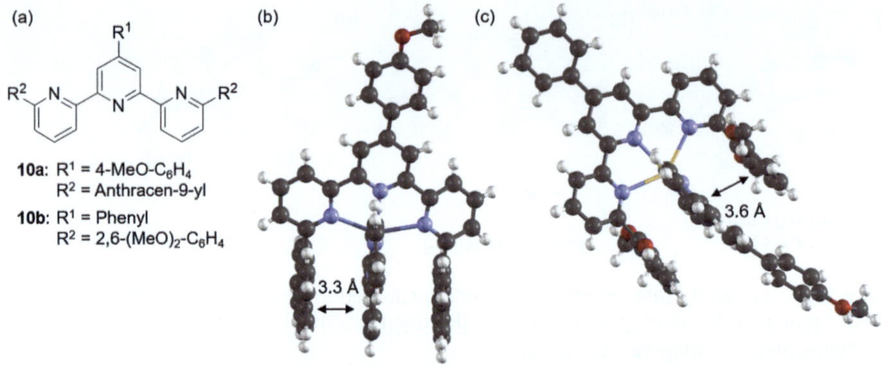

Figure 1.19 (a) Schematic representation of the 6,6″-disubstituted tpy ligands **10**. (b) Solid state structure of $[Zn(\mathbf{10a})(tpy)]^{2+}$. Adapted from [160]. (c) Solid-state structure of $[Cd(\mathbf{10b})(4\text{-MeOPh-tpy})]^{2+}$. Adapted from [162]. In panels (b) and (c), counterions and solvent molecules were omitted for clarity.

well as star-shaped materials were obtained [160, 161]. In a similar fashion, the heteroleptic complex $[Cd(\mathbf{10b})(L)]^{2+}$ was prepared (L: 4-MeOphenyl-tpy) [162]. Here, charge-transfer interactions between the electron-rich substituents and the electron-deficient tpy moiety accounted for the kinetic stability. Based on this complementary tpy pairing, 2D/3D metallo-supramolecular nanosized constructs were assembled.

Regarding the formation of square-planar complexes with Pd^{II}, Pt^{II}, or $Au^{I/III}$ ions, a chlorido complex is typically prepared first by coordinating the tpy ligand to the respective precursor salt (*e.g.*, K_2PtCl_4, $PdCl_2$, $HAuCl_4$, *etc.*) [39]. Subsequently, the chlorido ligand can be replaced by other co-ligands in the presence of Ag^I, as a trap for the liberated Cl^- anions. The alkynyl complexes represent a special class of complexes, in which an ethynyl-aryl derivative is attached *via* a metal-to-carbon bond. They are typically synthesized from the chlorido complexes by a Cu^I-catalyzed dehydrohalogenation in the presence of NEt_3; alternatively, one might also use the respective MeCN-complexes, whose loosely bound solvato ligands can readily be displaced under basic conditions (Figure 1.20) [32]. Such complexes have gained an enormous interest as photocatalysts, emitters for organic light-emitting diodes (OLEDs) or luminescent probes for bio-related applications (*i.e.*, sensing and imaging) [32, 39, 163, 164].

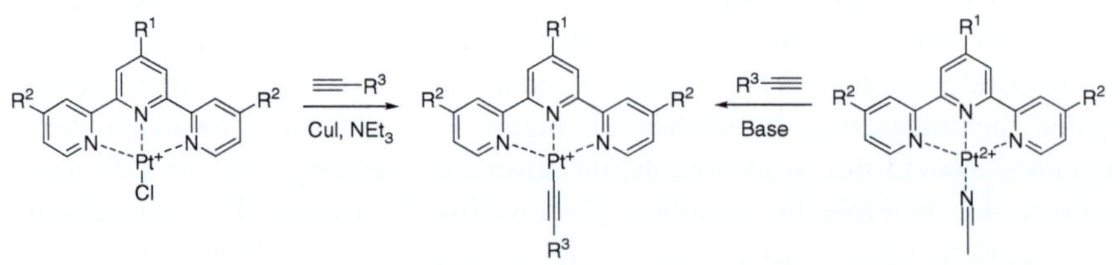

Figure 1.20 Schematic representation of the synthesis of Pt^{II}–tpy complexes bearing alkynyl ligands (R^1, R^2, and R^3 denote optional substituents; counterions were omitted for clarity).

1.4.2 Structural Analysis of Tpy Complexes

Regarding the structural characterization of tpy complexes, XRD analysis represents the method of choice if appropriate single crystals can be obtained. Accordingly, numerous solid-state structures of tpy ligands and their complexes/assemblies have been reported. A recent survey in the CCDC database, using the ConQuest 2024.3.0 software [165], provided 6277 hits for the "M(tpy)" fragment. The presumably first report on the solid-state structure of tpy complexes was provided by Corbridge and Cox in 1956 [166]. These authors utilized XRD to verify the assumed square-pyramidal coordination geometry of $M(tpy)Cl_2$ (M = Zn, Cd, or Cu; Figure 1.21).

However, XRD analysis is typically limited to mono- or oligonuclear complexes, but rather not for sophisticated metallo-supramolecular 2D/3D architectures or for polydisperse materials (macromolecules, coordination networks, *etc.*). This obstacle has been overcome by the exploitation of synchrotron XRD analysis of single crystals [167]. This method provides a significantly higher resolution than conventional XRD analysis. Thus, the structural analysis of complicated and macromolecular structures has been enabled (for selected examples, see Chapter 10). In the case of tpy-containing macromolecules, the analytical toolbox of polymer chemistry comes into play (*e.g.*, SEC, analytical ultracentrifugation, *etc.*) [11]. Mass spectrometry (MS), in particular the electrospray ionization (ESI) technique, represents a versatile tool to prove the identity of tpy complexes. ESI MS offers the main advantage that multiple charged ions can be detected, the coordination remains unaffected upon ionization (*i.e.*, disassembly-reassembly processes are typically absent) and causes only little fragmentation [168]. However, one can hardly deconvolute and analyze superimposing ions with different charges and shapes (but same *m/z* values) by conventional ESI MS. Newkome, Wesdemiotis *et al.* established a method, in which traveling-wave ion-mobility (TWIM) MS was coupled to gradient tandem MS (gMS^2) [169]. By doing so, fragments and intact assemblies of the same *m/z* ratio as well as isomeric architectures of the same self-assembly process can easily be

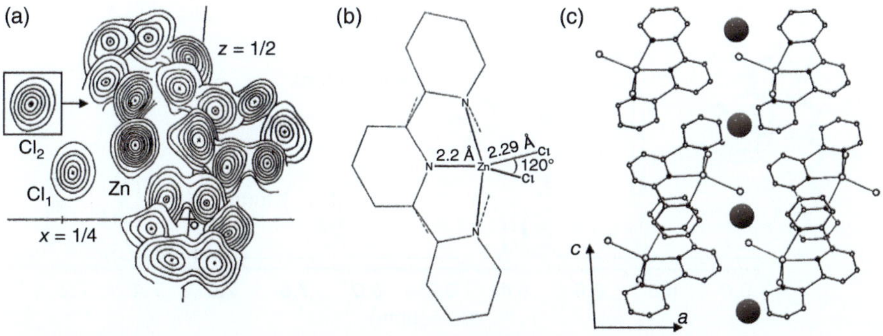

Figure 1.21 Reprint of the original XRD data of $Zn(tpy)Cl_2$. (a) Composite 3D Fourier diagram; (b) representation of the molecular structure including representative bond parameters; (c) representation of the packing in the crystal lattice (the grey spheres indicate voids) [166] / with permission of Royal Society of Chemistry.

differentiated. This strategy, which allows one to overcome the obstacles involved in the mass-spectrometric analysis of advanced metallo-supramolecular 2D/3D assemblies, is now utilized by numerous groups worldwide [23, 170, 171]. This field of research is discussed later in Chapter 10. Aside from these methods, which basically address the proof-of-identity, MS might also be used to determine the gas-phase stability of tpy *bis*-complexes, at least in a qualitative fashion. The applicability of this approach was shown by Schubert *et al.*, who analyzed a range of $[M(tpy)_2]^{2+}$ complexes (M denotes a divalent transition-metal ion) by the means of matrix-assisted laser desorption/ionization (MALDI) time-of-flight (ToF) MS. The thusly obtained order-of-stability was in line with the aforementioned results of calorimetric measurements and NMR titration studies (Section 1.3).

^{1}H-NMR spectroscopy represents the most versatile analytical tool, at least when dealing with tpy complexes of diamagnetic metal ions. The proton resonances experience characteristic shifts upon complexation; thus, the coordinated ligands can readily be distinguished from metal-free ones on the basis of their ^{1}H-NMR spectra. In octahedral *bis*-complexes, the tpy ligands are oriented almost perpendicular to each other. The protons in 6,6″-position of one tpy ligand are located above the ring plane of the central pyridine moiety of the second tpy ligand; as a result, these protons are shielded and their NMR signals are shifted upfield (Figure 1.22). Due to a change in the chemical environment upon complexation, the other signals are also moderately shifted.

If the complex contains a paramagnetic metal center, however, the situation becomes more complicated. Paramagnetic NMR is commonly associated with substantial signal broadening and, thus, the resolution will be too low to resolve couplings (sometimes signals in close proximity to the paramagnetic center might

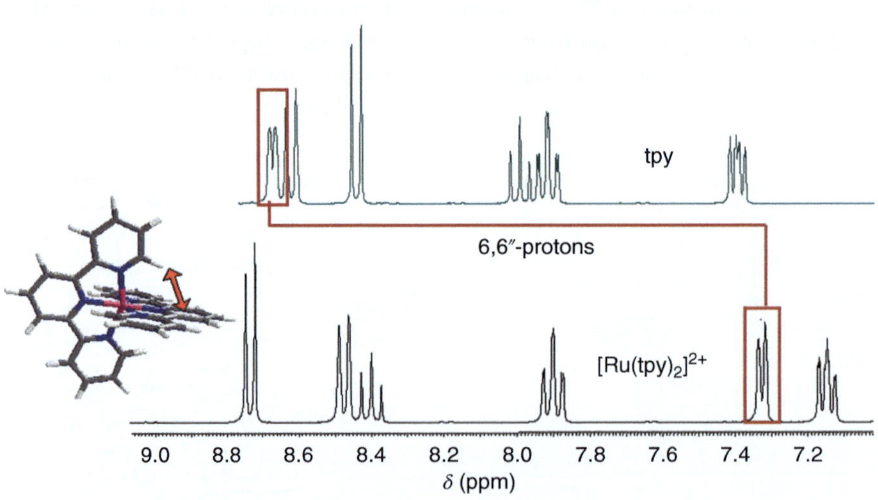

Figure 1.22 Comparison of the ^{1}H-NMR spectra of tpy (top) and its homoleptic Ru^{II} complex (bottom) as measured in CD_3CN. The upfield shift of the 6,6″-protons, which is due to the illustrated C–H···pyridine interaction in the complex, is highlighted [172] / with permission of John Wiley & Sons.

vanish at all) [173]. Besides this, the chemical-shift range is remarkably wide, even ranging over *ca.* 200 ppm in ^{1}H-NMR. Two factors contribute to the hyperfine shift (*i.e.*, the difference between the chemical shift of a proton in a diamagnetic environment *vs.* a paramagnetic one) – the so-called contact and pseudo-contact shifts. In the case of transition-metal complexes, the former is known to vanish quickly with increasing number of bonds since spin-density is mainly located on the metals' d-orbitals [174]. Whereas the latter effect is based on magnetic anisotropy with a distance and angular dependence. Finally, paramagnetism often entails short relaxation times (T_1); this, in turn, brings along high spectral acquisition rates. These aspects in concert make ^{1}H-NMR spectroscopy on paramagnetic complexes rather unalluring. However, the Co^{II} ion represents one notable exception since the spectra of its complexes, such as $[Co(tpy)_2]^{2+}$, appear well-resolved. This behavior is attributed to a hyperfine interaction between the ion's unpaired electrons and the ligand's protons; this coupling results in a considerable downfield shift of the signals – the so-called Knight's shift [175] – without affecting the signals' shapes [176]. According to the ligand-field theory, metals with d^6-electron configuration, such as the Co^{II} ions, might form high-spin as well as low-spin complexes (HS and LS, respectively). Constable *et al.* reported that one can distinguish between these two scenarios on the basis of the ^{1}H-NMR spectra (Figure 1.23): LS complexes, such as $[Co(tpy)_2]^{2+}$, exhibit their lowest resonance at *ca.* 100 ppm; whereas weak-field ligands give complexes with high-spin configuration and lower symmetry. For such complexes, the most downfield signal can be observed at *ca.* 255 ppm (*e.g.*, in $[Co(Br\text{-}tpy)_2]^{2+}$; Br-tpy: 4′-Br-tpy) [177, 178].

Beyond conventional ^{1}H-NMR spectroscopy, further versatile information can also be gained from NMR studies making use of various NMR-active nuclei that can be coordinated to the tpy ligand (*e.g.*, ^{7}Li, ^{31}P, ^{119}Sn, ^{195}Pt, ^{207}Pb, *etc.*). As listed in Figure 1.5, $Ru^{II/III}$ complexes of tpy ligands have been studied most intensively. Has $^{99/101}$Ru-NMR also been in this context? Despite its lower natural abundance and sensitivity, the ^{99}Ru nucleus is the more NMR-relevant one, due to the lower quadrupole moment (^{99}Ru, $I = 3/2$; ^{101}Ru: $I = 5/2$) [179, 180]. Elsevier *et al.* reported that the ^{99}Ru chemical shifts and signal line widths of complexes and organometallic

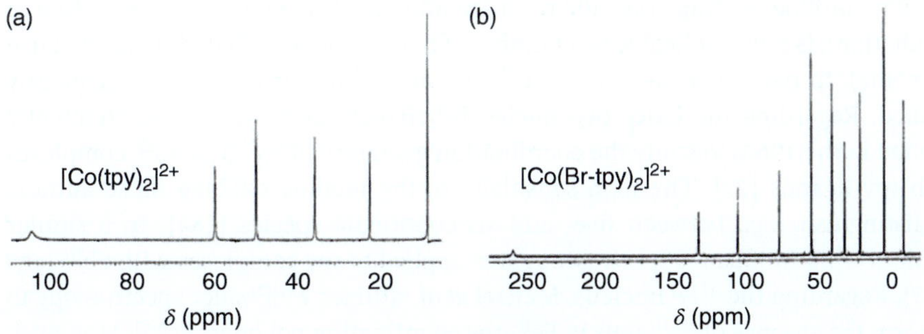

Figure 1.23 ^{1}H NMR spectra of $[Co(tpy)_2]^{2+}$ (a) and $[Co(Br\text{-}tpy)_2]^{2+}$ (b), which show the characteristic signal shifts for low-spin and high-spin complexes, respectively (spectra measured in CD_3CN at 250 MHz) [177] / with permission of Royal Society of Chemistry.

Table 1.2 ^{15}N-NMR chemical and coordination shifts of tpy and its homoleptic *bis*-complexes with various M^{II} ions (all values listed in ppm).

	$\delta_{N,N''}$	$\delta_{N'}$	$\Delta\delta_{N,N''}$ [c)]	$\Delta\delta_{N'}$ [c)]
tpy[a)]	−73.0	−82.8		
$[Fe(tpy)_2]^{2+}$[b)]	−134.6	−95.2	−61.6	−12.4
$[Ru(tpy)_2]^{2+}$[b)]	−140.1	−101.8	−67.1	−19.0
$[Os(tpy)_2]^{2+}$[b)]	−163.8	−117.3	−90.8	−34.5

a) Measured in d_6-DMSO.
b) Measured in D_2O.
c) $\Delta\delta$ represents the coordination shift, *i.e.*, the resonance difference between the respective complex and tpy.
[181] / with permission of John Wiley & Sons.

compounds strongly depended on the compounds' electronic and steric properties. Even though ^{99}Ru-NMR is suited for the analysis of Ru^{II} complexes, tpy complexes have not yet been studied with this technique. Attempts have also been made to study the interaction of tpy ligands with metal centers in solution by comparing the ^{15}N-NMR chemical shifts before and after coordination (the ^{15}N nucleus with a spin of ½ provides a chemical-shift range from 0 to 900 ppm) [181–183]. In one study, Pazderski *et al.* performed ^{15}N-NMR for the tpy *bis*-complexes of the divalent group-8 ions (Table 1.2) [181]. Complex formation resulted in a considerable shielding of the ligand's *N*-atoms. This effect was most pronounced for the *N*-atoms of the outer pyridine rings, whose ^{15}N resonances were shifted by 60–90 ppm; whereas the central pyridine moiety was less affected ($\Delta\delta$ of 10–35 ppm). In either case, the shielding effect increased in the order $Os^{II} > Ru^{II} > Fe^{II}$. However, this method is, due to the isotope's poor natural abundance (0.37% for ^{15}N), by far less sensitive when compared to other NMR experiments and, thus, hardly applicable as a routine experiment.

Finally, Mößbauer spectroscopy can be used to study the chemical environment of complexes. However, this spectroscopic technique, which is based on the so-called Mößbauer effect (*i.e.*, the resonant and recoil-free emission/absorption of γ-radiation), is only applicable to a number of nuclei (the so-called Mößbauer-active elements). In particular, the ^{57}Fe, ^{129}I, ^{119}Sn, and ^{121}Sb nuclei are most frequently studied. Regarding the latter two nuclei, Mößbauer spectroscopy was frequently applied in the 1960s to study the coordination geometry of Sn^{II} and Sn^{IV} complexes with tpy ligands [29]. The high sensitivity of the method enabled those authors to distinguish, *e.g.*, between five- and six-coordinate species [184]. In a similar fashion, ^{121}Sb Mößbauer spectroscopy was applied to tpy complexes with Sb^{III} ions [185]. Regarding the ^{57}Fe nucleus, Menzel *et al.* utilized Mößbauer spectroscopy to confirm the absence of Fe^{III} ions in Fe^{II}–tpy coordination polymers [186]. Moreover, this method was applied to determine the spin-states in homoleptic Fe^{II} and Fe^{III} complexes [187, 188]; Sarkar *et al.* noted that the spin-state of Fe^{II} *bis*-complexes plays a crucial role regarding their electrocatalytic performance in CO_2 reduction

[189]. Wieghard *et al.* used Mößbauer spectroscopy in combination with density functional theory (DFT) calculations to analyze the electronic structures of complexes within the electron-transfer series $[Fe(tpy)_2]^n$ (n = 3+, 2+, 1+, 0, 1−) [190]. This study corroborated the assumption that stepwise one-electron reductions down to the $[Fe(tpy)_2]^-$ species were all ligand-centered and did not affect the metal center in terms of charge and spin-state. This short excursion into the field of Mößbauer spectroscopy reveals that versatile information on tpy complexes can be gained from this method (*e.g.*, regarding spin-state or electronic structure); however, this technique is intrinsically limited to only a handful of nuclei.

1.5 Synthesis of Terpyridine Ligands – A Short Overview

As mentioned beforehand (Section 1.2), the parent tpy ligand was initially obtained from the oxidative dehydrogenation of pyridine in a disappointing yield of only *ca.* 4% [28]. Since then, numerous synthetic strategies have been followed to prepare tpy or its substituted derivatives. This topic was already covered in our previous books [144, 172] and has been summarized in various review articles [114, 191–193]. Therefore, we will discuss the synthesis aspects only briefly at this stage.

1.5.1 Synthesis of the Parent 2,2′:6′,2″-Terpyridine

The very low yield (*i.e.*, 55 g of tpy from a 15 kg batch of pyridine), which was encountered by Morgan and Burstall, called for significant improvement. However, Sasse's protocols to use Raney nickel, as the dehydrogenation catalyst, afforded a good selectivity for bpy, as the main product (tpy was only obtained in marginal amounts) [194, 195]. The same holds true for Burstall's attempts to cross-couple 2-bromopyridine and 6-bromo-bpy in an Ullmann-type fashion (here, bpy and qtpy were the main products) [54]. All these nondirected coupling reactions were inefficient with respect to tpy syntheses (yield <10%) and required enormous effort for product separation and purification [192]. This also holds true for the reaction of pyridine *N*-oxide with 3-methylpyrimidin-4(3*H*)-one in the presence of platinated Pd/C catalyst – this reaction also gave a complex product mixture, in which tpy was represented in *ca.* 13% yield.

Special credit goes to Kauffmann *et al.*, who presented the first applicable tpy synthesis [196]. The reaction between bpy and pyrid-2-yl lithium, as the nucleophile, at −40 °C gave tpy, as the exclusive product (40% yield). However, when the reaction was re-examined by Newkome *et al.* some years later, the chiral dimer **11**, which presumably resulted from an anionic [6 + 4]-cycloaddition, was observed as the by-product (Figure 1.24a). As one notable improvement to Kauffmann's protocol, Wakabayashi, Oae *et al.* introduced a sulfinyl-substituent in the 6-position of bpy, which could readily be replaced in an *ipso*-substitution (Figure 1.24a) [198].

With the emergence of the directed transition-metal catalyzed cross-coupling reactions, a substantial breakthrough in tpy synthesis was achieved [114]. To the best of our knowledge, the first such tpy synthesis *via* a Stille-type cross-coupling

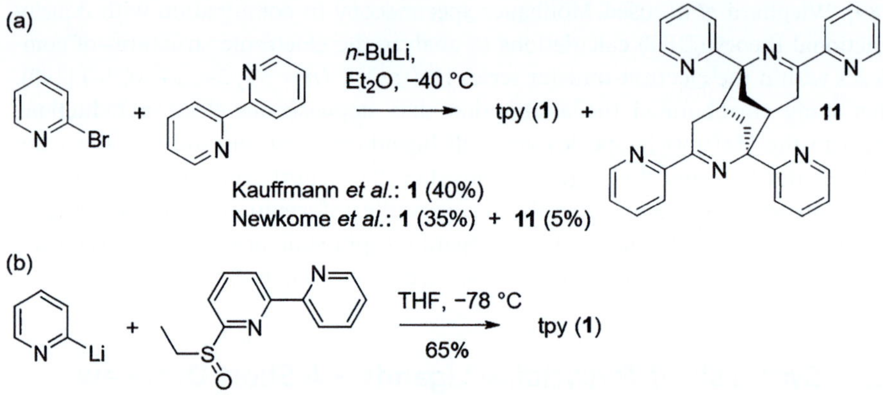

Figure 1.24 (a) Schematic representation of Kauffmann's tpy synthesis [196], which was revisited by Newkome *et al.* [197]. (b) Schematic representation of the tpy synthesis according to Wakabayashi, Oae *et al.* Adapted from [198].

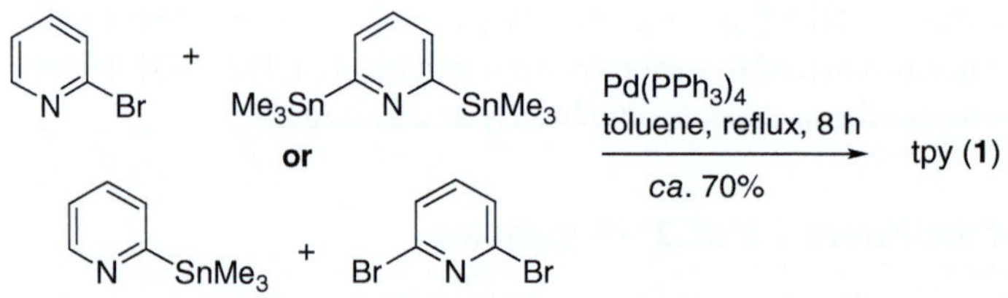

Figure 1.25 Schematic representation of Yamamoto's tpy synthesis. Adapted from [199].

was reported in 1996 by Yamamoto *et al.* (Figure 1.25) [199]. In order to overcome the intrinsic disadvantage of the Stille reaction, *i.e.*, the use of harmful organotin reagents, other Pd^0-catalyzed cross-coupling methods were also subsequently tested: Negishi [200], Suzuki-Miyaura [201], Hiyama [202], *etc.*

Besides the synthesis of tpy *via* coupling procedures, directed as well as nondirected ones, ring-assembly methods have also gained significant attention. As detailed in Section 1.5.3, pyridines can be prepared by condensation reactions, which use carbonyl derivatives as the starting materials. However, neither the original Chichibabin reaction nor the more general approaches according to Frank and Seven [203] or Kröhnke [204] are applicable for the synthesis of the parent molecule. Nonetheless, various ring-assembly methods have been published, in which at least one of the three pyridine rings is formed during the synthesis [114]. Firstly, Potts *et al.* disclosed that "ketene dithioacetals" represented versatile synthons for the synthesis of pyridine derivatives [205–207]. The optimized three-step synthesis afforded **1** in an overall yield of *ca.* 60% (Figure 1.26a). However, the last step, *i.e.*, the crucial borohydride reduction, is hardly applicable to a large-scale synthesis. A similar protocol was developed by Jameson and Guise, in which an enaminone served as the starting material (Figure 1.26b) [208]. The procedure is advantageous to Pott's one for several reasons: applicable on a multi-gram scale, avoidance of smelly sulfur compounds and absence of the uncomfortable reductive

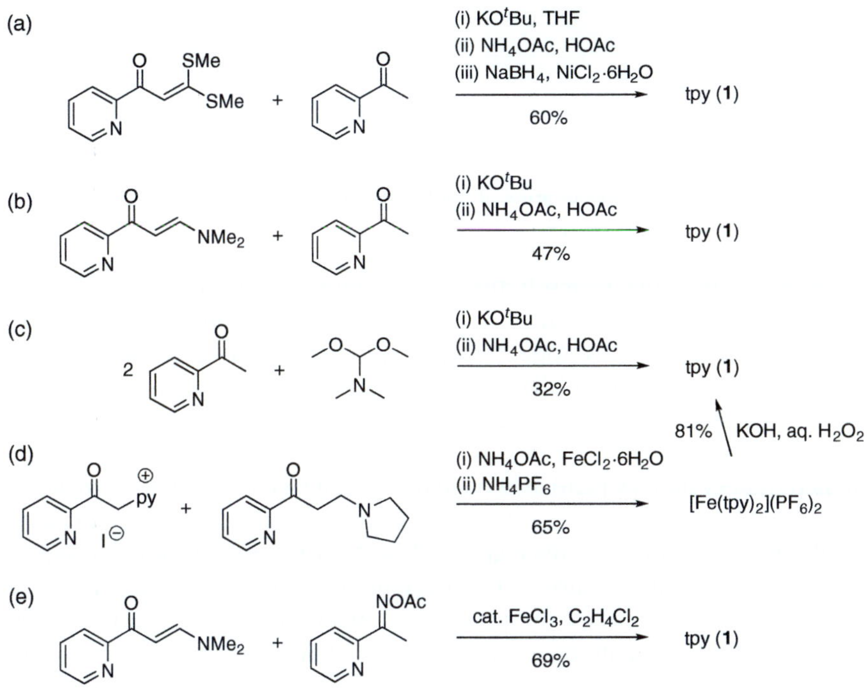

Figure 1.26 Schematic representation of various tpy syntheses, in which the central pyridine ring is assembled by condensation reactions. Adapted from [207–210].

elimination step. However, the reversibility of the initial Michael-addition step might give rise to the scrambling of substituents and, thus, by-products might be formed (*e.g.*, 2,2′:6′,4″-terpyridine) [211]. In 2006, Hanan *et al.* re-evaluated the aforementioned methods and established an easy-to-handle protocol to prepare **1** in a one-pot reaction under mild and benign conditions (Figure 1.26c) [209]. Constable *et al.* utilized a template synthesis to overcome one intrinsic limitation of the conventional Kröhnke-type synthesis, *i.e.*, the poor-yield synthesis of parent tpy (see Section 1.5.3). These authors used Fe^{II} ions to trap the *in situ* formed tpy ligand as its $[Fe(tpy)_2]^{2+}$ *bis*-complex; subsequently, tpy was liberated from the complex by oxidative hydrolysis (Figure 1.26d) [210]. This short overview on tpy synthesis does not claim to be exhaustive and research on pyridine synthesis is still ongoing. In this context, oxime acetates have emerged as versatile synthons for the transition-metal catalyzed formation of pyridine rings [212]. For example, Vidavalur *et al.* recently reported the Fe^{III}-catalyzed [3+4]-annulation between an oxime acetate and an enaminone, both prepared from 2-acetylpyridine; this highly modular protocol gave tpy in 69% yield (Figure 1.26e) [213]. Whereas the dehydrogenative coupling of pyridines in the presence of sodium dispersions, as the catalyst, afforded the bpy derivatives with high chemoselectivity (*i.e.*, the corresponding tpy derivatives were obtained in <10% yield) [214].

A fundamentally different strategy for the synthesis of **1** was reported by Pabst and Sauer (Figure 1.27) [215]. First, a 3,5-disubstituted 1,2,4-triazine derivative

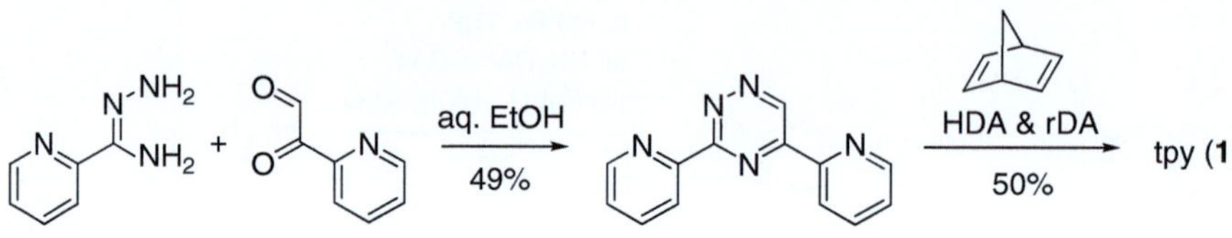

Figure 1.27 Schematic representation of the Papst-Sauer synthesis *via* a 1,2,4-triazine intermediate. Adapted from [215].

was obtained from the regioselective condensation of picolinamidrazone with 2-pyridylglyoxal. This intermediate, as a highly electron-deficient diene, reacted in a sequence of a hetero-Diels–Alder (HDA) and retro-Diels–Alder (rDA) reaction to afford **1** in 50% yield (norbornadiene was used as a masked acetylene equivalent).

Finally, tpy synthesis has also been targeted *via* the ring-assembly of both outer rings. In one example, Adrian *et al.* converted 2,6-diacetylpyridine by a sophisticated four-step strategy into tpy [216]. Although the overall yield of *ca.* 70% was relatively high, Adrian's protocol misses the straightforwardness, cost-efficiency, and experimental ease of the aforementioned approaches. Alternative strategies, also starting from 2,6-diacetylpyridine, relied on Chichibabin- or Kröhnke-type ring-closure (see Section 1.5.2). However, these strategies were only useful for the synthesis of tpy derivatives with substituents on the outer rings, but failed for the synthesis of parent **1** [217].

These various methods for the synthesis of unsubstituted tpy have also been applied to prepare a plethora of ring-substituted derivatives; however, a detailed listing of all the chemical structures, the wide range of reported reaction conditions, and discussion of practical aspects and yields goes far beyond the scope of this book. Basically, all the protocols rely on the fundamental concept mentioned in this section beforehand.

1.5.2 Synthesis and Modification of 4′-Chloro-2,2′:6′,2″-Terpyridine

Besides the parent tpy ligand, its 4′-chloro derivative (**12**, Cl-tpy) has also received widespread attention – this commercially available ligand is mentioned in >450 journal articles and >110 patents (SciFinder^{n} research on 1st April 2025). The ligand's prominent role results from the possibility to replace the chloro-substituent *via* an aromatic substitution reaction (S_NAr). For this purpose, a wide range of nucleophiles has already been used (*e.g.*, alcohols, amines, thiols, phosphines, carbanions, *etc.*) [218]. The synthesis of **12**, which was reported by Constable and Ward, involves the initial formation of a 1,3,5-triketone derivative, which is cyclized to a dihydropyridin-4(1*H*)-one species using NH_4OAc, as the source for NH_3; subsequent chlorination with PCl_5 in $POCl_3$ afforded **12** in 62% yield (Figure 1.28) [219]. This protocol from the early 1990s still represents the method of choice for the preparation of this particular compound.

As shown in numerous publications, the nucleophilic displacement of the chloro-substituents enables one to prepare 4′-substituted tpy derivatives. This strategy has *inter alia* been employed in the field of macromolecular chemistry:

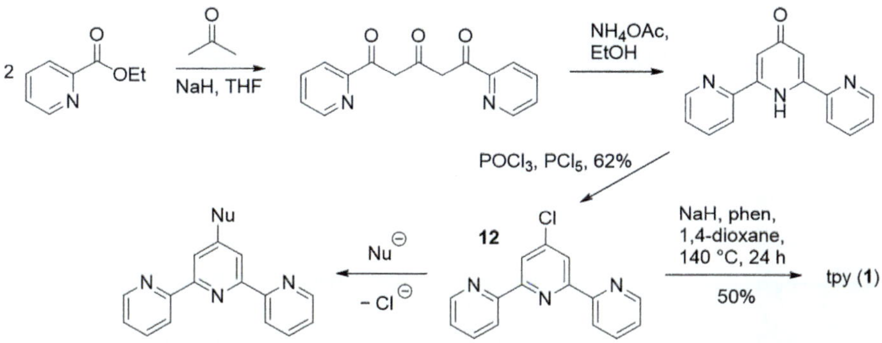

Figure 1.28 Schematic representation of Constable's synthesis of Cl-tpy (**12**). Adapted from [219] and its derivatization *via* nucleophilic or radical nucleophilic aromatic substitution.

The chain-end functionalization of preformed polymers paved the way toward metallo-supramolecular block copolymer architectures [220]. Moreover, polymers with tpy moieties within the side chains have been prepared from tpy-containing monomers; these, in turn, were synthesized from **12** according to Figure 1.28 [144, 172]. Such polymers are applied, for example, in the context of self-healing and shape-memory polymers, which are discussed in Chapter 9 [15, 143].

Recently, Studer *et al.* reported that aryl bromides and chlorides can be hydrodehalogenated by NaH in 1,4-dioxane using phen as the initiator for this radical nucleophilic aromatic substitution ($S_{RN}1$) [221]. In this context, also the reduction of **12** to afford tpy (**1**) in *ca.* 50% yield was shown (Figure 1.28). This $S_{NR}1$ methodology represents a notable extension of the preparative toolbox to synthesize tpy derivatives with unsubstituted central rings.

1.5.3 Synthesis of 4′-Aryl-2,2′:6′,2″-Terpyridines

The thermal cyclo-condensation of aldehydes and ammonia in the presence of an appropriate catalyst (*e.g.*, alumina) affords pyridine derivatives. Although this observation was already made by Lyubavin in 1873 [222, 223], this pyridine synthesis is today named after Chichibabin, who was the first to thoroughly analyze its outcome [224, 225]. Almost 20 years later, Chichibabin republished his results also in German [226]; this disclosure is commonly cited instead of the original paper from 1905, which is only available in Russian [227]. The primary protocol was revised in 1949 by Frank and Seven, who reacted an enolizable ketone with a non-enolizable aldehyde in the presence of a source for ammonia (*e.g.*, NH_4OAc). The reactions are typically performed in a protic environment (*e.g.*, acetic acid, ethanol, or acetamide). As outlined in Figure 1.29, a 1,5-diketone intermediate ("chalcone") is formed *via* a sequence of Claisen-Schmidt condensation and Michael addition. The subsequent cyclization with an NH_3 equivalent affords a 2,4,6-trisubsituted 1,4-dihydropyridine (DHP) derivative. This, in turn, is oxidized to the corresponding pyridine compound by air or by adding an external oxidant (the stability of the DHP strongly depends on the electronic nature of its substituents). As recently summarized by Shabalin,

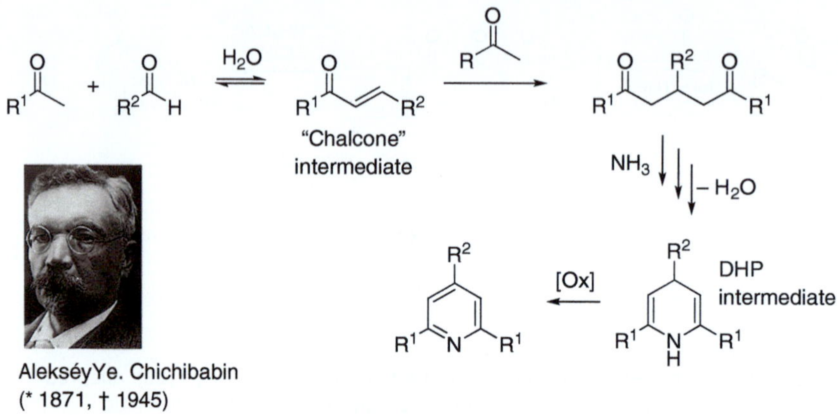

Figure 1.29 Schematic representation of the mechanism of Chichibabin's pyridine synthesis (R^1, R^2 = alkyl or aryl). [228] with permission of Royal Society of Chemistry. A photograph of A. Ye. Chichibabin is also shown [227] / with permission of John Wiley & Sons.

numerous variants to the original protocol are available; these include solvent-free or microwave-assisted reactions, the use of homogeneous or heterogeneous catalysts, *etc.* [228]. Furthermore, benzyl amines or oxime acetates have been utilized as alternative starting materials to the conventional carbonyl derivatives.

In order to circumvent one major limitation of the Chichibabin-type pyridine synthesis, which is the ability of the chalcone intermediate to act also as an internal oxidant (thereby significantly reducing the yield), Kröhnke suggested the use of a CH-acidic pyridinium salt in combination with a chalcone derivative (Figure 1.30) [204]. The Michael reaction of these two components affords an unsaturated 1,3-diketone, which, upon cyclization with NH_3, directly yields the corresponding pyridine (instead of the DHP). Thus, the crucial oxidation step becomes obsolete. Moreover, the use of preformed chalcones enabled this author to also synthesize unsymmetrical pyridine derivatives (*i.e.*, different substituents in 2,6-positions). The acylmethylpyridinium salts, such as 2-pyridacylpyridinium iodide, were obtained *via* an Ortoleva–King reaction of the respective ketone and iodine in pyridine [230–232]. In one application of this method, Kröhnke prepared 4′-aryl-substituted 2,2′:6′,2″-terpyridines, which are often referred to as the Kröhnke-type tpys; thus ascribing credit to his notable contributions to the field of pyridine synthesis [204].

Today, Kröhnke-type terpyridines are conveniently prepared by the condensation between 2-acetylpyridine and an aromatic aldehyde to directly (or sequentially *via* the chalcone intermediate) afford the respective 4-aryl-1,5-di(pyridine-2-yl)pentane-1,5-dione [233]. In this context, Raston *et al.* developed a solvent-free method, in which the 1,5-diketone was prepared by grinding in a mortar [234, 235]. Whereas various one-pot protocols were reported, in which ethanol or poly(ethylene glycol) (PEG) in combination with aqueous ammonia was used as the reaction medium [209, 236, 237]. Moreover, Tu *et al.* performed such a synthesis under microwave irradiation [238]. To rectify a common misunderstanding: Although Kröhnke-type

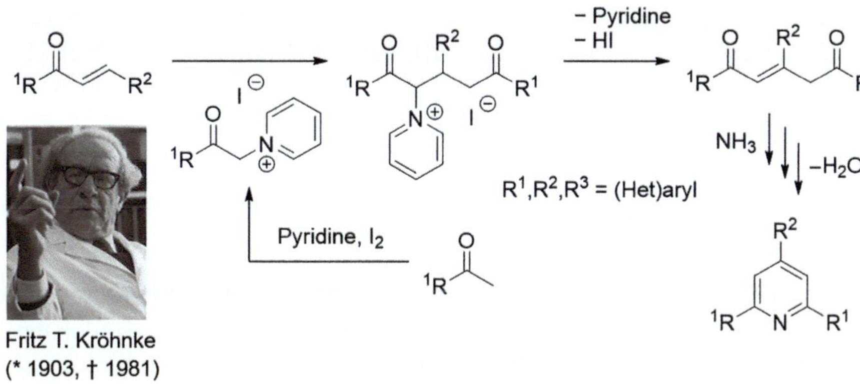

Figure 1.30 Schematic representation of the mechanism of Kröhnke's pyridine synthesis including the formation of the pyridinium-salt reactant (R^1, R^2 = alkyl or aryl) [204]. A photograph of F. T. Kröhnke is also shown [229] / with permission of John Wiley & Sons.

tpys are obtained *via* the aforementioned methods, these reactions are strictly speaking rather Chichibabin-type ones, but not Kröhnke-type (the latter requires the pyridinium-salt reagent).

However, one has to be aware that the tpy synthesis still has some shortcomings, which are due to the formation of by-products [233]. Already in the 1890s, von Kostanecki and Rossbach proposed the formation of a condensation product with a ketone-to-aldehyde stoichiometry of 3:2 (Figure 1.31a) [239]; surprisingly, the proof of structure for this triketone was provided only recently [240]. Such ketones often undergo an intramolecular aldol addition to afford penta-substituted cyclohexanol derivatives, as a diastereoisomeric mixture (Figure 1.31b); in some cases, these have been isolated even as the major products [233]. In addition to this, cyclohexane-1,3-diols might arise from the condensation of aromatic ketones and aldehydes in a 3:1 ratio and subsequent aldol-type ring closure (Figure 1.31c) [241]. Moreover, 6′-aryl-2,2′:4′,2″-terpyridines are formed if the chalcone intermediate

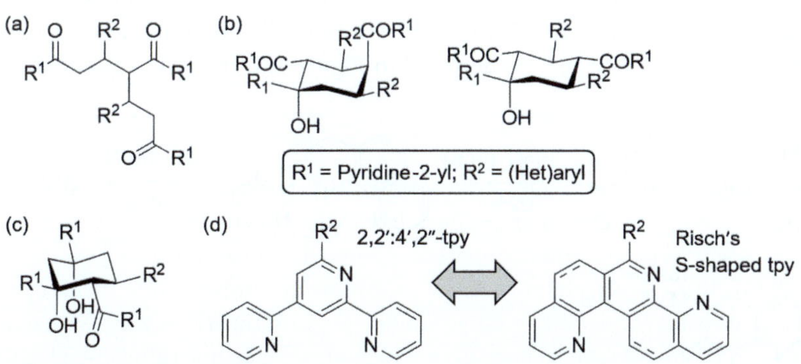

Figure 1.31 If tpy synthesis goes wrong ... Schematic representation of typical by-products of Chichibabin-type tpy synthesis: (a) Kostanecki's ketone, (b) cyclohexanols, (c) cyclohexandiols, and (d) 2,2′:4′,2″-terpyridines. Adapted from [233].

reacts with 2-acetylpyridine in an aldol-type fashion (instead of the usual Michael addition), followed by cyclization with an NH_3-equivalent (Figure 1.31d) [242]. The latter correspond to the so-called S-shaped tpys, which were introduced by Risch *et al.* [243–245].

1.6 Beyond the Classical Terpyridine Motif

As mentioned beforehand, the ligand's architecture has a profound impact on the complex formation as well as on its properties. In this respect, preorganization represents one important aspect. The DPA ligand, for example, represents tpy's dibenzannelated analogue (Figure 1.15, Table 1.1). Due to its rigid structure, this ligand formed complexes with higher K_a values (at least in those cases, where the size of the metal ion fitted into the ligand's rigid cleft) [130]. DPA is conveniently prepared *via* Friedländer condensation and subsequent dehydrogenation of intermediate **12** (Figure 1.32a) [246]. Thummel *et al.* also prepared ligand **13** as the tetrahydro derivative of DPA *via* an enamine route [247]. The same strategy was chosen for the synthesis of the substituted derivatives **14**, which are not accessible *via* conventional Chichibabin-type or Kröhnke-type reactions [248]. An alternative, highly modular procedure for the synthesis of **14** was developed by Risch *et al.* (Figure 1.32b); these authors used *in situ*-generated imines and preformed ternary iminium salts as the reactants in their one-pot, two-step protocol [244]. In none of these studies, K_a values for the complexation of metal ions were determined in order to elaborate the effect of rigidity.

The synthesis of tpy complexes typically proceeds in a straightforward fashion and affords the assemblies with mostly octahedral or square-planar geometry (Section 1.3). However, the ligand structure prevents the complexes from taking up the ideal geometry and, thus, considerable distortion occurs. This issue, which has been intensively discussed at the example to the Ru^{II} complexes, is highly detrimental regarding optoelectronic applications of such complexes [249]. Unlike $[Ru(bpy)_3]^{2+}$, whose *tris*-bidentate coordination environment allows for a non-distorted octahedral geometry, $[Ru(tpy)_2]^{2+}$ is practically non-emissive at room temperature (*i.e.*, 1150 *vs.* 0.25 ns). As depicted in Figure 1.33, emission quenching

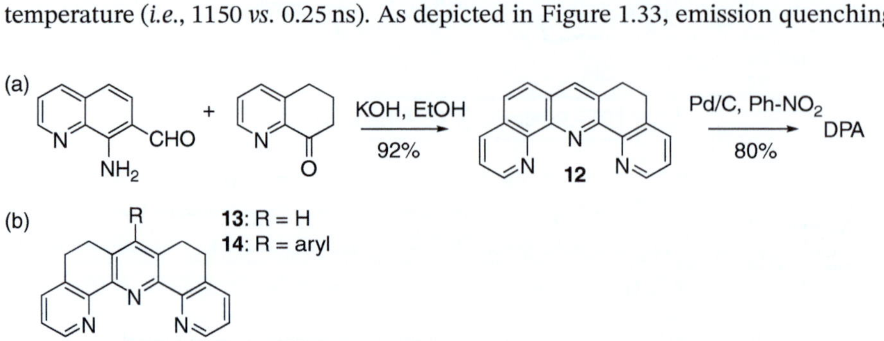

Figure 1.32 (a) Schematic representation of Thummel's DPA synthesis. Adapted from [246]. (b) Schematic representation of the semirigid ligands **13** and **14** [247, 248].

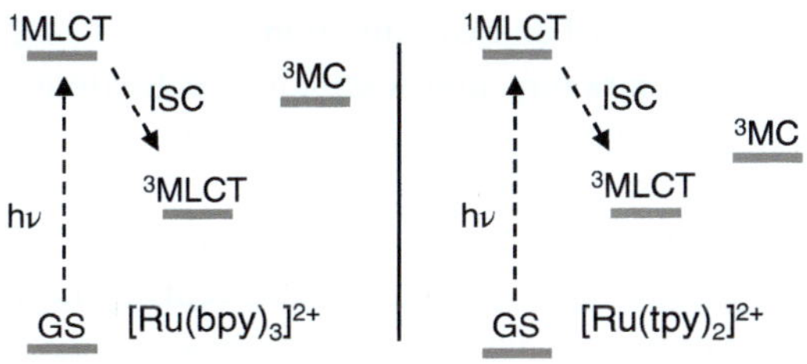

Figure 1.33 Comparison of the Jablonski diagrams of photophysically attractive $[Ru(bpy)_3]^{2+}$ (left) *vs.* the easily accessible $[Ru(tpy)_2]^{2+}$; only a simplified version of the diagram is shown for clarity [249] / with permission of Royal Society of Chemistry.

is due to the ligand's weak ligand-field strength, which decreases the energy of the metal-centered triplet state (^{3}MC). As a result, the energy gap between the metal-to-ligand charge-transfer triplet state (^{3}MLCT) and ^{3}MC becomes smaller, and the latter can be thermally populated; this, in turn, facilitates the non-radiative decay back to the ground state (GS).

To counteract this, some fundamentally different approaches have been followed: Introduction of electron-withdrawing or -donating substituents, increasing the ligand's planarity, increasing the ligand-field strength and decreasing the ligand's steric strain. The latter strategy affords complexes with almost ideal octahedral (or square-planar) geometry and, thus, significantly improved photophysical properties. Three representative examples for tridentate ligands, as "tpy mimics," which fulfil this requirement, are shown in Figure 1.34a. From these, the dqp ligand has attracted the most notable attention so far: Its modular synthesis also allows one to introduce substituents on the central ring [255] and, thus, the dqp moiety has already been incorporated into various architectures, including

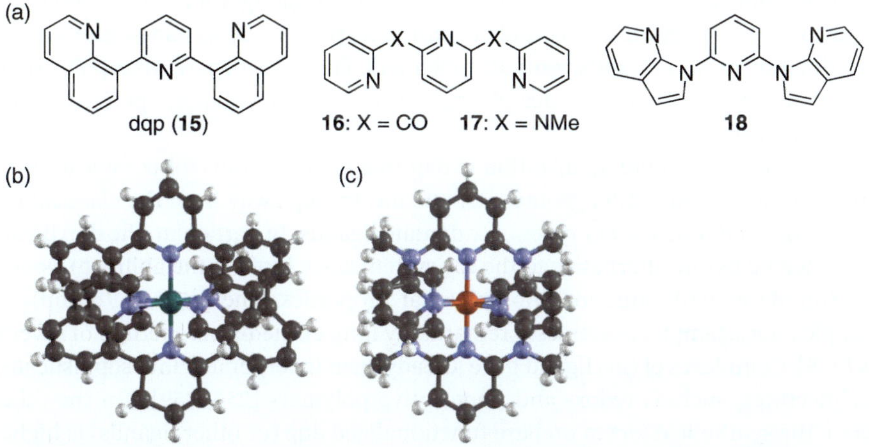

Figure 1.34 (a) Schematic representation of tridentate ligands **15–18** [250–253]. (b) Solid-state structure of *mer*-$[Ru(dqp)_2]^{2+}$. Adapted from [250]. (c) Solid-state structure of *mer*-$[Cr(\mathbf{17})_2]^{3+}$. Adapted from [254]. In panels (b) and (c), counterions and solvent molecules were omitted for clarity.

metallo-supramolecular block copolymers [256] and coordination polymers [257]. The $[Ru(dqp)_2]^{2+}$ complex exhibits a remarkably long excited-state lifetime of *ca.* 3 μs and good photo-stability [258]. However, when dealing with such more flexible ligands, one has to take the formation of coordination isomers into account; unlike for tpy, *bis*-tridentate coordination might lead to the *mer*-, *cis-fac* as well as *trans-fac* species (Figure 1.7). In general, the meridional one represents the thermodynamically favored isomer. Representatively, the solid-state structures of *mer*-$[Ru(dqp)_2]^{2+}$ and *mer*-$[Cr(\mathbf{17})_2]^{3+}$ are shown in Figure 1.34b,c [250, 254].

1.7 Conclusion

2,2′:6′,2″-Terpyridine was incidentally discovered in the 1930s, and its ability to form complexes with transition-metal ions was discovered right away [28]. Thus, the pioneering work by Morgan and Burstall paved the way to the wide field of tpy chemistry, which started to flourish in the 1980s when the photophysical and electrochemical properties of tpy complexes were discovered (Figure 1.1). Today, a wide range of strategies for the synthesis of tpy ligands is known, and tpy ligands have been combined with almost all elements in the periodic table (Figure 1.5). The coordination behavior of tpy ligands is widely understood: Complexes with octahedral or square-planar geometry represent the most prevalent ones (but many other coordination architectures are known). The stability of such complexes depends on the ligand's structure as well as on the coordinated metal ion (*i.e.*, its ionic radius, oxidation state, *etc.*). Regarding the characterization of tpy complexes, XRD analysis is favored, but only applicable if suited single crystals can be obtained. Besides, NMR spectroscopy and MS are commonly used.

As discussed in the subsequent chapters, tpy complexes have been used to assemble a plethora of novel and interesting compounds – ranging from discrete mononuclear complexes to dyads, triads, and oligonuclear species, 2D/3D nanoarchitectures, coordination networks, and eventually polymers. The range of potential applications is as wide and *inter alia* includes pharmaceuticals, photovoltaics, optoelectronics, materials science, *etc.*

However, if ligand preorganization is required or if one aims for complexes with non-distorted coordination geometries, one has to step away from the classical tpy framework and utilize other types of tridentate ligands. In particular, the dqp ligand has emerged as one alternative to the tpy system since it offers the ability to assemble complexes with superior photophysical properties. The parent $[Ru(dqp)_2]^{2+}$ complex, for example, possesses a remarkably long excited-state lifetime of several μs [258]. Complexes of this ligand have already been incorporated into sophisticated architectures, such as redox- and photoactive polymers [256, 259]. On the other hand, the synthetic effort to prepare functionalized dqp (or other ligands) is higher when compared to a related tpy synthesis. Thus, ligands from the tpy family still represent the more favorable ligands when aiming for the self-assembly of metallo-supramolecular materials – as long as photophysical aspects play only a minor role.

References

1 D. J. Cram, *Angew. Chem. Int. Ed.* 1988, 27, 1009–1020.
2 J.-M. Lehn, *Angew. Chem. Int. Ed.* 1988, 27, 89–112.
3 C. J. Pedersen, *Angew. Chem. Int. Ed.* 1988, 27, 1021–1027.
4 J.-M. Lehn, *Science* 1993, 260, 1762–1763.
5 U. S. Schubert, G. R. Newkome, A. Winter, *Supramolecular Polymers and Assemblies*, Wiley-VCH, Weinheim, 2021.
6 J. W. Steed, J. L. Atwood, *Supramolecular Chemistry*, John Wiley & Sons, Hoboken & Chichester, 2022.
7 B. L. Feringa, *Angew. Chem. Int. Ed.* 2017, 56, 11060–11078.
8 J.-P. Sauvage, *Angew. Chem. Int. Ed.* 2017, 56, 11080–11093.
9 J. F. Stoddart, *Angew. Chem. Int. Ed.* 2017, 56, 11094–11125.
10 V. A. Friese, D. G. Kurth, *Curr. Opin. Colloid Interface Sci.* 2009, 14, 81–93.
11 A. Winter, U. S. Schubert, *Chem. Soc. Rev.* 2016, 45, 5311–5357.
12 Y. Zhu, W. Zheng, W. Wang, H.-B. Yang, *Chem. Soc. Rev.* 2021, 50, 7395–7417.
13 D. G. Kurth, *Sci. Technol. Adv. Mater.* 2008, 9, 014103.
14 J.-C. Eloi, L. Chabanne, G. R. Whittell, I. Manners, *Mater. Today* 2008, 11, 28–36.
15 S. Götz, S. Zechel, M. D. Hager, G. R. Newkome, U. S. Schubert, *Prog. Polym. Sci.* 2021, 119, 101428.
16 G. R. Whittell, M. D. Hager, U. S. Schubert, I. Manners, *Nature Mater.* 2011, 10, 176–188.
17 F. Zeng, S. C. Zimmerman, *Chem. Rev.* 1997, 97, 1681–1712.
18 D. Astruc, E. Boisselier, C. Ornelas, *Chem. Rev.* 2010, 110, 1857–1959.
19 J. M. J. Fréchet, *Proc. Natl. Acad. Sci. U.S.A.* 2002, 99, 4782–4787.
20 G. R. Newkome, C. Shreiner, *Chem. Rev.* 2010, 110, 6338–6442.
21 J. E. M. Lewis, J. D. Crowley, *ChemPlusChem* 2020, 85, 815–827.
22 S. J. Dalgarno, N. P. Power, J. L. Atwood, *Coord. Chem. Rev.* 2008, 252, 825–841.
23 X. Yu, C. Guo, S. Lu, Z. Chen, H. Wang, X. Li, *Macromol. Rapid Commun.* 2022, 43, 2200004.
24 S. M. Elahi, M. Raizada, P. K. Sahu, S. Konar, *Chem. Eur. J.* 2021, 27, 5858–5870.
25 C. Wei, Y. He, X. Shi, Z. Song, *Coord. Chem. Rev.* 2019, 385, 1–19.
26 R. Sakamoto, K. Takada, X. Sun, T. Pal, T. Tsukamoto, E. J. H. Phua, A. Rapakousiou, K. Hoshiko, H. Nishihara, *Coord. Chem. Rev.* 2016, 320-321, 118–128.
27 H. Maeda, K. Takada, N. Fukui, S. Nagashima, H. Nishihara, *Coord. Chem. Rev.* 2022, 470, 214693.
28 G. T. Morgan, F. H. Burstall, *J. Chem. Soc. Res.* 1932, 20–30.
29 A. Winter, G. R. Newkome, U. S. Schubert, *Inorg. Chem. Front.* 2024, 11, 342–399.
30 D. Saccone, C. Magistris, N. Barbero, P. Quagliotto, C. Barolo, G. Viscardi, *Materials* 2016, 9, 137.
31 P. Liu, G. Shi, X. Chen, *Front. Chem.* 2020, 8, 592055.

32 K. M.-C. Wong, V. W.-W. Yam, *Coord. Chem. Rev.* 2007, 251, 2477–2488.
33 J.-P. Sauvage, J.-P. Collin, J.-C. Chambron, S. Guillerez, C. Coudret, V. Balzani, F. Barigelletti, L. De Cola, L. Flamigni, *Chem. Rev.* 1994, 94, 993–1019.
34 R. D. Costa, E. Ortí, H. J. Bolink, F. Monti, G. Accorsi, N. Armaroli, *Angew. Chem. Int. Ed.* 2012, 51, 8178–8211.
35 R. Sakamoto, K.-H. Wu, R. Matsuoka, H. Maeda, H. Nishihara, *Chemical Society Reviews* 2015, 44, 7698–7714.
36 S. Liu, P. Zhang, J. Fu, C. Wei, G. Cai, *Front. Energy Res.* 2021, 9, 620203.
37 M. Higuchi, *Polym. J.* 2009, 41, 511–520.
38 X. Yu, Y. Hu, C. Guo, Z. Chen, H. Wang, X. Li, *Supramol. Mater.* 2022, 1, 100017.
39 I. Eryazici, C. N. Moorefield, G. R. Newkome, *Chem. Rev.* 2008, 108, 1834–1895.
40 R. Musiol, P. Malecki, M. Pacholczyk, J. Mularski, *Expert Opin. Drug Discov.* 2022, 17, 259–271.
41 R. Abhijnakrishna, K. Magesh, A. Ayushi, S. Velmathi, *Coord. Chem. Rev.* 2023, 496, 215380.
42 A. Winter, M. Gottschaldt, G. R. Newkome, U. S. Schubert, *Curr. Top. Med. Chem.* 2012, 12, 158–175.
43 A. Winter, U. S. Schubert, *ChemCatChem* 2020, 12, 2890–2941.
44 A. Winter, G. R. Newkome, U. S. Schubert, *ChemCatChem* 2011, 3, 1384–1406.
45 Grace's Guited Ltd., Exeter, UK. *National Physical Laboratory*, https://www.gracesguide.co.uk/National_Physical_Laboratory (last accessed: 2025-04-01).
46 J. C. Irvine, *Obit. Notes Fellows R. Soc.* 1941, 3, 355–362.
47 E. A. Coulson, *Nature* 1956, 177, 210–211.
48 G. T. Morgan, *Chem. Ind. (London)* 1939, 665–673.
49 n/a, *Science* 1951, 113, 261–263.
50 C. E. Housecroft, E. C. Constable, *Molecules* 2021, 26, 3110.
51 L. Payen, L. Kletsch, T. Lapić, M. Wickleder, A. Klein, *Inorganics* 2023, 11, 174.
52 G. T. Morgan, F. H. Burstall, *J. Chem. Soc. Res.* 1934, 1498–1500.
53 G. Morgan, F. H. Burstall, *J. Chem. Soc. Res.* 1937, 1649–1655.
54 F. H. Burstall, *J. Chem. Soc. Res.* 1938, 1662–1672.
55 G. T. Morgan, F. H. Burstall, *Inorganic Chemistry: A Survey of Modern Developments*, W. Heffer & Sons, Cambridge, 1936.
56 W. W. Brandt, J. P. Wright, *J. Am. Chem. Soc.* 1954, 76, 3082–3083.
57 B. O. West, *J. Chem. Soc. Res.* 1954, 578–579.
58 F. Basolo, H. B. Gray, R. G. Pearson, *J. Am. Chem. Soc.* 1960, 82, 4200–4203.
59 F. P. Dwyer, E. C. Gyarfas, *J. Am. Chem. Soc.* 1954, 76, 6320–6321.
60 J. H. Baxendale, *Chem. Soc. Spec. Publ.* 1954, 40–46.
61 K. Nakamoto, *J. Phys. Chem.* 1960, 64, 1420–1425.
62 R. T. Pflaum, W. W. Brandt, *J. Am. Chem. Soc.* 1955, 77, 2019–2022.
63 G. J. Sutton, J. Aust, *Sci. Res., B: Biol. Sci.* 1951, 4, 651–654.
64 J. Leiter, J. L. Hartwell, J. S. Kahler, I. Kline, M. J. Shear, *J. Natl. Cancer Inst.* 1953, 14, 365–374.
65 F. Hein, W. Retter, *Ber. Dtsch. Chem. Ges.* 1928, 61, 1790–1791.
66 E. C. Constable, C. E. Housecroft, *Molecules* 2019, 24, 3951.

67 M. Pirani, *Sci. Nat.* 1930, 18, 97–101.
68 F. Blau, *Ber. Dtsch. Chem. Ges.* 1888, 21, 1077–1078.
69 Ceolas.net, P. Thornes. *100 Jahre OSRAM – Licht hat einen Namen*, http://ceolas.net/Docs/Osram_Geschichte.pdf (last accessed: 2025-04-01).
70 F. Blau, *Monatsh. Chem.* 1889, 10, 375–388.
71 H. Weidel, *Ber. Dtsch. Chem. Ges.* 1879, 12, 1989–2012.
72 A. Werner, Z. Anorg, *Chem.* 1893, 3, 267–330.
73 F. Blau, *Monatsh. Chem.* 1898, 19, 647–689.
74 F. Blau, Patent US 674754, 1901.
75 Y.-F. Shi, Z.-X. Yang, S. Ma, P.-L. Kang, C. Shang, P. Hu, Z.-P. Liu, *Engineering* 2023, 27, 70–83.
76 J. A. Keith, V. Vassilev-Galindo, B. Cheng, S. Chmiela, M. Gastegger, K.-R. Müller, A. Tkatchenko, *Chem. Rev.* 2021, 121, 9816–9872.
77 Z. Tu, T. Stuyver, C. W. Coley, *Chem. Sci.* 2023, 14, 226–244.
78 L. Yang, Q. Guo, L. Zhang, *Chem. Commun.* 2024, 60, 6977–6987.
79 M. Trobe, M. D. Burke, *Angew. Chem. Int. Ed.* 2018, 57, 4192–4214.
80 S. A. Biyani, Y. W. Moriuchi, D. H. Thompson, *Chem. Methods* 2021, 1, 323–339.
81 L. E. Horsburgh, A. P. Monkman, I. D. W. Samuel, *Synth Met.* 1999, 101, 113–114.
82 L. L. Patera, J. Amler, J. Repp, *Chemistry* 2022, 4, 112–117.
83 E. C. Constable, C. E. Housecroft, *Coord. Chem. Rev.* 2017, 350, 84–104.
84 E. C. Constable, C. E. Housecroft, *Aust. J. Chem.* 2020, 73, 390–398.
85 E. C. Constable, in *Advances in Inorganic Chemistry*, Vol. 30, (Ed.: H. J. Emeléus), Academic Press, London, 1986, pp. 69–121.
86 H. A. Favre, W. H. Powell, *Nomenclature of Organic Chemistry: IUPAC Recommendations and Preferred Names 2013*, RSC Publishing, Cambridge, 2014, pp. 1–129.
87 J. Rigaudy, S. P. Klesney, Eds, *Nomenclature of Organic Chemistry, IUPAC Recommendations and Preferred Names*, Pergamon Press, Oxford, 1979.
88 C. E. Housecroft, *Dalton Trans.* 2014, 43, 6594–6604.
89 C. E. Housecroft, *CrystEngComm* 2015, 17, 7461–7468.
90 C. E. Housecroft, E. C. Constable, *J. Inorg. Organomet. Polym. Mater.* 2018, 28, 414–427.
91 C. E. Housecroft, E. C. Constable, *Chimia* 2019, 73, 462.
92 Z. Kolarik, *Chem. Rev.* 2008, 108, 4208–4252.
93 D. Magnusson, B. Christiansen, M. R. S. Foreman Jr., A. Geist, J.-P. Glatz, R. Malmbeck, G. Modolo, D. Serrano-Purroy, C. Sorel, *Solvent Extr. Ion Exch.* 2009, 27, 97–106.
94 S. S. Chitnis, J. H. W. LaFortune, H. Cummings, L. L. Liu, R. Andrews, D. W. Stephan, *Organometallics* 2018, 37, 4540–4544.
95 S. S. Chitnis, F. Krischer, D. W. Stephan, *Chem. Eur. J.* 2018, 24, 6543–6546.
96 R. J. Andrews, S. S. Chitnis, D. W. Stephan, *Chem. Commun.* 2019, 55, 5599–5602.

97 A. K. Adcock, R. L. Ayscue, L. M. Breuer, C. P. Verwiel, A. C. Marwitz, J. A. Bertke, V. Vallet, F. Réal, K. E. Knope, *Dalton Trans.* 2020, 49, 11756–11771.
98 H. Hofmeier, E. Herdtweck, U. S. Schubert, Z. Anorg, *Allg. Chem.* 2004, 630, 683–688.
99 R. L. Ayscue III, V. Vallet, J. A. Bertke, F. Réal, K. E. Knope, *Inorg. Chem.* 2021, 60, 9727–9744.
100 B. Weber, *Coordination Chemistry: Basics and Current Trends*, Springer, Berlin & Heidelberg, 2023, pp. 47–72.
101 N. Pantalon Juraj, M. Krklec, T. Novosel, B. Perić, R. Vianello, S. Raić-Malić, S. I. Kirin, *Dalton Trans.* 2020, 49, 9002–9015.
102 G. H. Frost, F. A. Hart, C. E. Heath, M. B. Hursthouse, *J. Chem. Soc. D: Chem. Commun.* 1969, 1421–1422.
103 L. I. Semenova, A. N. Sobolev, B. W. Skelton, A. H. White, *Aust. J. Chem.* 1999, 52, 519–530.
104 D. Venkataraman, Y. Du, S. R. Wilson, K. A. Hirsch, P. Zhang, J. S. Moore, *J. Chem. Educ.* 1997, 74, 915.
105 The Cambridge Crystallographic Data Centre (CCDC), Cambridge, UK. *CCDC*, https://www.ccdc.cam.ac.uk/ (last accessed: 2025-04-01).
106 M. Grätzel, *J. Photochem. Photobiol. C: Photochem. Rev.* 2003, 4, 145–153.
107 J. J. Concepcion, J. W. Jurss, J. L. Templeton, T. J. Meyer, *J. Am. Chem. Soc.* 2008, 130, 16462–16463.
108 K. W. Jennette, S. J. Lippard, G. A. Vassiliiades, W. R. Bauer, *Proc. Natl. Acad. Sci. U.S.A.* 1974, 71, 3839–3843.
109 B. W. J. Harper, M. van Holst, J. Aldrich-Wright, in *Metallointercalators: Synthesis and Techniques to Probe Their Interactions with Biomolecules*, (Ed.: J. Aldrich-Wright), Springer, Vienna, 2011, pp. 101–128.
110 N. G. Connelly, T. Damhus, R. M. Hartshorn, A. T. Hutton, *Chem. Int.* 2005, 27, 25–26.
111 A. G. Blackman, *C. R. Chim.* 2005, 8, 107–119.
112 E. R. Civitello, P. S. Dragovich, T. B. Karpishin, S. G. Novick, G. Bierach, J. F. O'Connell, T. D. Westmoreland, *Inorg. Chem.* 1993, 32, 237–241.
113 K. Peng, A. Friedrich, U. Schatzschneider, *Chem. Commun.* 2019, 55, 8142–8145.
114 O. S. Taniya, D. S. Kopchuk, A. F. Khasanov, I. S. Kovalev, S. Santra, G. V. Zyryanov, A. Majee, V. N. Charushin, O. N. Chupakhin, *Coord. Chem. Rev.* 2021, 442, 213980.
115 A. Hergold-Brundić, Z. Popovic, D. Matkovic-Calogovic, *Acta Crystallogr. C* 1996, 52, 3154–3157.
116 T. N. Y. Hoang, T. Lathion, L. Guénée, E. Terazzi, C. Piguet, *Inorg. Chem.* 2012, 51, 8567–8575.
117 M. G. B. Drew, M. J. Hudson, P. B. Iveson, M. L. Russell, J.-O. Liljenzin, M. Sklberg, L. Spjuth, C. Madic, *J. Chem. Soc., Dalton Trans.* 1998, 2973–2980.
118 C. A. Bessel, R. F. See, D. L. Jameson, M. R. Churchill, K. J. Takeuchi, *J. Chem. Soc., Dalton Trans.* 1992, 3223–3228.

119 A. Winter, A. M. J. van den Berg, R. Hoogenboom, G. Kickelbick, U. S. Schubert, *Synthesis* 2006, 2873–2878.

120 L. M. Guard, J. L. Palma, W. P. Stratton, L. J. Allen, G. W. Brudvig, R. H. Crabtree, V. S. Batista, N. Hazari, *Dalton Trans.* 2012, 41, 8098–8110.

121 E. C. Constable, in *Advances in Inorganic Chemistry*, Vol. 79, (Eds.: R. van Eldik, C. D. Hubbard), Academic Press, London, 2022, pp. 109–149.

122 P. Thordarson, *Chem. Soc. Rev.* 2011, 40, 1305–1323.

123 M. Ringleb, L. Klähn, S. Zechel, U. S. Schubert, *ChemPlusChem* 2025, 90, e202400686.

124 F. Sha, T.-Y. Tai, M. A. Gaidimas, F. A. Son, O. K. Farha, *Langmuir* 2022, 38, 6771–6779.

125 W. R. Archer, M. D. Schulz, *Soft Matter* 2020, 16, 8760–8774.

126 M. W. Freyer, E. A. Lewis, *Methods Cell Biol.* 2008, 84, 79–113.

127 G. Schwarzenbach, *Helv. Chim. Acta* 1952, 35, 2344–2359.

128 R. Dobrawa, F. Würthner, *J. Polym. Sci., Part A: Polym. Chem.* 2005, 43, 4981–4995.

129 R. D. Hancock, *Chem. Soc. Rev.* 2013, 42, 1500–1524.

130 J. M. Hamilton, J. R. Whitehead, N. J. Williams, M. El Ojaimi, R. P. Thummel, R. D. Hancock, *Inorg. Chem.* 2011, 50, 3785–3790.

131 J. M. Hamilton, M. J. Anhorn, K. A. Oscarson, J. H. Reibenspies, R. D. Hancock, *Inorg. Chem.* 2011, 50, 2764–2770.

132 A. N. Carolan, A. E. Mroz, M. El Ojaimi, D. G. VanDerveer, R. P. Thummel, R. D. Hancock, *Inorg. Chem.* 2012, 51, 3007–3015.

133 R. G. Pearson, *J. Am. Chem. Soc.* 1963, 85, 3533–3539.

134 H. Mayr, M. Breugst, A. R. Ofial, *Angew. Chem. Int. Ed.* 2011, 50, 6470–6505.

135 S. Del Piero, P. Di Bernardo, R. Fedele, A. Melchior, P. Polese, M. Tolazzi, *Eur. J. Inorg. Chem.* 2006, 3738–3745.

136 G. E. Coates, S. I. E. Green, *J. Chem. Soc. Res.* 1962, 3340–3348.

137 Atomistic Simulation Group, Imperial College London, London, UK. *Database of Ionic Radii*, http://abulafia.mt.ic.ac.uk/shannon/ptable.php, (last accessed: 2025-04-01).

138 R. D. Shannon, *Acta Crystallogr. A* 1976, 32, 751–767.

139 R. Dobrawa, P. Ballester, C. R. Saha-Möller, F. Würthner, in *Metal-Containing and Metallosupramolecular Polymers and Materials*, Vol. 928, (Eds.: U. S. Schubert, G. R. Newkome, I. Manners), ACS Publishing, Washington, D.C., 2006, pp. 43–62.

140 H. M. N. H. Irving, R. J. P. Williams, *J. Chem. Soc. Res.* 1953, 3192–3210.

141 R. Hogg, R. G. Wilkins, *J. Chem. Soc. Res.* 1962, 341–350.

142 R. H. Holyer, C. D. Hubbard, S. F. A. Kettle, R. G. Wilkins, *Inorg. Chem.* 1966, 5, 622–625.

143 G. I. Dzhardimalieva, B. C. Yadav, S. Singh, I. E. Uflyand, *Dalton Trans.* 2020, 49, 3042–3087.

144 U. S. Schubert, A. Winter, G. R. Newkome, *Terpyridine-Based Materials*, Wiley-VCH, Weinheim, 2011.

145 D. L. Greene, D. M. P. Mingos, *Transit. Met. Chem.* 1991, 16, 71–72.

146 E. Gabano, M. Ravera, *Molecules* 2022, 27, 4249.

147 J. E. Beves, E. C. Constable, C. E. Housecroft, M. Neuburger, S. Schaffner, J. A. Zampese, *Inorg. Chem. Commun.* 2008, 11, 1006–1008.

148 S. Prusty, Y.-T. Chan, *Chem. Lett.* 2021, 50, 1202–1212.

149 M.-P. Santoni, A. K. Pal, G. S. Hanan, A. Proust, B. Hasenknopf, *Inorg. Chem. Commun.* 2011, 14, 399–402.

150 R. Ziessel, V. Grosshenny, M. Hissler, C. Stroh, *Inorg. Chem.* 2004, 43, 4262–4271.

151 B. Schulze, D. Escudero, C. Friebe, R. Siebert, H. Görls, S. Sinn, M. Thomas, S. Mai, J. Popp, B. Dietzek, L. González, U. S. Schubert, *Chem. Eur. J.* 2012, 18, 4010–4025.

152 J. Paul, S. Spey, H. Adams, J. A. Thomas, *Inorg. Chim. Acta* 2004, 357, 2827–2832.

153 J. A. G. Williams, A. J. Wilkinson, V. L. Whittle, *Dalton Trans.* 2008, 2081–2099.

154 C. Mugemana, P. Guillet, S. Hoeppener, U. S. Schubert, C.-A. Fustin, J.-F. Gohy, *Chem. Commun.* 2010, 46, 1296–1298.

155 J.-R. Jiménez, B. Doistau, M. Poncet, C. Piguet, *Coord. Chem. Rev.* 2021, 434, 213750.

156 E. C. Constable, C. E. Housecroft, M. Neuburger, J. Schönle, J. A. Zampese, *Dalton Trans.* 2014, 43, 7227–7235.

157 R.-A. Fallahpour, M. Neuburger, M. Zehnder, *New J. Chem.* 1999, 23, 53–61.

158 P. Zimmer, P. Müller, L. Burkhardt, R. Schepper, A. Neuba, J. Steube, F. Dietrich, U. Flörke, S. Mangold, M. Gerhards, M. Bauer, *Eur. J. Inorg. Chem.* 2017, 1504–1509.

159 P. Zimmer, L. Burkhardt, R. Schepper, K. Zheng, D. Gosztola, A. Neuba, U. Flörke, C. Wölper, R. Schoch, W. Gawelda, S. E. Canton, M. Bauer, *Eur. J. Inorg. Chem.* 2018, 5203–5214.

160 Y.-J. He, T.-H. Tu, M.-K. Su, C.-W. Yang, K. V. Kong, Y.-T. Chan, *J. Am. Chem. Soc.* 2017, 139, 4218–4224.

161 T. Bätz, I. Anufriev, C.-H. Pai, T. Schuett, I. Nischang, S. Zechel, M. D. Hager, Y.-T. Chan, U. S. Schubert, A. C. S. Appl, *Polym. Mater.* 2023, 5, 1354–1363.

162 S.-Y. Wang, J.-H. Fu, Y.-P. Liang, Y.-J. He, Y.-S. Chen, Y.-T. Chan, *J. Am. Chem. Soc.* 2016, 138, 3651–3654.

163 V. W.-W. Yam, A. S.-Y. Law, *Coord. Chem. Rev.* 2020, 414, 213298.

164 M.-C. Tang, A. K.-W. Chan, M.-Y. Chan, V. W.-W. Yam, *Top. Curr. Chem.* 2016, 374, 46.

165 I. J. Bruno, J. C. Cole, P. R. Edgington, M. Kessler, C. F. Macrae, P. McCabe, J. Pearson, R. Taylor, *Acta Crystallogr. B* 2002, 58, 389–397.

166 D. E. C. Corbridge, E. G. Cox, *J. Chem. Soc. Res.* 1956, 594–603.

167 W. Clegg, Philos., *Transact. Royal Soc. A: Math. Phys. Eng. Sci.* 2019, 377, 20180239.

168 P. Wang, G. R. Newkome, C. Wesdemiotis, *Int. J. Mass Spectrom.* 2006, 255-256, 86–92.

169 X. Li, Y.-T. Chan, G. R. Newkome, C. Wesdemiotis, *Anal. Chem.* 2011, 83, 1284–1290.

170 S. Chakraborty, G. R. Newkome, *Chem. Soc. Rev.* 2018, 47, 3991–4016.

171 J. Shi, M. Wang, *Chem. Asian J.* 2021, 16, 4037–4048.

172 U. S. Schubert, H. Hofmeier, G. R. Newkome, *Modern Terpyridine Chemistry*, Wiley-VCH, Weinheim, 2006.

173 A. J. Pell, G. Pintacuda, C. P. Grey, *Prog. Nucl. Magn. Reson. Spectrosc.* 2019, 111, 1–271.

174 J. Novotný, M. Sojka, S. Komorovsky, M. Nečas, R. Marek, *J. Am. Chem. Soc.* 2016, 138, 8432–8445.

175 W. D. Knight, *Phys. Rev.* 1949, 76, 1259–1260.

176 I. Bertini, C. Luchinat, *Coord. Chem. Rev.* 1996, 150, 1–292.

177 E. C. Constable, C. E. Housecroft, T. Kulke, C. Lazzarini, E. R. Schofield, Y. Zimmermann, *J. Chem. Soc., Dalton Trans.* 2001, 2864–2871.

178 H. S. Chow, E. C. Constable, C. E. Housecroft, K. J. Kulicke, Y. Tao, *Dalton Trans.* 2005, 236–237.

179 S. Gaemers, J. van Slageren, C. M. O'Connor, J. G. Vos, R. Hage, C. J. Elsevier, *Organometallics* 1999, 18, 5238–5244.

180 M. A. Fedotov, *J. Struct. Chem.* 2016, 57, 563–613.

181 L. Pazderski, T. Pawlak, J. Sitkowski, L. Kozerski, E. Szlyk, *Magn. Reson. Chem.* 2011, 49, 237–241.

182 L. Pazderski, *Ann. Rep. NMR Spectr.* 2020, 101, 151–284.

183 L. Pazderski, *Ann. Rep. NMR Spectr.* 2013, 80, 33–179.

184 N. W. G. Debye, E. Rosenberg, J. J. Zuckerman, *J. Am. Chem. Soc.* 1968, 90, 3234–3236.

185 G. Alonzo, M. Consiglio, F. Maggio, N. Bertazzi, *Inorg. Chim. Acta* 1988, 147, 217–220.

186 A. Lindner, M. Menzel, F. Renz, D. G. Kurth, A. F. Thünemann, *Hyperfine Interact.* 2005, 166, 465–468.

187 W. M. Reiff, *J. Am. Chem. Soc.* 1974, 96, 3829–3834.

188 H. Oshio, H. Spiering, V. Ksenofontov, F. Renz, P. Gütlich, *Inorg. Chem.* 2001, 40, 1143–1150.

189 S. Suhr, N. Schröter, M. Kleoff, N. Neuman, D. Hunger, R. Walter, C. Lücke, F. Stein, S. Demeshko, H. Liu, H.-U. Reissig, J. van Slageren, B. Sarkar, *Inorg. Chem.* 2023, 62, 6375–6386.

190 J. England, C. C. Scarborough, T. Weyhermüller, S. Sproules, K. Wieghardt, *Eur. J. Inorg. Chem.* 2012, 4605–4621.

191 R.-A. Fallahpour, *Synthesis* 2003, 0155–0184.

192 A. M. W. Cargill Thompson, *Coord. Chem. Rev.* 1997, 160, 1–52.

193 S. F. Kainat, M. B. Hawsawi, E. U. Mughal, N. Naeem, A. M. Almohyawi, H. M. Altass, E. M. Hussein, A. Sadiq, Z. Moussa, A. S. Abd-El-Aziz, S. A. Ahmed, *RSC Adv.* 2024, 14, 21464–21537.

194 G. M. Badger, W. H. F. Sasse, *J. Chem. Soc. Res.* 1956, 616–620.

195 W. H. F. Sasse, C. P. Whittle, *Aust. J. Chem.* 1963, 16, 31–33.

196 T. Kauffmann, J. König, A. Woltermann, *Chem. Ber.* 1976, 109, 3864–3868.

197 G. R. Newkome, D. C. Hager, F. R. Fronczek, *J. Chem. Soc., Chem. Commun.* 1981, 858–859.

198 J. Uenishi, T. Tanaka, S. Wakabayashi, S. Oae, H. Tsukube, *Tetrahedron Lett.* 1990, 31, 4625–4628.
199 Y. Yamamoto, T. Tanaka, M. Yagi, M. Inamoto, *Heterocycles* 1996, 42, 189–194.
200 Y.-Q. Fang, G. S. Hanan, *Synlett* 2003, 0852–0854.
201 G. D. Harzmann, M. Neuburger, M. Mayor, *Eur. J. Inorg. Chem.* 2013, 3334–3347.
202 S. Aroua, T. K. Todorova, P. Hommes, L.-M. Chamoreau, H.-U. Reissig, V. Mougel, M. Fontecave, *Inorg. Chem.* 2017, 56, 5930–5940.
203 R. L. Frank, R. P. Seven, *J. Am. Chem. Soc.* 1949, 71, 2629–2635.
204 F. Kröhnke, *Synthesis* 1976, 1–24.
205 K. T. Potts, M. J. Cipullo, P. Ralli, G. Theodoridis, *J. Am. Chem. Soc.* 1981, 103, 3585–3586.
206 K. T. Potts, M. J. Cipullo, P. Ralli, G. Theodoridis, *J. Org. Chem.* 1982, 47, 3027–3038.
207 K. T. Potts, P. Ralli, G. Theodoridis, P. Winslow, *Org. Synth.* 1986, 64, 189.
208 D. L. Jameson, L. E. Guise, *Tetrahedron Lett.* 1991, 32, 1999–2002.
209 M. W. Cooke, J. Wang, I. Theobald, G. S. Hanan, *Synth. Commun.* 2006, 36, 1721–1726.
210 E. C. Constable, M. D. Ward, S. Corr, *Inorg. Chim. Acta* 1988, 141, 201–203.
211 E. C. Constable, C. E. Housecroft, P. Kopecky, C. J. Martin, I. A. Wright, J. A. Zampese, H. J. Bolink, A. Pertegas, *Dalton Trans.* 2013, 42, 8086–8103.
212 R. L. Podugu, V. Bodala, K. Yettula, B. K. Karasala, S. Vidavalur, *Polycycl. Aromat. Compd.* 2024, 44, 4455–4466.
213 V. Bodala, R. L. Podugu, K. Yettula, P. Gollamudi, S. Vidavalur, S. Pulipaka, *Chem. Asian J.* 2023, 18, e202201004.
214 P. B. De, L. Ilies, K. Takai, S. Asako, *Synlett* 2024, 35, 459–463.
215 G. R. Pabst, J. Sauer, *Tetrahedron Lett.* 1998, 39, 6687–6690.
216 J. C. Adrian, L. Hassib, N. De Kimpe, M. Keppens, *Tetrahedron* 1998, 54, 2365–2370.
217 I. Sasaki, J.-C. Daran, G. G. A. Balavoine, *Synthesis* 1999, 815–820.
218 M. Heller, U. S. Schubert, *Eur. J. Org. Chem.* 2003, 947–961.
219 E. C. Constable, M. D. Ward, *J. Chem. Soc., Dalton Trans.* 1990, 1405–1409.
220 A. Winter, U. S. Schubert, in *New Polymeric Materials Based on Element-Blocks*, (Ed.: Y. Chujo), Springer, Singapore, 2019, pp. 307–346.
221 T. Hokamp, A. Dewanji, M. Lübbesmeyer, C. Mück-Lichtenfeld, E.-U. Würthwein, A. Studer, *Angew. Chem. Int. Ed.* 2017, 56, 13275–13278.
222 N. N. Lyubavin, Z. Russ, *Fiz.-Khim. O-vA.* 1873, 5, 99–109.
223 N. N. Lyubavin, Z. Russ, *Fiz.-Khim. O-vA.* 1873, 5, 339–347.
224 Z. Wang, Ed, *Comprehensive Organic Name Reactions and Reagents*, John Wiley & Sons, 2010, pp. 635–638.
225 A. Y. Chichibabin, Z. Russ, *Fiz.-Khim. O-vA.* 1905, 37, 1229–1236.
226 A. E. Tschitschibabin, *J. Prakt. Chem.* 1924, 107, 122–128.
227 D. E. Lewis, *Angew. Chem. Int. Ed.* 2017, 56, 9660–9668.
228 D. A. Shabalin, *Org. Biomol. Chem.* 2021, 19, 8184–8204.
229 H. Ahlbrecht, R. Huisgen, *Chem. Ber.* 1983, 116, I–XXVIII.

230 G. Ortoleva, *Gazz. Chim. Ital.* 1900, 30, 509–514.
231 L. C. King, *J. Am. Chem. Soc.* 1944, 66, 894–895.
232 F. Kröhnke, *Angew. Chem. Int. Ed.* 1963, 2, 225–238.
233 D. Rocco, C. E. Housecroft, E. C. Constable, *Molecules* 2019, 24, 1799.
234 G. W. V. Cave, C. L. Raston, *Chem. Commun.* 2000, 2199–2200.
235 G. W. V. Cave, C. L. Raston, *J. Chem. Soc., Perkin Trans.* 2001, 1, 3258–3264.
236 N. M. Smith, C. L. Raston, C. B. Smith, A. N. Sobolev, *Green Chem.* 2007, 9, 1185–1190.
237 J. Wang, G. S. Hanan, *Synlett* 2005, 1251–1254.
238 S. Tu, T. Li, F. Shi, Q. Wang, J. Zhang, J. Xu, X. Zhu, X. Zhang, S. Zhu, D. Shi, *Synthesis* 2005, 3045–3050.
239 S. von Kostanecki, G. Rossbach, *Ber. Dtsch. Chem. Ges.* 1896, 29, 1488–1494.
240 Z. Shan, X. Hu, L. Hu, X. Peng, *Helv. Chim. Acta* 2009, 92, 1102–1111.
241 A. M. W. Cargill Thompson, E. C. Constable, P. Harverson, D. Phillips, P. R. Raithby, H. R. Powell, M. D. Ward, *J. Chem. Res. S* 1995, 122–123.
242 E. Rajalakshmanan, V. Alexander, *Inorg. Chem.* 2007, 46, 6252–6260.
243 U. Westerwelle, N. Risch, *Tetrahedron Lett.* 1993, 34, 1775–1778.
244 D. Sielemann, A. Winter, U. Flörke, N. Risch, *Org. Biomol. Chem.* 2004, 2, 863–868.
245 R. Keuper, N. Risch, U. Flörke, H.-J. Haupt, *Liebigs Ann.* 1996, 705–715.
246 C.-Y. Hung, T.-L. Wang, Y. Jang, W. Y. Kim, R. H. Schmehl, R. P. Thummel, *Inorg. Chem.* 1996, 35, 5953–5956.
247 R. P. Thummel, Y. Jahng, *J. Org. Chem.* 1985, 50, 2407–2412.
248 V. Hegde, Y. Jahng, R. P. Thummel, *Tetrahedron Lett.* 1987, 28, 4023–4026.
249 E. A. Medlycott, G. S. Hanan, *Chem. Soc. Rev.* 2005, 34, 133–142.
250 M. Abrahamsson, M. Jäger, T. Österman, L. Eriksson, P. Persson, H.-C. Becker, O. Johansson, L. Hammarström, *J. Am. Chem. Soc.* 2006, 128, 12616–12617.
251 A. K. C. Mengel, C. Förster, A. Breivogel, K. Mack, J. R. Ochsmann, F. Laquai, V. Ksenofontov, K. Heinze, *Chem. Eur. J.* 2015, 21, 704–714.
252 S. Otto, M. Grabolle, C. Förster, C. Kreitner, U. Resch-Genger, K. Heinze, *Angew. Chem. Int. Ed.* 2015, 54, 11572–11576.
253 T. N. Y. Hoang, M. Humbert-Droz, T. Dutronc, L. Guénée, C. Besnard, C. Piguet, *Inorg. Chem.* 2013, 52, 5570–5580.
254 C. Förster, M. Dorn, T. Reuter, S. Otto, G. Davarci, T. Reich, L. Carrella, E. Rentschler, K. Heinze, *Inorganics* 2018, 6, 86.
255 T. Mede, M. Jäger, U. S. Schubert, *Inorg. Chem.* 2019, 58, 9822–9832.
256 T. Schlotthauer, R. Schroot, S. Glover, L. Hammarström, M. Jäger, U. S. Schubert, *Phys. Chem. Chem. Phys.* 2017, 19, 28572–28578.
257 S. Hennessey, C. S. Burke, R. González-Gómez, D. Sensharma, W. Tong, C. Amal, F. Kathalikkattil, W. Cucinotta, T. E. Schmitt, P. F. Keyes, *ChemPhotoChem* 2022, 6, e202100299.
258 M. Abrahamsson, M. Jäger, R. J. Kumar, T. Österman, P. Persson, H.-C. Becker, O. Johansson, L. Hammarström, *J. Am. Chem. Soc.* 2008, 130, 15533–15542.
259 J. Kübel, R. Schroot, M. Wächtler, U. S. Schubert, B. Dietzek, M. Jäger, *J. Phys. Chem. C* 2015, 119, 4742–4751.

2

Beyond 2,2′:6′,2″-Terpyridine – Coordination Chemistry of Its Constitutional Isomers

2.1 Introduction

When talking with undergraduate students in chemistry about "terpyridine ligands," the overwhelming majority of them will think only of 2,2′:6″,2″-terpyridine (tpy), as the archetypal and most prominent structure. Thereby, the possible 47 constitutional isomers, which are also "terpyridines" in the sense of meaning, that three pyridine rings are interconnected *via* two σ-bonds, are commonly overlooked [1, 2]. A literature survey on the parent structures, using SciFinder^{n}, reflects the extraordinary role of tpy in a scientific context; whereas for many other isomers, only scattered studies are found (Table 2.1).

Surprisingly, a significant number of references can be found for 3,2′:4′,3″-terpyridine, but its coordination chemistry has apparently not yet been considered. This contrariety can be rationalized by the fact that this particular isomer, referred to as nicotelline, occurs in tobacco [3] and tobacco smoke [4]. Thus, nicotelline can be distinguished from all the other isomers, since it represents a naturally occurring alkaloid. Since the fraction of nicotelline in tobacco smoke is considerably larger than in non-burned tobacco, it has been proposed that nicotelline is largely derived from the more abundant anatalline under oxidative conditions (*i.e.*, combustion or smoking; Figure 2.1) [6]. In recent years, nicotelline has been discussed in the context of medicinal issues to a considerable extent [7, 8]. After inhalative uptake, this compound is metabolized to its *N*-oxides; however, its toxicity has not yet been reliably quantified.

Housecroft and Constable noted that the synthesis of the various isomers is not always straightforward [1]. In the case of symmetrical isomers (*i.e.*, those isomers, whose outer pyridine rings are equivalent), cross-coupling or ring-assembly methods – similar to those of tpy (see Section 1.5.1) – might be used. The 3,2′:6′,3″- and 4,2′:6′,4″-isomers (Figure 2.2), for example, are available *via* the established condensation reactions according to Kröhnke [9] or Hanan *et al.* [10]. These approaches can also be followed for the synthesis of substituted derivatives, in particular with aryl moieties in 4′-position.

As already mentioned in Chapter 1, 2,2′:6′,2″-tpy typically acts as a chelating, tridentate ligand, which favorably forms octahedral or square-planar complexes with transition-metal ions. Klein *et al.* emphasized that such tridentate coordination

Terpyridines: Multifaceted Ligands for Diverse Applications, First Edition.
Ulrich S. Schubert, Andreas Winter, and George R. Newkome.

Table 2.1 Literature survey on the constitutional isomers of 2,2′:6′,2″-terpyridine (for their chemical structures, see Figures A.1–A.3 in Appendix A).

Terpyridine	Number of references[a)]	Commercially available	Coordination compounds
2,2′:6′,2″	1965	✓	✓
2,2′:6′,3″	3	✓	×
2,2′:6′,4″	8	✓	✓
2,2′:4′,2″	2	×	×
2,3′:5′,2″	2	✓	✓
2,2′:4′,4″	6	✓	✓
3,2′:4′,3″	47	✓	×
3,2′:6′,3″	15	✓	✓
3,2′:6′,4″	2	✓	×
3,3′:5′,3″	10	✓	✓
3,3′:5′,4″	1	✓	×
4,2′:4′,4″	2	✓	×
4,2′:6′,4″	16	✓	✓
4,3′:5′,4″	11	✓	✓
2,2′:3′,2″	16	✓	✓
2,2′:3′,3″	1	×	×
2,4′:3′,3″	1	×	×
2,3′:4′,2″	1	×	×
3,2′:3′,3″	1	✓	×
3,3′:4′,3″	1	✓	×
2,2′:5′,2″	16	✓	✓
2,2′:5″,3″	1	×	×
2,2′:5′,4″	6	✓	✓
2,2′:6′,3″	2	✓	×
3,2′:5′,3″	1	✓	×
3,2′:5′,4″	1	✓	×
3,3′:6′,4″	1	✓	×
4,2′:5′,4″	11	✓	✓

For the non-listed isomers, no references were found.

a) Number of scientific references (excluding patents), in which the respective parent structure without any substituents is mentioned (SciFinder^{n} search performed on April 1st, 2024).

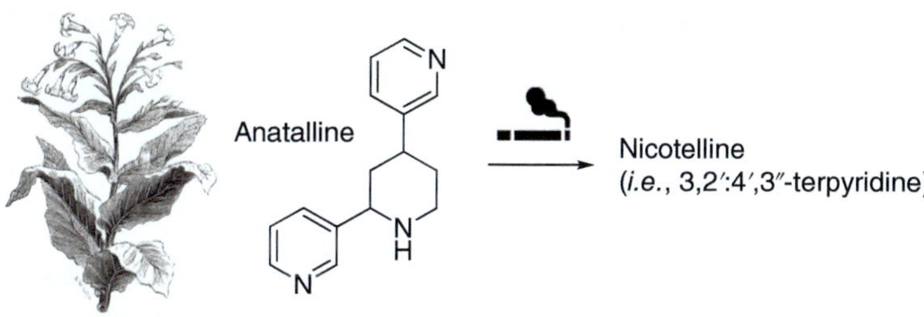

Figure 2.1 Schematic representation of the formation of nicotelline from anatalline upon tobacco combustion. An illustration of a tobacco plant is also shown. [5] / THE INTERNET ARCHIVE / Public Domain.

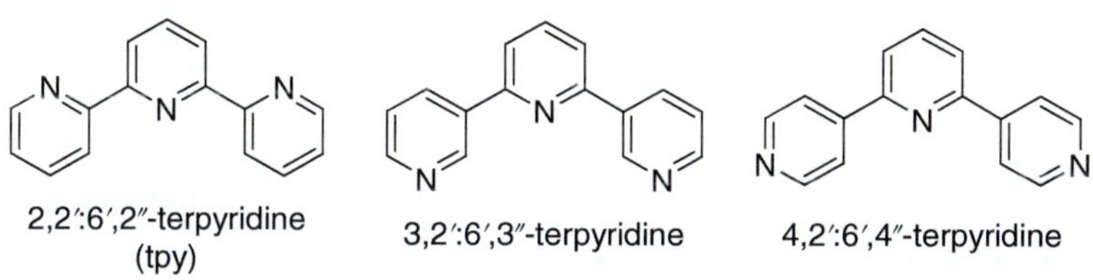

Figure 2.2 Schematic representation of three symmetrical terpyridines.

would also be possible for 20 of the other 47 isomers [11]; however, this implies that the mode of coordination will be different. The tpy ligand, which coordinates to metal centers *via* its three *N*-atoms, is a neutral N^N^N ligand; whereas in all other cases, C–H activation results in cyclometallation. In another extreme case, as for 4,3′:5′,4″-terpyridine, tridentate coordination will only be possible in a C^C^C-type fashion; *i.e.*, the three binding sites are formally carbanionic. In between these scenarios, a range of mixed binding motifs is therefore possible: C^N^N, N^C^N, C^C^N, and C^N^C. Examples for complexes, which contain some of these motifs, are presented in the subsequent Section 2.2. One has to be aware that tpy might also act as a cyclometalating ligand: Due to "rollover" cyclometallation, for example, a C^N^N motif might be obtained instead of the conventional N^N^N-type coordination (Figure 2.3) [12].

Besides this, terpyridine ligands bearing the 3,2′:6′,3″- or 4,2′:6′,4″-pattern have been used, as directing ligands, for the assembly of coordination polymers or networks [13–16]. In these assemblies, the central pyridine rings play only a spectator's role and their *N*-atom is typically not involved in coordination. This field of research,

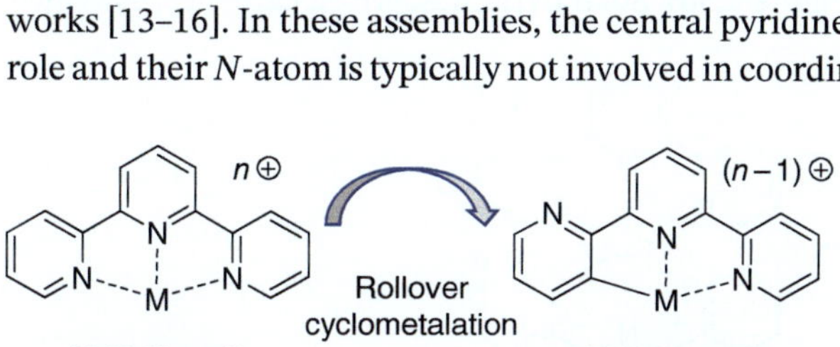

Figure 2.3 Schematic representation of the rollover cyclometallation in a tpy complex (further ligand and counterions are omitted for clarity).

which has been extensively reviewed by Housecroft and Constable, will be detailed in Section 2.3.

2.2 Transition-Metal Complexes of Terpyridine Isomers

As mentioned beforehand, the 48 constitutional isomers can be classified according to their (designated) denticity. According to Klein *et al.*, 21 isomers could theoretically act as tridentate ligands; these structures have to be distinguished from the remaining 27 structures, which are capable of binding at most in a bidentate fashion [11]. However, the synthesis of discrete complexes of these ligands has been sparsely targeted. In order to keep this overview comprehensive, only metal complexes derived from unsubstituted terpyridines are covered in this Section.

In line with their studies on cyclometalated Ru^{II} *bis*-tridentate complexes, Constable, Cargill Thompson *et al.* prepared the heteroleptic Ru^{II} complex **1**, which contained two different terpyridine ligands (Figure 2.4) [17]. A cyclometalated complex was obtained, as the major product, when the synthesis was performed in ethylene glycol; whereas the competing non-cyclometalated chlorido complex was favored in the presence of acetic acid, as the solvent. Unlike its homoleptic $[Ru(tpy)_2]^{2+}$ counterpart, **1** showed room-temperature emission in solution. Moreover, the photophysics of the complex was pH-dependent. However, this aspect, which might point toward a photoacid/photobase behavior of **1** was not analyzed any further [18]. Later, Kershaw Crook and Halcrow used this same Ru^{II} complex as a dopant for the modulation of the spin-crossover (SCO) properties of an Fe^{II} complex [19]. This behavior was attributed to the ability of **1** to intercalate into the crystal lattice of the Fe^{II} complex, thereby stabilizing the high-spin state of the latter with increasing content.

One further *mono*-cyclometalated Ru^{II} complex was reported by Tanaka *et al.*, who employed 4′-methylated 2,2′:4′,2″-terpyridine, as the cyclometalating ligand [20]. Complex **2**, whose solid-state structure is shown in Figure 2.5a, also contained a bidentate 1,8-naphthyridine and a CO ligand. These authors applied ^{13}C-NMR spectroscopy to analyze the nature of the metal-to-carbon bond: A strong low-field shift of the relevant carbon signal suggested a considerable contribution of a pyridylidene-type binding (Figure 2.5b). Such pyridine-derived carbenes, classified

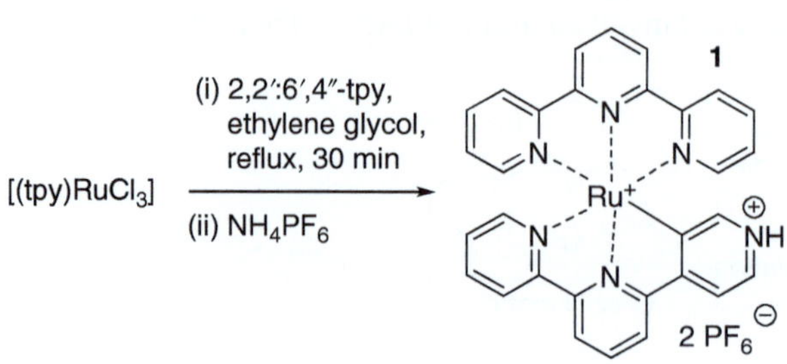

Figure 2.4 Schematic representation of the synthesis of the cyclometalated Ru^{II} complex **1**.

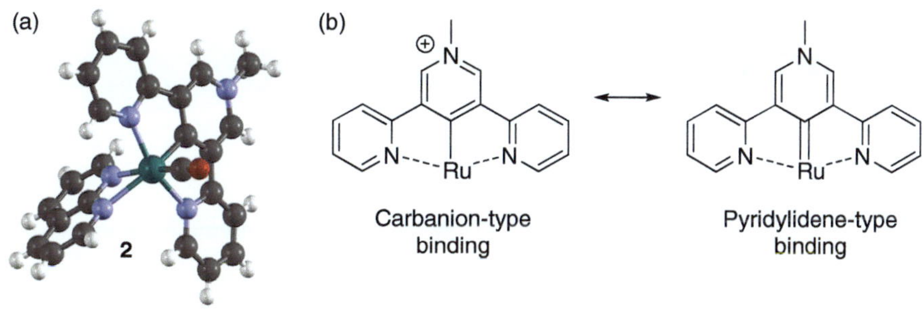

Figure 2.5 (a) Solid-state structure of Ru^{II} complex **2** (counterions and solvent molecules were omitted for clarity). Adapted from [20]. (b) Schematic representation of proposed mixing character of the metal-to-carbon bond in **2**.

as less-heteroatom stabilized carbenes, are already known since the mid-1970s, but their chemistry is still less developed than that of the related imidazole-derived *N*-heterocyclic carbenes (NHCs) or mesoionic carbenes [21].

More recently, Klein *et al.* prepared the square-planar complexes of Ni^{II}, Pd^{II}, and Pt^{II} ions with two terpyridine isomers, namely 2,3′:5′,2″- and 2,2′:4′,2″-terpyridine [11]. This study was performed in line with the authors' previous work on tridentate, *mono*-cyclometalating ligands, which afforded neutral complexes with divalent ions of the Ni-Pd-Pt triad (Figure 2.6a) [22]. Complexes of these types have attracted widespread interest due to their potential use as efficient triplet emitters [23]. In the case of 2,3′:5′,2″- and 2,2′:4′,2″-terpyridine, as the cyclometalating ligands, the full series of complexes was obtained (**3** and **4**, respectively; Figure 2.6b) [11]. Representatively, the solid-state structure of protonated **2c** is shown in Figure 2.7a. The electrochemical and photophysical analysis of complexes **3** and **4** revealed an increased π-acceptor and decreased σ-donor capability of the central cyclometalated

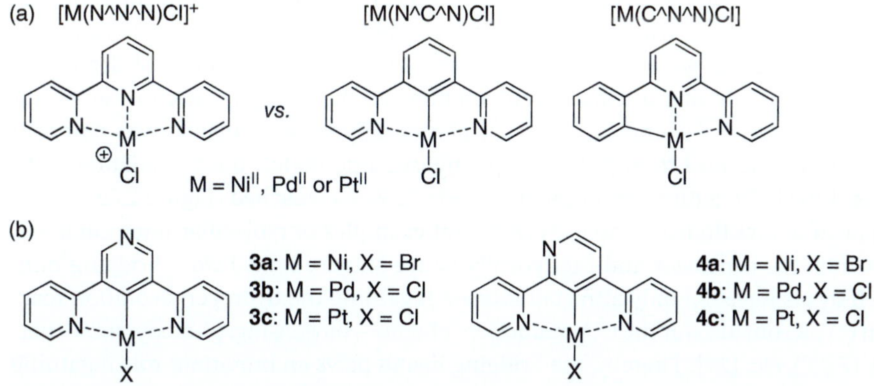

Figure 2.6 (a) Schematic representation of cationic M–tpy complexes in comparison to their neutral analogues with *mono*-cyclometalating N^C^N and C^N^N ligands. (b) Schematic representation of the *mono*-cyclometalated complexes **3** and **4**.

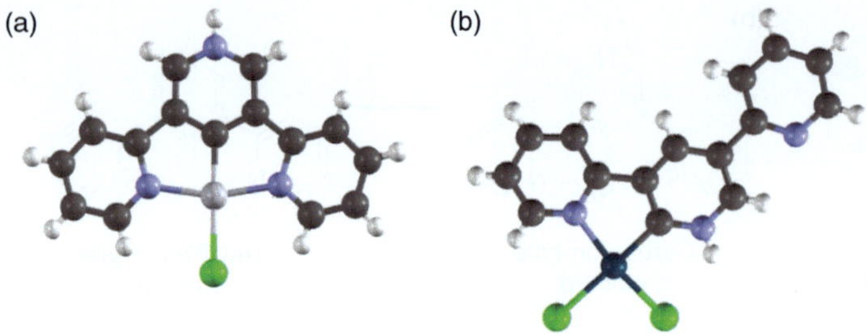

Figure 2.7 Solid-state structures of (a) [(**3c**)H]+ and (b) [(**5**)H] (counterions and solvent molecules were omitted for clarity). Adapted from [11].

pyridine ring compared to phenyl moiety in the prototypical [M(N^C^N)Cl] complexes (Figure 2.6a). Whereas the corresponding complexes with a C^N^N-type coordination were not isolated for 2,2′:6′,3″- or 2,2′:6′,4″-terpyridine. It was assumed that the required C–H activation was hampered once the ligand was pre-coordinated to the metal ion in a bidentate fashion. Moreover, in the case of 2,3′:5′,2″-terpyridine, the *mono*-cyclometalated complex **5** was formed with $PdCl_2$, in which the ligand acted as a bidentate C^N-type chelate. The unusual coordination mode was corroborated by single-crystal X-ray diffraction (XRD) analysis (Figure 2.7b). Related to this, the coordination of 2,2′:4′,2″-terpyridine to the dimeric $[Ir(ppy)_2Cl]_2$ precursor also gave a complex, in which the terpyridine derivative acted as a bidentate ligand – albeit in the conventional N^N-type manner (ppy: 2-phenylpyridinato) [24].

2,2′;4′,4″-Terpyridine as well as its linear 2,2′;5′,4″-isomer can be considered a 2,2′-bipyridine (bpy), which is equipped with a remote pyridine-4-yl moiety. Accordingly, complexes have been formed, in which appropriate precursor complexes have been coordinated to the bpy subunit [25–28]. In the case of the dianionic complex **6** (Figure 2.8a), a pH-dependent luminescence behavior, indicative of photobasic properties, was observed [25]. However, the pK_a and $pK_a{}^*$ values, representing the acidity constants in the ground and excited state, respectively, were not determined. Hanan, Elias *et al.* synthesized the Ir^{III} complexes **7** and **8**, as sensitizers, for the photocatalytic generation of H_2 (Figure 2.8b) [27, 28]. A series of dyads was assembled by binding a cobaloxime complex, as the catalytic site, to the vacant *N*-atom of the Ir^{III} complex. The best photocatalytic performance was observed for the dyad derived from **7** (C^N : 1-phenylisoquinolinato); upon irradiation with green LED light, a turnover number (TON) of 250 was reached (Figure 2.8c).

The aforementioned complexes represent examples of molecular photocatalysts, in which the sensitizer and catalytically active site are linked *via* a bridging unit. Such dyads have not only attracted interest regarding the hydrogen-evolution reaction (HER) but also for the CO_2 reduction, chemo–/biosensing, photodynamic therapy (PDT), *etc.* [29]. Thereby, the bridging ligand plays an important role regarding both the photocatalyst's stability and its overall activity [30]. For example, Vos *et al.* studied various dyads regarding their capability to act as a photocatalyst for the

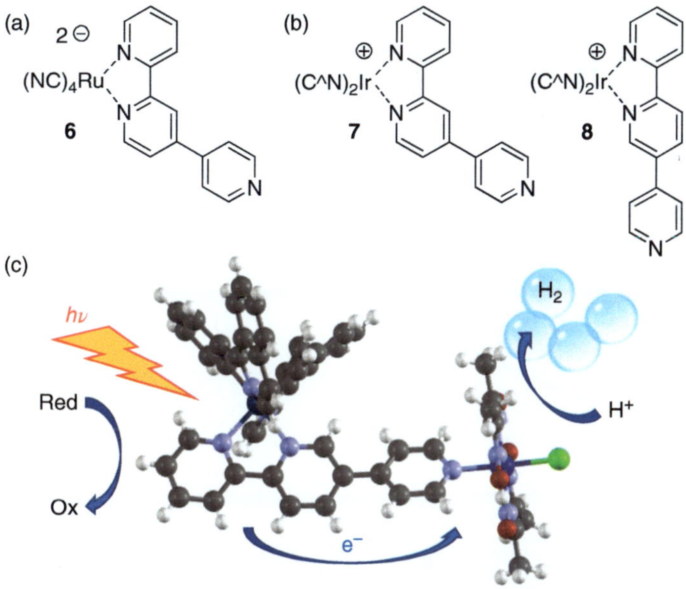

Figure 2.8 (a) Schematic representation of the Ru^{II} complex **6**. (b) Schematic representation of the Ir^{III} complexes **7** and **8**. (c) Schematic representation of the photocatalytic H_2 generation using Ir-Co dyads, as the photocatalyst (counterions were omitted for clarity). Adapted from [28].

HER [31, 32]. These dyads were comprised of a terpyridine ligand, as the bridging unit, whose bpy-like cleft was carrying the light-absorbing Ir^{III} or Ru^{II} complex; the catalytically active moiety, either a Pd^{II} or Pt^{II} center, was attached on the opposite site in a bidentate C^N-type fashion (**9** and **10**, Figure 2.9a). The drawn structures represent the proposed catalytically active species; however, the presence of μ-dichloro-bridged dimers could not be ruled out (at least for the solid state). All dyads were capable of producing H_2 upon irradiation with visible light. Thereby, the assemblies **9** were found to be the more potent photocatalysts due to the linear structure of the bridging ligand. As revealed by density-functional theory (DFT) calculations, the lowest unoccupied molecular orbital (LUMO) of **9c/d** was located on the terpyridine ligand [32]. Thus, the transport of electrons toward the Pd^{II}/Pt^{II} center occurred in an ultrafast and highly directional fashion [34]; whereas the location of the LUMO on the bpy ligands in **10c/d** was detrimental regarding the photocatalytic activity of these dyads.

In subsequent work, Browne, Vos *et al.* demonstrated that the photocatalytic activity could be further increased if the dyad's excited state is stabilized [33]. For this purpose, these authors introduced electron-withdrawing ester moieties on the ancillary ligand (**9e**, Figure 2.9a). This structural modification resulted in a remarkably increased TON for the HER from aqueous MeCN using NEt_3, as the sacrificial donor (*i.e.*, 80 for **9d** *vs.* 650 for **9e** after 6 h irradiation; Figure 2.9b). Moreover, the

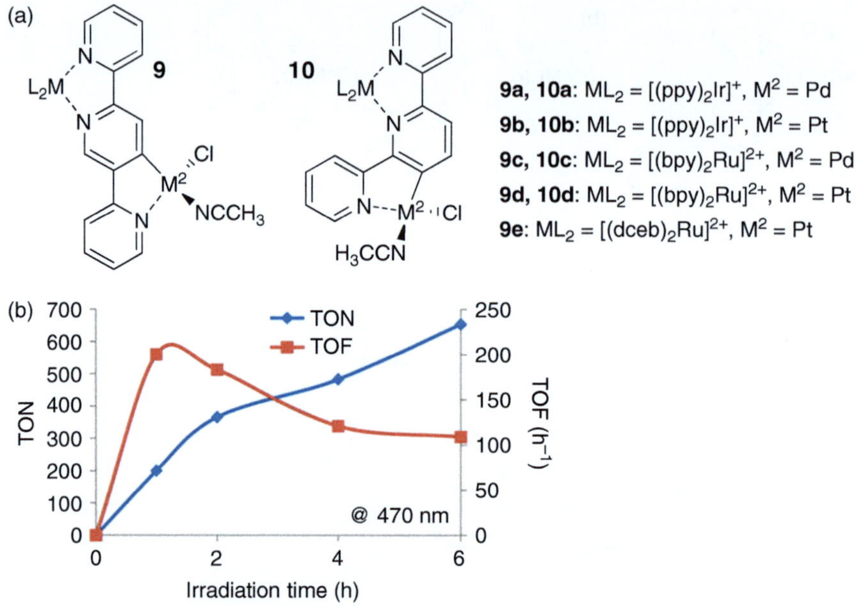

Figure 2.9 Schematic representation of the Ir/Ru–Pd/Pt photocatalysts **9** and **10** (dceb: 4,4′-di(carboxyethyl)-2,2′-bipyridine; the counterions were omitted for clarity). (b) TON and TOF values of dyad **9e** (solvent: aq. MeCN, donor: NEt_3) [33] / with permission of American Chemical Society.

maximum turnover frequency (TOF) was increased from 15 to 200 h^{-1} for **9d** and **9e**, respectively. This study elucidated that all organic ligands – the bridging as well as the sensitizer's ancillary ones – are important and that cautious structural design is required to develop dyads, as molecular catalysts, for the HER.

Slinker, Zysman-Coleman *et al.* targeted the use of cationic Ir^{III} complexes, as emitters, for light-emitting electrochemical cells (LEECs) [35]. Unlike for state-of-the-art OLEDS, whose multilayer architecture is typically fabricated by high-vacuum deposition under stringent conditions, LEECs represent single-layer devices, which can be fabricated in a rather straightforward fashion (see Section 7.4) [36]. Firstly, the deposition of the active material can be performed from solution *via* spin-coating or inkjet-printing technologies. Secondly, the operational mechanism relies on the presence of mobile ions within the active layer; thus, air-stable electrode materials can be used, and device encapsulation becomes obsolete. As one further benefit, thick and large-scale active-material layers can be fabricated, which reduces the overall error rate on the one hand and enables industrial fabrication on the other. The fundamentals of LEECs are detailed further in Section 7.4. As elaborated there, since the introduction of the LEEC concept in 1995 by Pei *et al.* [37], numerous compounds have been studied as LEEC active materials (*e.g.*, organic compounds, conjugated polymers, transition-metal complexes, *etc.*) [38]. Concerning the latter, cationic Ir^{III} complexes have *inter alia* been tested in the context of LEEC applications [39]. In one of those studies, the complexes **11** were tested (Figure 2.10a) [35].

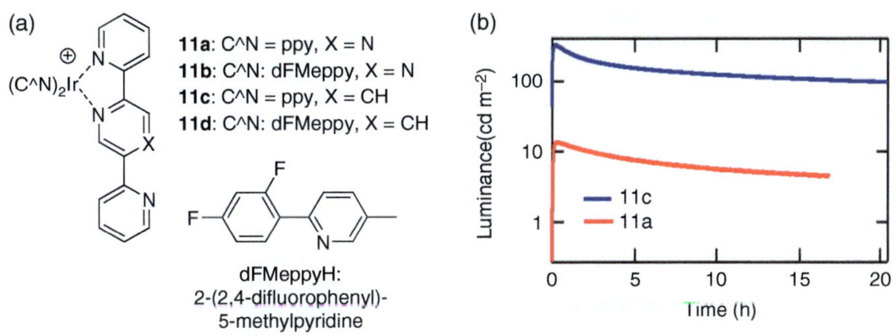

Figure 2.10 (a) Schematic representation of the Ir^{III} complexes **11** (the counterions were omitted for clarity). (b) Luminance *vs.* time for the LEECs with the configuration Al|LiF|**11**|PEDOT:PSS|ITO [PEDOT: poly(3,4-ethylenedioxythiophene), PSS: poly(styrene 4-sulfonate), ITO: indium-tin oxide] [35] / with permission of Royal Society of Chemistry.

All these complexes were highly luminescent with the emission color in solution ranging from red (**11a**) to orange (**11b/c**) and green (**11d**). However, when incorporated into LEEC devices, the fluorinated complexes **11b** and **11d** were not stable under operational conditions; whereas sustained emission was observed for **11a/c** over several hours (Figure 2.10b). The luminescent half-lives of these devices were 380 min and 250 min, respectively. Besides this example, various transition-metal complexes containing 2,2′:6′,2″-terpyridine derivatives, as bi- or tridentate ligands, have also been used in the context of LEECs. We will come back to this topic in Chapter 7.

Finally, Moore, Beves *et al.* prepared the heteroleptic Ru^{II} complex **12**, which contained 2,2′:5′,4″-terpyridine, acting as a bidentate bpy-type ligand [40]. Similar to some of the aforementioned complexes (*e.g.*, **6**; Figure 2.8a), **12** exhibited a pH-dependent UV/vis absorption and emission behavior (notably the $pK_a/pK_a{}^*$ values were not determined). As the most notable aspect of this work, the complex could be prepared in enantiopure form *via* a two-step, “chemistry on the complex” procedure. This strategy is of particular virtue to functionalize/modify preformed complexes of sufficient kinetic stability, *e.g.*, Ru^{II}–polypyridyl complexes [41]. Both the Λ- and Δ-isomers could be synthesized from the respective enantiopure precursor complex, according to Figure 2.11. Thereby, Λ and Δ refer to the optical isomers present in octahedral, *tris*-bidentate complexes (*i.e.*, the left- and right-handed enantiomers, respectively; Figure 2.11). Ru^{II}–polypyridyl complexes are known to bind ribonucleic acid (RNA) with good enantiomeric selectivity [42] and, thus, such chiral complexes might be used in a biomedical context, *e.g.*, antitumor treatment [43]. Moore, Beves *et al.* also pointed out that such chiral complexes could be used as supramolecular building blocks for the assembly of 1D/2D/3D architectures incorporating chiral sites [40]. This aspect, although not considering chirality, directly brings us to the topic of the subsequent Section 2.3, which is coordination polymers and networks.

As one further notable aspect, the coordination chemistry of 2,2′:4′,2″-tpy has not yet been explored, despite the available bidentate bpy substructure (Table 2.1). As

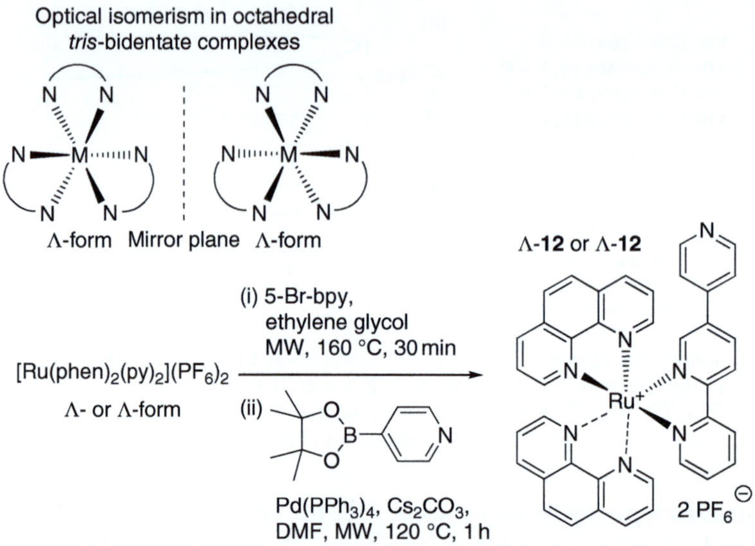

Figure 2.11 Schematic representation of the synthesis of the chiral Ru^{II} complex **12**. The origin of the optical isomerism in such *tris*-bidentate complexes is shown for clarity.

mentioned in Section 1.5.3, the 2,2′:4′,2″-motif resembles that of Risch's S-shaped tpy ligands; for the latter, the coordination chemistry has been studied to a significant extent [44–46].

2.3 Coordination Polymers and Networks Derived from Terpyridine Isomers

The aforementioned 2,2′:5′,4″-terpyridine ligand was used by Maekawa *et al.* for the formation of coordination polymers, whose solid-state structure depended on the applied reaction conditions, *i.e.*, the choice of the solvent and nature of the counterion [47]. Firstly, the self-assembly of $[Cu(CN)_4]X$ and this ligand in a MeOH/MeCN mixture afforded 1D coordination polymers whose topology was mainly governed by the counterion (*i.e.*, X = BF_4^- or PF_6^-). Firstly, the structure of coordination polymer $\{[Cu(2,2':5',4''\text{-tpy})(MeCN)](BF_4)\}_n$ (**13**) was comprised of tetrahedral $[Cu(NCMe)]^+$ sites, which were connected by ligand molecules in a head-to-tail fashion, thus forming a 1D zig-zag-type chain (Figure 2.12a,b). Whereas the related polymer $\{\{[Cu(2,2':5',4''\text{-tpy})(MeCN)](PF_6)\}_2{\cdot}H_2O\}_n$ (**14**) possessed a more complex topology: The unit cell contained two independent 1D zig-zag chains, which ran along the cell's *b*- and *c*-axes (Figure 2.12c). Unlike for **13**, ππ-stacking interactions between opposing ligand molecules were observed in **14**. Moreover, the PF_6^- counterions were surrounded by four opposite MeCN molecules to form two crossed zig-zag chains. Thus, the anions had a templating effect, which resulted in a fundamentally different structure of **14** when compared to that of **13**.

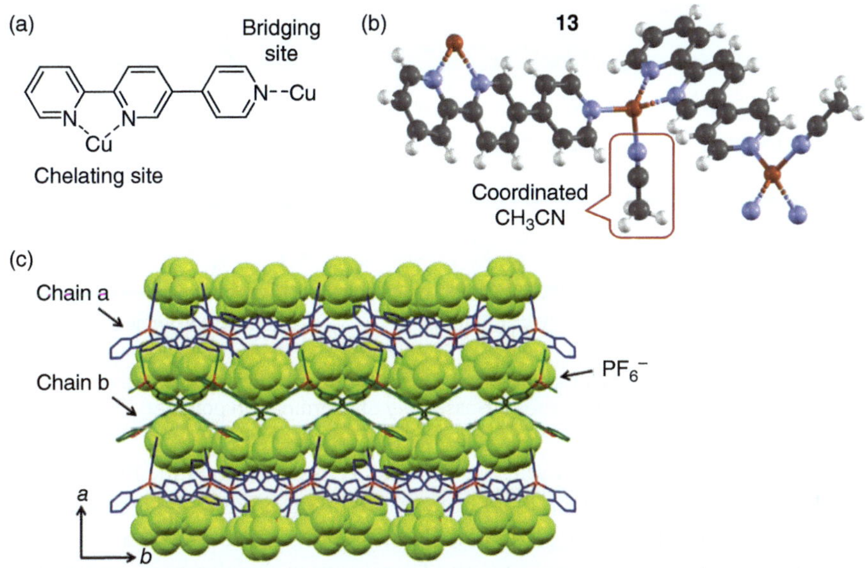

Figure 2.12 (a) Schematic representation of the two different binding sites of the 2,2′:5′,4″-terpyridine ligand. (b) Excerpt of the solid-state structure of **13** showing the head-to-tail binding of ligand to the Cu^I center (counterions and solvent molecules were omitted for clarity). Adapted from [47]. The figure was generated from the original data set using the Mercury 2022.3.0 software (CCDC deposition number 1942779; counterions and solvent molecules were omitted for clarity). (c) Representation of the crystal-packing structure of **14** (solvated H_2O molecules were omitted for clarity, the templating chains of PF_6^- anions were drawn by the space-filling model) [47] / with permission of Elsevier.

When the self-assembly was conducted in an acetone/MeCN mixture, a rhombic metallo-macrocycle, comprised of four Cu^I centers and four ligand molecules, was formed [47]. Furthermore, each macrocycle hosted an acetone molecule in its cavity, and two PF_6^- anions were each sandwiched between two macrocycles. Thus, an overall pillar-like packing was obtained. Finally, those authors demonstrated that the Cu^I ions, required for the self-assembly process, could be generated *in situ* by a comproportionating reaction between $Cu(NO_3)_2 \times 3H_2O$ and Cu tunings. The structure of the resulting coordination polymer was reminiscent of that of **13**: Single-crystal XRD analysis revealed a similar regioregular zig-zag chain with non-interacting anions.

The 2,2′:5′,4″-isomer represents one of the terpyridine structures, which provide divergent binding sites – a requirement to use them for the assembly of coordination polymers and networks. This also holds true for other isomers, which are summarized in Figure 2.13. Housecroft and Constable also noted that also substituted derivatives of these parent structures can be used for the same purpose [2]; however, they excluded pyridyl-substituted derivatives, which are rather quaterpyridines or even higher oligopyridines with their own coordination chemistry [48]. For the sake of clarity, these aspects are not considered within the scope of this section.

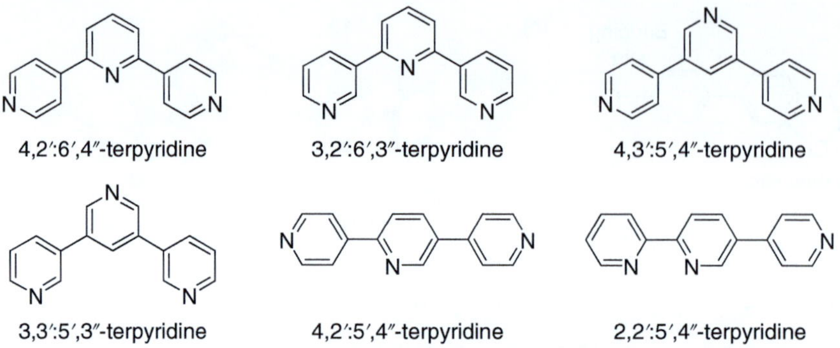

Figure 2.13 Schematic representation of terpyridine isomers, which provide "divergent" binding sites – one prerequisite for the self-assembly of coordination polymers or networks.

With respect to 3,3′:5′,3″-terpyridine, as a D_{2h}-symmetric ligand, Fujita *et al.* reported the self-assembly of metallo-supramolecular nanotubes with Pd(en)$(NO_3)_2$, as the precursor complex (en: 1,2-ethylenediamine) [49]. However, the assemblies, which were comprised of six metal sites and four ligand molecules, were only stable in the presence of biphenyl-4,4′-dicarbxylate, as a templating agent. The resulting host-guest structure was stabilized by ππ-stacking as well as CH–π interactions, as evidenced by single-crystal XRD analysis (Figure 2.14). In subsequent work, the scope of the nanotube syntheses was extended by using other *cis*-fixed Pd^{II} precursors, rod-like aromatic guests (*e.g.*, 1,4-dichlorobenzene, or *β*-naphthalene derivatives) as well as *bis*- and tetra*kis*-3,3′:5′,3″-terpyridine building blocks [50–52].

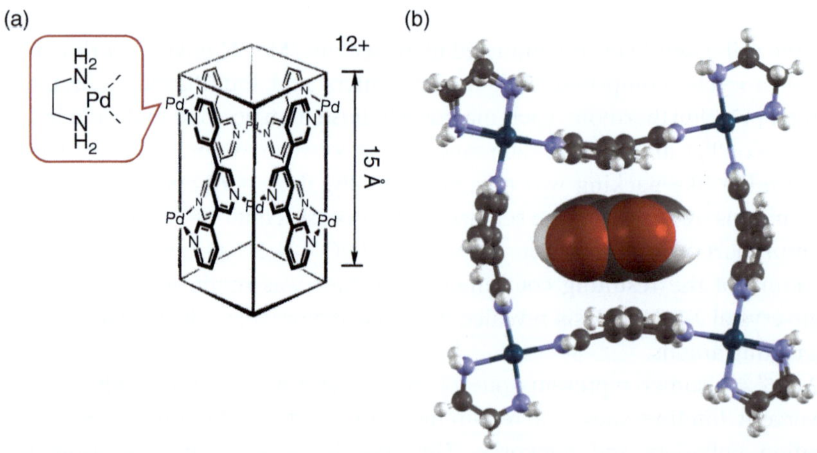

Figure 2.14 (a) Schematic representation of the nanotube architecture comprising six metal centers and four ligand molecules (the counterions were omitted for clarity) [49] / with permission of American Chemical Society. (b) Solid-state structure of the host-guest complex formed by the nanotube and biphenyl-4,4′-dicarboxylate (counterions and solvent molecules were omitted for clarity). The figure was generated from the original data set using the Mercury 2022.3.0 software (CCDC deposition number 1291830; counterions and solvent molecules were omitted for clarity). Adapted from [49].

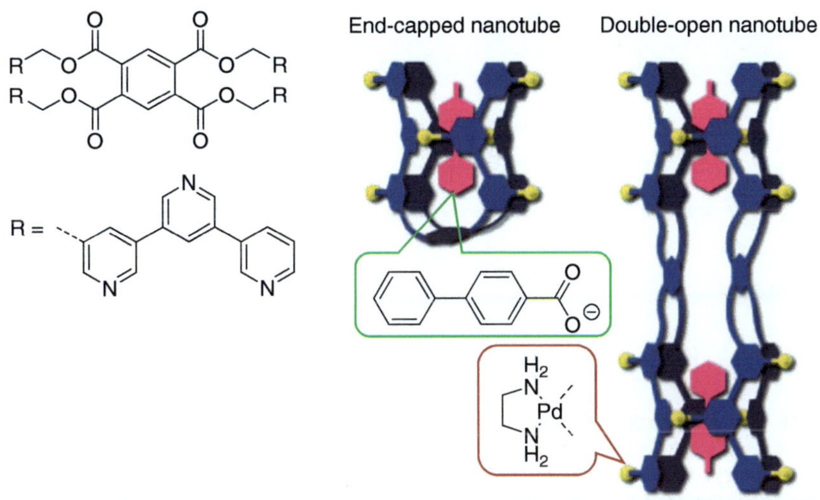

Figure 2.15 Schematic representation of a tetratopic 3,3′,5′,3″-terpyridine ligand, which afforded two different types of host-guest assemblies *via* a template-assisted self-assembly process [51, 52] / permission of Chemical Society of Japan.

The tetratopic building block, for example, afforded two different host-guest assemblies (Figure 2.15) [52]. As the major product, an end-capped tube comprised of one ligand molecule and six $[(en)Pd]^{2+}$ units was obtained. This assembly was capable of hosting unsymmetrical guest molecules, such as biphenyl-4-carboxylate, in an unidirectional fashion. Secondly, an extended double-open tube was formed, which hosted two such guests in a tail-to-tail orientation. The XRD analysis revealed that the large nanotube was comprised of two ligand scaffolds and 12 $[(en)Pd]^{2+}$ sites. The preference of the guest orientation was attributed to the binding of the nonpolar tail within the nanotube's pocket on the one hand and electrostatic repulsion between the nanotube and the polar head group on the other.

In similar fashion, 4,3′:5′,4″-terpyridine also gave guest-stabilized nanotubes with $[(en)Pd]^{2+}$, $[(bpy)Pd]^{2+}$, or $[(cxn)Pd]^{2+}$ moieties, as the nodes (cxn: 1,2-cyclohexyldiamine; Figure 2.16a) [53]. Owing to the ligand's D_{2h} symmetry, the derived assemblies possessed intrinsic C_2-chirality. The use of chiral Pd^{II} precursors, derived from (R, *R*′)-cxn or (*S*, *S*″)-cxn enabled the assembly of host-guest complexes with moderate diastereoselectivity. Nonetheless, these species could be distinguished by CD spectroscopy due to their opposite Cotton effects (CD: circular dichroism; Figure 2.16b).

A literature survey revealed that structurally defined coordination polymers/networks, solely based on the V-shaped 3,3′:5′,3″-terpyridine and transition-metal ions, have not yet been reported. Whereas a three-component self-assembly, which used aromatic dicarboxylic acids as auxiliary ligands, has recently been accomplished. The reaction of 3,3′:5′,3″-terpyridine, $Zn(NO_3)_2 \times 4H_2O$, and a benzenedicarboxylic acid under hydrothermal conditions afforded the

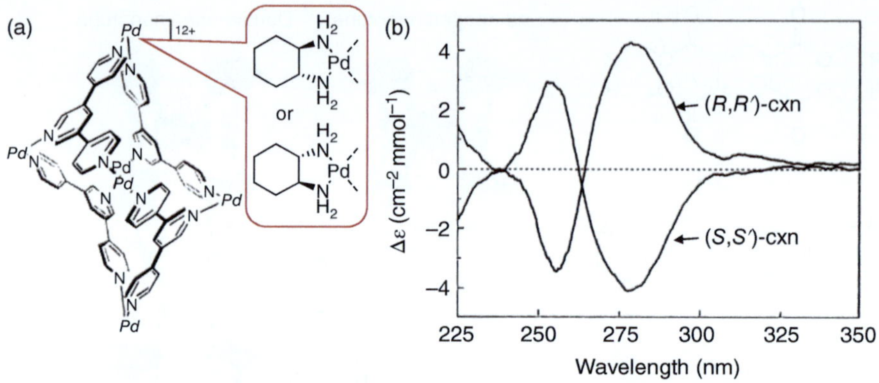

Figure 2.16 (a) Schematic representation of the nanotube, which was derived from 4,3′:5′,4″-terpyridine (the 1,4-dichlorobenzene guest molecule and the counterions were omitted for clarity). (b) CD spectra of the nanotubes, which contained enantiopure diamine ligands on the Pd^{II} nodes (0.5 mM in D_2O) [53] / with permission of OXFORD UNIVERSITY PRESS.

metal–organic frameworks (MOFs), whose structure was dependent on the geometry of the used dicarboxylic acid [54]. **MOF1**, which contained terephthalic acid (1,4-bdc) as the auxiliary ligand, possessed an unusual [3,3,8]-connected framework structure, which was established by the tritopic terpyridine ligand, individual Zn^{II} ions, $\{Zn_2(carboxylate)_2\}$ clusters, and dicarboxylate bridges (Figure 2.17a). The (3,5)-connected structure of **MOF2** resulted from the self-assembly of 3,3′:5′,3″-terpyridine with Zn^{II} ions and isophthalate bridges (Figure 2.17b). The occurrence of different structures and topologies was attributed to the different binding angles of the dicarboxylate ligands on one hand and different twist angles of the terpyridine ligand on the other. **MOF1** was utilized further for the sensing of nitro-containing compounds and (transition-)metal ions in water. In particular, the luminescence of **MOF1** was efficiently quenched by picric acid and Fe^{III}

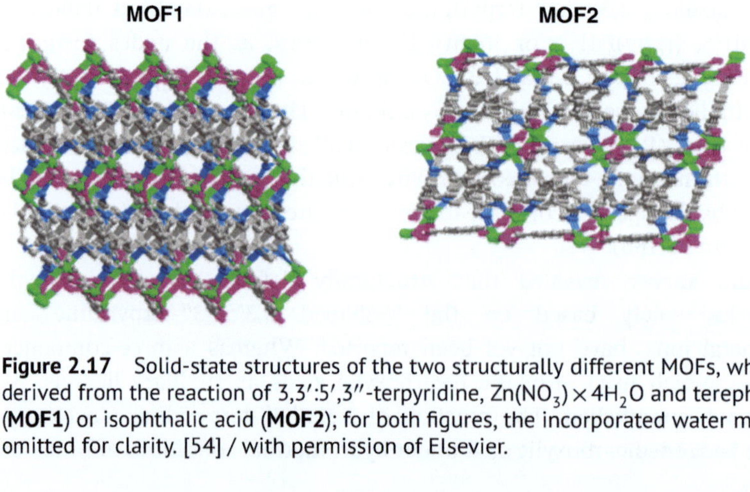

Figure 2.17 Solid-state structures of the two structurally different MOFs, which were derived from the reaction of 3,3′:5′,3″-terpyridine, $Zn(NO_3) \times 4H_2O$ and terephthalic acid (**MOF1**) or isophthalic acid (**MOF2**); for both figures, the incorporated water molecules were omitted for clarity. [54] / with permission of Elsevier.

ions. The lower detection limits for these analytes were 6.17×10^{-4} mmol and 1.55×10^{-3} mmol, respectively. These authors prepared **MOF1**-coated papers as luminescent test stripes to probe the presence of picric acid in aqueous solutions in a straightforward fashion.

The first coordination polymers based on 4,3′:5′,4″-terpyridine were reported by Champness, Schröder *et al.* in 2006 [55]. These authors obtained the coordination polymer $\{[CuI]_3(4,3':5',4''\text{-tpy})_2\}_n$ (**MOF3**) from the reaction of the terpyridine ligand and CuI under solvothermal conditions in *N*,*N*-dimethylformamide (DMF). Single-crystal XRD analysis revealed the presence of 2D bilayer sheets, in which the ligand's three *N*-atoms were involved in coordination. Firstly, the ligand assembled with rhomboid $(CuI)_2$ dimers into a ribbon-like structure. These ribbons, in turn, were linked by $(CuI)_2$ dimers *via* the ligands' vacant binding sites in the N^2-position to generate a 2D array of linked ribbons. The $(CuI)_2$ dimers were oriented in such a way that a bilayer structure was favored. Interpenetration of this motif by an analogous one gave rise to the overall bilayers-of-bilayers assembly (Figure 2.18). The solvothermal synthesis generated two further coordination polymers as by-products. The solid-state structure of one of these was clarified by XRD analysis: This assembly was comprised of 1D chains, in which the ligand molecules were coordinated to the Cu^I centers *via* their outer pyridine rings; the metal centers, in turn, adopted a trigonal-planar geometry.

Furthermore, Gen *et al.* obtained two Co^{II} coordination compounds from the hydrothermal self-assembly of 4,3′:5′,4″-terpyridine and $Co(NO_3)_2 \times 6H_2O$ in the presence of a dicarboxylic acid [56]. Firstly, a mononuclear Co^{II} complex was formed with *trans*-1,4-cyclohexanedicarboxylic acid (*trans*-chdc). In this octahedral complex, two ligand molecules were coordinated to the Co^{II} center in the axial positions *via* their central pyridine ring; the equatorial positions were occupied by H_2O ligands. The dicarboxylate anion represented the counterion. When using 4, 4′-oxydibenzoic acid instead, an infinite 3D structure was found that resulted from the interconnection of Co^{II}–tpy chains *via* hydrogen-bonding (*H*-bonding) interactions between coordinated water molecules and the auxiliary dicarboxylate.

4,3′:5′,4″-Terpyridine has also been used as a building block for the formation of various MOF-type architectures. The first such MOF was reported by Rosseinsky

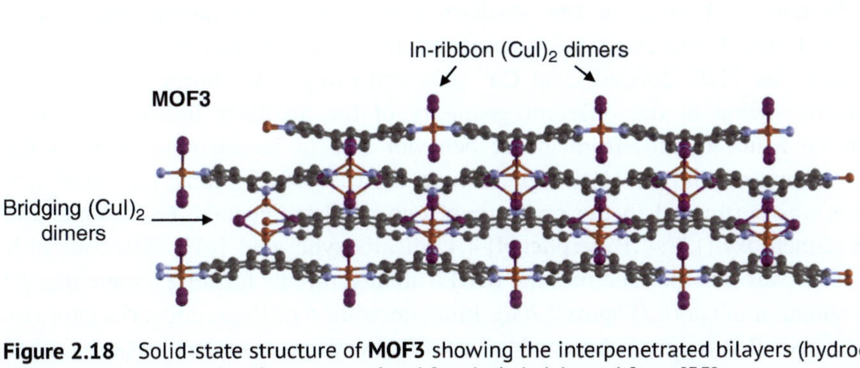

Figure 2.18 Solid-state structure of **MOF3** showing the interpenetrated bilayers (hydrogen atoms and solvent molecules were omitted for clarity). Adapted from [55].

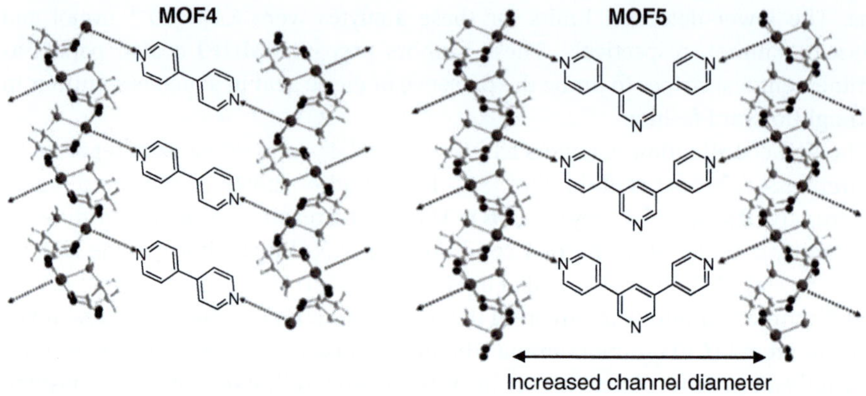

Figure 2.19 Schematic representation of the MOF, which were comprised of pillared Ni(L-asp) layers linked by organic bridges [57] / with permission of John Wiley & Sons.

et al., who were interested in nonporous materials based on natural amino acid [57]. Among the proteinogenic amino acids, L-aspartic acid (L-asp) represents a highly suited building block for MOF formation. The two carboxylate sites, in an appropriate distance from each other, can be used to assemble a network by linking the metal centers, as nodes; moreover, the NH_2-moieties can be used for further coordination of functional sites within the assembly [58]. The nanoporous framework **MOF4** was prepared by pillaring layers of octahedral Ni^{II}–(L-asp) complexes by 4,4′-bpy linkers (Figure 2.19); this framework exhibited chiral-sorption properties, *i.e.*, short-chain diols were chemisorbed within the channels with moderate enantioselectivity [59]. In order to manipulate the porosity, various linkers – as alternatives to 4,4′-bpy, were subsequently tested [57]. However, when using 4,3′:5′,4″-terpyridine in this context, the pillared framework **MOF5** was formed (Figure 2.19), whose solvent-accessible voids corresponded to *ca.* 31% of the overall framework volume. Since excess linker molecules were enclosed in the channels and could not be removed by solvent rinsing, an application for chemisorption became obsolete.

Further MOFs were obtained from the solvothermal reactions of Co^{II} or Ni^{II}, 4,3′:5′,4″-terpyridine and a dicarboxylic acid (*i.e.*, 1,4-bdc or *trans*-chdc) [60]. In the case of 1,4-bdc, as the auxiliary ligand, two isomorphous MOFs were obtained, which represented 3,4-connected threefold interpenetrating networks. Whereas the MOF derived from Co^{II} ions and *trans*-chdc afforded 4-connected network, owing to the different geometry of the auxiliary ligand. The latter material featured antiferromagnetic behavior due to interactions between the Co^{II} centers within the MOF structure. In the case of Zn^{II} ions, the MOF synthesis was performed in the presence of two auxiliary ligands, *i.e.*, 1,4-bdc and 2′,5′-dimethoxy-[1,1′:4′,1″-terphenyl]-4,4″-dicarboxylic acid [61]. However, only the latter was incorporated into the **MOF6** architecture by forming a wave-like 2D coordination network (Figure 2.20a). Interpenetration of these networks gave rise to the overall structure, whose voids were filled with guest molecules (*i.e.*, 1,4-bdc and water; Figure 2.20b).

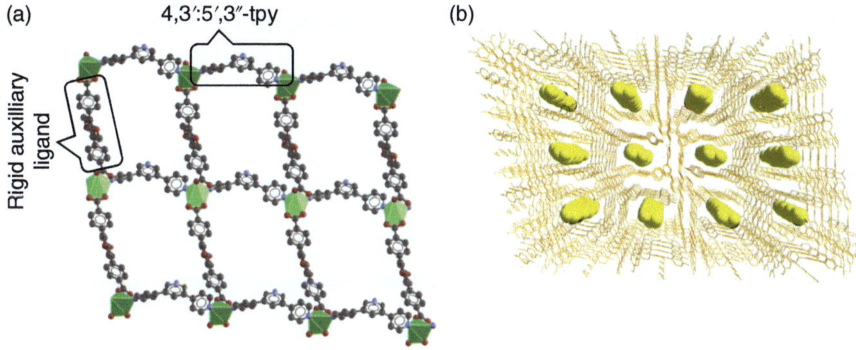

Figure 2.20 (a) Excerpt of the solid-state structure showing the 2D wave-like network structure formed by Zn^{II} ions, 4.3′:5′,4′-tpy and the rigid terphenyl-based linker (the green polyhedral represent the Zn^{II}-based nodes; hydrogen atoms, solvent molecules and non-coordinating ligand molecules were omitted for clarity). Adapted from [61]. (b) Schematic representation of **MOF6**, whose voids were filled with guest molecules (shown in yellow) [61] / with permission of Elsevier.

MOFs assembled from π-conjugated ligands and {MSCu} clusters (M = Mo or W) have attracted considerable interest regarding applications in nonlinear optics (NLO) or luminescence sensing [62]. Against this background, J. Zhang, C. Zhang *et al.* prepared the **MOF7**, which contained unusual 7-connected $\{WS_4Cu_4\}$ clusters, as the nodes [63]. Each Cu^I center was connected to three such clusters *via* bridging CN^- ligands; each tpy ligand coordinated to three cluster moieties *via* the three *N*-atoms. Thus, a [3,3,7]-connected framework structure was obtained (Figure 2.21a). **MOF7** showed a strong NLO response – the material's hyperpolarizability (γ) was by a factor of 10–100 times higher than those of related MOFs or coordination polymers, which contained $\{MoS_4Cu_4\}^{2+}$ or $\{MoS_4Cu_4\}^{2+}$ clusters (Figure 2.21b). Similar to the luminescence quenching of **MOF1**, **MOF7** also exhibited a high sensitivity toward the presence of 2,4,6-trinitrophenol. Sensing of this analyte in water was possible down to the detection limit of *ca.* 0.8 μmol; whereas other nitro-containing compounds were detected with significantly less sensitivity (Figure 2.21c).

Recently, Pang, Ma *et al.* prepared a POM-encapsulated MOF, which exhibited a remarkably high activity in both the electrocatalytic HER and the photocatalytic dye degradation (POM: polyoxometalate) [64]. As the most notable aspect of this work, these authors observed a partial ligand fragmentation/rearrangement of the used 4,3′:5′,4″-terpyridine under the applied hydrothermal reaction conditions. It was assumed that the 3,4′-bpy and 4,3′:5′,3″:5″,4″-quaterpyridine ligands, which were identified in the resultant structure of **MOF8**, resulted from a radical reaction, as outlined in Figure 2.22a. The MOF structure was comprised of edge-sharing triangular and rhombic channels; the POM moieties, as charge-compensating guests, were encapsulated within the latter (Figure 2.22b). Despite the presence of the large $\{SiW_{12}O_{40}\}^{4-}$ clusters within the MOF, the structure provided sufficient voids in which chemical reactions might proceed (*ca.* 18.6% of the overall crystal volume). Firstly, composites were prepared by mixing **MOF8** with nickel foam or Pd/C-loaded

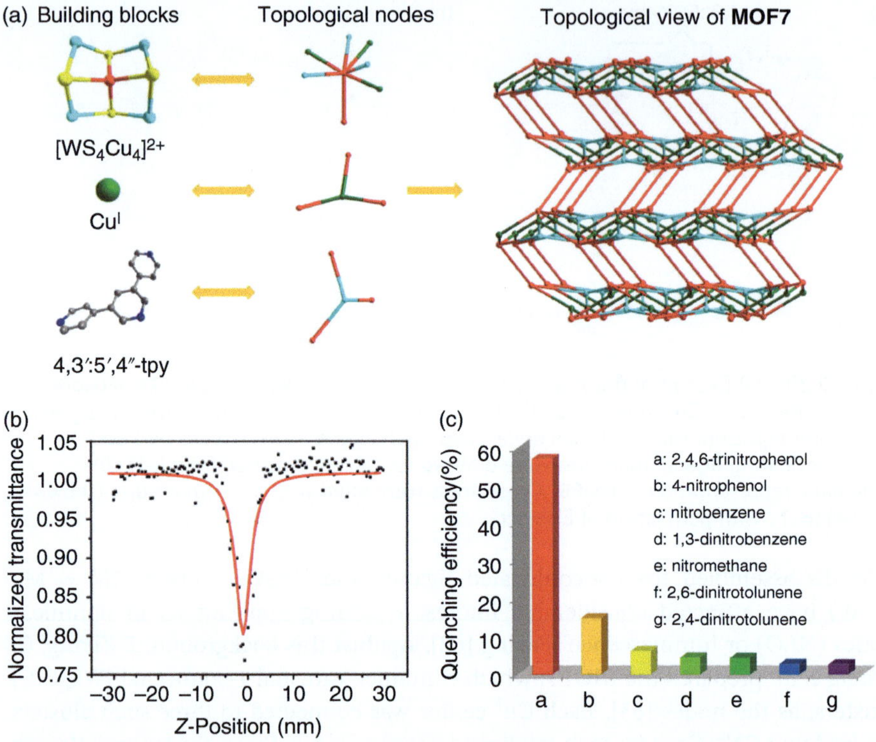

Figure 2.21 (a) Schematic representation of the molecular building blocks (left) and their topological nodes (middle), which generate the topological structure of **MOF7** (right, view along the *a*-axis). (b) NLO absorption of **MOF7** in a *Z*-scan experiment. (c) Comparison of the quenching efficiency of various nitro-containing compounds (1 mmol in H_2O) [63] / with permission of American Chemical Society.

nickel foam as well as carbon black. In particular, the composites, which contained 50 wt% of **MOF8** were highly active in the electrocatalytic HER. Moreover, **MOF8** was used as a recyclable photocatalyst for the degradation of dyes (*e.g.*, rhodamine-B; Figure 2.22c). In subsequent work, these same authors assembled one further MOF, which incorporated Ni^{II} ions and Keggin-type $\{PW_{12}O_{40}\}^{3-}$ polyanions [65]. Unlike the previous example, the formation of the twofold interpenetrating MOF occurred without noticeable ligand fragmentation/rearrangement. This material featured the same bifunctional catalytic activity as **MOF8**.

Unlike the previously discussed isomers, which acted as ditopic or tritopic vectorial ligands, 2,2′:6′,4″-terpyridine represents a derivative of 2,2′-bpy, which carries one further pyridine-4-yl donor in its 6-position. This ligand has been utilized for the formation of some transition-metal complexes, including cyclometalated ones (Section 2.2); whereas its use as a building block for the assembly of coordination polymers/networks has only been sparsely explored. Besides the dinuclear $[Ag_2(2,2',6',4''\text{-tpy})_2]^{2+}$ complexes with BF_4^-, ClO_4^-, or NO_3^- counterions, a zig-zag-shaped coordination polymer, in which the ligand molecules were oriented in a head-to-tail fashion, was reported by Beauchamp and Loeb [66]. In these assemblies, the Ag^I centers were coordinated in a trigonal-planar fashion. More

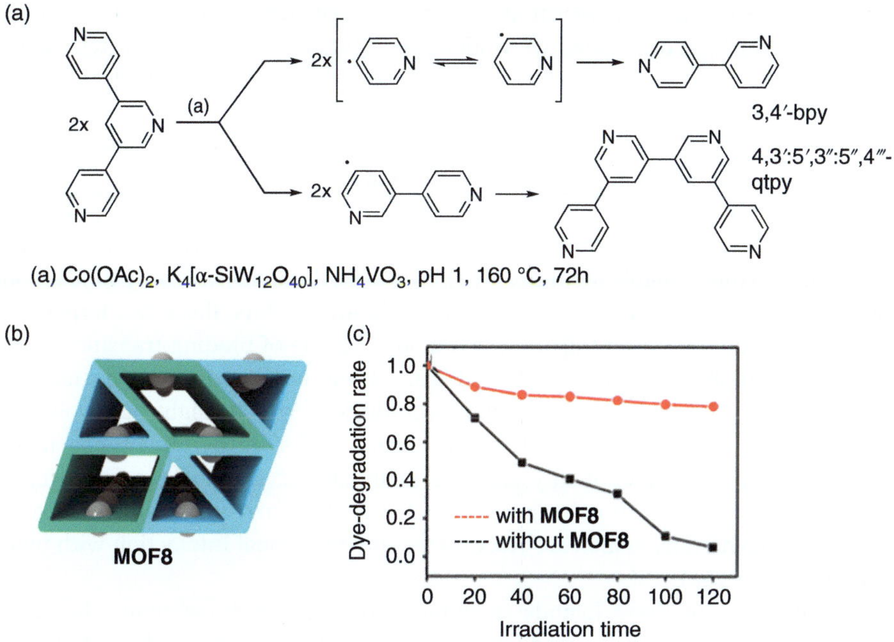

Figure 2.22 (a) Schematic representation of the proposed mechanism for the observed ligand fragmentation/rearrangement under the applied hydrothermal reaction conditions. (b) Schematic representation of the structure of **MOF8**. (c) Plot of the dye-degradation rate *vs.* time (aq. solution of rhodamine-B) [64] / with permission of American Chemical Society.

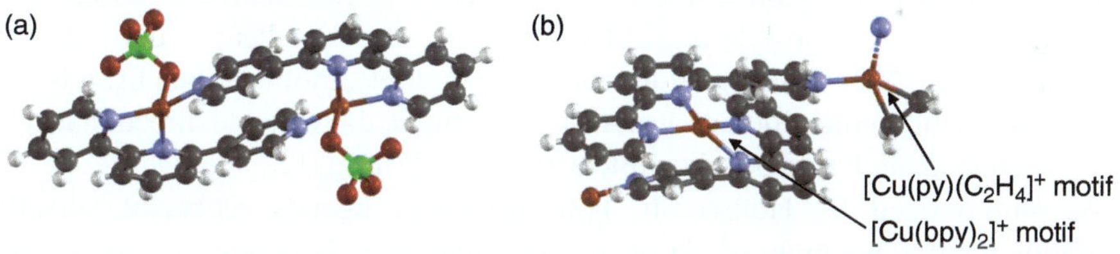

Figure 2.23 (a) Solid-state structure of the dimeric Cu^I complex with coordinated ClO_4^- anions. (b) Excerpt from the solid-state structure of the helical Cu^I coordination polymer showing the two different coordination motifs. In both figures, non-coordinating anions and solvent molecules were omitted for clariyt). Adapted from [67].

recently, Maekawa *et al.* studied the coordination behavior of Cu^I ion with this same terpyridine ligand [67]. This work was directly related to the coordination compounds derived from Cu^I ions and the linear 2,2′:5′,4″-terpyridine (Figure 2.12) [47]. Firstly, two dimeric $[Cu_2(2,2′:6′,4″\text{-tpy})]^{2+}$ complexes were obtained, which differed in the counterion's coordination behavior. In the case of non-coordinating BF_4^-, the complex's rectangular structure resembled that of the analogous Ag^I species. Whereas the ClO_4^- anions were coordinated to the Cu^I centers in a monodentate fashion in order to saturate the coordination spheres (Figure 2.23a).

However, the self-assembly $[Cu(MeCN)_4](PF_6)$ with 2,2′:6′,4″-tpy in a $MeOH/CHCl_3$ solvent mixture and under an atmosphere of ethene gave a 1D coordination polymer of helical shape. In this structure, two distinctly different types of coordination modes were identified (Figure 2.23b). Firstly, tetrahedral Cu^I complexes were formed by the coordination of two ligand molecules in a head-to-head fashion (*i.e.*, *via* their 2,2′-bpy binding sites). In the second set of complexes, two ligands were bound to the Cu^I centers *via* their vacant tails and one ethene molecule was bound in a side-on fashion.

For both 4,2′:6′,4″- and 3,2′:6′,3″-terpyridine complexes, in which coordination occurs *via* the central pyridine ring, has, to the best of our knowledge, not yet been published in the available literature [1, 2]. However, this analysis is only valid as long as one ignores the possibility of hypodentate binding. Thus, these two terpyridine isomers virtually represent ligands, which are capable of binding transition-metal ions in a ditopic, *bis*-monodentate fashion. As one notable distinction, these ligands differ in the number of accessible conformers, which might be involved in coordination. Owing to its substitution pattern, 4,2′:6′,4″-terpyridine is considered a ligand with fixed directionality since this is not altered, if one of the outer rings is rotated around a C—C bond; whereas the 3,2′:6′,3″-isomer might adopt three different conformations, which are relevant for the directional interaction with metal ions (Figure 2.24a).

When either of these ligands is “dimerized” *via* the 4′-positions, the corresponding *bis*-terpyridine ligands are obtained (Figure 2.24b). These telechelic ligands represent the counterparts to the manifold of ditopic 2,2′:6′,2″-terpyridine ligands, which are widely used in the field of metallo-supramolecular chemistry [68]. Their coordination polymers are used, for example, as active materials in electrochromic devices [69, 70]. The term “telechelic” is borrowed from polymer science and refers to ligands, which are comprised of two identical coordination moieties linked *via* a flexible or rigid spacer. The telechelic ligands derived from 4,2′:6′,2″- and 3,2′:6′,3″-terpyridine represent tetratopic, monodentate ligands. For the latter, owing to the different available conformers, the variable directionality in coordination again has to be taken into account (*vide infra*) [2].

As summarized by Housecroft, both divergent ligands represent versatile scaffolds for the assembly of 2D or 3D coordination polymers [14]. Firstly, both ligands were used for the assembly of metallo-macrocycles, in which the ring topology was determined by the ligand’s directionality [71]. The self-assembly with the dinuclear precursor **13** afforded a rhomboid and an overlapped double rhomboidal architecture with 4,2′:6′,4″-tpy and 3,2′:6′,3″-tpy, respectively (Figure 2.25a). However, the presumed structures were only concluded from multinuclear magnetic resonance (NMR) spectroscopy and electrospray ionization (ESI) mass spectrometry (MS). The idea of coordination-driven self-assembly was also followed by Yang, Stang *et al.*, who performed the self-assembly of **13** and 4,2′:6′,4″-tpy in the presence of one further linear/angular ditopic or tripodal tritopic ligand. According to NMR and ESI MS analyses, these reactions afforded complex mixtures of architectures including rhomboid, triangular, and triangular bipyramidal species. Fujita *et al.* studied the two-component self-assembly of 4,2′:6′,4″-tpy with Pd^{II} ions [72]. Due to the ligand’s geometry, *i.e.*, the bend angle of 112°, kinetically stable Pd_8L_{16} and Pd_9L_{19} frameworks

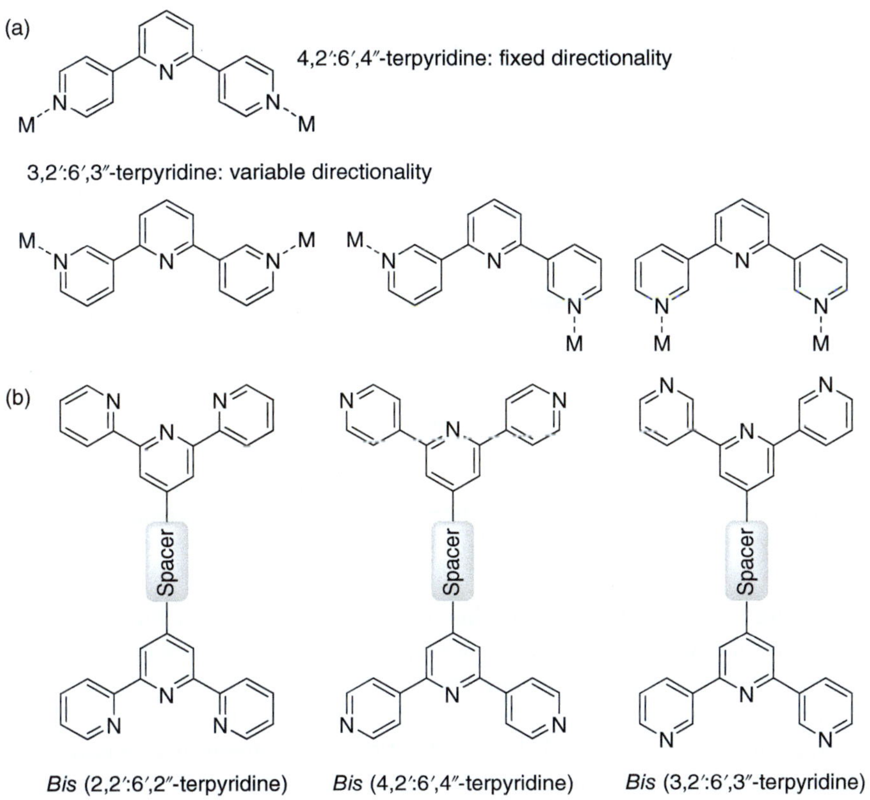

Figure 2.24 (a) Schematic representation of the potential binding modes of 4,2′:6′,2″- and 3,2′:6′,2″-terpyridine. (b) Schematic representation of the telechelic *bis*-terpyridine ligands derived from the 2,2′:6′,2″- 4,2′:6′,2″- and 3,2′:6′,3″-isomers.

were obtained (Figure 2.25b). Single-crystal XRD analysis of the Pd_9L_{19} assembly revealed the proposed D_{3h} symmetry with almost distortion-free Pd^{II} complexes (*i.e.*, the *cis*-N-Pd-N angles deviated only by ±0.2° from the ideal 360° angle).

Beforehand, González Garmendia *et al.* prepared coordination compounds of 4,2″:6′,4″-tpy with various transition-metal chlorides: $MnCl_2$, $CoCl_2$, $NiCl_2$, $CuCl_2$, and $ZnCl_2$ [73]. Only in the latter case, single crystals suited for XRD analysis were obtained. In the 1D coordination polymer, whose chains were not interacting with each other, the $ZnCl_2$ sites were linked *via* the ligand's 4,4″-positions and the Zn^{II} ions adopted a highly distorted tetrahedral coordination geometry. According to Housecroft and Constable, further coordination polymers based on unsubstituted 4,2″:6′,4″-tpy have not been described in the available literature until recently. Powel, Neville *et al.* synthesized a geometrically strained Hofmann-type MOF, in which both penta- and hexa-coordinate Fe^{II} centers were incorporated [74]. Such MOFs are of particular interest due to their remarkable SCO properties [75, 76]. Conventionally, the assemblies are comprised of $[Fe^{II}(M^{II}(CN)_4)]$ or $[Fe^{II}(M^{I}(CN)_2)_2]$ layers, which are linked *via* ditopic *N*-heteroaromatic ligands (M^{II} = Pt, Pd, or

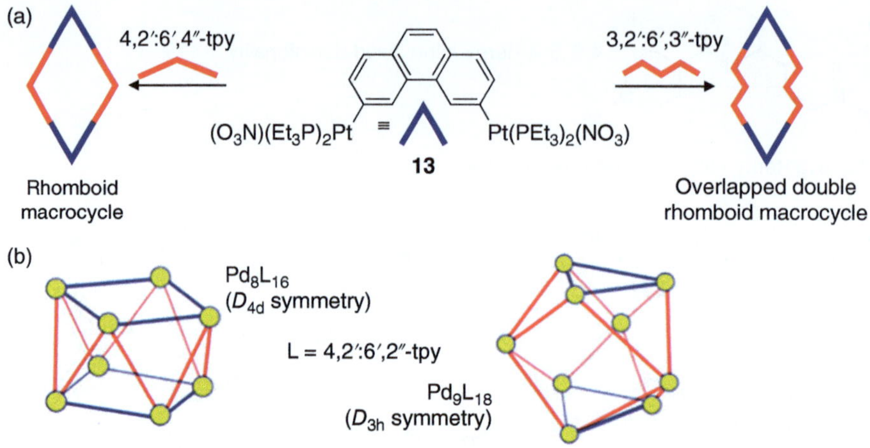

Figure 2.25 (a) Schematic representation of the formation of metallo-macrocycles *via* coordination-driven self-assembly. Adapted from [71]. (b) Schematic representation of Fujita's frameworks Pd_8L_{16} and Pd_9L_{18} [72] / with permission of John Wiley & Sons.

Ni; M^I = Ag or Au). If the bridging ligand is linear geometry, a highly regular, pillared-layer topology is obtained. However, angular 4,2″:6′,4″-tpy prevented the formation of such a regular topology and stabilized penta-coordinated Fe^{II} sites by "framework frustration" (*i.e.*, steric hindrance) and *H*-bonding [74]. It was noted that these sites mimicked the coordination environment of Fe^{II}–porphyrin complexes; thus, a potential use for O_2 binding/activation or CO removal was suggested.

The various ring-assembly methods, as already introduced in Section 1.5, enable the synthesis of 4′-substituted terpyridines. Although this strategy is most commonly used for 2,2′:6′,2″-terpyridines, it can also be applied to its 4,2′:6′,4″- and 3,2″:6′,3″-counterparts. The coordination chemistry of these derivatives has been studied to a significant extent [1, 2, 14]. Firstly, some fundamental lessons can be learned from the coordination of Zn^{II} ions by divergent 4,2′:6′,4″-tpy ligands, which give coordination polymers or macrocycles, depending on the choice of the counterion and the 4′-substituent [13]. When using halide salts, the formation of 1D coordination polymers, which are assembled around one crystallographic screw axis is favored. Thus, the self-assembly of 4′-substituted ligands, as achiral building blocks, provides a chiral polymer chain; whereas crystallization results in right-handed (*P*) and left-handed (*M*) helices being present in the lattice in equal numbers. Such coordination polymers are referred to as racemic or heterochiral ones. On the other hand, homochiral polymers are a conglomerate comprised of enantiomerically pure crystals. For both scenarios, examples have been reported in the literature and, in many cases, one type of polymer is formed alongside the other. Detrimentally, the prediction of the packing mode, as a function of the 4′-substituent, is still not possible. For example, ligand **14a** gave a polymer with a very short helical pitch of *ca.* 7.8 Å (Figure 2.26a/b) [77]. This was due to the interdigitation of neighboring chains of identical handedness. These assemblies

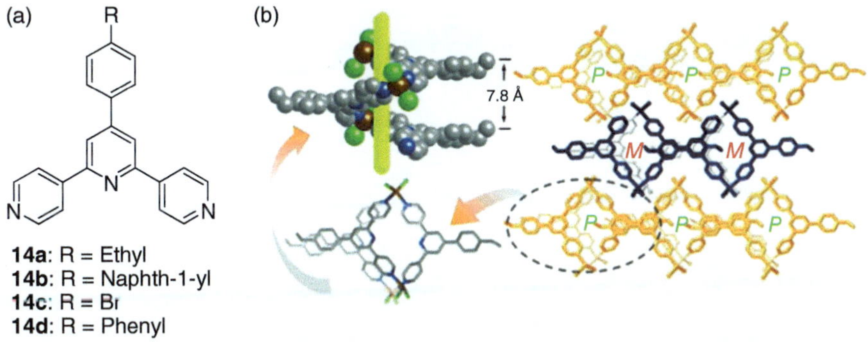

Figure 2.26 (a) Schematic representation of the 4′-substituted 4,2′:6′,4″-terpyridine ligands **14**. (b) Representations of solid state-structure the helical coordination polymer of Zn^{II} ions and ligand **14a**: Left: Side and top views of a helices; right: packing of the homochiral layers [77] / with permission of John Wiley & Sons.

are stabilized *via* ππ-stacking interactions involving the phenylpyridine moieties of adjacent *M*- or *P*-configured chains. The crystal lattice was comprised of alternating homochiral layers parallel to the crystal's *ab*-plane. Whereas the aforementioned racemic coordination polymer derived from $ZnCl_2$ and parent 4′,2′:6′,4″-tpy featured the association of helices of the same handedness *via* ππ-stacking of the pyridine rings in neighboring chains. This mode of interaction gave homochiral 2D sheets [73]. Further examples for hetero- or homochiral polymers based on 4′-substituted 4,2′:6′,4″-tpy ligands can be found in Housecroft's review on this topic [14].

In the case of extended aryl groups in 4′-position, the self-assembly of 4,2′:6′,4″-tpy ligands favors the formation of hexameric metallo-macrocycles. It was assumed that cyclic species were interlocked by ππ-stacking to afford tube-like arrays, thus suppressing the assembly into helical 1D chains [13]. Reprehensively, the tubular stack of macrocycles, derived from ligand **14b** (Figure 2.26a) and Zn^{II} ions is shown in Figure 2.27a [77]. This structure was capable of hosting C_{60} molecules in an unusual fashion: The fullerenes were located in the cavity of a macrocycle and stabilized via ππ-stacking by in-total six naphthyl moieties of the neighboring two macrocycles (three from each). Thus, the host-guest structure featured a highly ordered array in which every second macrocycle along the array hosted a C_{60} guest (Figure 2.27b).

Besides the self-assembly with Zn^{II}–halide salts, 4,2′:6′,2″-terpyridines were also combined with $Zn(OAc)_2$. Zn^{II} ions are well-known for their ability to afford $\{Zn_2(\mu\text{-}O_2CR)_4\}$ clusters with carboxylate anions (R: alkyl). This so-called "paddle-wheel" motif represents a versatile building block, as a node, in coordination polymers or MOFs, which offers a high degree of directionality [79]. Due to its available binding sites in opposing axial positions, incoming ligands will be oriented in a strictly linear fashion and, thus, enforce the formation of 1D coordination polymers when combined with a directional ditopic ligand, such as 4,2′:6′,2″-terpyridine or its derivatives (Figure 2.28a) [14]. Unlike helical coordination polymers, derived from the halide salts, the paddle-wheel motif

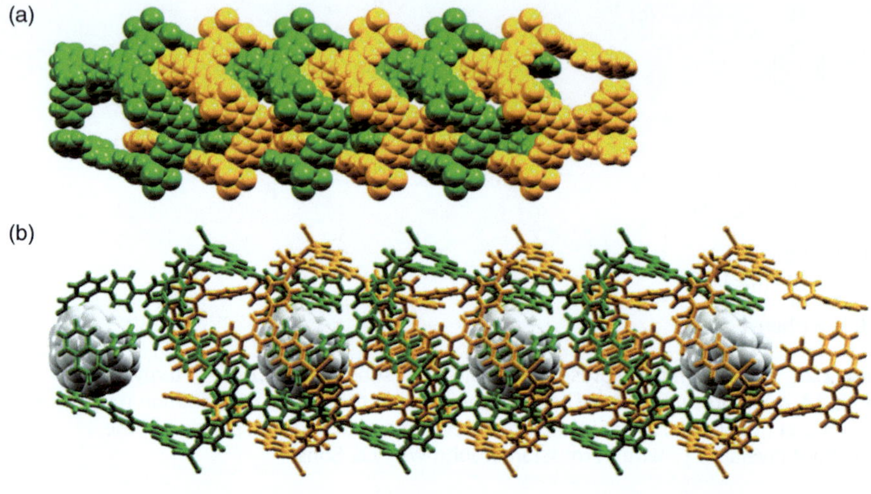

Figure 2.27 (a) Solid-state structure of the tubular assembly of $[Zn(\mathbf{14b})]_6$ metallo-macrocycles. (b) Solid-state structure of the host-guest assembly formed by the inclusion of C_{60} molecules into the tubular structure. In both representations, counterions and solvent molecules were omitted for clarity [13, 78] / with permission of Royal Society of Chemistry.

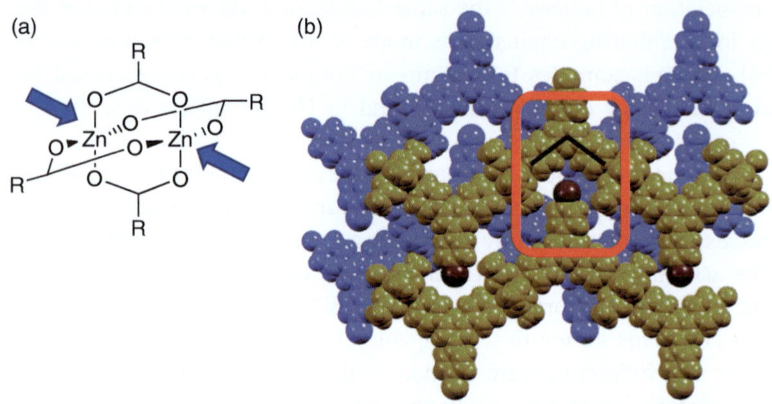

Figure 2.28 (a) Schematic representation of the Zn^{II} paddle-wheel motif (the arrows indicate the vacant axial binding sites). (b) Excerpt from the solid-state structure of the polymer derived from ligand **14c** (Figure 2.26) showing the packing of chains in a face-to-face fashion; thereby, the 4′-substituents are located in the V-shaped cleft ligands of the adjacent chain [80] / with permission Royal Society of Chemistry.

results in a planar zig-zag chain, in which the terpyridine moieties of adjacent chains are stacked in face-to-face fashion to give flat sheets. Thereby, the 4′-substituents along one polymer chain are located within the V-shaped clefts of the respective adjacent chain (the clefts offer sufficient space to accommodate even large aromatic substituents, *e.g.*, biphenyl). As one example, the coordination polymer derived from 4′-bromophenyl-substuted ligand **14c** and $Zn(OAc)_2$ is highlighted here; the packing mode, as determined by XRD analysis, is shown in Figure 2.28b [80].

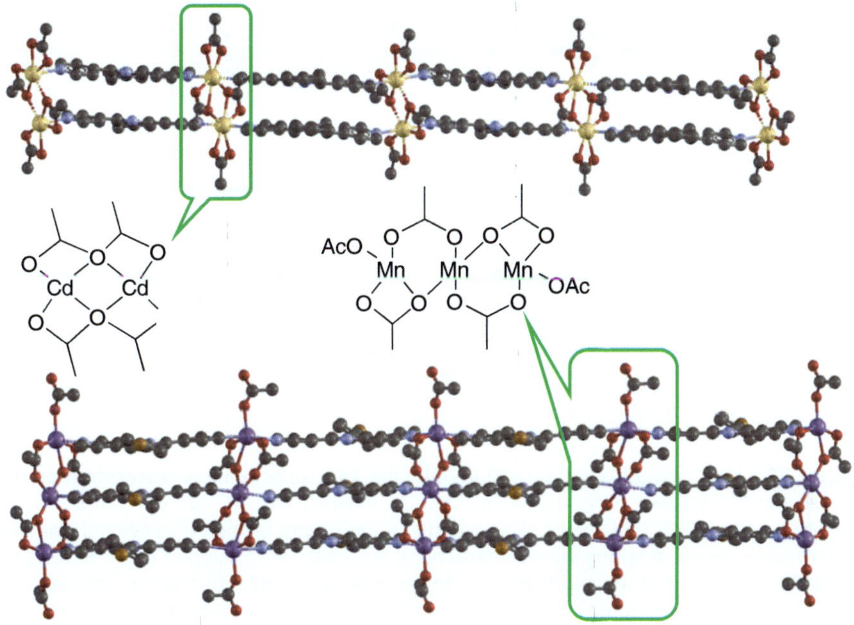

Figure 2.29 Excerpts from the solid-state structures of the coordination polymers derived from $Cd(OAc)_2$/**14d** (top) or $Mn(OAc)_2$/**14c** (bottom). The difference in the number of strands is due to the number of metal centers per connection cluster; each metal center coordinated two tpy ligands in axial positions (hydrogen atoms and solvent molecules were omitted for clarity). Adapted from [80, 81].

Owing to the isostructural nature of the corresponding Cu^{II}-based $\{Cu_2(\mu\text{-}O_2CCH_3)_4\}$ cluster, the coordination polymers incorporating this paddle-wheel motif resembled their Zn^{II}-containing analogs in terms of crystal packing [81]. Whereas in the case of Cd^{II}, a planar cluster motif was observed, which was capable of binding four pyridyl rings *via* the two metal centers in axial positions (Cd^{II} permits higher coordination numbers than Zn^{II}) [81]. Thus, a double-stranded, ladder-type coordination polymer was obtained with ligand **14d**, in which the 4,2′:6′,4″-tpy moieties were stacked in a face-to-face fashion (Figure 2.29). Likewise, the trinuclear Mn^{II} cluster was able to bind six ligand molecules in a highly direct fashion and a triple-stranded coordination polymer with pronounced intrachain $\pi\pi$-stacking interactions was obtained (Figure 2.29) [80]. This study revealed that the structure of the polymer could be controlled *via* the coordination chemistry of the connecting clusters, whereas the influence of the 4′-substituent was only marginal.

The fact that the Cd^{II} ions favor higher coordination numbers was utilized further for the synthesis of other assemblies with higher dimensionality. For example, the self-assembly of $Cd(NO_3)_2 \times 4H_2O$ and ligand **14e** (Figure 2.30a) afforded a coordination network in which hexa-coordinated Cd^{II} centers were present as nodes [82]. Each center coordinated two NO_3^- anions and four ligand molecules *via* one of its outer pyridine rings. This mode of interaction resulted in a highly regular (4,4)-network of undulating chains, which intersected at the Cd^{II}

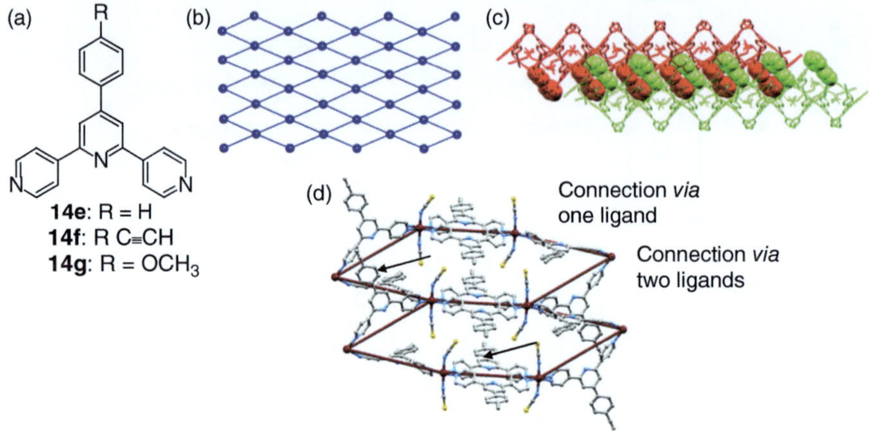

Figure 2.30 (a) Schematic representation of ligand **14e**. (b) Schematic representation of the (4,4)-net of $\{[Cd(\mathbf{14e})_2(ONO_2)_2]\cdot MeOH\cdot CHCl_3\}_n$, in which the Cd^{II} centers as the connection points [82]. (c) Excerpt of the solid-state structure of the Cd^{II}/**14e** coordination polymer showing the stacking *via* ππ-interactions [82]/with permission of Royal Society of Chemistry. (d) Excerpt of the solid-state structure of the Cd^{II}/**14f** coordination polymer showing the (6,3)-net topology [83] / with permission of Royal Society of Chemistry.

centers (Figure 2.30b). These sheets, in turn, were densely stapled *via* ππ-stacking interactions involving the phenyl substituents (Figure 2.30c). Topologically similar (4,4)-nets were also obtained when 4,2′:6′,2″-tpy ligands were combined with $Cd(SCN)_2$; however, this applies only to those ligands, which contained a non-coordinating substituent in 4′-position [84]. The substituent had an influence on the geometry of the rhombuses as well as on the packing behavior. In the case of 4′-*tert*-butyl-4,2′:6′,2″-tpy, as the bridging ligand, a double-layered structure was observed; thereby, the hydrophobic substituents coated the (4,4)-nets. Thus, the crystal structure was comprised of layers with alternating wide and close spacings. Remarkably, a fundamentally different topology was observed for the assembly derived from ligand **14f** [83]. In this case, a (6,3)-net was formed, in which the Cd^{II} nodes were connected by either one or two ligand molecules (Figure 2.30d). It was assumed that *H*-bonding interactions between the ethynyl moieties and SCN^- ligands in adjacent sheets accounted for the deviation from the usual assembly mode.

Finally, Konar *et al.* assembled related (4,4)-nets using ligand **14g** (Figure 2.30) and Co^{II} or Fe^{II} ions [85]. Owing to different conformations of the ligand in the 2D lattice (*i.e.*, planar with Co^{II} *vs.* twisted for Fe^{II}), different 3D packing modes were obtained. These authors analyzed the magnetic properties of the materials. Measurements of the magnetic susceptibility showed a field-induced slow magnetic relaxation of the magnetically isolated Co^{II} centers. Such a behavior is typical for single-molecule magnets, but occurs sparsely in 2D coordination polymers.

In related studies, Constable, Housecroft *et al.* studied the coordination behavior of the 4′-substituted 3,2′:6′,3″-tpy ligands **15** with $Co(SCN)_2$ (Figure 2.31a) [86, 87]. Depending on the length of the ligand's alkoxy tail, different assembly modes were observed. This behavior was attributed to the ability of 3,2′:6′,3″-tpy derivatives to

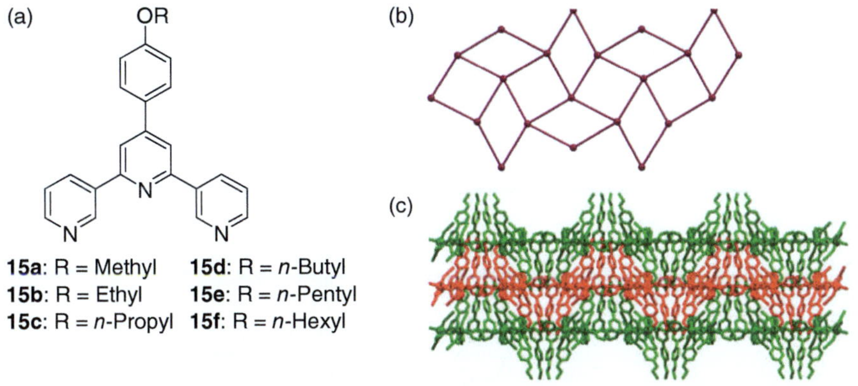

Figure 2.31 (a) Schematic representation of the 4′-substituted 3,2′:6′,3″-terpyridines **15**. (b) Schematic representation of the (4,4)-net, which was observed for the coordination polymers Cd^{II}/**15b**, Cd^{II}/**15c** and Cd^{II}/**15d**. (c) Excerpt from the solid structure of the coordination polymers Cd^{II}/**15b** showing the packing of the cone-shaped (4,4)-nets [87] / with permission of MDPI / CC BY 4.0.

respond to the steric requirements of substituents by changing their conformation and, thus, the directionality of coordination (Figure 2.24a) [2]. It was shown, for example, that the 4-alkoxyphenyl moieties of ligands **15b–d** favored the formation of cone-shaped (4,4)-nets instead of the regular rhomboid motif (Figure 2.31b) [87]. In the crystal lattice, the substituents of one layer were accommodated in the cavities of cone-shaped cavities of the neighboring one. Whereas the rhomboid (4,4)-motif was obtained with ligands **15a** and **15e–f**. However, a difference in the arrangement of the sheets was identified between **15a** on the one hand and **15e–f** on the other. In the former case, eclipsed networks were obtained; whereas the latter two ligands afforded an ABAB-type layering of sheets, in which the substituents were located in the cavities of the adjacent sheets (Figure 2.31c). Moreover, the layer-to-layer distance was larger in the assemblies derived from **15e–f** when compared to those of **15b–d** (Figure 2.31).

The first study devoted to the use of tetratopic, telechelic ligands, *i.e.*, *bis*(4,2′:6′,4″-terpyridine)s and *bis*(3,2′:6′,3″-terpyridine)s, in the field of coordination chemistry was performed by Yoshida *et al.* [88]. Specifically, these authors combined ligands **16** and **17** with *bis*(β-diketonato) Co^{II} complexes (Figure 2.32). In the case of the angular ligand **16**, triply interpenetrating networks were obtained. Housecroft *et al.* first employed the rigid ligand **18**, but then – for the benefit of solubility – opted for the linear ligands **19**, as building blocks for the formation of coordination polymers (Figure 2.32) [89].

Firstly, coordination polymers were assembled using Zn^{II}–halide salts and ligands **19a** and **19d–f** (Figure 2.32) [89–91]. The resulting coordination sheets were comprised of (4,4)-motifs, in which the Zn^{II} centers and ligand molecules represented the linkages and nodes, respectively (Figure 2.33a). The methoxyphenyl substituents were alternatingly located above and below the layers' mean planes. Owing to its shortness, the alkoxy chain of **19a** had only a marginal influence on the

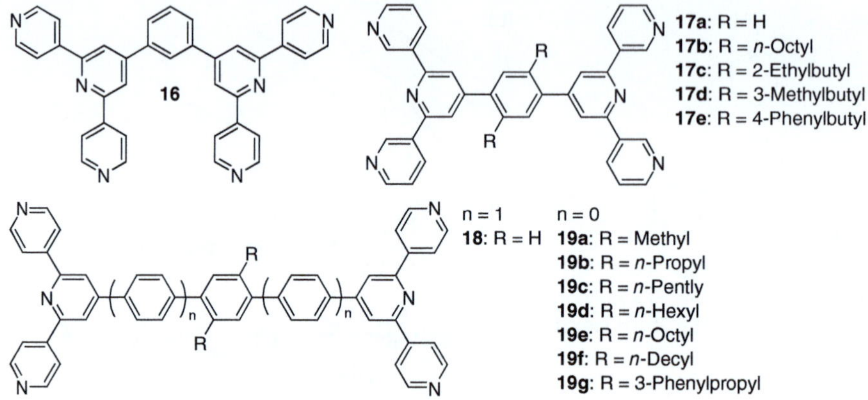

Figure 2.32 Schematic representation of the telechelic ligands **16–19**, which contained ditopic 4,2′:6′,2″- or 3,2′:6′,2″-terpyridine moieties.

overall topology of the coordination compound, *i.e.*, the packing of sheets occurred almost undisturbed by ππ-stacking of 4,2′:6′,4″-tpy moieties of neighboring layers in a face-to-face fashion [90]. Whereas in the case of Zn^{II}/**19e**, the long *n*-octyloxy chains in extended conformations reached through the sheets, thus giving rise to a 2D → 2D parallel interpenetrated assembly (Figure 2.33b) [89, 90]. In subsequent work, Housecroft *et al.* showed that this assembly mode was rather sustained since the same topology was also found for the coordination polymers prepared from **19d** and **19f** [91]. However, the pendant phenyl moieties of **19g** had a profound influence on the ligand's self-assembly with Zn^{II} ions. Unexpectedly, the crystal lattice was comprised of twofold interpenetrating ($6^4{\cdot}8^2$) nbo nets (nbo: niobium-oxide type; Figure 2.33c). Explaining the nomenclature rules of such MOF-type assemblies goes far beyond the scope of this Chapter and, thus, the reader is referred to a comprehensive review by Li, Zhang, Lu, *et al.* for further reading [92]. In short, the ($6^4{\cdot}8^2$) nbo notation means that adjacent four-connecting nodes are all perpendicular to one another. The crystals ($R\bar{3}$ space group) featured hexagonal channels, which ran in parallel along the *c*-axis; the overall solvent-accessible void space was *ca.* 65% of the total crystal's volume (Figure 2.33c) [91]. The entire framework was stabilized by ππ-stacking interactions involving the ligands' phenyl moieties, which lay above ligand sites of the neighboring sheets. This ππ-stacking brought the sheets into close proximity to each other and, eventually, generated the aforementioned channels.

In line with the studies on the self-assembly of ditopic 4,2′:6′,4″-tpy ligands in the presence of thioyanide salts (*vide supra*), Housecroft *et al.* also reported the formation of coordination compounds of the tetratopic ligands **19b/c** with $Co(SCN)_2$ [93, 94]. In both cases, networks of ($6^5{\cdot}8$) cds structure were formed, *i.e.*, 50% of the neighboring 4-connecting nodes were oriented perpendicular to one another (Figure 2.34). In this motif, both components – the Cd^{II} ions and ligand molecules – represented 4-connecting nodes. Analysis of the cell parameters revealed that the 3D structure was capable of accommodating the alkoxy chains, independent of their actual lengths.

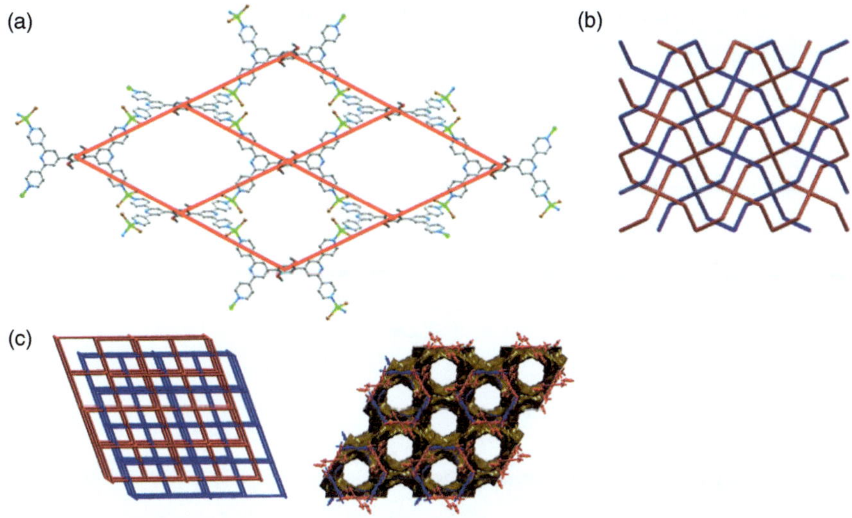

Figure 2.33 (a) Excerpt from the solid-state structure of the Zn^{II}/19a coordination polymer showing the formation of a (4,4)-motif, in which the ligand molecules acted as nodes (hydrogen atoms, counterions, and solvent molecules were omitted for clarity) [90]. (b) Simplified schematic representation of the topology in 2D → 2D parallel interpenetrating sheets (for clarity, the two layers are shown in different colorations) [90]/with permission of Elsevier. (c) Left: Simplified schematic representation of the topology of the interpenetrated nbo nets in the Zn^{II}/**19g** coordination polymer. Right: Excerpt from the solid-state structure of this same material showing the void space in the hexagonal channels [91] / with permission of Royal Society of Chemistry.

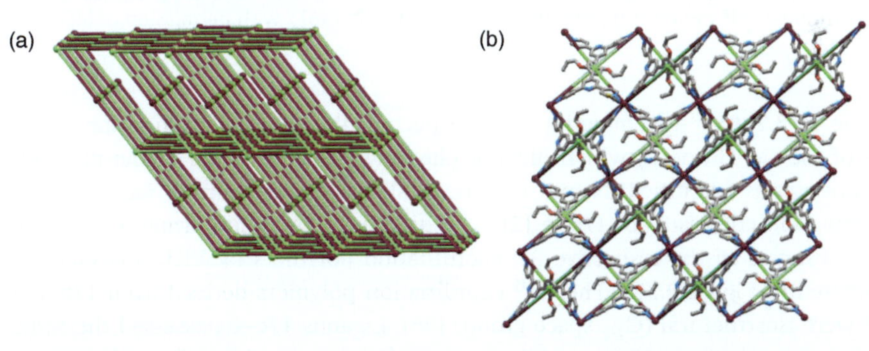

Figure 2.34 (a) Simplified schematic representation of the $(6^5 \cdot 8)$ cds network structure. (b) Excerpt from the solid-state structure of the Co^{II}/**19b** coordination polymer (hydrogen atoms and solvent molecules were omitted for clarity); for comparison, the topological motif of the structure is overlaid [93] / CC BY 4.0.

When the tetratopic ligands **17b–e** were used in combination with $Co(SCN)_2$, the situation was complicated, owing to the different possible ligand conformers, which might be involved in coordination. All these ligands afforded 3D coordination networks, but with distinct differences regarding structure and topology [94–96]. Moreover, conformational freedom is also given regarding the rotation

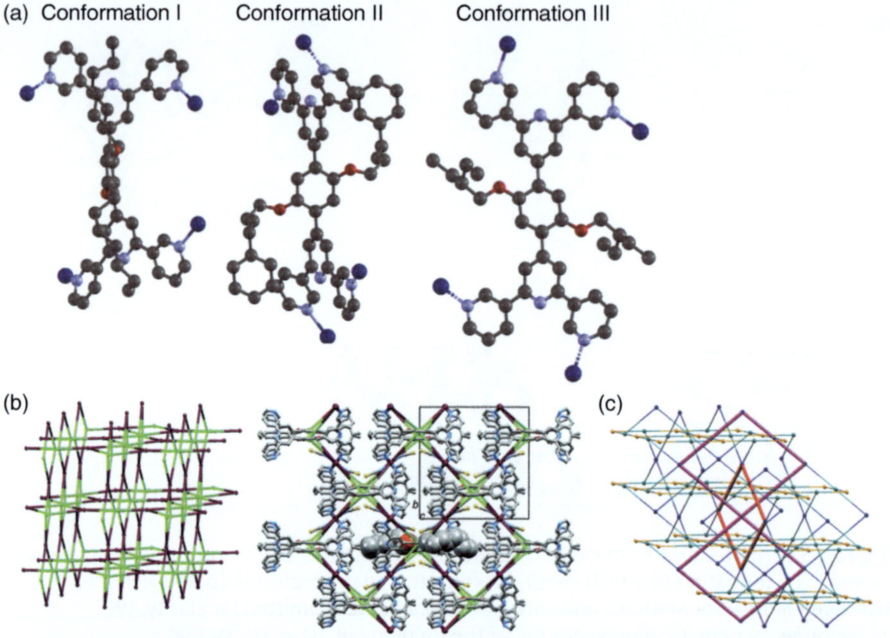

Figure 2.35 (a) Excerpts from the solid state structures of the $Co(SCN)_2$/**17b**, $Co(SCN)_2$/**17e** and $Co(SCN)_2$/**17c** showing the conformations of the respective ligands (hydrogen atoms were omitted for clarity). Adapted from [2, 94, 95]. (b) Left: Schematic representation of the ($4^2 \cdot 8^4$) lvt net, as present in $Co(SCN)_2$/**17b**. Right: Excerpt from the solid-state structure of this same coordination polymer; the topological structure is superimposed for comparison [95] / with permission of Royal Society of Chemistry. (c) Simplified schematic representation of the trinodal self-penetrating topology of $Co(SCN)_2$/**17c** [96] / with permission of MDPI / CC BY 4.0.

of one of 3,2′:6′,3″-tpy moieties with respect to the other. This rotation about one of the C—C bonds, which did not play any significant role in the previous examples, was identified as one important structure-determining factor in the coordination chemistry of **17b–e** [2]. The different ligand conformations, which were observed in the respective 3D coordination polymers by XRD analysis, are compared in Figure 2.35a. The Co^{II} coordination polymers derived from **17c** and **17d** were isostructural ($C_{2/c}$ space group) [96]. Ligands **17c–e** possessed the same conformation in their 3D assemblies with $Co(SCN)_2$, in which the 3,2′:6′,2″-tpy moieties were almost coplanar, but not in-plane with the central phenyl ring (conformer II in Figure 2.35a). Whereas ligand **17b** adopted a conformation, which enabled the coordination of the planar Co^{II} nodes in an almost perfect tetrahedral topology (conformer I in Figure 2.35a). This arrangement resulted in a binodal ($4^2 \cdot 8^4$) lvt structure (Figure 2.35b,c). Thereby, the ligands' side chains reached into the lattice's cavities and interacted with the 3,2′:6′,3′-tpy moieties *via* CH···π binding. Whereas a ($6^5 \cdot 8$) cds network structure was obtained with ligand **17e** (cds: cadmium sulfate). This solid-state structure resembled those which were also found for coordination polymers Co^{II}/**19b** and Co^{II}/**19c** (Figure 2.34). This is somewhat

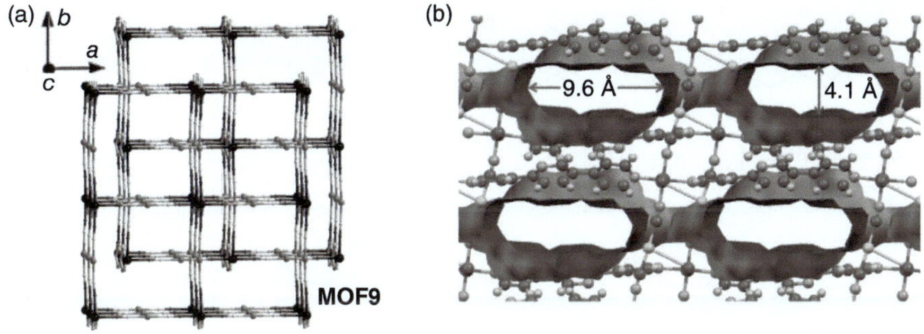

Figure 2.36 (a) Simplified schematic representation of the 3D-interpenetrated structure of **MOF9** (view along the *c*-axis). (b) Excerpt from the solid-state structure of **MOF9** showing its 1D channel (view along the *b*-axis) [97] / with permission of John Wiley & Sons.

remarkable since ligand **17e** provided a different directionality in coordination than **19b/c** and possessed the bulky phenyl moieties in the side chains. Unlike the phenyl groups of **19g**, these groups did not participate in the self-assembly process of **17e**. Finally, ligands **17c** and **17d**, which possessed branched side chains, both afforded trinodal self-penetrating nets of unusual $(6^2{\cdot}84)(6^4{\cdot}8^2)(6^5{\cdot}8)^2$ topology [96]. In this arrangement, the network contained four 4-connecting nodes; the net structure was trinodal, due to the topological equivalence of two ligand nodes.

The 4,2′:5′,4″-terpyridine ligand represents the final isomer to be mentioned in this Section. This ligand possesses the same vectoral properties as 4,4′-bpy, *i.e.*, it can be seen as a linear ditopic spacer, in which the central pyridine moiety is not involved in coordination (Figure 2.13). This feature is comparable to the behavior of the previously discussed 4,2′:6′,2″- and 3,2′:6′,3″-isomers. The presumably first coordination compound of this ligand was reported by Ni, Tong *et al.* [97]. Unlike the aforementioned Hofmann-type MOF, which contained the angular 4,2″;6′,4″-tpy ligand [74], linear 4,2′:5′,4″-tpy afforded **MOF9** with a regular topology upon self-assembly with Fe^{II} and $[Au(CN)_2]^-$ ions. **MOF9** possessed a 3D structure of interpenetrated networks, which generated 1D channels accessible for solvent molecules (Figure 2.36). This MOF exhibited a solvent-dependent SCO behavior with abrupt, stepwise, and hysteretic features. It was assumed that the protic solvents of varying polarity established *H*-bonding interactions with the ligands' vacant binding sites within the channels. These interactions were believed to contribute to the lattice cooperativity, which, in turn, influenced the material's SCO behavior.

A Zn^{II}-based MOF was prepared from $Zn(OAc_2)$, 4,2′:5′,4″-tpy and 5-hydroxyisophthalic acid, as the ancillary ligand [98]. The 1D chains of Zn^{II} ions and tpy ligands were interlinked by the dicarboxylate on the one hand and *H*-bonding interactions on the other. As a result, a highly regular structure with 3D channels was obtained. This environment was used for the sensing of metal ions, which impacted the MOF's luminescence. In particular, Fe^{III} and Cu^{II} ions were capable of quenching the emission almost quantitatively. The efficiency of quenching was attributed to the presence of the two binding sites, vacant pyridyl moieties, and OH-groups, on the

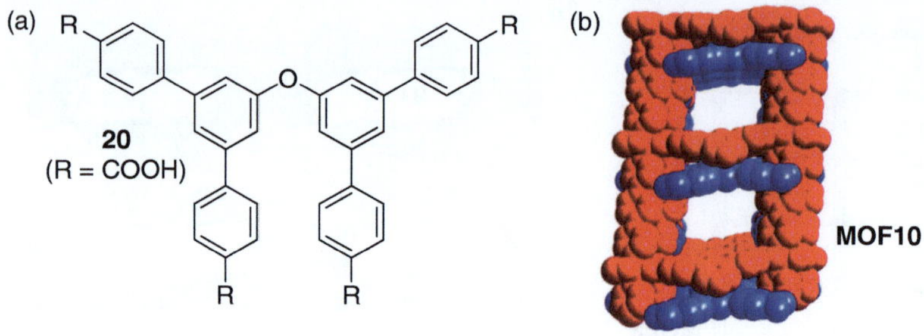

Figure 2.37 (a) Schematic representation of the tetracarboxylic acid **20**. (b) Simplified representation of the **MOF10** structure showing the twofold interpenetrating framework [99] / with permission of Royal Society of Chemistry.

channels' surfaces. For a couple of years, the use of 4,2′:5′,4″-tpy, as a building block for MOFs has been flourishing. In one of these studies, Gao and Zheng prepared pillar-layered MOFs, as designated CO_2 absorbents [99]. One of these MOFs was comprised of Co^{II} ions, 4,2′:5′,4″-tpy and the tetracarboxylic acid **20** (Figure 2.37a). The structure of **MOF10** consisted of a twofold interpenetrating 3D framework (Figure 2.37b) and possessed 1D channels, which were used for the low-pressure CO_2 capture at room temperature, *i.e.*, under ambient conditions. However, **MOF10** revealed only a moderate CO_2 capability for storage (*ca.* 33 cm^3 g^{-1}), despite its relatively large solvent-accessible volume of >40%. One further MOF, which also had a pillared architecture, was reported by Mudassir, Jin *et al.*; these authors used a Cd^{II} salt and 1,3,5-benzenetricarboxylic acid, instead of a Co^{II} salt and **20** [100]. This MOF possessed a 6-node pcu topology with the assigned $(4^{12}{\cdot}6^3)$ point symbol. The assembly was tested as a fluorescent sensor for agricultural pesticides. In particular, pymetrozine, representing a toxic and cancerogenic insecticide, efficiently quenched the MOF's emission. The detection limit for this environmental pollutant was as low as 86 ppm. The efficiency in quenching was attributed to energy-/electron-transfer processes with the framework's channels, which were facilitated by the accessible binding sites.

The tripodal, semirigid tricarboxylic acid **21** (Figure 2.38a) represents a versatile component for MOF synthesis, due to its flexibility and shape [101]. **MOF11**, which was prepared from this auxiliary ligand, 4,2′:5′,4″-tpy and Zn^{II} ions, featured a twofold interpenetrating topology with rectangular channels (Figure 2.38b) [102]. This particular MOF was employed as a photocatalyst for the degradation of antibiotics in aqueous solution. It was proposed that the antibiotics were bound in the channels *via* electrostatic interactions and that their degradation was initiated by the photo-generation of hydroxyl radicals. This study revealed a particularly good recognition of nitrofurazone by **MOF11** and, thus, this antibiotic was degraded more efficiently than dimetridazole and oxytetracycline (Figure 2.38c). Related to this, Kumar, Guo, Jin *et al.* obtained an MOF from the reaction of hydrothermal synthesis of $Zn(OAc)_2$, 4,2′:5′,4″-tpy and 1,3,5-benzenetricarboxylic acid under hydrothermal conditions [103]. Paddle-wheel Zn^{II} moieties (see Figure 2.28) were linked by the organic ligands to afford an environmentally stable MOF, which displayed intense

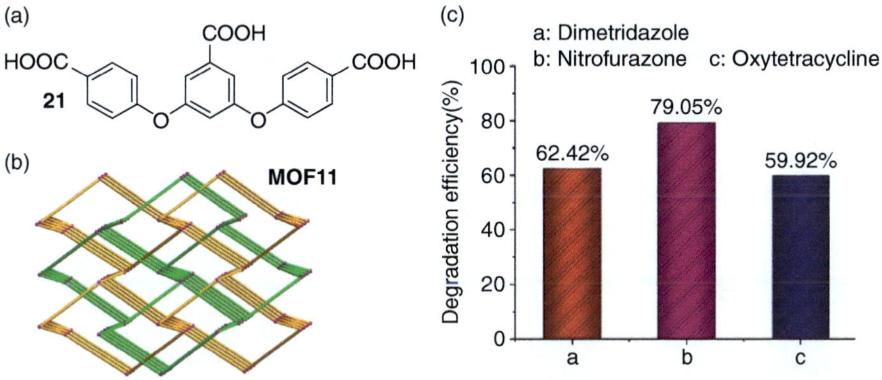

Figure 2.38 (a) Schematic representation of the tripodal auxiliary ligand **21**. (b) Simplified schematic representation of the twofold interpenetrating 3D topology of **MOF11**. (c) Comparison of the efficiency of the photocatalytic degradation of some antibiotics [102] / with permission of Elsevier.

green emission in aqueous media. This MOF was also used for the recognition of antibiotics. Unlike the previous example, sensing due to emission quenching was the focus of interest. From the in-total 16 analytes, sulfasalazine induced a rapid and almost quantitative quenching (lower detection limit of 1.65 ppm), whereas the other antibiotics (including the three examples from the aforementioned degradation study) were sensed less efficiently. The MOF could be reactivated by rinsing with EtOH and used in repetitive sensing experiments without loss of activity and structural changes (according to powder XRD analysis). According to DFT calculation and Hirshfeld surface analysis, emission quenching was due to charge-transfer processes on one hand and competitive absorption by the analyte on the other. When comparing the results of the former two studies, one can conclude that the photophysical detection of analytes by MOFs depends on both the chemical structure of the former and the topology of the latter. Thus, non-covalent analyte binding in the MOF channels as well as matching of the frontier orbitals have also to be considered (Figure 2.38).

The photocatalytic degradation of organic dyes represents one further important application of MOFs in the context of water preparation [104]. In one recent example, Muddassir, Zhong, Wang, Jin *et al.* reported a Cd^{II}-based coordination compound, which was comprised of $\pi\pi$-stacked 1D double chains (4,2′:5′,4″-tpy and 4-nitrobenzoic acid, as the ligands) [105]. Upon irradiation with UV light, rhodamine-B was degraded with high efficiency (*i.e.*, >95%), whereas the degradation efficiency for methyl blue, methyl violet, and methyl orange was significantly lower (*i.e.*, 29–59%). As for the photocatalytic degradation of antibiotics, the superoxide radical anion and hydroxyl radical were identified by radical-trapping experiments as the active species occurring during the degradation process. Whereas methyl blue was degraded with high efficiency over methyl violet and acid-chrome blue, when **MOF12** was used, as the photocatalyst (Figure 2.39b) [106]. This MOF possessed a 3D binodal (3,4)-connected network structure, which

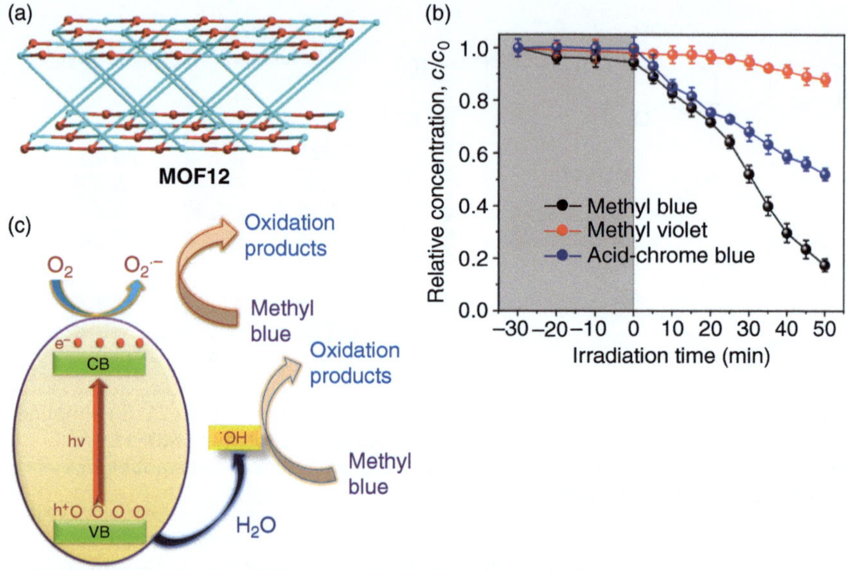

Figure 2.39 (a) Simplified schematic representation of the topology of **MOF12**. (b) Comparison of the efficiency of the dye degradation using **MOF12**, as the photocatalyst. (c) Schematic representation of the proposed mechanism of the photocatalytic dye degradation [106] / with permission of Elsevier.

was formed by Zn^{II} ions, 4,2′:5′,4″-tpy, and 5-aminoisophthalic acid, as the auxiliary ligand (Figure 2.39a). It was assumed that the dye molecules were adsorbed in the voids of the **MOF12** structure, where the photocatalytic degradation occurred. As depicted in Figure 2.39c, electrons are excited into the MOF's conduction band, leaving behind holes in the valence band. The thusly formed charge carriers contribute to the degradation by forming $O_2^{\cdot-}$ and $HO^{\cdot-}$ radicals, respectively. However, the related MOF, which contained 5-nitroisophthalic acid, represented a

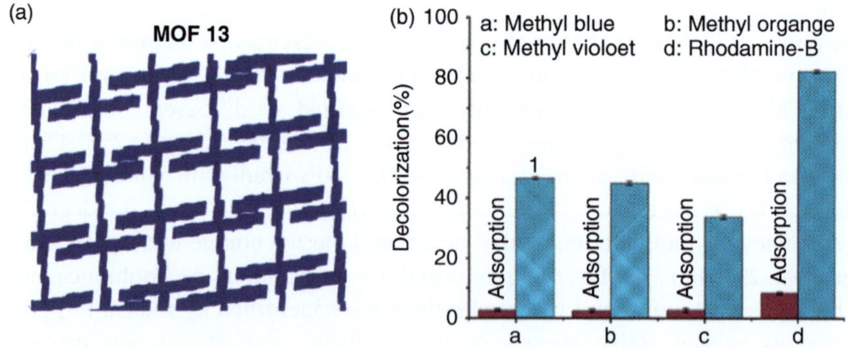

Figure 2.40 (a) Simplified schematic representation of the topology of **MOF13**. (b) Comparison of the photocatalytic decoloration efficiency of **MOF13** against various organic dyes [108] / with permission of Elsevier.

potent photocatalyst for the degradation of methyl orange [107]. It was concluded that the alteration of the substituent on the auxiliary ligand (*i.e.*, changing from $-NH_2$ to $-NO_2$) had an influence on both the MOF's molecular and electronic structures.

Finally, **MOF13** was capable of photocatalyzing the degradation of rhodamine in a highly efficient fashion (*i.e.*, decoloration rate of *ca.* 82%; Figure 2.40) [108]. This MOF contained 4-nitrobenzoic acid, as the auxiliary ligand, and was, thus, reminiscent of the aforementioned Cd^{II}-containing MOF [105]. Single-crystal XRD analysis revealed that the structure was comprised of linear 1D chains, in which the Zn^{II} ions were coordinated in tetragonal geometry [108]. The chains formed a 2D framework due to an interplay of *H*-bonding and ππ-stacking interactions.

2.4 Concluding Remarks

The term "terpyridine isomer game" was coined by Housecroft and Constable to describe the multifaceted coordination chemistry, which is based on the various constitutional isomers of the famous 2,2′:6′,2″-terpyridine ligand [2]. From the in-total 47 isomers, only a handful of examples have been used for the formation of metal complexes or assemblies of higher dimensionality, such as linear coordination polymers, 2D coordination networks, or MOFs. Other isomers have only sparsely been employed in the context of coordination chemistry (if at all), and some isomers have not even been synthesized yet.

Up to now, mononuclear complexes of non-2,2′:6′,2″-tpy ligands have only been studied to a limited extent. The only systematic study in this respect was performed by Klein *et al.* in 2023 [11]. This work indicates the scope and limitations when aiming for the synthesis of new emitters or catalysts based on transition-metal ions and nonclassical tpy ligands.

Whereas the use of these ligands for the assembly of 2D or 3D coordination compounds has quite recently gained momentum. This field of research was pioneered by Fujita *et al.* as well as Constable, Housecroft *et al.*, who were interested in metallo-supramolecular macrocycles and nanocages [51] on the one hand and coordination polymers on the other [1, 2]. Metallo-nanocages of defined architecture can be obtained *via* metallo-supramolecular self-sorting of cautiously designed ligands. This topic represents a currently highly active field of research, in particular with respect to sophisticated oligotopic 2,2′:6′,2″-tpy ligands [109]. The elegant approaches to assembling advanced 2D and 3D architectures by transition-metal ion complexation are discussed separately in Chapter 10.

Regarding 2D coordination polymers and MOFs, various tpy isomers have already been utilized – alone or in combination with carboxylates, as auxiliary ligands. First of all, the proper choice of molecular components enables crystal engineering, *i.e.*, the assembly of solid-state structures with variable topologies. However, the involved self-assembly processes are far from being fully understood, and unexpected structural motifs might even be obtained [1, 2]. Therefore, fundamental studies are still devoted to understanding how a ligand's molecular structure influences the topology of its coordination compound with metal ions. Nonetheless, practical use of terpyridine-based MOFs has also been proposed. The range of potential applications ranges from single-molecule magnets [74, 97] to the sensing and/or degradation of organic pollutants in water [102, 105, 106, 108].

Therefore, the "terpyridine isomer game" is definitely not yet over. Quite the contrary, one might expect substantial progress in this field, for example, with respect to photophysically or electrochemically active assemblies: emitters, electrochromic materials, catalysts, *etc.* One might assume further that artificial intelligence, *i.e.*, computer-aided crystal engineering, will help control and eventually understand the self-assembly of the various terpyridine ligands [110].

References

1 C. E. Housecroft, E. C. Constable, *Molecules* 2021, 26, 3110.

2 C. E. Housecroft, E. C. Constable, *Chem. Commun.* 2020, 56, 10786–10794.

3 E. Noga, *Fach. Mitt. Tabakregie* 1914, 1 & 2.

4 F. Kuffner, *Fach. Mitt. Tabakregie* 1956, 18–19.

5 J. W. Orr, Harper's Magazine 1855, 11, LXI - June.

6 P. Jacob III, M. L. Goniewicz, C. M. Havel, S. F. Schick, N. L. Benowitz, *Chem. Res. Toxicol.* 2013, 26, 1615–1631.

7 A. Díez-Izquierdo, P. Cassanello-Peñarroya, C. Lidón-Moyano, N. Matilla-Santander, A. Balaguer, J. M. Martínez-Sánchez, *Environ. Res.* 2018, 167, 341–371.

8 N. J. Aquilina, C. M. Havel, P. Cheung, R. M. Harrison, K.-F. Ho, N. L. Benowitz, P. Jacob III, *Environ. Int.* 2021, 150, 106417.

9 F. Kröhnke, *Synthesis* 1976, 1–24.

10 J. Wang, G. S. Hanan, *Synlett* 2005, 1251–1254.

11 L. Payen, L. Kletsch, T. Lapić, M. Wickleder, A. Klein, *Inorganics* 2023, 11, 174.

12 B. Butschke, H. Schwarz, *Chem. Sci.* 2012, 3, 308–326.

13 C. E. Housecroft, *Dalton Trans.* 2014, 43, 6594–6604.

14 C. E. Housecroft, *CrystEngComm* 2015, 17, 7461–7468.

15 C. E. Housecroft, E. C. Constable, *J. Inorg. Organomet. Polym. Mater.* 2018, 28, 414–427.

16 C. E. Housecroft, E. C. Constable, *Chimia* 2019, 73, 462.

17 E. C. Constable, A. M. W. Cargill Thompson, J. Cherryman, T. Liddiment, *Inorg. Chim. Acta* 1995, 235, 165–171.

18 O. Bysewski, M. Sittig, A. Winter, B. Dietzek-Ivanšić, U. S. Schubert, *Coord. Chem. Rev.* 2024, 498, 215441.

19 L. J. Kershaw Cook, M. A. Halcrow, *Polyhedron* 2015, 87, 91–97.

20 T.-a. Koizumi, T. Tomon, K. Tanaka, *Bull. Chem. Soc. Jpn.* 2003, 76, 1969–1975.

21 Á. Vivancos, C. Segarra, M. Albrecht, *Chem. Rev.* 2018, 118, 9493–9586.

22 L. Kletsch, R. Jordan, A. S. Köcher, S. Buss, C. A. Strassert, A. Klein, *Molecules* 2021, 26, 5051.

23 C. Cebrián, M. Mauro, *Beilstein J. Org. Chem.* 2018, 14, 1459–1481.

24 E. C. Constable, C. E. Housecroft, P. Kopecky, C. J. Martin, I. A. Wright, J. A. Zampese, H. J. Bolink, A. Pertegas, *Dalton Trans.* 2013, 42, 8086–18103.

25 S. Encinas, A. F. Morales, F. Barigelletti, A. M. Barthram, C. M. White, S. M. Couchman, J. C. Jeffery, M. D. Ward, D. C. Grills, M. W. George, *J. Chem. Soc. Dalton Trans.* 2001, 22, 3312–3319.

26 T. Rawling, C. Austin, F. Buchholz, S. B. Colbran, A. M. McDonagh, *Inorg. Chem.* 2009, 48, 3215–3227.

27 C. Lentz, O. Schott, T. Auvray, G. S. Hanan, B. Elias, *Dalton Trans.* 2019, 48, 15567–15576.

28 C. Lentz, O. Schott, T. Auvray, G. Hanan, B. Elias, *Inorg. Chem.* 2017, 56, 10875–10881.

29 G. Li, D. Zhu, X. Wang, Z. Su, M. R. Bryce, *Chem. Soc. Rev.* 2020, 49, 765–838.

30 M. Schulz, M. Karnahl, M. Schwalbe, J. G. Vos, *Coord. Chem. Rev.* 2012, 256, 1682–1705.

31 S. Soman, G. Singh Bindra, A. Paul, R. Groarke, J. C. Manton, F. M. Connaughton, M. Schulz, D. Dini, C. Long, M. T. Pryce, J. G. Vos, *Dalton Trans.* 2012, 41, 12678–12680.

32 G. Singh Bindra, M. Schulz, A. Paul, R. Groarke, S. Soman, J. L. Inglis, W. R. Browne, M. G. Pfeffer, S. Rau, B. J. MacLean, M. T. Pryce, J. G. Vos, *Dalton Trans.* 2012, 41, 13050–13059.

33 T. Kowacs, L. O'Reilly, Q. Pan, A. Huijser, P. Lang, S. Rau, W. R. Browne, M. T. Pryce, J. G. Vos, *Inorg. Chem.* 2016, 55, 2685–2690.

34 Q. Pan, F. Mecozzi, J. P. Korterik, D. Sharma, J. L. Herek, J. G. Vos, W. R. Browne, A. Huijser, *J. Phys. Chem. C* 2014, 118, 20799–20806.

35 K. Hasan, L. Donato, Y. Shen, J. D. Slinker, E. Zysman-Colman, *Dalton Trans.* 2014, 43, 13672–13682.

36 S. B. Meier, D. Tordera, A. Pertegás, C. Roldán-Carmona, E. Ortí, H. J. Bolink, *Mater. Today* 2014, 17, 217–223.

37 Q. Pei, G. Yu, C. Zhang, Y. Yang, A. J. Heeger, *Science* 1995, 269, 1086–1088.

38 K. Youssef, Y. Li, S. O'Keeffe, L. Li, Q. Pei, *Adv. Funct. Mater.* 2020, 30, 1909102.

39 A. F. Henwood, E. Zysman-Colman, in *Iridium(III) in Optoelectronic and Photonics Applications*, (Ed.: E. Zysman-Coleman), John Wiley & Sons Ltd., Chichester, 2017, pp. 275–357.

40 H. Iranmanesh, K. S. A. Arachchige, M. Bhadbhade, W. A. Donald, J. Y. Liew, K. T. C. Liu, E. T. Luis, E. G. Moore, J. R. Price, H. Yan, J. Yang, J. E. Beves, *Inorg. Chem.* 2016, 55, 12737–12751.

41 T. Mede, M. Jäger, U. S. Schubert, *Chem. Soc. Rev.* 2018, 47, 7577–7627.

42 Y. Li, X. Liu, L. Tan, *Dyes Pigments* 2021, 192, 109406.

43 W. Zhang, Y. Sun, J. Wang, X. Ding, E. Yang, L. L. Martin, D. Sun, *J. Inorg. Biochem.* 2021, 216, 111339.

44 R. Keuper, N. Risch, *Z. Naturforsch. B* 1995, 50, 1115–1120.

45 A. Winter, C. Ulbricht, E. Holder, N. Risch, U. S. Schubert, *Aust. J. Chem.* 2006, 59, 773–782.

46 E. Altuntaş, A. Winter, A. Baumgaertel, R. M. Paulus, C. Ulbricht, A. C. Crecelius, N. Risch, U. S. Schubert, *J. Mass Spectrom.* 2012, 47, 34–40.

47 M. Maekawa, T. Hayashi, K. Sugimoto, T. Okubo, T. Kuroda-Sowa, *Inorg. Chim. Acta* 2019, 497, 119088.

48 E. C. Constable, in *Advances in Inorganic Chemistry*, Vol. 30, (Ed.: H. J. Emeléus), Academic Press, London, 1986, pp. 69–121.

49 M. Aoyagi, K. Biradha, M. Fujita, *J. Am. Chem. Soc.* 1999, 121, 7457–7458.

50 M. Yoshizawa, T. Kusukawa, M. Fujita, S. Sakamoto, K. Yamaguchi, *J. Am. Chem. Soc.* 2001, 123, 10454–10459.

51 M. Tominaga, M. Fujita, *Bull. Chem. Soc. Jpn.* 2007, 80, 1473–1482.

52 S. Tashiro, M. Tominaga, T. Kusukawa, M. Kawano, S. Sakamoto, K. Yamaguchi, M. Fujita, *Angew. Chem. Int. Ed.* 2003, 42, 3267–3270.

53 Y. Kubota, K. Biradha, M. Fujita, S. Sakamoto, K. Yamaguchi, *Bull. Chem. Soc. Jpn.* 2002, 75, 559–565.

54 J. Zhang, Q. Qiu, Q. Xiang, S. Ren, C. Zhang, *J. Solid State Chem.* 2021, 294, 121849.

55 F. Thébault, S. A. Barnett, A. J. Blake, C. Wilson, N. R. Champness, M. Schröder, *Inorg. Chem.* 2006, 45, 6179–6187.

56 C.-L. Zhanh, H.-Y. Wang, H.-G. Zheng, *Chinese J. Inorg. Chem.* 2016, 32, 859–863.

57 J. Perez Barrio, J.-N. Rebilly, B. Carter, D. Bradshaw, J. Bacsa, A. Y. Ganin, H. Park, A. Trewin, R. Vaidhyanathan, A. I. Cooper, J. E. Warren, M. J. Rosseinsky, *Chem. Eur. J.* 2008, 14, 4521–4532.

58 R. Chen, F. Zhou, B. Sheng, Z. Zhang, Q. Yang, Y. Yang, Q. Ren, Z. Bao, *ACS Sustain. Chem. Eng.* 2022, 10, 11330–11337.

59 R. Vaidhyanathan, D. Bradshaw, J.-N. Rebilly, J. P. Barrio, J. A. Gould, N. G. Berry, M. J. Rosseinsky, *Angew. Chem. Int. Ed.* 2006, 45, 6495–6499.

60 T. Wang, L. Qin, C. Zhang, H. Zheng, *RSC Adv.* 2015, 5, 64514–64519.

61 C.-L. Zhang, H. Xu, X.-W. Song, Y. Meng, J.-J. Chen, *Inorg. Chem. Commun.* 2017, 84, 229–233.

62 Z.-Z. Lu, R. Zhang, Z.-R. Pan, Y.-Z. Li, Z.-J. Guo, H.-G. Zheng, *Chem. Eur. J.* 2012, 18, 2812–2824.

63 J. Zhang, Q. Xiang, Y. Zhu, J. Yang, Y. Song, C. Zhang, *Cryst. Growth Des.* 2021, 21, 3225–3233.

64 Z. Zhang, C. J. Gómez-García, Q. Wu, J. Xin, H. Pang, H. Ma, D. Chai, S. Li, C. Zhao, *Inorg. Chem.* 2022, 61, 11830–11836.

65 Z. Zhuanfang, F. Enze, Z. Shuangqi, W. Zhaojun, Z. Wenzhi, Z. Ming, D. GuoHua, *J. Solid State Chem.* 2023, 323, 124066.

66 D. A. Beauchamp, S. J. Loeb, *Supramol. Chem.* 2005, 17, 617–622.

67 M. Maekawa, M. Yabuta, K. Sugimoto, T. Okubo, T. Kuroda-Sowa, *Inorg. Chim. Acta* 2021, 528, 120628.

68 A. Winter, U. S. Schubert, *Chem. Soc. Rev.* 2016, 45, 5311–5357.

69 M. Higuchi, *Polym. J.* 2009, 41, 511–520.

70 M. Higuchi, *J. Syn. Org. Chem. Jpn.* 2023, 81, 1089–1095.

71 J.-J. Kim, O. K. Kwak, J. K. Park, K. S. Min, Y.-J. Yoon, *J. Incl. Phenom. Macrocycl. Chem.* 2014, 79, 495–502.

72 D. Fujita, H. Yokoyama, Y. Ueda, S. Sato, M. Fujita, *Angew. Chem. Int. Ed.* 2015, 54, 155–158.

73 M. Barquín, J. Cancela, M. J. González Garmendia, J. Quintanilla, U. Amador, *Polyhedron* 1998, 17, 2373–2378.

74 L. Xu, Z. Xie, K. A. Zenere, J. K. Clegg, E. Kenny, N. J. Rijs, G. N. L. Jameson, C. J. Kepert, B. J. Powell, S. M. Neville, *Dalton Trans.* 2022, 51, 9596–9600.

75 K. Senthil Kumar, M. Ruben, *Coord. Chem. Rev.* 2017, 346, 176–205.

76 M. C. Muñoz, J. A. Real, *Coord. Chem. Rev.* 2011, 255, 2068–2093.

77 X.-Z. Li, M. Li, Z. Li, J.-Z. Hou, X.-C. Huang, D. Li, *Angew. Chem. Int. Ed.* 2008, 47, 6371–6374.

78 E. C. Constable, C. E. Housecroft, S. Vujovic, J. A. Zampese, *CrystEngComm* 2014, 16, 328–338.

79 M. Köberl, M. Cokoja, W. A. Herrmann, F. E. Kühn, *Dalton Trans.* 2011, 40, 6834–6859.

80 E. C. Constable, G. Zhang, E. Coronado, C. E. Housecroft, M. Neuburger, *CrystEngComm* 2010, 12, 2139–2145.

81 E. C. Constable, C. E. Housecroft, M. Neuburger, J. Schönle, S. Vujovic, J. A. Zampese, *Polyhedron* 2013, 60, 120–129.

82 E. C. Constable, G. Zhang, C. E. Housecroft, M. Neuburger, J. A. Zampese, *CrystEngComm* 2009, 11, 2279–2281.

83 E. C. Constable, C. E. Housecroft, M. Neuburger, S. Vujovic, J. A. Zampese, G. Zhang, *CrystEngComm* 2012, 14, 3554–3563.

84 Y. M. Klein, A. Prescimone, E. C. Constable, C. E. Housecroft, *Polyhedron* 2016, 103, 58–65.

85 A. K. Mondal, S. Khatua, K. Tomar, S. Konar, *Eur. J. Inorg. Chem.* 2016, 3545–3552.

86 D. Rocco, A. Prescimone, Y. M. Klein, D. J. Gawryluk, E. C. Constable, C. E. Housecroft, *Polymers* 2019, 11, 1224.

87 D. Rocco, A. Prescimone, E. C. Constable, C. E. Housecroft, *Molecules* 2020, 25, 1663.

88 J. Yoshida, S.-i. Nishikiori, H. Yuge, *J. Coord. Chem.* 2013, 66, 2191–2200.

89 E. C. Constable, C. E. Housecroft, S. Vujovic, J. A. Zampese, *CrystEngComm* 2014, 16, 3494–3497.

90 S. Vujovic, E. C. Constable, C. E. Housecroft, C. D. Morris, M. Neuburger, A. Prescimone, *Polyhedron* 2015, 92, 77–83.

91 Y. M. Klein, A. Prescimone, M. Neuburger, E. C. Constable, C. E. Housecroft, *CrystEngComm* 2017, 19, 2894–2902.

92 D.-S. Li, Y.-P. Wu, J. Zhao, J. Zhang, J. Y. Lu, *Coord. Chem. Rev.* 2014, 261, 1–27.

93 Y. M. Klein, A. Prescimone, E. C. Constable, C. E. Housecroft, *Materials* 2017, 10, 728.

94 Y. M. Klein, A. Prescimone, M. Karpacheva, E. C. Constable, C. E. Housecroft, *Polymers* 2018, 10, 1369.

95 M. Y. Klein, E. C. Constable, C. E. Housecroft, A. Prescimone, *CrystEngComm* 2015, 17, 2070–2073.
96 G. Manfroni, A. Prescimone, S. R. Batten, Y. M. Klein, D. J. Gawryluk, E. C. Constable, C. E. Housecroft, *Crystals* 2019, 9, 529.
97 J.-Y. Li, Y.-C. Chen, Z.-M. Zhang, W. Liu, Z.-P. Ni, M.-L. Tong, *Chem. Eur. J.* 2015, 21, 1645–1651.
98 J.-C. Jin, R.-L. Guo, W.-Y. Zhang, C. Jiang, Y.-Y. Wang, *J. Solid State Chem.* 2016, 243, 253–258.
99 X.-J. Gao, H.-G. Zheng, *Dalton Trans.* 2021, 50, 9310–9316.
100 C. Han, J. Li, G. Li, A. Kumar, M. Muddassir, J. Jin, *Inorg. Chem. Commun.* 2021, 123, 108296.
101 D.-D. Yang, L.-P. Lu, M.-L. Zhu, *Dalton Trans.* 2019, 48, 10220–10234.
102 J.-C. Jin, M. Ray, B. Wu, Z.-J. Zhou, X. Wang, M. Muddassir, A. Mohanty, *J. Solid State Chem.* 2023, 327, 124283.
103 G.-L. Wang, Y. Lu, R. Wang, D. Srivastava, A. Kumar, H. Sakiyama, M. Muddassir, J. Guo, J.-C. Jin, *Polyhedron* 2022, 227, 116141.
104 M. S. Khan, Y. Li, D.-S. Li, J. Qiu, X. Xu, H. Y. Yang, *Nanoscale Adv.* 2023, 5, 6318–6348.
105 J. Zhao, Z. Dang, M. Muddassir, S. Raza, A. Zhong, X. Wang, J. Jin, *Molecules* 2023, 28, 6848.
106 J. Zhang, L. Peng, G. Li, A. Kushwaha, M. Sudheer, X. Muddassir, A. Wang, J.-C. J. Kumar, *J. Solid State Chem.* 2023, 326, 124220.
107 L.-L. Bao, Y. Li, Z. Xi, X.-Y. Wang, M. Afzal, A. Alarifi, D. Srivastava, O. Prakash, A. Kumar, J.-C. Jin, *J. Mol. Struct.* 2023, 1292, 136103.
108 S.-Q. Li, T. Verma, K.-H. Sun, X.-H. Zou, A. Mohanty, M. Muddassir, M. Ray, J. Jin, *J. Solid State Chem.* 2023, 327, 124239.
109 S. Chakraborty, G. R. Newkome, *Chem. Soc. Rev.* 2018, 47, 3991–4016.
110 E. V. Alexandrov, A. P. Shevchenko, N. A. Nekrasova, V. A. Blatov, *Russ. Chem. Rev.* 2022, 91, RCR5032.

3

Terpyridine Ligands and Main-Group Elements: An Often-Overlooked Field of Research*

3.1 Introduction

As already explained in Chapter 1, the coordination chemistry of 2,2′:6′,2″-terpyridine is dominated by the transition-metal and rare-earth-metal ions. Some of these complexes have found notable use, for example, as sensitizers in dye-sensitized solar cells (DSSCs), water-oxidation catalysts or metallo-drugs (*i.e.*, alternatives to cisplatin) [1]. Moreover, such complexes have been incorporated as a functional unit into metallo-supramolecular assemblies: polymers, dendrimers, nanostructures, *etc.* In particular, *bis*-tridentate complexes of octahedral geometry play an outstanding role in this respect. Concerning polymers, the complexes might play a structural role, *i.e.*, they link different polymer chains to afford block copolymers [2–5]. The same strategy has also been applied to prepare dendrimers, which contain tpy complexes as the core, branching units, and/or terminal moieties [6, 7]. The next level of complicity was reached when the metal-to-tpy linkage was utilized to assemble metallo-macrocycles and, eventually, sophisticated geometric figures [8]. This concept, which relies on the metallo-supramolecular self-assembly and self-sorting of organic ligands and transition-metal ions, affords nanosized architectures, as kinetically and/or thermodynamically stable species, which can be characterized by nuclear magnetic resonance (NMR) spectroscopy, mass spectrometry (MS), and microscopy imaging [9–11]. This fascinating aspect of modern terpyridine chemistry is discussed later in Chapter 10. Besides this, the metal-to-terpyridine linkage in polymers might also represent the property-giving moiety. In this context, electrochromic [12–14] as well as self-healing and shape-memory polymers [15, 16] have to be mentioned. Regarding mononuclear tpy complexes, numerous studies have been devoted to the development of metallodrugs [17–19], catalysts for organic or energy-related transformations [20, 21], sensors, sensitizers [22], *etc.*

However, most of the systematic overviews on these topics overlook or disregard the use of complexes/assemblies with main-group ions. Nonetheless, some notable examples for the use can be found in recent literature: Catalytic hydrodefluorination

* Parts of this chapter have previously been published elsewhere: A. Winter, G. R. Newkome, U S. Schubert, The chemistry of the s- and p-block elements with 2,2′:6′,2″-terpyridine ligands, *Inorg. Chem. Front.* **2024**, *11*, 342–399. Source: The content is reproduced with kind permission; © 2024 The Royal Society of Chemistry.

Terpyridines: Multifaceted Ligands for Diverse Applications, First Edition.
Ulrich S. Schubert, Andreas Winter, and George R. Newkome.

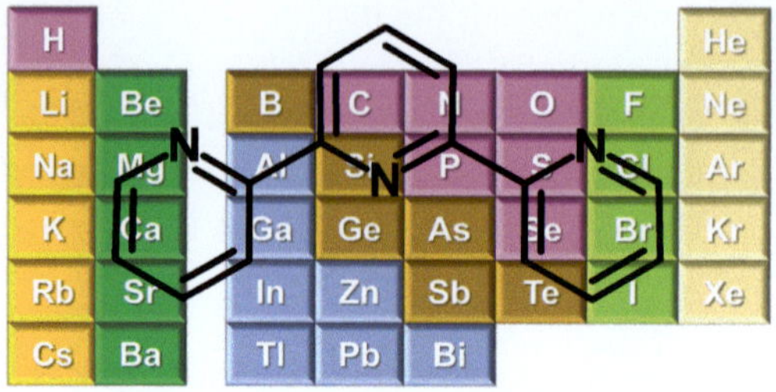

Figure 3.1 2,2′:6′,2″-Terpyridine ligands in the context of the chemistry of main-group elements.

by formal P^{III} or P^{V} species [23], or cytotoxic organotin complexes [24], Bi^{III}-based solid-state emitters [25], *etc.* Moreover, tpy ligands have successfully been tested, as an additive, in perovskite-type materials for solar cell applications [26]. This aspect represents a topic of its own and is, thus, excluded at this stage. This particular application is detailed in Chapter 6. In the following sections, the use of tpy ligands in the field of main-group chemistry is summarized in order to draw attention to this often overlooked field of research and to demonstrate the multifaceted chemistry related to this topic (Figure 3.1). The main aspects and group-/element-specific applications will be presented in groups according to the periodic table of elements (PTE). Some overlapping topics (*e.g.*, catalysis, 2D/3D coordination polymers, *etc.*) are presented in the second part of this chapter. However, the overview provided herein does not claim to be exhaustive. A full coverage of this topic was presented in a recent review [1].

3.2 Terpyridine Complexes of Alkali-Metal Ions

Before dealing with the complexes of the alkali-metal ions in more detail, the reader is reminded of the fundamental considerations of binding according to Pearson's HSAB concept (HSAB: hard and soft acids and bases; see Section 1.3) [27]. The group-1 cations, which represent hard Lewis acids, favor the interaction with hard *O*-based donors, such as ether moieties, over intermediate ones, such as pyridine derivatives (see Table 1.1). When compared to pyridyl complexes with transition-metal ions, one stabilizing factor is absent for alkali-metal ions. Back bonding from the metal to the ligand's π^*-orbital is disabled and, thus, σ-donation by the organic ligand represents the indicator for the efficiency of coordination. The capability for σ-donation correlates with the electronegativity of the donor site. Since the sp^2-hybridized *N*-atom of pyridines is less electronegative than the sp^3 *O*-centers, as present in crown ethers (*i.e.*, 3.2 *vs.* 3.44), pyridine is an intrinsically weaker ligand for electropositive alkali-metal ions. Thus, the complexes, which are only based on electrostatic forces, have to be considered kinetically unstable (whereas crown-ether complexes exhibit significantly higher stabilities) [28].

Constable pointed to the consequences of electrostatic binding as the relevant interaction in complexes of the alkali-metal ions [29]. This author referred to the considerations made by Poonia *et al.* already in the late 1970s [30, 31]. In a simplified view, such complexes can be considered positively charged cations, which are surrounded by negative point charges, which represent the ligands' donor sites. In such a scenario, one might simply use the Coulomb equation (Eq. 3.1) to calculate the electrostatic force (F_C) between the oppositely charged species (q_1 and q_2). In this equation, ε_0 and r represent the vacuum permittivity (8.85×10^{-12} F m^{-1}) and the point-to-point distance between the charged species, respectively. Notably, F_C is, according to the International System of Units (SI), a type of energy with the derived unit of Joule.

$$F_C = \frac{q_1 \cdot q_2}{k_C \cdot r^2} \text{ with } k_C = \frac{1}{4\pi \cdot \varepsilon_0} \tag{3.1}$$

The value of F_C depends only on two parameters: (i) the magnitude of the species' charges and (ii) the distance between them. Geometric considerations, on the other hand, become irrelevant since the spatial distribution of the negative charges around a positively charged center is not reflected by Eq. (1). In other words, a complex, whose metal center coordinates four ligands, might adopt two different geometries, in which the ligands are in equal distance to the metal center – the square-planar and tetrahedral one. According to Eq. (1), one will not be able to distinguish between these since their total energy of metal-to-ligand interactions will be equal. Thus, the ligands themselves will enforce the geometry of the complexes *via* their chemical structures as well as electric repulsion and steric constraints.

One has to keep all of these fundamental aspects in mind when dealing with polypyridyl complexes of alkali-metal ions. Naturally, all this applies also to the analogous complexes based on earth-alkaline-metal ions, as relatively hard Lewis acids [29].

To the best of our knowledge, Nakamura was the first to perform a systematic study on tpy complexes with alkali-metal ions. This author reacted the parent ligand with the respective metal in various polar-aprotic solvents [32]. This reaction, which was a single-electron transfer (SET) reduction, afforded the tpy radical anion ($tpy^{\bullet-}$). The formation of singly reduced oligopyridyl ligands, *i.e.*, 2,2′-bipyridine (bpy) and tpy, has been studied to a significant extent [33–35]. In one of these studies, Wieghardt *et al.* showed that the "zero-valent" complexes $[Fe(tpy)_2]$ and $[Ru(tpy)_2]$ were indeed comprised of divalent d^6-cations and two $tpy^{\bullet-}$ ligands [35]. In one early study on the complexes of alkali-metal ions with the $bpy^{\bullet-}$ ligand, electron paramagnetic resonance (EPR) spectroscopy revealed that the complexes were either loose ion pairs or possessed "covalent" metal-to-ligand bonds (Figure 3.2a) [36]. Depending on nature of the metal ion, either of these scenarios was observed. In Nakamura's work, UV/vis absorption spectroscopy revealed that the properties of the solvent and the cation's size had a significant influence on the product distribution in solution [32]. The presence of loose ion pairs was assumed in *N*,*N*-dimethylformamide (DMF) solution since $tpy^{\bullet-}$ represented the only detectable species (absorption band at *ca.* 573 nm). Whereas, in ethereal solvents (*e.g.*, tetrahydrofuran [THF]), mixtures of complexes **1a** and **1b** were observed; these two species possess distinct absorption maxima at 610 and 560 nm, respectively. The large Rb^I and Cs^I cations favored the coordination in the "hypodentate" $\kappa^2 N^1,N^2$ mode, owing to a mismatch of their size with the binding pocket of the tridentate ligand; accordingly, the smaller alkali-metal ions preferred

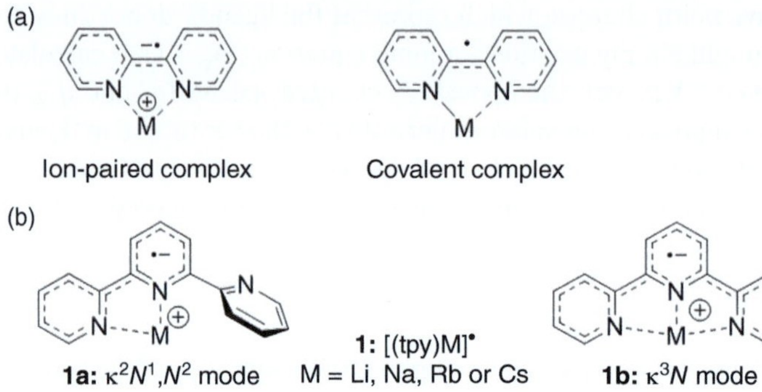

Figure 3.2 (a) Schematic representation of ion-paired and covalent complexes of the bpy•− ligand. (b) Schematic representation of the two proposed structures of complexes **1**.

the "conventional" tridentate κ³*N* binding (Figure 3.2b). It was assumed that the respective coordination spheres were saturated by solvent molecules.

Satterfield and Brodbelt analyzed the gas-phase stability of metal-oligopyridyl complexes by electrospray ionization (ESI) MS [37]. This approach allows for the comparison of the binding strength of related complexes, based on the measured threshold activation voltages. The molecular polarizability is commonly accepted as one indicator for a ligand's affinity to metal ions; this value can readily be estimated *via* Miller's incremental approach [38]. Although the polarizability of tpy is higher than that of bpy (*i.e.*, tpy: 27.6 $Å^3$ *vs.* bpy: 18.7 $Å^3$), very low threshold activation voltages were observed for the $[Li(tpy)_2]^+$ and $[Na(tpy)_2]^+$ complexes (the other *bis*-complexes were not observed at all) [37]. It was assumed that the destabilization of the *bis*-complexes was due to the initial formation of stable $[M(tpy)]^+$ species, which disfavored the binding of a second tpy ligand for electronic and steric constraints. These findings are in line with those made beforehand by Vögtle *et al.*, who performed ^{7}Li- and ^{23}Na-NMR titration studies to quantify the formation of $[Li(tpy)]^+$ and $[Na(tpy)]^+$ from the respective thiocyanate salts and tpy in solution [39]. Remarkably, the combination of $(NH_4)(SCN)$ and tpy afforded an assembly of 1:2 stoichiometry, which contradicted Grimm's hydride-displacement law [39]. This early hypothesis, which was initially used to describe bioisosterism, supposes that certain chemical moieties might be capable of mimicking others; in this concept, NH_4^+ and Na^+ are isosteric to each other [40]. However, this analogy did not apply with respect to the observed coordination behavior [39].

The stability of tpy complexes can be increased by suppressing competitive binding of solvent molecules, in particular when representing strong *O*-donors. In order to apply this concept for the complexation of Li^+ ions by tpy ligands, van Eldik *et al.* selected ionic liquids, such as non-Lewis-basic solvents [41]. Solvents (and other molecules) can be classified according to their ability to act as Lewis bases (donors) or Lewis acids (acceptors). Gutmann *et al.* developed a method to quantify this behavior based on NMR titration and/or calorimetric experiments [42]. The donor number (DN) is defined as the negative enthalpy value of formation of a 1:1 adduct between a Lewis-basic molecule and $SbCl_5$, as the standard Lewis acid, in dilute 1,2-dichloroethane solution. By definition, the DN of this non-coordinating solvent is equal to zero. Analogous considerations can also be made to determine the acceptor number (AN) of a Lewis-acidic molecule. In this case, Et_3PO represents the reference Lewis base. The AN can then be deduced from the change in the ^{31}P NMR chemical shift of the phosphine oxide upon interaction with the Lewis acid. This method is referred to as the Gutmann–Becket method [43]. Shamsipur *et al.* determined the stability constants (K_1) for complexes $[Li(tpy)]^+$ and $[Na(tpy)]^+$ in $MeNO_2$ solution: The lgK_1 value for the latter was significantly lower than for the former (*i.e.*, lgK_1 = 1.67 *vs.* 3.24) [44]. On the contrary, lgK_1 values could not be determined in solvents with higher DN. This reflects the general trend that the lower the lgK_1 (or lgβ_2) values will be, the better the solvent solvates/shields the cation. In this respect, $MeNO_2$ (DN = 2.7) and certain ionic liquids, *e.g.*, (emim)(NTf_2) and (emim)(ClO_4), represent versatile solvents of low DN when aiming for high stability constants (emim: *N*-ethyl-*N′*-methylimidazolium; Tf^- trifluoromethylsulfonyl) [45, 46]. Van Eldik *et al.* applied ^{7}Li-NMR spectroscopy to monitor the complexation of Li^+ by tpy in such solutions [41]. Although the quadrupole moment of ^{7}Li is relatively high (spin of 3/2), this nucleus is highly suited for this purpose, owing to its natural abundance (*ca.* 92.5%) and gyromagnetic ratio ($103{,}962 \times 10^7$ rad·s^{-1} T^{-1}) [47]. The ^{7}Li-NMR titration experiments confirmed the formation of $[Li(tpy)_2]^+$ in both ionic liquids as well as in $MeNO_2$ (Figure 3.3a) [41]. Stability constants (lgβ_{21}) in the range from 4.68 in (emim)(ClO_4) to 5.80 in $MeNO_2$ were extracted from the titration data (Figure 3.3a). However, the *bis*-complex could not yet be isolated. Whereas crystallization from (enim)(ClO_4) afforded a dimeric μ-perchlorate-bridged complex (Figure 3.3b); likewise, a 1D coordination polymer was isolated from the (enim)(NTf_2) solution. In both cases, the penta-coordinated Li^+ ions possessed an overall "N_3O_2" coordination environment. It was suggested that charge neutralization, due to close ion pairing, as well as ππ-stacking interactions, precluded the anticipated crystallization of the *bis*-complex.

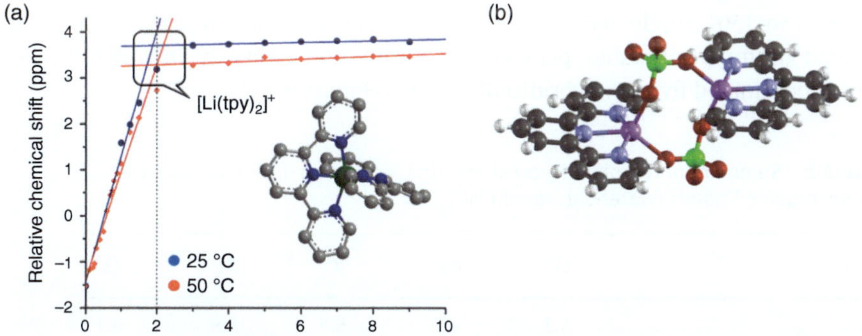

Figure 3.3 (a) Results of the ^{7}Li-NMR titration experiment in (emim)(NTf_2) at two different temperatures. The simulated structure of the proposed $[Li(tpy)_2]^+$ complex is also shown (hydrogen atoms were omitted for clarity). Adapted from [41]. (b) Solid-state structure of the dimeric $[(tpy)Li(ClO_4)]_2$ complex. Adapted from [41].

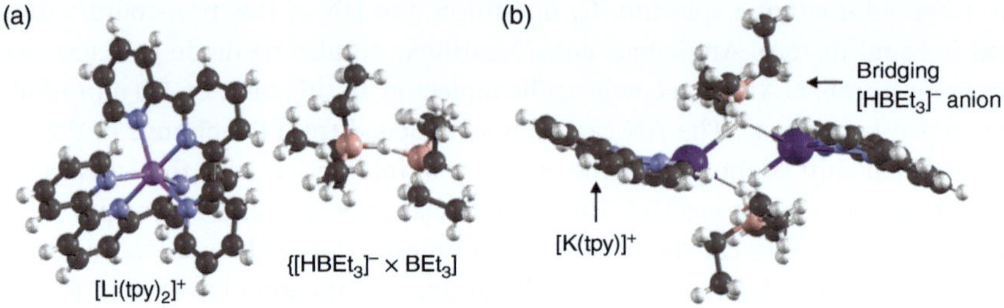

Figure 3.4 (a) Solid-state structure of $[Li(tpy)_2]\{[HBEt_3]^- \times BEt_3\}$. (b) Solid-state structure of the dimeric $\{[K(tpy)][HBEt_3]\}_2$ complex. Adapted from [48].

Eventually, a complex containing the $[Li(tpy)_2]^+$ cation was obtained in 22% yield from the reaction of tpy with $Li[HBEt_3]$ (Figure 3.4a) [48]. Whereas $Na[HBEt_3]$ and $K[HBEt_3]$ afforded dimeric complexes, in which the coplanar $[M(tpy)]^+$ fragments were bridged *via* two $[HBEt_3]^-$ moieties (M = Na or K; Figure 3.4a,b). The dimeric Na^+ complex represented a notable catalyst for the hydroboration of epoxides to give Markovnikov-type alcohols.

Not only the parent tpy ligand but also some structural variants were used for the complexation of group-1 ions. For example, Bell *et al.* reported the ligands **2** and **3** (Figure 3.5a), which contained a tridentate tpy subunit. In particular, the rigidified derivatives **2a/b** showed a high capability of binding the picrate salts in a water-saturated $CHCl_3$ solution [49]. The stability constants were remarkably high when compared to those determined for 2,3-naphtho-18-crown-6, as the macrocyclic reference, but did not show any size-selectivity for binding (Table 3.1). Ligand **3** provided lower stability constants than **2**; thus, it was concluded that preorganization on the one hand and the presence of carbonyl groups as additional donor moieties on the other accounted for the increased $\lg K_s$ values for **2a/b** when compared to **3**. Constable *et al.* prepared various macrocyclic ligands, which contained polypyridyl subunits [29, 51]. In one of these studies, Li^+ ions were complexed by macrocycle **6** (Figure 3.5b) [50]. Single-crystal X-ray diffraction (XRD) analysis of the 1:1 complex revealed an untypical planar, pentagonal coordination geometry, in which the Li^+ center coordinated five sp^2-hybridized *N*-atoms (Figure 3.5b).

Table 3.1 Summary of the binding constants ($\lg K_s$) of group-1 picrate salts with various organic ligands (water-saturated $CHCl_3$, 25 °C).

	Li^+	Na^+	K^+	Rb^+	Cs^+
2a	8.3	9.7	9.5	8.9	8.1
2b	5.2	4.7	4.7	4.3	4.1
3	<3	<8	<3	<3	<3
2,3-naphtho-18-crown-6	4.4	6.1	7.9	7.1	6.1

[49]/American Chemical Society.

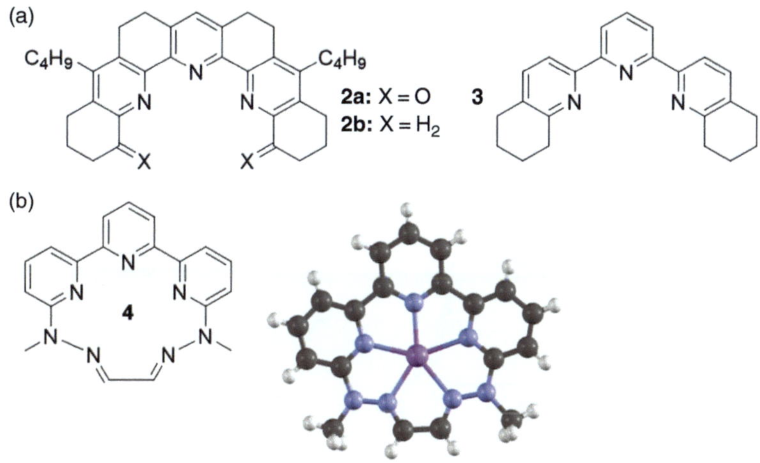

Figure 3.5 (a) Schematic representation of Bell's molecular clefts **2** and **3**. Adapted from [49]. (b) Schematic representation of Constable's macrocyclic host **6**; the solid-state structure of the [Li(**6**)]$^+$ cation is also shown (anion and solvent molecules are omitted for clarity). Adapted from [50].

3.3 Terpyridine Complexes of Alkaline-Earth-Metal Ions (Group 2)

According to Pearson's HSAB concept, the divalent cations of the alkaline-earth metals also represent hard Lewis acids [27] and, thus, similar fundamental aspects apply regarding their coordination chemistry (Section 3.2). In particular, the lighter homologs (*i.e.*, Be^{II}, Mg^{II}, and Ca^{II}) represent reasonably hard Lewis acids and, thus, their coordination compounds with polypyridyl-type ligands, as intermediate Lewis bases, should exhibit a certain degree of lability (at least in solvents with high DNs). Such behavior is detrimental for preparative chemistry (*i.e.*, the synthesis of solution-stable complexes). However, lability/reversibility in binding is utilized in polymer science to create stimuli-responsive and self-healing materials [15, 16]. For this purpose, a wide range of polymers, containing tpy moieties in the side chains, have already been combined with transition-metal ions (*e.g.*, Fe^{II}, Mn^{II}, Zn^{II}, Cd^{II}, *etc.*; see Chapter 9) [52]. Whereas ions of the s-block elements have not yet been used for this application. Overall, the number of available studies dealing with the synthesis and structural characterization of complexes with the alkaline-earth-metal ions is relatively small.

The specific implications of the small size and pronounced Lewis acidity of the Be^{II} ion on its coordination chemistry have already been discussed in Section 1.2 [53]. Scattered studies in the 1960s have dealt with the coordination of the Be^{II} ion with some bidentate polypyridyl ligands [*i.e.*, bpy or 1,10-phenanthroline (phen)] [53–55]. For example, [Be(bpy)Me_2] was obtained from the reaction of $BeMe_2$ with bpy; whereas the transmetalation of Li(bpy)$_2$ with $BeCl_2$ afforded the green-colored [Be(bpy)$_2$] complex. This "zero-valent" complex possessed paramagnetic properties due to the presence of the two bpy$^{\bullet-}$ ligands, as revealed by detailed EPR studies [33].

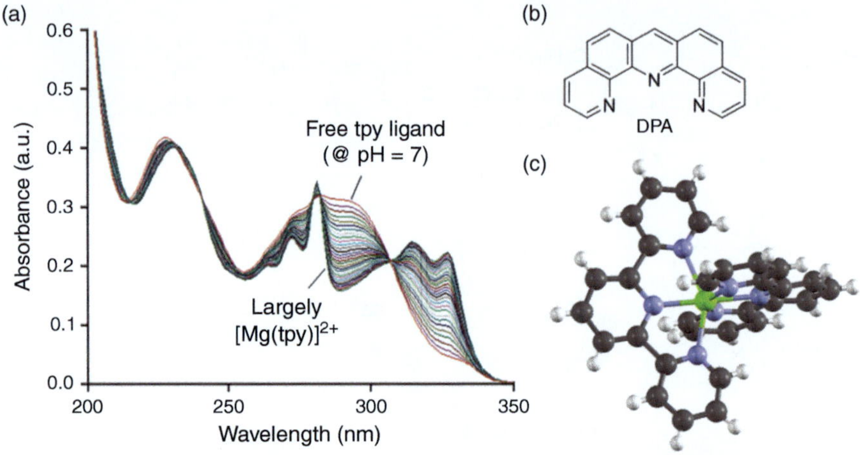

Figure 3.6 (a) Evolution of the UV/vis absorption spectrum of tpy (2×10^{-5} M in pH-neutral water) upon titration with Mg^{II} ions; complex formation was concluded from the appearance of sharp signals at *ca.* 275 nm [58] / with permission from American Chemical Society. (b) Schematic representation of the DPA ligand. (c) Solid-state structure of $[Mg(tpy)_2](ClO_4)_2$ (anions and solvent molecules were omitted for clarity). Adapted from [60].

To the best of our knowledge, analogous experiments using tpy, as the chelating ligand, have not yet been performed [53]. Certainly, further restrictions arise from safety concerns, which are related to the severe chronic beryllium disease (CBD) and generally retard research on Be^{II} coordination chemistry [56].

The fundamental aspects of the complexation behavior of the other alkaline-earth-metal ions have already been discussed in Section 1.3. As pointed out there, Thummel, Hancock *et al.* provided a thorough analysis taking also the matching/mismatching of the metal ions with tpy's binding pocket into account [57–59]. In all cases, the $\lg K_1$ values for complexation were derived from UV/vis titration experiments [58]. These authors used the ligand's $\pi \rightarrow \pi^*$ transitions as the reliable marker to monitor the formation of the $[M(tpy)]^{2+}$ *mono*-complexes in an aqueous environment (M = Mg, Ca, Sr or Ba; Figure 3.6a). To briefly recap the most relevant aspects: firstly, owing to the tridentate chelation by the tpy ligand, the $\lg K_1$ values were higher than those of the coordination of bidentate bpy; secondly, the transition-metal ions, as moderate Lewis acids, afforded an even stronger coordination, as expressed by the increased $\lg K_1$ values; and finally, preorganized ligands, such as dipyrido[4,3-*b*;5,6-*b*]acridine (DPA, Figure 3.6b), afforded the highest $\lg K_1$ values – at least in the case of Ca^{II} and Sr^{II} ions, whose medium-sized ionic radii matched the size of the ligand's cleft. Accordingly, the gain in stabilization was lower for the smaller and the larger ions (*i.e.*, Mg^{II} and Ba^{II}, respectively). Noteworthy, at least in the case of the $[Mg(tpy)]^{2+}$ *mono*-complex, the determined $\lg K_1$ value matched with that reported by Emara *et al.* beforehand (*i.e.*, 0.74 vs. 0.77) [61, 62].

The first study on tpy complexes of the alkaline-earth-metal ions was performed by Brown *et al.* already in the 1960s [55], thus predating the aforementioned report on the reaction of alkali-metal ions with tpy (Section 3.2) [32]. As the main commonality of these reports, the formation of radical anions by one-electron reduction of tpy was proposed. Brown *et al.* used the respective metal-mercury amalgam to generate the $tpy^{\bullet-}$ species, which formed "zero-valent" $[M(tpy)_2]^{2\bullet}$ diradical complexes of S_4-symmetry (M = Mg^{II}, Ca^{II}, Ba^{II}) [55]. This proposal was based on the experimental and simulated EPR spectra, but further insight into the molecular structure of these complexes was not provided by those authors.

Constable *et al.* obtained $[Mg(tpy)_2](ClO_4)_2$ from the straightforward self-assembly of $Mg(ClO_4)_2$ and tpy in the binary $MeCN/CHCl_3$ solvent mixture [60]. Unlike all previous attempts to structurally characterize such complexes [1], structural analysis in solution and the solid state was performed in this case. The solid-state structure of the *bis*-tridentate complex, which is shown in Figure 3.6c, revealed a typical distorted octahedral geometry, which is well-known for tpy complexes with transition-metal ions [60]. Furthermore, some mixed-ligand complexes containing Mg^{II} or, for the first time, Ca^{II} ions were prepared in solvents with high DNs [63]. For example, the dimeric $[CaCl_2(tpy)(DMF)]_2$ complex was obtained, whose Ca^{II} centers were coordinated in a pentagonal-bipyramidal fashion. The metal centers were linked *via* μ-dichloro bridges – a motive, which can *inter alia* be found in the dimers of *bis*-cyclometalated Ir^{III} complexes [64]. In the case of Ca^{II} and Sr^{II}, the remarkable *tris*-tridentate $[M(tpy)_3]X_2$ complexes were also obtained (M: Ca or Sr; X = Br or I) [65]. Whereas owing to steric constraints, such high coordination numbers (CNs) were not observed for the Ba^{II} ion. Instead of the *tris*-tridentate complexes, $[Ba(tpy)_2]I_2$ and $[Ba(tpy)](ClO_4)_2$ were isolated from the respective Ba^{II} salts and tpy.

One interesting potential application was proposed by Batista, Hizari *et al.*, who synthesized the Mg^{II} *mono*-complexes **5** from the respective ligands and preformed Grignard reagents (Figure 3.7a) [66]. Thereby, a remarkable selectivity was observed since the dihalide complexes **5** were formed almost exclusively. It was assumed that these complexes, owing to their poor solubility in the used ethereal solvents (*i.e.*, Et_2O or THF), were withdrawn from the reaction equilibrium, thus suppressing the formation of other species (*i.e.*, complexes with coordinated $MgXR^3$ moieties). According to density-functional theory (DFT) calculations, ligand pre-association *via* $\pi\pi$-stacking interactions represented the essential process, and the subsequent binding of MgX_2 was thermodynamically more favored than that of $MgXR^3$ or $MgR^3{}_2$. These authors claimed that the addition of tpy ligands to a Grignard reagent in solution might allow one to shift the Schlenk equilibrium (Figure 3.7b) [67] and, thereby, modulate the reagent's reactivity: Chemistry textbooks commonly teach that MgR_2 species are more reactive than the respective RMgX ones. However, this proposal has not yet proven its applicability in practice. One might speculate that similar considerations might also apply for the tpy complexes of "heavy" Grignard reagents, *i.e.*, their Ca^{II}-containing analogs [68].

McLuckey *et al.* developed remarkable protocols for the analysis of lipids by mass spectrometric techniques [69]. These authors used gas-phase ion/ion chemistry to convert one type of lipid ion into another by reaction with an appropriate reagent ion/complex. In other words, the gas-phase reactions turn lipid cations into anions (or *vice versa*), which can be analyzed, *e.g.*, by collision-induced dissociation (CID).

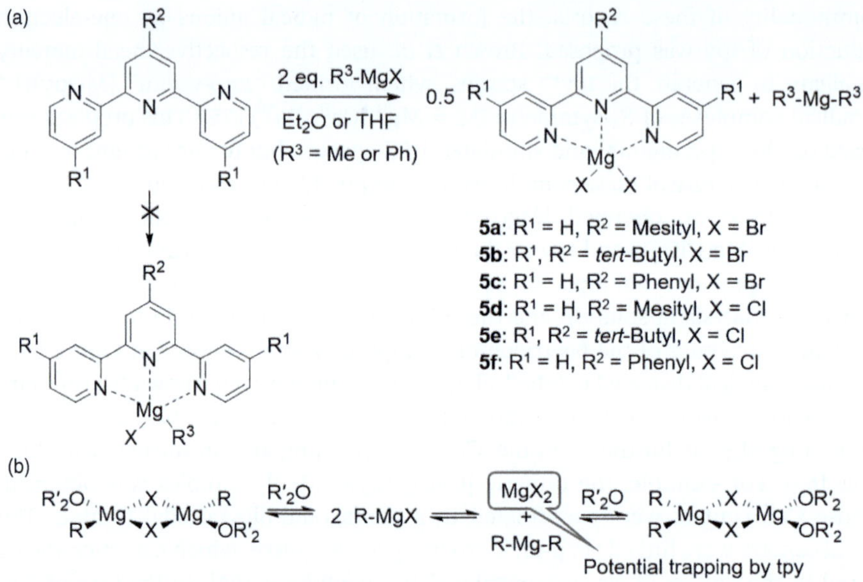

Figure 3.7 (a) Synthesis of complexes **5** from tpy ligands and Grignard reagents. Adapted from [66]. (b) Schematic representation of the simplified Schlenk equilibrium. Adapted from [67].

In particular, $[Mg(tpy)_2]^{2+}$ was identified as a versatile reagent for the generation of such charge-converted complexes. Its applicability was demonstrated first in the example of glycosphingolipids (GSLs), specifically *lyso*-glycosphingolipids (*lyso*-GSLs) and cerebrosides [70]. Diastereomeric mixtures of these GSLs are typically found in mammalian cells, due to the co-occurrence of glucose and galactose carbohydrate head groups (Figure 3.8a). The isomeric resolution of such mixtures is impossible with most analytical tools [71]; however, the charge-inverted complexes, which resulted from the reaction of deprotonated GSLs and $[Mg(tpy)_2]^{2+}$ (Figure 3.8b), could be distinguished when subjected to CID [70]. Due to the different interaction mode between the $[Mg(tpy)]^{+}$ cation and the deprotonated GSL diastereomers, the resulting mass spectra revealed distinct differences in the signal intensities (Figure 3.8c). Thus, GlcSL and GalSL could be differentiated and even quantified, based on tandem MS. This highly accurate and reproducible method was subsequently applied to other substrates regarding structure elucidation and/or differentiation/separation of isomers: Phosphoinositides (essential signaling phospholipids in mammalian cellular membranes) [72], cerebrosides (a sphingolipid subclass) [73], and fatty acids [74, 75]. One might assume that the scope of this protocol can be extended further to other biomolecules, as analytes, on the one hand and other regent complexes on the other [69].

Besides these examples for mono- and binuclear complexes, alkaline-earth-metal ions and tpy ligands have also been combined to afford 2D/3D coordination polymers, from which some materials featured notable properties. For clarity reasons, this topic is discussed separately in Section 3.9.

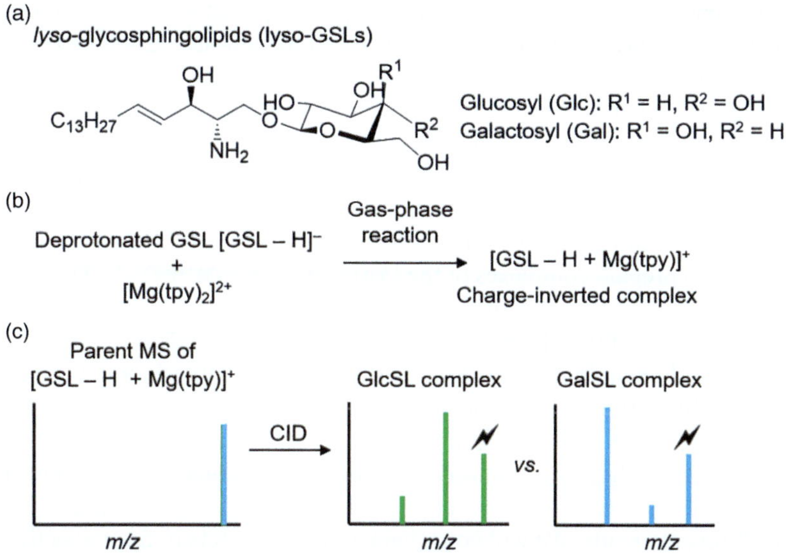

Figure 3.8 (a) Schematic representation of *lyso*-glycosphingolipids with glucose or galactose head groups. (b) Schematic representation of the formation of charge-inverted complex cations in a gas-phase reaction. (c) Representation of the differentiation between glucose- and galactose-containing GSL by tandem MS [70] / with permission from American Chemical Society.

3.4 Terpyridine Complexes of Boron-Group Ions

Research in the field of "boron-terpyridine" chemistry has up until now attracted very little attention, and the number of publications referring to this is still exceedingly small. This contrasts the relatively intense work on tpy complexes with the heavier homologs within the triel/boron group. In this respect, literature reports can even be traced back to the 1960s. Whereas a SciFinder^{n} survey revealed that the first concrete study, in which an electron-deficient B^{III} center was used for the coordination of tpy, was only published in 2000. In continuation of their research on "chelated fluoroboron cations" [76], Hartman and Shoemaker prepared a series of $[(N^\wedge N)BF_2]^+$ complexes, in which *N*-heterocyclic ligands were coordinated to a highly Lewis-acidic B^{III} center in a ditopic fashion (N^N = bpy, phen, or tpy) [77]. As shown by ^{19}F- and ^{11}B-NMR spectroscopy, the $[(phen)BF_2]^+$ complex was formed efficiently, owing to the ligand's rigidity. Whereas the coordination of bpy afforded a mixture of non-chelated *mono-* and *bis*-adducts. On the other hand, the even more flexible tpy ligand readily underwent ligand-exchange reactions with precursors $BF_3{\cdot}OEt_2$, $BF_nCl_{3-n}{\cdot}$isox, and $BF_nBr_{3-n}{\cdot}$isox, to give the respective complexes, in which the tpy ligand was always coordinated in the hypodentate κ^2N^1,N^2 mode (n = 3, 2, 1 or 0; isox: isoxazole). However, all of the $[(tpy)BX_2]^+$ adducts were obtained in low yields (X = F, Cl or Br) and their solid-state structures were not clarified. Unexpectedly, the otherwise labile py ligands of precursor $[(py)_2BF_2](PF_6)$ could not be substituted by tpy, despite the assumed favoring chelate effect [77, 78].

Besides this systematic study, very few other B^{III}–tpy complexes were found in the available literature. The reader is referred to our recent review article, where these compounds are mentioned [1].

As mentioned beforehand, the first examples for tpy complexes with the heavier triel ions were already mentioned in the 1960s. The reaction of $In(ClO_4)_3$ with tpy in EtOH afforded an ionic compound, which was assumed to be $[In(tpy)_2](ClO_4)_3$; however, this assumption was based only on infrared (IR) spectroscopic studies [79]. Nonetheless, the applied reaction conditions were versatile and established as the standard method to prepare complexes of the $[M(tpy)]X_3$ and $[M(tpy)_2]X_3$ families (M = Al, Ga, In or Tl; X = monovalent anion).

When dealing with thallium chemistry, one has to be aware of relativistic effects, which stabilize the element in its low oxidation state [80, 81]. Thus, thallium chemistry is distinctly different from that of its lighter triel homologs. As one result, Tl is widely considered a "relativistic" alkali metal and its chemistry is often similar to that of potassium (note that the Tl^{I} and K^+ ions have almost identical ionic radii) [82]. Likewise, relativistic effects are also of importance for lead and bismuth, as the subsequent 6th-row elements. We will come back to these aspects in the respective parts of this chapter, where the coordination chemistry of these elements is discussed (see Sections 3.5 and 3.6). The coordination chemistry of Pb^{II} and Bi^{III} ions with polypyridyl ligands, specifically tpy ligands, is well established; whereas Tl^{I} complexes with such ligands have attracted only little attention [83] and the related Tl^{III} complexes are more common [1].

In one early contribution, Walton prepared coordination compounds from thallium(III) halides and tpy [84]. According to elemental analysis, the combination of $TlCl_3$ with tpy afforded two different compounds, whose relative compositions were $TlCl_3(tpy)$ and $(TlCl_3)_2(tpy)$. The TlI_3 precursor salt, on the other hand, provided $[(TlI_3)_2(tpy)]$ as a single product. Conductivity and molar-mass measurements suggested that all complexes existed as $[TlX_2(tpy)]^+$ cations and $[TlX_4]^-$ anions in MeCN solution (X = Cl or I); however, it remained unclarified whether this also applied for the solid state. In subsequent work, Walton analyzed the full series of crystalline samples of $[TlX_2(tpy)][TlX_4]$ (X = Cl, Br or I) by low-frequency vibrational spectroscopy and X-ray photoelectron spectroscopy (XPS) to confirm the proposed salt-like structure of these complexes [85, 86]. It was impossible to distinguish the two different Tl^{III} centers in these complexes by XPS [86] and, based on vibrational spectroscopy, an ionic structure, such as the assumed $[Tl(tpy)_2][TlX_6]$ motif, had to be ruled out (X = Cl or Br) [85]. Indeed, $[Tl(tpy)Cl_3]$ possessed a pseudo-octahedral geometry with three distinctly different M—Cl bonds and, thus, was isomorphous to its homologs (*i.e.*, the corresponding *mono*-complexes with Al^{III}, Ga^{III}, or In^{III} ions) [87]. These complexes were prepared by Carty *et al.*, who provided the first systematic study of complexes of the $[M(tpy)]X_3$ family by correlating the results from single-crystal XRD analysis with those from vibrational spectroscopy [87, 88]. These complexes were poorly conductive when dissolved in aprotic solvents (*e.g.*, MeCN or $MeNO_2$); thus, the degree of dissociation of halide ligands was only marginal; whereas the corresponding *bis*-complexes behaved as 1:3 electrolytes, indicative for the formation of dissociated complex cations and anions in these solvents [87, 89]. Timoshkin *et al.* directed to a masking effect of the

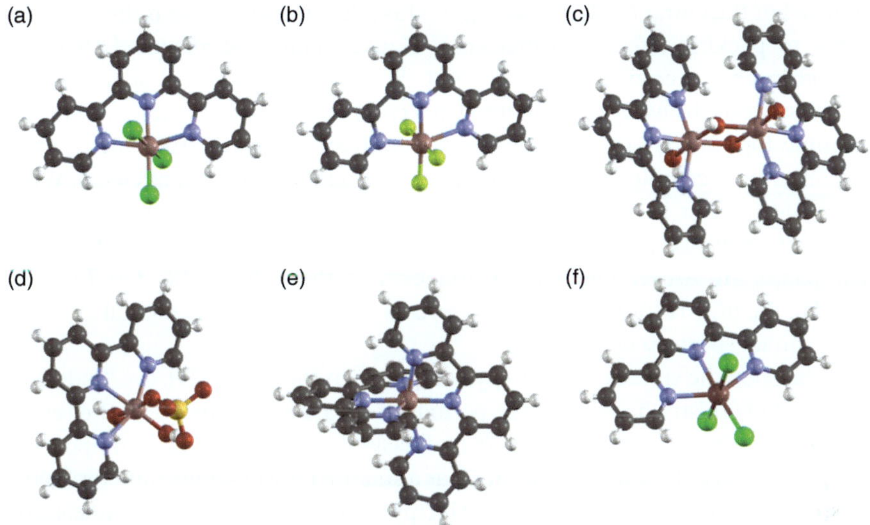

Figure 3.9 Solid-state structures of (a) [Ga(tpy)Cl_3]. Adapted from [90], (b) [Al(tpy)F_3]. Adapted from [91], (c) [Ga(tpy)(H_2O)]$_2$(μ-OH)$_2$(NO_4)$_4$. Adapted from [92], (d) [Ga(tpy)(OH)(SO_4)]. Adapted from [93], (e) $[Ga(tpy)_2]^{3+}$. Adapted from [94], and (f) [In(tpy)Cl_3]. Adapted from [95]. In all figures, non-coordinating ligands, counterions and solvent molecules were omitted for clarity.

tpy ligand on the Lewis acidity of $GaCl_3$ and $GaBr_3$ [90]. The solid-state structure of [Ga(tpy)Cl_3] is depicted in Figure 3.9a. It was concluded from single-crystal XRD analyses that [Ga(tpy)Cl_3] and [Ga(tpy)Br_3] possessed similar Ga^{III}-based Lewis acidities in the solid state; whereas the gas-phase Lewis acidity of the parent $GeCl_3$ is significantly higher than that of $GeBr_3$ [96].

Since the aforementioned general method for the synthesis of complexes [M(tpy)]X_3 was not applicable for the fluorido complexes, Reid *et al.* developed an alternative strategy in which the complexation was performed under forcing hydrothermal reaction conditions [91]. The solid-state structures of the resulting *mer*-[M(tpy)]F_3·3H_2O complexes closely resembled those of the analogous complexes with chlorido or bromido ligands (M = Al, Ga or In; Figure 3.9b). Whereas the packing in the crystal lattice was governed by extensive hydrogen-bonding (*H*-bonding) interactions involving coordinated F^- anions and water molecules. Upon preparing the related complexes [M(tpy)](OTf)$_3$ (M = Al, Ga or In), Reid *et al.* also observed the formation of the dimeric μ-dihydroxy-bridged complex [In(tpy)(OTf)$_2$(OH)]$_2$[OTf]$_2$ [97]. Notably, similar dimers were also obtained with other triel(III) ions and/or bridging ligands, such as fluoride (Figure 3.9c) [91, 92], but not sulfate (Figure 3.9d) [93]. Reid *et al.* mentioned one further aspect, which is the characterization of complexes [M(tpy)]X_3 by heteronuclear NMR spectroscopy [97]. When dealing with such low-symmetry complexes, fast spin relaxation, as enabled by the relevant quadrupolar nuclei (*i.e.*, ^{27}Al: $I = 5/2$, ^{71}Ga: $I = 3/2$, ^{115}In $I = 9/2$), excluded the recording of meaningful NMR spectra.

The $[Ga(tpy)_2]^{3+}$ *bis*-complex exhibited a remarkable single-crystal structure: The achiral complex, whose solid-state structure is shown in Figure 3.9e, was crystallized in the chiral $P2_12_12_1$ space group [94]. It was assumed that the chirality was due to packing effects. Furthermore, Hauser, Piquet *et al.* noted that the limitation in characterizing paramagnetic Cr^{III} complexes by NMR spectroscopy could be overcome by using isostructural Ga^{III} complexes. Thus, $[Ga(tpy)_2]^{3+}$ served as a mimic for $[Cr(tpy)_2]^{3+}$ and enabled the indirect NMR spectroscopic characterization of such complexes (undisturbed by the quadrupolar ^{71}Ga nucleus). With the growing interest in photoactive *bis*-tridentate Cr^{III} complexes, this approach might be of relevance in this field of research [98]. Reports of other triel(III)–tpy *bis*-complexes are exceedingly rare. Complexes of the $[Al(tpy)_2]X_3$ family were reported in the mid-1970s but only characterized superficially by elemental analysis and vibrational spectroscopy [99].

After the sporadic studies on In^{III}–tpy complexes in the 1960s [79], the "indium–terpyridine" chemistry has seen some periods of renaissance. Firstly, Fischer *et al.* were interested in the synthesis of safe and easy-to-handle precursor compounds for the generation of indium nitride (InN), as a small-bandgap semiconductor [100]. In this context, the *tris*-azido complex $[In(tpy)](N_3)_3$ was *inter alia* synthesized. The idea was to tame the highly explosive $In(N_3)_3$ by the coordination of a tpy ligand to the In^{III} center; however, $[In(tpy)](N_3)_3$ was prone to decomposition in THF solution and/or presence of air and could, thus, not be isolated in its pure form. Research in this respect was discontinued since the controlled detonation of related complexes, such as $[In(py)_3](N_3)_3$, only afforded metallic indium instead of the anticipated semiconductor material. As the main limitation of InN fabrication *via* chemical-vapor deposition, the material decomposes into metallic In and N_2 at temperatures >500 °C, owing to its metastable character [101]. Renewed attention in combining tpy with $InCl_3$ and $InCl_5$ resulted from their potential use as perovskite-type materials. As detailed in Chapter 6, perovskite-type solar cells represent one emerging photovoltaic technology, which combines fabrication costs with high power-conversion efficiency (PCE) [102]. In addition to this, perovskite-structured materials, such as $CsPbX_3$ nanocrystals (X = Cl, Br or I), are highly photoluminescent with very narrow emission-band widths [103]. Among others, organic–inorganic hybrid, lead-free metal halides, which are stable under ambient conditions, represent one alternative to the established Pb^{II}-containing systems [104]. In this broader context, (tpy)$InCl_3$ (Figure 3.9f) and (tpy)$InCl_5$ as well as $[Ga(tpy)]_2PbI_5$ were studied [95, 105]. For further reading on this topic, reference to Section 6.5 is given at this point.

Besides the aforementioned range of *mono*-complexes, which mainly contained halide ligands, examples have also been reported in which tpy has been combined with organometallic moieties, *i.e.*, fragments containing a metal-to-carbon bond. In this respect, tpy was coordinated to the unusual organoindium(III) species $MeInCl_2$ and $(C_6F_5)InCl_2$; however, the resulting products were only analyzed by molar-mass measurements and IR studies [106, 107]. Organoindium compounds have found widespread use in organic synthesis since such reagents combine relative stability and good functional-group tolerance with the ability to apply them under mild reaction conditions [108]. Ligand-stabilized derivatives have *inter alia* been employed, but tpy has (not yet) been considered in this context. The analogous compounds,

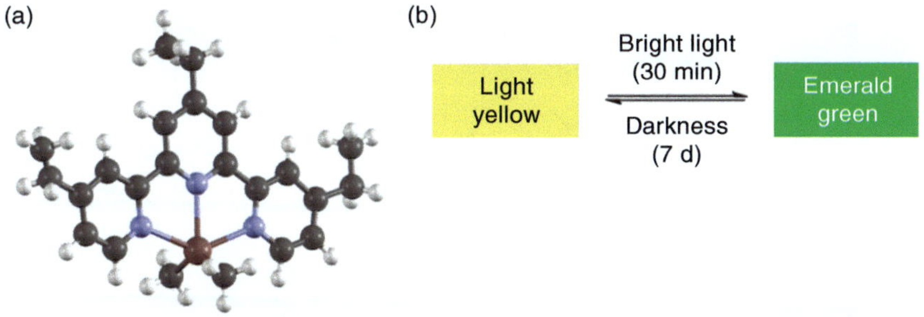

Figure 3.10 (a) Solid-state structure of $[In(L)Me_2](NO_3)$ (L = 4,4′4″-Et_3-tpy; counterions and solvent molecules were omitted for clarity). (b) Representation of the reversible solid-state photochromism of [In(L)MeCl] (L = tpy or 4,4′,4″-Et_3-tpy). Adapted from [109].

in which tpy ligands (*i.e.*, the parent ligands or their 4,4′,4″-triethyl derivative) were coordinated to a formal $[InMe_2]^+$ or $[TlMe_2]^+$ fragment, were reported by Canty *et al.* [109–111]. The products were characterized by NMR spectroscopy and single-crystal XRD analysis (Figure 3.10a). Furthermore, ESI MS measurements revealed the high stability of the covalent metal-to-carbon and coordinative metal-to-nitrogen bonds, which persisted independently from the applied ion-source energy [110]. The complexes derived from $InMe_2Cl$ possessed photochromic properties in the solid state [109]. Upon exposure to bright light, the initially pale-yellow crystals adopted an emerald-green coloration within 30 min; the original coloration was slowly restored in the dark within a few days (Figure 3.10b). It was shown that this phenomenon only occurred at the crystals' surfaces, presumably due to a light-induced change of the binding mode and, thus, different absorption behavior. It was assumed that the tpy ligand switched between a bidentate κ^2N^1,N^2 (light) and tridentate κ^3N (dark) coordination. It has to be emphasized that this type of photochromism is fundamentally different from the more common case, where photochromism of complexes is due to structural changes within the photoactive ligands [112].

Zhang, Zheng, Dub *et al.* intended to prepare complex **6**, in which a singly reduced tpy ligand was coordinated to the organoaluminum fragment $[Al^iBu_2]^+$ [113]. The study was performed in the context of using Al^{III} complexes with "non-innocent" ligands as catalysts for organic transformations [114]. The term "non-innocent" was initially coined by Jørgensen to describe the oxidation-state ambiguity of redox-active ligands in their complexes with redox-active transition-metal ions [115]. However, the reaction of 4′-phenyl-tpy with Al^iBu_3 afforded **6** only as a by-product and dearomatized derivatives, such as **7**, were formed instead (either by radical alkylation of **6** or *via* an undisclosed pathway; Figure 3.11a) [113]. The structures of these complexes were *inter alia* verified by single-crystal XRD analysis (Figure 3.11b,c). In order to explore the generalizability of this observation, other tpy ligands and organoaluminum precursors were screened. However, in the case of $Al(CH_2SiMe_3)_3$, an unusual zwitterionic complex was obtained, which serendipitously represented a versatile hydroboration catalyst. We will come back to this in Section 3.8 when dealing with catalytic applications.

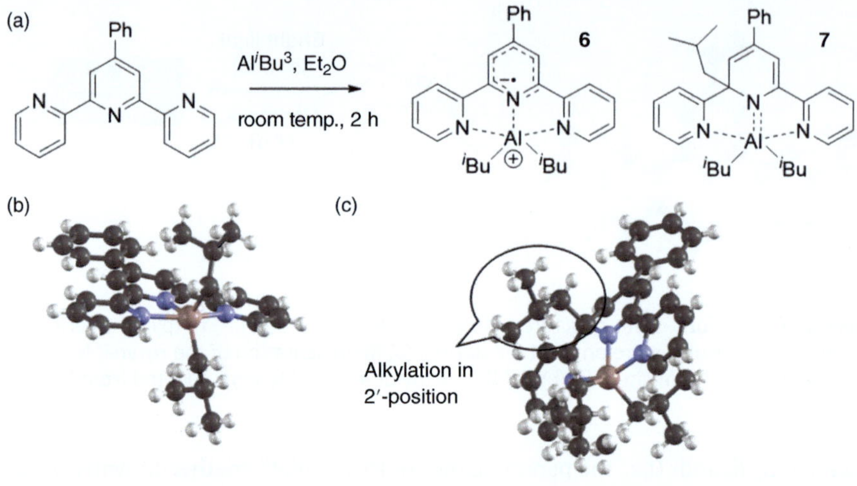

Figure 3.11 (a) Schematic representation of the synthesis of complexes **6** and **7**. (b) Solid-state structure of **6**. Adapted from [113]. (c) Solid-state structure of **7**. Adapted from [113].

The same holds true for one further family of complexes (**10–12**; Figure 3.12a), which all contained an uncommon metal-to-metal bond. Takaya and Iwasawa synthesized these complexes *via* a two-step procedure, in which the *mono*-complexes **8** or **9** were formed first [116–118]. These complexes resembled the aforementioned [M(tpy)]Cl_3 complexes (M = Al, Ga, In, Tl). However, owing to steric constraints, the metal centers were only penta-coordinated (Figure 3.12a). Representatively, the solid-state structures of the Al^{III}-containing complexes **8a** and **10a** are depicted in Figure 3.12b [116]. Remarkably, the triel element in **8** and **9** acted as a two-electron (*i.e.*, σ-acceptor or *Z*-type) moiety to coordinate a transition-metal ion, despite their intrinsic Lewis acidity. Typically, this interaction, which occurs within a (mainly) tridentate ligand framework, is supported by flanking auxiliary ligands (*e.g.*, phosphines) [119]. Besides the Pd^{II} complexes **10**, these authors also formed the Rh^{III} and Ir^{III} complexes **11** and **12** *via* complexation-induced reduction of the triel(III) centers by Rh^{I} or Ir^{I} species, respectively (Figure 3.12a) [117, 118]. This process resulted in inversed formal oxidation states, *i.e.*, formation of a ligand-stabilized Ga^{I} species (in the case of **12**) [118]. As detailed in Section 3.8, some of these complexes were utilized as catalysts in hydrosilylation reactions.

3.5 Terpyridine Complexes of Carbon-Group Ions

Unlike for Sn^{IV} and Pb^{II}, whose coordination chemistry with polypyridyl ligands is well explored, the complexes of the lighter homologs of the carbon group, *i.e.*, Si^{IV} or Ge^{IV}, are by far less explored [120]. Before turning the attention to the combination of these ions with tpy ligands, a short digression concerning carbon is made. Naturally, carbenium ions will not form complexes with polypyridyl ligands, but the *N*-alkylation of pyridine derivatives affords pyridinium salts,

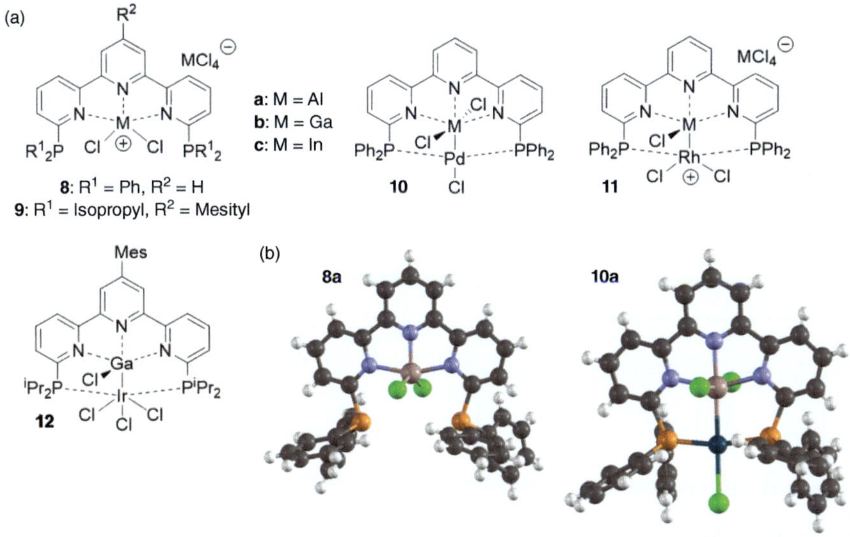

Figure 3.12 (a) Schematic representation of complexes **8–12**. (b) Solid-state structures of **8a** (left) and **10a** (right); In both figures, counterions and solvent molecules were omitted for clarity. Adapted from [116].

which can be considered adducts of a formal carbenium ion, at least in the broadest sense [1]. In this respect, paraquat or methyl viologen (MV^{2+}, *i.e.*, *N*,*N*′-dimethyl-4,4′-bipyridinium dichloride) is certainly the most famous representative [121]. Likewise, *N*-alkylated terpyridines have also been prepared, but these derivatives have not received any significant acknowledgment. A *N*,*N*′-dialkylated tpy derivative, as a paraquat mimic, was already reported in 1969 [122]. Aiming for the fabrication of antimicrobial fabrics, a threefold *N*-methylated tpy derivative, which was bearing a polymerizable group, was synthesized and used further to prepare a lignin-based composite coating [123]. Finally, Yang and Li proposed the use of a redox-active electrolyte, composed of the *N*,*N*″-dimethylterpyridinium dication and the S^{2-}/S_x^{2-} redox couple, for solar-cell applications [124].

Concerning the formal Si^{IV} ion, two contributions have to be mentioned within the scope of this section. Firstly, Schmedake *et al.* prepared $[Si(tpy)_2](PF_6)_4$ from SiI_4 and tpy under forcing conditions (*i.e.*, pyridine, 125 °C, 3 h, sealed ampoule), followed by anion metathesis [125]. These reaction conditions were required since this complex was not formed from $SiCl_4$ and tpy in CS_2 solution – a protocol that was applicable for the synthesis of $[Ge(tpy)_2]Cl_4$ and $[Sn(tpy)_2]Cl_4$ [28]. Noteworthy, the neutral complex $Si(bpy)_3$ was already reported in 1963 [126]; whereas the oxidized complex, $[Si(bpy)_3]^{2+}$ was obtained later from the reaction of SiI_4 and bpy in the melt [127]. Cyclic voltammetry (CV) revealed a remarkable electrochemistry of $[Si(tpy)_2](PF_6)_4$, *i.e.*, four reversible one-electron reduction steps were observed (Figure 3.13a). Detailed EPR and magnetochemical studies on $Si(bpy)_3$ indicated that the added electrons were located in the ligands' π^* orbitals, *i.e.*, ligand-based reduction was favored over the metal-based one [128, 129]. These reductions occurred at significantly more positive potentials

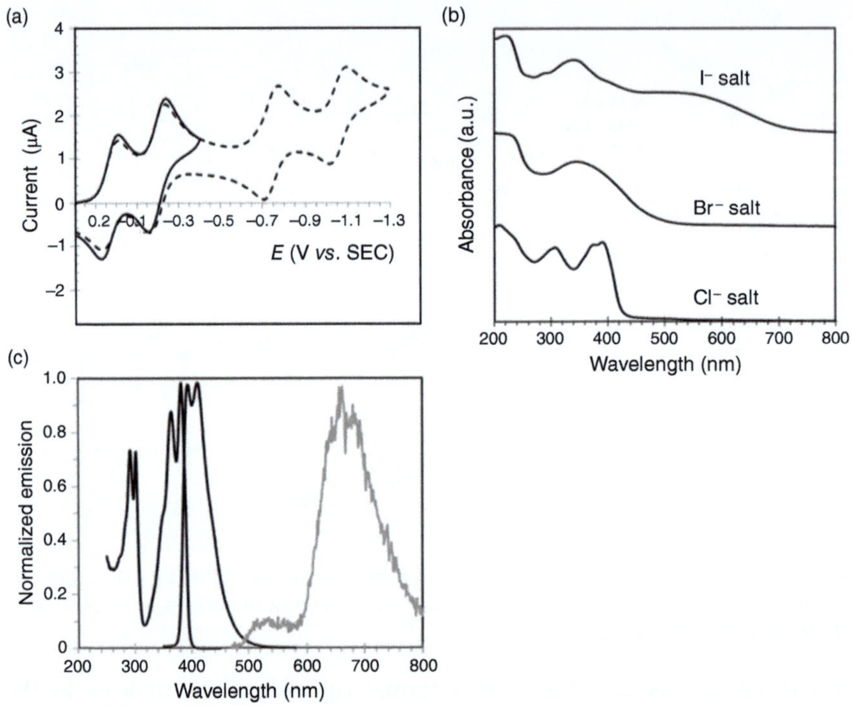

Figure 3.13 (a) CV curves of $[Si(tpy)_2](PF_6)_4$ covering two (solid line) and four reduction steps (dashed line). The measurements were performed in MeCN with 0.1 M $TBAPF_6$ and a Pt reference electrode at the scan rate of $200\,mV\,s^{-1}$. (b) Solid-state UV/vis absorption spectra of $[Si(tpy)_2]^{4+}$ halide salts (thin films on quartz plates). (c) Excitation and emission spectra of $[Si(tpy)_2](PF_6)_4$ in MeCN solution (black curves) as well as emission spectrum of $[Si(tpy)_2]Br_4$ in the solid state at 77 K (gray curve) [125] / with permission of Elsevier.

than the metal-centered reductions of $[Ru(tpy)_2](PF_6)_2$, as the reference, under otherwise similar experimental conditions. At very low potential (*i.e.*, −1.66 V *vs.* SCE [saturated calomel electrode]), a fifth reduction step afforded the $[Si(tpy)_2]^-$ anion; whereas the sixth reduction step, which was present for $[Si(bpy)_3](PF_6)_4$, was absent within the available solvent-limited potential window. Owing to the absence of d-electrons, the UV/vis absorption spectrum of $[Si(tpy)_2](PF_6)_4$ only featured strong bands in the near-UV regime, which originated from ligand-centered (LC) $\pi \rightarrow \pi^*$ transitions. In the solid state, the halide salts of the complex, *i.e.*, $[Si(tpy)_2]Br_4$ and $[Si(tpy)_2]I_4$, were colored (Figure 3.13b), presumably due to charge-transfer (CT) processes from a halide's orbital to a tpy-centered π^* orbital. For $[Si(tpy)_2]I_4$, the absorption profile was almost panchromatic, *i.e.*, it covered the full vis regime of the absorption spectrum. Complex $[Si(tpy)_2](PF_6)_4$ also featured blue emission upon excitation with UV light, which arose from the aforementioned $\pi \rightarrow \pi^*$ transitions (Figure 3.13c). Whereas the halide salts represented only poor solid-state emitters, their weak emission at low temperature was attributed to the CT between the halide anions and tpy ligands. In subsequent work, these same authors showed that Si^{IV}-polypyridyl complexes could also be used as electrochromic dyes, precisely

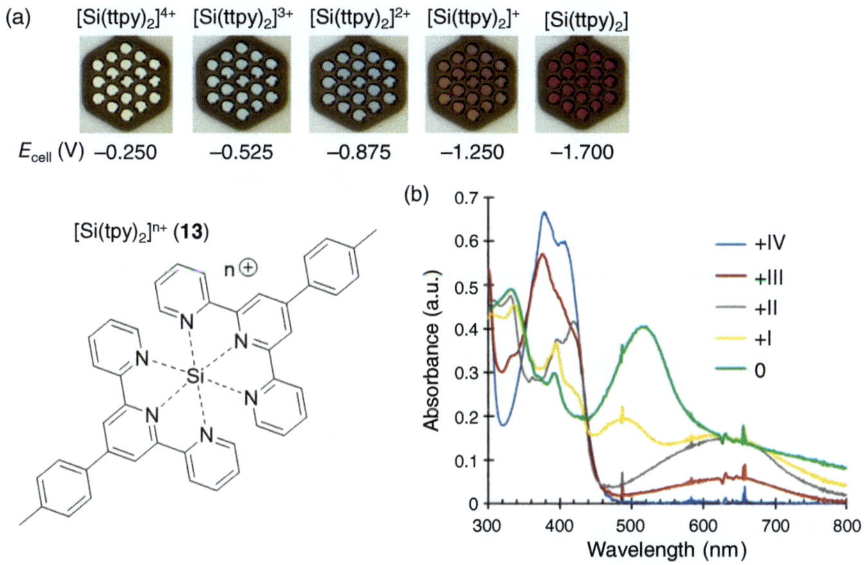

Figure 3.14 Schematic representation of $[Si(ttpy)_2]^{4+}$ (**13**) as polychromatic cathode colorant (the counterions are omitted for clarity). (a) Photos of the electrochromic window of the dye in MeCN solution; the electrochemical cell contained a Pt honeycomb working electrode. (b) UV/vis absorption spectra of **13** at the different redox states. [130] / MDPI / CC BY 4.0.

as cathodic colorants. In MeCN solution, the complex **13** (Figure 3.14) displayed polychromatic behavior: Its coloration – ranging from yellow (+IV state) to magenta (+I and charge-neutral state) – could be addressed by applying a proper cell voltage (Figure 3.14a). Further insight into nature of the involved ligand-based reduction steps was gained from time-dependent and broken-symmetry DFT calculations as well as spectroelectrochemical measurements (Figure 3.14b).

Besides these examples for homoleptic Si^{IV}–terpyridine *bis*-complexes, Hermannsdorfer and Driess reported some *mono*-complexes, in which tpy was coordinated to Lewis-superacidic cations of the types R_2Si^{2+} and RSi^{3+} (R denotes an alkyl or aryl group) [131]. Complexes **14** and **15** (Figure 3.15a) could be isolated and characterized *inter alia* by NMR and single-crystal XRD analysis [131]. Originally, superacids were defined as acids whose acidity is even greater than that of pure H_2SO_4 and, thus, their Hammet acidity function (H_0) is smaller than −12 [133]. Whereas the contemporary perception uses a unified absolute pH scale, which is independent of the phase. The proton's chemical potential in the respective medium represents the measure and, for a superacid, it must be higher than in pure H_2SO_4 [134]. Similar considerations have also been made for "Lewis superacids": Molecules, whose gas-phase Lewis acidity is higher than that of SbF_5, are referred to as "Lewis superacids" [135]. Basically, this concept can also be applied to the condensed phase. However, the highly reactive Lewis-superacidic species have to be stabilized in solution by appropriate ligands. Greb provided a thorough analysis of the concept of Lewis superacidity, thereby taking the common scaling methods,

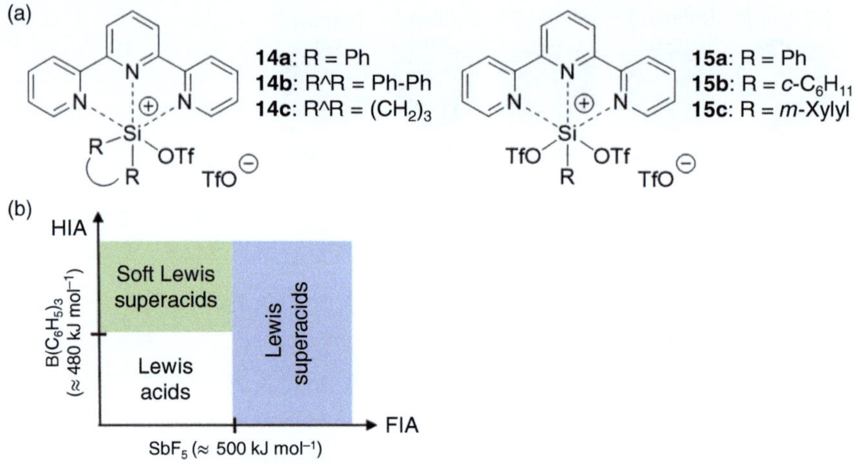

Figure 3.15 (a) Schematic representation of complexes **14** and **15**. Adapted from [131]. (b) Classification of Lewis acids according to hydride-ion affinity (HIA) and fluoride-ion affinity (FIA). Adapted from [132].

i.e., the hydride-ion affinity and fluoride-ion affinity (HIA and FIA, respectively; Figure 3.15b) into account [132]. Compared to the gas-phase FIAs of cations R_2Si^{2+} and RSi^{3+}, the FIAs of the corresponding complexes **17** and **18** in solution were significantly lower [131]. Hence, stabilization by the chelating tpy ligand came at the expense of Lewis superacidity. For example, the calculated FIA of the Ph_2Si^{2+} cation in solution was 673 kJ mol^{-1}; whereas a solution-based FIA of only 263 kJ mol^{-1} was estimated for **14a**. The FIA of complexes **15** in solution was even higher than that of **14** (*e.g.*, 563 kJ mol^{-1} in the case of **15a**). Complexes **14** and **15** represented good acceptors for both fluoride as well as hydride ions. This behavior was utilized, *e.g.*, in hydrodefluorination reactions (see Section 3.8).

Examples for coordination compounds of the formal Ge^{IV} cation with tpy are also rare. Already in 1965, Fergusson *et al.* performed the first study in this respect [136]. These authors obtained a complex of 1:1 stoichiometry from the reaction of GeI_4 with tpy in an aromatic solvent. Based on the measurements of the molar conductivity in nitrobenzene (24.2 cm Ω^{-1} mol^{-1}), dissociation into $[(tpy)GeI_3]^+$ and I^- ions was assumed. Further evidence for the molecular structure of this assembly was not provided. Whereas the reaction of $GeCl_4$ or $GeBr_4$ with tpy under the same experimental conditions did not afford similar complexes. As mentioned beforehand, CS_2 represents an appropriate solvent to overcome this restriction [28]. Various attempts have also been made to prepare structurally defined coordination compounds with GeF_4. Firstly, elemental analysis corroborated the formation of a compound with the composition "$(GeF_4)_3(tpy)_2$" when the $GeF_4(MeCN)_2$ precursor was reacted with tpy in CH_2Cl_2 [120]. IR studies suggested that the coordination compound was of oligomeric structure in which hexa-coordinated Ge^{IV} centers were bound to tpy ligands in the hypodentate κ^2- and κ^1-modes. Unlike the former compound, which could not be characterized further due to its poor solubility,

$[Ge(tpy)F_3](OTf)$ possessed increased solubility in organic solvents and could be analyzed *inter alia* by ^{19}F-NMR spectroscopy; these measurements gave evidence for the existence of the $[Ge(tpy)F_3]^+$ cation [137]. The binding of tpy to small-cage metallocarborane precursors has also been studied. In particular, Hosmane *et al.* reported various such coordination compounds [138]; these compounds are presented separately in Section 3.10.

Unlike for the lighter congeners from the carbon group, the coordination chemistry of tin with tpy ligands has been studied to a significant extent – this holds true for the formal Sn^{IV} and Sn^{II} ions [1, 28]. Once more, it was Fergusson *et al.* who pioneered this field of research [136]. The Sn^{IV} halides revealed remarkable differences in their coordination behavior. $SnCl_4$ and $SnBr_4$ gave complexes of 2:1 stoichiometry (according to elemental analysis); based on conductivity measurements in $MeNO_2$, the molecular formula $[Sn(tpy)X_3]_2(SnX_6)$ was proposed (X = Cl or Br). Whereas SnI_4 afforded a coordination compound, which behaved as a typical 1:1 electrolyte; thus, the molecular formula $[Sn(tpy)I_3]I$ was assumed. The difference in the anion's structure was attributed to the intrinsically poor stability of the $(SnI_6)^{2-}$ species. In the same context, the coordination chemistry of tin(IV)organyls was also explored. Complexes of similar structure were also obtained from mono- and dialkyltin(IV) halides, but not from the corresponding trialkyltin(IV) halides. However, the proposed structure of the $SnMe_2Cl_2$/tpy complex was revised when Einstein and Penfold disclosed the solid-state structure by XRD analysis. The molecular structure of complex **16** was composed of the highly distorted octahedral $[Sn(tpy)MeCl_2]^+$ cations and penta-coordinate $[SnMe_2Cl_3]^-$ anions (Figure 3.16a) [139, 141, 142].

The discrepancy between the originally proposed structure of **16** and the revised one unfolds the necessity of reliable analytical techniques to gain insight into the molecular structure of the compounds. Regarding the solid-state structure, this is given by XRD analysis, as long as suited single crystals are obtained. Therefore, it is advantageous to have other independent methods at hand (*e.g.*, NMR spectroscopy, MS, *etc.*). As discussed in Section 1.4.2, Mößbauer spectroscopy represents one

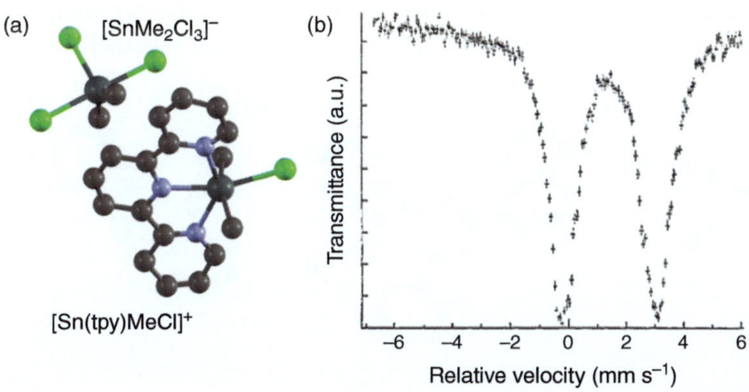

Figure 3.16 (a) Solid-state structure of complex **16**. Adapted from [139]. (b) ^{119}Sn Mößbauer spectrum of complex **16** (recorded *vs.* a $Ba^{119}SnO_3$ source) [140] / with permission of American Chemical Society.

further versatile technique to probe the chemical structure of complexes. The presence of Mößbauer-active elements represents the prerequisite for such measurements. Such elements feature the so-called Mößbauer effect, *i.e.*, the resonant and recoil-free emission/absorption of γ-radiation. Among others, the ^{119}Sn nucleus is highly active in this respect and, consequently, Sn^{II} and Sn^{IV} compounds can be analyzed by Mößbauer spectroscopy. As one notable recent innovation, Barone *et al.* established DFT calculations of the electric-field gradient at the ^{119}Sn nucleus as a versatile method to probe compounds also from a theoretical point of view. Thus, the correlation of experimental ^{119}Sn Mößbauer spectra with the predicted ones was enabled [143].

Zuckerman *et al.* were the first to apply this method to a wide range of tpy complexes with Sn^{IV} and Sn^{II} centers [140, 144, 145]. The ^{119}Sn Mößbauer studies confirmed, *e.g.*, the presence of hexa-coordinate cations and penta-coordinate anions in various complexes (*e.g.*, **16**), due to the distinct isomer shift and quadrupole splitting (Figure 3.16b) [140]. The reaction of **16** with Li(TCNQ) provided a stable compound, whose blue coloration was attributed to CT processes in the electron-paired dimeric $(TCNQ)_2^{2-}$ anions, which accompanied the $[Sn(tpy)MeCl_2]^+$ cations (TCNQ: 7,7,8,8-tetracyanoquinodimethane) [146]. The results of EPR measurements ruled out the presence of $TCNQ^{\cdot-}$ radical anions. Whereas the chloride anions of the non-chelated Sn^{IV} precursors were displaced to afford $R_3Sn(TCQN^{\cdot\cdot})$ (R = Me, *n*-Pr or *n*-Bu) or $Me_2Sn(TCQN^{\cdot})_2$, as persistent (di)radicals. Wassef and Hessin observed an unusual ligand-transfer reaction upon dealing with the coordination chemistry of Ph_2SnCl_2 [147]. The 1:1 complex, which had resulted from the binding of tpy, was capable of transferring the tpy ligand to a Hg^{II} center, thus liberating Ph_2SnCl_2. The directionality of this process was presumably due to the Hg^{II} cation's affinity to the ligand, which was higher than that of the Sn^{IV} center (the β_{21} value for forming the Hg^{II}–tpy *bis*-complex is *ca.* 23 [148]). Furthermore, an Sn^{IV} center with a high CN of seven was observed in $[Sn(tpy)Me_2](SCN)_2$ [149]. Thereby, the Me-groups occupied the *trans*-axial positions of the pentagonal-bipyramidal coordination motif (Figure 3.17a). Penta-coordination of the Sn^{IV} was corroborated further by ^{119}Sn Mößbauer, IR and ^{119}Sn NMR measurements of some $[Sn(tpy)R_2](SCN)_2$ complexes (R = Me, *n*-Bu, Ph or vinyl) [152–154]. These measurements revealed that the

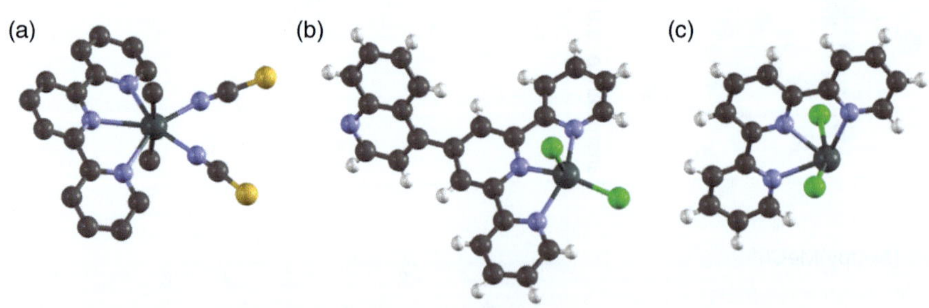

Figure 3.17 (a) Solid-state structure of $[Sn(tpy)Me_2](SCN)_2$ (the positions of the hydrogen atoms were not resolved). Adapted from [149]. (b) Solid-state structure of **17**. Adapted from [150]. (c) Solid-state structure of $[Sn(tpy)Cl_2]$. Adapted from [151]. In all three figures, solvent molecules were omitted for clarity.

relatively short tin-to-carbon bonds in $[Sn(tpy)Ph_2](SCN)_2$ were of double-bond character, due to the occurrence of a $^+C_6H_5 = Sn^-$ resonance structure [154].

Authors also paid attention to the coordination chemistry of the formal Sn^{II} ion. Recently, Momeni *et al.* reported the low-valent complex **17**, which was prepared by the coordination of the 4′-(4-quinolin-yl)-tpy ligand to $SnCl_2$ (Figure 3.17b) [150]. Like in the complex, which was derived from the parent tpy ligand, the Sn^{II} center was penta-coordinated and the ligands adopted a distorted square-pyramidal arrangement (Figure 3.17c) [151, 155]. The Lewis-basic character of these complexes was due to the metal's lone pair of electrons, which occupied one equatorial position. This positioning offered sufficient space for the interaction with Lewis-acidic molecules. $GeCl_2$ and $SnCl_2$ have been found to undergo auto-ionization in the presence of chelating Lewis bases [156]. However, this behavior was not observed for the $SnCl_2$/tpy couple. Auto-ionization to afford $[Sn(tpy)Cl][SnCl_3]$ or $[Sn(tpy)_2][SnCl_4]$ was presumably not possible due to insufficient electronic stabilization and steric hindrance, respectively. As one of the rare examples, where such a behavior occurred in the presence of a tpy ligand, Yasuda *et al.* reported the formation of $[Fe(tpy)_2][FeCl_4]$ upon refluxing an equimolar mixture of $FeCl_2$ and tpy in THF [157].

In coordination chemistry, a template reaction is a ligand-based reaction that involves (at least) two adjacent coordination sites. In many cases, the coordinated metal helps the ligand to adopt the required geometry to enable, *e.g.*, macrocyclization. If the metal center is absent, ligand pre-organization is no longer given and, thus, other products might be obtained. As one prominent example, the formation of crown ethers, which is templated by alkali-metal ions, has to be highlighted [158]. Constable *et al.* used this concept for the templated synthesis of macrocycles [159]. If the template reaction between ligand **18** and glyoxal was conducted in MeOH with Ni^{II}, the corresponding metallized macrocycles $[Ni(\mathbf{19})]^{2+}$ were obtained (Figure 3.18a). However, a different behavior was observed when the reaction was performed in the presence of Me_2SnCl_2. In the absence of glyoxal, the self-assembly of ligand **18a** with Me_2SnCl_2 gave the solution-stable complex $[Sn(\mathbf{18a})Me_2]^{2+}$. Whereas, under the conditions of the template reaction, the "transient template effect" came into play, and the demetalated macrocycle **19a** was formed. Upon macrocyclization, the ligand's binding pocket was contracted. The resulting mismatch in size of the $[SnMe_2]^{2+}$ fragment with the available binding pocket caused its release once the macrocycle was formed. The release of $[SnMe]^{2+}$ was followed by its hydrolysis into SnO_2, thus withdrawing the species entirely from the reaction mixture. On the other hand, the size of the Ni^{II} ion matched with the cycle's diameter and, thus, remained coordinated. The characterization of $[Sn(\mathbf{18a})Me_2]^{2+}$ using NMR spectroscopy and single-crystal XRD analysis revealed the typical hepta-coordinate complex geometry, in which the axial positions were occupied by the Me-groups (Figure 3.18b) [159, 161, 162]. In one further example, Constable *et al.* intended to use the transient template effect of $[Me_2Sn]^{2+}$ also for the synthesis of macrocycles **20** [160]: Unexpectedly, a Sn^{II}-induced rearrangement instead afforded the ring-contracted **21** (Figure 3.18c). Owing to their poor kinetic stability in solution, the proposed $[Sn(\mathbf{21})Me_2]^{2+}$ intermediates were not isolated or characterized.

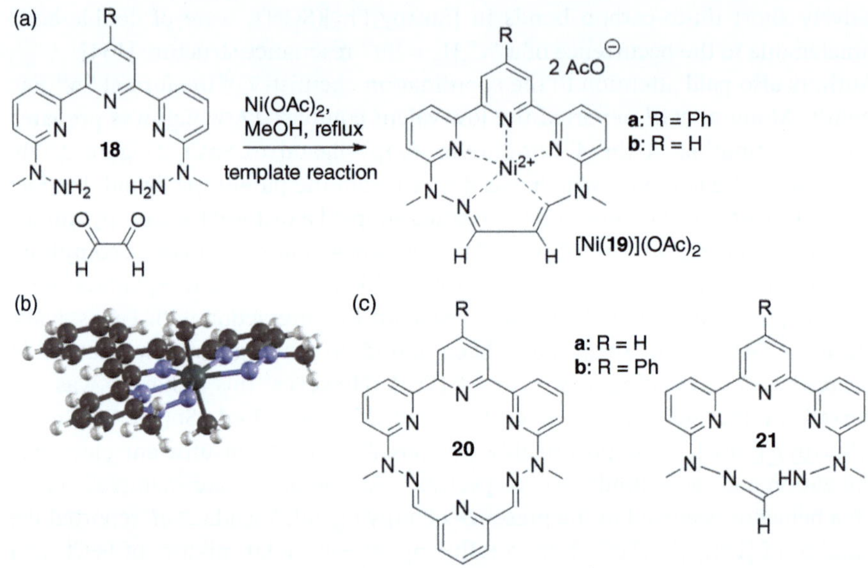

Figure 3.18 (a) Schematic representation of the template reaction to give [Ni(**19**)](OAc)$_2$. Adapted from [159]. (b) Solid-state structure of the [Sn(**18a**)Me$_2$]$^{2+}$ cation (counterions and solvent molecules were omitted for clarity). Adapted from [159]. (c) Schematic representation of targeted and obtained macrocycles **20** and **21**, respectively .

Although the vast majority of studies devoted to Sn^{IV}–tpy complexes have dealt with fundamental aspects, practical applications in a biological/medicinal context have also been proposed [24, 163]. It is well known that terpyridines and their transition-metal complexes possess biomedical relevance (*e.g.*, antimicrobial or antitumor agents, biosensors, labels for bioimaging, *etc.*) [17–19, 164, 165]. The field of research, which can be traced back (at least) to Lippard's seminal work from 1974 on the binding of Pt^{II}–tpy complexes to DNA [166], is summarized in Chapter 4. Therein, the relevant tpy complexes of main-group elements (MGEs) will be mentioned briefly.

Lead represents the heaviest naturally occurring element of the carbon group and, unlike its lighter homologs, the +II oxidation state is favored over the +IV one. This exception is due to relativistic effects and parallels the behavior of thallium and bismuth, as lead's neighbors in the PTE [80, 81]. As one consequence, the coordination chemistry of Pb^{II} is similar to that of the late transition-metal ions and differs from that of the aforementioned tetravalent ions of the Si-to-Sn triad. According to Thummel. Hancock *et al.*, the K_a values for the binding tpy are similar for Pb^{II} and Cd^{II} (*i.e.*, $\lg K_1 = 6.0$ *vs.* 6.2), but lower than that of Ni^{II}, whose tpy complexes are considered kinetically rather stable ($\lg K_1 = 10.7$; see Table 1.2) [167]. However, this general trend contradicts the observation that a tpy-functionalized polystyrene (PS) resin represented an efficient absorbent for the extraction of Pb^{II} ions from pH-neutral aqueous solutions [168]. Mubarak *et al.* reported that the capability for the metal-ion uptake followed the unexpected order: $Pb^{II} > Cu^{II} > Zn^{II} > Ni^{II}$.

As one part of this systematic study on the coordination chemistry of the tetryl(IV) halides, Fergusson *et al.* also reacted Ph_2PbX_2 with tpy (X = Cl, Br or I) [136]. In this

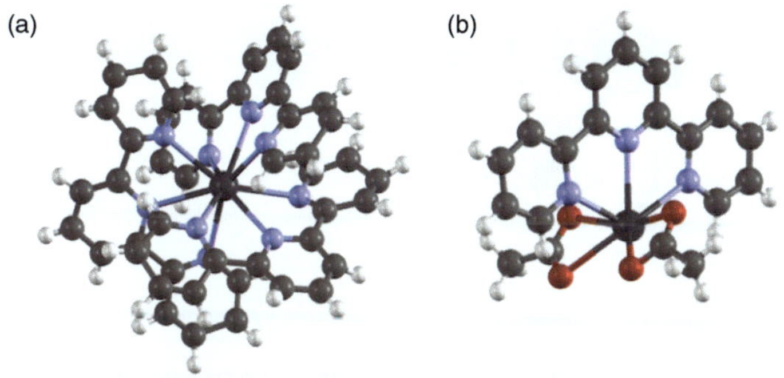

Figure 3.19 (a) Solid-state structure of $[Pb(tpy)_3](ClO_4)_2$. Adapted from [170]. (b) Solid-state structure of $[Pb(tpy)(OAc)_2]$. Adapted from [171]. In both figures, non-ccordinating counterionsand solvent molecules were omitted for clarity.

primary study, complexes of the general formula $[Pb(tpy)Ph_2X_2]$ were obtained, which barely dissociated in solution and, thus, afforded poor molecular conductivities. The different structure of these complexes (compared to those of the related Sn^{IV} species, *vide supra*) was due to the metal's increased Lewis acidity on one hand and the tolerance of high CNs in its complexes on the other [169]. This behavior is expressed to its best by the formation of the *tris*-tridentate complex $[Pb(tpy)_3]^{2+}$ with the very high CN of nine (the ClO_4^- and BF_4^- salts were prepared; Figure 3.19a) [170, 172]. High CNs were also observed in the solid-state structures of other Pb^{II}–tpy complexes. For example, the combination of halide salts PbX_2 with tpy in 1:1 ratios provided binuclear (X = Cl or I) or mononuclear complexes (X = Br) with CNs ≥6 [173]. Whereas $Pb(SCN)_2$, $Pb(ClO_4)_2$, $Pb(NO_3)_2$, $Pb(NO_2)_2$, and $Pb(SiF_6)_2$ all gave anion-bridged 1D coordination polymers under the applied self-assembly conditions (see Section 3.9) [172–175]. In the case of $Pb(OAc)_2$, no such extended structures were formed. Morsali isolated the *mono*-nuclear complex $[Pb(tpy)(OAc)_2]$, in which the acetate anions were each coordinated in a bidentate fashion (Figure 3.19b) [171]. Notably, this author also used ^{207}Pb-NMR spectroscopy to characterize the complex in solution. This particular heteronuclear NMR technique is of importance, due to the properties of the ^{207}Pb nucleus, which combines natural abundance (*ca.* 22%) with beneficial characteristics, *i.e.*, applicable spin (1/2) and gyromagnetic ratio (5.6264×10^7 rad T^{-1} s^{-1}) [176]. Moreover, the chemical shift of Pb^{II} compounds is highly sensitive to the chemical environment and signals might appear over a wide chemical-shift range (*ca.* 11,500 ppm). Conventional ^{207}Pb-NMR or ^{207}Pb-^1H correlated spectroscopy (COSY) is widely used to study Pb^{II} coordination compounds in the solid state or in solution.

In addition to the aforementioned examples, where the parent tpy ligand was used for complexation, 4′-substituted derivatives were combined with Pb^{II} ions to provide (in most cases) coordination compounds of higher dimensionality (see Section 3.9) [177–181]. Despite the intrinsically high (cyto)toxicity of lead compounds, Pb^{II} complexes with tpy ligands have even been considered for biomedical applications (in line with their Sn^{IV}-containing analogs) [182, 183]. For further reading on this topic, reference to Chapter 4 is given here.

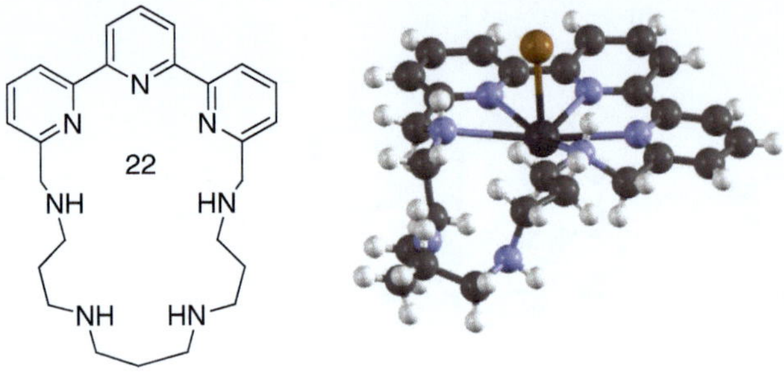

Figure 3.20 Schematic representation of the macrocycle **22** and solid structure of its Pb^{II} complex having one coordinated Br^- ion in axial position (non-coordinating counterions and solvent molecules were omitted for clarity). Adapted from [184].

Bencini, Pina *et al.* prepared the macrocyclic ligand **22**, which was similar to Constable's systems (Figure 3.18); however, these authors did not use the templating effect with transition-metal ions for their synthesis [184]. pH-Titration studies revealed that **22** was capable of undergoing five successive protonation steps involving the four secondary amine sites and, finally, the tpy moiety. The binding of Cd^{II} or Pb^{II} ions afforded the corresponding mononuclear complexes, in which the two most remote *N*-atoms opposite to the tpy site were not involved in coordination (the solid-state structure of [Pb(**22**)Br](ClO_4) is depicted in Figure 3.20). The analogous $[Cd(\mathbf{22})(H_2O)_2]^{2+}$ complex was emissive in solution, though only in the lower pH regime. It was assumed that protonation of the non-coordinating *N*-atoms blocked the pathways for intramolecular quenching. The $[Pb(\mathbf{22})Br]^+$ cation did not show such a behavior in aqueous solution.

In metallo-supramolecular chemistry, the combination of Pb^{II} ions with polypyridyl ligands has been used for the assembly of coordination compounds, whose architecture depended on the folding of the oligotopic ligands upon metal-ion coordination. This field of research was pioneered by Lehn *et al.*, who prepared a plethora of metallo-supramolecular constructs using Pb^{II} ions: twisted/helical assemblies [185], grid-like arrays [186], zippers [187], tweezers [188], *etc.* The aspect of rack/grid formation, *inter alia*, by the reversible coordination of Pb^{II} ions, was already one subject in our previous textbooks [189, 190]. Admittedly, in most of these studies, the sophisticated ligands did not contain tpy sub-units in the narrower sense of meaning. At this stage, only one example is highlighted, in which the tpy-centered ligands **23** afforded single-stranded helices upon binding of two Pb^{II} ions (Figure 3.21a) [191]. This study was performed in the broader context to mimic the self-organization of natural biomacromolecules (*e.g.*, DNA or peptides). In Lehn's artificial systems, the molecular strands are reversibly folded/unfolded *via* the pH-induced binding and release of metal ions (Figure 3.21b). The utmost importance of the ligand design was emphasized by Hanton *et al.* [192]. These authors showed that one has to pay attention to the number of donor sites as well as their distance within the ligand strand when aiming for the assembly of single-strand helices with Pb^{II} ions. Changing one of these parameters will result in a structural mismatch and hamper (or even hinder) the self-assembly process.

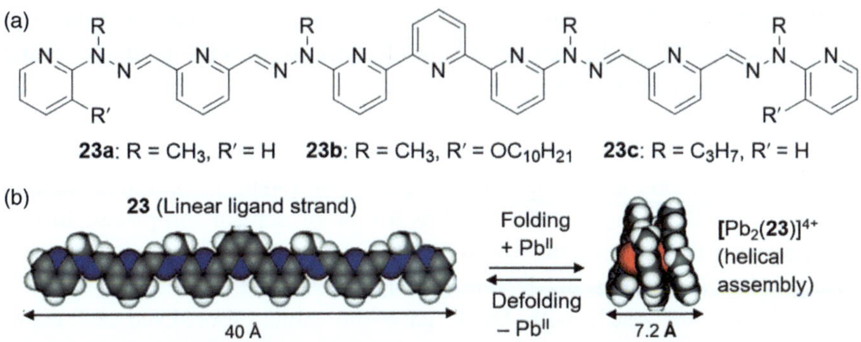

Figure 3.21 (a) Schematic representation of Lehn's concept of reversible metallo-supramolecular folding/unfolding. (b) Schematic representation of linear ligands **31**. (c) Schematic representation of the reversible assembly of **31** to give the corresponding helical, binuclear complexes; the lengths of the compounds were estimated by molecular modelling (MM) and derived from single-crystal XRD analysis, respectively. [191] / with permission of Royal Society of Chemistry.

Noteworthy, this consideration is of equal importance concerning the formation of metallo-supramolecular constructs, which are discussed later in Chapter 10.

At the end of this section, the use of Pb^{II} ions to assemble coordination polymers is worth mentioning. In this regard, a wide range of architectures has already been reported: 1D coordination polymers [172–175, 193], 2D coordination nanosheets (CONASHs) [194] as well as 3D coordination networks or metal–organic frameworks (MOFs) [195–197]. These assemblies are dealt with separately in Section 3.9.

In today's perception, the use of lead is undoubtedly associated with perovskite solar cells [198, 199]. This emerging class of photovoltaic device uses perovskite-type materials, as the light absorber; organic–inorganic hybrid materials, such as $MAPbX_3$, are *inter alia* used for this purpose (MA: methylammonium; X = Br or I). In order to manipulate the material's crystallinity and, thus, its properties, additives have been explored. At this point, the polypyridyl ligands, such as tpy, came into play [26, 200, 201]. As mentioned beforehand (see Section 3.4), the use of tpy in the context of photovoltaic applications represents a topic of its own and, accordingly, is treated later in Chapter 6.

3.6 Terpyridine Complexes of Pnictide Ions

Back in the 1970s, Brodie and Wilkins performed the first study on the coordination chemistry of various pnictogen(III) halides with polypyridyl ligands, including tpy [202]. The thusly obtained complexes were analyzed by the techniques available by that time: IR spectroscopy, elemental analysis, and conductivity measurements. $BiCl_3$ afforded a complex, whose properties were distinctly different from those of the complexes derived from AsX_3 or SbX_3 (X = Cl, Br, I). The unique role of the Bi^{III} ions is again due to relativistic effects [80, 81] and, unlike for As^{III} or Sb^{III}, a substantial coordination chemistry with tpy is known (*vide infra*). The tpy complexes of AsX_3 and SbX_3 almost uniformly contained the $[M(tpy)X_2]^+$ cation

(M = As or Sb; X = Cl, Br or I); whereas some variations in the nature of the anion were observed [202]. In the case of Sb^{III}, preference to form the penta-coordinate $[SbX_5]^{2-}$ anion was found. Likewise, the reaction between $AsCl_3$ and tpy provided $[As(tpy)Cl_2]_2(AsCl_5)$, whose ions both possessed a penta-coordinate geometry. However, in the case of $AsBr_3$, a complex was formed, which was composed of $[As(tpy)Br_2]^{2+}$ cations and simple Br^- anions. The $[AsBr_5]^{2-}$ anion is virtually unknown and will, owing to its instability, not be present in complexes. Instead, the complexes either contain the simple Br^- anion or even hypervalent anions (*e.g.*, $[As_2Br_8]^{2-}$, $[As_2Br_9]^{3-}$, $[As_3Br_{12}]^{3-}$, *etc.*) [203].

The reduction of $AsCl_3$ with $SnCl_3$ gave the labile, short-lived As^I cation. However, this species can be generated *in situ* and trapped by coordination. For this purpose, chelating ligands with sp^2-hybridized *N*-atoms might be used. For example, this was achieved with bpy [204] as well as some tridentate ligands [205], but not reported yet for tpy.

Similar to the aforementioned complexes of Sn^{II}/Sn^{IV} ions, which can be analyzed by ^{119}Sn Mößbauer spectroscopy (see Section 3.5), the corresponding Sb^{III} compounds are also highly susceptible to this method. Indeed, the ^{121}Sb nucleus is one of the three most relevant Mößbauer-active isotopes (aside ^{57}Fe and ^{119}Sn). The characteristics of ^{121}Sb enable the probing of the electronic structure of Sb compounds and allow one to accurately conclude on their molecular structure [206]. For example, IR and ^{121}Sb Mößbauer spectroscopic analysis of the complex $[Sb(tpy)F_3]$ suggested a hepta-coordinate geometry and stereochemical activity of the metalloid's lone pair [207]. The solid-state structure of this complex, which was disclosed later, confirmed the assumed distorted pentagonal-bipyramidal coordination geometry, in which two fluorido ligands together with the tpy ligand were in the basal plane and the apical positions were occupied by the third F^- ligand and the lone pair (Figure 3.22) [208, 209].

Owing to the quadrupole of the ^{121}Sb (I = 5/2) and ^{123}Sb nuclei (I = 72), one further notable technique, *i.e.*, nuclear quadrupole resonance (NQR) spectroscopy, is predestined for getting insight into bond lengths between an Sb center and its neighbors [210, 211]. NQR spectroscopy is commonly named "zero-field NMR": This method does not rely on an external field, as it is based only on the interaction between the quadrupole and the electric-field gradient at the nucleus. Using Semin's approach, one can calculate the coupling constant for each nucleus [212]. Bryce *et al.* utilized NQR spectroscopy to analyze the binding strength of polypyridyl ligands in complexes of SbF_3 and $SbCl_3$ [213]. For all complexes, the NQR coupling

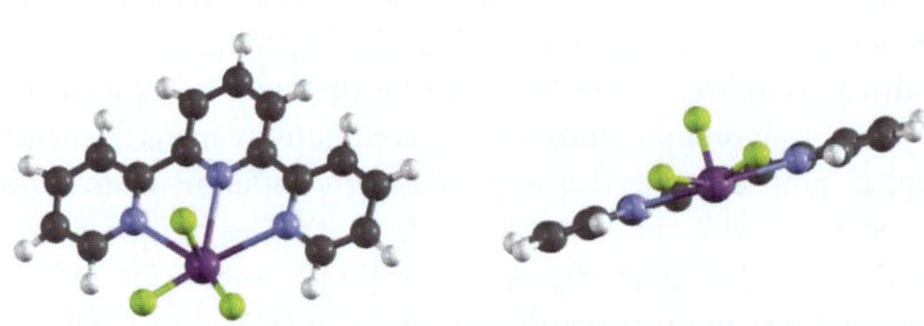

Figure 3.22 Two different views of the solid-state structure of $[Sb(tpy)F_3]$ (solvent molecules were omitted for clarity). Adapted from [208].

constant was smaller than in the respective precursor; furthermore, a gradual decrease of the coupling constant with increasing number of *N*-donor sites became apparent. It was thus concluded that Sb^{III} was strongly bound to chelating bidentate or tridentate ligands.

To the best of our knowledge, further studies on the coordination chemistry of the formal As^{III} or Sb^{III} ions with tpy ligands are (not yet) available. But what about the other two relevant pnictogen(III) centers, *i.e.*, the formal P^{III} and Bi^{III} ions? Considering the coordination chemistry of the former, as the acceptor, appears odd at first glance, since trivalent phosphorus compounds are traditionally ranked as nucleophiles and Lewis bases, but not as Lewis acids. However, Lewis-acidic phosphines have gained attention due to their ability to promote CH-activation [214]. In order to evoke such reactivity, the electrophilic P^{III} center has to be stabilized by strong π- and/or σ-donor ligands (*e.g.*, carbenes, pyridines, *etc.*). Against this background, Stephan *et al.* prepared the air-stable complexes **24** from $PPhCl_2$ and tpy on the gram-scale under mild conditions (Figure 3.23a) [23]. In an alternative approach, Hirano, Mura *et al.* reacted phenylphosphinic acid and tpy in the presence of triflic anhydride (Tf_2O) to obtain (**24**)$(OTf)_2$ [216]. Single-crystal XRD analysis revealed a "seesaw" geometry of the tetra-coordinate P^{III} center with the lone pair being stereochemically active. As shown in Figure 3.23b, one anion of (**24**)$(OTf)_2$ bound axially to the P^{III} center in a bidentate fashion [23]. The Lewis acidity of cation **24**$^{2+}$ was utilized to catalyze various reactions, *e.g.*, the hydrodefluorination of alkyl fluorides (see Section 3.8).

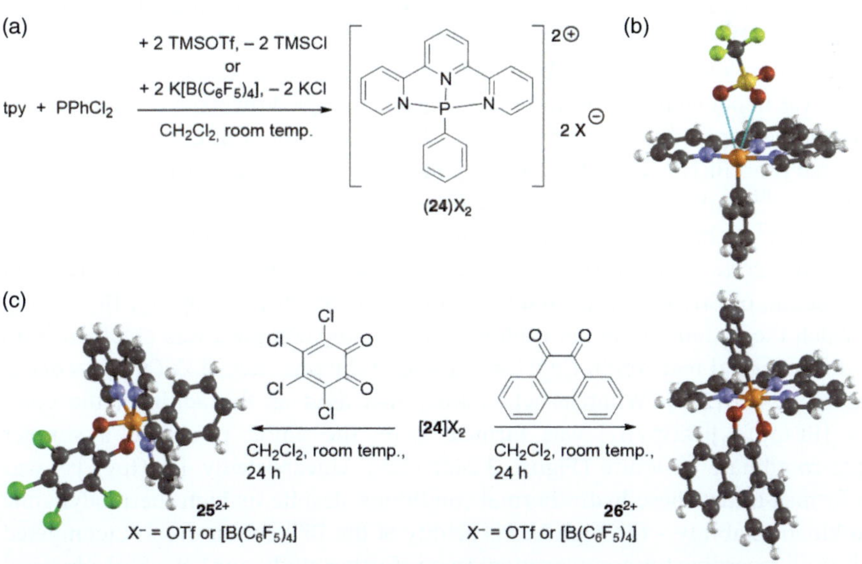

Figure 3.23 (a) Schematic representation of the synthesis of complexes (**24**)X_2. (b) Solid-state structure of complex (**24**)$(OTf)_2$. Adapted from [23]. (c) Schematic representation of the synthesis of complexes (**25**)X_2 and (**26**)X_2; the solid-state structures of the respective cations are also shown. Adapted from [215]. In figures in panels (b) and (c), (non-coordinating counterions and solvent molecules were omitted for clarity).

The complexes (**25**)X_2 and (**26**)X_2 were prepared from the parent complexes (**24**)X_2 by oxidation with an *ortho*-quinone (*i.e.*, *ortho*-chloranil or 9,10-phenanthrenequinone). These complexes contained a formal P^V dication, which featured a hexa-coordinate, pseudo-octahedral coordination sphere (Figure 3.23c) [215]. The fundamentals of the Gutmann–Becket method to estimate a compound's Lewis acidity have been explained beforehand (see Section 3.2) [42]. When applied to dications **25**$^{2+}$ and **26**$^{2+}$, a weak Lewis-acidic character became apparent, despite the actually saturated coordination sphere [215]: The gauged AN for these complexes was *ca.* 12, thus significantly lower than for classical Lewis acids [*e.g.*, $B(OMe)_3$: AN = 23]. However, DFT calculations indicated that the observed Lewis acidity was carbon-based and, thus, of different origin than in complex **24**$^{2+}$ or its Si^{IV}-based relatives **14** and **15** (Figure 3.15). The lowest unoccupied molecular orbital (LUMO) of cations **25**$^{2+}$ and **26**$^{2+}$ was located on the central pyridine ring of the tpy ligand (most pronounced in its 4′-position). Despite this fundamental difference, both complexes likewise represented catalysts for the hydrodefluorination reaction (see Section 3.8).

As mentioned beforehand, the reaction of $BiCl_3$ with tpy afforded a complex, whose properties significantly differed from those of the complexes derived from $AsCl_3$ and $SbCl_3$ [202]. Complex $[Bi(tpy)Cl_3]$ was barely soluble in polar-aprotic solvents and possessed a high degree of covalency, as expressed by the low molar conductivity in solution. The different behavior was attributed to the size of the Bi^{III} ion. Unlike the As^{III} and Sb^{III} ions, the very large Bi^{III} is able to exceed the CN of 5; compared to the size of the As^{III} ion, hexa-coordinate Bi^{III} is *ca.* 2.25 times larger [217]. Whereas the ionic radii of Bi^{III} and La^{III} are similar (at least for the CNs 6 and 8); therefore, the coordination chemistry of Bi^{III} often parallels that of the lanthanide(III) ions [218]. Raston *et al.* reported the first solid-state structure of a Bi^{III}–tpy complex, specifically $[Bi(tpy)(S_2CNEt_2)I_2]$ [219]. In this complex, the Bi^{III} center was hepta-coordinated and the I^- ligands occupied the axial positions of the distorted pentagonal-bipyramid (Figure 3.24a). Moreover, a dimeric complex was prepared from $Bi(NO_3)_3 \times 5\,H_2O$ and tpy in hot dimethylsulfoxide (DMSO); in this case, the Bi^{III} centers were even nona-coordinate [222]. This example highlights the ability of Bi^{III} to handle very high CNs up to 10 (similar to the Ln^{III} ions).

zur Loye *et al.* explored the influence of transition-metal salts, as additives, on the reaction of BiI_3 with tpy. Firstly, the unusual complex $[Bi(tpy)_2I_2][Bi_2I_7(tpy)]$, in which the cation and anion each were a Bi^{III}–tpy complex, was obtained from the solvothermal reaction in EtOH/water and in the presence of $Zn(NO_3)_2 \times 6H_2O$ (Figure 3.24b) [220]. Whereas when RuI_3 was used as the additive, the complex $[Bi(tpy)_2I_2][Bi(tpy)I_4]$ was formed; here, the anion possessed a simpler hepta-coordinate structure (Figure 3.24c) [221]. Unexpectedly, $[Ru(tpy)_2]^{2+}$ was not formed under these hydrothermal conditions, despite its high thermodynamic and kinetic stability – the high Lewis acidity of the Bi^{III} center even outcompeted Ru^{III}/Ru^{II} regarding tpy complexation. In one further study, zur Loye *et al.* obtained complexes, which were composed of $[M(tpy)_2]^{2+}$ cations (M = Fe^{II}, Co^{II} or Zn^{II}) and iodobismuthate anions [223, 224]. The $[Bi_2I_8]^{2-}$ anion was observed only when the reaction involved preformed Fe^{II} or Co^{II} *bis*-complexes [223]; whereas, in the *in situ* formation of Co^{II} or Zn^{II}, *bis*-complexes afforded the highly symmetric

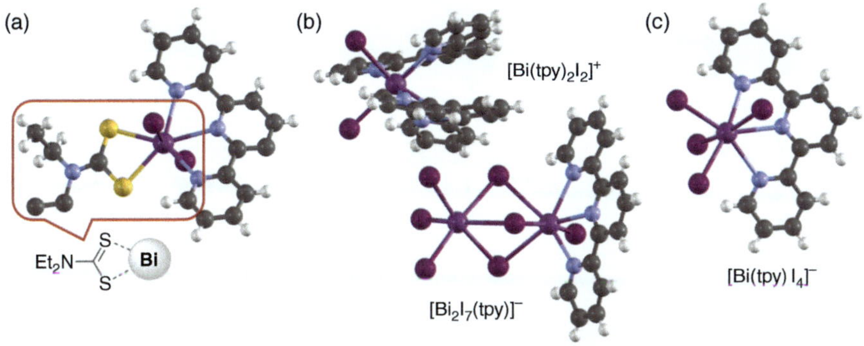

Figure 3.24 (a) Solid-state structure of [Bi(tpy)(S_2CNEt$_3$)I_2]. Adapted from [219]. (b) Solid-state structure of the [Bi(tpy)$_2$$I_2$] cation and the [$Bi_2I_7$(tpy)] anion. Adapted from [220]. (c) Solid-state structure of the [Bi(tpy)I_4]$^-$ anion. Adapted from [221]. In all figures, solvent molecules were omitted for clarity.

[Bi_2I_9]$^-$ anion [224]. In the latter case, the crystal lattice also contained isolated I^- ions, which ensured the balancing of charges. Similar to the parent BiI_3, complexes [M(tpy)$_2$][B_2I_9]I displayed thermochromic behavior. Thermochromism was due to changes of the interatomic distances in the anions and not the result of structural changes or even phase transitions upon heating/cooling. Finally, if the complexation was conducted in aqueous HI (but absence of transition-metal salts), a material was obtained, whose crystal lattice featured isolated [BiI_6]$^-$ octahedra, which were surrounded by highly regular stacks of [Htpy]$^+$ cations [225]. However, the charge-carrier mobility in this perovskite-type material was poor and, thus, excluded any further consideration in the context of photovoltaic applications (see Chapter 6).

Recently, Knope *et al.* disclosed that Bi^{III}–tpy complexes represent solid-state emitters, whose emission color could be adjusted over a broad range *via* the choice of the halide anion, the carboxylic-acid additive and/or the content of the Eu^{III} salt, as the dopant. For example, the dinuclear complex [Bi_2Cl_6(tpy)$_2$], whose solid-state structure is shown in Figure 3.25a, was obtained from the solvothermal reaction of $BiCl_3$ and tpy in the presence of dipicolinic acid, as the additive [226]. The origin of the material's long-lived yellow-green emission was assigned to ligand-to-ligand CT and ligand-to-metal CT processes (LLCT and LMCT, respectively). Furthermore, the solid-state structures of the complexes [Bi_2Cl_4(tpy)$_2$(κ^2-TC)$_2$] and [BiCl(tpy)(κ^2-TC)$_2$] were also reported (TC: thiophene-2-carboxylate; Figure 3.25b,c) [25]. A thorough discussion of this class of tpy-containing emitters is provided in Section 7.7.

3.7 The Far Right Site at a Glance: Combination of Tpy Ligands with Elements from Groups 16–18

Compared to most of the other elements in the PTE, the coordination chemistry of the chalcogens and halogens with polypyridyl ligands – specifically tpy – is rather

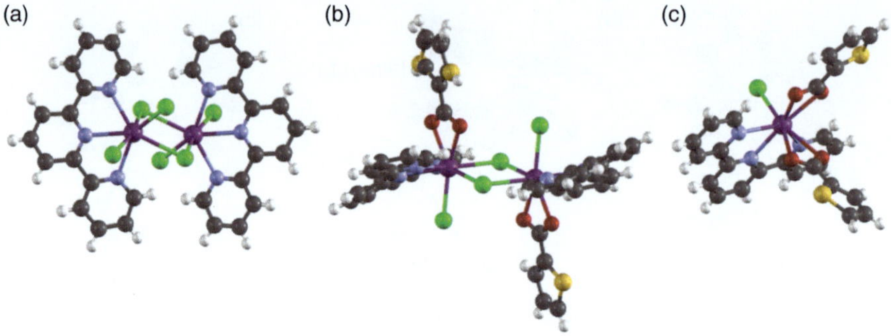

Figure 3.25 Solid-state structures of (a) $[Bi_2(tpy)_2Cl_6]$, (b) $[Bi_2Cl_4(tpy)_2(\kappa^2\text{-TC})_2]$, and (c) $[BiCl(tpy)(\kappa^2\text{-TC})_2]$ (solvent molecules were omitted for clarity). Adapted from [25, 226].

underdeveloped. This shortcoming can be rationalized when considering that such a coordination requires a drastic change of roles: the intrinsically Lewis-basic species need to act as electron-pair acceptors, *i.e.*, as Lewis acids [1]. The topic becomes even more exotic when one thinks of associates between (heavy) noble-gas ions and tpy ligands. Nonetheless, it was shown that such "coordination" is basically possible when using electron-poor $[RXe]^+$ cations (R = F or C_6F_5) [227, 228]. However, the complexes with py derivatives were only stable at low temperatures and decomposed upon melting. Stabilization of the complexes by using chelating bpy or tpy ligands was not yet attempted.

Unlike for tpy, bpy "complexes" with formal $X^{III/V}$ (X = Br or I) ions have been sparsely reported. For example, a 1:1 adduct between BrF_5 and bpy, which was stable up to room temperature, was obtained [229]. However, the actual structure of this compound could not be clarified by NMR spectroscopy. Besides this, the complex $[I(bpy)Cl_2][ICl_4]$, which contained a chelated $[ICl_2]^+$ ion, was reported by Wilson, Dutton *et al.* [230]. In solution, this complex represented a versatile reagent for the oxidative chlorination of organic and organometallic compounds. *N*-Halogenated tpy derivatives have only been reported for fluorine as the most electronegative element. In the only available study in this respect, Umemoto *et al.* prepared **27** as a bench-stable compound by reacting tpy with an F_2/N_2 mixture in the presence of SbF_5, as the scavenger for F^- ions (Figure 3.26a) [231]. Although not demonstrated with **27**, *N*-fluorinated polypyridines represented highly reactive reagents for the controlled electrophilic fluorination of CH-acidic substrates under mild conditions (*e.g.*, 1,3-dicarbonyl derivatives, electron-rich aromatics, *etc.*). Unlike most of the molecules presented in the previous sections, the *tris*-fluorinated tpy **27** represents an organic molecule with highly polar, yet covalent, N—F bonds.

The same holds true for the various *N*-oxides, which can be deduced from tpy (Figure 3.24b). Basically, pyridine *N*-oxides are prepared by the treatment of the respective pyridine with a peroxy acid. However, this method, which can be found in organic chemistry books, was not applicable for the synthesis of the N^2-oxide and the N^1,N^2-dioxide due to the absence of steric control [232, 233]. However, when opting for a step-wise protocol involving a sequence of *N*-oxidation,

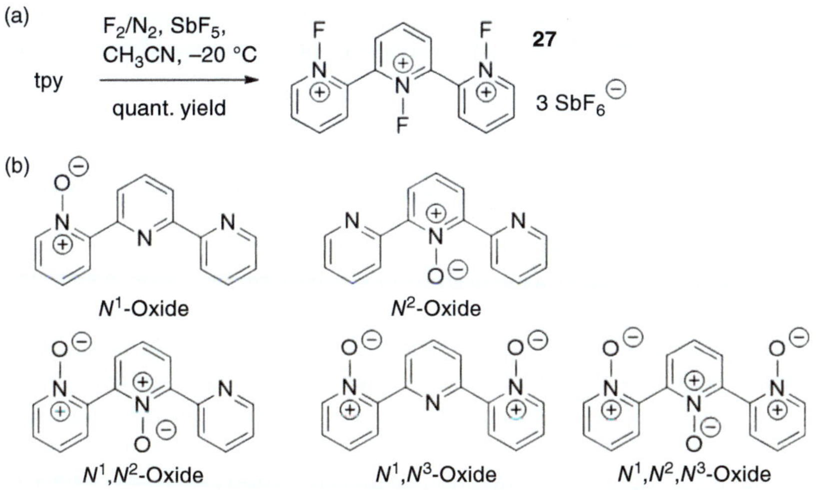

Figure 3.26 (a) Schematic representation of the synthesis of **27**. (b) Schematic representation of the various tpy-*N*-oxides.

cross-coupling and *N*-oxidation, this limitation was overcome [234]. Finally, the series was completed by Fallahpour and Neuburger, who conducted a stepwise synthesis of the two missing members [234]. Since then, a broad range of tpy ligands has been converted into *N*-oxides. The compounds have been utilized as organo-catalysts for various organic reactions (see Section 3.8). Moreover, complexes with transition-metal [235] or rare-earth-metal ions [236, 237] have been reported. The latter were characterized by UV/vis absorption and emission spectroscopy; in the case of Eu^{III}, the solid-state structures of the complexes with the N^2-oxide, the N^1,N^3-dioxide, and the N^1,N^2,N^3-trioxide were solved by single-crystal XRD analysis. Unlike the former two ligands, which gave homoleptic *tris*-complexes with nona-coordinate Eu^{III} centers, steric constraints prevent this coordination mode in the case of the trioxide ligand. Instead, a *tris*-complex with an octa-coordinate metal center was obtained (Figure 3.27).

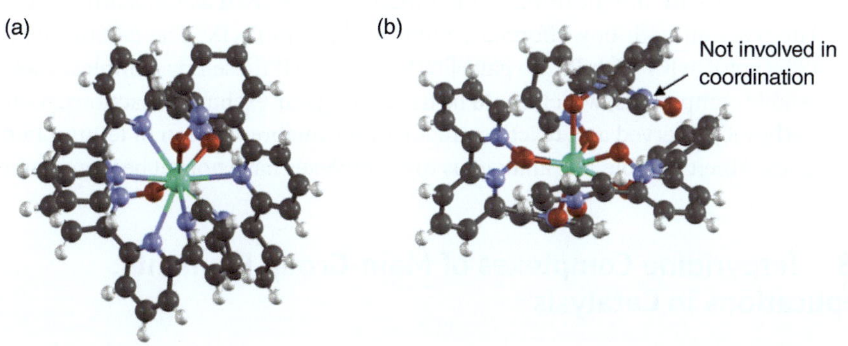

Figure 3.27 Solid-state structure of the homoleptic Eu^{III} complexes with (a) tpy-N^2-oxide and (b) tpy-N^1,N^2,N^3-trioxide (counterions and solvent molecules were omitted for clarity). Adapted from [237].

And what about the other chalcogenides? Owing to a mismatch in electronegativity, the analogous *N*-sulfides of tpy (or other polypyridines) are not accessible. Whereas the highly Lewis-acidic $[SF_3]^+$ cation was capable of coordinating py or phen to afford square-pyramidal complexes, whose S—N bonds possessed a substantial degree of covalency [238]. Flexible chelates, such as bpy or tpy were not considered in this study, but one might assume that tpy would only act as a bidentate κ^2N^1,N^2-type ligand in combination with $[SF_3]^+$.

Finally, attention is briefly turned to Se and Te. To the best of our knowledge, the only available study dealing with the reaction of EX_4 with tpy was performed in 1967 (E = Se or Te, X = Cl or Br) [239]. Adducts of 1:1 stoichiometry were obtained when reacting $SeCl_4$ or $TeBr_4$ with tpy. Owing to their poor solubility and the nonavailability of suitable single crystals, neither of these compounds was characterized further. However, in the case of $TeCl_4$, two different complexes were isolated, which both behaved as a 1:1 electrolyte in solution. Measurements of the molar conductivity corroborated that the composition of one was $[Te(tpy)Cl_3]Cl$ and $[Te(tpy)Cl_2][TeCl_6]$ of the other. The latter complex featured molecular conductivity of $122\,\Omega^{-1}\,cm^{-2}$ in DMF; IR spectroscopic analysis of a solid sample gave evidence for the proposed $[TeCl_6]^{2-}$ anion. Why was research in this field discontinued after a single study? Dutton and Ragogna analyzed the coordination chemistry of the Se and Te centers [240]. It was emphasized that the +IV species are highly susceptible to reduction to the +II state if a two-electron donor is present. In the case of the tetrahalides EX_4 (E = Se or Te; X = Cl, Br or I), reduction might already occur in the presence of neutral ligands (*e.g.*, phosphines, amines, *N*-heterocyclic carbenes, *etc.*) [241]. The respective dihalogen X_2 is released upon reduction; the X_2 might provoke side reactions and, thus, removal of side products and product isolation is hampered. When aiming for coordination compounds of the E^{II} species, the use of the respective preformed or even *in situ*-generated SeX_2 or TeX_2 is recommended (X = Cl or Br) [1]. However, these precursors are highly unstable, and attempts have been made to stabilize them or tame their reactivity. This can *inter alia* be achieved by the coordination of chelating Lewis bases, *e.g.*, polypyridines [242]. Ragogna *et al.* used freshly prepared SeX_2 to prepare the respective complexes $[Se(bpy)X_2]$ (X = Cl or Br). Whereas, in the case of tellurium, the tetrahalides were used as the starting materials. The reaction with bpy afforded complexes $[Te(bpy)X_3]X$. The careful reduction of $[Te(bpy)Cl_3]Cl$ with Ph_3Sb gave $[Te(bpy)Cl_2]$. All these bpy complexes were bench-stable compounds, which could be used for ligand-exchange reactions; moreover, $[Te(bpy)Cl_2]$ served as a reactive precursor for the preparation of tellurophene derivatives. Albeit feasible, the analogous tpy complexes have not yet been prepared.

3.8 Terpyridine Complexes of Main-Group Elements: Applications in Catalysis

Terpyridine ligands and their complexes with transition-metal ions are widely used as catalysts for numerous reactions. The wide range of catalytic applications not only includes organic chemistry, *i.e.*, functional-group transformations, but also

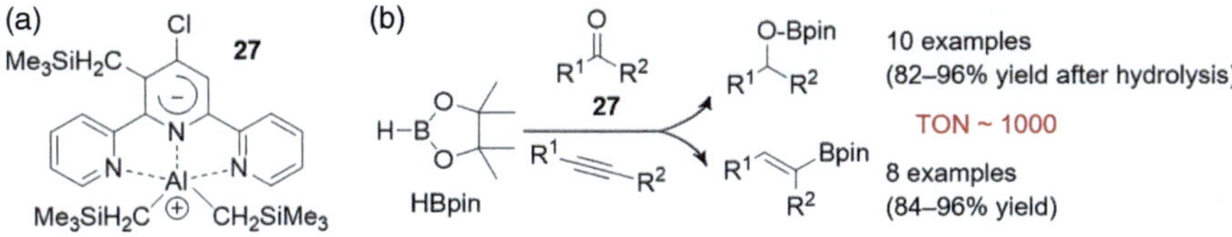

Figure 3.28 (a) Schematic representation of the betainic complex **27**. (c) Schematic representation of hydroboration reactions, which were catalyzed by **27**. Adapted from [113].

energy-related purposes: hydrogen-evolution reduction (HER), oxygen-evolution reaction (OER), CO_2 reduction, *etc.* (see Chapter 8) [20, 21, 243, 244]. Whereas the use of tpy complexes, which were derived from MGEs, represents a relatively young field of research, which was – with one exception – not even covered in the available reviews on "terpyridine catalysis."

This exception represents the presumably first example for stable tpy complexes of main-group metals, which was used, as the (pre)catalyst, in an organic reaction. As mentioned beforehand (Section 3.4; Figure 3.11), during their studies on the alkylative complexation of *tris*-alkylaluminium compounds, Zhang, Zheng, Dub *et al.*, by chance isolated the 3′-alkylated Al^{III} complex **27** (Figure 3.28a) [113]. Similar to the previously reported V^{III} complexes [245], this betainic Meisenheimer-type complex acted as a pre-catalyst in the hydroboration of C=O and C≡C bonds. The hydroboration of both types of substrates occurred in very high yields; in total, 10 carbonyl and 8 alkynyl derivatives were screened (Figure 3.28b) [113]. The turnover numbers (TONs) for these reactions were remarkably high (*i.e.*, up to ~1,000); thus, **27** outperformed all other Al^{III}-containing hydroboration (pre)catalysts reported by that time. Mechanistic studies suggested that Al^{III} hydride species were involved in the hydroboration; however, participation of the dearomatized "non-innocent" tpy ligand, as the hydride donor, could not fully be ruled out. Such an active role of "non-innocent" ligands in catalysis was first proposed by Chirik and Wieghardt [246]; today, this concept represents a versatile tool to control catalytic reactions beyond the stabilization of the metal centers in the various oxidation states, which occur in course of a catalytic cycle [114].

One further catalytic application is related to the unusual complexes **10–12**, in which an initially triel(III) center is involved in coordination (Figure 3.12). As elaborated by Nakajima, Takaya, Hashimoto *et al.*, complexes of this kind feature remarkable properties due to their unique *Z*-type metal-to-triel bonding (Figure 3.29a) [247]. Such complexes have *inter alia* been used for the activation of H—H, C—H, or C—F bonds or to catalyze certain reactions, *e.g.*, hydrosilylation of unreactive substrates. Firstly, complexes **10**, which contained a formal Pd^0 center due to the inversion of oxidation states, were applied for the synthesis of silyl formate from CO_2 under mild conditions (Figure 3.29b) [116]. Owing to the strong *trans*-effect of the electropositive Al-center, the Pd—Cl bond of **10a** was weakened and, thus, the catalytic activity was increased. Accordingly, the hydrosilylation of CO_2 occurred almost quantitatively and with a remarkably high turnover frequency (TOF) of 19,300 h^{-1}. Whereas **10b** and **10c** were by far inferior catalysts (*i.e.*, <10%

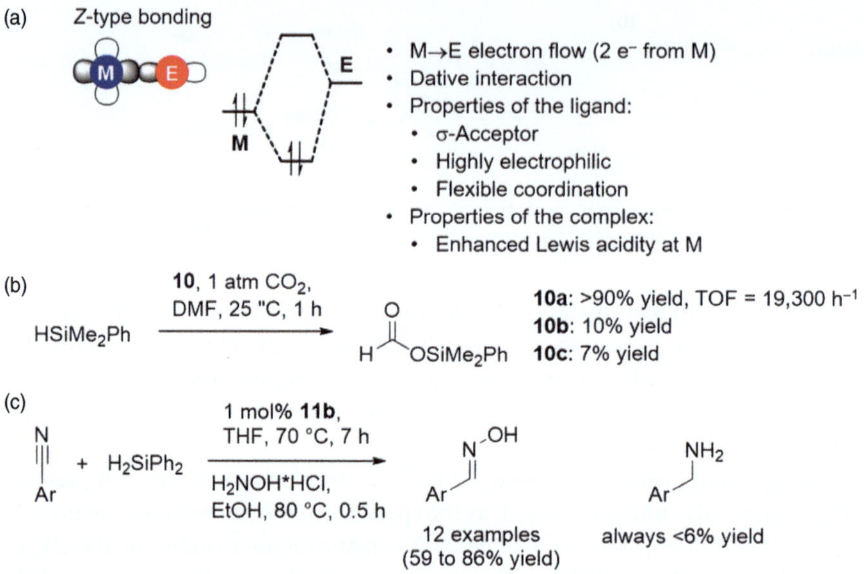

Figure 3.29 (a) Description of Z-type bonding in complexes **10–12**. Adapted from [247]. (b) Schematic representation of the hydrosilylation of CO_2. Adapted from [116]. (c) Schematic representation of the hydrosilylation of aromatic nitriles. Adapted from [117].

yield in the same test reaction). The diminished catalytic activity was attributed to the more covalent, charge-balanced bonding scenario in these two complexes.

Furthermore, these same authors also studied the reactivity of complexes **11** and **12** (Figure 3.12) [117, 118]. The latter, which possessed a formal Ga^{I}-to-Ir^{III} bond, featured a remarkable stability: Ligand-exchange reactions or the two-electron reduction by KC_8 were performed on the Ir^{III} center without affecting the coordinated metallo-ligand framework [118]. However, complex **12** was not tested regarding its catalytic activity. The related Rh^{III} complexes **11**, on the other hand, were used to catalyze the hydrosilylation of aromatic nitriles to afford the corresponding oximes (after work-up with $NH_2OH \times H_2O$; Figure 3.29c). In particular, complex **11b** was highly suited for this purpose since the reactions gave the desired products in good to high yields and with excellent chemoselectivity (*i.e.*, the corresponding primary amines were formed only in marginal amounts).

As mentioned in Section 3.5, complexes **14** and **15** (Figure 3.15) possessed high Lewis acidities, which were used for the abstraction of fluoride or hydride ions [131]. For example, this behavior enabled the abstraction of F^- from the SbF_6^- anion on the one hand and H^- from $HSiEt_3$ on the other. This reactivity was employed in the catalytic hydrodefluorination of *mono*-fluorinated (cyclo)alkanes. 1-Fluoroadamantane was selected as the test substrate. The best results in terms of substrate conversion and isolated product yield were obtained with catalysts **15a** and **15b** (in both cases >99%, Figure 3.15).

Even though the **24**$^{2+}$ cation possessed a hypervalent structure at the formal P^{III} ion (Figure 3.23), its electrophilicity was sufficiently high to also catalyze the

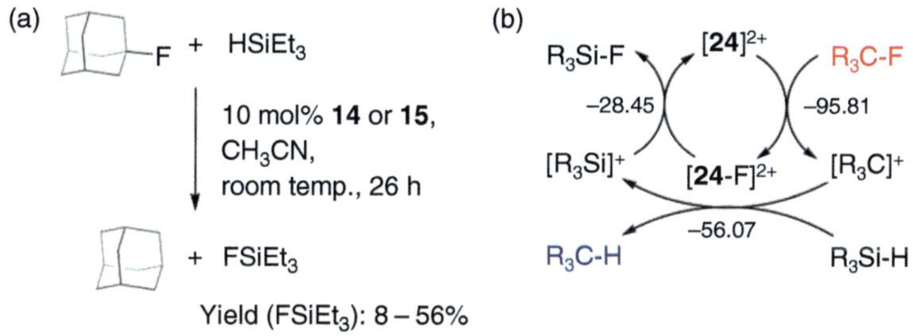

Figure 3.30 (a) Schematic representation of the hydrodefluorination of 1-fluoroadamantane, which was catalyzed by **14** or **15**. Adapted from [131]. (b) Schematic representation of the proposed mechanism of the hydrodefluorination of alkyl fluorides, catalyzed by $\mathbf{24}^{2+}$; for each step of the catalytic cycles, the calculated ΔG value (in kJ mol^{-1}), as an indicator for the thermodynamic driving force, is also given. Adapted from [23].

hydrodefluorination of alkyl fluorides [23]. Similar to the previous example, $HSiEt_3$ was used as the sacrificial hydride donor. At the low catalyst loading of 5–10 mol%, a wide range of primary, secondary, or tertiary alkyl substrates were converted into the corresponding alkanes. Notably, the CF_3 groups of various substrates were fully converted into CH_3 groups, even at room temperature. Although the catalyst could readily be prepared on the gram-scale, the convenient *in situ* generation of $\mathbf{24}^{2+}$ for catalytic applications was also demonstrated. Although the Lewis acidity of the P^V complexes **25** and **26** (Figure 3.23c) was of a different origin than that of **24**, *i.e.*, carbon-based *vs.* phosphorus-based, hydride abstraction from $HSiEt_3$ was possible [215]. Consequently, **25** and **26** also mediated the hydrodefluorination of alkyl fluorides, but presumably *via* a mechanistic pathway different from the one shown in Figure 3.30b.

In extension to the former application, the utilization of $\mathbf{24}^{2+}$, as a Lewis-acidic catalyst, for the allylation of C—F bonds was further shown [248]. The cross-coupling, in which two sp^3-hybridized centers were connected, features a good substrate scope: Sterically hindered as well as nonactivated substrates were converted at 35–100 °C under an atmosphere of air (*e.g.*, 1-fluoroadamantane and 1-fluoropentane, respectively; Figure 3.31a). Although being thermodynamically unfavored, Stephan *et al.* found evidence for a partial ligand dissociation under the applied reaction conditions (Figure 3.31b); it was proposed that this hemi-labile nature of the tpy ligand accounted for the observed catalytic activity. Detailed heteronuclear NMR experiments corroborated the suggested mechanistic pathway (Figure 3.31c). Moreover, Stephan *et al.* used (**24**)$[B(C_6F_5)_4]_2$ as the catalyst for the hydrosilylation of C=O and C=C double bonds (Figure 3.32) [249]. Ketones and aldehydes were reduced to the corresponding silyl ethers or hydrocarbons, depending on the applied substrate-to-reagent ratio. The hydrosilylation of olefins proceeded with the anticipated *anti*-Markovnikov regioselectivity. In both applications, the low catalyst loading of 5 mol% was used.

Within the context of this Section, the use of chiral tpy-*N*-trioxides also has to be mentioned. Naturally, the *N*-oxides represent organic derivatives of tpy and not their complexes, but for the sake of completeness, they are included in this Section. For

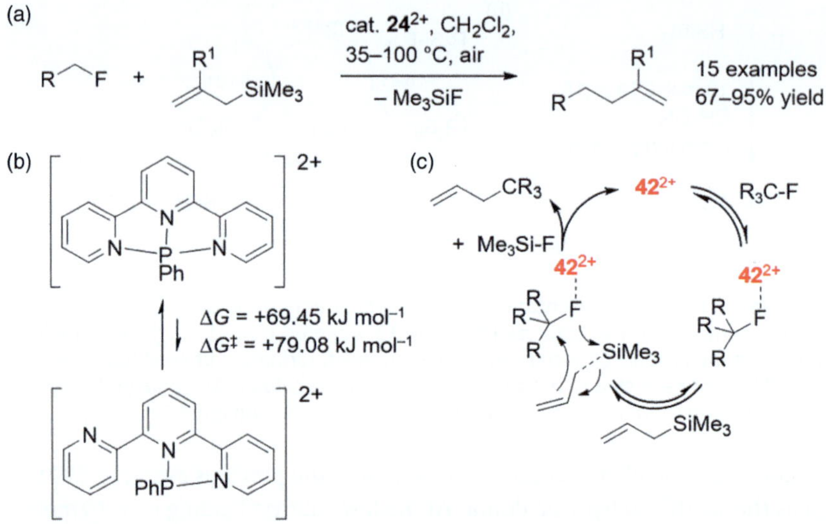

Figure 3.31 (a) Schematic representation of the catalytic allylation of C—F bonds. (b) Schematic representation of the reversible dissociation of one pyridine ring of **24**$^{2+}$. (c) Schematic representation of the proposed mechanism of the allylation reaction. Adapted from [248].

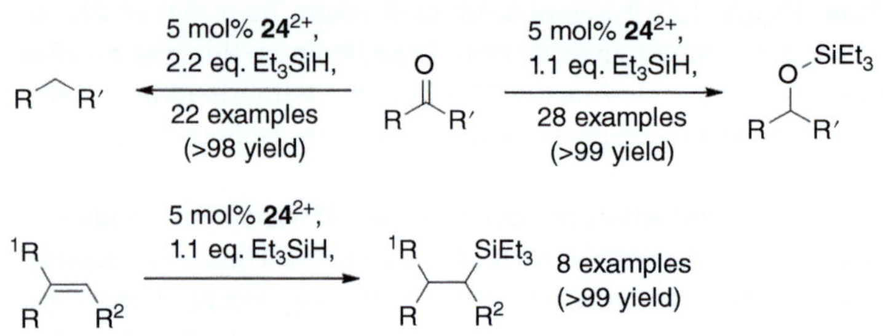

Figure 3.32 Schematic representation of the catalytic hydrosilylation of C=O and C=C bonds. Adapted from [249].

example, the *N*-trioxides **28** were prepared from the respective chiral tpy derivatives by oxidation with *m*-chloroperoxybenzoic acid (MCPBA; Figure 3.33a) [250]. The *N*-oxides were employed as Lewis-basic catalysts for the asymmetric allylation of aldehydes using allyltrichlorosilane (Figure 3.33b). The reaction afforded the homoallylic alcohols in high yields (up to 97%) and moderate enantioselectivity (up to 86%). The best results were obtained when electron-poor aldehydes were reacted at 0 °C. The use of chiral terpyridines as well as their *N*-oxides and transition-metal complexes in the context of asymmetric catalysis has been detailed in our previous textbook on tpy chemistry; we refer to this for further reading on this specific topic [190].

Aside from these catalytic applications of tpy complexes of MGEs, the stoichiometric use of such complexes in organic reactions has to be recapped very briefly. As

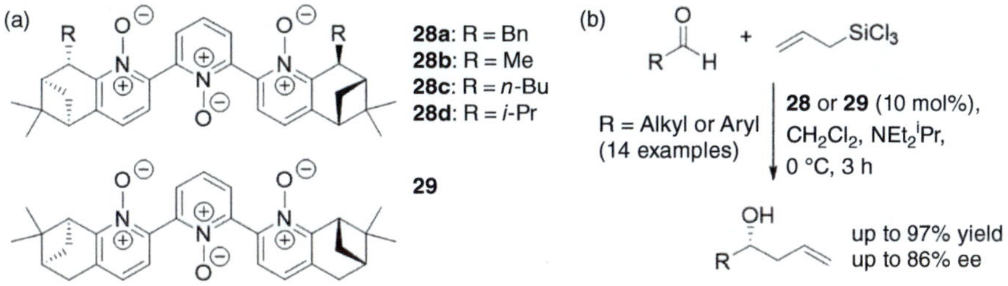

Figure 3.33 (a) Schematic representation of the chiral tpy-*N*-trioxides 28 and 29. (b) Schematic representation of the asymmetric allylation of aldehydes. Adapted from [250].

detailed in the previous sections 3.3 and 3.7, tpy ligands could be used to modulate the reactivity of Grignard reagents [66] and their *N*-fluorides might represent powerful reagents for the oxidative fluorination of CH-activated compounds [231].

3.9 Assembly of 1D, 2D, and 3D Coordination Polymers

In Chapter 2, we paid particular attention to the use of terpyridine ligands in the directed self-assembly of 2D/3D coordination compounds. Specifically, constitutional isomers of 2,2′:6′,2″-tpy were used in this context [251, 252]. Ligands containing the conventional 2,2′:6′,2″-motif have also been widely employed to prepare 1D coordination polymers [2], 2D CONASHs [194] or 3D networks/MOFs [253]. In most cases, these solid-state materials have been assembled from transition-metal ions, and the use of MGEs for this purpose has largely been overlooked. In this section, a short overview of this topic is provided.

For example, the combination of Pb^{II} ions with tpy was used to prepare 1D coordination polymers. Unlike its lighter carbon-group homologs, Pb^{II} possess a rich coordination chemistry, which is often similar to that of the divalent first-row transition-metal ions (see Section 3.5) [254]. As mentioned beforehand, the reaction of $Pb(SCN)_2$, $Pb(ClO_4)_2$, $Pb(NO_3)_2$, $Pb(NO_2)_2$, or $Pb(SiF_6)_2$ with tpy provided anion-bridged 1D coordination polymers [172–175]. In the case of $Pb(SCN)_2$, the material was composed of $[Pd(tpy)]^{2+}$ moieties, which were linked *via* the SCN^- anions; the average distance between the hepta-coordinate Pd^{II} centers was *ca.* 6.02 Å (Figure 3.34a) [174]. Whereas the polymer chain of $[Pb(tpy)(OH_2)](SiF_6)$ featured the seldom case, where the $[SiF_6]^-$ anion is coordinated to a Pb^{II} center; a substantial degree of packing *via* ππ-stacking interactions of the tpy moieties was observed (Figure 3.34b) [172].

Heterometallic coordination compounds containing Pb^{II} and Au^{I} centers in the main chain were reported by López-de-Luzuriaga, Olmos *et al.* [193]. Depending on the stoichiometry of the starting materials (*i.e.*, $[Au_2Ag_2(C_6F_5)_4(OEt_2)_2]_n$, $PbCl_2$, and tpy), either the metallopolymer **30** (1:1:1 ratio) or the binuclear complex **31** (0.5:1:1 ratio) was obtained (Figure 3.35a). In polymer **30**, the chain was formed by aurophilic interactions between the trinuclear Au^{I}–Pt^{II}–Au^{I} subunits. When **30** and **31** were combined in the (2/*n*)-to-2 ratio, the decanuclear complex **32** was

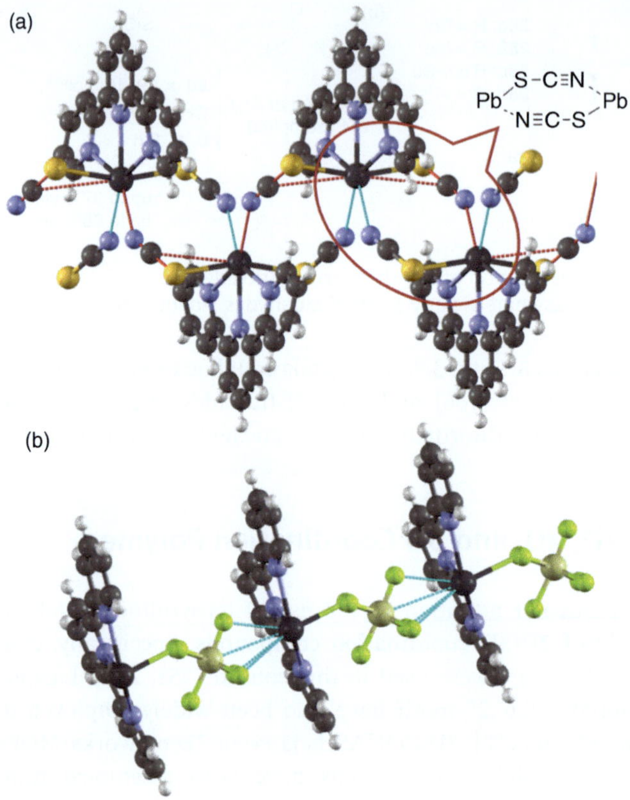

Figure 3.34 (a) An excerpt of the solid-state structure of the 1D coordination polymer $[Pb(tpy)(SCN)_2]_n$. Adapted from [174]. (b) Excerpt of the solid-state structure of $\{[Pb(tpy)(OH_2)](SiF_6)\}_n$. Adapted from[172]. In both figures, disordered atoms, non-coordinating counterions and solvent molecules were omitted for clarity.

formed. All three compounds represented solid-state emitters, whose emission resulted from metal–metal-to-ligand CT (MMLCT) transitions. With increasing strength of the metallophilic interactions, the emission maximum was red-shifted: the emission maxima of **30** and **32** were located at 670 nm and 650 nm, respectively (Figure 3.35b); whereas **31** was an orange emitter (λ_{em} = 595 nm).

Nanosheets are defined as 2D polymers with nanoscale thicknesses and graphene is presumably the most acknowledged representative [194]. Unlike covalent nanosheets, CONASHs are assembled from molecular components *via* metal-to-ligand complexation, *e.g.*, at the liquid-to-liquid or gas-to-liquid interface. Such materials are of interest regarding potential applications *inter alia* in electrochromic devices or for energy storage (see Chapter 5). In one study, Sakamoto, Schlüter *et al.* prepared such CONASHs by the air/water interfacial reaction of divalent metal ions (*i.e.*, Fe^{II}, Co^{II}, Ni^{II}, Zn^{II}, or Pb^{II}) and hexa *kis*- or *tris*-tpy ligands (**33** and **34**, Figure 3.36a,b) [255, 256]. The free-standing, μm^2-sized CONASHs were transferred onto TEM-grids and analyzed by microscopy imaging, specifically scanning-electron microscopy (SEM), transmission-electron microscopy (TEM),

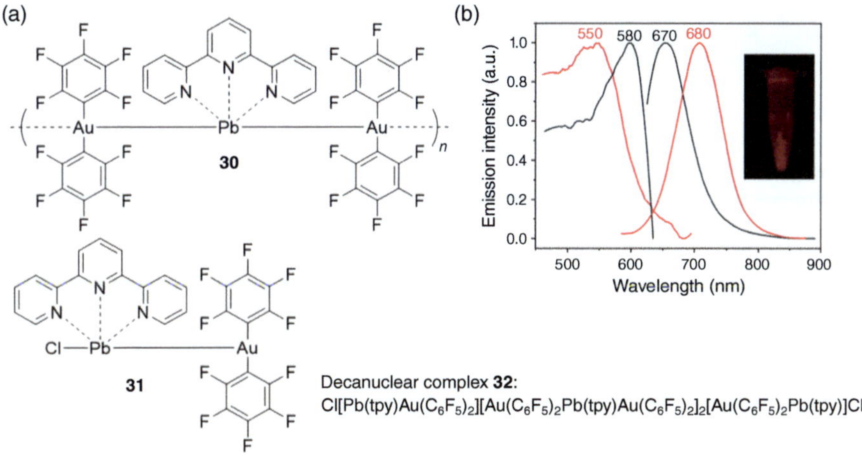

Figure 3.35 (a) Schematic representation of the metallopolymer **30** and the binuclear complex **31**; the condensed formula of the decanuclear complex **32** is also shown. (b) Solid-state excitation and emission spectra of **30** at room temperature (black curves) and 77 K (red curves) [193] / with permission of John Wiley & Sons.

and tapping-mode atomic-force microscopy (TM-AFM). The homogeneous sheets span the grids' holes with dimensions of $20 \times 20\,\mu m^2$ (Figure 3.36c). An average CONASH height of *ca.* 1.4 nm was measured; thus, the TM-AFM was slightly overestimating the calculated value of 0.8 nm (Figure 3.36c). Unlike for the analogous Fe^{II}-based systems, the lower kinetic stability and the absence of redox activity excluded any practical further use of the Pb^{II}-containing materials [194].

Anisotropic materials are of widespread importance due to their remarkable properties, which include piezo- or ferroelectricity, nonlinear optical behavior, and birefringence, just to name but a few. Anisotropy might arise from the molecular or crystallographic level. Optical birefringence, for example, will occur if the difference between the maximum refractive indices of a crystal in two perpendicular directions is unequal to zero [257]. Materials, which show such behavior might be used in polarizing prisms and retarder plates, liquid-crystal displays, or for medicinal diagnostics. Coordination polymers represent one prominent class of materials, which is discussed in this context [258]. The parent tpy ligand intrinsically combines a large polarizability (*i.e.*, 27.6 $Å^3$; see Section 3.2) with a large structural anisotropy [259]. However, in the solid state, anisotropy is partially cancelled since tpy crystallizes *inter alia* in an unfavorable herringbone-like motif (Figure 3.37a). Thus, these crystals are barely birefringent [261, 262]. Leznoff *et al.* combined tpy with divalent metal ions and the highly anisotropic anions $[Au(CN)_2]^-$, $[Ag(CN)_2]^-$ or $[Pt(CN)_4]^{2-}$ to obtain coordination polymers, which were both birefringent and emissive [195, 260]. Firstly, the divalent alkaline-earth-metal ions were used [260]. The solid-state structure of complex $[Sr(tpy)(H_2O)][Au(CN)_2]_2$ was comprised of infinite chains, in which the Sr^{II} centers were connected *via* $[Au(CN)_2]^-$ bridging units. The chains, in turn, were ordered into a 2D network due to aurophilic interchain interactions. These non-covalent interactions also influenced the material's

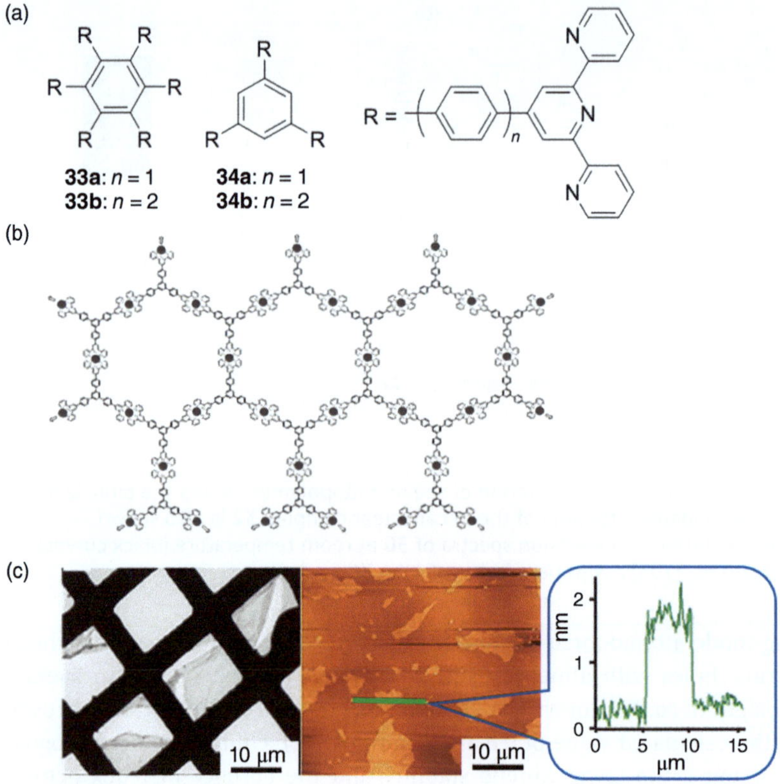

Figure 3.36 (a) Schematic representation of the ligands **33** and **34**. (b) Schematic representation of the molecular structure of the CONASH formed from **34a** and a divalent metal ion. (c) Representative SEM (left), TEM image (middle) of the CONASH **33a**/Pb^{II}, after its transfer onto a TEM grid; the TM-AFM height profile of this same CONASH is also shown [255] / with permission of John Wiley & Sons.

photophysical properties, which were distinctly different from those of the related compound [Sr(tpy)][$Pt(CN)_4$]. The occurrence of only a single emission maximum was attributed to the absence of metallophilic interactions in the solid-state structure of [Sr(tpy)][$Pt(CN)_4$]. Leznoff *et al.* prepared the isostructural series of complexes [M(tpy)][$Pt(CN)_4$] (M = Ca, Sr or Ba) *via* a solvothermal crystallization approach. In all cases, the hepta-coordinate metal centers were bridged by the square-planar $[Pt(CN)_4]^{2-}$ anions to afford a highly regular 3D coordination network. Owing to its mismatch in size and limitation in the CNs, the small Mg^{II} cation could not adopt such a coordination motif. Whereas the 1D chains of [Mg(tpy)(H_2O)][$Pt(CN)_4$] interacted *via* *H*-bonding to establish a 3D network. Independent of the particular 3D topology, all coordination polymers were highly birefringent, *i.e.*, their Δn values ranged from *ca.* 0.31 to *ca.* 0.46. These values become particularly meaningful when compared to calcite. All coordination compounds were significantly more birefringent than this common reference material ($\Delta n = 0.172$) [263]. The overall largest birefringence was observed for [Sr(tpy)][$Pt(CN)_4$] [260]. Single-crystal XRD

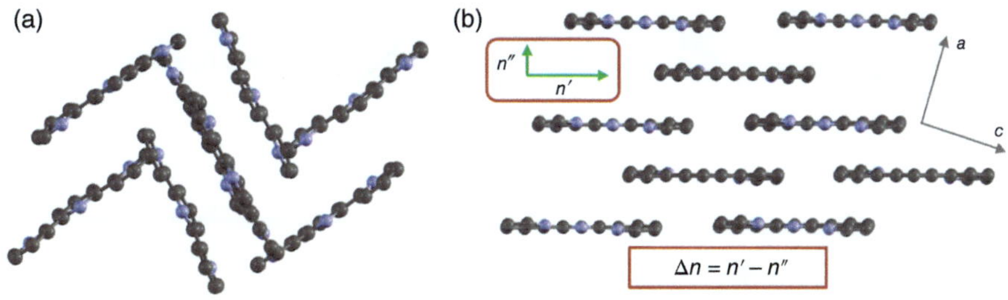

Figure 3.37 (a) Solid-state structure of the monoclinic polymorph showing the herringbone-type arrangement of tpy molecules. (b) Solid-state structure of $[Sr(tpy)][Pt(CN)_4]$ with the view normal to the primary crystallographic (010)-face (for clarity only the tpy moieties are shown); n' and n'' denote the refractive indices in the respective directions; the green arrows scale their magnitudes. in both figures, hydrogen atoms were omitted for clarity. Adapted from [260].

analysis revealed a strictly coplanar arrangement of the tpy ligands within the crystal lattice. This specific orientation was beneficial to obtain a very high polarizability anisotropy and refractive index in this direction; accordingly, the refractive index measured in the perpendicular direction was very low (the refractive indices are denoted as n' and n'' in Figure 3.37b).

Likewise, the coordination polymers $[M(tpy)[M'(CN)_2]_2$ were highly birefringent with Δn values more than twice as high as that of calcite (M = Mn or Pb; M′ = Au or Ag) [195]. Unlike the former examples, their solid-state structures were composed of $[M(tpy)]^{2+}$ cations, which were linked *via* the $[M'(CN)_2]^-$ anions. Thereby, the tpy moieties were ordered in a strictly face-to-face fashion, thus affording a highly regular network architecture (Figure 3.38a). The interplay of the anisotropic tpy ligands and the anions accounted for the observed high optical birefringence. On the other hand, the influence of the divalent metal (*i.e.*, its polarizability) on the optical properties was negligible: for $[Pb(tpy)][Au(CN)_2]$ and $[Mn(tpy)][Au(CN)_2]$, the Δn values differed only by 0.008. The coordination compounds, which contained the monovalent $[Ag[CN)_2]^-$ anion, were also highly birefringent, but were prone to degradation upon prolonged illumination. In subsequent work, Leznoff *et al.* elaborated the influence of the presence of polarizable substituents on the tpy ligand on the birefringence of the coordination polymers assembled thereof [265]. When using 4′-bromo-substituted tpy as the ligand, coordination polymers with increased birefringence were obtained (with both Pb^{II} and Mn^{II} ions). Kroeker *et al.* demonstrated that solid-state ^{207}Pb-NMR spectroscopy represented a versatile tool to analyze the relationship between a material's structural/electronic anisotropy and its birefringence (Figure 3.38b) [264]. The study revealed that the electron lone pair of the Pb^{II} centers became stereochemically more active with increasing ligand basicity. This, in turn, resulted in an alteration of the metal's coordination sphere, as expressed by a greater variation in the Pb—N bond lengths and broader ppm spans (Ω) in the ^{207}Pb-NMR spectra. The tpy ligands became less basic if substituted in 4′-position with a halide atom. As a result, the effect of the coordinated metal's lone pair of electrons was repelled and ^{207}Pb-NMR spectroscopy provided signals at lower frequencies and smaller Ω values.

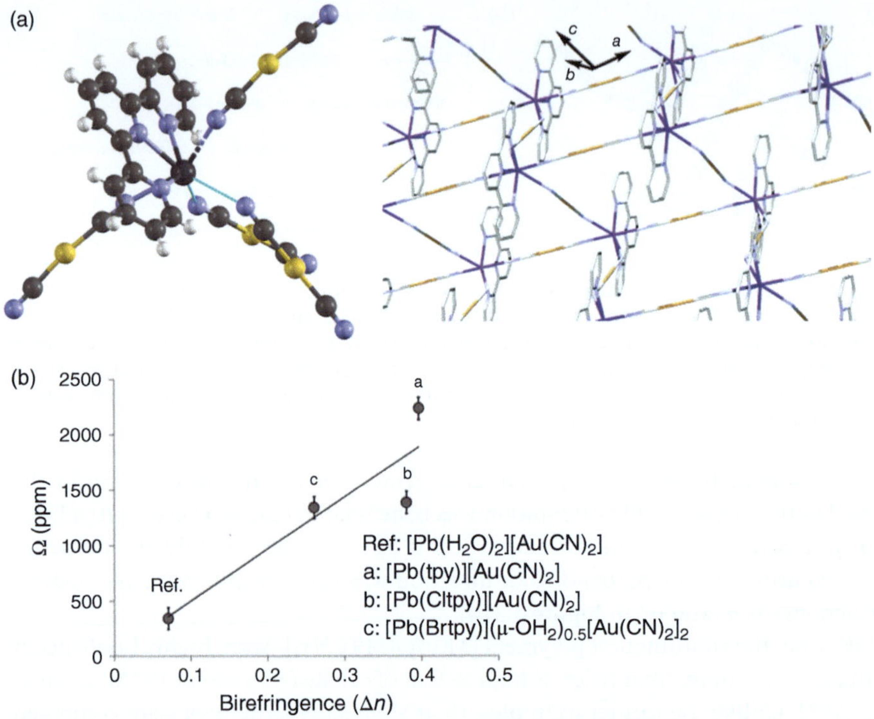

Figure 3.38 (a) Excerpts from the solid-state structure of coordination polymer $[Pb(tpy)][Au(CN)_2]$ (solvent molecules were omitted for clarity). Adapted from [195]. (b) Correlation between the structural anisotropy (*i.e.*, the measured birefringence) and the solid-state ^{207}Pb NMR ppm span [264] / with permission of John Wiley & Sons.

The coordination polymer $[Pb(tpy)Br_2[Hg(CN)_2]_{0.5}$ displayed both a high Δn value as well as ligand-based emission [266]. Finally, the square-planar anion $[Pt(SCN)_4]^{2-}$ was combined with various complex cations to afford coordination polymers, such as $[Pb(tpy)][Pt(SCN)_4]$ [267]. Unlike the former examples, birefringence was absent, but a pressure-dependency of the emission was observed. Specifically, the band of the metal-centered (d_{Pt}–d_{Pt}) luminescence ($\lambda_{em} \approx 690$ nm) was shifted to higher energy at high pressure (by $23\,cm^{-1}\,kbar^{-1}$). This behavior, which is typical for coordination compounds containing the $[Pt(SCN)_4]^{2-}$ anion, was due to the compression of the complex's metal-to-ligand bonds.

To the end of this section, the use of functionalized tpy ligands for the assembly of MOFs is presented. MOFs, representing crystalline micro-/mesoporous hybrid materials, have found numerous applications: gas storage or separation, catalysis, energy storage, sensing, water treatment, *etc.* [268]. In most MOFs, which are based on tpy ligands, the framework is formed with transition-metal ions [253]. However, few examples have been reported where main-group-metal ions have been used for this purpose.

In one study, the *bis*-sulfonated tpy ligand **35** (Figure 3.39a) was used for the MOF formation with Mg^{II} and Ca^{II} on the one hand and Pb^{II} on the other [269]. In

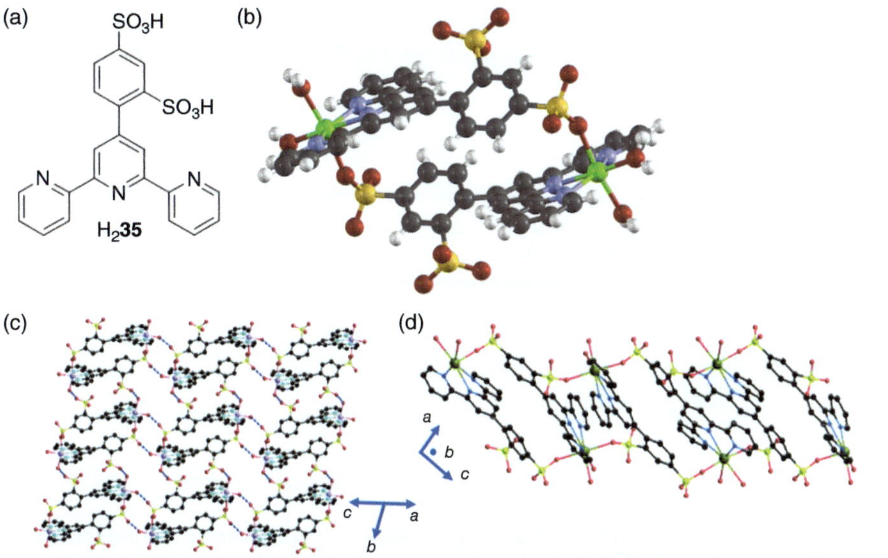

Figure 3.39 (a) Schematic representation of tpy ligand [H_2**35**]. (b) Solid-state structure of the binuclear [Mg(**35**)(H_2O)$_2$] subunit. Adapted from [269]. (c) Excerpt from the solid-state structure of [Mg(**35**)(H_2O)^2]·2H_2O showing the 2D network based on *H*-bonding. (d) Excerpt from the solid-state structure of {[Ca(**35**)(H_2O)$_2$]·1.5H_2O}$_n$. In panels (c) and (d), the hydrogen atoms were omitted for clarity [269] / with permission of Royal Society of Chemistry.

all cases, a binuclear [M_2(**35**)$_2$] coordination motif represented the central building block, which was assembled into different topologies *via* *H*-bonding between coordinated H_2O molecules and the ligands' sulfonate moieties (Figure 3.39b). The differences in the size of their ionic radii and number of coordinated water molecules accounted for the different structures observed for [Mg(**35**)(H_2O)$_2$]·2H_2O and {[Ca(**35**)(H_2O)$_2$]·1.5H_2O}$_n$. The hydrothermal reaction in the presence of Mg^{II} ions afforded a layered framework. The layers were comprised [Mg_2(**35**)$_2$] of subunits, interconnected *via* *H*-bonding (Figure 3.39c). Whereas the Ca^{II}-containing material possessed a highly regular 3D crystal structure, due to interchain *H*-bonding of the 1D coordination polymer [Ca_2(**25**)$_2$(H_2O)$_2$] (Figure 3.39d). Unlike for [Mg(**35**)(H_2O)$_2$]·2H_2O, which was barely luminescent, the ligand-based emission was mostly retained in {[Ca(**35**)(H_2O)$_2$]·1.5H_2O}$_n$. Whereas the two Pb^{II}-containing frameworks {[Pb(**35**)(H_2O)]·2H_2O}$_n$ and {[Pb(**35**)]·0.5H_2O} showed a fundamentally different emission behavior in that the albeit weak emission was attributed to metal-to-ligand CT (MLCT) processes within the solid-state materials. The topology of {[Pb(**35**)(H_2O)]·2H_2O}$_n$ resembled that of the Ca^{II}-containing MOF. The solid-state structure of {[Pb(**35**)]·0.5H_2O} was comprised of 2D coordination polymers with interlayer *H*-bonding.

In one further study, the carboxylated ligands **36a** and **36b** (Figure 3.40a) were combined with Pb^{II} ions to afford blue-emitting coordination polymers [196]. Owing to the ligands' different substitution patterns, different 3D topologies were obtained. Firstly, ligand **36a** formed a 2D layered structure, which featured square pores of

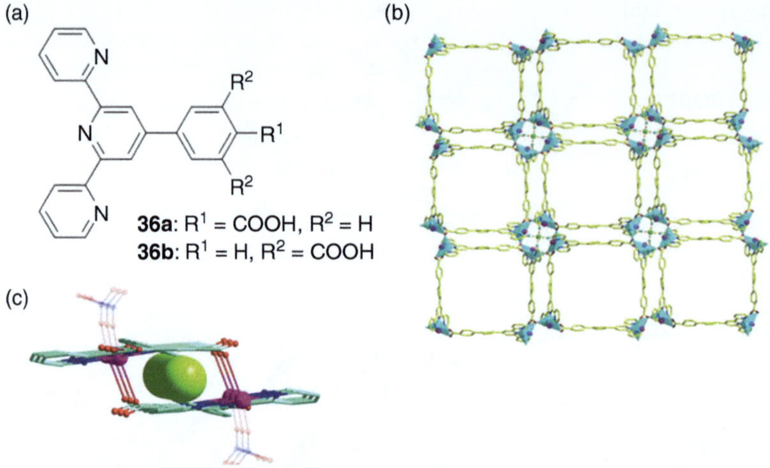

Figure 3.40 (a) Schematic representation of carboxylated ligands **36**. (b) Schematic representation of the 2D layer structure of $[Pb_4(\mathbf{36a})_4Cl]_n$. (c) Schematic representation of the open channels in $[Pb(\mathbf{36b})(NO_3)_2]_n$. (d) Comparison of the solid-state emission of ligands **36** and their respective assemblies with Pb^{II} ions [196] / with permission of Elsevier.

18.35 × 18.35 Å^2 size (Figure 3.40b); whereas a 1D coordination polymer with an open channel was obtained with ligand **36b** (Figure 3.40c). In both cases, the highly porous MOF resulted from interlayer ππ-stacking interactions between ligand moieties of adjacent layers/chains. The materials' intense solid-state emission was due to an interplay of intraligand CT and LMCT processes.

Recently, Cahill *et al.* reported the formation of an MOF comprised of Pb^{II} ions, uranyl cations, tpy, and an auxiliary ligand (*i.e.*, 1,2,3-benzenetricarboxylic acid or terephthalic acid) [197]. Owing to its hardness (according to Pearson's HSAB concept; see Section 1.3 [27]), the $[UO_2]^{2+}$ cation coordinated the carboxylate groups of the auxiliary ligands, whereas the Pb^{II} ions were chelated by the tpy ligands. The 3D structure was formed by Pb^{II}–carboxylate as well as Pb^{II}–oxo interactions. In particular, the latter had a profound impact on the MOFs' photophysical properties: the stronger these interactions were, the more expressed was the luminescence quenching of the uranyl-based MOF. Lustig and Li pointed out that actinide-containing MOFs have attracted interest due to their remarkable photophysical properties [270]. Unlike their lanthanide-containing counterparts, which require sensitization to activate f–f transitions, actinide-based MOFs are emissive simply due to the available HOMO-to-LUMO transitions. In the case of $[UO_2]^{2+}$ nodes within MOF architectures, the emission is sensitive toward other species (*i.e.*, transition-metal ions or organic molecules) within the framework or as guests [271].

3.10 Terpyridine Ligands and (Metallo)carboranes

Finally, we will allude to a topic, which is a bit exotic, but deserves coverage within the scope of this chapter: the combination of terpyridines with (metallo)carboranes.

In this respect, one has to distinguish between carboranes, which have been equipped with tpy ligands on the one hand and metallocarboranes to which tpy ligands have been coordinated.

Polyhedral carboranes of the general formula $C_xB_yH_z$ were first reported in 1950 [272], but have gained attention due to their remarkable electronic and photophysical properties more recently. Carboranes are classified as 3D σ-electron delocalized, carbon–boron clusters [273] and one has to distinguish between *closo*-, *nido*-, *arachno*-, and *hypho*-type structures. The derivatives, whose deltahedra are complete, are referred to as *closo*-carboranes; whereas, in the other types, at least one vertex is missing. Dicarba-*closo*-dodecaborane is certainly the most prominent carborane and is known as the "*closo*-carborane" or, simply, "carborane." This structure is comprised of two carbon and ten boron centers, which are located at the vertices of an icosahedron. Thereby, three constitutional isomers can be written, in which the carbon atoms occupy the 1,2-, 1,7- or 1,12-positions of the framework (in analogy to disubstituted benzenes, these isomers are also named the *ortho*-, *meta*-, or *para*-form; Figure 3.41a). The electron-withdrawing character of *closo*-1,2-carborane is due to the highly polarizable σ-aromatic character, *i.e.*, delocalization of skeletal electrons over the 3-center-2-electron bonds. For this reason, this molecule is widely considered an "inorganic benzene." When an electron-rich chromophore is attached to *ortho*-carborane *via* its adjacent carbon centers, donor–acceptor (DA) arrays are obtained. In these DA systems, intramolecular CT (ICT) from the π-conjugated chromophore to the carborane's antibonding orbital of the C—C bond results in emissive excited states. In order to capitalize these characteristics, *ortho*-carborane has been incorporated into a wide range of functional materials (*e.g.*, polymers) [274].

Kokado and Chujo prepared the tpy-equipped carborane derivative **37** (Figure 3.41a) and assembled it into an organo-soluble coordination polymer by complexation with Zn^{II} ions [275]. The intense blue emission of the polymer

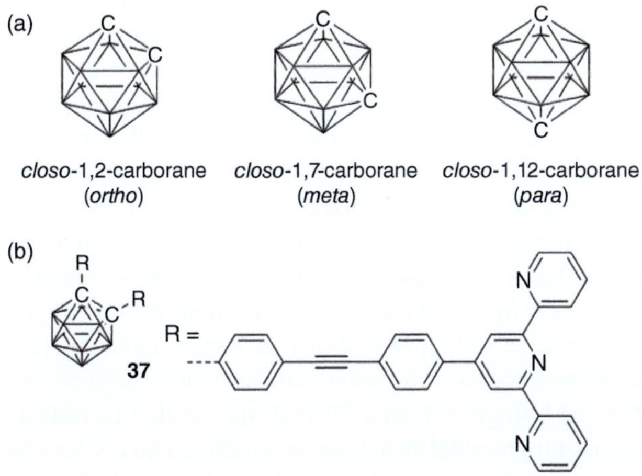

Figure 3.41 (a) Schematic representation of the three constitutional isomers of dicarba-*closo*-dodecaborane. (b) Schematic representation of the tpy-equipped *ortho*-carborane derivative **37**.

was due to an intraligand CT, a typical behavior in metallopolymers, based on $[Zn(tpy)_2]^{2+}$ complexes. A wide range of such materials has been prepared and studied regarding their photophysical properties and their use, *e.g.*, in polymer light-emitting diodes, has been discussed (see Chapter 7) [276]. Moreover, Wesdemiotis, Newkome *et al.* analyzed the metallo-macrocyclization of **37** in the presence of Fe^{II} or Zn^{II} ions [277]. In the former case, a triangular 3:3 assembly, as characteristic of 60°-based *bis*-tpy ligands, was obtained. Whereas the more labile Zn^{II}–tpy complexation resulted in an equilibrium between the enthalpically favored triangle and a cyclic dimer, which represented the entropically favored species. This study was performed in the context of using multitopic tpy ligands for the formation of advanced metallo-supramolecular constructs (see Chapter 10) [9].

Research on metallacarboranes emerged with the discovery of the 11-vertex *nido*-dicarbollide dianion, $[C_2B_9H_{11}]^{2-}$, which is able to bind metal centers to its open face in a η^5-type fashion *via* its delocalized electrons [278]. In this respect, the dianion resembles the coordination behavior of cyclopentadienyl anions; specifically, the electron-donor abilities and steric requirements of $[C_2B_9H_{11}]^{2-}$ and the pentamethylcyclopentadienyl anion are almost identical [279]. Likewise, two different binding modes can be distinguished, of which the first one is reminiscent of the organometallic half-sandwich complexes [280, 281]. In this case, the binding of a metal center to a *nido*-carborane affords the corresponding *closo*-metallacarborane, in which the metal occupies one of the vertices. Notably, such *closo*-metallacarboranes can be formed with the metal ions from the s-, p-, d-, or f-block. Secondly, the counterpart to a full-sandwich complex (*e.g.*, ferrocene) is obtained if the metal centers occupy a common vertex of two carborane moieties. Such compounds are referred to as *commo*-metallacarboranes. Such coordination chemistry is not limited to the larger carborane dianions, such as $[C_2B_9H_{11}]^{2-}$, but also small-sized cluster dianions show this behavior. For example, Hosmane *et al.* focused on metallocarboranes derived from the four-vertex dianion $[R_2C_2B_4H_4]^{2-}$ (R = H or alkyl/silyl substituent) [280]. These authors paid particular attention to the Lewis acidity of the grafted metal center, which was used for the coordination of *inter alia* polypyridyl ligands (*e.g.*, bpy, tpy, *etc.*) [280]. Within the scope of this section, only the coordination of tpy to metallacarboranes derived from MGEs will be presented. However, in the case of the *closo*-gallacarborane and *closo*-germacarborane, the structures of the corresponding complexes could not be verified by XRD analysis [282, 283]. Single-crystal XRD could only be performed for the *closo*-metallacarboranes and the adducts with bpy. Thus, characterization of the materials relied only on IR and NMR spectroscopy as well as elemental analysis. However, it was assumed that the tpy ligands were bound in a similar fashion as bpy (exemplarily, the proposed structure of adduct **38** is shown in Figure 3.42a [283]).

In the case of the *closo*-stannacarboranes, the solid-state structures of their tpy complexes were reported [284–286]. Single-crystal XRD analysis revealed a considerable distortion of the pentagonal-bipyramidal structure, in which the Sn center was displaced from the centroid position above the carborane cluster toward the "boron side" (Figure 3.42b). Semi-empirical quantum calculations, *i.e.*, the modified neglect of diatomic overlap (MNDO) and Fenske–Hall methods suggested that this slip was due to orbital overlap as well as steric constraints.

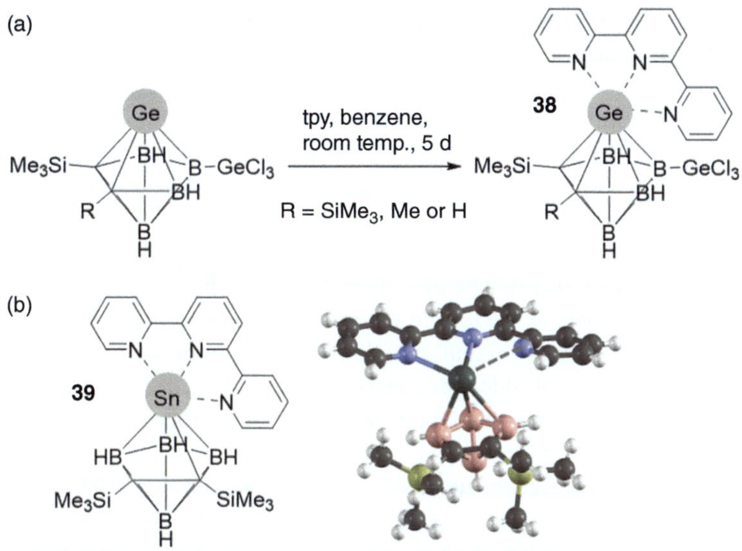

Figure 3.42 (a) Schematic representation of the synthesis of **38** from substituted *closo*-germacarboranes and tpy. Adapted from [283]. (b) Schematic representation of the tpy-complex of closo-stannacarborane **39**, the solid-state structure of this compound is also shown. Adapted from [284].

3.11 Concluding Remarks

When thinking of the chemistry of the MGEs with tpy, one might first think that this topic is hardly extensive. However, a closer inspection reveals that a manifold of structures has already been reported and, in some cases, notable applications for various fields of research can be deduced. Recently, a comprehensive review on this specific topic was provided by us [1]. Before this, Constable and Housecroft already paid attention to the "unexpected structural diversity" of complexes, which can be assembled from MGE ions and tpy ligands (as well as other polypyridyl ligands) [287]. This range of reported solid-state structures includes mono- and dinuclear complexes as well as 1D coordination polymers, 2D CONASHs and 3D MOFs. Thereby, the topology of the crystal structures is often influenced by non-covalent interactions between metal ions, ligand molecules, and counterions. Taking this into account, cautious selection of the single components [*i.e.*, the MGE ions, tpy ligands, auxiliary ligands (*e.g.*, carboxylates), and charge-compensating ions] enables one to engineer crystals with tailor-made properties. For example, highly birefringent coordination polymers, which even out-competed calcite, as the benchmark crystal, were obtained from carefully selected metal ions, tpy, and cyanometallate anions (see Section 3.9) [195].

Further notable applications of tpy complexes with MGE ions can be found in areas of interest, which have not been covered in this chapter. Owing to their importance, these topics will be discussed in Chapter 4 (Biomedical Applications), Chapter 6 (Photovoltaic Applications), and Chapter 7 (Optoelectronic Applications). Representatively, the modification of perovskite-type materials for solar-cell applications [26, 95] or the invention of Bi^{III}-based solid-state emitters [25] is highlighted at this stage.

The complexation of Mg^{II} ions is worth mentioning for two reasons: Firstly, the $[Mg(tpy)_2]^{2+}$ complex represents a versatile gas-phase reagent, which has been utilized to improve the MS analysis of certain biomolecules (*e.g.*, fatty acids) [75]. The Mg^{II}–tpy interaction could also be employed to modulate the reactivity of Grignard reagents for organic synthesis [66]. The use of tpy complexes, as catalysts, in organic synthesis represents a highly active field of research that has been extensively reviewed earlier [20, 21]. The already rich toolbox has been expanded significantly by the tpy complex of the formal P^{III} ions. This complex was reported to be a useful catalyst for various organic reactions (*e.g.*, hydrodefluorination of alkyl fluorides, hydrosilylation of aldehydes/ketones, olefins, *etc.*) [23, 249]. The same holds true for the highly Lewis-acidic tpy-masked R_2Si^{2+} cations, which also catalyzed the hydrodefluorination reaction [131].

To sum up this chapter, research in the field of MGE–tpy complexes has been continuously ongoing for more than 50 years. However, research in this field has been eclipsed by the more prominent work on tpy complexes with transition-metal ions. This is justifiable since remarkable applications have been derived from these studies (*e.g.*, Ru^{II}–tpy complexes for photovoltaic applications; see Chapter 6). This is also wrongful by the same time since notable progress in various fields of research has recently been made (*vide supra*). Could MGE–tpy complexes replace their analogs based on transition-metal ions in certain applications? This is one central question in today's applied coordination chemistry, in particular when considering the use of complexes based on heavy transition-metal ions, whose natural abundance is limited, and use is related to high (environmental) costs. In particular, the first-row transition metal ions are considered for replacing, *e.g.*, Ru^{II} or Ir^{III} centers in sensitizers or emitters [288]. Given that the performance in the field of interest is accordingly improved (*e.g.*, organic synthesis, biomedicine, photochemistry/photophysics, optics, photovoltaics, *etc.*), MGE–tpy systems can certainly represent alternatives in future applied science. For these reasons, the authors encourage intensified work on MGE–tpy systems.

References

1 A. Winter, G. R. Newkome, U. S. Schubert, *Inorg. Chem. Front.* 2024, 11, 342–399.

2 A. Winter, U. S. Schubert, *Chem. Soc. Rev.* 2016, 45, 5311–5357.

3 C.-A. Fustin, P. Guillet, U. S. Schubert, J.-F. Gohy, *Adv. Mater.* 2007, 19, 1665–1673.

4 A. Winter, U. S. Schubert, in *New Polymeric Materials Based on Element-Blocks*, (Ed.: Y. Chujo), Springer, Singapore, 2019, pp. 307–346.

5 J. Zhou, G. R. Whittell, I. Manners, *Macromolecules* 2014, 47, 3529–3543.

6 S.-H. Hwang, C. D. Shreiner, C. N. Moorefield, G. R. Newkome, *New J. Chem.* 2007, 31, 1192–1217.

7 U. S. Schubert, A. Winter, G. R. Newkome, in *Ruthenium-Containing Polymers*, (Eds.: U. S. Schubert, A. Winter, G. R. Newkome), Springer International Publishing, Cham, 2021, pp. 275–336.

8 C. N. Moorefield, A. Schultz, G. R. Newkome, *Braz. J. Pharm. Sci.* 2013, 49, 67–84.

9 S. Chakraborty, G. R. Newkome, *Chem. Soc. Rev.* 2018, 47, 3991–4016.

10 X. Yu, C. Guo, S. Lu, Z. Chen, H. Wang, X. Li, *Macromol. Rapid Commun.* 2022, 43, 2200004.

11 H. Wang, C. Guo, X. Li, *CCS Chem.* 2022, 4, 785–808.

12 M. Higuchi, *Polym. J.* 2009, 41, 511–520.

13 M. Higuchi, *Bull. Jpn. Soc. Coord. Chem.* 2022, 79, 68–77.

14 M. Higuchi, *J. Syn. Org. Chem. Jpn.* 2023, 81, 1089–1095.

15 G. I. Dzhardimalieva, B. C. Yadav, S. Singh, I. E. Uflyand, *Dalton Trans.* 2020, 49, 3042–3087.

16 C.-H. Li, J.-L. Zuo, *Adv. Mater.* 2020, 32, 1903762.

17 R. Abhijnakrishna, K. Magesh, A. Ayushi, S. Velmathi, *Coord. Chem. Rev.* 2023, 496, 215380.

18 I. Eryazici, C. N. Moorefield, G. R. Newkome, *Chem. Rev.* 2008, 108, 1834–1895.

19 A. Winter, M. Gottschaldt, G. R. Newkome, U. S. Schubert, *Curr. Top. Med. Chem.* 2012, 12, 158–175.

20 A. Winter, G. R. Newkome, U. S. Schubert, *ChemCatChem* 2011, 3, 1384–1406.

21 A. Winter, U. S. Schubert, *ChemCatChem* 2020, 12, 2890–2941.

22 N. Tomar, A. Agrawal, V. S. Dhaka, P. K. Surolia, *Sol. Energy* 2020, 207, 59–76.

23 S. S. Chitnis, F. Krischer, D. W. Stephan, *Chem. Eur. J.* 2018, 24, 6543–6546.

24 P.-F. Shi, Q. Jiang, H.-C. Duan, D.-Q. Wang, *Chin. Chem. Lett.* 2014, 25, 586–588.

25 R. L. Ayscue III, V. Vallet, J. A. Bertke, F. Réal, K. E. Knope, *Inorg. Chem.* 2021, 60, 9727–9744.

26 N. T. P. Hartono, M. Ani Najeeb, Z. Li, P. W. Nega, C. A. Fleming, X. Sun, E. M. Chan, A. Abate, A. J. Norquist, J. Schrier, T. Buonassisi, *Cryst. Growth Des.* 2022, 22, 5424–5431.

27 R. G. Pearson, *J. Am. Chem. Soc.* 1963, 85, 3533–3539.

28 E. C. Constable, in *Advances in Inorganic Chemistry*, Vol. 30, (Ed.: H. J. Emeléus), Academic Press, London, 1986, pp. 69–121.

29 E. C. Constable, in *Advances in Inorganic Chemistry*, Vol. 79, (Eds.: R. van Eldik, C. D. Hubbard), Academic Press, London, 2022, pp. 109–149.

30 S. K. Sarad, V. W. Bhagwat, N. S. Poonia, *Res. J. Sci. (Univ. Indore)* 1977, 4, 27–30.

31 N. S. Poonia, A. V. Bajaj, *Chem. Rev.* 1979, 79, 389–445.

32 K. Nakamura, *Bull. Chem. Soc. Jpn.* 1972, 45, 1943–1943.

33 W. Kaim, *Chem. Ber.* 1981, 114, 3789–3800.

34 M. Wang, J. England, T. Weyhermüller, K. Wieghardt, *Eur. J. Inorg. Chem.* 2015, 2015, 1511–1523.

35 J. England, C. C. Scarborough, T. Weyhermüller, S. Sproules, K. Wieghardt, *Eur. J. Inorg. Chem.* 2012, 2012, 4605–4621.
36 J. dos Santos Veiga, W. L. Reynolds, J. R. Bolton, *J. Chem. Phys.* 1966, 44, 2214–2216.
37 M. Satterfield, J. S. Brodbelt, *Inorg. Chem.* 2001, 40, 5393–5400.
38 K. J. Miller, *J. Am. Chem. Soc.* 1990, 112, 8533–8542.
39 F. Vögtle, W. M. Müller, W. Raßhofer, *Isr. J. Chem.* 1979, 18, 246–248.
40 H. G. Grimm, *Z. Elektrochem. Angew. Phys. Chem.* 1925, 31, 474–480.
41 K. Pokorny, M. Schmeisser, F. Hampel, A. Zahl, R. Puchta, R. van Eldik, *Inorg. Chem.* 2013, 52, 13167–13178.
42 V. Gutmann, *The Donor-Acceptor Approach to Molecular Interactions*, Springer, New York, 1978.
43 P. Erdmann, L. Greb, *Angew. Chem. Int. Ed.* 2022, 61, e202114550.
44 T. Madrakian, A. Afkhami, J. Ghasemi, M. Shamsipur, *Polyhedron* 1996, 15, 3647–3652.
45 M. Schmeisser, P. Illner, R. Puchta, A. Zahl, R. van Eldik, *Chem. Eur. J.* 2012, 18, 10969–10982.
46 *Gutmann Acceptor and Donor Numbers*, http://www.stenutz.eu/chem/solv21.php?sort=5 (last accessed: 2025-04-01).
47 *Lithium-7 NMR*, https://chem.ch.huji.ac.il/nmr/techniques/1d/row2/li.html (last accessed: 2025-01-15).
48 G. Zhang, H. Zeng, S. Zheng, M. C. Neary, P. A. Dub, *iScience* 2022, 25, 105119.
49 T. W. Bell, P. J. Cragg, A. Firestone, A. D. I. Kwok, J. Liu, R. Ludwig, A. Sodoma, *J. Org. Chem.* 1998, 63, 2232–2243.
50 E. C. Constable, M. J. Doyle, J. Healy, P. R. Raithby, *J. Chem. Soc. Chem. Commun.* 1988, 1988, 1262–1264.
51 A. Haque, R. Ilmi, I. J. Al-Busaidi, M. S. Khan, *Coord. Chem. Rev.* 2017, 350, 320–339.
52 S. Götz, S. Zechel, M. D. Hager, G. R. Newkome, U. S. Schubert, *Prog. Polym. Sci.* 2021, 119, 101428.
53 G. E. Coates, S. I. E. Green, *J. Chem. Soc. Res.* 1962, 1962, 3340–3348.
54 G. E. Coates, *Proc. Chem. Soc.* 1961, 1961, 376–376.
55 I. M. Brown, S. I. Weissman, L. C. Snyder, *J. Chem. Phys.* 1965, 42, 1105–1111.
56 T. M. McCleskey, V. Buchner, R. W. Field, B. L. Scott, *Rev. Environ. Health* 2009, 24, 75–116.
57 A. N. Carolan, A. E. Mroz, M. El Ojaimi, D. G. VanDerveer, R. P. Thummel, R. D. Hancock, *Inorg. Chem.* 2012, 51, 3007–3015.
58 J. M. Hamilton, M. J. Anhorn, K. A. Oscarson, J. H. Reibenspies, R. D. Hancock, *Inorg. Chem.* 2011, 50, 2764–2770.
59 J. M. Hamilton, J. R. Whitehead, N. J. Williams, M. El Ojaimi, R. P. Thummel, R. D. Hancock, *Inorg. Chem.* 2011, 50, 3785–3790.
60 E. C. Constable, J. Healy, M. G. B. Drew, *Polyhedron* 1991, 10, 1883–1887.
61 M. M. Emara, C.-T. Lin, *J. Indian Chem. Soc.* 1980, 57, 876–883.
62 M. M. Emara, C.-T. Lin, G. Atkinson, *Bull. Soc. Chim. Fr.* 1980, 1980, 173–180.

63 A. F. Waters, A. H. White, *Aust. J. Chem.* 1996, 49, 147–154.

64 J. C. Deaton, F. N. Castellano, in *Iridium(III) in Optoelectronic and Photonics Applications*, (Ed.: E. Zysman-Coleman), John Wiley & Sons Ltd., Chichester, 2017, pp. 1–69.

65 B. W. Skelton, A. F. Waters, A. H. White, *Aust. J. Chem.* 1996, 49, 137–146.

66 L. M. Guard, J. L. Palma, W. P. Stratton, L. J. Allen, G. W. Brudvig, R. H. Crabtree, V. S. Batista, N. Hazari, *Dalton Trans.* 2012, 41, 8098–8110.

67 W. Schlenk, W. Schlenk Jr., *Ber. Dtsch. Chem. Ges.* 1929, 62, 920–924.

68 M. Westerhausen, *Coord. Chem. Rev.* 2008, 252, 1516–1531.

69 H.-C. Chao, S. A. McLuckey, *Trends Anal. Chem.* 2023, 158, 116852.

70 H.-C. Chao, S. A. McLuckey, *Anal. Chem.* 2020, 92, 13387–13395.

71 X. Han, R. W. Gross, *Mass Spectrom. Rev.* 2005, 24, 367–412.

72 K. C. Fabijanczuk, H.-C. Chao, J. L. Fischer, S. A. McLuckey, *Analyst* 2022, 147, 5000–5010.

73 H.-C. Chao, S. A. McLuckey, *Anal. Chem.* 2021, 93, 7332–7340.

74 C. E. Randolph, C. H. Beveridge, S. Iyer, S. J. Blanksby, S. A. McLuckey, G. Chopra, *J. Am. Soc. Mass Spectrom.* 2022, 33, 2156–2164.

75 S. T. Nsiah, K. C. Fabijanczuk, S. A. McLuckey, *Rapid Commun. Mass Spectrom.* 2024, 38, e9741.

76 J. A. W. Shoemaker, J. S. Hartman, *Can. J. Chem.* 1999, 77, 1856–1868.

77 J. S. Hartman, J. A. W. Shoemaker, *Polyhedron* 2000, 19, 165–176.

78 J. S. Hartman, J. A. W. Shoemaker, A. F. Janzen, P. J. Ragogna, W. R. Szerminski, *J. Fluor. Chem.* 2003, 119, 125–139.

79 A. J. Carty, D. G. Tuck, *J. Chem. Soc. Res.* 1964, 1964, 6012–6017.

80 P. Pyykkö, *Annu. Rev. Phys. Chem.* 2012, 63, 45–64.

81 J. S. Thayer, *J. Chem. Educ.* 2005, 82, 1721.

82 A. Askarinejad, A. A. Torabi, A. Morsali, *Z. Naturforsch. B* 2006, 61, 565–569.

83 K. Akhbari, A. Morsali, *Coord. Chem. Rev.* 2010, 254, 1977–2006.

84 R. A. Walton, *Inorg. Chem.* 1968, 7, 640–648.

85 R. A. Walton, *J. Inorg. Nucl. Chem.* 1970, 32, 2875–2884.

86 R. A. Walton, *J. Inorg. Nucl. Chem.* 1977, 39, 549–550.

87 G. Beran, K. Dymock, H. A. Patel, A. J. Carty, P. M. Boorman, *Inorg. Chem.* 1972, 11, 896–898.

88 G. Beran, A. J. Carty, H. A. Patel, G. J. Palenik, *J. Chem. Soc. D Chem. Commun.* 1970, 1970, 222–223.

89 R. A. Walton, *Inorg. Nucl. Chem. Lett.* 1976, 12, 767–771.

90 I. V. Kazakov, M. Bodensteiner, A. Y. Timoshkin, *Acta Crystallogr. C* 2014, 70, 312–314.

91 R. Bhalla, W. Levason, S. K. Luthra, G. McRobbie, F. M. Monzittu, J. Palmer, G. Reid, G. Sanderson, W. Zhang, *Dalton Trans.* 2015, 44, 9569–9580.

92 G. S. Papaefstathiou, A. Sofetis, C. P. Raptopoulou, A. Terzis, G. A. Spyroulias, T. F. Zafiropoulos, *J. Mol. Struct.* 2007, 837, 5–14.

93 A. Sofetis, G. S. Papaefstathiou, A. Terzis, C. P. Raptopoulou, T. F. Zafiropoulos, *Z. Naturforsch. B* 2004, 59, 291–297.

94 D. Zare, B. Doistau, H. Nozary, C. Besnard, L. Guénée, Y. Suffren, A.-L. Pelé, A. Hauser, C. Piguet, *Dalton Trans.* 2017, 46, 8992–9009.
95 Y. Zhang, S. Yuan, Y. Yuan, Y. Bao, Q. Ran, E. Liu, J. Fan, W. Li, *Adv. Opt. Mater.* 2022, 10, 2102041.
96 A. Y. Timoshkin, M. Bodensteiner, T. N. Sevastianova, A. S. Lisovenko, E. I. Davydova, M. Scheer, C. Graßl, A. V. Butlak, *Inorg. Chem.* 2012, 51, 11602–11611.
97 K. R. Cairns, W. Levason, G. Reid, W. Zhang, *Polyhedron* 2021, 207, 115367.
98 P. A. Scattergood, in *Organometallic Chemistry*, Vol. 43, (Eds.: N. J. Patmore, P. I. P. Elliott), The Royal Society of Chemistry, Croydon, 2020.
99 P. L. Bellavance, E. R. Corey, J. Y. Corey, G. W. Hey, *Inorg. Chem.* 1977, 16, 462–467.
100 R. A. Fischer, H. Sussek, H. Parala, H. Pritzkow, *J. Organomet. Chem.* 1999, 592, 205–211.
101 S. V. Ivanov, T. V. Shubina, T. A. Komissarova, V. N. Jmerik, *J. Cryst. Growth* 2014, 403, 83–89.
102 S. Liu, V. P. Biju, Y. Qi, W. Chen, Z. Liu, *NPG Asia Mater.* 2023, 15, 27.
103 M. I. Saidaminov, J. Almutlaq, S. Sarmah, I. Dursun, A. A. Zhumekenov, R. Begum, J. Pan, N. Cho, O. F. Mohammed, O. M. Bakr, *ACS Energy Lett.* 2016, 1, 840–845.
104 R. Wang, J. Wang, S. Tan, Y. Duan, Z.-K. Wang, Y. Yang, *Trends Chem.* 2019, 1, 368–379.
105 Y. Xian, S. Yuan, Y. Zhang, Y. Yuan, Y. Zhang, J. Fan, W. Li, *J. Phys. Chem. Lett.* 2021, 12, 5571–5579.
106 H. C. Clark, A. L. Pickard, *J. Organomet. Chem.* 1968, 13, 61–71.
107 G. B. Deacon, J. C. Parrott, *Aust. J. Chem.* 1974, 27, 2547–2555.
108 Z.-L. Shen, S.-Y. Wang, Y.-K. Chok, Y.-H. Xu, T.-P. Loh, *Chem. Rev.* 2013, 113, 271–401.
109 A. J. Canty, L. A. Titcombe, B. W. Skelton, A. H. White, *J. Chem. Soc. Dalton Trans.* 1988, 1988, 35–45.
110 A. J. Canty, R. Colton, I. M. Thomas, *J. Organomet. Chem.* 1993, 455, 283–289.
111 A. J. Canty, K. Mills, B. W. Skelton, A. H. White, *J. Chem. Soc. Dalton Trans.* 1986, 1986, 939–945.
112 Y. Hasegawa, T. Nakagawa, T. Kawai, *Coord. Chem. Rev.* 2010, 254, 2643–2651.
113 G. Zhang, J. Wu, H. Zeng, M. C. Neary, M. Devany, S. Zheng, P. A. Dub, *ACS Catal.* 2019, 9, 874–884.
114 V. Lyaskovskyy, B. de Bruin, *ACS Catal.* 2012, 2, 270–279.
115 C. K. Jørgensen, *Coord. Chem. Rev.* 1966, 1, 164–178.
116 J. Takaya, N. Iwasawa, *J. Am. Chem. Soc.* 2017, 139, 6074–6077.
117 J. Takaya, K. Ogawa, R. Nakaya, N. Iwasawa, *ACS Catal.* 2020, 10, 12223–12228.
118 N. Saito, J. Takaya, N. Iwasawa, *Angew. Chem. Int. Ed.* 2019, 58, 9998–10002.
119 A. Amgoune, D. Bourissou, *Chem. Commun.* 2011, 47, 859–871.
120 F. Cheng, M. F. Davis, A. L. Hector, W. Levason, G. Reid, M. Webster, W. Zhang, *Eur. J. Inorg. Chem.* 2007, 2007, 4897–4905.

121 J. Ding, C. Zheng, L. Wang, C. Lu, B. Zhang, Y. Chen, M. Li, G. Zhai, X. Zhuang, *J. Mater. Chem. A* 2019, 7, 23337–23360.
122 J. E. Dickeson, L. A. Summers, *Experientia* 1969, 25, 1247–1248.
123 S. Kumaran, E. Oh, S. Han, H.-J. Choi, *Nano Lett.* 2021, 21, 5422–5429.
124 X. Yang, L. Li, Patent CN-102024569, 2011.
125 B. Suthar, A. Aldongarov, I. S. Irgibaeva, M. Moazzen, B. T. Donovan-Merkert, J. W. Merkert, T. A. Schmedake, *Polyhedron* 2012, 31, 754–758.
126 S. Herzog, F. Krebs, *Naturwissenschaften* 1963, 50, 330–331.
127 D. Kummer, K. E. Gaisser, J. Seifert, R. Wagner, *Z. Anorg. Allg. Chem.* 1979, 459, 145–156.
128 P. M. Broudy, A. D. Berry, B. B. Wayland, A. G. MacDiarmid, *J. Am. Chem. Soc.* 1972, 94, 7577–7579.
129 E. Wulf, S. Herzog, *Z. Anorg. Allg. Chem.* 1972, 387, 81–90.
130 D. M. Peloquin, A. N. Bimukhanov, A. A. Aldongarov, J. W. Merkert, B. T. Donovan-Merkert, T. A. Schmedake, *Molecules* 2022, 27, 8521.
131 A. Hermannsdorfer, M. Driess, *Angew. Chem. Int. Ed.* 2020, 59, 23132–23136.
132 L. Greb, *Chem. Eur. J.* 2018, 24, 17881–17896.
133 N. F. Hall, J. B. Conant, *J. Am. Chem. Soc.* 1927, 49, 3047–3061.
134 D. Himmel, S. K. Goll, I. Leito, I. Krossing, *Angew. Chem. Int. Ed.* 2010, 49, 6885–6888.
135 L. O. Müller, D. Himmel, J. Stauffer, G. Steinfeld, J. Slattery, G. Santiso-Quiñones, V. Brecht, I. Krossing, *Angew. Chem. Int. Ed.* 2008, 47, 7659–7663.
136 J. E. Fergusson, W. R. Roper, C. J. Wilkins, *J. Chem. Soc. Res.* 1965, 1965, 3716–3720.
137 M. S. Woodward, R. P. King, R. D. Bannister, J. Grigg, G. McRobbie, W. Levason, G. Reid, *Inorganics* 2022, 10, 107.
138 N. S. Hosmane, J. A. Maguire, *Comment. Inorg. Chem.* 2005, 26, 183–215.
139 F. W. B. Einstein, B. R. Penfold, *J. Chem. Soc. Part A* 1968, 1968, 3019–3024.
140 N. W. G. Debye, E. Rosenberg, J. J. Zuckerman, *J. Am. Chem. Soc.* 1968, 90, 3234–3236.
141 F. W. B. Einstein, B. R. Penfold, *Chem. Commun.* 1966, 1966, 780–781.
142 F. W. B. Einstein, B. R. Penfold, *Acta Crystallogr.* 1966, 20, 924–926.
143 G. Barone, A. Silvestri, G. Ruisi, G. La Manna, *Chem. Eur. J.* 2005, 11, 6185–6191.
144 D. L. Tomaja, J. J. Zuckerman, *Synth. React. Inorg. Met. Org. Chem.* 1976, 6, 323–335.
145 W. A. Gustavson, L. M. Principe, W. Z. M. Rhee, J. J. Zuckerman, *Inorg. Chem.* 1981, 20, 3460–3468.
146 A. B. Cornwell, P. G. Harrison, J. A. Richards, *J. Organomet. Chem.* 1977, 140, 273–287.
147 M. A. Wassef, S. Hessin, *Commun. Fac. Sci. Univ. Ankara Ser. B Chim.* 1981, 27, 161–169.
148 S. Uritis, R. P. Thummel, H.-S. Lee, R. D. Hancock, *Inorg. Chim. Acta* 2022, 530, 120670.

149 D. V. Naik, W. R. Scheidt, *Inorg. Chem.* 1973, 12, 272–276.

150 B. Z. Momeni, S. F. Hosseini, J. Janczak, *J. Mol. Struct.* 2023, 1279, 135038.

151 S. J. Archer, K. R. Koch, L. R. Nassimbeni, *J. Crystallogr. Spectrosc. Res.* 1986, 16, 449–458.

152 J. C. May, C. Curran, *J. Organomet. Chem.* 1972, 39, 289–294.

153 J. Otera, T. Hinoishi, R. Okawara, *J. Organomet. Chem.* 1980, 202, C93–C94.

154 D. V. Naik, C. Curran, *J. Organomet. Chem.* 1974, 81, 177–185.

155 M. A. Doskey, C. Curran, *Inorg. Chim. Acta* 1969, 3, 169–173.

156 A. P. Singh, H. W. Roesky, E. Carl, D. Stalke, J.-P. Demers, A. Lange, *J. Am. Chem. Soc.* 2012, 134, 4998–5003.

157 Y. Nakayama, Y. Baba, H. Yasuda, K. Kawakita, N. Ueyama, *Macromolecules* 2003, 36, 7953–7958.

158 F. Nicoli, M. Baroncini, S. Silvi, J. Groppi, A. Credi, *Org. Chem. Front.* 2021, 8, 5531–5549.

159 E. C. Constable, J. Lewis, M. C. Liptrot, P. R. Raithby, *J. Chem. Soc. Dalton Trans.* 1984, 1984, 2177–2185.

160 E. C. Constable, J. M. Holmes, *Polyhedron* 1988, 7, 2531–2536.

161 E. C. Constable, F. K. Khan, J. Lewis, M. C. Liptrot, P. R. Raithby, *J. Chem. Soc. Dalton Trans.* 1985, 333–335.

162 E. C. Constable, J. M. Holmes, R. C. S. McQueen, *J. Chem. Soc. Dalton Trans.* 1987, 1987, 5–8.

163 B. Z. Momeni, V. Noroozi, *Monatsh. Chem.* 2017, 148, 893–900.

164 R. Musiol, P. Malecki, M. Pacholczyk, J. Mularski, *Expert Opin. Drug Discov.* 2022, 17, 259–271.

165 S. D. Cummings, *Coord. Chem. Rev.* 2009, 253, 1495–1516.

166 K. W. Jennette, S. J. Lippard, G. A. Vassiliades, W. R. Bauer, *Proc. Natl. Acad. Sci. USA* 1974, 71, 3839–3843.

167 G. M. Cockrell, G. Zhang, D. G. VanDerveer, R. P. Thummel, R. D. Hancock, *J. Am. Chem. Soc.* 2008, 130, 1420–1430.

168 H. A. Saadeh, E. A. A. Shairah, N. Charef, M. S. Mubarak, *J. Appl. Polym. Sci.* 2012, 124, 2717–2724.

169 E. S. Claudio, H. A. Godwin, J. S. Magyar, in *Progress in Inorganic Chemistry*, Vol. 31, (Ed.: K. D. Karlin), John Wiley & Sons Inc., Hoboken, 2002, pp. 1–144.

170 D. L. Kepert, J. M. Patrick, B. W. Skelton, A. H. White, *Aust. J. Chem.* 1988, 41, 157–158.

171 A. Morsali, *Z. Naturforsch. B* 2004, 59, 1039–1044.

172 J. Burt, W. Grantham, W. Levason, M. E. Light, G. Reid, *Polyhedron* 2015, 85, 530–536.

173 L. M. Engelhardt, J. M. Harrowfield, H. Miyamae, J. M. Patrick, B. W. Skelton, A. A. Soudi, A. H. White, *Aust. J. Chem.* 1996, 49, 1135–1146.

174 L. M. Engelhardt, B. M. Furphy, J. M. Harrowfield, J. M. Patrick, A. H. White, *Inorg. Chem.* 1989, 28, 1410–1413.

175 A. J. Blake, C. Caltagirone, V. Lippolis, M. Shamsipur, *Acta Crystallogr. C* 2009, 65, m33–m36.

176 E. S. Claudio, M. A. ter Horst, C. E. Forde, C. L. Stern, M. K. Zart, H. A. Godwin, *Inorg. Chem.* 2000, 39, 1391–1397.
177 N. Noshiranzadeh, A. Ramazani, A. Morsali, A. D. Hunter, M. Zeller, *Inorg. Chem. Commun.* 2007, 10, 738–742.
178 N. Noshiranzadeh, A. Ramazani, A. Morsali, A. D. Hunter, M. Zeller, *Inorg. Chim. Acta* 2007, 360, 3603–3609.
179 T.-T. Li, Y. Zheng, Q.-Y. Chen, S.-R. Zheng, *J. Coord. Chem.* 2016, 69, 966–975.
180 L. Saghatforoush, F. Marandi, I. Pantenburg, G. Meyer, *Z. Anorg. Allg. Chem.* 2009, 635, 1523–1526.
181 F. Marandi, L. Saghatforoush, I. Pantenburg, G. Meyer, *J. Mol. Struct.* 2009, 938, 277–282.
182 L. Saghatforoush, K. Adil, E. Şahin, S. Babaei, S. J. Musevi, *J. Coord. Chem.* 2011, 64, 4421–4433.
183 L. Saghatforoush, E. Sahin, S. Babaei, A. Bakhtiari, A. Nasimian, Ö. Çelik, Z. Zabihollahi, *J. Coord. Chem.* 2014, 67, 1463–1477.
184 C. Bazzicalupi, A. Bencini, A. Bianchi, L. Borsari, A. Danesi, C. Giorgi, P. Mariani, F. Pina, S. Santarelli, B. Valtancoli, *Dalton Trans.* 2006, 2006, 5743–5752.
185 J.-M. Lehn, *Angew. Chem. Int. Ed.* 1990, 29, 1304–1319.
186 M. Ruben, J. Rojo, F. J. Romero-Salguero, L. H. Uppadine, J.-M. Lehn, *Angew. Chem. Int. Ed.* 2004, 43, 3644–3662.
187 M. Barboiu, E. Petit, G. Vaughan, *Chem. Eur. J.* 2004, 10, 2263–2270.
188 B. Doistau, C. Rossi-Gendron, A. Tron, N. D. McClenaghan, L.-M. Chamoreau, B. Hasenknopf, G. Vives, *Dalton Trans.* 2015, 44, 8543–8551.
189 U. S. Schubert, H. Hofmeier, G. R. Newkome, *Modern Terpyridine Chemistry*, Wiley-VCH, Weinheim, 2006.
190 U. S. Schubert, A. Winter, G. R. Newkome, *Terpyridine-Based Materials*, Wiley-VCH, Weinheim, 2011.
191 A.-M. Stadler, N. Kyritsakas, J.-M. Lehn, *Chem. Commun.* 2004, 2004, 2024–2025.
192 M. J. Lobo, S. C. Moratti, L. R. Hanton, *Chem. Asian J.* 2019, 14, 1184–1193.
193 J. M. López-de-Luzuriaga, M. Monge, S. Moreno, M. E. Olmos, M. Rodríguez-Castillo, *Angew. Chem. Int. Ed.* 2021, 60, 640–644.
194 R. Sakamoto, K. Takada, X. Sun, T. Pal, T. Tsukamoto, E. J. H. Phua, A. Rapakousiou, K. Hoshiko, H. Nishihara, *Coord. Chem. Rev.* 2016, 320–321, 118–128.
195 M. J. Katz, H. Kaluarachchi, R. J. Batchelor, A. A. Bokov, Z.-G. Ye, D. B. Leznoff, *Angew. Chem. Int. Ed.* 2007, 46, 8804–8807.
196 E.-J. Gao, J. Chen, Y. Hui, S. Wu, T. Zhang, D. Song, D. Liu, M.-C. Zhu, *Inorg. Chem. Commun.* 2019, 107, 107488.
197 D. M. Brager, A. D. Nicholas, M. H. Schofield, C. L. Cahill, *Inorg. Chem.* 2021, 60, 17186–17200.
198 J. Y. Kim, J.-W. Lee, H. S. Jung, H. Shin, N.-G. Park, *Chem. Rev.* 2020, 120, 7867–7918.
199 N.-G. Park, *Mater. Today* 2015, 18, 65–72.

200 Y. Yuan, Y. Xian, Y. Long, Y. Zhang, N. U. Rahman, Y. Zhang, J. Fan, W. Li, *RSC Adv.* 2021, 11, 24816–24821.

201 J. Chen, S.-G. Kim, X. Ren, H. S. Jung, N.-G. Park, *J. Mater. Chem. A* 2019, 7, 4977–4987.

202 A. M. Brodie, C. J. Wilkins, *Inorg. Chim. Acta* 1974, 8, 13–16.

203 J. Kaub, W. S. Sheldrick, *Z. Naturforsch. B* 1984, 39, 1257–1261.

204 B. D. Ellis, C. L. B. Macdonald, *Inorg. Chim. Acta* 2007, 360, 329–344.

205 G. Reeske, A. H. Cowley, *Chem. Commun.* 2006, 16, 1784–1786.

206 J. G. Stevens, in *Chemical Mössbauer Spectroscopy*, (Ed.: R. H. Herber), Springer U.S, Boston, 1984, pp. 319–342.

207 G. Alonzo, M. Consiglio, F. Maggio, N. Bertazzi, *Inorg. Chim. Acta* 1988, 147, 217–220.

208 L. P. Battaglia, A. B. Corradi, G. Pelosi, A. Cantoni, G. Alonzo, N. Bertazzi, *J. Chem. Soc. Dalton Trans.* 1991, 1991, 3153–3155.

209 S. L. Benjamin, J. Burt, W. Levason, G. Reid, M. Webster, *J. Fluor. Chem.* 2012, 135, 108–113.

210 J. Seliger, in *Encyclopedia of Spectroscopy and Spectrometry*, Third Edition, (Eds.: J. C. Lindon, G. E. Tranter, D. W. Koppenaal), Academic Press, Oxford, 2017, pp. 447–455.

211 O. K. Poleshchuk, J. N. Latosińska, M. Latosińska, in *Encyclopedia of Spectroscopy and Spectrometry*, Third Edition, (Eds.: J. C. Lindon, G. E. Tranter, D. W. Koppenaal), Academic Press, Oxford, 2017, pp. 432–446.

212 G. K. Semin, *Russ. J. Phys. Chem.* 2007, 81, 38–46.

213 C. Leroy, R. Johannson, D. L. Bryce, *J. Phys. Chem. A* 2019, 123, 1030–1043.

214 N. Burford, P. J. Ragogna, *J. Chem. Soc. Dalton Trans.* 2002, 2002, 4307–4315.

215 A. E. Waked, S. S. Chitnis, D. W. Stephan, *Chem. Commun.* 2019, 55, 8971–8974.

216 K. Nishimura, K. Hirano, M. Miura, *Org. Lett.* 2020, 22, 3185–3189.

217 *Database of Ionic Radii*, http://abulafia.mt.ic.ac.uk/shannon/ptable.php, (last accessed: 2025-04-01).

218 M. A. Katkova, G. Y. Zhigulin, R. V. Rumyantcev, G. S. Zabrodina, V. R. Shayapov, M. N. Sokolov, S. Y. Ketkov, *Molecules* 2020, 25, 4379.

219 C. L. Raston, G. L. Rowbottom, A. H. White, *J. Chem. Soc. Dalton Trans.* 1981, 1981, 1383–1388.

220 M. A. Tershansy, A. M. Goforth, M. D. Smith, H.-C. zur Loye, *J. Chem. Crystallogr.* 2008, 38, 453–459.

221 K. M. Lewis, J. Kelley, L. Peterson, M. D. Smith, R. C. Severance, S. A. Vaughn, H.-C. zur Loye, *J. Chem. Crystallogr.* 2010, 40, 867–871.

222 P. C. Junk, L. M. Louis, *Z. Anorg. Allg. Chem.* 2000, 626, 556–559.

223 M. A. Tershansy, A. M. Goforth, L. Peterson, M. C. Burns, M. D. Smith, H.-C. zur Loye, *Solid State Sci.* 2007, 9, 895–906.

224 A. M. Goforth, M. A. Tershansy, M. D. Smith, L. Peterson Jr., J. G. Kelley, W. J. I. DeBenedetti, H.-C. zur Loye, *J. Am. Chem. Soc.* 2011, 133, 603–612.

225 M. S. Ozório, W. X. C. Oliveira, J. F. R. V. Silveira, A. F. Nogueira, J. L. F. Da Silva, *Mater. Adv.* 2020, 1, 3439–3448.

226 A. K. Adcock, R. L. Ayscue, L. M. Breuer, C. P. Verwiel, A. C. Marwitz, J. A. Bertke, V. Vallet, F. Réal, K. E. Knope, *Dalton Trans.* 2020, 49, 11756–11771.

227 A. A. A. Emara, G. J. Schrobilgen, *J. Chem. Soc. Chem. Commun.* 1988, 1988, 257–259.

228 H. J. Frohn, T. Schroer, G. Henkel, *Z. Naturforsch. B* 1995, 50, 1799–1810.

229 W. Breuer, H. J. Frohn, *Z. Anorg. Allg. Chem.* 1992, 611, 85–91.

230 A. J. M. Shaw, R. Corbo, D. J. D. Wilson, J. L. Dutton, *Dalton Trans.* 2015, 44, 15083–15087.

231 T. Umemoto, M. Nagayoshi, K. Adachi, G. Tomizawa, *J. Org. Chem.* 1998, 63, 3379–3385.

232 F. H. Case, *J. Org. Chem.* 1962, 27, 640–641.

233 R. P. Thummel, Y. Jahng, *J. Org. Chem.* 1985, 50, 3635–3636.

234 R.-A. Fallahpour, M. Neuburger, *Eur. J. Org. Chem.* 2001, 2001, 1853–1856.

235 W. M. Reiff, W. A. Baker Jr., *Inorg. Chem.* 1970, 9, 570–576.

236 A. Musumeci, R. P. Bonomo, V. Cucinotta, A. Seminara, *Inorg. Chim. Acta* 1982, 59, 133–140.

237 A. J. Amoroso, M. W. Burrows, R. Haigh, M. Hatcher, M. Jones, U. Kynast, K. M. A. Malik, D. Sendor, *Dalton Trans.* 2007, 2007, 1630–1638.

238 D. Turnbull, P. Chaudhary, P. Hazendonk, S. D. Wetmore, M. Gerken, *Inorg. Chem.* 2021, 60, 3893–3901.

239 D. A. Couch, P. S. Elmes, J. E. Fergusson, M. L. Greenfield, C. J. Wilkins, *J. Chem. Soc. A* 1967, 1967, 1813–1817.

240 J. L. Dutton, P. J. Ragogna, *Inorg. Chem.* 2009, 48, 1722–1730.

241 J. L. Dutton, R. Tabeshi, M. C. Jennings, A. J. Lough, P. J. Ragogna, *Inorg. Chem.* 2007, 46, 8594–8602.

242 J. L. Dutton, G. J. Farrar, M. J. Sgro, T. L. Battista, P. J. Ragogna, *Chem. Eur. J.* 2009, 15, 10263–10271.

243 C. Wei, Y. He, X. Shi, Z. Song, *Coord. Chem. Rev.* 2019, 385, 1–19.

244 G. Chelucci, R. P. Thummel, *Chem. Rev.* 2002, 102, 3129–3170.

245 G. Zhang, J. Wu, S. Zheng, M. C. Neary, J. Mao, M. Flores, R. J. Trovitch, P. A. Dub, *J. Am. Chem. Soc.* 2019, 141, 15230–15239.

246 P. J. Chirik, K. Wieghardt, *Science* 2010, 327, 794–795.

247 T. Komuro, Y. Nakajima, J. Takaya, H. Hashimoto, *Coord. Chem. Rev.* 2022, 473, 214837.

248 S. S. Chitnis, J. H. W. LaFortune, H. Cummings, L. L. Liu, R. Andrews, D. W. Stephan, *Organometallics* 2018, 37, 4540–4544.

249 R. J. Andrews, S. S. Chitnis, D. W. Stephan, *Chem. Commun.* 2019, 55, 5599–5602.

250 W.-L. Wong, C.-S. Lee, H.-K. Leung, H.-L. Kwong, *Org. Biomol. Chem.* 2004, 2, 1967–1969.

251 C. E. Housecroft, E. C. Constable, *Chem. Commun.* 2020, 56, 10786–10794.

252 C. E. Housecroft, E. C. Constable, *Molecules* 2021, 26, 3110.

253 S. M. Elahi, M. Raizada, P. K. Sahu, S. Konar, *Chem. Eur. J.* 2021, 27, 5858–5870.

254 E. S. Claudio, H. A. Godwin, J. S. Magyar, in *Progress in Inorganic Chemistry*, Vol. 51, (Ed.: K. D. Karlin), John Wiley & Sons, Inc., Hoboken, 2002, pp. 1–144.

255 Z. Zheng, C. S. Ruiz-Vargas, T. Bauer, A. Rossi, P. Payamyar, A. Schütz, A. Stemmer, J. Sakamoto, A. D. Schlüter, *Macromol. Rapid Commun.* 2013, 34, 1670–1680.

256 T. Bauer, Z. Zheng, A. Renn, R. Enning, A. Stemmer, J. Sakamoto, A. D. Schlüter, *Angew. Chem. Int. Ed.* 2011, 50, 7879–7884.

257 Nikon-MicroscopyU, *Principles of Birefringence*, https://www.microscopyu.com/techniques/polarized-light/principles-of-birefringence (last accessed: 2025-04-01).

258 W. L. Leong, J. J. Vittal, *Chem. Rev.* 2011, 111, 688–764.

259 S. B. Bulgarevich, D. V. Bren, D. Y. Movshovich, S. E. Filippov, E. P. Olekhnovich, I. V. Korobka, *Russ. J. Gen. Chem.* 2002, 72, 1446–1452.

260 J. R. Thompson, K. A. S. Goodman-Rendall, D. B. Leznoff, *Polyhedron* 2016, 108, 93–99.

261 K. F. Bowes, I. P. Clark, J. M. Cole, M. Gourlay, A. M. E. Griffin, M. F. Mahon, L. Ooi, A. W. Parker, P. R. Raithby, H. A. Sparkes, M. Towrie, *CrystEngComm* 2005, 7, 269–275.

262 C. A. Bessel, R. F. See, D. L. Jameson, M. R. Churchill, K. J. Takeuchi, *J. Chem. Soc. Dalton Trans.* 1992, 1992, 3223–3228.

263 G. Ulian, G. Valdrè, *Sci. Rep.* 2022, 12, 299.

264 B. J. Greer, V. K. Michaelis, M. J. Katz, D. B. Leznoff, G. Schreckenbach, S. Kroeker, *Chem. Eur. J.* 2011, 17, 3609–3618.

265 M. J. Katz, D. B. Leznoff, *J. Am. Chem. Soc.* 2009, 131, 18435–18444.

266 D. Guan, J. R. Thompson, D. B. Leznoff, *Can. J. Chem.* 2018, 96, 226–234.

267 Y. Suffren, M. Kobayashi, J. S. Ovens, A. Rodrigue-Witchel, C. Genre, K. Sakai, C. Reber, D. B. Leznoff, *Eur. J. Inorg. Chem.* 2017, 2017, 2865–2875.

268 V. F. Yusuf, N. I. Malek, S. K. Kailasa, *ACS Omega* 2022, 7, 44507–44531.

269 S.-R. Zheng, R.-L. Chen, T. Xie, Z.-M. Liu, X.-L. Wen, X.-Y. Chen, J. Fan, W.-G. Zhang, *CrystEngComm* 2014, 16, 4029–4037.

270 W. P. Lustig, J. Li, *Coord. Chem. Rev.* 2018, 373, 116–147.

271 K. Lv, S. Fichter, M. Gu, J. März, M. Schmidt, *Coord. Chem. Rev.* 2021, 446, 214011.

272 R. N. Grimes, in *Carboranes*, 2nd Edition, (Ed.: R. N. Grimes), Academic Press, Oxford, 2011, pp. 1–6.

273 R. N. Grimes, in *Carboranes*, 3rd Edition, (Ed.: R. N. Grimes), Academic Press, Boston, 2016, pp. 1–5.

274 X. Zhang, L. M. Rendina, M. Müllner, *ACS Polym. Au* 2024, 4, 7–33.

275 K. Kokado, Y. Chujo, *Dalton Trans.* 2011, 40, 1919–1923.

276 C.-L. Ho, W.-Y. Wong, *Coord. Chem. Rev.* 2011, 255, 2469–2502.

277 J. M. Ludlow III, M. Tominaga, Y. Chujo, A. Schultz, X. Lu, T. Xie, K. Guo, C. N. Moorefield, C. Wesdemiotis, G. R. Newkome, *Dalton Trans.* 2014, 43, 9604–9611.

278 M. F. Hawthorne, D. C. Young, P. A. Wegner, *J. Am. Chem. Soc.* 1965, 87, 1818–1819.
279 T. P. Hanusa, *Polyhedron* 1982, 1, 663–665.
280 N. S. Hosmane, J. A. Maguire, *Organometallics* 2005, 24, 1356–1389.
281 N. S. Hosmane, J. A. Maguire, *Eur. J. Inorg. Chem.* 2003, 2003, 3989–3999.
282 N. S. Hosmane, A. K. Saxena, K.-J. Lu, J. A. Maguire, H. Zhang, Y. Wang, C. J. Thomas, D. Zhu, B. R. Grover, H. Isom, A. H. Cowley, *Organometallics* 1995, 14, 5104–5118.
283 N. S. Hosmane, J. Yang, K.-J. Lu, H. Zhang, U. Siriwardane, M. S. Islam, J. L. C. Thomas, J. A. Maguire, *Organometallics* 1998, 17, 2784–2796.
284 U. Siriwardane, J. A. Maguire, J. J. Banewicz, N. S. Hosmane, *Organometallics* 1989, 8, 2792–2800.
285 U. Siriwardane, N. S. Hosmane, *Acta Crystallogr. C* 1988, 44, 1572–1576.
286 N. S. Hosmane, L. Jia, H. Zhang, J. A. Maguire, *Organometallics* 1994, 13, 1411–1423.
287 E. C. Constable, C. E. Housecroft, *Aust. J. Chem.* 2020, 73, 390–398.
288 O. S. Wenger, *J. Am. Chem. Soc.* 2018, 140, 13522–13533.

4

The Bio-Medicinal Relevance of Terpyridine Ligands and Their Metal Complexes

4.1 Introduction

The pharmaceutical use of metals (or metal compounds) can unequivocally be traced back to the ancient civilizations of Mesopotamia, Egypt, India, and China. Thus, the important role of metal ions in biological processes was already known (though not understood) in these early days [1]. Today, research on metal ions or complexes in biological systems represents one notable subset of modern bioinorganic chemistry [2]. Their administration into (human) biological systems has found widespread use for the treatment of diseases (*i.e.*, therapy), imaging (*i.e.*, diagnostic), and the combination of both (*i.e.*, theranostic). Various metallodrugs are available for either of these applications.

Ehrlich *et al.* discovered the antimicrobial activity of a mixture of 3-amino-4-hydroxyphenyl-As^{III} compounds [3]. Specifically, a high activity against syphilis was disclosed; thus, the harmful “mercury curses” became obsolete. Already in 1910, the medical usage was commercialized as Salvarsan™ by Farbwerke Hoechst and its introduction to the market is widely defined as the origin of modern chemotherapy. Indeed, in Ehrlich’s laboratory, the first systematic screening of (metallo)drugs for the targeted treatment of infectious diseases was performed. Ehrlich believed that such infections could efficiently be fought by selective drugs, which kill the invading microorganisms without affecting the human host. This idea of “Magic Bullets” called for the extensive screening of potential drugs [4]. Although Salvarsan, as the primary metallodrug, was widely used in human medicine, its composition largely remains unclear; however, there is strong evidence for the presence of cyclic As_3 and As_5 species [5]. Together with Hg- and Bi-containing formulations, Salvarsan remained the standard medicine for the treatment of syphilis until it was eventually outcompeted by penicillin after World War II [6].

However, *cis*-$PtCl_2(NH_3)_2$ (cisplatin, Platinol; Figure 4.1) represents the brightest star among the metallodrugs. This compound was discovered in 1965 by serendipity, when Rosenberg *et al.* studied the effect of electrolysis products of Pt electrodes on *Escherichia coli* [7]. The cell division was inhibited by *cis*-$PtCl_2(NH_3)_2$, which was formed *in situ* by an electrode reaction. In subsequent work, the antitumor activity of cisplatin was also discovered [8]. This groundbreaking discovery paved the way to intensified research on anticancer metallodrugs and marks the birth of medicinal

Terpyridines: Multifaceted Ligands for Diverse Applications, First Edition.
Ulrich S. Schubert, Andreas Winter, and George R. Newkome.

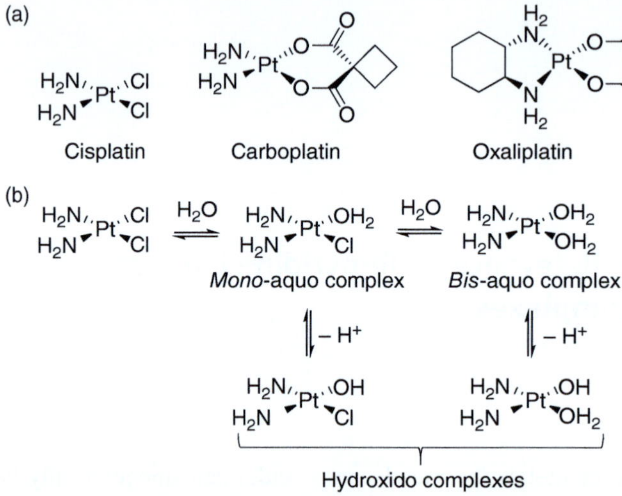

Figure 4.1 (a) Schematic representation of three important Pt^{II} anticancer drugs: Cisplatin, carboplatin, and oxaliplatin. (b) Schematic representation of the hydrolysis equilibria of cisplatin.

inorganic chemistry [9]. Eventually, cisplatin was approved by the FDA for cancer treatment in 1978 (FDA: Food and Drug Administration).

In the last decades, metallodrugs have widely been tested *inter alia* as anticancer, antimicrobial, or antiviral agents as well as diagnostic compounds [1]. However, when compared to the enormous number of organic or biological drugs, which are now available in the market, only a small minority of metal-based therapeutics have been translated so far into clinical use [10]. A histogram showing the metallodrugs per element, which have been approved in the United States and/or European Union until 2020, is shown in Figure 4.2a [11]. Furthermore, the timeline of platinodrug development at the example of cisplatin, carboplatin, and oxaliplatin, as the three most prominent representatives, is displayed in Figure 4.2b.

When talking about metallodrugs, one has to take the various electronic and structural features into account, which might contribute to their therapeutic activity: Oxidation state and coordination geometry of the metal center, number and type of ligands, *etc.* Thus, their activity might be due to ligand-substitution reactions, redox reactions involving the metal center and/or the ligands, or catalytic cycles [11]. Notably, metallodrugs mostly represent prodrugs, which require preliminary activation either during transport in the body or at the actual target site. The activation step might also involve ligand-displacement or redox reactions. In some cases, activation can be achieved by an external stimulus (*e.g.*, by irradiation or heat) once the metallodrug has reached its target. Hydrolysis, *i.e.*, the displacement of weakly bound σ-donor-type ligands by water molecules, represents a prominent activation pathway for transition-metal complexes, as metallodrugs. The exchange rate strongly depends on the nature of the metal ion and the leaving ligand and might range from nanoseconds to years. Also, this aspect has to be carefully considered in metallodrug design.

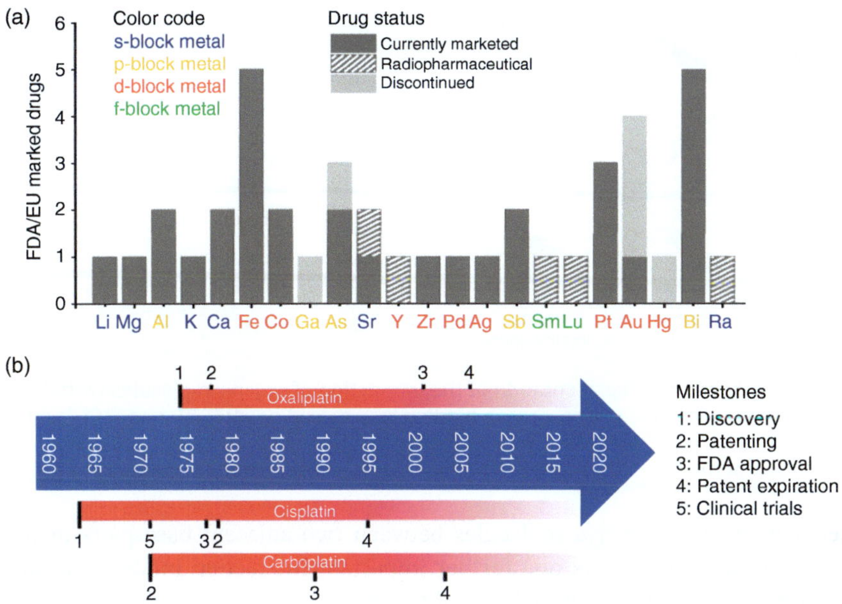

Figure 4.2 (a) The metallodrugs, which were approved in the United States (by the Food and Drug Administration) and/or European Union until 2020, are listed according to the used metal center (different formulations of the same active metallodrug as well as pharmaceuticals, where the metal only represents the counterion, are not counted; imaging/diagnostic agents, food supplements, agents used as anesthetics, and in implants are also excluded) [11]. (b) Representation of the timeline reflecting the development of the three common Pt^{II}-based anticancer drugs [11] / Royal Society of Chemistry / CC BY 3.0.

In the case of cisplatin, activation proceeds *via* hydrolysis, *i.e.*, the step-wise substitution of the chloride ligands (Figure 4.1b) [12]. In the extracellular environment, where the concentration of Cl^- ions is high (*i.e.*, > 100 mM), ligand substitution is suppressed. Whereas, inside cells, the Cl^- concentration is significantly lower (*e.g.*, *ca.* 23 mM and *ca.* 4 mM in the cytoplasm and nucleus, respectively) and, thus, the aquation of cisplatin is fast [13]. The thusly formed aquo species readily bind to the guanine (G) and adenine (A) bases of deoxyribonucleic acid (DNA). It is assumed that the less reactive hydroxido complexes are also involved in DNA binding. One has to be aware that only *ca.* 1% of the intracellular cisplatin [14], thus only a minor fraction, is available for the targeted processes, *i.e.*, cell cycle arrest and apoptosis [15]. Most of the cisplatin is lost by side reactions, such as the binding to cysteine's thiolate group, which is commonly present in proteins or peptides (*e.g.*, glutathione). These detoxification processes represent a considerable drawback, in particular when dealing with the so-called resistant cancer types [16]. Off-target side effects are also often caused by these intracellular reactions [17]. The 2nd-generation platinodrugs, such as carboplatin or oxaliplatin (Figure 4.1a), are more inert toward hydrolysis, owing to their chelating *bis*-carboxylate ligands; these drugs are less nephrotoxic than cisplatin [18].

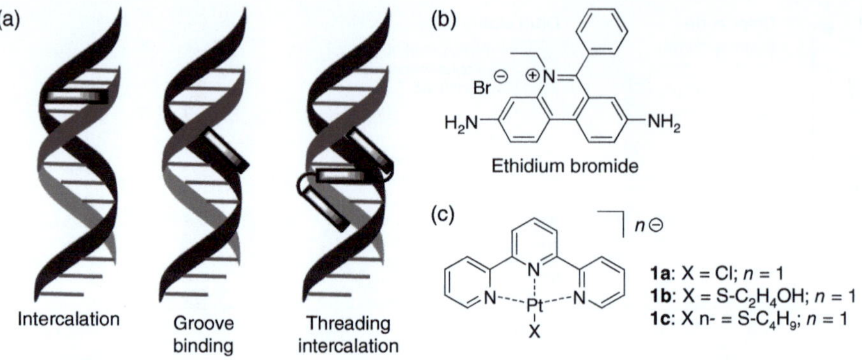

Figure 4.3 (a) Schematic representation of the interaction of organic molecules with the DNA double helix *via* intercalation, groove binding, and threading intercalation [19] / with permission of Elsevier. (b) Schematic representation of ethidium bromide. (c) Schematic representation of the Pt^{II}–tpy complexes **1** (counterions were omitted for clarity).

Intercalation of planar dye molecules between two adjacent base pairs of the DNA double helix represents one further important interaction mode. Basically, interaction of molecules with DNA might occur by (i) electrostatic interaction with the negatively charged sugar-phosphate backbone, (ii) interaction with the major or minor groove, (iii) intercalation between base pairs of the major or minor groove, and (iv) threading intercalation (Figure 4.3a) [19]. Intercalation was first described by Lerman at the example of acridine derivatives [20] and has, since then, been reported for a plethora of organic molecules [19]. Due to their extended π-conjugated system, these molecules can intercalate between the base pairs; however, with increasing steric hindrance, the other binding modes become more favorable. For example, intercalation of ethidium bromide (EthBr), as one prominent intercalating dye, might occur from the major or minor groove; in general, the intercalation *via* the former affords the less stable adducts [21].

Intercalation has a profound impact on the properties of DNA, as reflected by pronounced shifts of the UV/vis absorption bands, induced circular dichroism (CD), increased length, viscosity, and melting temperature (T_m), altered degree of supercoiling (in the case of closed circular duplex DNA) as well as decreased sedimentation coefficients [22]. Thus, DNA intercalation can be monitored and analyzed in a straightforward fashion by a wide range of methods.

4.2 Medicinal Relevance of Terpyridine Ligands and Complexes

In the 1970s, 2,2′:6′,2″-terpyridine (tpy) ligands eventually came into play. Although the parent square-planar complex $[Pt(tpy)Cl]^+$ was already reported by Morgan and Burstall in 1934 [23], Lippard *et al.* were the first who proposed that such complexes also represented potential antitumor drugs. Owing to their flat, disk-shaped geometries, Pt^{II}–tpy complexes also intercalate into DNA, thus resembling the behavior of conventional organic intercalators [24, 25]. In the first available studies in this field, complex **1b** (Figure 4.3c) was chosen over other Pt^{II}–tpy complexes as the intercalator to prevent covalent binding to the DNA strands. Unlike for complex **1a**, the Pt—X bond of **1b** is hydrolytically stable and, thus, the formation of reactive aquo species, as observed in the case of cisplatin, was suppressed. Likewise, the related complex **1c** was studied by McFayden *et al.* [26]. Without going too much into the details, two different binding modes were identified: Besides intercalation between the base pairs, **1b** and **1c** also interacted electrostatically with the DNA backbone. In the following, numerous Pt^{II}–tpy complexes have been reported and studied regarding their capability to act as an intercalator and the cytotoxicity against various cancer cell lines has been evaluated. A detailed overview of this research was provided by Cummings as well as Newkome *et al.* and the reader is referred to these review articles [27, 28]. In this context, also the medicinal relevance of the related Pd^{II}–tpy and Au^{III}–tpy complexes was summarized [28]. It has to be noted that scattered studies on the biological activity of tpy complexes were already performed even before Lippard's work. For example, Koch *et al.* reported the inhibition of acetylcholinesterase and toxicity for mice; the activity of the complexes $[M(tpy)_2]^{2+}$ was ascribed to electrostatic interaction of the cationic complexes with the enzyme's negatively charged sites (M = Zn, Fe, Ni, Ru, Os) [29]. Likewise, $[Ru(tpy)_2]^{2+}$ was screened for its ability to block the oxidation activity of liver mitochondrial enzyme systems; however, this complex was outcompeted by its $[Ru(phen)_3]^{2+}$ congener, which exhibited a higher and more general activity in this context (phen: 1,10-phenanthroline) [30].

In one further pioneering work on the antitumor activity of tpy complexes, Reedijk *et al.* showed that *mer*-$Ru(tpy)Cl_3$ strongly interacted with DNA and possessed activity against L1210 leukemia cells, which was in between those of cisplatin and carboplatin [31]. However, this complex acted not as an intercalator but reacted with DNA to afford *ca.* 2% of interstrand cross-links. Complexes from the Ru^{II}–polypyridyl family have attracted considerable interest due to their ability to (covalently) bind to DNA and to act as reactive probes for the oxidative fragmentation of DNA strands [32]. The latter property is not limited to Ru^{II} complexes – also other complexes, which contain redox-active metal centers (*e.g.*, Cu^{I} or Fe^{II}), are able to form reactive oxygen species (ROS), as agents for DNA cleavage. For example, the trinuclear Cu^{II} complex **2** (Figure 4.4a) displayed a high efficiency in cleaving DNA within the cells, even in the absence of co-oxidants (*e.g.*, peroxides) [33]. Its cytotoxic effect outcompeted that of cisplatin, in particular against a cisplatin-resistant leukemia-cell line (*i.e.*, MOLT-4; Figure 4.4b). It has to be noted at this stage, that the biological activity of tpy *bis*-complexes, specifically the acetylcholinesterase inhibition, was already studied in the mid-1950s [29]. It was shown that complexes of Fe^{II}, Ru^{II}, or Os^{II} ions were less potent inhibitors, but more toxic than their bpy- or phen-based counterparts.

Many therapeutically and diagnostically interesting tpy complexes contain heavy metal ions (*e.g.*, Pt^{II}, Ru^{II}, Ir^{III}, *etc.*) [28, 34]. However, when dealing with such complexes in the context of medicinal applications, one has to be aware of two notable limitations, which are the natural abundance of the metals on the one hand and the intrinsic toxicity of their compounds on the other [35]. In recent years, numerous

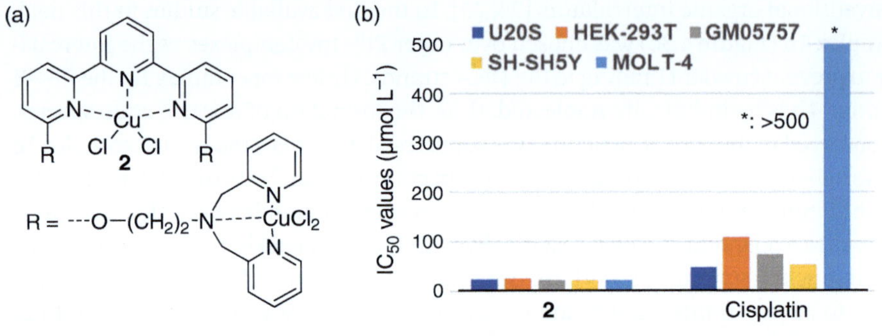

Figure 4.4 (a) Schematic representation of the trinuclear Cu^{II} complex **2**, (b) Comparison of the cellular toxicity of **2** and cisplatin against various cancer cell lines. Adapted from [33].

tpy complexes of the first-row transition-metal ions have been reported and their biomedical activity has been disclosed. Besides anticancer activity, antibacterial or antifungal properties have also been found [34]. Research on such complexes is highly attractive, due to the combination of high natural abundance and reduced intrinsic toxicity of the metal ions with the remarkably high biomedical activity. These aspects are of importance in view of the development of efficient low-cost metallodrugs [35].

In line with our previous review on this topic [35] and the aspects covered in the second edition of this textbook [36]. The focus of this chapter is mainly on tpy complexes of the first-row transition metal ions. However, before discussing the biomedicinal relevance of these complexes, attention has to be paid briefly to the tpy ligands themselves. The presumably first study on the biological activity of tpy ligands is commonly overlooked. Already in 1953, Leiter *et al.* tested parent tpy, amongst numerous other (hetero)aromatic compounds, regarding their ability to damage the murine Sarcoma-37 tumor [37]. This study revealed only marginal toxicity of tpy; whereas 2,2′-bipyridine (bpy) and 1,10-phenanthroline (phen) were by far more potent. The first detailed report, which showed that tpy ligands possess substantial anticancer activity, was published by Lee *et al.*, who showed that tpy ligands, equipped with (het)aryl substituents in 4′-position, were highly cytotoxic toward several human tumor-cell lines (at least when compared to doxorubicin, as the reference) [38]. Whereas the corresponding 2,2′:6′,3″- and 2,2′:6′,4″-tpy derivatives represented potent topoisomerase-I inhibitors. Thus, it was concluded that antitumor activity was not primary due to the inhibition of topoisomerase-I and that different mechanisms were involved, depending on the "pattern" of the tpy moiety. More recently, Zych, Musiol *et al.* screened the anticancer activity of various 4′-[4-(1*H*-1,2,3-triazol-4-yl)phenyl]-tpy derivatives (**3**, Figure 4.5a) [39]. All these compounds revealed anticancer activity at the nM level against five different cell lines. Moreover, the high photoluminescence of **3** allowed the visualization of their uptake and localization within the cells by fluorescence-microscopy imaging (Figure 4.5b). The antiproliferative activity was ascribed to interactions with DNA by electrostatic binding and intercalation (at least in the case of halogenated **3e** and **3f**).

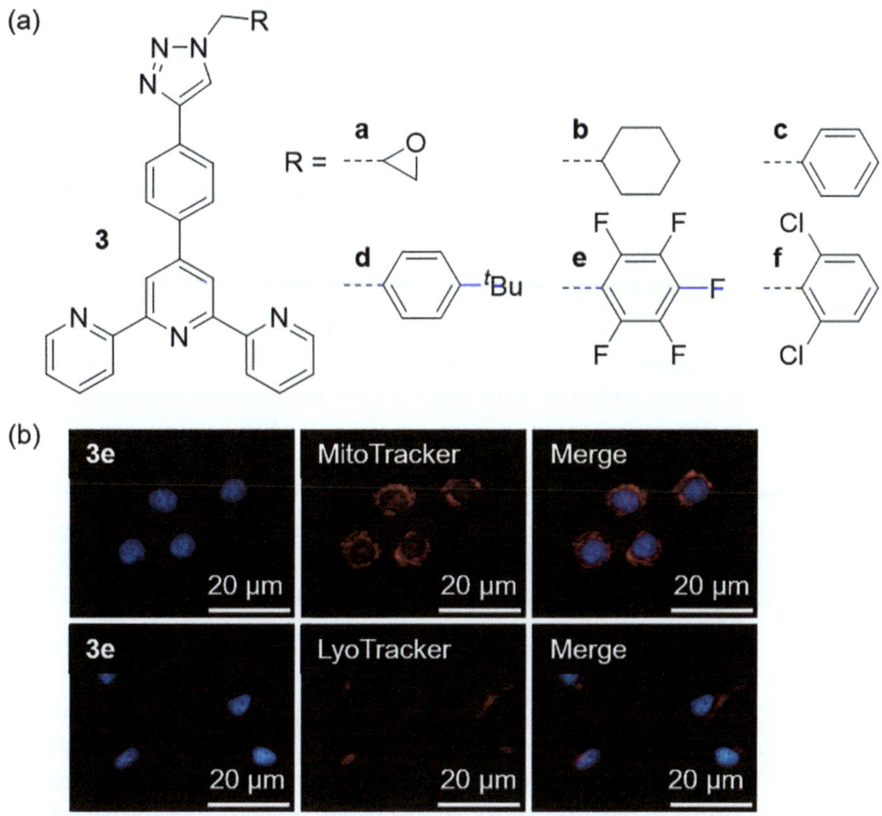

Figure 4.5 (a) Schematic representation of the tpy derivatives **3**. (b) Fluorescence-microscopy images of MCF-7 cells, which were incubated with **3f** (25 μM) for 2 h at 37 °C. The images show the co-localization of **3f** with organelle-specific trackers (top row: mitochondria, bottom row: lysosomes) [39] / with permission of John Wiley & Sons.

Recently, Musiol *et al.* provided a thorough evaluation of the available literature dealing with the biological, specifically antitumor, activity of metal-free tpy ligands [40]. Some important lessons can be learned from this survey. Firstly, the mode of action is not yet fully understood and depends on the particular ligand structure – steric and electronic factors significantly affect the interaction with DNA. Moreover, the biological activity of ligands might differ from those of their transition-metal complexes and complexes of 3d and 4d/5d metal ions often behave differently. This observation has been ascribed to the kinetic lability of the former, *i.e.*, dissociation into metal ions and free ligand molecules might occur under physiological conditions. This behavior complicates (or even precludes) the interpretation of the observed experimental results. Whereas tpy complexes of heavy transition-metal ions (*e.g.*, Pt^{II}, Ru^{II}, *etc.*) are kinetically stable in solution and, thus, anticancer activity can be ascribed to the complex itself. Thus, we fully subscribe to the proposed view that every study on the biological activity of tpy complexes should always include control experiments with the metal-free ligands.

4.3 Biological Activity of Tpy Complexes of First-Row Transition-Metal Ions

Following up on our earlier review in this field [35], the biological studies on tpy complexes of the first-row transition metal ions will be discussed in the following. With the exception of the two lightest representatives, *i.e.*, Sc and Ti, such studies have been performed with all other metals. However, in the case of Cr^{III}, no follow-up studies on the binding of Cr^{III}–tpy complexes to DNA or on its nuclease activity have been made [34, 41]. This is a bit surprising when taking the enormous progress in Cr^{III}–polypyridyl complexes for catalytic [42, 43] or optoelectronic applications into account [44]. Moreover, DNA fragmentation in the presence of co-reagents has been reported for other types of Cr^{III}–polypyridyl complexes, excluding Cr^{III}–tpy species [45].

For the sake of clarity, the various transition-metal complexes will be discussed per incorporated metal ions. It is emphasized that general structure-property relationships can hardly be deduced since the discussed complexes differ strongly in chemical structure (*i.e.*, homoleptic *bis*-complexes *vs.* *mono*-complexes with more or less labile ancillary ligands), hydrolytic stability, and redox-activity.

4.3.1 Tpy Complexes of Vanadium Ions

In 2017, Hong *et al.* reported the pentacoordinate V^V complex $[V(O)_2(tpy)](ClO_4)$ (**4**, Figure 4.6a), which exhibited antitumor activity against various cancer cell lines: HepG2, HeLa, A-549, BEL-7402, and HOS [46]. Remarkably, the cytotoxicity of **4** against these cell lines was even higher than that of cisplatin, as the reference (Figure 4.7). Moreover, healthy human hepatic cells (LO2) were less affected by **4**; thus showing a high selectivity, which is desirable when aiming for the use of such metallodrugs for cancer treatment. Mechanistic studies suggested that the formation of ROS, which caused DNA breakage, and alteration of the mitochondrial membrane potential (MMP) represented the main factors in provoking apoptosis in BEL-7402 cells. Cell cycle analyses revealed that suppression of the cell development at the G0/G1 phase also accounted for the antitumor activity.

The formation of ROS by V^V–peroxo complexes was already shown beforehand. Specifically, Kwong *et al.* reported the cleavage of DNA strands *inter alia* by $[VO(O_2)_2(tpy)(H_2O)](ClO_4)$ [47]. However, unlike for the previous example, photoactivation, *i.e.*, the liberation of singlet oxygen by photolysis according to the proposed mechanism, was required (Figure 4.6b). These authors proposed that a V^{III} intermediate was formed, which eventually decomposed to afford the $[VO_3]^-$ anion and uncomplexed tpy. However, the "non-innocent" role of the coordinated tpy ligands, *i.e.*, the temporary formation of a radical-anion species, could not fully be ruled out. Presumably due to the strong chelation by the tpy ligand, the DNA-cleavage efficiency of $[VO(O_2)_2(tpy)(H_2O)](ClO_4)$ was only moderate (*i.e.*, *ca.* 25%). Whereas the corresponding complexes, which contained bidentate polypyridyl ligands (*e.g.*, phen or bpy), performed with efficiencies >50%.

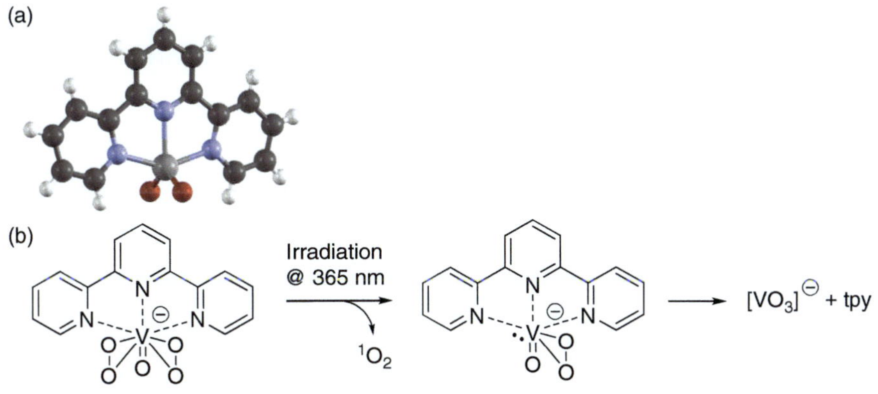

Figure 4.6 (a) Solid-state structure of complex **4** (counterions and solvent molecules were omitted for clarity). Adapted from [46]. (b) Schematic representation of the photoinduced generation of $^1O_2$ from the $[V(O_2)_2O(tpy)]^-$ anion (counterions were omitted for clarity). Adapted from [47].

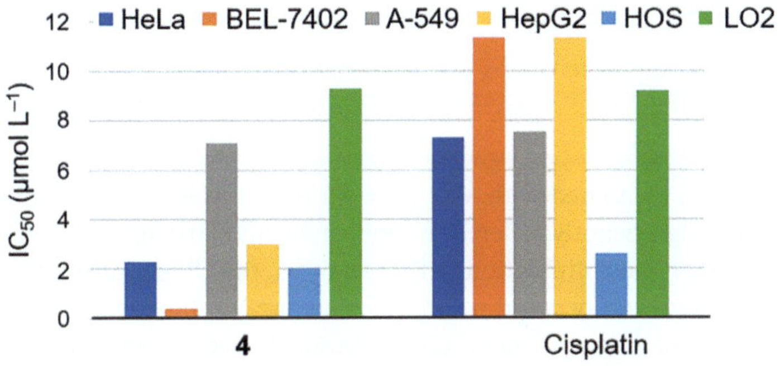

Figure 4.7 Comparison of the IC_{50} values (μM) of complex **4** against various cancer cell lines. Adapted from [46].

Further V-containing complexes with DNA-cleavage activity were reported by Chakravarty *et al.* [48]. Unlike for the previous cases, complexes **5** contained oxyvanadium(IV) centers, to which a tpy and catechol ligand were each coordinated (Figure 4.8a). The glycosylated derivatives (**5c** and **5d**) were readily taken up by cancer cells; this uptake occurred with high selectivity even if a high glucose concentration was offered in the cell medium. Due to their intense fluorescence, the localization of the complexes within the cells could be monitored by confocal microscopy. In particular, complex **5d** showed a preferential localization within the cells' cytosol and, to a minor extent, the nuclei. Furthermore, these complexes displayed strong ligand-to-metal charge-transfer (LMCT) absorption bands at *ca.* 500 nm. This behavior is of particular interest regarding their use in the context of photodynamic therapy (PDT). Indeed, the complexes showed cytotoxicity against HeLa and HepG2 cancer cell lines and DNA-cleavage activity, but only if irradiated with visible light of various wavelengths (Figure 4.8b).

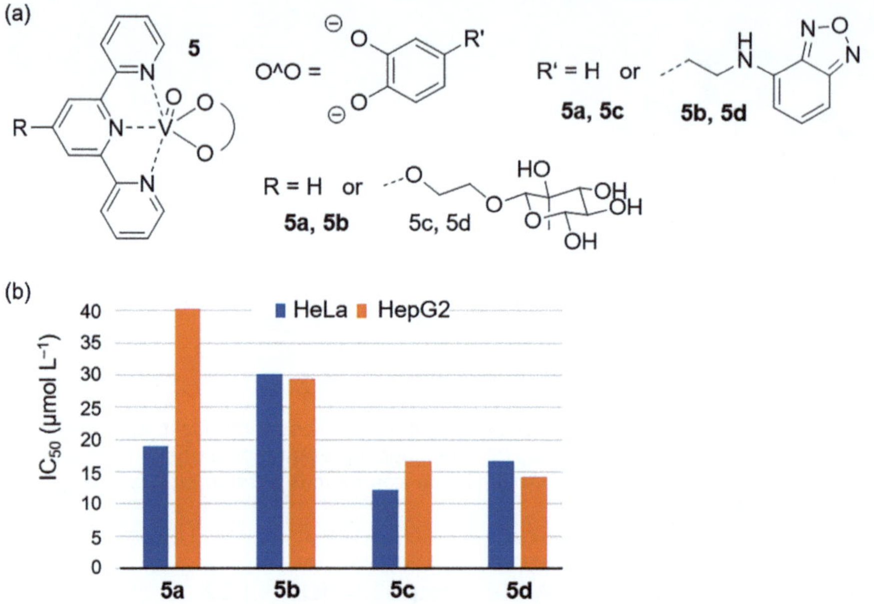

Figure 4.8 (a) Schematic representation of the oxyvanadium (IV) complexes **5**. (b) Summary of the IC_{50} (μM) values of these same complexes upon irradiation with visible light. Adapted from [48].

The 6,6″-tpy ligand was combined with VX_3 (X = Cl or Br) to afford the complexes **6a** and **6b**, which significantly differed in their molecular structure [49]. As confirmed by single-crystal X-ray diffraction (XRD) analysis, the Cl^- anions of the V^{IV}–oxido complex **6a** were closely coordinated, whereas the V^V-dioxido complex **6b** possessed ionic character (Figure 4.9a,b). UV/vis absorption/emission titration studies suggested that the complexes acted as intercalators; the K_a values for the binding to DNA were $4.4 \times 10^5\ M^{-1}$ and $1.0 \times 10^6\ M^{-1}$ for **6a** and **6b**, respectively. Thus, the molecular structure of the complex (and indirectly the size of the halide anion) had an influence on the intercalator strength. Cytotoxicity studies and cell cycle analysis by flow cytometry further corroborated this finding. Remarkably, **6a** and **6b** represented the first tpy complexes, which were reasonably cytotoxic against the T-47D cancer cell line (Figure 4.9c). Whereas the metal-free ligand showed no detectable activity against the tested cancer cell lines. Moreover, the cytotoxicity of **6a/b** against healthy cell lines, *e.g.*, GM07492, was lower than that of cisplatin. Owing to their good solubility in water, ionic V^V–dioxo complexes with tpy ligands might be promising candidates for studies in the field of anticancer metallodrugs.

4.3.2 Tpy Complexes of Manganese Ions

Metalloenzymes play a vital role in biological processes. For example, manganese superoxide dismutase (MnSOD) and catalase represent Mn-containing enzymes, which contribute to the body's antioxidant defense system [50]. Detailed knowledge of the mode of action *inter alia* of MnSOD has paved the way toward the

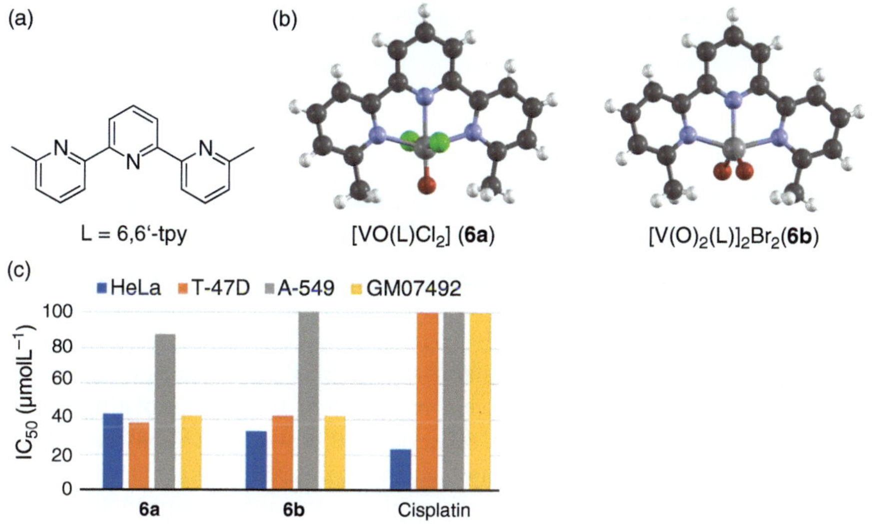

Figure 4.9 (a) Schematic representation of 6,6″-dimethyl-tpy. (b) Solid-state structures of complexes **6a** and **6b** (hydrogen atoms, counterions, and solvent molecules were omitted for clarity). Adapted from [49]. (c) Summary of the IC_{50} (μM) values of **6**. Adapted from [49].

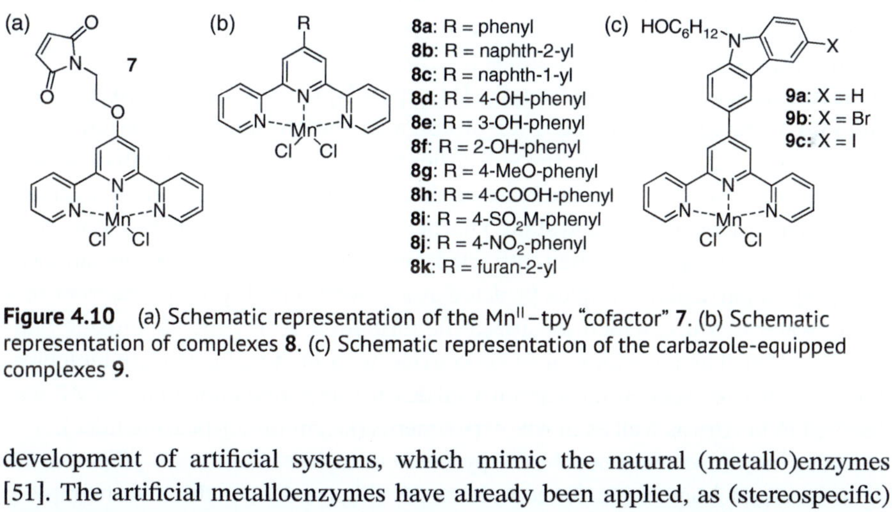

Figure 4.10 (a) Schematic representation of the Mn^{II}–tpy "cofactor" **7**. (b) Schematic representation of complexes **8**. (c) Schematic representation of the carbazole-equipped complexes **9**.

development of artificial systems, which mimic the natural (metallo)enzymes [51]. The artificial metalloenzymes have already been applied, as (stereospecific) catalysts for a plethora of organic reactions. Against this background, Lewis *et al.* prepared such systems by the conjugation of the Mn^{II}–tpy cofactor to protein scaffolds *via* the reaction of the maleimide moiety of **7** with the proteins' thiol groups (Figure 4.10a) [52]. The resulting artificial metalloenzymes revealed a very good oxygenation capability, which was used for both benzylic oxygenation as well as olefin epoxidation. This reactivity resembled that of Crabtree's catalyst, which contained an $Mn(\mu\text{-}O)_2Mn$ reactive center [42, 43, 53].

Complexes from the $[Mn(tpy)Cl_2]$ family were also exploited by Pan, Ma *et al.*, who focused on the anticancer activity of complexes **8** (Figure 4.10b) [54]. All these

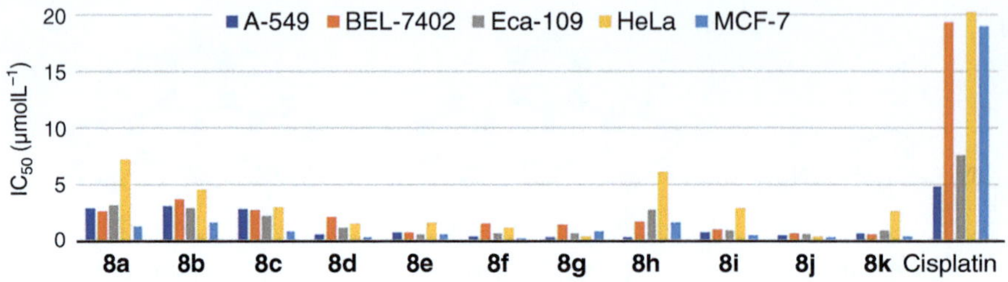

Figure 4.11 Comparison of the IC_{50} values (μM) of complexes **8** against various cancer cell lines. Adapted from [54].

complexes exhibited *in vitro* antiproliferative activities against various common human carcinoma-cell lines, which were higher than those of cisplatin, as the reference drug (Figure 4.11). Complex **8g** performed best against A-549 and HeLa cells; whereas **8k**, **8j**, and **8f** were most active against BEL-7402, Eca-109, and MCF-7, respectively. Within the series of complexes, the nitro-equipped derivative **8j** featured the overall highest cytotoxicity. Spectrophotometric experiments suggested that the strong interaction of the complexes to DNA was due to intercalation, which involved – according to molecular-modelling (MM) studies – ππ-stacking and hydrogen-bonding (*H*-bonding) interactions. In particular, the latter, which was enabled by the *O*-containing ligands, provided very strong interactions and DNA deformation. Thus, these complexes possessed higher antitumor potential than their derivatives missing *O*-atoms (*i.e.*, **8a–8c**).

One further set of complexes from the [Mn(tpy)Cl_2] family was reported by Zhang *et al.* [55]. Complexes **9** showed three-proton aggregation-induced emission (AIE), which could be modulated *via* the substituent X (Figure 4.10c). Basically, the AIE phenomenon is related to the increase of a material's luminescence upon aggregation in a bad solvent or the solid state. This behavior was observed first by Tang *et al.* in 2001 [56] and has, since then, attracted enormous interest for various applications [57]. In particular, complex **9c** acted as a powerful multiphoton absorber in a DMSO/water mixture, which sensitized the formation of $^1O_2$ and superoxide radicals (O_2^-) [55]. The formation of ROS occurred with high yield, causing biological damage with a deep penetration depth. Cell death, due to reduction of the MMP, was observed in *in vitro* as well as *in vivo* experiments (*i.e.*, in tumor-bearing mice).

Besides the aforementioned Mn^{II}–tpy *mono*-complexes, Tan *et al.* prepared the highly charged homoleptic *bis*-complexes **10** (Figure 4.12), which were rapidly taken up into the nuclei of living cells and "switched on" the nucleolar ribonucleic acid (RNA) [58]. The specific recognition of ribosomal RNA in living cells by highly hydrophilic tpy derivatives was reported by Tan *et al.* beforehand [59]. In this context, the tpy ligand of complex **10a** was identified as a powerful two-photon fluorescent probe for the staining of intracellular nucleolar RNA. Due to the strong electrostatic interaction of complexes **10** with the RNA, their fluorescence was intensified, in particular for **10a** [58]. This behavior was capitalized to visualize the nucleolar ultrastructure using super-resolution imaging methods, *i.e.*, stimulated emission depletion (STED) nanoscopy. This method is of high

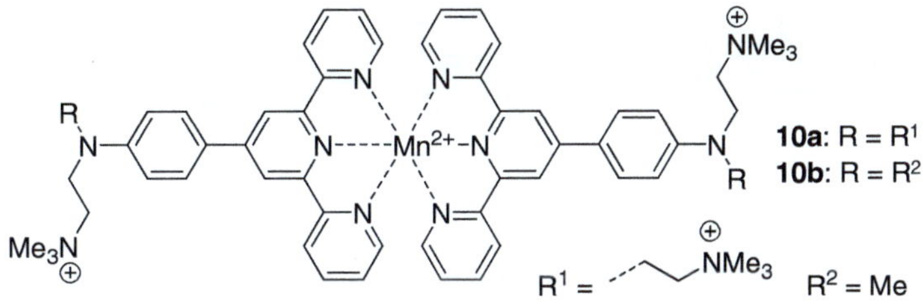

Figure 4.12 Schematic representation of the homoleptic *bis*-complexes **10** (counterions were omitted for clarity).

virtue for the visualization of subcellular structures, but the probes must fulfill two premises – biocompatibility and photostability [60]. Many organometallic compounds and transition-metal complexes withstand the prolonged exposure to the STED laser and, thus, allow for the intricate assembly of high-resolution 3D structures. Tan *et al.* employed complex **10a** to study the RNA-rich compartments of the nuclei, *i.e.*, the dense fibrillar regions and fibrillar centers, by STED nanoscopy. Moreover, the administration of **10a** allowed for an analysis of the nucleolar RNA dynamics and its interaction with ribosomes upon protein translation. These results represent significant achievements in the field of bio-imaging using tpy complexes of a first-row transition-metal ion, as the fluorescent probes.

Not only have the Mn^{II}–tpy complexes been discussed regarding their use for bio-imaging or anticancer treatment but also the Mn^{I} complexes **11**, as CO-releasing molecules (CORMs), were studied by Mansour and Radacki [61]. The delivery of CO by, *e.g.*, carbonyl complexes, to cells and tissues represents a versatile therapeutic alternative to conventional gas-inhalation protocols, which bear the risk of CO poisoning [62]. The main challenges related to the development of practicably useful CORMs are their stability in the dark, solubility in water or biocompatible media as well as leaving behind non-harmful residues. PhotoCORMs are a notable subclass, which are able to release CO only upon irradiation; thereby, the use of visible light is desirable to achieve a deeper and less harmful penetration of light [63]. In this context, the mono- and dinuclear complexes were considered antimicrobial PhotoCORMs (**11–13** and **14**; Figure 4.13) [61]. Owing to the strong binding of the CO ligands, the tpy ligands were coordinated only in the hypodentate $\kappa^2 N^1,N^2$ mode (see Section 1.3). Without going too much into detail, complexes **11** and **14c** were synthesized from the azido species **12** and **14b**, respectively, *via* the so-called iClick reaction, *i.e.*, the catalyst-free 1,3-dipolar cycloaddition reaction between an azido ligand and electron-deficient alkyne derivative. Unlike the bromide complexes (*i.e.*, **11** and **14a**), which possessed poor solubility and were prone to decomposition in many organic solvents, the corresponding triazolato species (*i.e.*, **13** and **14c**) provided sufficient solubility and stability. Thus, these complexes were tested further as PhotoCORMs. In particular, complex **13b**, when photoexcited at 468 or 525 nm, featured a high antimicrobial activity against *Escherichia coli*, *Candida albicans*, and *Cryptococcus neoformans*. Unlike all other complexes and their metal-free tpy ligands, complex **13b** was not cytotoxic to a

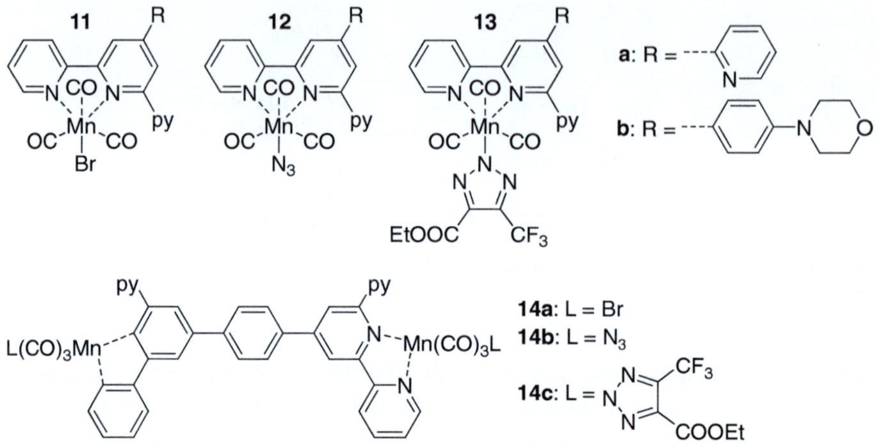

Figure 4.13 Schematic representation of the mono- and dinuclear Mn^{I} complexes **11–14**.

normal human cell line (*i.e.*, the embryonic kidney cells HEK-293). Apparently, the cytotoxicity of the morpholino-equipped ligand was erased upon coordination of the Mn^{I} center. Beneficially, the compatibility of all compounds with human red blood cells was high.

4.3.3 Tpy Complexes of Iron Ions

As already discussed in our previous review on this topic, Fe^{II} complexes are basically suited as anticancer agents [35]. Beyond doubt, the Fe-containing glycopeptide bleomycin, which is used in anticancer chemotherapy, is the best-known representative [64]. However, very few Fe^{II}–polypyridyl complexes, being able to oxidatively cleave DNA in absence of further oxidants, have been reported [65]. On the other hand, complexes from the Fe^{II}–tpy family have not been utilized in such a context for a longer period of time. The first examples were published after the aforementioned review.

The homoleptic Fe^{II} complexes **15** (Figure 4.14a) were studied regarding their photo-cytotoxicity; these complexes interacted with DNA, partially by intercalation [67]. Upon irradiation with green light, photo-cleavage occurred in the presence of the pyrene-containing complex **15c**. Thus, this complex represented a potential candidate for application in PDT. Moreover, **15b** and **15c** featured light-induced cytotoxicity against HeLa cells, whereas the complexes were noncytotoxic in the dark. Mechanistic studies revealed that the killing of HeLa cells was due to the formation of ROS inside the cells. In one further application, **15**b and **15**c were employed for *in vitro* bio-imaging: Confocal microscopy imaging and dual-staining studies, using propidium iodide, suggested that the complexes were preferentially located in the cells' nuclei.

Photo-cytotoxicity against HeLa cells was also reported for the *mono*-cyclometalated Fe^{II} complex **16**, which contained a BODIPY moiety, as intramolecular sensitizer (BODIPY: boron-dipyrromethene; Figure 4.15a) [68]. Unlike the

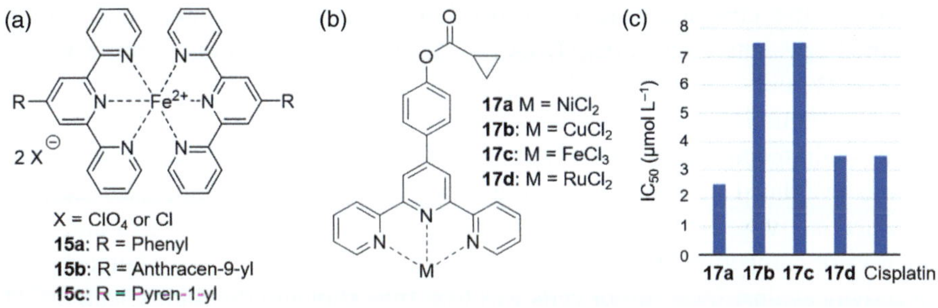

Figure 4.14 (a) Schematic representation of the Fe^{II} complexes **15** (counterions were omitted for clarity). (b) Schematic representation of the *mono*-complexes **17**. (c) Summary of the IC_{50} (µM) values of complexes **17**. Adapted from [66].

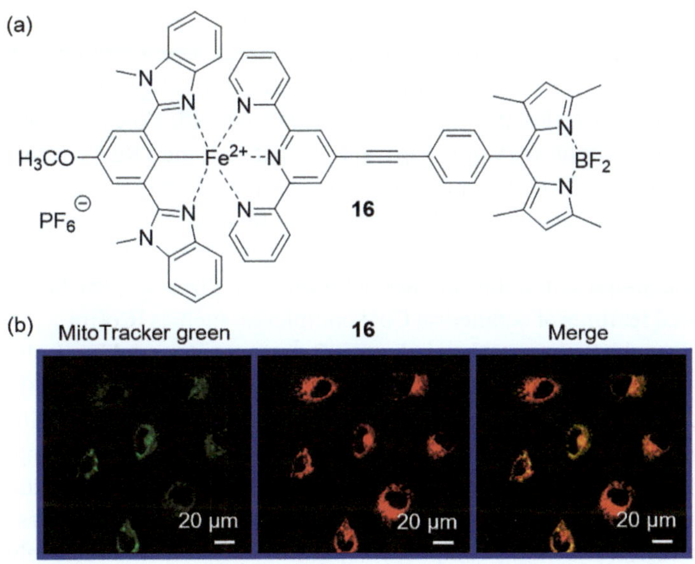

Figure 4.15 (a) Schematic representation of the *mono*-cyclometalated Fe^{II} complex **16**. (b) Visualization of the cellular localization of **16** in HeLa cells after incubation at 37 °C for 1 h; staining with the mitochondria-selective dye "MitoTracker green" was performed for comparison [68] / with permission of John Wiley & Sons.

previous complexes, this complex possessed a high selectivity for entering the mitochondria (Figure 4.15b). The complex was only cytotoxic upon irradiation with visible light (λ_{ex} = 500 nm), but was not in the dark. At this wavelength, the formation of ROS and, thus, activity against HeLa cells was most efficient (IC_{50} = 7.12 ± 0.2). This complex was also active against the non-tumorigenic MRC-5 cells, but only to a moderate extent (IC_{50} = 55.28 ± 0.2). This complex possessed a remarkable potential for application in mitochondria-targeted PDT.

Very recently, Pandya and Sivaramakrishna compared the binding affinity of *mono*-complexes **17** (Figure 4.14b) toward DNA and bovine serum albumin (BSA) [66]. All complexes were able to bind to these biomacromolecules. The lowest binding energies were determined for complexes **17c** and **17d**, which contained trivalent metal ions (DNA: −9.62 and − 8.28 kcal mol^{-1}, respectively; BSA: −8.23 and − 10.05 kcal mol^{-1}). It was concluded on the basis of UV/vis absorption/emission titration studies that groove binding rather than intercalation was the main mode of interaction. The cytotoxicity of the complexes against MCF-7 cell was evaluated *in vitro*. Unlike the metal-free ligand, all complexes showed anticancer activity. Thereby, the best performance was observed for **17a** and **17d**, which were competitive to cisplatin, as the reference (Figure 4.14c). Remarkably, the activity against L929 tumor cells was low, thus showing the high selectivity of the potential metallodrugs.

4.3.4 Tpy Complexes of Cobalt Ions

It is well-known that Co-containing complexes represent promising metallo(pro)drugs, *e.g.*, as hypoxic selective agents [69]. In one recent example, a Co^{III} complex acted as a tyrosine-kinase inhibitor of the epidermal growth-factor receptor [70]. Moreover, DNA cleavage has also been realized with Co complexes; the performance of these complexes was comparable to that of related Ru^{II} systems [35]. Against this background, homoleptic Co^{III}–tpy complexes were studied, as potential PDT agents, by Nair *et al.* [71, 72]. These studies revealed that these complexes interacted with DNA mainly by intercalation and that their DNA-cleavage activity arose from the photoexcited state of the Co^{III} centers. At this stage, one has to be reminded of special features of octahedral Co^{III} complexes, such as $[Co(tpy)_2]^{3+}$, which are kinetically stable; whereas the analogous Co^{II} complexes are kinetically labile and are prone to ligand-exchange reactions in solution (see Section 1.3) [73].

Furthermore, the Co^{II} complexes **18** and **19** (Figure 4.16), which were formed with the commercially available 4′-chloro-tpy (Cl-tpy) ligand, were reported by Zhang, Cheng *et al.* [74]. Notably, the formation of **18a** and **19** could be controlled *via* the applied metal-to-ligand ratio, *i.e.*, dinuclear **19** was obtained when using an excess of Co^{II} ions. Unlike the related Cu^{II} complex **20** (Figure 4.16), which was barely cytotoxic, complexes **18** and **19** exhibited distinct anticancer activity, although with

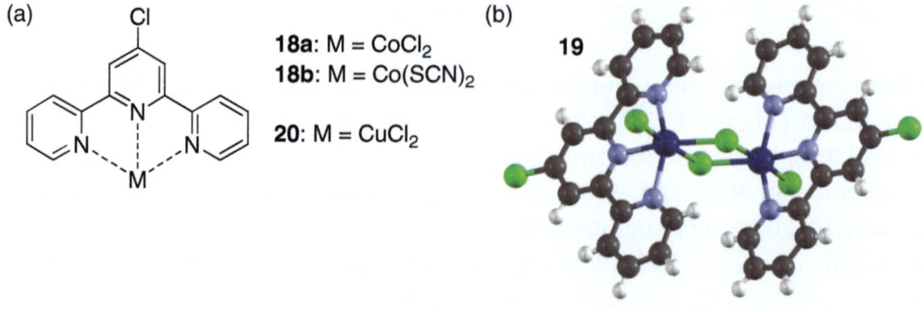

Figure 4.16 (a) Schematic representation of complexes **18** and **20**. (b) Solid-state structure of dinuclear **19** (counterions and solvent molecules were omitted for clarity). Adapted from [74].

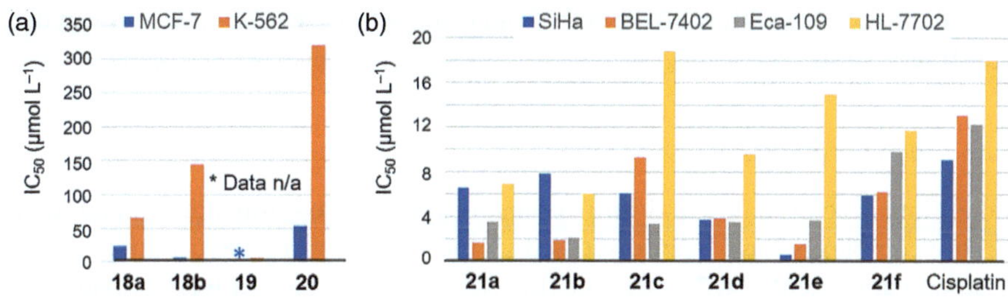

Figure 4.17 (a) Summary of the IC_{50} (μM) values of **18–20**. Adapted from [74]. (b) Summary of the IC_{50} (μM) values of **21**. Adapted from [75].

different selectivity [74]. As shown in Figure 4.17a, the highest cytotoxicity against MCF-7 cancer cells was found for complex **18b**; whereas **19** performed best against human leukemia cells (*i.e.*, K-562). For comparison, the previously reported Cu^{II} complex **20** was by far less cytotoxic. This is a bit surprising since Cu^{II}–tpy complexes typically display substantial anticancer activity (see Section 4.3.6).

4.3.5 Tpy Complexes of Nickel Ions

To the best of our knowledge, Ni^{II}–tpy complexes have first been utilized in the context of cytotoxicity studies very recently. Current interest in Ni complexes is due to the metal's low price and broad abundance, at least when compared to its heavier homologs Pd and Pt. Moreover, complexes of these metals are often less soluble than their Ni-based counterparts. Accordingly, Ni complexes have successfully been used as catalysts for the water-splitting or CO_2-reduction reaction [76]; moreover, their use as emitters represents one further active field of research [77].

Having these recent achievements in mind, Xue, Liu, Ma *et al.* prepared the homoleptic *bis*-complexes **21** (Figure 4.18) and tested their antiproliferative activities against various cancer cell lines: SiHa, BEL-7402, ECA-109, and HL-7702 [75]. Compared to cisplatin, which was used in control experiments, all complexes exhibited a superior cytotoxicity against SiHa, BEL-7420, and Eca-109 cells (Figure 4.17b). Moreover, complex **21a** featured a high inhibition selectivity toward SiHa cells, whilst being less toxic to the HL-7702 cell line. Such behavior is of relevance regarding the development of cell-line specific metallodrug, in particular for the human cervical squamous carcinoma. Spectroscopic studies, supported by molecular-docking simulation, suggested that the complexes interacted with

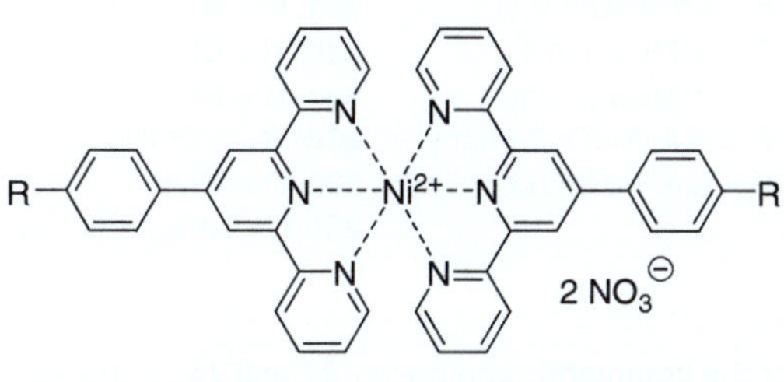

21a: R = H
21b: R = OMe
21c: R = Me
21d: R = F
21e: R = Br
21f: R = I

Figure 4.18 Schematic representation of the Ni^{II} complexes **21**.

DNA *via* embedding with contributions from electrostatic, ππ-stacking, and van der Waals interactions.

4.3.6 Tpy Complexes of Copper Ions

As mentioned beforehand, and unlike most of the first-row transition-metal ions, Cu^{II}–tpy complexes have been discussed in the context of biological applications to a significant extent [34, 35, 78]. The first studies in this respect were provided by Nair *et al.*, who reported the apoptotic activity of mixed-ligand, penta-coordinate complexes from the $[Cu(tpy)(phen)]^{2+}$ family (phen: 1,10-phenanthroline) [79]. Contemporaneously, Rao *et al.* showed that Cu^{II}–tpy complexes were able to cleave plasmid DNA and provide antiproliferative activity against various cancer cell lines [80]. Since then, a broad range of Cu^{II}–tpy complexes has been prepared and evaluated for the biological activity [34, 78]. These complexes can be classified according to their ligands: homoleptic Cu^{II}–tpy *bis*-complexes, Cu^{II}–tpy with a bidentate polypyridyl ligand, Cu^{II}–tpy *mono*-complexes with (pseudo-)halogens or *O*-donors, as the ancillary ligands, as well as oligonuclear complexes.

4.3.6.1 Hexa- and Penta-Coordinate Cu^{II}–Polypyridyl Complexes

The homoleptic Cu^{II} complex **22a** (Figure 4.19) was already reported by Rao *et al.* in 2011 [80]. This complex was identified as a broad-band anticancer agent, which was highly active against various types of cancer cell lines. Thereby, the antiproliferative performance of **22a** was higher than that of carboplatin, which was used as the reference in these experiments. Mechanistic studies revealed that **22a** interacted with DNA by intercalation and groove binding. It was shown further that the complex acted as a redox mediator for thc clcavage of plasmid DNA. A similar behavior in terms of nuclease activity in the presence of H_2O_2 was also reported for complexes **22b-e** (Figure 4.19) [81, 82]. These complexes were similarly cytotoxic as cisplatin against various cancerous cell lines, but did not affect the non-tumorigenic human dermal fibroblast (NHDF) cell line (Figure 4.20a). Experimental and computational binding studies revealed that the complexes acted as intercalators for DNA and were also able to bind to BSA. Follow-up studies on **22b** showed that the complex went inside the cytoplasm and was accumulated in the cells' nuclei [84]. These authors also paid attention to the mechanism of cell death, which occurred by apoptosis *via* a ROS-mediated mitochondrial pathway.

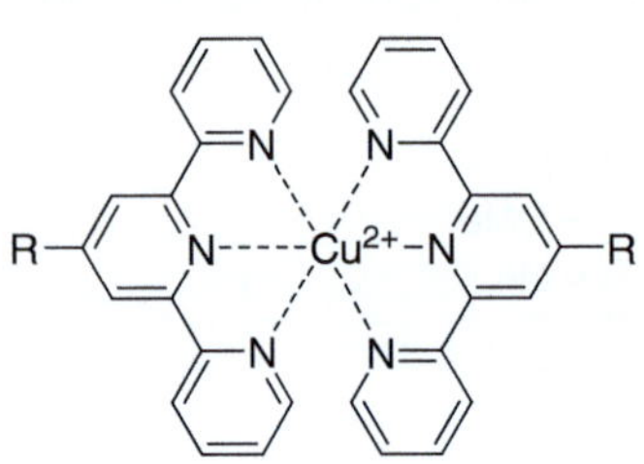

22a: R = anthracen-9-yl
22b: R = 4-Me-phenyl
22c: R = 4-MeO-phenyl
22d: R = 3,4-(MeO=2-phenyl
22e: R = 3,4,5-(MeO)3-phenyl

23a: R = H
23b: R = Cl
23c: R = Br
23d: R = COOH
23e: R = OMe
23f: R = NMe_2

Figure 4.19 Schematic representation of the homoleptic complexes **22** and **23** (counterions were omitted for clarity).

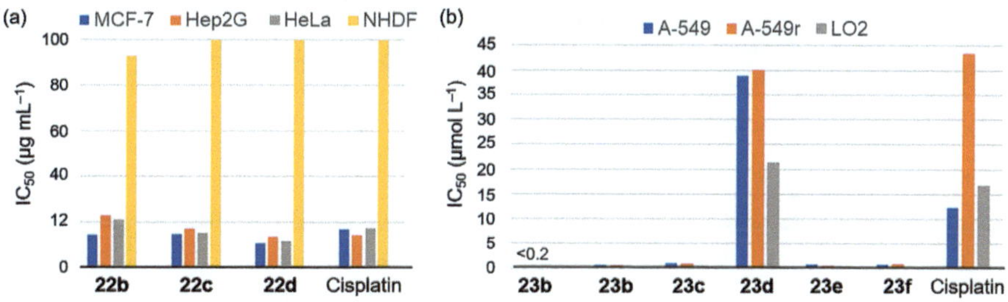

Figure 4.20 (a) Summary of the IC_{50} values of **22**. Adapted from [81]. (b) Summary of the IC_{50} [µM] values of **23**. Adapted from [83].

In vivo experiments with female Swiss albino mice were performed and the tumor growth of Ehrlich ascites carcinoma was inhibited upon administration of **22b**. In a similar fashion, the homoleptic complexes **23** (Figure 4.19) were also screened as anticancer metallodrugs [83]. Except for COOH-substituted **23d**, all complex displayed admirable cytotoxicity in the nM range; whereas cisplatin was an order of magnitude less toxic against the tested cancer cell lines (Figure 4.20b). Due to their poor solubility in an aqueous environment, the corresponding [Cu(R-tpy)Cl_2] *mono*-complexes could not be analyzed regarding their biological activity.

There are plenty of examples for anticancer studies on penta-coordinated Cu^{II} complexes, which contain a tridentate tpy ligand in combination with a bidentate one (*e.g.*, bpy or phen derivatives) [78]. Firstly, the mixed-ligand complexes **24** (Figure 4.21a), which were studied in line with the homoleptic complex **22e**, acted as DNA intercalators and also showed nuclease activity [81]. However, a clear structure–property relationship could not be provided: Although **24a** was more cytotoxic than **24b**, its binding to DNA was weaker. Thus, a multi-target mechanism, which could not be clarified, was postulated to rationalize the observed behavior. In one further study, Pathak, Balamurali *et al.* elaborated the effect of the bidentate ligand on the cytotoxicity of complexes **25**; in this work, dipyrido[3,2-*d*:2′,3′-*f*]quinoxaline (dpq) and dipyrido[3,2-*a* : 2′,3′-*c*]phenazine (dppz) were *inter alia* used (Figure 4.21b) [86]. These bidentate ligands are prominent DNA intercalators and many biologically active complexes bearing either of these ligands have already been prepared [87]. With increasing size of the π-conjugated system, the strength of the interaction with DNA *via* intercalation increased; accordingly, **25d** represented the strongest intercalator and this complex exhibited the highest cytotoxicity within the studied series of complexes (*i.e.*, IC_{50} of 23.8 µg mL^{-1} against HepG2 cells) [86]. Albeit, the anticancer activity of complexes **25** was only moderate when compared to other designated metallodrugs, which are presented in this chapter. Due to its dppz ligand, complex [Cu(tpy)(dppz)]$^{2+}$ (**26**) also efficiently intercalated into DNA [88]. The *in vitro* cytotoxicity study revealed an IC_{50} of 4.57 µM against MCF-7, as the representative cancer cell line.

In continuation of their work on the biological activity of complexes from the [Cu(tpy)(phen)]$^{2+}$ family, Nair *et al.* reported complexes **27** (Figure 4.21c), which both represented good DNA intercalators [89]. The benzimidazol-2-yl moiety of **27b** enabled *H*-bonding with the nucleobases and, thus, its binding to DNA was stronger

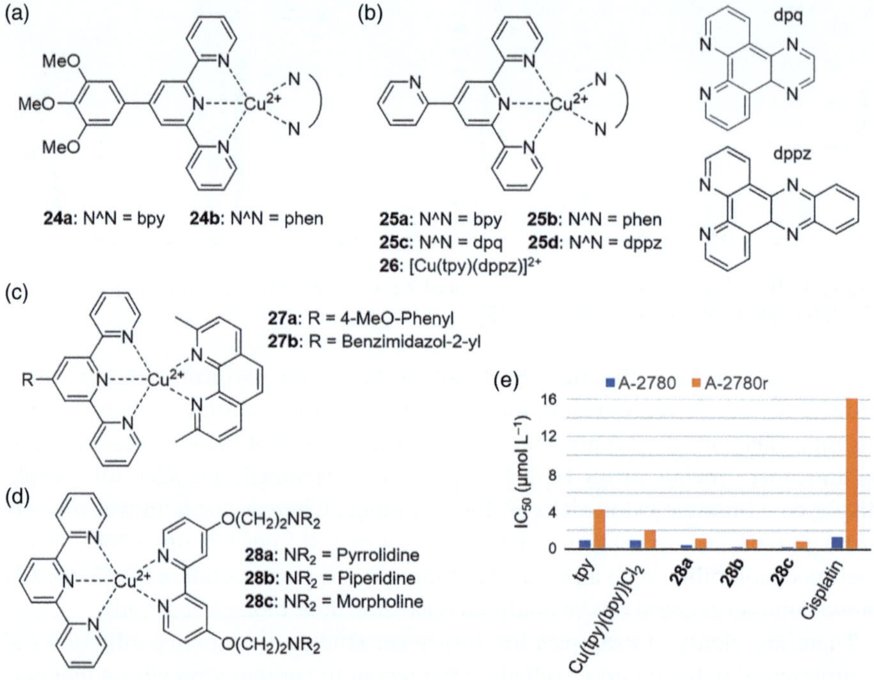

Figure 4.21 (a–d) Schematic representation of the penta-coordinated Cu^{II} complexes **24–28** (counterions were omitted for clarity). (e) Summary of the IC_{50} [μM] values of **28**. Adapted from [85].

than that of **27a**. It was shown further that both complexes were able to induce hydrolytic DNA cleavage in electrophoresis experiments. Moreover, cytotoxicity against the cancerous breast cancer cell line MCF-7 was observed, but the non-cancerous African green monkey's kidney cells (VERO) were hardly affected. Such selectivity, in particular in the case of **27b**, is of great importance for finding potent antiproliferative agents for breast cancer treatment. Cell cycle studies gave further insight into the mechanism of cell death: the complexes induced apoptosis *via* a mitochondrial-mediated pathway. As the last example to be mentioned in the context of mixed-ligand Cu^{II}–polypyridyl complexes, Garra *et al.* presented the complexes **28**, which contained cyclic tertiary amines as the lateral substituents (Figure 4.21d) [85]. These substituents were chosen to improve the electrostatic interaction between the complexes and the DNA's phosphodiester backbone, thereby increasing the nuclease reactivity. Remarkably, the complexes efficiently cleaved the double strands of plasmid DNA – even in the absence of an oxidizing or reducing agent. Moreover, the cytotoxicity against ovarian cancer cells (A-2780 and A-2780r) was higher than that of cisplatin, as the reference metallodrug (Figure 4.21e).

It has been concluded that DNA is targeted by all penta-coordinate Cu^{II}–polypyridyl complexes [78]. Thereby, the higher biological activity is obtained if rigid bidentate ligands are used since these represent better intercalators than bpy-based ones. Whereas the influence of substituents on the tpy ligand is less predictable and still less understood.

4.3.6.2 Cu^{II}–Tpy *Mono*-Complexes with "Inorganic" Ligands

The Cu^{II}–tpy *mono*-complexes, which carry charge-compensating "inorganic" ligands (*e.g.*, halides, pseudo-halides, nitrate, *etc.*), represent the second class of complexes to deal with in this section. As mentioned beforehand, the dichloride *mono*-complexes, which corresponded to the *bis*-complexes **23** (Figure 4.19), could not be tested regarding their biological activity due to their poor solubility [90]. Contrary to this, the *mono*-complexes **29** (Figure 4.22a), which contained Kröhnke-type tpy ligands, were more soluble and, thus, their anticancer activity was evaluated [90]. Throughout the series of complexes, significantly increased antiproliferative activity (compared to that of cisplatin) against various human carcinoma cell lines was observed in *in vitro* experiments (Figure 4.22b). However, due to the only small differences of the deduced IC_{50} values, a clear structure–property relationship could not be established. Spectrophotometric results indicated strong interactions with DNA, mainly by intercalation, which gave rise to considerable changes in the DNA's conformation. This, in turn, disturbed DNA replication and transcription. According to computational studies, the interaction was due to an interplay of non-covalent forces: $\pi\pi$-stacking, van-der-Waals, and hydrophobic interactions as well as *H*-bonding. The latter was identified as the major contribution for **29c–f** and **29k**. In particular, **29k** revealed the lowest binding energy for the interaction with biomacromolecules (*i.e.*, DNA, oligonucleotides, etc.).

Moreover, Chao *et al.* analyzed how the position and number of substituents on the tpy ligand influenced the anticancer activity. For this purpose, the DNA binding and cytotoxicity of **29f** was compared to that of **29l,m** [91]. This study revealed that

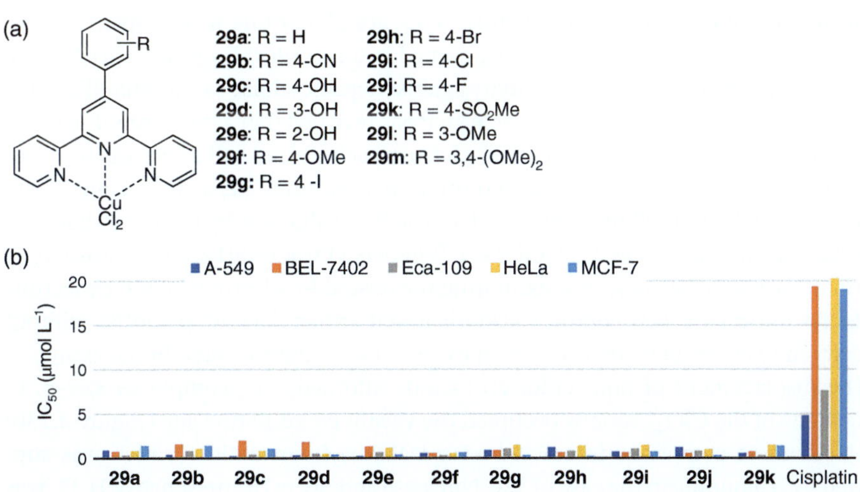

Figure 4.22 (a) Schematic representation of the Cu^{II}–tpy *mono*-complexes **29**. (b) Summary of the cytotoxic activity of complexes **29** against various cancer cell lines. Adapted from [90].

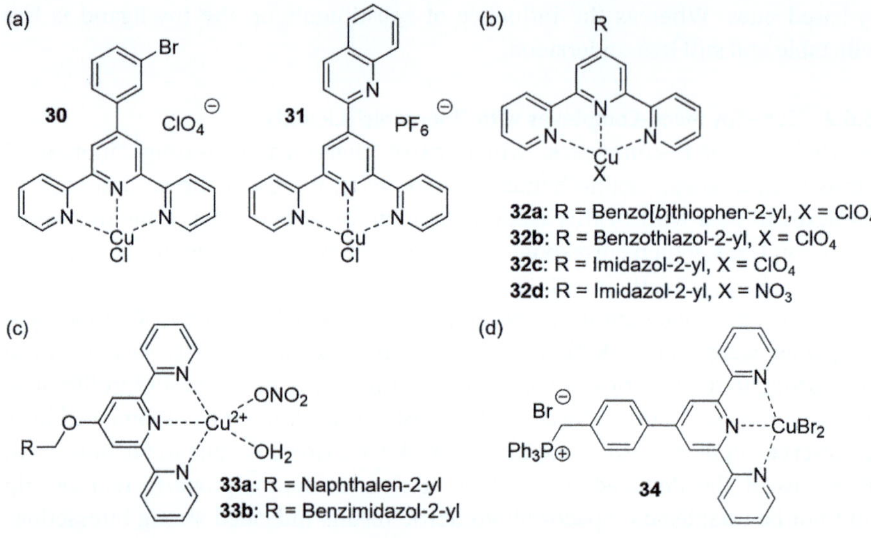

Figure 4.23 Schematic representation of the Cu^{II}–tpy *mono*-complexes **30–34** (counterions partially omitted for clarity).

strong binding to DNA does not necessarily imply high biological activity. Although the interaction of **29f** with DNA was the strongest, its nuclease activity, which was attributed to the generation of ROS in the presence of glutathione, was the lowest. Nonetheless, all three complexes were more potent metallodrugs than cisplatin against the three tested cancer cell lines (*i.e.*, HeLa, HepG2, and BEL-7402).

Attempts have also been made to modulate the biological activity by replacing one of the coordinated Cl^- anions by a non-coordinating one (*e.g.*, ClO_4^-, PF_6^-, BF_4^-). This substitution resulted in a change of the coordination geometry: Instead of the distorted trigonal-bipyramidal structure, a square-planar one was adopted. Representatively, the complexes **30** and **31** are named in this context (Figure 4.23a) [92, 93]. Complex **30** featured a very high activity against hepatocellular carcinoma BEL-7402 cells (IC_{50} of *ca.* 0.90 μM) [92]. Cell cycle analysis revealed that activity was due to the blocking of the cell cycle in the G0/G1 phase, which caused cell apoptosis by mitochondrial dysfunction. In the case of quinoline-substituted **31**, distinct cytotoxicity against the HCT-116 cell line was detected (the IC_{50} value was by far lower than that offered by cisplatin: 0.21 *vs.* 15.4 μM) [93]. Whereas **31** was only barely cytotoxicity in normal dermal fibroblasts, thus exhibiting the desirable selectivity to kill cancerous cells. As important, this complex also displayed antiangiogenic potential without toxicity in *in vivo* experiments using an ex-ovo chorioallantoic membrane assay.

The displacement of both chlorido ligands afforded, *e.g.*, complexes **32a–c**, in which one of the ClO_4^- anions occupied the vacant coordination site (Figure 4.23b) [94]. All three complexes interacted with DNA by intercalation, which was supported by *H*-bonding in the case of **32c**. Nuclease activity in the presence of H_2O_2 was also reported. Thereby, the oxidative stress due to the formed ROS affected cancerous cells (*e.g.*, Hep2G and CAL-51) much more strongly than healthy cells. Since **32d** possessed a similar binding strength and nuclease activity than **32c**, it was concluded

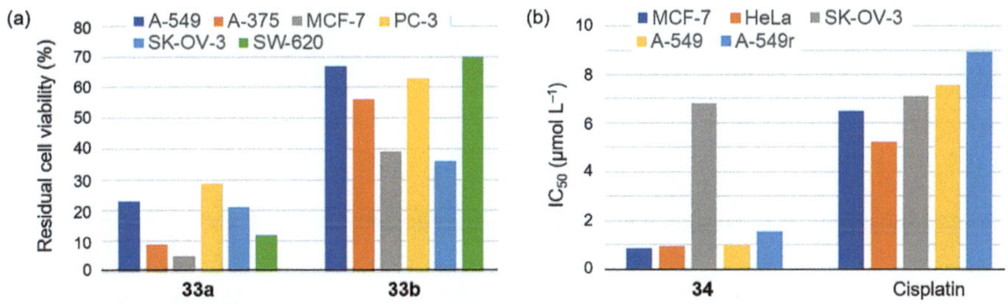

Figure 4.24 (a) Summary of the viability of cancer cells (%) after an incubation time of 24 h in the presence of **33**. Adapted from [95]. (b) Summary of the IC_{50} [μM] values of **34**. Adapted from [96].

that the counterion (coordinated or non-coordinated) had only a marginal influence of the anticancer performance of these complexes; whereas this behavior is mainly governed by the structure of the tpy ligand, *i.e.*, its substituent in 4′-position [78]. Representatively, the nitrate complexes **33** (Figure 4.23c) have to be highlighted in this context. Despite their (at first glance) similar chemical structure, their biological activity differed significantly: Naphthyl-equipped **33a** was the stronger intercalator than **33b** and was also more cytotoxic (Figure 4.24a) [95]. However, this behavior could not be correlated to the nuclease activity, which was higher for **33b**. In other words, metallodrugs, which are capable of cleaving DNA, must not necessarily be harmful to cells. Without going further into the details here, other Cu^{II}–tpy *mono*-complexes with nitrate, as the counterion, have also been reported [34, 78].

Finally, Zhen, Wang *et al.* reported the Cu^{II} complex **34**, which contained a targeting and lipophilic phosphonium moiety (Figure 4.23d), which enabled the crossing of the cytoplasmic and mitochondrial membranes of tumor cells [96]. Detailed mechanistic studies corroborated that apoptosis of cancer cells occurred *via* a mitochondrial pathway involving the collapse of the MMP, generation of ROS, Bax mitochondrial aggregation as well as release of cytochrome-c. This behavior accounted for the remarkably high cytotoxicity of **34** against various types of cancer cells (Figure 4.24b). It was emphasized that such mitochondria-selective complexes might possess a superior anticancer effect than conventional Pt^{II}-based metallodrugs (*e.g.*, cisplatin).

4.3.6.3 Cu^{II}–Tpy *Mono*-Complexes with Oxygen-Donor Ligands

In order to modulate the biological activity of Cu^{II}–tpy *mono*-complexes, Ma, Fernandes *et al.* explored the use of anionic oxygen-donor ligands, specifically aromatic sulfonates and carboxylates (**35** and **36**, Figure 4.25a) [97]. Cytotoxicity tests revealed that all four complexes caused apoptosis in HepG2 and HCT-116 cell lines; thereby, the cationic complex **36** featured the highest activity (Figure 4.25b). Beneficially, the cytotoxicity against noncancerous cell lines, *i.e.*, normal human fibroblasts (NHFs), was significantly lower. As one further interesting aspect, the complexes intercalated into DNA and induced the breakdown of the DNA double-strand helix. Since this occurred in the absence of an oxidant, a hydrolytic pathway for DNA cleavage was assumed.

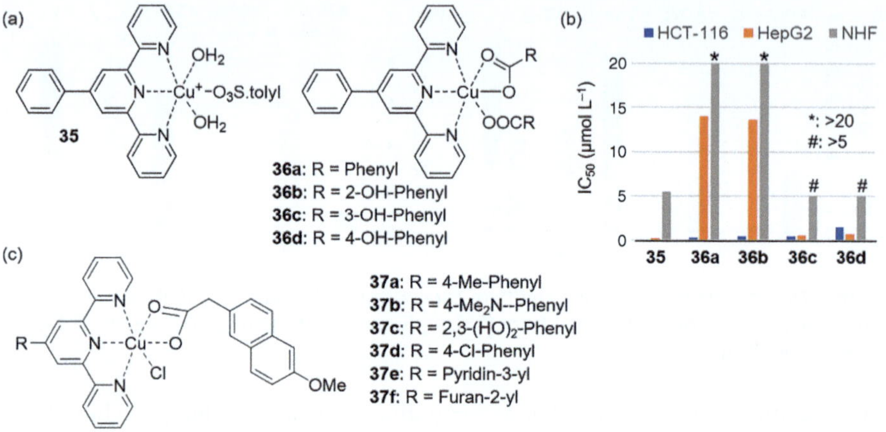

Figure 4.25 (a) Schematic representation of the Cu^{II} complexes **35** and **36** (counterions were partially omitted for clarity). (b) Summary of the IC_{50} [μM] values of **35** and **36**. Adapted from [97]. (c) Schematic representation of the Cu^{II} complexes **37**.

In one further study on complexes with carboxylates, as the ancillary ligand, Rahiman *et al.* prepared the complexes **37**, which varied in the substituent in 4′-position (Figure 4.25c) [98]. In this study, naproxen, a nonsteroidal anti-inflammatory drug, was used as the co-ligand. From the series of complexes, three examples (*i.e.*, **37a**, **37e**, and **37f**) were analyzed further regarding their binding to DNA and antiproliferative activity. These complexes interacted with DNA by groove binding. The strongest interaction was observed for **37e**, presumably due to *H*-bonding *via* the pyridyl substituent. This complex also provided the highest cytotoxicity against the MCF-7 cell line and outcompeted cisplatin in this respect. According to mechanistic studies, cytotoxicity was caused by mitochondrial-mediated and caspase-dependent apoptosis.

Patel *et al.* reported two series of complexes in which fluoroquinolone antibiotics, namely norfloxacin or ciprofloxacin, were used as bidentate *O*-donor ligands (**38** and **39**, respectively: Figure 4.26) [99, 100]. Firstly, the antibacterial activity of **38** against specific $Gram^{(+ve)}$ and $Gram^{(-ve)}$ microorganisms was screened (Figure 4.27) [99]. All complexes exhibited a significantly higher antimicrobial activity than $CuCl_2$

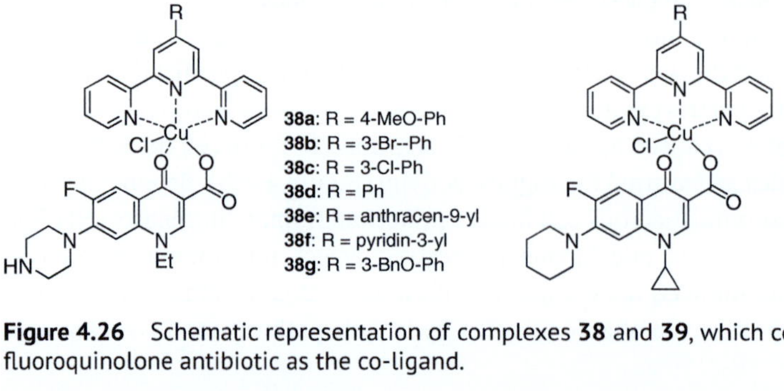

Figure 4.26 Schematic representation of complexes **38** and **39**, which contained a fluoroquinolone antibiotic as the co-ligand.

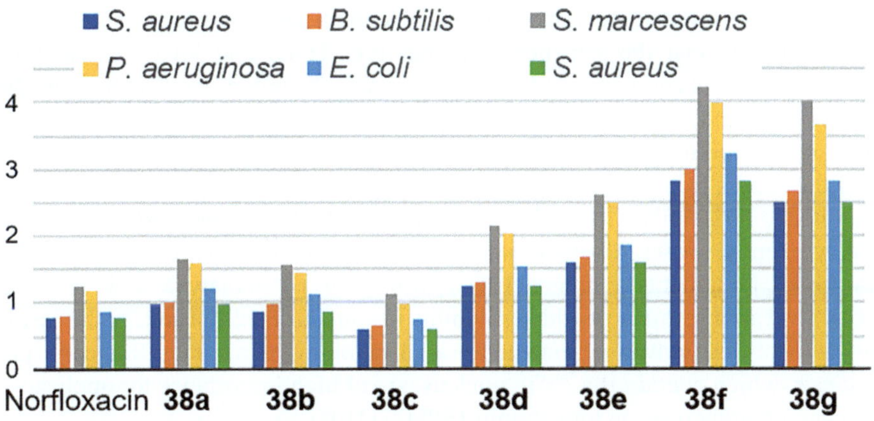

Figure 4.27 Summary of the IC_{50} [μM] values of **38** (*Staphylococcus aureus: S. aureus; Bacillus subtilis: B. subtilis; Serratia marcescens: S. marcescens, Escherichia coli: E. coli; Pseudomonas aeruginosa: P. aeruginosa*). Adapted from [99].

alone. Remarkably, the activity of **38c** was even higher than that of the antibiotic itself. Binding studies revealed that the complexes acted as DNA intercalators; the highest binding affinity was found for **38c**. This complex also displayed the highest cytotoxicity and SOD-like activity within the series, *i.e.*, the ability to scavenge oxygen radicals under physiological conditions (SOD: superoxide dismutase). Similar studies were also performed with complexes **39** [100]. In this case, the order of activity was **39a** > **39b** > **39c** > **39d**. This held true for the antibacterial activity, binding to DNA, nuclease activity as well as scavenging of oxygen radicals.

Banerjee, Saikia, Deka, Hussain *et al.* studied complexes, which contained *O,O*-donor ligands: plumbagin (**40a/d**), chrysin (**40b/e**), and curcumin (**40c/f**, Figure 4.28) [101]. These co-ligands were chosen due to their use as anticancer agents in traditional Chinese medicine. Presumably, due to the presence of

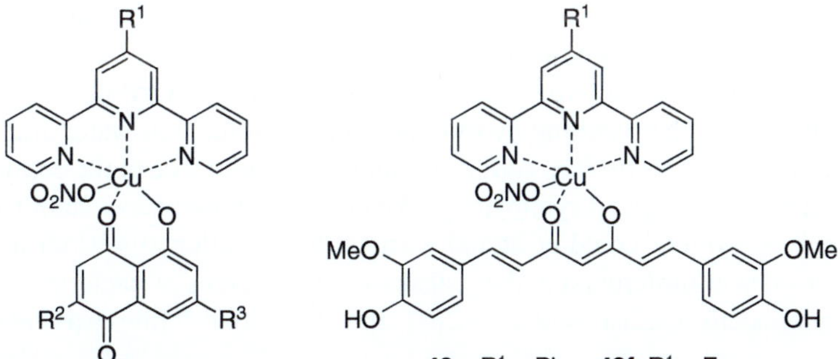

40a: R¹ = Ph, R² = Me, R³ = H
40b: R¹ = Ph, R² = Ph, R³ = OH
40a: R¹ = Fc, R² = Me, R³ = H
40b: R¹ = Fc, R² = Ph, R³ = OH

40c: R¹ = Ph **40f**: R¹ = Fc

Figure 4.28 Schematic representation of complexes **40**.

the redox-active ferrocene (Fc) moiety, the complexes **40d–f** were significantly more active against the screened cell lines (*i.e.*, HeLa and MCF-7) than their phenyl-containing counterparts **40a–c** or the organic ligands alone. The complex **40f** was emissive and, thus, its uptake could be monitored by fluorescence-microscopy imaging. The preferential location in the mitochondria of HeLa cells, in concert with the results from mechanistic studies, suggested that complexes **40** also acted *via* mitochondrial apoptosis, which was caused by the formation of ROS.

Very recently, Gasser, Alberto *et al.* reported on a dinuclear complex, in which two $[Cu(tpy)Cl_2]$ fragments were linked *via* a $[M(\eta^6\text{-arene})_2]^+$ moiety (M = Re or ^{99m}Tc) [102]. The cationic linker provided solubility in aqueous medium and, in the case of the radio-labeled derivative, enabled the application in theranostics. Radiopharmaceuticals, which contain the ^{99m}Tc nucleus, are of high importance for application in single-photon emission tomography (SPECT) [103].

4.3.6.4 Miscellaneous Cu^{II}–Tpy *Mono*-Complexes Including Oligonuclear Ones

As mentioned beforehand, the trinuclear Cu^{II} complex **2** (Figure 4.4) featured a remarkably high cytotoxicity and nuclease activity [33]. This work highlights the use of oligonuclear complexes in the context of biological studies. Further examples for similar assemblies were summarized by Gil-Moles and Gimeno in their recent review [78]. Beyond this, the coordination of Cu^{II}–tpy *mono*-complexes to biomacromolecules has also been reported.

Hassan *et al.* prepared a hybrid material in which Cu^{II}–tpy *mono*-complexes were bound to carboxymethyl cellulose (CMC) *via* the biopolymer's carboxy moieties [104]. Besides the coordination of Cu^{II} complexes along the CMC chains (Figure 4.29a), metallo-supramolecular linking *via* COOH groups of two different CMC strands was also observed. Thus, a partially cross-linked material was obtained. This water-soluble material was not only studied regarding its physical and mechanical properties but also the antimicrobial activity was the focus of interest. At the minimum concentration of 6–8 $mg\,L^{-1}$, high activity against various Gram-positive bacteria was observed; in all cases, the microbial growth was inhibited to at least 90%. Such antimicrobial hybrid materials might be used for the packing of foods, as pharmaceuticals or medicinal tissues. Related to this, Hassan *et al.* also grafted Cu^{II}–tpy complexes to pectin, as one further biopolymer (Figure 4.29b) [105]. The resulting hybrid material was soluble in water and combined the properties of pectin (*i.e.*, acting as emulsifier) with those of the complexes (*i.e.*, antimicrobial activity and cytotoxicity). An oil-in-water emulsion, stabilized by the "metallobiopolymer," could be stored without deterioration at 4 °C for at least 3 months. Besides antimicrobial activity against various types of bacterial strains, moderate cytotoxicity against HepG2, MCF-7, and HCT-116 cell lines was detected.

4.3.7 Tpy Complexes of Zinc Ions

Complexes with Zn^{II} ions have gained significant attention, as designated metallodrugs, for a number of reasons. Firstly, Zn^{II} ions are essential for humans and play a prominent physiological role; supplementation of Zn^{II} is generally considered

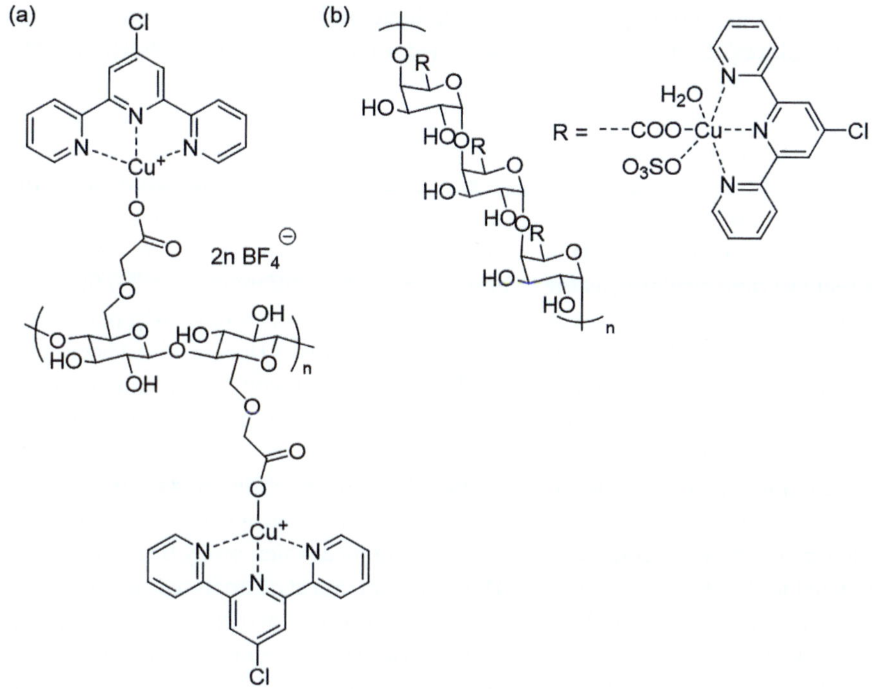

Figure 4.29 Schematic representation of CMC (a) and pectin (b), to which Cu^{II}–tpy complexes have been coordinated. For clarity, for each, only one representative structure is shown.

beneficial in cancer treatment. Due to restricted side effects, even at higher doses, Zn^{II}-based drugs are advantageous over those that contain metal ions, which are handled less well by human physiology [106]. For comparison, many other transition-metal ions (*e.g.*, Fe^{II}, Cu^{II}, Ru^{II}, Pt^{II}, Hg^{II}, *etc.*) are less biocompatible or even toxic at high doses or upon prolonged administration. Next, the mode of action of Zn^{II}-based anticancer drugs is fundamentally different from that of common Pt^{II}-based systems or those metallodrugs, which contain redox-active metal centers [107]. Due to the widespread use of Zn^{II}–porphyrin derivatives in the context of PDT, the role of Zn^{II}-containing photosensitizers in biological systems is largely understood [108]. Finally, the Lewis acidity and fast ligand-exchange processes facilitate *inter alia* the catalytic hydrolysis and DNA cleavage, which represent prerequisites to function as anticancer drugs [109, 110]. Taking these considerations into account, one can understand the enormous interest in research on Zn^{II}-containing compounds/complexes for application in biomedicine. In this context, the Zn^{II} complexes, which contain nitrogen-donor ligands, such as those from the polypyridyl family, have recently been summarized by Porchia *et al.* [111]. These authors pointed to the potential anticancer activity of Zn^{II}–tpy complexes, which has also been elaborated in our previous review on this topic [35].

In one recent study, Pan *et al.* focused on the antiproliferative activity of the Zn^{II}–tpy *mono*-complexes **41** (Figure 4.30) [112]. Thereby, the substituents were varied to modulate the lipophilicity on one hand and steric demand on the other. The complexes' fluorescence was quenched upon interaction with DNA. MM studies suggested that the strong interaction was due to intercalation. This, in turn, induced conformational changes and prevented replication and transcription. Notably, the highest preference for intercalation was found for the ATGC base sequences of the DNA strands, where ππ-stacking and *H*-bonding were most efficient. The complexes were tested against four cell lines regarding their anticancer activity. The complexes were more cytotoxic throughout (*i.e.*, IC_{50} values in the low μM regime) than cisplatin, as the reference, and subtle differences arising from the lateral substituents were observed (Figure 4.31a). Notably, complex **41g** performed best against both MCF-7 and Eca-109 cells. Derivative **41k** was also tested as a fluorescent probe for the detection of HClO in an aqueous environment [114]. Fluorescence quenching was presumably due to the oxidation of the *S*-center, thus affording nonfluorescent **41j**. The sensing of HClO occurred within a few seconds and with a low detection limit of 0.16 mM. The sensing ability of **41k** was further used for bio-imaging, *i.e.*, the detection of exogenous HClO in HeLa cells. The depletion of emissions could be monitored by confocal-microscopy imaging.

Related to the study on complexes **41**, Chen, Pan, Ma *et al.* analyzed the influence of halogen substituents and ligands on the antitumor activity of complexes **42** (Figure 4.30) [115]. This study revealed a distinct dependency on the nature of the halogen in terms of electronegativity (EN) and lipophilicity: The complexes, which contained bromido ligands, were generally more lipophilic than their iodido counterparts. Whereas the halogen's EN influenced the antiproliferative activity of the complexes. The highest K_a values for binding to DNA were found for the F-substituted complexes, which also exhibited the highest antitumor activity against MCF-7 and BEL-7420 cells; whereas the lowest K_a values and cytotoxicity were determined for the iodo-substituted derivatives.

Interestingly, the parent complex $[Zn(tpy)Cl_2]$ also featured significant cytotoxicity against MDA-MB-231 and HCT-116 after incubation for 72 and 24 h, respectively [116]. The anticancer activity of this complex was higher than that of $[Zn(en)Cl_2]$ (en: ethylenediamine). This difference was ascribed to the solution-stable square-pyramidal structure of $[Zn(tpy)Cl_2]$, which strongly interacted with guanosine-5′-monophosphate, as the model substrate, and could also

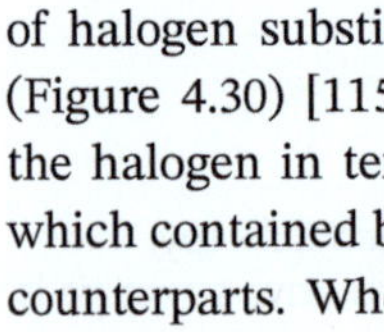

41a: R = H
41b: R = 4-Me
41c: R = 4-OMe
41d: R = 4-Ph
41e: R = 4-(4-Me-Ph)
41f: R = 4-OH
41g: R = 3-OI
41h: R = 2-OH
41i: R = 4-COOH
41j: R = 4-OSO_2Me
41k: R = 4-SMe
(41a–41k: X = Cl)

42a: R = 4-F, X = Br
42b: R = 4-F, X = I
42c: R = 4-Cl, X = Br
42d: R = 4-Cl, X = I
42e: R = 4-Br, X = Br
42f: R = 4-Br, X = I
42g: R = 4-I, X = Br
42h: R = 4-I, X = I

43a: R = 4-OH, X = Br
43b: R = 3-OH, X = B
43c: R = 2-OH, X = Br
43d: R = 4-Cl, X = I
43e: R = 3-OH, X = I
43f: R = 2-OHr, X = I

Figure 4.30 Schematic representation of the Zn^{II}–tpy *mono*-complexes **41–43**.

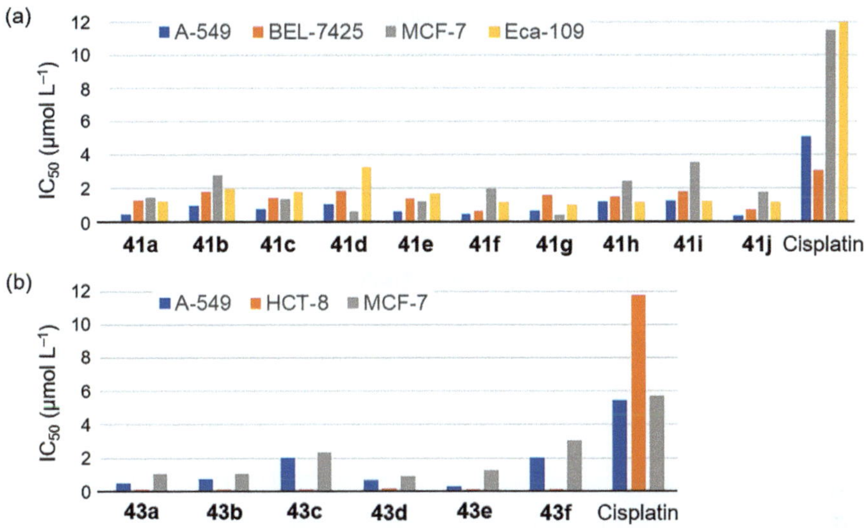

Figure 4.31 (a) Summary of the cytotoxic activity of (a) complexes **41** and (b) complexes **43** against various cancer cell lines. Adapted from [112, 113].

potentially act as intercalator; whereas tetrahedral [Zn(en)Cl_2] was not stable in solution and the interaction with the target molecule was weak.

The square-planar coordination geometry was also observed for complexes **43** (Figure 4.30) [113]. In all cases, a higher cytotoxicity, compared to that of cisplatin, against various cancerous cell lines was observed (Figure 4.31b). This high antitumor activity was mainly ascribed to the presence of the OH moieties, which facilitated *H*-bonding to biomolecules. This assumption was further corroborated by molecular-docking simulations, which suggest that binding occurred spontaneously, at least in the case of hemoglobin proteins, *via* an interplay of *H*-bonding, van der Waals and ππ-stacking interactions.

The Zn^{II} complex **44** (Figure 4.32a), to which lipoic acid, as an antioxidant, was conjugated, strongly interacted with DNA by groove binding and intercalation [117]. However, due to the absence of redox activity, this complex did not show nuclease activity and only very poor cytotoxicity. For comparison, the analogous Cu^{II} complex efficiently cleaved DNA and was cytotoxic against MCF-7 and HeLa cells.

Busto, Jalón, Manzano, García *et al.* targeted the interaction of Zn^{II} complexes with G-quadruplex (G4) DNA [118]. The development of anticancer drugs, which selectively interact with this guanine-rich domain, represents a highly active field of research [119]. To the best of our knowledge, the first Zn^{II}–tpy complex, which selectively bound to G4 DNA, was reported in 2016 (**45**, Figure 4.32a) [120]. This complex was cytotoxic against BEL-7404, SK-OV-3, MGC80-3, and T-24 cell lines with IC_{50} values in the low μM regime. In line with this study, the complexes **46–48** (Figure 4.32c) were tested regarding their G4 DNA affinity and cytotoxicity [118]. The chloride complexes were more active against SW-480 cells, thus showing the important influence of the co-ligand. Whereas the structure of the tpy ligand played a crucial role regarding the interaction with G4 DNA, in particular, human telomeric

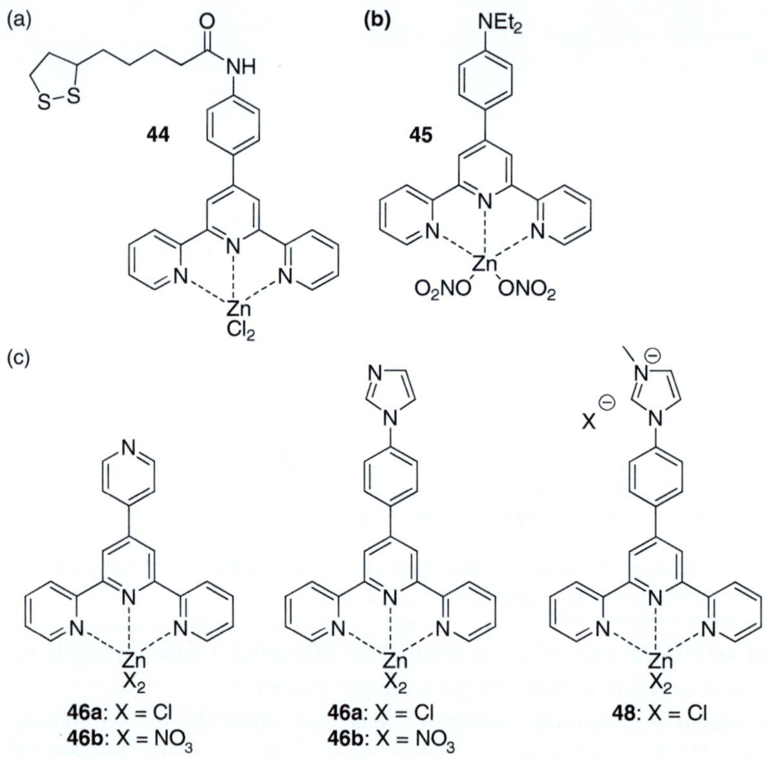

Figure 4.32 (a–c) Schematic representation of the Zn^{II}–tpy mono-complexes **44–48**.

G4 (Tel22). Presumably, due to additional positive charge, the strongest interaction was observed for the imidazolium derivative **48**. The cellular internalization revealed that complex **48** was mainly located in the nuclei of the SW-480 cells.

Similar to the aforementioned Cu^{II} complexes, the use of carboxylates as the co-ligands has also been explored. For example, the imadol-2-yl-subsituted complex **49** (Figure 4.33) revealed an intercalative binding mode with DNA [121]. Unlike the corresponding Cu^{II} complex, which was able to oxidatively cleave DNA, **49** cleaved DNA hydrolytically. This behavior accounted for the observed cytotoxicity against A-549 lung-cancer cells (IC_{50} = 5.83 μM). Moreover, this complex induced considerable conformational changes upon binding to proteins, such as BSA. More recently, Zhou, Pan, Chen, Ma *et al.* reported the complexes **50**, which contained benzoate co-ligands (Figure 4.33) [122]. In order to modulate the complexes' photo-physical and biological properties, the substituents on the Kröhnke-type tpy ligands were varied. The complexes were photoluminescent in the solid state and in DMSO solution. Since the UV/vis absorption/emission characteristics were retained over several days, a high kinetic stability in solution was assumed. Their cytotoxicity was screened *in vitro* against three different cell lines. As summarized in Figure 4.34a, all complexes outcompeted cisplatin in terms of antiproliferative properties. In particular, **50c** revealed the highest activity, thus showing once more the remarkable

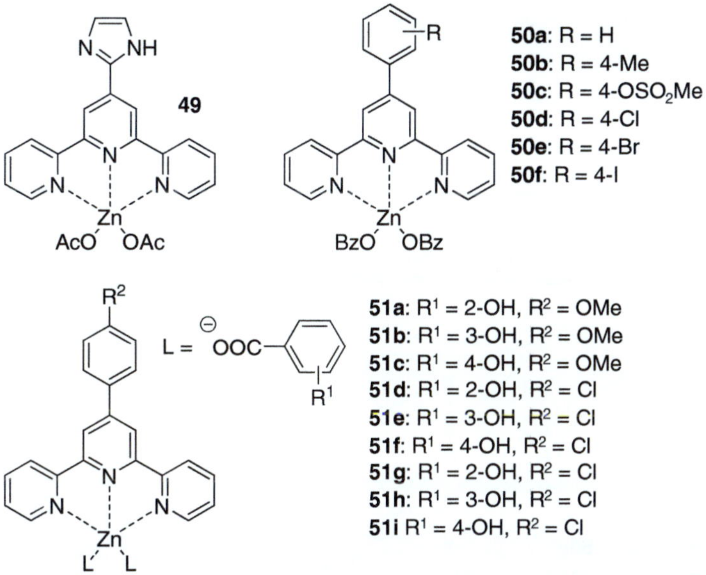

Figure 4.33 Schematic representation of the Zn^{II} complexes **49–51**, which contained carboxylates as the co-ligands (counterions were partially omitted for clarity).

influence of the mesylate substituent. Contrary to the aforementioned trend, the iodo-substituted complex **50f** was more potent against A-549 and MCF-7 cell lines than **50d/e**. It was assumed that the benzoate co-ligands had a regulatory effect on the cytotoxicity, which overcame the influence of EN. The growth-inhibitory effects of complexes **50** on the A549 cell line are shown in Figure 4.34b. The binding of **50** to different DNA motives was analyzed thoroughly by UV/vis absorption/emission titration, CD spectroscopy, and molecular-docking simulations. These experiments suggested that the complexes intercalated into the DNA bases, mainly driven by ππ-stacking interactions, whereas in the case of G4 DNA, electrostatic binding was prevalent. This study represents a significant achievement in understanding the DNA binding and activity of Zn^{II}–tpy complexes to DNA and, thus, development of potent anticancer metallodrugs.

In one follow-up study, these same authors reported that the complexes **51**, which contained hydroxybenzoate co-ligands, provided antiproliferative and antibacterial activity (Figure 4.33) [123]. The complexes exhibited a very high cytotoxicity against three different cancerous cell lines (*i.e.*, Eca-109, BEL-7402, and MCF-7), but were less harmful to human normal liver cells (HL-7702; Figure 4.35). The binding behavior toward double-stranded and G4 DNA was similar to that of complexes **50** (*vide supra*). However, in the case of the former, a distinct preference of the complexes to intercalate between GCGC base pairs was observed. The activity against various bacterial strains was also screened. The best antimicrobial performance was registered for **51b** against *S. aureus*; however, the minimum bacteriostatic concentration of 20 μg mL^{-1} was significantly lower than that of enrofloxacin, as a common bactericidal agent (*i.e.*, 0.125 μg mL^{-1}).

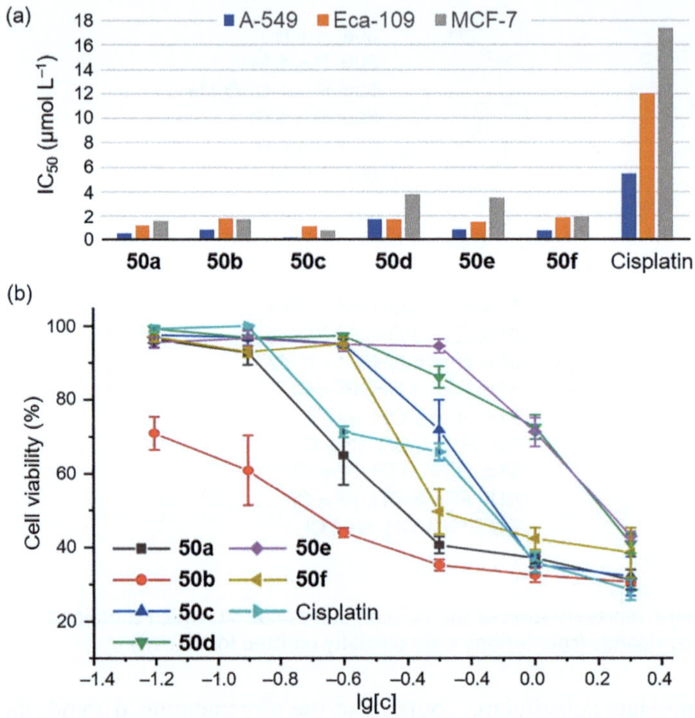

Figure 4.34 (a) Summary of the IC_{50} [μM] values of **50**. Adapted from [122]. (b) Representation of the growth-inhibitory effects of complexes **50** on the A549 cell line [122] / with permission of Springer Nature.

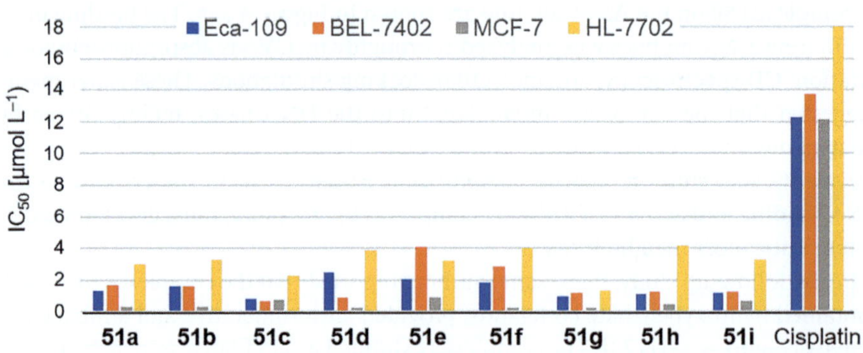

Figure 4.35 Summary of the cytotoxic activity of complexes **51** against various cancer cell lines. Adapted from [123].

Zn^{II}–tpy complexes have also been widely used for bio-imaging [34]. Confocal-microscopy imaging is commonly used to detect the emission of lumiphores, which have been taken up by cells. The complexes, which are used in this context, have to be photoactive (*i.e.*, bright emission of visible light upon excitation) and photostable in the physiological environment. In order to employ such complexes solely for diagnostic and biological purposes, low cytotoxicity might be beneficial.

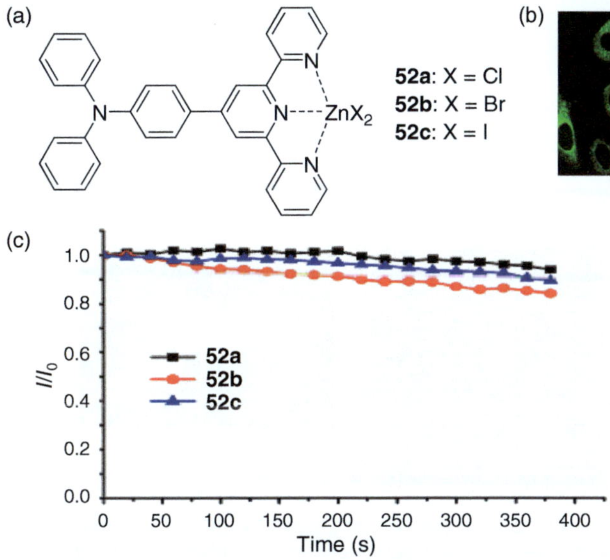

Figure 4.36 (a) Schematic representation of complexes **52**. (b) Confocal-microscopy image of living HeLa cells after incubation with **52c** for 30 min. (c) Representation of the photostability of complexes **52** [124] / with permission of Elsevier.

In this respect, the fluorescent TPA-substituted Zn^{II} complexes **52** are worth mentioning (TPA: triphenylamine; Figure 4.36a) [124]. These complexes were barely cytotoxic against various cancer cell lines, *i.e.*, the residual cell viability was >80%. However, they displayed good membrane permeability and photostability, even upon prolonged illumination (Figure 4.3c). Fluorescence microscopy revealed that the complexes were rapidly taken up and located in the cytosolic space of living HeLa cells (Figure 4.3c). These results highlight the potential Zn^{II}–tpy complexes for application in long-term as well as real-time bio-imaging.

As mentioned beforehand in the context of the Mn^{II} complex **10a** (Figure 4.12), STED nanoscopy represents an important technique to visualize the ultrastructure of cells with super-resolution [125]. X. Tian, Y. Tian *et al.* prepared the homoleptic Zn^{II} complexes **53**, which were brightly fluorescent (Figure 4.37a) [126]. Specific binding of **53a** to mitochondrial DNA by intercalation was observed in *in vitro* and *in vivo* experiments. This behavior, in concert with its photophysical properties, **53a** was used for the visualization of mtDNA in living cells by STED nanoscopy. The achieved super-resolution, which was by far superior to that provided by confocal-microscopy imaging, revealed the distribution of mtDNA within the mitochondrial cristae and inner matrix (Figure 4.37b).

The previous two examples have been selected to highlight the enormous potential of Zn^{II}–tpy complexes for use in *in vitro*/*in vivo* bio-imaging. Further examples in this respect can be found in Velmathi's review on this topic [34]. Although the details on the modes of action are not yet fully clear, cautious structural design allows one to employ Zn^{II}–tpy metallodrugs for therapeutic and/or diagnostic purposes.

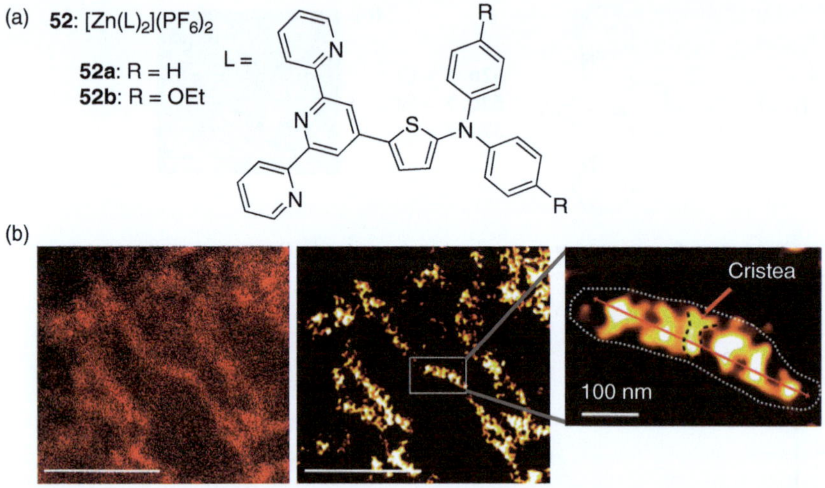

Figure 4.37 (a) Schematic representation of the Zn^{II} complexes **52**. (b) Left and middle: Confocal and STED microscopy images of living HepG2 cells after incubation with **52a** (unknown scaling); right: enlargement of the STED microscopy image, showing a single mitochondrion [126] / with permission of Royal Society of Chemistry.

4.4 Concluding Remarks

Terpyridine complexes have been intensively studied regarding their biological activity. This field of research emerged in 1970s when Lippard *et al.* showed that square-planar Pt^{II}–tpy complexes could act as DNA intercalators, similar to organic molecules, such as EthBr [24, 25]. This seminal work paved the way for intense research on such complexes, as metallodrugs, for use in therapy and/or diagnostics. Initially, complexes of Pt^{II} and Ru^{II} ions were the focus of interest [28, 127]. The latter also exhibit nuclease activity, which is due to the redox-active nature of the metal ion [128]. This property, which can be induced photochemically, is of significant importance in PDT [129]. However, due to their poor natural abundance, the use of metallodrugs *inter alia* based on Pt^{II} or Ru^{II} ions is cost-intensive and, presumptively, not future-oriented. This aspect runs like a thread through this book and is addressed over and over again in the subsequent chapters. Consequently, metallodrugs, which contain earth-abundant metal ions, mainly of the 3d-block metals, have gained significant importance in recent years [34, 35, 40].

Following up on the corresponding Chapter 7 in the 2nd edition of this textbook [36], the biological activity of tpy complexes of the 1st-row transition-metal ion has been evaluated. In particular, the complexes of Cu^{II} and Zn^{II} ions have received considerable recognition over the last decade. It has been shown, for example, that the selectivity in anticancer treatment can be modulated *via* substituents on the tpy ligand, the choice of the metal ion and auxiliary ligand(s) as well as the counterions. Most tpy complexes interact with DNA by intercalation and their cytotoxicity levels are in the low μM to nM regime. In the case of redox-active metal centers, such as Cu^{II} or Fe^{II}, the formation of ROS accounted for their increased nuclease activity.

Moreover, luminescent tpy complexes, in particular the Zn^{II}-containing ones, have been used for cell imaging. In this respect, the imaging of living cells by STED nanoscopy achieved super-resolution, which was by far superior to that offered by confocal microscopy imaging.

Admittedly, cell imaging has also been performed with tpy complexes of Ln^{III} or Ir^{III} ions [34, 130], but the discussion of these imaging agents, which also featured (light-induced) anticancer activity, is beyond the scope of this chapter. The same also holds true for the complexes of p-block ions, which have been explored to some extent. So far, the complexes of Sn^{IV} or Pb^{II} ions cannot compete with the other tpy-based metallodrugs in terms of anticancer activity and selectivity (*i.e.*, non-toxicity against healthy cells) [131]. Very recently, In^{III} complexes have been screened, for the first time, regarding their applicability as antifungal and anticancer agents [132].

This issue, *i.e.*, the dependence of anticancer activity and selectivity on the structure of the tpy-based metallodrug, is still largely unclear. Thus, the structure–property relationships need to be elucidated in future work and computational chemistry will certainly play a significant role in this research. The same holds true for the next steps toward the practical use of such metallodrugs, which are detailed *in vivo* studies on promising candidates. Moreover, other aspects of biological activity, *i.e.*, antibacterial and antimicrobial properties, should not be disregarded. In this respect, promising results have already been achieved, *e.g.*, with metal-free ligands [133] and Cu^{II}–tpy complexes [100, 104].

References

1 K. D. Mjos, C. Orvig, *Chem. Rev.* 2014, 114, 4540–4563.
2 K. H. Thompson, C. Orvig, *Science* 2003, 300, 936–939.
3 P. Ehrlich, A. Bertheim, *Ber. Dtsch. Chem. Ges.* 1912, 45, 756–766.
4 P. Ehrlich, in *The Collected Papers of Paul Ehrlich*, (Ed.: F. Himmelweit), Pergamon, 1960, pp. 106–117.
5 N. C. Lloyd, H. W. Morgan, B. K. Nicholson, R. S. Ronimus, *Angew. Chem. Int. Ed.* 2005, 44, 941–944.
6 C. Ros-Vivancos, M. González-Hernández, J. F. Navarro-Gracia, J. Sánchez-Payá, A. González-Torga, J. Portilla-Sogorb, *Rev. Esp. Quim.* 2018, 31, 485–492.
7 B. Rosenberg, L. Van Camp, T. Krigas, *Nature* 1965, 205, 698–699.
8 B. Rosenberg, L. Vancamp, J. E. Trosko, V. H. Mansour, *Nature* 1969, 222, 385–386.
9 R. A. Alderden, M. D. Hall, T. W. Hambley, *J. Chem. Educ.* 2006, 83, 728.
10 T. W. Hambley, *Science* 2007, 318, 1392–1393.
11 E. J. Anthony, E. M. Bolitho, H. E. Bridgewater, O. W. L. Carter, J. M. Donnelly, C. Imberti, E. C. Lant, F. Lermyte, R. J. Needham, M. Palau, P. J. Sadler, H. Shi, F.-X. Wang, W.-Y. Zhang, Z. Zhang, *Chem. Sci.* 2020, 11, 12888–12917.
12 J. K.-C. Lau, B. Ensing, *Physi. Chem. Chem. Phys.* 2010, 12, 10348–10355.

13 A. M. Pizarro, A. Habtemariam, P. J. Sadler, in *Medicinal Organometallic Chemistry*, (Eds.: G. Jaouen, N. Metzler-Nolte), Springer, Berlin & Heidelberg, 2010, pp. 21–56.

14 A. Mandic, J. Hansson, S. Linder, M. C. Shoshan, *J. Biol. Chem.* 2003, 278, 9100–9106.

15 Z. H. Siddik, *Oncogene* 2003, 22, 7265–7279.

16 H. H. W. Chen, M. T. Kuo, *Met.-Based Drugs* 2010, 2010, 430939.

17 F. Arnesano, M. Losacco, G. Natile, in *Binding, Transport and Storage of Metal Ions in Biological Cells*, (Eds.: W. Maret, A. Wedd), The Royal Society of Chemistry, Cambridge, 2014, pp. 429–460.

18 F. Martinez, G. Deray, M. Dubois, H. Beaufils, C. Jacquiaud, R. Bourbouze, M. Benhmida, M.-C. Jaudon, C. Jacobs, *Anti-Cancer Drugs* 1993, 4, 85–90.

19 A. Rescifina, C. Zagni, M. G. Varrica, V. Pistarà, A. Corsaro, *Eur. J. Med. Chem.* 2014, 74, 95–115.

20 L. S. Lerman, *J. Mol. Biol.* 1961, 3, 18-IN14.

21 R. Galindo-Murillo, T. E. Cheatham III, *Nucleic Acids Res.* 2021, 49, 3735–3747.

22 E. C. Long, J. K. Barton, *Acc. Chem. Res.* 1990, 23, 271–273.

23 G. T. Morgan, F. H. Burstall, *J. Chem. Soc. Res.* 1934, 1934, 1498–1500.

24 K. W. Jennette, S. J. Lippard, G. A. Vassiliades, W. R. Bauer, *Proc. Natl. Acad. Sci. U.S.A.* 1974, 71, 3839–3843.

25 S. J. Lippard, *Acc. Chem. Res.* 1978, 11, 211–217.

26 W. D. McFadyen, L. P. G. Wakelin, I. A. G. Roos, B. L. Hillcoat, *Biochem. J.* 1987, 242, 177–183.

27 S. D. Cummings, *Coord. Chem. Rev.* 2009, 253, 449–478.

28 I. Eryazici, C. N. Moorefield, G. R. Newkome, *Chem. Rev.* 2008, 108, 1834–1895.

29 J. H. Koch, E. C. Gyarfas, F. Dwyer, *Aust. J. Biol. Sci.* 1956, 9, 371–381.

30 J. H. Koch, C. H. Gallagher, *Biochem. Pharmacol.* 1960, 3, 231–243.

31 P. M. van Vliet, S. M. S. Toekimin, J. G. Haasnoot, J. Reedijk, O. Nováková, O. Vrána, V. Brabec, *Inorg. Chim. Acta* 1995, 231, 57–64.

32 V. Brabec, O. Nováková, *Drug Resist. Updat.* 2006, 9, 111–122.

33 K. Suntharalingam, D. J. Hunt, A. A. Duarte, A. J. P. White, D. J. Mann, R. Vilar, *Chem. Eur. J.* 2012, 18, 15133–15141.

34 R. Abhijnakrishna, K. Magesh, A. Ayushi, S. Velmathi, *Coord. Chem. Rev.* 2023, 496, 215380.

35 A. Winter, M. Gottschaldt, G. R. Newkome, U. S. Schubert, *Curr. Top. Med. Chem.* 2012, 12, 158–175.

36 U. S. Schubert, A. Winter, G. R. Newkome, *Terpyridine-Based Materials*, Wiley-VCH, Weinheim, 2011.

37 J. Leiter, J. L. Hartwell, J. S. Kahler, I. Kline, M. J. Shear, *J. Natl. Cancer Inst.* 1953, 14, 365–374.

38 L.-X. Zhao, T. S. Kim, S.-H. Ahn, T.-H. Kim, E.-k. Kim, W.-J. Cho, H. Choi, C.-S. Lee, J.-A. Kim, T. C. Jeong, C.-j. Chang, E.-S. Lee, *Bioorg. Med. Chem. Lett.* 2001, 11, 2659–2662.

39 D. Zych, A. Slodek, S. Krompiec, K. Malarz, A. Mrozek-Wilczkiewicz, R. Musiol, *ChemistrySelect* 2018, 3, 7009–7017.

40 R. Musiol, P. Malecki, M. Pacholczyk, J. Mularski, *Expert Opin. Drug Discov.* 2022, 17, 259–271.

41 U. Dasmahapatra, B. Maiti, M. M. Alam, K. Chanda, *Eur. J. Med. Chem.* 2024, 275, 116603.

42 A. Winter, G. R. Newkome, U. S. Schubert, *ChemCatChem* 2011, 3, 1384–1406.

43 A. Winter, U. S. Schubert, *ChemCatChem* 2020, 12, 2890–2941.

44 C. Förster, K. Heinze, *Chem. Soc. Rev.* 2020, 49, 1057–1070.

45 B. Babu, T. Nandhini, V. G. Vaidyanathan, B. U. Nair, *Inorg. Chem. Commun.* 2016, 73, 124–128.

46 X.-L. Hong, L.-J. Liu, W.-G. Lu, X.-B. Wang, *Transit. Met. Chem.* 2017, 42, 459–467.

47 D. W. J. Kwong, O. Y. Chan, R. N. S. Wong, S. M. Musser, L. Vaca, S. I. Chan, *Inorg. Chem.* 1997, 36, 1276–1277.

48 B. Banik, K. Somyajit, G. Nagaraju, A. R. Chakravarty, *RSC Adv.* 2014, 4, 40120–40131.

49 M. A. Fik, A. Gorczyński, M. Kubicki, Z. Hnatejko, A. Wadas, P. J. Kulesza, A. Lewińska, M. Giel-Pietraszuk, E. Wyszko, V. Patroniak, *Polyhedron* 2015, 97, 83–93.

50 A. K. Holley, V. Bakthavatchalu, J. M. Velez-Roman, D. K. St. Clair, *Int. J. Mol. Sci.* 2011, 12, 7114–7162.

51 F. Schwizer, Y. Okamoto, T. Heinisch, Y. Gu, M. M. Pellizzoni, V. Lebrun, R. Reuter, V. Köhler, J. C. Lewis, T. R. Ward, *Chem. Rev.* 2018, 118, 142–231.

52 C. Zhang, P. Srivastava, K. Ellis-Guardiola, J. C. Lewis, *Tetrahedron* 2014, 70, 4245–4249.

53 S. Das, C. D. Incarvito, R. H. Crabtree, G. W. Brudvig, *Science* 2006, 312, 1941–1943.

54 J. Li, M. Chen, J. Jiang, J. Huang, H. Chen, L. Pan, D. S. Nesterov, Z. Ma, A. J. L. Pombeiro, *Int. J. Mol. Sci.* 2023, 24, 3903.

55 L. Liu, T. Zhu, C. Cai, L. Wang, S. Li, X. Lian, Y. Feng, Y. Tian, Q. Zhang, *Dalton Trans.* 2022, 51, 16915–16920.

56 J. Luo, Z. Xie, J. W. Y. Lam, L. Cheng, H. Chen, C. Qiu, H. S. Kwok, X. Zhan, Y. Liu, D. Zhu, B. Z. Tang, *Chem. Commun.* 2001, 2001, 1740–1741.

57 Q. Xia, Y. Zhang, Y. Li, Y. Li, Y. Li, Z. Feng, X. Fan, J. Qian, H. Lin, *Aggregate* 2022, 3, e152.

58 J. Liu, Q. Gu, W. Du, Z. Feng, Q. Zhang, Y. Tian, K. Luo, Q. Gong, X. Tian, *Biosens. Bioelectron.* 2022, 203, 114058.

59 W. Du, H. Wang, Y. Zhu, X. Tian, M. Zhang, Q. Zhang, S. C. De Souza, A. Wang, H. Zhou, Z. Zhang, J. Wu, Y. Tian, A. C. S. Appl, *Mater. Interfaces* 2017, 9, 31424–31432.

60 Y. Xu, R. Xu, Z. Wang, Y. Zhou, Q. Shen, W. Ji, D. Dang, L. Meng, B. Z. Tang, *Chem. Soc. Rev.* 2021, 50, 667–690.

61 A. M. Mansour, K. Radacki, *Inorg. Chim. Acta* 2020, 511, 119806.

62 H.-I. Choi, A. Zeb, M.-S. Kim, I. Rana, N. Khan, O. S. Qureshi, C.-W. Lim, J.-S. Park, Z. Gao, H.-J. Maeng, J.-K. Kim, *J. Control. Release* 2022, 350, 652–667.

63 U. Schatzschneider, *Inorg. Chim. Acta* 2011, 374, 19–23.

64 M. Froudarakis, E. Hatzimichael, L. Kyriazopoulou, K. Lagos, P. Pappas, A. G. Tzakos, V. Karavasilis, D. Daliani, C. Papandreou, E. Briasoulis, *Crit. Rev. Oncol. Hematol.* 2013, 87, 90–100.

65 E. L.-M. Wong, G.-S. Fang, C.-M. Che, N. Zhu, *Chem. Commun.* 2005, 2005, 4578–4580.

66 C. Pandya, A. Sivaramakrishna, *Int. J. Biol. Macromol.* 2024, 274, 132792.

67 U. Basu, I. Khan, D. Koley, S. Saha, P. Kondaiah, A. R. Chakravarty, *J. Inorg. Biochem.* 2012, 116, 77–87.

68 L. Tabrizi, *Appl. Organomet. Chem.* 2018, 32, e4161.

69 P. J. Blower, J. R. Dilworth, R. I. Maurer, G. D. Mullen, C. A. Reynolds, Y. Zheng, *J. Inorg. Biochem.* 2001, 85, 15–22.

70 M. Mathuber, H. Schueffl, O. Dömötör, C. Karnthaler, É. A. Enyedy, P. Heffeter, B. K. Keppler, C. R. Kowol, *Inorg. Chem.* 2020, 59, 17794–17810.

71 R. Indumathy, M. Kanthimathi, T. Weyhermuller, B. U. Nair, *Polyhedron* 2008, 27, 3443–3450.

72 R. Indumathy, S. Radhika, M. Kanthimathi, T. Weyhermuller, B. Unni Nair, *J. Inorg. Biochem.* 2007, 101, 434–443.

73 I. M. Henderson, R. C. Hayward, *J. Mater. Chem.* 2012, 22, 21366–21369.

74 G. Zhang, J. Tan, Y. Z. Zhang, C. Ta, S. Sanchez, S.-Y. Cheng, J. A. Golen, A. L. Rheingold, *Inorg. Chim. Acta* 2015, 435, 147–152.

75 X. Huang, B. Wang, D. Sun, M. Chen, X. Xue, H. Liu, Y. Zhou, Z. Ma, *J. Inorg. Biochem.* 2024, 256, 112554.

76 J.-W. Wang, W.-J. Liu, D.-C. Zhong, T.-B. Lu, *Coord. Chem. Rev.* 2019, 378, 237–261.

77 Y.-S. Wong, M.-C. Tang, M. Ng, V. W.-W. Yam, *J. Am. Chem. Soc.* 2020, 142, 7638–7646.

78 M. Gil-Moles, M. Concepción Gimeno, *ChemMedChem* 2024, 19, e202300645.

79 S. Rajalakshmi, T. Weyhermüller, A. J. Freddy, H. R. Vasanthi, B. U. Nair, *Eur. J. Med. Chem.* 2011, 46, 608–617.

80 A. Kumar, J. Prakash Chinta, A. Kumar Ajay, M. Kumar Bhat, C. P. Rao, *Dalton Trans.* 2011, 40, 10865–10872.

81 D. Mahendiran, R. S. Kumar, V. Viswanathan, D. Velmurugan, A. K. Rahiman, *Dalton Trans.* 2016, 45, 7794–7814.

82 S. Jain, K. Bhar, S. Kumar, S. Bandyopadhyaya, S. Tapryal, C. C. Mandal, A. K. Sharma, *Dalton Trans.* 2020, 49, 4100–4113.

83 J. Karges, K. Xiong, O. Blacque, H. Chao, G. Gasser, *Inorg. Chim. Acta* 2021, 516, 120137.

84 D. Mahendiran, R. S. Kumar, V. Viswanathan, D. Velmurugan, A. K. Rahiman, *J. Biol. Inorg. Chem.* 2017, 22, 1109–1122.

85 S. Gama, I. Rodrigues, F. Marques, E. Palma, I. Correia, M. F. N. N. Carvalho, J. C. Pessoa, A. Cruz, S. Mendo, I. C. Santos, F. Mendes, I. Santos, A. Paulo, *RSC Adv.* 2014, 4, 61363–61377.

86 K. Tummalapalli, C. S. Vasavi, P. Munusami, M. Pathak, M. M. Balamurali, *Int. J. Biol. Macromol.* 2017, 95, 1254–1266.
87 F. Jia, S. Wang, Y. Man, P. Kumar, B. Liu, *Molecules* 2019, 24, 769.
88 K. Abdi, H. Hadadzadeh, M. Weil, M. Salimi, *Polyhedron* 2012, 31, 638–648.
89 S. Rajalakshmi, T. Weyhermüller, M. Dinesh, B. U. Nair, *J. Inorg. Biochem.* 2012, 117, 48–59.
90 J. Li, H. Yan, Z. Wang, R. Liu, B. Luo, D. Yang, H. Chen, L. Pan, Z. Ma, *Dalton Trans.* 2021, 50, 8243–8257.
91 J.-W. Liang, Y. Wang, K.-J. Du, G.-Y. Li, R.-L. Guan, L.-N. Ji, H. Chao, *J. Inorg. Biochem.* 2014, 141, 17–27.
92 Y.-Q. Gu, Y.-J. Zhong, M.-Q. Hu, H.-Q. Li, K. Yang, Q. Dong, H. Liang, Z.-F. Chen, *Dalton Trans.* 2022, 51, 1968–1978.
93 K. Choroba, B. Machura, A. Szlapa-Kula, J. G. Malecki, L. Raposo, C. Roma-Rodrigues, S. Cordeiro, P. V. Baptista, A. R. Fernandes, *Eur. J. Med. Chem.* 2021, 218, 113404.
94 V. M. Manikandamathavan, M. Thangaraj, T. Weyhermuller, R. P. Parameswari, V. Punitha, N. N. Murthy, B. U. Nair, *Eur. J. Med. Chem.* 2017, 135, 434–446.
95 J. Grau, R. F. Brissos, J. Salinas-Uber, A. B. Caballero, A. Caubet, O. Roubeau, L. Korrodi-Gregório, R. Pérez-Tomás, P. Gamez, *Dalton Trans.* 2015, 44, 16061–16072.
96 J. Shao, M. Li, Z. Guo, C. Jin, F. Zhang, C. Ou, Y. Xie, S. Tan, Z. Wang, S. Zheng, X. Wang, *Cell Commun. Signal.* 2019, 17, 149.
97 Z. Ma, B. Zhang, M. F. C. Guedes da Silva, J. Silva, A. S. Mendo, P. V. Baptista, A. R. Fernandes, A. J. L. Pombeiro, *Dalton Trans.* 2016, 45, 5339–5355.
98 D. Mahendiran, P. Gurumoorthy, K. Gunasekaran, R. Senthil Kumar, A. K. Rahiman, *New J. Chem.* 2015, 39, 7895–7911.
99 M. N. Patel, H. N. Joshi, C. R. Patel, *Polyhedron* 2012, 40, 159–167.
100 M. N. Patel, P. A. Dosi, B. S. Bhatt, *Med. Chem. Res.* 2012, 21, 2723–2733.
101 B. Deka, A. Bhattacharyya, S. Mukherjee, T. Sarkar, K. Soni, S. Banerjee, K. K. Saikia, S. Deka, A. Hussain, *Appl. Organomet. Chem.* 2018, 32, e4287.
102 J. Csucker, M. Scarpi-Luttenauer, P. Mesdom, S. Hidalgo, O. Blacque, G. Gasser, R. Alberto, *Inorg. Chem.* 2024, 63, 18154–18161.
103 A. Duatti, *Nucl. Med. Biol.* 2021, 92, 202–216.
104 E. A. Hassan, M. L. Hassan, C. N. Moorefield, G. R. Newkome, *Carbohydr. Polym.* 2015, 116, 2–8.
105 E. A. Hassan, W. S. Abou Elseoud, M. T. Abo-Elfadl, M. L. Hassan, *Carbohydr. Polym.* 2021, 268, 118230.
106 A. S. Prasad, F. W. J. Beck, D. C. Snell, O. Kucuk, *Nut. Cancer* 2009, 61, 879–887.
107 B. S. Mendiguchia, D. Pucci, T. F. Mastropietro, M. Ghedini, A. Crispini, *Dalton Trans.* 2013, 42, 6768–6774.
108 L. B. Josefsen, R. W. Boyle, *Met.-Based Drugs* 2008, 2008, 276109.
109 J. Qian, L. Wang, W. Gu, X. Liu, J. Tian, S. Yan, *Dalton Trans.* 2011, 40, 5617–5624.
110 E. Boseggia, M. Gatos, L. Lucatello, F. Mancin, S. Moro, M. Palumbo, C. Sissi, P. Tecilla, U. Tonellato, G. Zagotto, *J. Am. Chem. Soc.* 2004, 126, 4543–4549.

111 M. Porchia, M. Pellei, F. Del Bello, C. Santini, *Molecules* 2020, 25, 5814.

112 J. Li, R. Liu, J. Jiang, X. Liang, L. Huang, G. Huang, H. Chen, L. Pan, Z. Ma, *Molecules* 2019, 24, 4519.

113 L. Huang, R. Liu, J. Li, X. Liang, Q. Lan, X. Shi, L. Pan, H. Chen, Z. Ma, *J. Inorg. Biochem.* 2019, 201, 110790.

114 Y. Pan, Y. Yan, Y. Li, X.-W. Gao, D. Chao, *New J. Chem.* 2019, 43, 15120–15125.

115 X. Liang, J. Jiang, X. Xue, L. Huang, X. Ding, D. Nong, H. Chen, L. Pan, Z. Ma, *Dalton Trans.* 2019, 48, 10488–10504.

116 T. V. Soldatović, E. Selimović, B. Šmit, D. Ašanin, N. S. Planojević, S. D. Marković, R. Puchta, B. M. Alzoubi, *J. Coord. Chem.* 2019, 72, 690–706.

117 K. Velugula, A. Kumar, J. P. Chinta, *Inorg. Chim. Acta* 2020, 507, 119596.

118 N. Busto, M. C. Carrión, S. Montanaro, B. Díaz de Greñu, T. Biver, F. A. Jalón, B. R. Manzano, B. García, *Dalton Trans.* 2020, 49, 13372–13385.

119 A. Kellett, Z. Molphy, C. Slator, V. McKee, N. P. Farrell, *Chem. Soc. Rev.* 2019, 48, 971–988.

120 H.-H. Zou, J.-G. Wei, X.-H. Qin, S.-G. Mo, Q.-P. Qin, Y.-C. Liu, F.-P. Liang, Y.-L. Zhang, Z.-F. Chen, *MedChemComm* 2016, 7, 1132–1137.

121 V. M. Manikandamathavan, T. Weyhermüller, R. P. Parameswari, M. Sathishkumar, V. Subramanian, B. U. Nair, *Dalton Trans.* 2014, 43, 13018–13031.

122 J. Jiang, J. Li, C. Liu, R. Liu, X. Liang, Y. Zhou, L. Pan, H. Chen, Z. Ma, *J. Biol. Inorg. Chem.* 2020, 25, 311–324.

123 N. Zhang, D. Sun, X. Guan, B. Wang, Z. Wang, M. Chen, X. Xue, H. Liu, Z. Ma, *Inorg. Chim. Acta* 2024, 565, 121975.

124 H. Wang, F. Cai, D. Feng, L. Zhou, D. Li, Y. Wei, Z. Feng, J. Zhang, J. He, Y. Wu, *J. Mol. Struct.* 2019, 1194, 157–162.

125 P. Bianchini, C. Peres, M. Oneto, S. Galiani, G. Vicidomini, A. Diaspro, *Cell Tissue Res.* 2015, 360, 143–150.

126 Y. Shen, T. Shao, B. Fang, W. Du, M. Zhang, J. Liu, T. Liu, X. Tian, Q. Zhang, A. Wang, J. Yang, J. Wu, Y. Tian, *Chem. Commun.* 2018, 54, 11288–11291.

127 I. Kostova, *Curr. Med. Chem.* 2006, 13, 1085–1107.

128 M. J. Clarke, *Coord. Chem. Rev.* 2003, 236, 209–233.

129 J. Liu, C. Zhang, T. W. Rees, L. Ke, L. Ji, H. Chao, *Coord. Chem. Rev.* 2018, 363, 17–28.

130 X. Yu, C. Guo, S. Lu, Z. Chen, H. Wang, X. Li, *Macromol. Rapid Commun.* 2022, 43, 2200004.

131 A. Winter, G. R. Newkome, U. S. Schubert, *Inorg. Chem. Front.* 2024, 11, 342–399.

132 X. Chen, Q. Zhang, A. Kaswer, H. Kim, M. Ahmed, M. C. Neary, S.-Y. Cheng, G. Zhang, J. M. Rauceo, *Dalton Trans.* 2025, 54, 4656–4663.

133 S. Behera, P. P. Dash, A. K. Bishoyi, K. Dash, P. Mohanty, C. R. Sahoo, R. N. Padhy, M. Mishra, B. N. Ghosh, S. Sahoo, B. R. Jali, *J. Biomol. Struc. Dyn.* 2023, 41, 11274–11285.

5

Terpyridine Complexes for Energy-Storage Applications

5.1 Introduction

The current century, commonly referred to as the "digital age," is characterized by the societies' insatiable demand for energy (Figure 5.1). The ongoing increase in demand can, however, only be met if reliable solutions not only for the generation of electric energy, but also for its storage, are made available. These targets become of utmost importance against the background of today's most relevant challenge, *i.e.*, facing climate change with sustained solutions for the production of electricity (see Chapter 6) and its storage. In other words, a significant share of the global energy demand has to be obtained from renewable sources, and its storage has to be enabled in an eco-friendly fashion. The EU's long-term strategy, for example, targets net-zero greenhouse gas emissions by 2050 (EU: European Union; Figure 5.1) [2]. Regarding energy storage, the respective technology needs to meet the requirements of the respective application: for example, lithium-ion batteries (LIBs) are hardly suited to fulfil the needs of a large-scale storage within an electrical power grid, whereas e-mobility calls for transportable, ideally light-weight devices.

Various energy-storage technologies have already been developed. They can be classified according to the type of technology and/or the capacity/power values, which these technologies are capable of reaching [3]. One has to distinguish between mechanical (*e.g.*, pumped hydropower, flywheels), electromagnetic (*e.g.*, capacitors), chemical (*e.g.*, batteries), biological, and thermal energy storage (Figure 5.2) [5]. Most of these technologies were developed to guarantee stationary, large-scale energy storage. Batteries [6], fuel cells [7], and supercapacitors (SCs) [8, 9], on the other hand, are of importance regarding the "mobile" storage of energy. The wide range of potential applications includes active radio frequency identification (aRFID) tags, integrated-circuit smart cards, mobile sensor systems, smart clothes, and packages, as well as an Internet of Things (IoT), (hybrid) electric vehicles (EV and HEV, respectively), *etc.* [10].

In 1936, archaeologists excavated some exceptional artifacts in Khujut Rabuah, Iraq (Figure 5.3a). These asphalt-sealed ceramic jars, which contained a cylindrical copper sheet and a rod of iron, have been assigned to the Parthian (150 BC to 223 AD) or subsequent Sasanian era (224–650 AD). The composition of the items provoked an intense debate concerning their historical context. Their use as an early galvanic

Terpyridines: Multifaceted Ligands for Diverse Applications, First Edition.
Ulrich S. Schubert, Andreas Winter, and George R. Newkome.

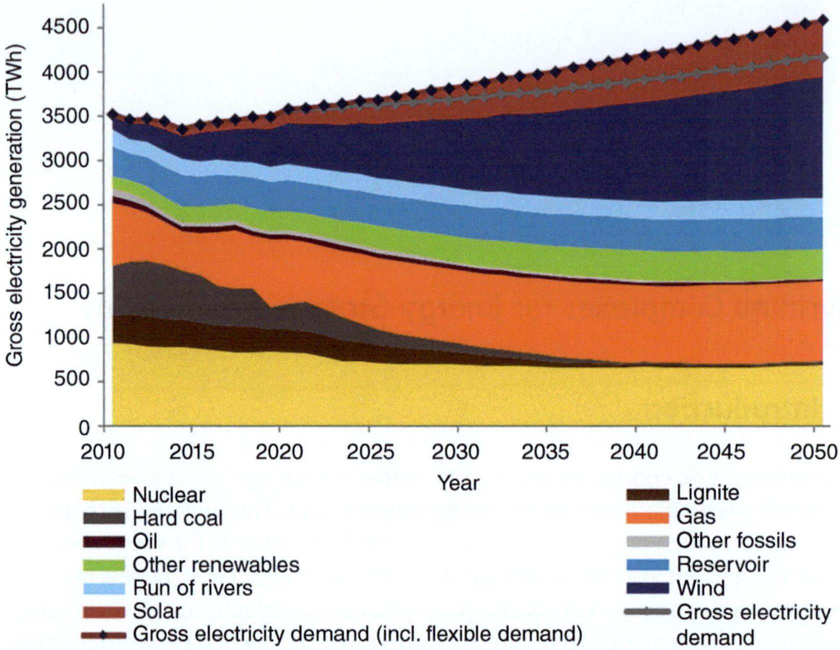

Figure 5.1 Gross electricity generation per generating technologies and gross electricity demand in Europe (*i.e.*, European Union 27 incl. the United Kingdom, Norway, and Switzerland). Adapted from [1].

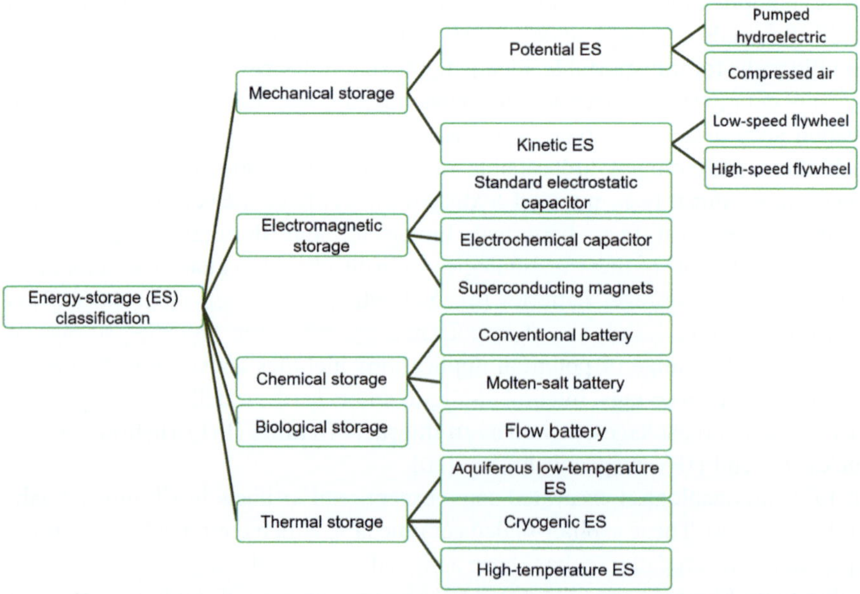

Figure 5.2 Classification of energy-storage (ES) technologies. Adapted from [4].

cell, which could have afforded a potential of *ca.* 0.8 V, was *inter alia* suggested. Accordingly, the name "Baghdad battery" was coined for this apparatus, which might have contained juice or vinegar as the electrolyte. However, the artifacts still represent a unique find and their actual meaning remains unclear. Thus, the suggested use in the context of electroplating (*i.e.*, the electrochemical gold coating of metal items) is only speculative. Unfortunately, the original items were lost in 2003 in the ravages of war, *i.e.*, the US-led invasion of Iraq and the subsequent civil war.

Far later, the term "battery" was coined by Benjamin Franklin, who in 1748 compared multiple, simultaneously acting "Leyden jars" to an artillery battery [15]. The so-called "Leyden jars," which were invented in the mid-1740s by von Kleist and van Musschenbroek, independently from each other, represent early days of capacitors, capable of storing high-voltage electric charge (Figure 5.3b). They are of historic importance since they were used in many early experiments in electricity and enabled the fundamental understanding of electrostatics. Luigi Galvani discovered the "animal electricity," which resulted in an intense controversy with Alessandro Volta regarding its origin [16]. Today, Galvani is recognized as the father of electrophysiology, whereas Volta is acknowledged for demonstrating the first electrochemical cell, the so-called "voltaic pile" (Figure 5.3c) [17]. Although the underlying chemical reactions were not understood, the device provided a steady current, at least for a reasonable length of time. Various variants of Volta's original

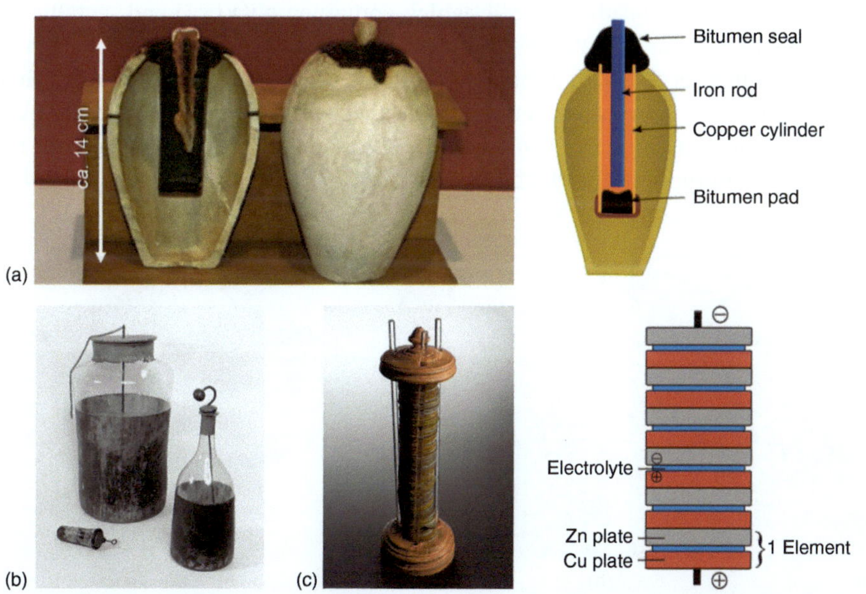

Figure 5.3 (a) Photographs (left) and schematic representation of the "Baghdad battery" [11] / with permission from Iraqi Civil Society Solidarity Initiative. (b) Photograph of three differently sized "Leyden jars" [12] / Welcome Collection gallery / Wikimedia Commons / CC BY 4.0. (c) Photograph (left) and schematic representation of a voltaic pile [13, 14] / Welcome Collection gallery / Wikimedia Commons / CC BY 4.0.

device were developed in subsequent years. In particular, the so-called Daniell cell, which was reported in 1836, became the most renowned one. The device featured a longer operational lifetime than Volta's pile and provided a potential of *ca.* 1.1 V. Accordingly, it became the durable power source for telecommunication and household applications (*i.e.*, door bells) [18]. Despite their commercial success for almost 100 years, some shortcomings of these 1st-generation "wet cells" are worth mentioning: The handling of corrosive and hazardous electrolytes in fragile glassware apparatuses was often dangerous and prevented any "portable" applications. Thus, these batteries were intrinsically limited to stationary use.

This limitation was overcome when the more convenient dry-cell batteries were invented. In this context, Leclanché and Gassner performed pioneering work. Their zinc-carbon cells afforded potentials of 1.4 and 1.5 V, respectively [15]. In particular, Gassner's cell, which was patented *inter alia* in Germany (1886) and the United States (1887), became commercially successful [19, 20]. This dry cell was comprised of an MnO_2 cathode, which was dipped into the $NH_4Cl/ZnCl_2$ electrolyte paste; the surrounding Zn shell acted as the anode. The next notable achievement in battery technology was the step from single-use (primary) devices to rechargeable (secondary) ones. In 1859, Gaston Planté developed a lead-acid battery, which could be recharged by applying an appropriate reverse current (Figure 5.4a/b) [21]. This secondary battery can certainly be considered the most successful secondary battery invented so far; the fundamental design can still be found in today's lead-acid automotive batteries (Figure 5.4c).

Likewise, the nickel-cadmium cell, which contained NiO(OH) and metallic Cd as the electrodes, was invented by Jungner in 1899 [23]. This alkaline cell, which entered the market in 1910, afforded a significantly higher energy density than the lead-acid battery and was chemically and mechanically more robust; however, the production costs were by far higher. As one improvement, Edison invented the nickel-iron battery in 1901, which also provided a high energy density, relatively

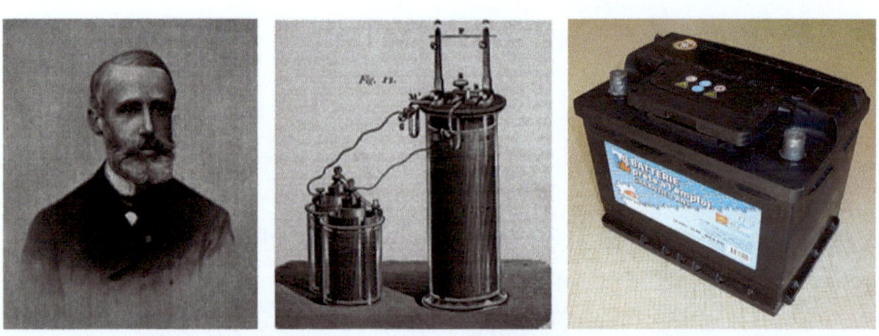

(a) (b) (c)

Figure 5.4 (a) Photograph of G. Planté (1834–1889) [21] / with permission of Elsevier. (b) Representation of Planté's secondary cell from 1872 [the rolled Pb plates (60 cm × 20 cm × 1 mm) were charged by two Bunsen cells] [21] / with permission from Wikimedia Commons. (c) Photograph of a commercial 12 V lead-acid automotive battery [22] / with permission of Cjp24 / Wikimedia Commons / CC BY-SA 4.0.

short recharging times, and reduced production costs; however, these batteries suffered from their poor performance at low temperatures [24]. As the next notable milestone, the alkaline-manganese battery was developed by Urry in the late 1950s [15]. This battery design, which contained an MnO_2 cathode, a powdered Zn anode, and an alkaline electrolyte, was superior to the aforementioned zinc-carbon cells in terms of shelf lifetime, while providing similar potential. The nickel-metal hydride (NiMH) batteries represent a consumer-grade alternative to their nickel-hydrogen predecessors, which were developed in the 1970s as an energy-storage subsystem in commercial satellites [25]. The NiMH batteries reveal improved lifetimes and are, due to the absence of toxic Cd, environmentally friendlier.

Eventually, secondary LIBs were brought to the market in the 1990s [26, 27]. As already noted by Lewis in 1913, lithium metal is the alkali metal with the lowest density, greatest electrochemical potential, and thus the optimum energy-to-weight ratio [28]. Moreover, its low molecular weight and small ionic radius are highly beneficial for fast diffusion processes [29]. Thus, Li is predestinated for use as battery material in electronic devices (Figure 5.5a) [26]. Indeed, secondary LIBs represent the state-of-the-art technology to be incorporated into mobile devices: mobile phones, notebooks, vehicles, *etc.* (Figure 5.5b). The success story of LIB technology was written *inter alia* by J.B. Goodenough, M.S. Whittingham and A. Yoshino, who were honored with the *Chemistry Nobel Prize* in 2019 for their fundamental work in this field [30].

The LIB technology features distinguished benefits (*e.g.*, low cell-fabrication costs, high voltages, increased power densities), but also some serious downside aspects have to be addressed. The use of LIB is associated with severe safety issues: Battery fires or explosions might arise from unproper charging, overheating, or damage [31]. Damage to the basically non-flexible devices might result from mechanical treatment and thus, LIBs are hardly suited for the fabrication of thin and flexible batteries. Moreover, large battery packs require a cautious temperature control to ensure safe operational conditions and prevent serious hazards, *e.g.*, fires or explosions [32, 33].

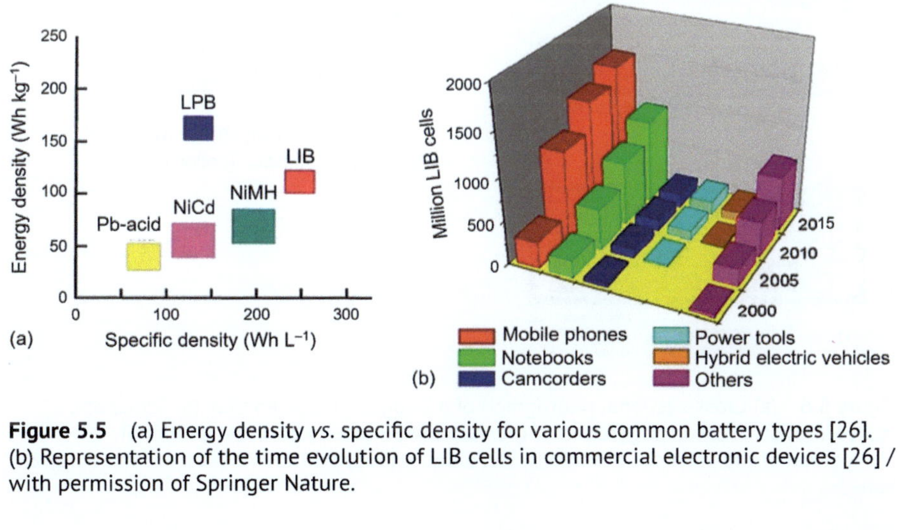

Figure 5.5 (a) Energy density *vs.* specific density for various common battery types [26]. (b) Representation of the time evolution of LIB cells in commercial electronic devices [26] / with permission of Springer Nature.

These aspects limit the applicability of LIBs, in particular in view of smart clothing or packaging, where flexible thin-film batteries are desired, and other demanding applications, such as grid stabilization and on-site storage [34]. Accordingly, other battery concepts have been considered to bypass these limitations.

Redox-active polymers, which can be derived from organic raw materials, represent one promising alternative approach [10, 34–37]. For example, organic polymers allow the fabrication of flexible devices, and their use is almost independent of the fluctuating commodity prices of rare metals. Moreover, polymers can be synthesized in a wide range of structures and topologies, thus facilitating the preparation of tailor-made materials for energy-storage applications. Particular reference is given to a recent textbook on this cutting-edge topic [38]. Notably, the first polymer-containing batteries were already developed in the late 1980s [39]. These batteries, representing the 1st generation of lithium-polymer batteries (LPBs), utilized conducting polymers as the active materials. The rechargeable 3 V coin cell, which was commercialized in 1987 by Bridgestone/Seiko, contained polyaniline (PAni) and $LiBF_4$, as the cathode and electrolyte material, respectively (Figure 5.6a) [40]. Whereas the Varta/BASF coin cell utilized partially oxidized (*i.e.*, p-doped) polypyrrole (PPy) as the cathode material (Figure 5.6b) [41]. Miller emphasized that these LPBs afforded lower energy than the aforementioned Ni-Cd batteries (representing the most common battery type by that time), but were superior in terms of shelf lifetime in the charged state and eco-friendliness [39]. However, the Varta/BASF battery never went beyond pilot production and the sale of the Bridgestone/Seiko coin cells was discontinued after only five years. The main reason for this was the development of a competitive product by Valence Technologies, Inc. (Figure 5.6c), which contained a Li^+-conductive polymer instead of an intrinsically conductive one. This battery could reliably be recharged up to

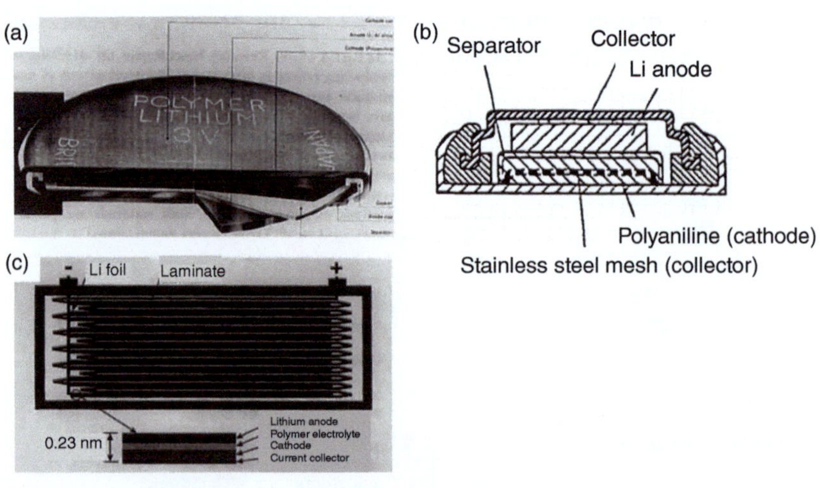

Figure 5.6 (a) Cross-sectional photograph of a Bridgestone/Seiko LPB. (b) Schematic representation of the configuration of the Varta/BASF battery. (c) Cross-sectional photograph of the Valence Technology, Inc. battery [39] / with permission of John Wiley & Sons.

300 times without significant loss of efficiency. The shelf lifetime, as expressed by the loss of charge per month, was by far superior to that of Ni-Cd batteries (*i.e.*, 0.5% *vs*. 15%) [42]. This type of LPB became commercially successful, with $\$10^8$ worth of this battery being ordered by Motorola in 1992 [43].

Since then, polymer-based batteries have experienced a tremendous evolution with a wide range of battery systems having already been realized – ranging from conventional coin cells to thin-film batteries and scalable solutions (*e.g.*, solution-based redox-flow cells). Also, the nature of the actually redox-active moiety within the polymer architecture has been widely varied. Metal-containing polymers, *i.e.*, polymers, which are equipped with redox-active metal complexes (or organometallic moieties), represent one notable subclass of redox polymers. Among others, tpy complexes of Fe^{II} or Co^{II} ions have been incorporated. Such complexes are capable of storing energy by utilizing their highly reversible +II/+III redox couple. In the context of energy storage, the specific capacity (C_{spec}) of the redox-active component represents one relevant parameter. This value refers to the amount of storable charge with respect to the mass of the respective active material. Thus, C_{spec} is related to the material's molar mass in an inversely proportional fashion. For this reason, the use of the light-weight (and earth-abundant) 3d-metal ions is preferred. Since the molar mass of Ru is about twice as high as that of Fe, the C_{spec} of the $Fe^{II/III}$ redox couple is particularly favorable for practical applications.

In Sections 5.3–5.5, the use of tpy complexes in energy-storage applications will be discussed. These applications range from solid-state batteries (*e.g.*, LIBs and related technologies) to SCs and redox-flow batteries (RFBs). However, before doing so, the so-called performance parameters will shortly be introduced (Section 5.2). These values are of importance in order to evaluate whether a designated active material is suited for battery applications or not; moreover, the performance parameters are also of importance regarding the performance of a whole battery setup [38, 44, 45].

5.2 Battery Performance Parameters

A comprehensive review, which discussed the characterization techniques relevant in this context of energy-storage devices, was published by Friebe and Schubert in 2015 [45]. Referring to this, a brief overview of the so-called performance parameters is given at this stage (thereby, details concerning the electrochemical characterization of materials are excluded).

The cell voltage of an energy-storage device (V in V), *i.e.*, its working voltage, can readily be determined from the voltage profile of a charge/discharge experiment (Figure 5.7). However, plateau regions are only observed in the ideal case, whereas, in real cases, a substantial change of the voltage in course of charging or discharging will be observed; such a "sloping" voltage requires the determination of an arithmetic average [44]. The voltage efficiency (VE or η_V in %) corresponds to the ratio between the measured discharging and charging voltage. The ideal cell, whose discharging and charging occur at the same voltage, would possess a VE of 100%. A cell's maximum voltage, *i.e.*, the theoretical voltage (V_{theo} in V), can simply be calculated

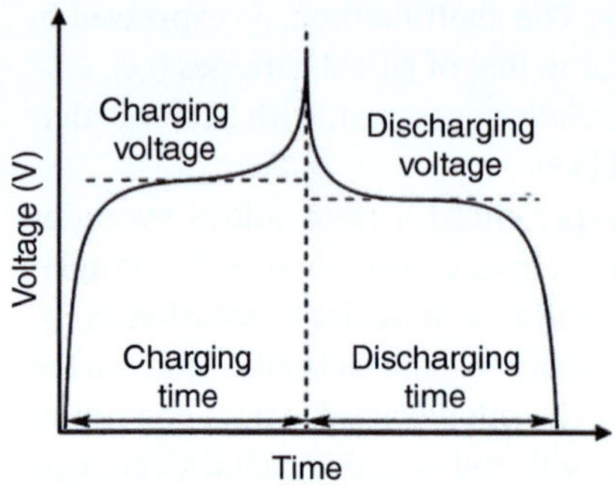

Figure 5.7 Idealized representation of a charging/discharging curve [45] / with permission of John Wiley & Sons.

from the redox potentials (E) of the employed active materials (Eq. (5.1)). The redox potentials, in turn, are accessible *via* electrochemical measurements, such as cyclic voltammetry (CV) or differential pulse voltammetry (DPV) [38].

$$V_{\text{theo}} = E_{\text{cathode}} - E_{\text{anode}} \tag{5.1}$$

The cell capacity (C in Ah) corresponds to the amount of charge that the device can store. This value depends on the number of redox sites and the number of transferred electrons per redox site on the one hand and the accessibility of redox sites within the electrode on the other. The theoretical capacity (C_{theo} in Ah), however, takes only the first two parameters into account. Thus, C will always be decreased due to limitations in the charge-carrier transport to the electrodes.

Furthermore, the specific capacity (C_{spec} in mAh g^{-1}) relates the amount of storable charge to the mass of the respective active material and, thus, also to the molar mass of actual redox site (n: number of electrons transferred per redox reaction; F: Faraday constant; M: molar mass of the redox-active unit; Eq. (5.2)). This value becomes of particular relevance regarding applications, which require both a high capacity and a low weight (*e.g.*, mobile and/or flexible devices).

$$C_{\text{spec}} = \frac{nF}{M} = \frac{n \cdot 96485}{M} \text{As g}^{-1} = \frac{n \cdot 26801}{M} \text{mAh g}^{-1} \tag{5.2}$$

If the C_{spec} values for the anode and cathode material are known, one can use Eq. (5.3) to calculate C_{spec} also for a whole battery cell. As a result of (ir)reversible capacity losses, the experimental values for the discharging and charging process will differ to some extent. This deviation is expressed by the coulombic efficiency (CE or η_{C} in %), which is simply the ratio between the discharging and charging capacities.

$$C^{-1}_{\text{spec,cell}} = C^{-1}_{\text{spec,anode}} + C^{-1}_{\text{spec,cathode}} \tag{5.3}$$

The cell voltage and capacity can be used to calculate one further important parameter, which is the storable energy (E in Wh; Eq. (5.4)). In this respect, one has to distinguish between the theoretical (E_{theo} in Wh) and the specific energy (E_{spec} in mWh g^{-1}). These values can be determined using the respective capacity values, *i.e.*, C_{theo} and C_{spec}. Regarding applications, where the active material's volume is of importance (*e.g.*, RFBs), the energy density is typically used (E_{dens} in Wh L^{-1}).

Table 5.1 Summary of recent benchmark values for lithium-ion and organic batteries [10, 47]. Specific capacities and energies are listed for the combined anode–cathode system.

Battery type	V (V)	C_{spec} (mAh g^{-1})	E_{spec} (mWh g^{-1})	Rate capability	Cyclability
LIB	≤4.5	≤200	≤650	≤5C	1000–2000
LPB	≤4.0	≤200[a)]	≤400[a)]	≤20C	≤1500
Polymer	≤1.3	≤50	≤90	≤320C	≤250

a) Values refer to a lithium-metal anode.

The energy efficiency (EE or η_W in %) corresponds to the aforementioned voltage and CEs.

$$E = V \cdot C \tag{5.4}$$

The rate capability of an energy-storage device describes its performance at different charging/discharging currents. Commonly, this value corresponds to the highest current density that affords a capacity, which is similar to those obtained at significantly lower current densities. The so-called *C*-rate, *i.e.*, the current (I), which is needed for the cell charging within 1 h, is usually used to express the current density (Eq. (5.5)). For example, cell charging with 1C or 60C corresponds to the full charging within an hour or a minute, respectively. One has to be aware that at high *C*-rates, a portion of the energy is transformed into heat, which lowers the capacity and might even result in material degradation [46].

$$C - \text{rate} = \frac{I_{\text{appl}}}{I_{\text{1h}}} \tag{5.5}$$

The cell cyclability, *i.e.*, its capacity retention over a certain number of discharging/charging cycles, and shelf lifetime or self-discharge rate (*i.e.*, the capacity loss of the fully charged cell with time) represent further important performance parameters. A comparison of benchmark values for various battery types is provided in Table 5.1 [10, 44, 47].

5.3 Terpyridine Complexes in Metal-Ion and Related Batteries

In 1985, the first LIB, which contained $LiCoO_2$ and graphite electrodes, was developed by Yoshino *et al.* from the Asahi Kasei Corp. [48] and subsequently commercialized by the SONY Corp. for use in cell phones and camcorders (Figure 5.8) [50]. Since then, the secondary LIB technology experienced a tremendous evolution and such devices are commonly used in daily-life applications, due to their high energy densities, high CEs, long shelf lifetimes and the ability to adjust the cell potential *via* a wide range of available electrode materials [51]. The last decade witnessed the emergence of next-generation LIBs to satisfy the need for high charge

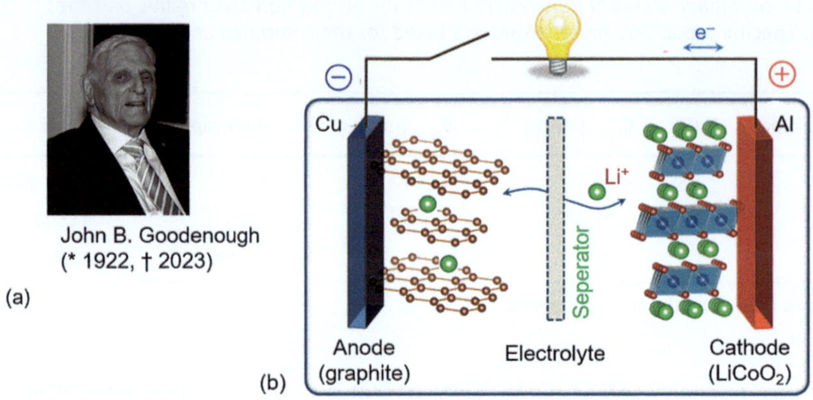

Figure 5.8 (a) Photograph of John B. Goodenough / US Embassy Sweden / Wikimedia Commons / CC BY 2.0. (b) Schematic representation of the first LIB (configuration: Al|$LiCoO_2$|Li^+ electrolyte|graphite|Cu) [50] / with permission of American Chemical Society.

capacities (typically >200 mAh g^{-1}) and power densities, which are required for (hybrid) electric vehicles, aerospace and grid applications, autonomous electric devices, *etc.* [50, 52–56]. These requirements cannot be met by the conventional LIB cathode materials (*e.g.*, $LiCoO_2$, $LiFePO_4$, and $LiMn_2O_4$), whose energy-storage mechanism essentially involves the intercalation of Li^+ ions [57, 58]. Conventionally, the electrolyte is comprised of a Li^+ salt (*e.g.*, $LiPF_6$) and an organic solvent of polar-aprotic character, mostly carbonates [59]. In order to overcome the shortcomings of these electrolytes, which are *inter alia* related to long-term stability and safety issues, solid-state polymer electrolytes have been developed [60]. These materials are favorable due to their low flammability, excellent processability, good flexibility, high safety levels, and superior thermal stability [61].

In order to overcome the shortcomings of conventional LIBs, numerous battery solutions, which are capable of providing high energy densities, have been developed: Li-rich (*e.g.*, $Li_{1.2}Mn_{0.54}Ni_{0.13}Co_{0.13}O_2$) and Ni-rich transition-metal oxides (*e.g.*, $LiNi_{0.8}Co_{0.15}Al_{0.05}O_2$ or $Li[Ni_{0.8}Co_{0.1}Mn_{0.1}]O_2$) [62] as well as high-voltage spinels; moreover, the range of alternative concepts includes Li-S, Li-O_2, Li-air, organic/polymer electrode [63], Li-CO_2 as well as solid-state batteries [51]. The timeline of innovations in LIB technology, starting from the first demonstration of the LIB concept in the 1970s, is given in Figure 5.9. The major advantage of these energy-storage devices is the remarkably high charge capacities in concert with lower production costs. A standard LIB affords an energy density of *ca.* 150 Wh kg^{-1}, whereas theoretical energy densities of 2600 and 11,400 Wh kg^{-1} can be expected from Li-S and LiO_2 batteries, respectively [26]. Moreover, they are considered eco-friendlier and less toxic when compared to LIBs, which contain a $LiCoO_2$ cathode.

Despite all recent innovations, commercial LIBs are still based on inorganic materials. This holds true for the cathode (*e.g.*, $LiCoO_2$, $LiNiO_2$, $LiMn_2O_4$, $LiFePO_4$, $LiNi_xCo_yMn_{1-x-y}O_2$, *etc.*) as well as for the anode (*e.g.*, Li metal, carbon- or

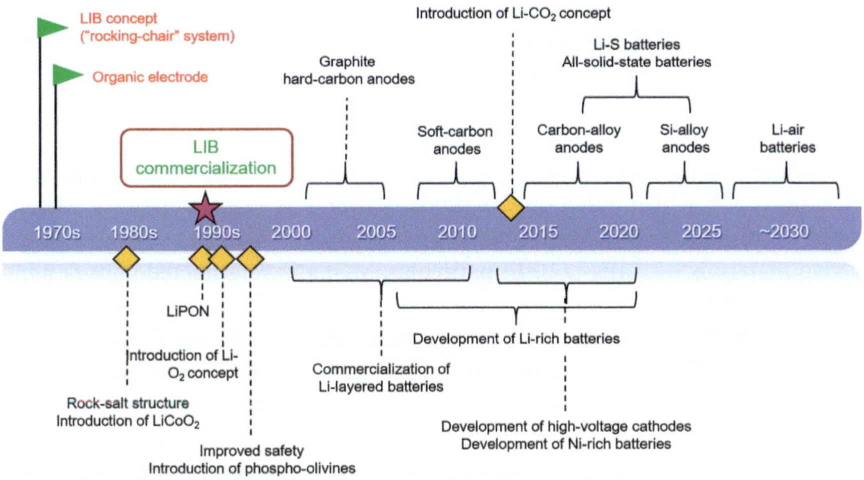

Figure 5.9 Overview of the historical evolution of the LIB technology and advances therein [51] / with permission of Royal Society of Chemistry.

silicon-based materials, *etc.*). Owing to the high production costs and the relatively low theoretical E_{spec} values, various alternative electrode materials have been explored. Thereby, the combination of eco-friendliness, low fabrication costs, hazardous-free operation, and high energy density represents the "golden goal" of this research [63–65]. In this context, organic active materials have attracted an enormous interest since such materials can be prepared, at least in theory, from natural resources and *via* established organic chemistry. Moreover, organic or polymeric materials allow for a thin and flexible electrode design. The redox potential of the organic materials can be varied over a wide range (due to the diversity of possible structures) and might involve multi-electron reactions [66]. Besides small organic molecules and polymers [38], covalent organic frameworks (COFs) [67], and metal–organic frameworks (MOFs) [68] have already been utilized. Regarding organic/macromolecular materials, conductive polymers [69, 70], organo-sulfur compounds [71], organic free radicals [72, 73], and π-conjugated carbonyl derivatives (*e.g.*, quinones, carboxylates, *etc.*) [74] are most commonly used.

Carboxylates are particularly interesting active materials since they offer high capacities, low synthesis costs as well as fast redox kinetics; detrimentally, they are prone to dissolution in aprotic electrolytes and their electronic conductivity is rather low [74]. In 2009, Tarascon *et al.* reported the first LIB, which contained carboxylates, as the anode material [75]. Specifically, the dilithium salts of all-*trans*-muconic acid and terephthalic acid were evaluated in this context. These materials provided initial specific capacities of 205 mAh g^{-1} and 300 mAh g^{-1}, which were – after continuous charging/discharging at 1C – retained to 61% (80 cycles) and 78% (50 cycles), respectively. It was assumed that irreversible structural changes of the material and/or its dissolution in the electrolyte accounted for the observed capacity fading. In the following, a wide range of π-conjugated carboxylates was studied regarding their energy-storage capability. As summarized by Schubert *et al.*, aromatic derivatives were particularly suited to increase the electrodes' charging/discharging cyclability [74]. From these, the tetralithium salt of perylen-3,4,9,10-tetra*kis*-carboxylic acid afforded an initial capacity of 200 mAh g^{-1}; this value was retained to 98% after 100 charging/discharging cycles [76]. The corresponding sodium-ion battery (SIB) provided a decent reversible capacity of *ca.* 100 mAh g^{-1}, which faded only marginally over 100 cycles. Moreover, various types of coordination compounds, based on carboxylate ligands, have been tested as active materials for LIBs [77]. In particular, MOFs have attracted considerable attention [78–80].

In line with these studies, Liu *et al.* prepared the homoleptic Fe^{II} complex **1** (Figure 5.10a), which provided multiple redox-active sites (*i.e.*, the metal center and the four carboxylate moieties attached to the π-conjugated tpy scaffold) as well as high chemical stability and mitigated solubility in aprotic electrolyte systems [81]. When utilized as the anode material, the corresponding LIB afforded a high discharge capacity of 650 mAh g^{-1} after electrochemical cycling for 300 times at the current density of 100 mA g^{-1} (Figure 5.10b). The complex displayed a particularly good cyclability with high CE also at the high current density of 5000 mA g^{-1}. Under these experimental conditions, a discharge capacity of *ca.* 218 mAh g^{-1} was obtained after 1800 charging/discharging cycles.

Very recently, a different electrode design was suggested by Yoo *et al.* [82]. These authors prepared a binder-free cathode material *via* the straightforward in-cell electropolymerization of Co^{II} complex **2** (Figure 5.11a). The resulting metallopolymer possessed a bipolar structure, due to the incorporated *p*-type (*i.e.*, triphenylamine and Co^{II} center) and *n*-type subunits (*i.e.*, tpy scaffold). This particular structure accounted for the observed high specific capacity and energy density when used in a so-called dual-ion battery (DIB). DIBs represent one promising alternative to alkali-ion batteries (*e.g.*, LIBs or SIBs) due to their high working voltage, material

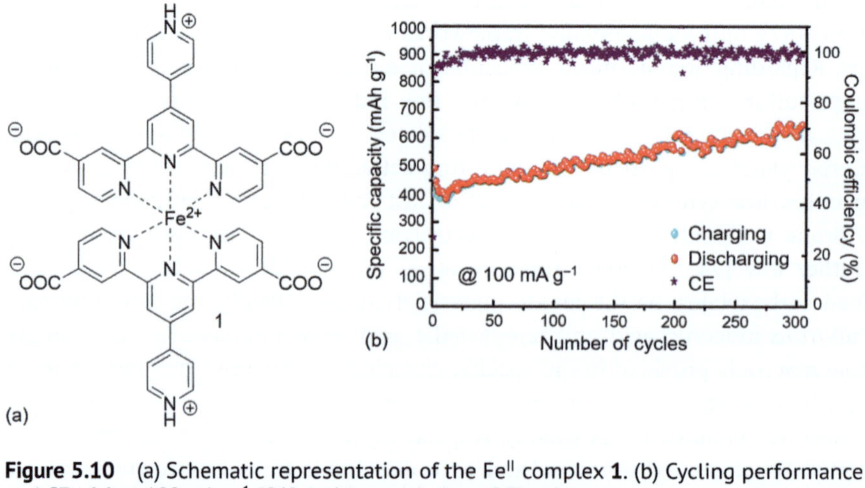

Figure 5.10 (a) Schematic representation of the Fe^{II} complex **1**. (b) Cycling performance and CE of **1** at 100 mA g^{-1} [81] / with permission of Elsevier.

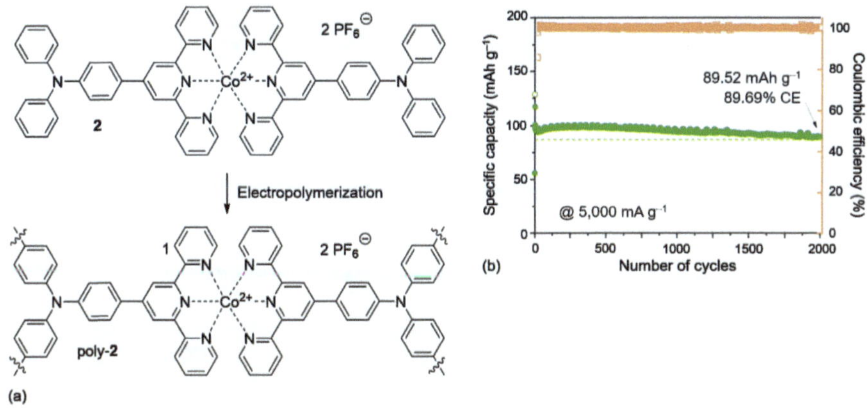

Figure 5.11 (a) Schematic representation of the preparation of poly-**2** by the in-cell electropolymerization of **2**. (b) Cycling performance and CE of poly-**2** at 5000 mA g^{-1} [82] / with permission of Elsevier.

availability, low fabrication cost, and superior operational safety [83]. The working principle of DIBs relies on the intercalation of cations and anions into the anode and cathode, respectively, during the charging process; upon discharging, these species are released back into the electrolyte. The poly-**2** cathode delivered a high specific capacity of *ca.* 192 mAh g^{-1} at 50 mA g^{-1} and revealed a very good rate capability [82]. Moreover, the initial capacity was retained to >89% even after 2000 charging/discharging cycles at 5000 mA g^{-1} (Figure 5.11b). Noteworthy, the strategy to deposit a redox-active metallopolymer film onto an electrode surface was already followed in 2013 by Nishihara *et al.* [84]. These authors fabricated a polymer film, which contained $[Fe(tpy)_2]^{2+}$ moieties on the surface of a glassy-carbon electrode. As a proof of concept, the energy-storage capability of this film was demonstrated: Rapid charging/discharging occurred with high capacity retention (*i.e.*, the initial capacity was retained to 95% after 3000 cycles at 10 V s^{-1}). Regrettably, a functional battery containing this metallopolymer electrode was not assembled.

In related work, Meng, Wong *et al.* assembled free-standing coordination nanosheets (CONASHs) *via* the liquid/liquid interfacial reaction between Co^{II} ions and a rigid tetra*kis*-tpy ligand (Figure 5.12) [85]. CONASHs are polymer networks whose 2D structure results from metal-to-ligand coordinative interactions. Such materials have attracted considerable interest due to their photophysical and electrochemical properties, which result from the interplay of the organic ligands and the metal ions; these properties can be utilized for various purposes, *inter alia* energy storage [86]. Similar to aforementioned poly-**2**, the Co^{II}-containing CONASH was tested as a cathode material for DIBs [85]. This DIB revealed a good cycle lifetime and rate performance, whereas the obtained maximum capacity stood behind that reported for poly-**2** (*vide supra*). COFs, as the covalent counterparts to CONASHs are still superior in terms of electro-activity and ease of synthesis (at least when taking the possibility for structural and functional tunability into account). Such 2D polymers have been employed as the active material in various

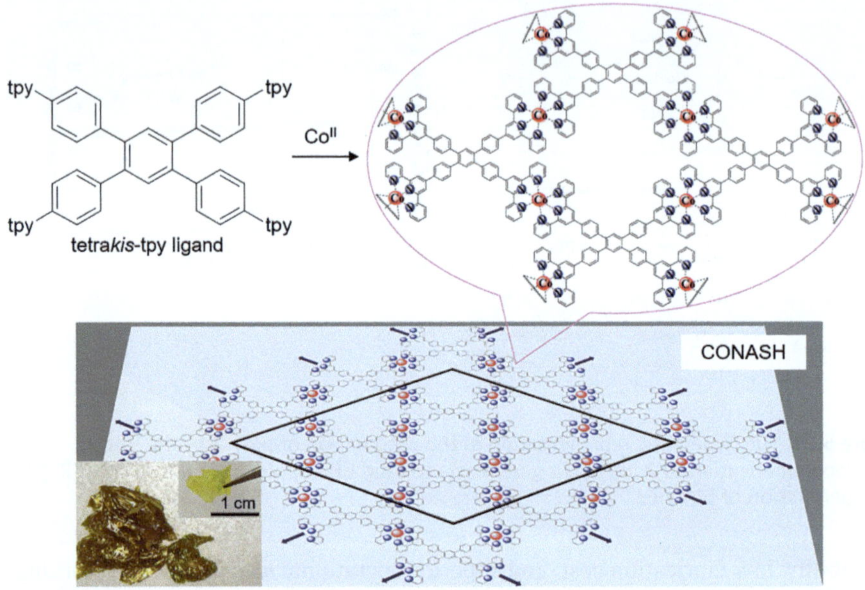

Figure 5.12 Schematic representation of the formation of a Co^{II}–tpy CONASH. Representative photographs of the air-dried CONASH and of the free-standing CONASH are also shown [85] / with permission of John Wiley & Sons.

battery types: LIBs, SIBs, potassium-ion batteries (PIBs), aqueous zinc batteries, lithium-sulfur batteries (LSBs), *etc.* [67].

Besides Co^{II}-containing polymers, also the parent $[Co(tpy)_2]^{2+}$ complex has been used in the context of energy-storage application – even though in a different context, *i.e.*, as a redox mediator in Li-O_2 batteries [87]. Li-O_2 (or Li-air) batteries have attracted an enormous interest due to the highest theoretical specific energy of *ca.* 3500 Wh kg^{-1}, which by far outcompetes the energy densities accessible with the contemporary LIB technology (*e.g.*, 544 Wh kg^{-1} for $LiFePO_4$ or 1013 Wh kg^{-1} for $LiCoO_2$) [88, 89]. Thus, Li-O_2 batteries, as depicted in Figure 5.13a, are considered for applications where high specific energies are required, *e.g.*, in electric vehicles or the military field [90]. The high specific energy results from the net cell reaction between lithium and oxygen, *i.e.*, the two-step formation of Li_2O_2, as the solid discharge product (Figure 5.13b), which proceeds with 2.96 V *vs.* Li^+/Li. In particular, the low molar mass of gaseous O_2 (compared to those of the transition-metal oxides used in LIBs) accounts for high gravimetric energy density of the Li-O_2 system. However, contemporary Li-O_2 batteries possess some severe shortcomings, such as the poor efficiency of the charging process [90]. This is mainly due to the insulating nature of Li_2O_2, as the discharge product (*ca.* 10^{-12} S cm^{-1}) [91, 92] and the languid kinetics of the oxygen-evolution reaction (OER). Thus, charging of Li-O_2 batteries typically occurs with high overpotentials. Detrimentally, high charging voltages might bring electrolyte degradation or electrode corrosion along. Thus, the cycling efficiency and lifetime of Li-O_2 cells are rather poor [88]. Wu, Chen

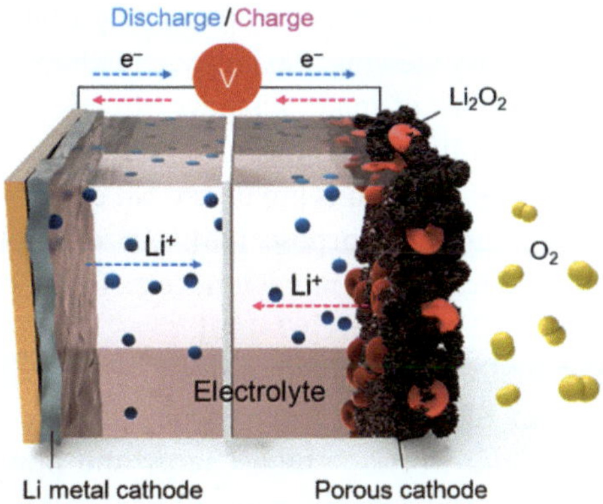

(a)

$$O_2 + Li^+ + e^- \longrightarrow LiO_2 \quad (1)$$

$$LiO_2 + Li^+ + e^- \longrightarrow Li_2O_2 \quad (2)$$

$$O_2 + 2\,Li^+ + 2\,e^- \longrightarrow Li_2O_2 \quad (3)$$

$$2\,LiO_2 \longrightarrow Li_2O_2 + O_2 \quad (4)$$

(b) Cell discharge reactions

$$M \longrightarrow M^+ + e^- \quad (1)$$

$$2\,M^+ + Li_2O_2 \longrightarrow 2\,M + O_2 + 2Li^+ \quad (2)$$

(c) Cell charge reactions

Figure 5.13 (a) Schematic representation of a Li-O_2 battery; configurations with aqueous, aprotic, hybrid, or solid-state electrolytes are known [88] / with permission of American Chemical Society. (b) Summary of the relevant discharge reactions: initial O_2 reduction (1), subsequent LiO_2 reduction (2), net cell discharge reaction (3), and LiO_2 disproportionation (4). (c) Summary of the cell charging in the presence of a redox mediator: mediator oxidation (1) and OER (2).

et al. elaborated that the performance of such cells can be improved significantly by using a redox mediator, which is present in the electrolyte [90]. The particular role of the mediator is to increase the electron-transfer rate between Li_2O_2 and the electrode in the course of the charging process. Thereby, the mediator is first oxidized; subsequently, Li^+ ions are released into the electrolyte in course of the OER (thereby, the mediator is re-reduced). Taking this two-step mechanism into account (Figure 5.13c), the cell charging potential will, in fair approximation, depend on the mediator's redox potential, which should be greater than 2.96 V of the net cell reaction. However, when using $[Co(tpy)_2]^{2+}$, as the redox mediator, only moderate cell charging with a O_2-recovery rate of 25% and considerable electrolyte degradation were observed, whereas tetrathiafulvalene (TTF), as a common organic mediator, achieved O_2-recovery rates up to 32% under comparable experimental conditions whilst the electrolyte was degraded to a smaller extent [87]. Notably, Co^{II}–polypyridyl complexes, *e.g.*, $[Co(tpy)_2]^{2+}$, have also been employed as redox mediators in dye-sensitized solar cells (DSSCs) [93]. Unlike their use in Li-O_2 batteries, the performance of DSSCs could be improved significantly when using such complexes instead of the commonly used I^-/I_3^- electrolyte system (see Section 6.3).

5.4 Terpyridine Complexes in Supercapacitors

The phenomenon of electrochromism was discussed first by Platt, who defined electrochromic (EC) materials as dyes whose apparent color might reversibly be changed under a strong electric field [94]. Since 2000, research in EC materials has been intensified in order to use them as stimuli-responsive dyes in numerous applications: wearable and portable electronics, electronic papers and billboards, see-through displays, *etc.* [95]. It has been noted that the outstanding properties of these materials (*i.e.*, low power consumption, easy viewing, flexibility, stretchability, *etc.*) accounted for their success. Numerous types of materials have already been evaluated regarding the EC performance. Metallo-supramolecular polymers, *i.e.*, coordination compounds, in which redox-active metal ions are linked *via* chelating organic ligands, have *inter alia* been used for this purpose [96]. For example, Higuchi *et al.* employed tpy-based ligands and transition-metal ions (*e.g.*, Fe^{II}, Co^{II}, Ru^{II}) for the assembly of robust and efficient EC materials [97–100].

In 2020, this same group reported the first example for the use of such polymer films as active electrode materials in SCs [101]. This particular type of energy-storage system contains the best of two worlds – high energy and power densities. Whereas only one or the other can be realized with conventional batteries or capacitors. Although the boundaries are rather diffuse, energy-storage devices can be categorized according to their specific energy (in Wh kg^{-1}) and specific power (W kg^{-1}); typically, this categorization is summarized in a so-called Ragone plot [102]. The representative plot (Figure 5.14) shows that conventional batteries offer high specific energies, but their specific power is intrinsically low (owing to the reaction kinetics and mass-transfer processes involved in charging/discharging) [104]; whereas capacitors feature high specific power and low specific energy. This characteristic is a result of the involved electrostatic processes, which proceed in a fast and highly reversible fashion, *i.e.*, the electrode absorbs ions from the electrolyte (charging) and releases them back again (discharging). However, SCs, often also named ultracapacitors, represent the link between secondary (*i.e.*, rechargeable) batteries and the aforementioned capacitors [105]. Advantageously, SCs are superior in terms of maximum capacitance values and possess much lower voltage limits than the former. In general, SC devices can store significantly more energy when compared to electrolytic capacitors (*i.e.*, 10–100 times more per volume/mass unit).

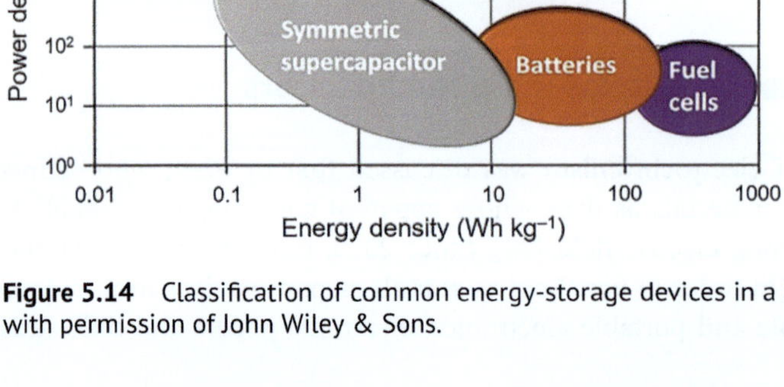

Figure 5.14 Classification of common energy-storage devices in a Ragone plot [103] / with permission of John Wiley & Sons.

As one further benefit, SCs can be charged/discharged at much faster rates and their cycle lifetime is remarkably longer than those of secondary batteries.

The history of capacitors can be traced back to the 1740s, when von Kleist and van Musschenbroek invented early condensers, which were able to store large quantities of electricity on their plates (the so-called "Leyden jars", see Section 5.1) [106]. For more details on the development of capacitor technology, reference is given at this stage to a comprehensive overview by Ho *et al.* [107]. Eventually, electric double-layer capacitors (EDLCs) were patented in 1957 by Becher; this inventor developed a "low-voltage electrolytic capacitor with porous carbon electrodes" [108]. However, the physical background, *i.e.*, the double-layer mechanism, was not clarified at that time and, thus, Becker could not yet rationalize his observation: "It is not known exactly what is taking place in the component if it is used for energy storage, but it leads to an extremely high capacity." Nonetheless, this invention marks the origin of what is known as the SC technology and the charge-mechanism of (super)capacitors is largely understood [109, 110]. The basic configuration of an EDLC is shown in Figure 5.15a. In such a device, the current collectors are coated with a porous, carbonaceous material and kept in an electrolyte/salt solution; two such half-cells are connected *via* a mesoporous separator. The working principle is basically the same as for a conventional electrostatic capacitor, in which a dielectric material is sandwiched between two conductive plates (Figure 5.15b). If a potential is applied between the electrodes, the supporting electrolyte-salt ions will diffuse toward the electrodes' surface to balance the accumulating charges. The applied charge is stored in the thusly formed electrical double layer at the electrodes and is solely of physical (*i.e.*, electrostatic) nature. Whereas the energy storage in the secondary batteries relies on electrochemical processes due to faradaic charge-transfer processes and typically involves changes in the active materials' chemical structures [34]. Although these energy-storage mechanisms appear fundamentally different, similarities between them have been emphasized. According to Frontana-Uribe *et al.*, ion diffusion represents one central aspect in the charge-storage process. Ion migration is due to an electric field between the electrodes on the one hand and electron injection into the electrodes on the other [105].

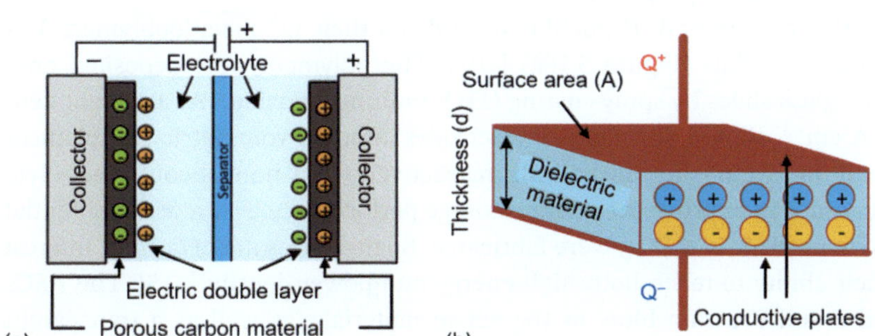

Figure 5.15 Schematic representation of the configuration of an EDLC (a) and a conventional electrostatic capacitor (b) [111] / CC BY 4.0.

As mentioned before, the capacitance (C in F, 1 F = 1 As V^{-1}) represents the most relevant performance parameter of a capacitor. This value can be calculated according to Eq. (5.6) in which A and d are the surface of the conductive plates and the thickness of the dielectric material, respectively [ε_0: vacuum permittivity (8.854×10^{-12} F m^{-1}), ε: permittivity of the medium between the plates] [112]. Accordingly, one will only obtain high capacitance values if the capacitor's plates have high surface areas, the separation between them is small, and/or the permittivity of the dielectric material is high.

$$C = \frac{A\varepsilon\varepsilon_0}{d} \tag{5.6}$$

Regarding practical energy-storage applications, the devices' maximum energy (E_{max}) also has to be considered. The breakdown voltage (U_d) represents the capacitor's limit (beyond this value, dielectric breakdown will occur). E_{max} can be determined according to Eq. (5.7). Although not expressed by this equation, the electrolyte's electrochemical stability represents one further limiting parameter [113].

$$E_{max} = \frac{CV}{2} = \frac{\varepsilon\varepsilon_0 \ AdU_d}{2} \tag{5.7}$$

But what is the relevance of redox-active materials (*e.g.*, inorganic oxides, organic polymers, *etc.*) in the context of capacitors? As said, EDLCs operate in the absence of such materials since the conductive carbon materials with their high surface areas already guarantee high charge-storage efficiencies [114]. The charge-storage concept has been developed further to combine the electrostatic charge storage with the faradaic one. To reach this goal, the traditional carbonaceous active materials have been replaced by (or modified with) redox-active ones. The resulting hybrid devices, which are commonly referred to as pseudocapacitors, typically feature a superior charge-storage capability [115, 116]. Fast charging/discharging rates, as known from capacitors, are combined with the benefits of faradaic charge-storage systems, *i.e.*, the stable voltage output, to afford significantly increased capacitance, the pseudocapacitance. As summarized elsewhere [38], mainly conductive polymers have been utilized, as the active materials, in the context of SC technologies. However, other types of materials, *e.g.*, coordination compounds, have also been employed for this same purpose [117].

Against this background, Higuchi *et al.* utilized their EC metallopolymers **3** as SC electrode materials (Figure 5.16a) [101]. The polymers were deposited onto ITO-coated glass slides by spray coating (ITO: indium-tin oxide). At a current density of 2 A cm^{-3}, the thusly obtained electrodes afforded volumetric capacitances of 62.6 and 98.5 F cm^{-3} for **3a** and **3b**, respectively, in a nonaqueous electrolyte system. In order to improve the energy-storage performance over a wider potential range, asymmetric SCs (ASCs) were fabricated. Such devices are of current interest due to their ability to reach both high energy and power density [115]. The ASCs contained **3** and Prussian blue, as the active materials, as well as a transparent polymeric ion gel, as the electrolyte (Figure 5.16b) [101]. The maximum volumetric energy densities of these ASCs (*i.e.*, 10–18 mWh cm^{-3}) were higher than those of Li thin-film batteries and the observed power densities (*ca.* 7 W cm^{-3}) were

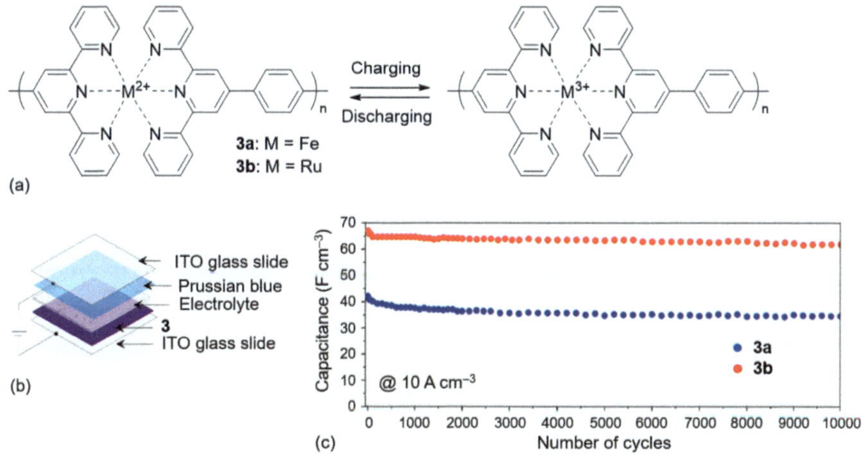

Figure 5.16 (a) Schematic representation of the charging and discharging of metallopolymers **3** (counterions were omitted for clarity). (b) Schematic representation of the configuration of an ASC, which contained **3** and Prussian blue, as the active materials. (c) Cycling stability of these same ASCs [101] / with permission of American Chemical Society.

competitive with those of conventional electrolyte capacitors. The Ru^{II}-containing polymer **3b** delivered a higher capacitance than its Fe^{II}-based counterpart **3a**. Both polymers represented highly robust active materials, which could be cycled up to 10,000 times without significant loss of capacitance (Figure 5.16c). The fully charged device, which contained **3b**, was able to light up an LED bulb for up to 5 min.

One further 1D metallopolymer, based on the $[Fe(tpy)_2]^{2+}$ linkage, was reported by Chakraborty *et al.* [118]. Similar to Higuchi's materials, polymer **4** also exhibited EC properties, *i.e.*, the material could reversibly be switched between the purple-colored reduced state and the yellow-/orange-colored oxidized one, as well as charge-storage capability (Figure 5.17a/b). EC SCs are of particular interest since such "dual performer" devices offer the advantage that the charge condition can be read out *via* the EC color indicator. When fabricated into a solid-state SC, which contained a $LiClO_4$-based gel electrolyte, average volumetric energy and power densities of 13 mWh cm^{-3} and 5 W cm^{-3} were obtained (Figure 5.17c). The SC operated with a high CE of *ca.* 83%; the initial capacitance was retained to 73% after 10,000 galvanostatic charge/discharge cycles (*i.e.*, charging at 2.5 V, discharging at −1.0 V). The fully charged device was able to illuminate a red LED bulb for up to 6 min (Figure 5.17d).

Not only linear coordination polymers but also CONASHs were evaluated as active materials for SC applications. For example, Qiu, Han, Zhuang *et al.* prepared a CONASH by the self-assembly of Fe^{II} ions and a rigid *tris*-tpy ligand at the liquid/liquid interface [119]. The resulting ultrathin CONASH (thickness of *ca.* 300 nm) was transferred onto a Si wafer to give an on-chip micro-SC, which achieved remarkably high areal capacitance of 1.25 mF cm^{-2} (at 50 mV s^{-1}) as well as a volumetric energy density of 5.8 mWh cm^{-3}. Moreover, the on-chip CONASH exhibited good alternating current (AC) line-filtering properties (*i.e.*, −71.3° at

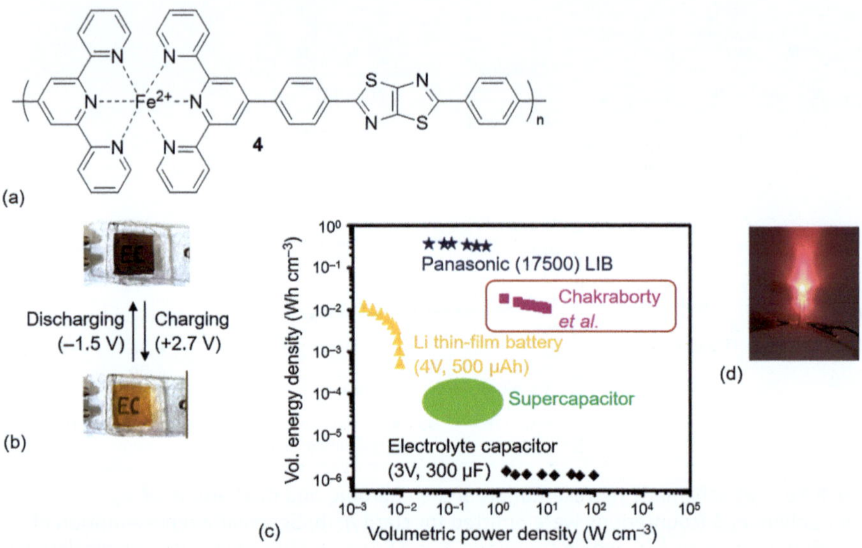

Figure 5.17 (a) Schematic representation of the Fe^{II}-containing metallopolymer **4** (counterions were omitted for clarity). (b) Photographs of the EC thin-film device in the reduced (discharged) and oxidized (charged) state. (c) Ragone plot of Chakraborty's SC in comparison to other energy-storage devices. (d) Photograph of a red LED, which was illuminated by the fully charged device. Reproduced from [118] / with permission of Elsevier.

120 Hz) and a short resistance-capacitance time of only 0.06 ms [electrolyte: LiCl in polyvinylalcohol (PVA)]. In related work, Liu *et al.* prepared a CONASH from Fe^{II} ions and a star-shaped ligand, whose three tpy arms were attached to a benzene core *via* flexible linkers [120]. Aiming for a potential application in smart windows, these authors assembled a dual-functional ASC, which contained the CONAH and Prussian blue, as the EC and charge-storage layer, respectively. This device provided a reasonably good coloration contrast of 64%, short coloring time of *ca.* 2 s as well as a high coloration efficiency of >440 $cm^2\ C^{-1}$. Regarding its energy-storage capability, a specific capacitance of 29.12 $F\ g^{-1}$, at a current density of 0.4 $A\ g^{-1}$, was obtained.

In one further study, Shankar, Ajayaghosh *et al.* compared the capacitive behavior of three EC CONASHs [121]. These CONASHs were prepared from Fe^{II} ions and the respective tetra*kis*-tpy ligands **5** (Figure 5.18a) *via* a straightforward self-assembly in solution. The symmetrical two-electron SCs, which contained these CONASHs, as the active materials, provided volumetric capacitances up to 866.2 $F\ cm^{-3}$ at the current density of 0.25 $A\ g^{-1}$. The charging/discharging kinetics of the SCs as well as their volumetric capacitances and energy densities correlated with the CONASH's electrical conductivities (*i.e.*, the best overall performance was observed for the SC, which contained the Fe^{II}–**5b** CONASH; Figure 5.18b/c). In either case, a particularly good cycle lifetime was found, *i.e.*, the initial capacitance remained stable over at least 3000 charging/discharging cycles at the current density of 1.0 $A\ g^{-1}$.

Jena and Choudhury followed the same strategy, *i.e.*, using $[Fe(tpy)_2]^{2+}$-based CONASHs, as EC-active materials for SCs [122]. However, these authors went one

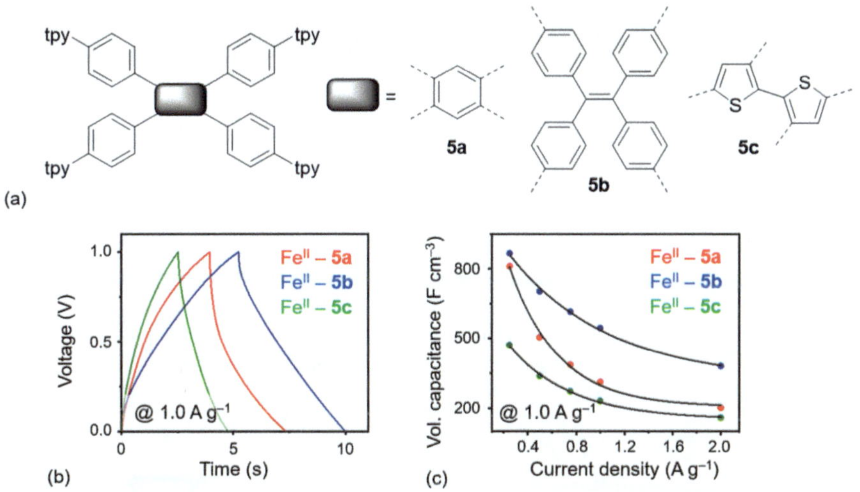

Figure 5.18 (a) Schematic representation of the tetra*kis*-tpy ligands, which were used for the formation of Fe^{II} CONASHs. (b) Galvanostatic charging/discharging curves of the SCs. (c) Comparison of the volumetric capacitances of the SCs at different current densities [121] / with permission of American Chemical Society.

step further and coupled their SC to a commercial Si solar cell, thus demonstrating the storage of solar energy as well as its on-demand use. In this study, three different *tris*-tpy ligands were used for the CONASH formation (**6**, Figure 5.19). According to a structure–property evaluation, ligand **6a** gave the most promising CONASH due to its 3D structure and porosity on the one hand and the stabilization of the Fe^{III} species in the charged state by a high density of present counter anions on the other. The five-layered device, driven by solar light, featured electrochromism with fast color-switching kinetics (*i.e.*, 1.2 and 2 s for the charging and bleaching, respectively), high optical contrast ($\Delta T = 47\%$), high open-circuit charge-retention time ($t_{50\%} = 193$ min) as well as a remarkably high stability over 10,000 cycles. The energy-storage capability was employed to energize some applications, *e.g.*, a 3.5 V LED, a 1.7 V digital thermo-hygrometer, and a 3 V multimeter (Figure 5.19b). For these purposes, five such devices were connected in series. Although the operational times remained short, the proof-of-concept study showed that hybrid solar SCs can be realized on the basis of metallopolymers as the active materials.

A fundamentally different strategy to fabricate a tpy-containing electrode for SC applications was followed by Xu, Dai *et al.*, who opted for a metal-loaded composite electrode instead of the aforementioned 1D/2D metallopolymers [123]. PAni represents one of the most prominent conductive polymers used in energy-storage applications – either as a conducting agent or the actual active material [124]. Regarding SCs, this polymer has attracted widespread interest due to various oxidation states, which enable a tunable pseudocapacitive performance. Detrimentally, pristine PAni exhibits a poor cycling stability, which is due to considerable volumetric changes and irreversible backbone degradation in course of the charging/discharging process. Many strategies have been followed to

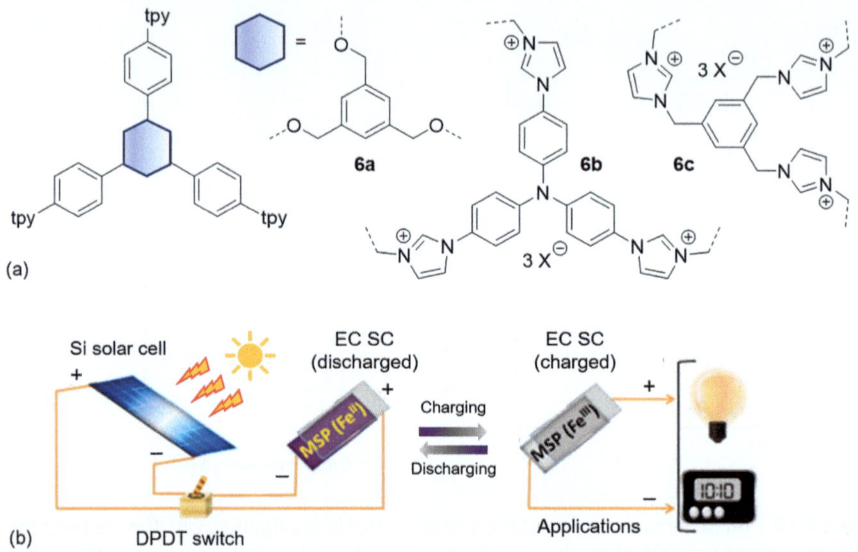

Figure 5.19 (a) Schematic representation of the *tris*-tpy ligands, which were used for the formation of Fe^{II} CONASHs (X = Br or OAc). (b) Schematic representation of the coupled device, which enabled the harvesting of solar light, energy storage, and the on-demand delivery of electricity. Reproduced from [122] / with permission of Elsevier.

counteract this. In this respect, the use of carbon materials, in particular graphene, represents a versatile approach to improve the cycling stability. Graphene/PAni composites not only feature sufficient flexibility to withstand structural damage but the conductivity of PAni is also largely increased [125]. Moreover, the doping of PAni with first-row transition-metal ions is an effective method to increase the cycle stability [126]. *Inter alia*, the adsorption of Cr^{III} ions on a carbon/PAni composite resulted in an enhanced capacitance [127]. Taking up this idea to obtain high-performance SC electrode materials, Xu, Dai *et al.* fabricated a Cr^{III}-loaded composite material *via* a multistep procedure [123]. In detail, graphene oxide (GO) nanosheets were first treated with 1,4-diaminobenzene to introduce active sites for further functionalization. Subsequently, tpy-equipped PAni nanofibers were grown on the GO surface by the *in situ* polymerization of aniline in the presence of an NH_2-functionalized tpy derivative. This protocol was initially elaborated by Xiong, Cheng *et al.*, who aimed for the improvement of PAni's EC properties by loading with metal ions [128]. The thusly obtained composite was loaded with Cr^{III} ions, followed by reaction with ascorbic acid to obtain a reduced GO (rGO) surface [123]. The resulting composite (named rGPT-Cr) provided a significantly higher specific capacitance than other PAni/graphene composite electrodes (*i.e.*, 1499 $F\,g^{-1}$ at 1 $A\,g^{-1}$ in 1 M H_2SO_4, as the electrolyte). As shown in Figure 5.20a,b, the capacitance performance of rGPT-Cr was superior to those of the reference materials, which were not loaded with Cr^{III} ions (GPT) or did not contain tpy moieties (rGP-Cr). An ASC was assembled, which contained rGPT-Cr and activated carbon as the cathode and anode, respectively, as well as 1 M H_2SO_4 in PVA, as the

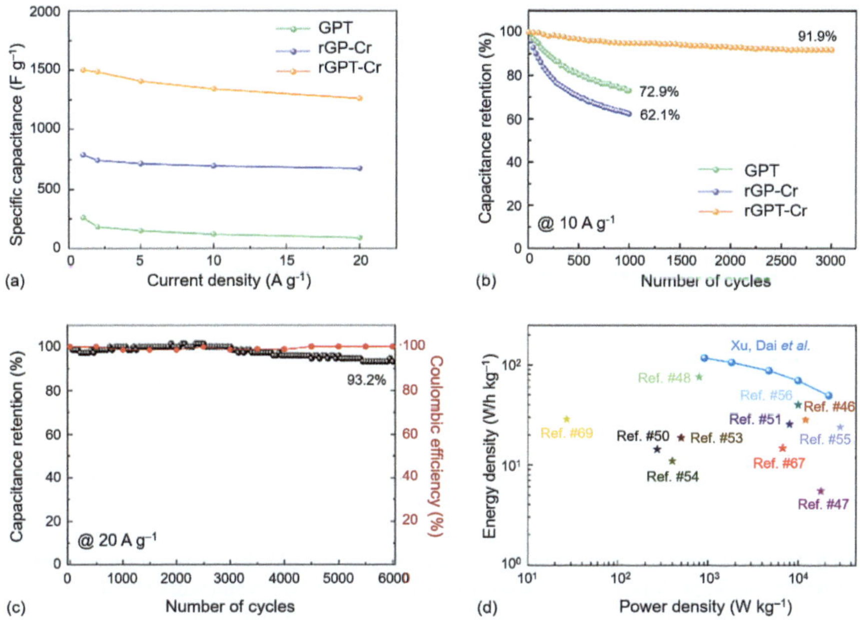

Figure 5.20 (a) Rate performance of the three different composite electrodes. (b) Cycling stability of these same electrodes. (c) Cycling stability of the as-prepared ASC. (d) Ragone plot of the as-prepared ASC and various related devices (the reference numbers refer to those in the original work) [123] / with permission of Elsevier.

gel electrolyte. This device possessed a very good cycling stability with the initial capacitance being retained to >92% after 6000 cycles at 20 A g^{-1} (Figure 5.20c). Moreover, a maximum energy density of 117.5 Wh kg^{-1} and a power density of 915.0 W kg^{-1} were achieved, thus outcompeting many related ASCs (Figure 5.20d).

MOFs, which are formed by transition-metal ions and organic ligands, represent an important class of materials. Their porosity and channels can be tuned *via* cautious selection of the building blocks. MOFs are commonly classified according to their components and structures [129]. Thereby, their abbreviation often includes the place of finding. The porous structure and large specific surface area of MOFs have already been capitalized for numerous purposes: Gas-storage, waste-water treatment, medicine (*i.e.*, drug delivery), energy, *etc.* However, when aiming for the use of MOFs in the context of energy-storage applications, specifically SCs, their intrinsically poor conductivity represents a limiting factor. This can be counteracted, *e.g.*, by mixing the MOF with a conductive material (*e.g.*, graphene or polymers) [130]. Besides this, electrical conductivity and/or redox activity can be introduced by the proper choice of the inorganic and organic MOF building blocks [131–133]. In one common approach, the electrochemically inert Zr^{IV}-based nodes are substituted by redox-active ones (*i.e.*, transition-metal ions with partially filled d-shells: Co^{II}, Cu^{II}, Ni^{II}, *etc.*). Moreover, the use of carboxylate ligands with extended π-conjugated systems instead of the traditional ones (*e.g.*, terephthalic acid, *etc.*) is beneficial for the charge transport within the MOF [134].

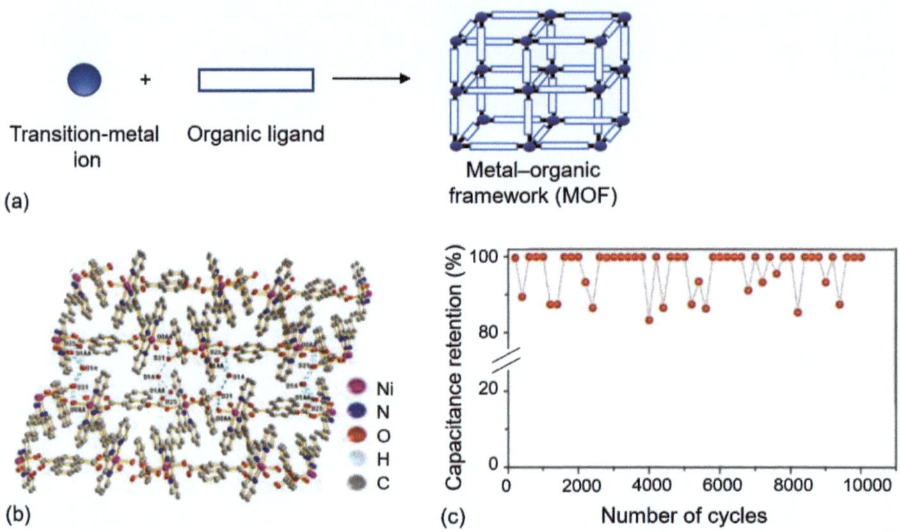

Figure 5.21 (a) Schematic representation of a MOF synthesis [129] / with permission of American Chemical Society. (b) Solid-state structure of the HU17 framework [135]. (c) Cycling stability of the HU17 electrode [135] / with permission of Elsevier.

In line with studies, Ma *et al.* assembled a MOF by the hydrothermal reaction of $Ni(NO_3)_2$, 1,4-*bis*(2,2′:6′,2″-terpyridin-4′-yl)benzene and 1,4-naphthalenedicarboxylic acid (Figure 5.21a) [135]. The resulting MOF, referred to as HU17, was comprised of zigzag chains, which were interconnected *via H*-bonding interactions (Figure 5.21b). Notably, this MOF represents one of the scare examples, where photo- or redox-active tpy complexes have been incorporated into a MOF framework. Some structures, which contained Ru^{II}–tpy or Fe^{II}–tpy *bis*-complexes have been discussed in the context of photocatalytic applications [136–138]. Besides this, nonclassical tpy ligands have been utilized as linkers within MOF architectures (these structures are dealt with in Chapter 2) [139]. Ma's Ni^{II}-based MOF was employed as the electrode active material in an SC, which featured a specific capacitance of *ca.* 333 $F\,g^{-1}$ at the scanning rate of 1 $mV\,s^{-1}$ and a good rate performance [135]. Moreover, the device showed an exceptionally good cycling stability with efficient capacitance retention over 10,000 cycles (Figure 5.21c).

5.5 Terpyridine Complexes in Redox-Flow Batteries

Unlike for the aforementioned battery/capacitor systems, whose working principle basically relies on the storage of energy in/at the electrodes, the energy of RFBs is stored in the electrolyte. The history of RFBs can at least be traced back to 1879, when Doyle's patent on a "galvanic battery" was filed [140]. This device, which was comprised of multiple refillable cells (Figure 5.22a), is commonly considered the precursor to modern Zn/halogen flow cells, which were proposed as power source for electric vehicles almost a century later (Figure 5.22b) [142]. However, recharging

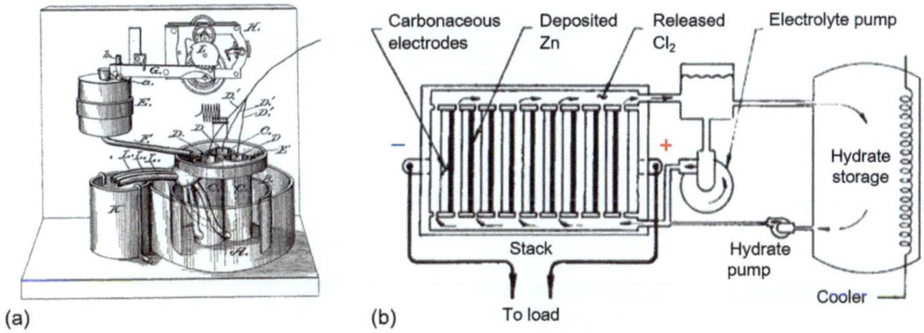

Figure 5.22 (a) Reproduction of the original drawing of Doyle's Zn-Br_2 flow battery; A: spill enclosure (dielectric container), B: cylindrical Zn anode, C: porous dielectric jars/separators (3 are shown), D: porous electron-conducting cathode (*e.g.*, carbon) coated on the separators' inner surfaces, D′: electric wires to the cathodes, E: container with a catholyte liquid, E′: a funnel for the catholyte fluid, F: catholyte delivery tube, L: overflow tubes, K: container for discharged fluid. (b) Schematic representation of the Zn/chlorine-hydrate flow cell (24 such cells with 40 V and 8 A were used to power a modified 1971 Vega hatchback car) [141] / with permission of IOP Publishing.

of the original cell was not claimed (but would have been feasible by connecting pumps) [141].

In the 1940s, Kangro and Pieper elaborated – independently from each other – the concept to store electrical energy in solution further; unlike Doyle, these researchers focused on reversible redox couples of certain transition metals and constructed rechargeable flow-cell prototypes [143]. The first functional RFB, which used $Cr^{III/IV}$ and $Ti^{III/IV}$ redox couples, was patented by Kangro in 1949 [144]. Numerous alternatives were tested in subsequent years until the all-vanadium RFB (VRFB) was eventually invented by Rychcik and Skyllas-Kazacos [145, 146]. Beneficially, VRFBs, which conventionally contain H_2SO_4, as the electrolyte, are capable to use in-total four of vanadium's oxidation states for the half-cell reactions (Figure 5.23).

The working principle of a VRFB is shown in Figure 5.24 (this fundamental configuration not only applies to all RFBs, independent from the actually used active materials but also to fuel cells [149]). The active materials of RFBs are dissolved in an electrolyte and stored in reservoirs, one for each half-cell. The

$$VO_2^+ + 2\,H^+ + e^- \rightarrow VO^{2+} + H_2O \qquad E^0 = +1.00\text{ V}$$

$$V^{3+} + e^- \rightarrow V^{2+} \qquad E^0 = -0.26\text{ V}$$

Figure 5.23 Half-cell reactions of a VRFB. The photograph shows solutions of "vanadium sulfate" in its four different oxidation states [147] / Walter Tau / Wikimedia Commons / CC BY-SA 4.0.

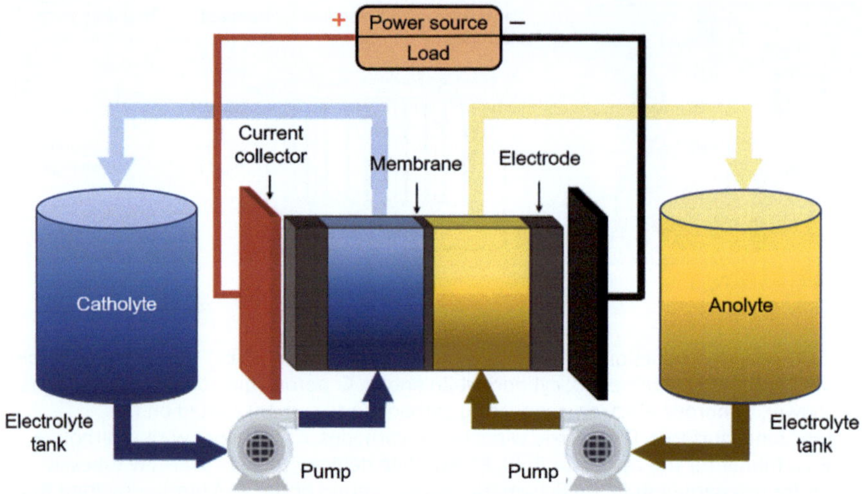

Figure 5.24 Schematic representation of a redox-flow battery [148] / MDPI / CC BY 4.0.

electrolyte solutions, commonly named catholyte (posolyte) and anolyte (negolyte), are pumped into the galvanic cell, where the redox reactions (*i.e.*, charging and discharging) proceed [150].

The membrane, which separates the half-cells from each other, represents one critical RFB component; its efficiency is crucial regarding both the performance and cycle lifetime. According to Epps III *et al.*, one has to take numerous criteria into account in order to select a proper membrane [151]. These include conductivity and selectivity, as well as stability (mechanically and chemically), cost-efficiency, and sustainability. Conductivity and selectivity represent competing phenomena, which have to be cautiously balanced. The migration of charge-balancing ions through the membrane has to be enabled. In other words, high ionic conductivity is desirable for reaching high VE values and, thus, high power densities [152]. However, the membrane should simultaneously exclude the crossover of active species in order to suppress CE-reducing self-discharge processes [153]. As mentioned beforehand, the storable energy of an energy-storage device is defined as the product of its capacity and voltage (Eq. (5.4); Section 5.2; the same relation also applies for the EE, which is the product of CE and VE. Thus, two crucial membrane properties – the selectivity and ionic conductivity – are respected [151]. Moreover, the membrane must possess sufficient mechanic stability to withstand the *in-operando* condition, *i.e.*, the pressure and flow rates. By the same time, membrane degradation by the harsh chemical environment should be minimized. This issue is of particular relevance regarding the commonly used RFB electrolytes, which are often highly corrosive liquids (*e.g.*, H_2SO_4 in VRFBs) [154].

Nonetheless, VRFBs have evolved to become the leading technology, which targets the large-scale energy storage by attaching it to power plants or electrical grids. The largest VRFB battery located in Germany was installed in Pfinztal. This energy-storage device, coupled to a 2 MW wind turbine, is prospected for a power output of 2 MW and a capacity of 20 MWh [155]. In 2022, the worldwide largest VRFB began operating in Dailan, China; this peak-shaving energy park is reported to provide an initial capacity of 400 MWh and a 100-MW power output [156]. Besides these grid applications, the scalability of the technology has also been adapted for residential RFBs. For example, a German manufacturer offers RFBs, which can be coupled to domestic photovoltaic systems (capacity of 4 kW, peak power of 5 kW) [157].

The main advantages of VRFBs over solid-state batteries, which rely on the LIB technology, are their remarkably long cycle lifetime and time durability (*i.e.*, >15,000 cycles; 20–30 years) in concert with low costs per MWh, which are as low as a few tens of €/$ cents [158]. However, one also has to consider the shortcomings of vanadium-based electrolytes, which use highly corrosive electrolytes, dependency on commodity prices, and toxicity [159]. Related RFBs, such as the vanadium-bromide or vanadium-oxygen cells have been developed in order to increase the energy density [160]. The former are capable to provide a voltage of 1.3 V and energy densities in the range of 35–70 Wh L^{-1}; however, the evolution of hazardous bromine vapors represents a considerable drawback [161]. In V-O_2 cells, the positive half-cell electrolyte and electrode are replaced simply by air, thus reducing the material costs and increasing the specific energy up to 40 Wh kg^{-1} [162]. In order to fully refrain from the use of vanadium in RFBs, research has been intensified to identify cheaper and less environmentally problematic active materials. In this respect, the zinc-iron RFB represents a notable invention. This type of RFB, which uses two cheap, earth-abundant, and nontoxic transition metals, is competitive with VRFBs in terms of energy density and efficiency as well as cycle lifetime [163, 164]. Moreover, the estimated capital costs of Zn-Fe RFBs are remarkably low when compared to those of other RFB technologies or conventional batteries (*e.g.*, Ni-Cd or Pb-acid; Figure 5.25) [165].

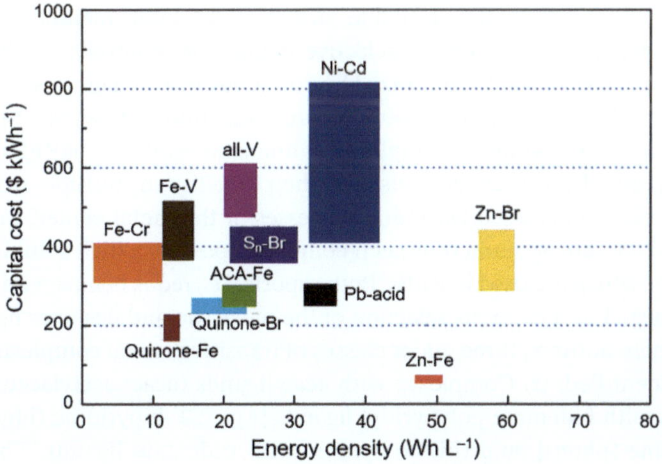

Figure 5.25 Comparison of various RFB and conventional battery technologies in terms of cost and energy density (ACA: alloxazine 7/8-carboxylic acid) [165] / with permission of John Wiley & Sons.

Next, the various polysulfide-containing RFBs are worth mentioning. These PSRFBs capitalize on the wide abundance of sulfur and its lowest cost per unit charge among the available reactive materials (PSRFB: polysulfide RFB) [166]. The high solubility of polysulfide anions, as the involved redox-active species, in aqueous and organic solvents is beneficial when aiming for RFBs with high energy densities. In particular, S_n-Br RFBs have attracted widespread interest due to the favorable combination of low chemical cost of the active materials with high theoretical energy densities. These PSRFBs, which rely on the Na_2S_2/Na_2S_4 and $NaBr_3/NaBr$ redox couples, are considered to meet the requirements of grid storage.

One further example for a metal-free RFB was reported in 2014 by Aziz *et al.* [167]. These authors used a water-soluble 9,10-anthraquinone derivative, as the redox active moiety, in combination with the Br_2/Br^- redox couple [167]. This quinone-bromide RFB reached a maximum galvanic power density of 600 mW cm^{-2} at a current density of 1.3 A cm^{-2}; charging/discharging occurred with an average capacity retention of 99% over at least 10 cycles. Later, the hazardous Br_2/Br^- redox couple was replaced by the safer and nonvolatile ferrocyanide/ferricyanide one [168]. In recent years, a wide range of other organic compounds and transition-metal complexes was tested as active materials for RFB applications [169–171]. In one example, an all-organic RFB was assembled, which operated with 2,2,6,6-tetramethylpiperidinyl-1-oxyl (TEMPO) and *N*-methylphthalimide, as the catholyte and anolyte compounds, respectively (electrolyte: $NaClO_4$ in MeCN) [172].

As the final notable milestone to be mentioned in this overview, the development of a polymer-based RFB (pRFB) has to be highlighted. Schubert *et al.* utilized redox polymers, *i.e.*, polymers, which contained redox-active moieties in their side chains, as the active materials [173]. This all-pRFB provided an energy density of 10 Wh L^{-1}, a current density of >100 mA cm^{-2}, and an exceptionally long cycle lifetime: The initial capacity was retained to *ca.* 80% after 10,000 charging/discharging cycles 20 mA cm^{-2}. Moreover, the pRFB contained aqueous NaCl, as the environmentally benign electrolyte, and commercially available size-selective filter membranes instead of the expensive, typically used ion-selective membrane materials. In the following, the concept of pRFBs has been utilized by numerous research groups and a wide range of redox polymers has already been incorporated into pRFBs [34, 38].

Regarding the use of transition-metal complexes in nonaqueous RFBs (naRFBs), Hogue and Toghill pointed out their benefits. *i.e.*, their stability in multiple oxidation states, which can be accessed *via* redox processes of the metal center, the ligands, or both [174]. Moreover, numerous such complexes possess a high solubility in organic solvents, which are used in RFBs. Both aspects, *i.e.*, redox behavior and solubility, can be adjusted by a cautious selection of the metal ion and designer ligands. According to these authors, three major classes of transition-metal complexes for naRFBs can be identified: (i) Complexes with acac ligands (acac: acetylacetonate), (ii) complexes with bidentate polypyridyl ligands [*e.g.*, 2,2′-bipyridine (bpy) or 1,10-phenanthroline (phen)], and (iii) complexes with tridentate ligands. The latter class includes ligands from the tpy family as well as their counterparts, in which at least one of the outer pyridine rings has been replaced (preferentially) by an azole one.

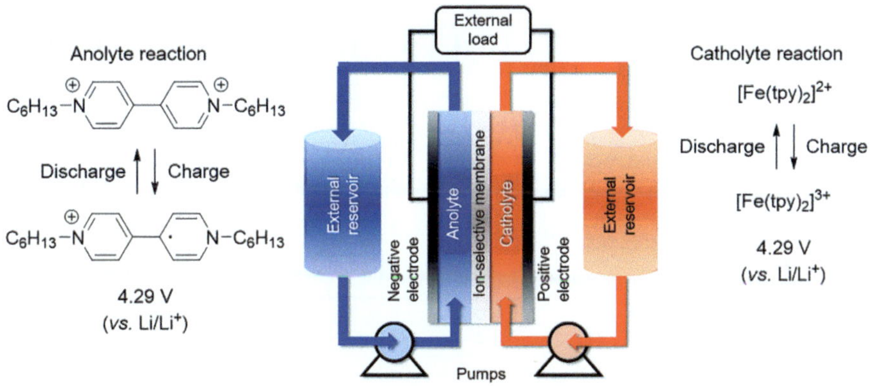

Figure 5.26 Schematic representation of the configuration of the naRFB, which used a mixture of ethylene carbonate and dimethyl carbonate, as the solvent. The relevant half-cell reactions are also shown (counterions are omitted for clarity) [176] / Royal Society of Chemistry / CC BY 3.0.

To the best of our knowledge, the idea to use tpy complexes, specifically $[Fe(tpy)_2]^{2+}$, in the context of naRFBs was taken up first by Ikeda *et al.* [175]. The complex was evaluated as the catholyte compound in bmim(Tf_2N), as the ionic-liquid solvent (bmim: 1-butyl-3-methylimidazolium; Tf: trifluoromethylsulfonyl). The catholyte possessed a good cycling stability, as expressed by continuous CV scanning over 100 cycles at 100 mV s^{-1}. However, the poor solubility of the complex in the solvent disabled any practicable use. On the other hand, $[Fe(tpy)_2](BF_4)_2$ was sufficiently soluble in carbonate-based electrolytes, which enabled its use as catholyte for naRFBs [176]. The naRFB, which used a methyl viologen (MV^{2+}) derivative, as the anolyte, and a mixture of ethylene carbonate and dimethyl carbonate, as the solvent, afforded a high discharge voltage of *ca.* 1.4 V as well as CE and EE of >98% and >84, respectively, over 50 cycles (Figure 5.26).

In related work, Armstrong and Toghill compared the electrochemistry of the corresponding $[Co(tpy)_2]^{2+}$ complex to those of complexes $[Co(\mathbf{7})_2]^{2+}$ (Figure 5.27a) [177]. All complexes showed reversible Co^{II}/Co^{III} and Co^{I}/Co^{II} redox processes (Figure 5.27b). Thus, symmetric RFBs were assembled, in which these redox systems were used in the posolyte and negolyte half-cells, respectively [electrolyte: $(TBA)PF_6$ in MeCN; TBA: tetrabutylammonium]. Due to the weaker σ-donor/π-acceptor character of the 3,5-dimethylpyrazole moiety, the Co^{II}/Co^{III} redox potential was increased, thus raising the cell potential of the symmetrical RFBs from 1.06 V for $[Co(tpy)_2]^{2+}$ sequentially to 1.81 V for $[Co(\mathbf{7b})_2]^{2+}$, whereas the presence of the electron-poor 1,2,4-triazole moiety in $[Co(\mathbf{7c})_2]^{2+}$ was reflected by positive shift of both redox couples; this, in turn, resulted in a cell potential of 1.50 V (Table 5.2). The RFBs operated reliably with good to high efficiencies at least over 10 charging/discharging cycles (Figure 5.27c,d). Owing to the intrinsically low solubility of the complexes, the overall energy density was limited to 2.58–12.80 Wh L^{-1} (the highest energy density was obtained for the most soluble complex $[Co(\mathbf{7b})_2]^{2+}$).

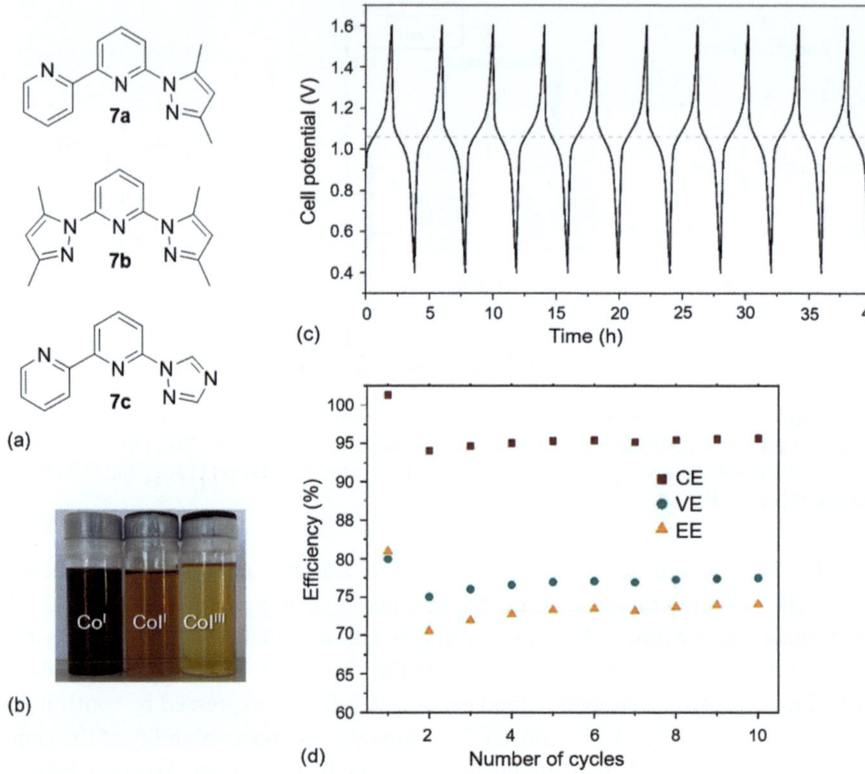

Figure 5.27 Schematic representation of the tridentate ligands **7**. (b) Photographs of the solutions of the Co–tpy *bis*-complex in its various oxidation states. (c) Continuous charging/discharging of the symmetrical RFB; which contained $[Co(tpy)_2]^{2+}$, as the active material. (d) Cycling efficiencies of this same RFB [177] / with permission of Elsevier.

Table 5.2 Summary of the performance parameters of bis-tridentate Co^{II} complexes in symmetrical RFBs.

Complex	ΔE (V)$^{a)}$	Charge/discharge time (h)$^{b)}$	CE (%)$^{c)}$	VE (%)$^{c)}$	EE (%)$^{c)}$
$[Co(tpy)_2]^{2+}$	1.06	2.14 (2.03)	95.2	76.8	73.0
$[Co(\mathbf{7a})_2]^{2+}$	1.38	2.84 (2.84)	99.8	81.2	80.8
$[Co(\mathbf{7b})_2]^{2+}$	1.81	2.84 (2.55)	89.7	70.3	63.1
$[Co(\mathbf{7c})_2]^{2+}$	1.50	2.90 (2.73)	94.2	77.0	72.5

a) Difference between the two half-cell potentials [CVs measured in MeCN containing 0.1 M $TBA(PF_6)$ at a scan rate of 50 mV s^{-1}, Au electrode, potentials reported *vs.* Fc/Fc^+].
b) Mean values; the discharge times are given in brackets.
c) Efficiency values averaged over 9 out of 10 cycles (2–10); for all cell test: 1 mM complex in MeCN containing 0.5 M $TBA(PF_6)$ at ±4.4 μA cm^{-2}.
[177] / with permission of Elsevier.

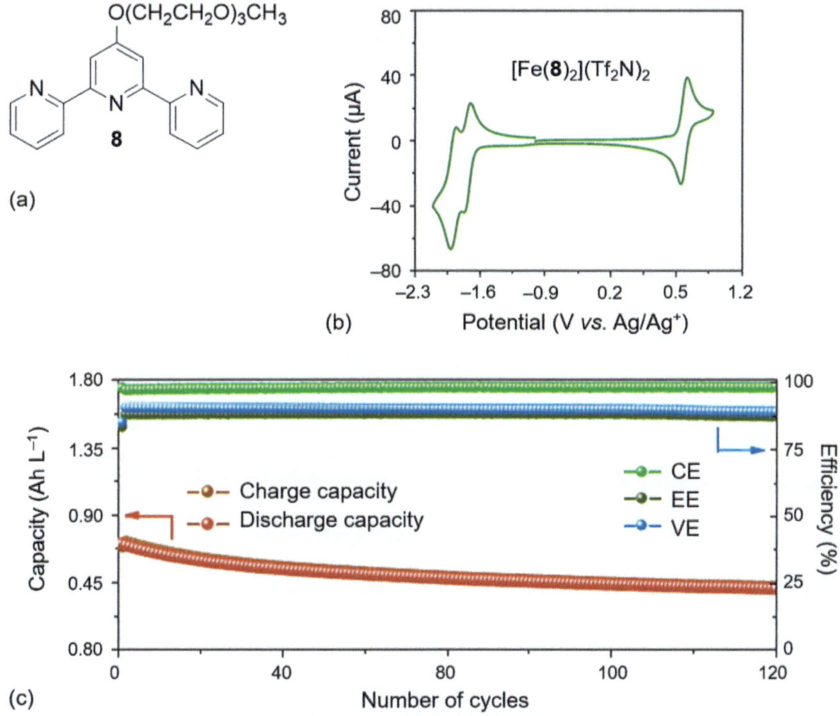

Figure 5.28 (a) Schematic representation of the oligoether-substituted tpy ligand **8**. (b) CV of a 2 mM solution of $[Fe(\mathbf{8})_2](Tf_2N)_2$ in a 0.5 M $TBA(PF_6)$/MeCN solution. (c) Cell performance of a RFB with 0.04 M $[Fe(\mathbf{8})_2][Tf_2N]_2$ in 0.5 M $TBA(Tf_2N)$/MeCN solution [178] / with permission of Elsevier.

In order to improve the solubility, Zhao *et al.* synthesized the 4′-substituted ty ligand **8** (Figure 5.28a) [178]. Due to the short oligoether tails, the corresponding complexes possessed improved solubility in MeCN and were analyzed regarding their applicability, as active material, in symmetrical RFBs. All complexes $[M(\mathbf{8})_2]^{2+}$ (M = Mn; Fe, Co) and $[Cr(\mathbf{8})_2]^{3+}$ exhibited multiple, reversible redox processes. The complex $[Fe(\mathbf{8})_2](Tf_2N)$ was selected for use in an RFB, due its superior solubility and auspicious electrochemistry (Figure 5.28b). The symmetrical RFB, which contained this complex as catholyte and anolyte active material, afforded a high cell voltage of 2.3 V. The RFB operated with high efficiency (*i.e.*, CE of 97%, EE of 88%) and promising capacity retention (*i.e.*, *ca.* 60% after 160 charging/discharging cycles) at the current density of 15 mA cm^{-2} (Figure 5.28c).

What about tpy complexes as catholyte active materials for aqueous RFBs? When dealing with aqueous electrolytes, one has to be aware of the narrow electrochemical potential window of *ca.* 1.23 V, which can be utilized to operate the aqueous RFB (aRFB). Thus, the maximum ΔE value, which is the difference between the potentials of the catholyte and anolyte redox reaction, is relatively small and limits the accessible energy density [179]. High energy densities can only be reached, if the predominant side reactions, which are the OER and hydrogen-evolution

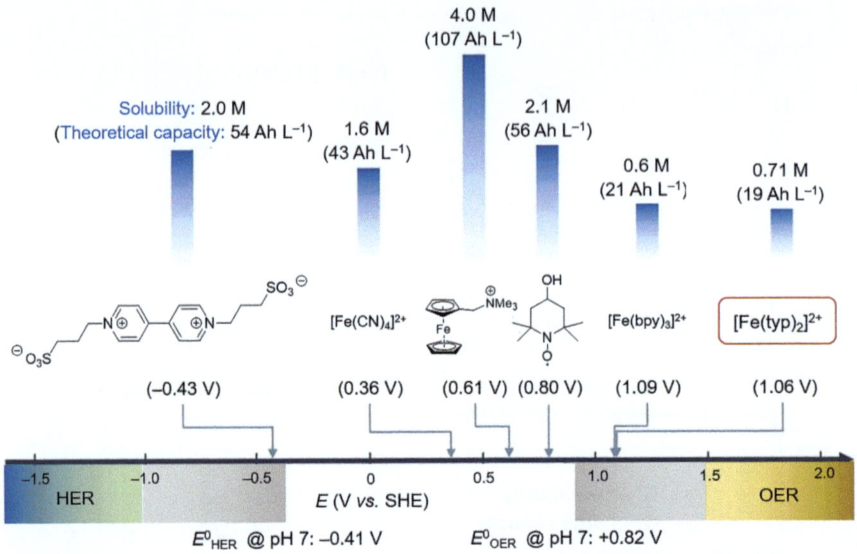

Figure 5.29 Overview over various commonly used active materials in aRFBs and archetypal Fe^{II}–polypyridyl complexes. Their redox potentials (*vs.* SHE), solubility in water as well as the theoretical energy density are listed. The gray regions represent the potential ranges, where the HER and OER are thermodynamically allowed under pH-neutral conditions. The colored regimes represent the potential ranges for the kinetic HER and OER at glassy-carbon electrodes (electrolyte: 0.5 M NaCl in water) [179] / with permission of AIP Publishing LLC.

reaction (HER), are suppressed. This becomes of particular importance when using high-voltage catholytes and/or low-voltage anolytes (Figure 5.29).

As shown in Figure 5.29, Fe^{II}–polypyridyl complexes, such as $[Fe(bpy)_3]^{2+}$ or $[Fe(tpy)_2]^{2+}$ possess relatively high redox potentials when compared to ferrocene (Fc) derivatives and, thus, their use as catholytes might already compete with the thermodynamically enabled HER [180]. Nonetheless, aRFBs, which use water-soluble complexes from the $[Fe(bpy)_3]^{2+}$-family, as catholytes, have been realized [181]. These aRFBs, which used a methyl viologen (MV^{2+}) derivative, as the anolyte, provided a high cell voltage of 1.3 V and maximum power density of 120 mW cm^{-2} under pH-neutral condition. The remarkably high cycling stability was expressed by the low capacity fading of 0.07% per day over 35 days of continuous cycling. Regarding high energy densities, the solubility of the complex becomes of importance. Unlike for naRFBs, the use of complexes with PF_6^- or BF_4^- anions in aRFBs is unfavorable, whereas solubility in water can be guaranteed when opting for chloride or sulfonate anions. Moreover, according to Carnelley's rule, solubility in aqueous media can be improved by a symmetry-breaking design of the molecular structure [182]. This can be achieved, *e.g.*, by using asymmetric ligands or mixed-ligand, *i.e.*, heteroleptic complexes. Substituents on the ligand can not only be used to modulate the solubility in water, they have also an impact on the voltage *via* their electron-donating/-withdrawing properties (Figure 5.30) [180]. This strategy is applicable to both families of complexes, Fe^{II} *tris*-bidentate as well

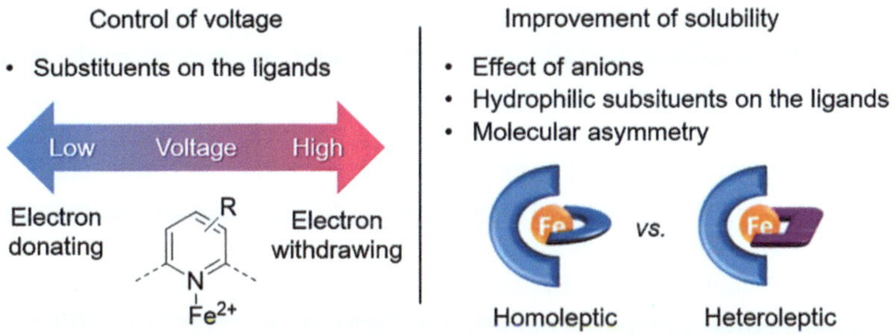

Figure 5.30 Summary of the design strategy for Fe^{II}–polypyridyl catholytes for aRFBs with improved voltage and solubility [179] / with permission of AIP Publishing LLC.

as *bis*-tridentate ones. However, $[Fe(tpy)_2]^{2+}$ (or its derivatives) have not yet been implemented, as catholyte active material, in aRFBs.

5.6 Concluding Remarks

In order to improve the performance of energy-storage devices, numerous electrode and electrolyte materials have been explored in the last few decades. Regarding electrode materials, research aims for the improvement of the energy-storage capability on the one hand and sustainability on the other. It is intended to overcome the shortcomings of, *e.g.*, LIBs by eco-friendlier solutions that eventually fully avoid the use of Li compounds. This goal can be reached by using organic/polymeric active materials or by alternative metals (or their organo-metal compounds), which are by far more abundant than Li. These aspects have been discussed in several reviews on this topic [10, 183, 184]. In the same context, alternative electrolyte systems for solid-state batteries have been developed [185, 186].

For all three major classes of energy-storage systems – metal-ion (and related) batteries, SCs and RFBs – tpy complexes have been tested, as active materials. When dealing with such active materials, one has to take their solubility in the electrolyte into account. This property is beneficial for RFBs, in which the redox-active complexes are dissolved in the catholyte. This concept has been realized for nonaqueous and aqueous RFBs. So far, Fe^{II}–tpy and Co^{II}–tpy complexes have mainly been used. However, when dealing with such complexes, as aRFB catholytes, one has to be aware of the competing electrochemical HER, which becomes relevant, if the catholyte's redox potential is high, as in the case of $[Fe(tpy)_2]^{2+}$ [180]. Moreover, the specific capacity (C_{spec}) of such complexes is low, owing to the molar mass of the redox-inactive tpy ligands. In the case of the complex cation $[Fe(tpy)_2]^{2+}$, the C_{spec} value is only 51.30 mAh g^{-1}.

This shortcoming also holds true for other battery concepts, which capitalize on the redox behavior of tpy complexes. Additionally, the complexes have to be prevented from dissolving into the electrolyte, thus reducing the capacity fading and shorting the cycle lifetime. For this purpose, the tpy complexes have been incorporated into polymer networks by self-assembly [85] or electropolymerization strategies [82]. The resulting metallopolymers featured an exceptionally good long-term performance in metal-ion batteries (MIBs), as expressed by capacity retention of *ca.* 90% over some thousands of charging/discharging cycles. Similarly, linear or 2D metallopolymers have been employed, as active materials, in SCs. These SCs could be charged/discharged over 10,000 cycles with almost quantitative capacitance retention. The fully charged devices have been used to successfully power small electric devices, *e.g.*, LED lamps. The EC SCs represent one notable type of device, due to their dual performance. If the active material features different colorations in the charged and discharged state, the state-of-charging can simply be read out *via* the EC color indicator (this implies, that a transparent electrode substrate is used). Functional EC SCs were fabricated using Fe^{II}–tpy or Ru^{II}–tpy metallopolymers. In the case of the former, the purple-colored reduced state could easily be distinguished from the oxidized one, which possessed a yellow-orange coloration [118].

To conclude, tpy complexes of redox-active transition-metal ions feature a good performance, when used, as the active material, in energy-storage devices. However, the moderate specific capacity represents a considerable shortcoming, in particular when compared to the wide range of organic compounds, which have been presented in recent years. Whereas the combination of energy storage and electrochromism affords EC SCs, which can be fabricated as flexible and transportable as well as large-scale and stationary devices. In this emerging application, polymers, which incorporate $[Fe(tpy)_2]^{2+}$ complexes, might be of particular interest, due to the well compromised specific capacitance, long-term operational stability, and good coloration efficiency [187].

References

1 EU Energy Outlook 2060, https://blog.energybrainpool.com/en/eu-energy-outlook-2060-how-will-the-european-electricity-market-develop-over-the-next-37-years-2/ (last accessed: 2025-04-01).

2 EU Climate Action: 2050. *Long-Term Strategy*, https://climate.ec.europa.eu/eu-action/climate-strategies-targets/2050-long-term-strategy_en (last accessed: 2025-04-01).

3 D. O. Akinyele, R. K. Rayudu, *Sustain. Energy Technol. Assess.* 2014, 8, 74–91.

4 D. Fernandes, F. Pitié, G. Cáceres, J. Baeyens, *Energy* 2012, 39, 246–257.

5 S. Hameer, J. L. van Niekerk, *Int. J. Energy Res.* 2015, 39, 1179–1195.

6 M. Armand, J. M. Tarascon, *Nature* 2008, 451, 652–657.

7 M. A. Abdelkareem, K. Elsaid, T. Wilberforce, M. Kamil, E. T. Sayed, A. Olabi, *Sci. Total Environ.* 2021, 752, 141803.

8 A. González, E. Goikolea, J. A. Barrena, R. Mysyk, *Renew. Sustain. Energy Rev.* 2016, 58, 1189–1206.

9 G. Wang, L. Zhang, J. Zhang, *Chem. Soc. Rev.* 2012, 41, 797–828.

10 S. Münch, A. Wild, C. Friebe, B. Häupler, T. Janoschka, U. S. Schubert, *Chem. Rev.* 2016, 116, 9438–9484.

11 *Who Stole the Mysterious Baghdad Battery?*, https://www.iraqicivilsociety.org/archives/4946 (last accessed: 2025-04-01).
12 Welcome Images, *Apparatus used by Galvani – three Leyden jars*, https://commons.wikimedia.org/wiki/File:Apparatus_used_by_Galvani_-_three_Leyden_jars_Wellcome_L0011369.jpg (last accessed: 2025-04-01).
13 *Voltaic pile*, https://commons.wikimedia.org/wiki/File:Voltaic_pile,_Europe,_1800-1899_Wellcome_L0057740.jpg (last accessed: 2025-04-01).
14 *Voltaic pile*, https://commons.wikimedia.org/wiki/File:Voltaic_pile.png (last accessed: 2025-04-01).
15 *History and Timeline of the Battery*, https://www.thoughtco.com/battery-timeline-1991340 (last accessed: 2025-04-01).
16 M. Bresadola, *Brain Res. Bull.* 1998, 46, 367–380.
17 E. C. Smith, *Nature* 1945, 155, 473–474.
18 *History of Technology*, https://www.mpoweruk.com/history.htm#daniell (last accessed: 2025-04-01).
19 C. Gassner Jr., Patent DE-37758, Germany, 1886.
20 C. Gassner Jr., Patent US-373064A, USA, 1887.
21 P. Kurzweil, *J. Power Sources* 2010, 195, 4424–4434.
22 *Lead-acid automotive battery*, https://commons.wikimedia.org/wiki/File:Lead-acid_automotive_battery,_55_Ah.jpg (last accessed: 2025-04-01).
23 C. Jeyaseelan, A. Jain, P. Khurana, D. Kumar, S. Thatai, in *Rechargeable Batteries*, (Eds.: R. Boddula, R. Pothu, A. M. Asiri), John Wiley & Sons, Hoboken, 2020, pp. 177–194.
24 A. E. Kennelly, *Sci. Am.* 1901, 51, 21260–21261.
25 P. Ruetschi, F. Meli, J. Desilvestro, *J. Power Sources* 1995, 57, 85–91.
26 B. Scrosati, *J. Solid State Electrochem.* 2011, 15, 1623–1630.
27 C. A. Vincent, *Solid State Ionics* 2000, 134, 159–167.
28 G. N. Lewis, F. G. Keyes, *J. Am. Chem. Soc.* 1913, 35, 340–344.
29 M. Winter, B. Barnett, K. Xu, *Chem. Rev.* 2018, 118, 11433–11456.
30 *Press release of the Nobel Committee*, https://www.nobelprize.org/prizes/chemistry/2019/press-release/ (last accessed: 2025-04-01).
31 Y. Chen, Y. Kang, Y. Zhao, L. Wang, J. Liu, Y. Li, Z. Liang, X. He, X. Li, N. Tavajohi, B. Li, *J. Energy Chem.* 2021, 59, 83–99.
32 C. Menale, F. D'Annibale, B. Mazzarotta, R. Bubbico, *Energy* 2019, 182, 57–71.
33 D. Lu, S. Lin, S. Hu, W. Cui, T. Fang, A. Iqbal, Z. Zhang, W. Peng, *J. Solid State Electrochem.* 2021, 25, 315–325.
34 P. Rohland, E. Schröter, O. Nolte, G. R. Newkome, M. D. Hager, U. S. Schubert, *Prog. Polym. Sci.* 2022, 125, 101474.
35 K. Oyaizu, H. Nishide, *Adv. Mater.* 2009, 21, 2339–2344.
36 J. Lopez, D. G. Mackanic, Y. Cui, Z. Bao, *Nat. Rev. Mater.* 2019, 4, 312–330.
37 A. Saal, T. Hagemann, U. S. Schubert, *Adv. Energy Mater.* 2021, 11, 2001984.
38 U. S. Schubert, A. Winter, G. R. Newkome, *An Introduction to Redox Polymers for Energy-Storage Applications*, Wiley-VCH, Weinheim, 2023.
39 J. S. Miller, *Adv. Mater.* 1993, 5, 671–676.
40 T. Matsunaga, H. Daifuku, T. Nakajima, T. Kawagoe, *Polym. Adv. Technol.* 1990, 1, 33–39.

41 D. Naegele, R. Bittihn, *Solid State Ionics* 1988, 28–30, 983–989.
42 W. E. Sheeline, *Fortune* 1992, 126, 72.
43 C. McCoy, *Wall Street J.* 1992, 220, B1.
44 C. Friebe, A. Lex-Balducci, U. S. Schubert, *ChemSusChem* 2019, 12, 4093–4115.
45 C. Friebe, U. S. Schubert, *Adv. Energy Mater.* 2015, 5, 1500858.
46 S. Ma, M. Jiang, P. Tao, C. Song, J. Wu, J. Wang, T. Deng, W. Shang, *Prog. Nat. Sci. Mater. Int.* 2018, 28, 653–666.
47 N. Nitta, F. Wu, J. T. Lee, G. Yushin, *Mater. Today* 2015, 18, 252–264.
48 A. Yoshino, K. Sanechika, T. Nakajima, Patent JP-1989293, 1985.
49 B. John, *Goodenough*, https://en.wikipedia.org/wiki/John_B._Goodenough (last accessed: 2025-04-01).
50 J. B. Goodenough, K.-S. Park, *J. Am. Chem. Soc.* 2013, 135, 1167–1176.
51 T. Kim, W. Song, D.-Y. Son, L. K. Ono, Y. Qi, *J. Mater. Chem. A* 2019, 7, 2942–2964.
52 D. Larcher, J.-M. Tarascon, *Nat. Chem.* 2015, 7, 19–29.
53 B. V. Ratnakumar, M. C. Smart, A. Kindler, H. A. Frank, R. C. Ewell, S. Surampudi, *J. Power Sources* 2003, 119–121, 906–910.
54 M. Arbabzadeh, J. X. Johnson, G. A. Keoleian, P. G. Rasmussen, L. T. Thompson, *Environ. Sci. Technol.* 2016, 50, 1046–1055.
55 M. A. Pellow, C. J. M. Emmott, C. J. Barnhart, S. M. Benson, *Energy Environ. Sci.* 2015, 8, 1938–1952.
56 A. Bilich, K. Langham, R. Geyer, L. Goyal, J. Hansen, A. Krishnan, J. Bergesen, P. Sinha, *Environ. Sci. Technol.* 2017, 51, 1043–1052.
57 W. Liu, P. Oh, X. Liu, M.-J. Lee, W. Cho, S. Chae, Y. Kim, J. Cho, *Angew. Chem. Int. Ed.* 2015, 54, 4440–4457.
58 A. Manthiram, J. C. Knight, S.-T. Myung, S.-M. Oh, Y.-K. Sun, *Adv. Energy Mater.* 2016, 6, 1501010.
59 S. Kainat, J. Anwer, A. Hamid, N. Gull, S. M. Khan, *Mater. Chem. Phys.* 2024, 313, 128796.
60 H. Wang, L. Sheng, G. Yasin, L. Wang, H. Xu, X. He, *Energy Storage Mater.* 2020, 33, 188–215.
61 Y. An, X. Han, Y. Liu, A. Azhar, J. Na, A. K. Nanjundan, S. Wang, J. Yu, Y. Yamauchi, *Small* 2022, 18, 2103617.
62 M. A. Kebede, *Curr. Opin. Electrochem.* 2023, 39, 101261.
63 B. Esser, F. Dolhem, M. Becuwe, P. Poizot, A. Vlad, D. Brandell, *J. Power Sources* 2021, 482, 228814.
64 N.-S. Choi, Z. Chen, S. A. Freunberger, X. Ji, Y.-K. Sun, K. Amine, G. Yushin, L. F. Nazar, J. Cho, P. G. Bruce, *Angew. Chem. Int. Ed.* 2012, 51, 9994–10024.
65 V. Etacheri, R. Marom, R. Elazari, G. Salitra, D. Aurbach, *Energy Environ. Sci.* 2011, 4, 3243–3262.
66 Y. Liang, Z. Tao, J. Chen, *Adv. Energy Mater.* 2012, 2, 742–769.
67 T. Sun, J. Xie, W. Guo, D.-S. Li, Q. Zhang, *Adv. Energy Mater.* 2020, 10, 1904199.
68 R. Mehek, N. Iqbal, T. Noor, M. Z. B. Amjad, G. Ali, K. Vignarooban, M. A. Khan, *RSC Adv.* 2021, 11, 29247–29266.
69 P. Sengodu, A. D. Deshmukh, *RSC Adv.* 2015, 5, 42109–42130.

70 E. Umeshbabu, B. Zheng, Y. Yang, *Electrochem. Energy Rev.* 2019, 2, 199–230.

71 Z. Shadike, S. Tan, Q.-C. Wang, R. Lin, E. Hu, D. Qu, X.-Q. Yang, *Mater. Horiz.* 2021, 8, 471–500.

72 T. Janoschka, M. D. Hager, U. S. Schubert, *Adv. Mater.* 2012, 24, 6397–6409.

73 C. Friebe, U. S. Schubert, in *Electrochemical Energy Storage: Next Generation Battery Concepts*, (Ed.: R.-A. Eichel), Springer Int. Publ., Cham, 2019, pp. 65–99.

74 B. Häupler, A. Wild, U. S. Schubert, *Adv. Energy Mater.* 2015, 5, 1402034.

75 M. Armand, S. Grugeon, H. Vezin, S. Laruelle, P. Ribière, P. Poizot, J. M. Tarascon, *Nat. Mater.* 2009, 8, 120–125.

76 R. R. Zhao, Y. L. Cao, X. P. Ai, H. X. Yang, *J. Electroanal. Chem.* 2013, 688, 93–97.

77 J. Liu, D. Xie, W. Shi, P. Cheng, *Chem. Soc. Rev.* 2020, 49, 1624–1642.

78 M. Shen, H. Ma, *Coord. Chem. Rev.* 2022, 470, 214715.

79 L. Zhang, H. Liu, W. Shi, P. Cheng, *Coord. Chem. Rev.* 2019, 388, 293–309.

80 R. Zhao, Z. Liang, R. Zou, Q. Xu, *Joule* 2018, 2, 2235–2259.

81 H. Zhu, S. Liao, R. Bian, B. Su, X. Ding, M. Li, S. Ge, H. Zhang, Q. Liu, *Inorg. Chim. Acta* 2022, 535, 120848.

82 C.-x. Zhang, X.-h. Chen, W.-s. Zhang, Y. Wang, S.-l. Mei, Y.-W. Zhong, C.-J. Yao, *Chem. Eng. J.* 2024, 483, 149198.

83 L. Zhang, H. Wang, X. Zhang, Y. Tang, *Adv. Funct. Mater.* 2021, 31, 2010958.

84 K.-H. Wu, H. Maeda, T. Kambe, K. Hoshiko, E. J. H. Phua, R. Sakamoto, H. Nishihara, *Dalton Trans.* 2013, 42, 15877–15880.

85 Y. Liu, W. Deng, Z. Meng, W.-Y. Wong, *Small* 2020, 16, 1905204.

86 R. Sakamoto, K. Takada, X. Sun, T. Pal, T. Tsukamoto, E. J. H. Phua, A. Rapakousiou, K. Hoshiko, H. Nishihara, *Coord. Chem. Rev.* 2016, 320-321, 118–128.

87 K. P. C. Yao, J. T. Frith, S. Y. Sayed, F. Bardé, J. R. Owen, Y. Shao-Horn, N. Garcia-Araez, *J. Phys. Chem. C* 2016, 120, 16290–16297.

88 W.-J. Kwak, D. Rosy, C. Sharon, H. Xia, L. R. Kim, P. G. Johnson, L. F. Bruce, Y.-K. Nazar, A. A. Sun, M. Frimer, S. A. Noked, D. A. Freunberger, *Chem. Rev.* 2020, 120, 6626–6683.

89 Y.-C. Lu, B. M. Gallant, D. G. Kwabi, J. R. Harding, R. R. Mitchell, M. S. Whittingham, Y. Shao-Horn, *Energy Environ. Sci.* 2013, 6, 750–768.

90 X. Shen, S. Zhang, Y. Wu, Y. Chen, *ChemSusChem* 2019, 12, 104–114.

91 O. Gerbig, R. Merkle, J. Maier, *Adv. Mater.* 2013, 25, 3129–3133.

92 V. Viswanathan, K. S. Thygesen, J. S. Hummelshøj, J. K. Nørskov, G. Girishkumar, B. D. McCloskey, A. C. Luntz, *J. Chem. Phys.* 2011, 135, 214704.

93 M. Freitag, G. Boschloo, *Curr. Opin. Electrochem.* 2017, 2, 111–119.

94 J. R. Platt, *J. Chem. Phys.* 1961, 34, 862–863.

95 C. Gu, A.-B. Jia, Y.-M. Zhang, S. X.-A. Zhang, *Chem. Rev.* 2022, 122, 14679–14721.

96 S. Liu, P. Zhang, J. Fu, C. Wei, G. Cai, *Front. Energy Res.* 2021, 9, 620203.

97 M. Higuchi, *Polym. J.* 2009, 41, 511–520.

98 C.-W. Hu, T. Sato, J. Zhang, S. Moriyama, M. Higuchi, *J. Mater. Chem. C* 2013, 1, 3408–3413.

99 M. Higuchi, *Bull. Jpn. Soc. Coord. Chem.* 2022, 79, 68–77.
100 M. Higuchi, *J. Syn. Org. Chem. Jpn.* 2023, 81, 1089–1095.
101 S. Mondal, T. Yoshida, S. Maji, K. Ariga, M. Higuchi, *ACS Appl. Mater. Interfaces* 2020, 12, 16342–16349.
102 I. Beyers, A. Bensmann, R. Hanke-Rauschenbach, *J. Energy Storage* 2023, 73, 109097.
103 M. L. Divya, S. Natarajan, Y.-S. Lee, V. Aravindan, *ChemSusChem* 2019, 12, 4353–4382.
104 T. Christen, *J. Energy Storage* 2020, 27, 101084.
105 J. G. Ibanez, M. E. Rincón, S. Gutierrez-Granados, M. H. Chahma, O. A. Jaramillo-Quintero, B. A. Frontana-Uribe, *Chem. Rev.* 2018, 118, 4731–4816.
106 E. J. Houston, *Electricity in Every-day Life*, P. F. Collier & Son, New York, 1905, pp. 71–84.
107 J. Ho, T. R. Jow, S. Boggs, *IEEE Electr. Insul. Mag.* 2010, 26, 20–25.
108 H. I. Becker, US-2800616, USA, 1957.
109 M. Salanne, B. Rotenberg, K. Naoi, K. Kaneko, P.-L. Taberna, C. P. Grey, B. S. Dunn, P. Simon, *Nat. Energy* 2016, 1, 16070.
110 K. Xu, H. Shao, Z. Lin, C. Merlet, G. Feng, J. Zhu, P. Simon, *Energy Environ. Mater.* 2020, 3, 235–246.
111 M. E. Şahin, F. Blaabjerg, A. Sangwongwanich, *Energies* 2022, 15, 674.
112 B. E. Conway, V. I. Birss, J. Wojtowicz, *J. Power Sources* 1997, 66, 1–14.
113 Y. Kumar, S. Rawal, B. Joshi, S. A. Hashmi, *J. Solid State Electrochem.* 2019, 23, 667–692.
114 Y. Wang, Y. Song, Y. Xia, *Chem. Soc. Rev.* 2016, 45, 5925–5950.
115 D. P. Dubal, O. Ayyad, V. Ruiz, P. Gómez-Romero, *Chem. Soc. Rev.* 2015, 44, 1777–1790.
116 D. Cericola, R. Kötz, *Electrochim. Acta* 2012, 72, 1–17.
117 S. Zheng, H. Xue, H. Pang, *Coord. Chem. Rev.* 2018, 373, 2–21.
118 S. Halder, C. Chakraborty, *Sol. Energy Mater. Sol. Cells* 2023, 254, 112288.
119 H. Wang, F. Qiu, C. Lu, J. Zhu, C. Ke, S. Han, X. Zhuang, *Polymers* 2021, 13, 1002.
120 H. Chen, J. Xing, W. Wang, X. Li, M. Shu, P. Gao, Y. Pan, J. Liu, *Sol. Energy Mater. Sol. Cells* 2022, 248, 111967.
121 I. Mukkatt, A. P. Mohanachandran, A. Nirmala, D. Patra, P. A. Sukumaran, R. S. Pillai, R. B. Rakhi, S. Shankar, A. Ajayaghosh, *ACS Appl. Mater. Interfaces* 2022, 14, 31900–31910.
122 S. R. Jena, J. Choudhury, *Sol. Energy Mater. Sol. Cells* 2022, 239, 111660.
123 H. Xie, J. Dai, L. Luo, L. Xia, X. Wang, C. Yang, Y. Xu, B. Zeng, L. Dai, *J. Energy Storage* 2022, 52, 104965.
124 A. Eftekhari, L. Li, Y. Yang, *J. Power Sources* 2017, 347, 86–107.
125 Z. Huang, L. Li, Y. Wang, C. Zhang, T. Liu, *Compos. Commun.* 2018, 8, 83–91.
126 M. M. Rahman, T. Mahtab, M. Z. B. Mukhlish, M. O. Faruk, M. M. Rahman, *Polym. Bull.* 2021, 78, 5379–5397.
127 S. Acharya, S. Sahoo, S. Sonal, J. H. Lee, B. K. Mishra, G. C. Nayak, *Compos. Part B Eng.* 2020, 193, 107913.

128 S. Xiong, J. Zhang, R. Wang, B. Wu, J. Chu, X. Wang, R. Zhang, M. Gong, Z. Li, M. Qu, Z. Chen, *ChemistrySelect* 2019, 4, 14343–14350.

129 V. F. Yusuf, N. I. Malek, S. K. Kailasa, *ACS Omega* 2022, 7, 44507–44531.

130 P. Zhang, B. Y. Guan, L. Yu, X. W. Lou, *Angew. Chem. Int. Ed.* 2017, 56, 7141–7145.

131 L. S. Xie, G. Skorupskii, M. Dincă, *Chem. Rev.* 2020, 120, 8536–8580.

132 J. Calbo, M. J. Golomb, A. Walsh, *J. Mater. Chem. A* 2019, 7, 16571–16597.

133 X. Fu, B. Ding, D. D'Alessandro, *Coord. Chem. Rev.* 2023, 475, 214814.

134 S.-J. Shin, J. W. Gittins, C. J. Balhatchet, A. Walsh, A. C. Forse, *Adv. Funct. Mater.* 2024, 34, 2308497.

135 H. Feng, W. Li, C. Ma, H. He, *Res. Appl. Mater. Sci.* 2022, 4, 42–46.

136 T. Toyao, M. Saito, S. Dohshi, K. Mochizuki, M. Iwata, H. Higashimura, Y. Horiuchi, M. Matsuoka, *Chem. Commun.* 2014, 50, 6779–6781.

137 M. Elcheikh Mahmoud, H. Audi, A. Assoud, T. H. Ghaddar, M. Hmadeh, *J. Am. Chem. Soc.* 2019, 141, 7115–7121.

138 G. Zhang, S. Zheng, M. C. Neary, *RSC Adv.* 2023, 13, 2225–2232.

139 S. M. Elahi, M. Raizada, P. K. Sahu, S. Konar, *Chem. Eur. J.* 2021, 27, 5858–5870.

140 J. Doyle, US-224404, USA, 1879.

141 Y. V. Tolmachev, *J. Electrochem. Soc.* 2023, 170, 030505.

142 C. J. Amato, *International Automotive Engineering Congress and Exposition*, SAE International, Warrendale, PA, 1973, p. 11.

143 W. Kangro, H. Pieper, *Electrochim. Acta* 1962, 7, 435–448.

144 W. Kangro, DE-914264C, Germany, 1949.

145 M. Rychcik, M. Skyllas-Kazacos, *J. Power Sources* 1987, 19, 45–54.

146 M. Rychcik, M. Skyllas-Kazacos, *J. Power Sources* 1988, 22, 59–67.

147 *Vanadium redox battery*, https://en.wikipedia.org/wiki/Vanadium_redox_battery (last accessed: 2025-04-01).

148 A. Clemente, G. A. Ramos, R. Costa-Castelló, *Electronics* 2020, 9, 1567.

149 L. Carrette, K. A. Friedrich, U. Stimming, *Fuel Cells* 2001, 1, 5–39.

150 E. Sánchez-Díez, E. Ventosa, M. Guarnieri, A. Trovò, C. Flox, R. Marcilla, F. Soavi, P. Mazur, E. Aranzabe, R. Ferret, *J. Power Sources* 2021, 481, 228804.

151 C. A. Machado, G. O. Brown, R. Yang, T. E. Hopkins, J. G. Pribyl, T. H. Epps III, *ACS Energy Lett.* 2021, 6, 158–176.

152 Q. Dai, Z. Liu, L. Huang, C. Wang, Y. Zhao, Q. Fu, A. Zheng, H. Zhang, X. Li, *Nat. Commun.* 2020, 11, 13.

153 S. Jiang, S. Lu, Y. Xiang, S. P. Jiang, *Adv. Sustain. Syst.* 2019, 3, 1900020.

154 X.-Z. Yuan, C. Song, A. Platt, N. Zhao, H. Wang, H. Li, K. Fatih, D. Jang, *Int. J. Energy Res.* 2019, 43, 6599–6638.

155 *Project "RedoxWind"*, https://www.ict.fraunhofer.de/en/comp/ae/rw.html (last accessed: 2025-04-01).

156 *World's largest flow battery connected to the grid in China*, https://newatlas.com/energy/worlds-largest-flow-battery-grid-china/ (last accessed: 2025-04-01).

157 *German manufacturer unveils 10 kWh residential redox flow battery*, https://www.pv-magazine.com/2023/05/25/german-manufacturer-unveils-10-kwh-residential-redox-flow-battery/ (last accessed: 2025-04-01).
158 M. Guarnieri, P. Mattavelli, G. Petrone, G. Spagnuolo, *IEEE Ind. Electron. Mag.* 2016, 10, 20–31.
159 A. Trovò, G. Marini, W. Zamboni, S. D. Sessa, *Electronics* 1844, 2023, 12.
160 P. Alotto, M. Guarnieri, F. Moro, *Renew. Sustain. Energy Rev.* 2014, 29, 325–335.
161 M. Skyllas-Kazacos, G. Kazacos, G. Poon, H. Verseema, *Int. J. Energy Res.* 2010, 34, 182–189.
162 C. Menictas, M. Skyllas-Kazacos, *J. Appl. Electrochem.* 2011, 41, 1223–1232.
163 K. Gong, X. Ma, K. M. Conforti, K. J. Kuttler, J. B. Grunewald, K. L. Yeager, M. Z. Bazant, S. Gu, Y. Yan, *Energy Environ. Sci.* 2015, 8, 2941–2945.
164 H. Zhang, C. Sun, M. Ge, *Batteries* 2022, 8, 202.
165 C. Xie, Y. Duan, W. Xu, H. Zhang, X. Li, *Angew. Chem. Int. Ed.* 2017, 56, 14953–14957.
166 S. Zhang, W. Guo, F. Yang, P. Zheng, R. Qiao, Z. Li, *Batter. Supercaps* 2019, 2, 627–637.
167 B. Huskinson, M. P. Marshak, C. Suh, S. Er, M. R. Gerhardt, C. J. Galvin, X. Chen, A. Aspuru-Guzik, R. G. Gordon, M. J. Aziz, *Nature* 2014, 505, 195–198.
168 K. Lin, Q. Chen, M. R. Gerhardt, L. Tong, S. B. Kim, L. Eisenach, A. W. Valle, D. Hardee, R. G. Gordon, M. J. Aziz, M. P. Marshak, *Science* 2015, 349, 1529–1532.
169 J. Winsberg, T. Hagemann, T. Janoschka, M. D. Hager, U. S. Schubert, *Angew. Chem. Int. Ed.* 2017, 56, 686–711.
170 J. Noack, N. Roznyatovskaya, T. Herr, P. Fischer, *Angew. Chem. Int. Ed.* 2015, 54, 9776–9809.
171 W. Wang, Q. Luo, B. Li, X. Wei, L. Li, Z. Yang, *Adv. Funct. Mater.* 2013, 23, 970–986.
172 Z. Li, S. Li, S. Liu, K. Huang, D. Fang, F. Wang, S. Peng, *Electrochem. Solid-State Lett.* 2011, 14, A171.
173 T. Janoschka, N. Martin, U. Martin, C. Friebe, S. Morgenstern, H. Hiller, M. D. Hager, U. S. Schubert, *Nature* 2015, 527, 78–81.
174 R. W. Hogue, K. E. Toghill, *Curr. Opin. Electrochem.* 2019, 18, 37–45.
175 K. Teramoto, T. Nishide, S. Okumura, K. Takao, Y. Ikeda, *Electrochemistry* 2014, 82, 566–572.
176 C. Chen, S. Zhang, Y. Zhu, Y. Qian, Z. Niu, J. Ye, Y. Zhao, X. Zhang, *RSC Adv.* 2018, 8, 18762–18770.
177 C. G. Armstrong, K. E. Toghill, *J. Power Sources* 2017, 349, 121–129.
178 B. Liu, C. W. Tang, W. Wei, C. Zhang, G. Jia, T. Zhao, *Energy Storage Mater.* 2023, 60, 102808.
179 A. Okazawa, T. Kakuchi, K. Kawai, M. Okubo, *APL Mater.* 2023, 11, 110901.
180 Y. W. D. Chen, K. S. V. Santhanam, A. J. Bard, *J. Electrochem. Soc.* 1981, 128, 1460.
181 J. Gao, K. Amini, T. Y. George, Y. Jing, T. Tsukamoto, D. Xi, R. G. Gordon, M. J. Aziz, *Adv. Energy Mater.* 2022, 12, 2202444.

182 R. Pinal, *Org. Biomol. Chem.* 2004, 2, 2692–2699.
183 J. Kim, Y. Kim, J. Yoo, G. Kwon, Y. Ko, K. Kang, *Nat. Rev. Mater.* 2023, 8, 54–70.
184 Z. Yu, Y. Wang, Z. Luo, *Ind. Eng. Chem. Res.* 2024, 63, 9619–9630.
185 L. Long, S. Wang, M. Xiao, Y. Meng, *J. Mater. Chem. A* 2016, 4, 10038–10069.
186 X. Zhao, C. Wang, H. Liu, Y. Liang, L.-Z. Fan, *Batter. Supercaps* 2023, 6, e202200502.
187 T. G. Yun, X. Chen, J. Y. Cheong, *Batter. Supercaps* 2023, 6, e202200454.

6

Terpyridine Ligands and Complexes in the Context of Photovoltaic Applications

6.1 Introduction

Today's society demands an ever-increasing energy supply, which brings the natural resources in terms of fossil fuels rapidly to depletion. In the 1970s, experts predicted the worldwide oil reserves to be used up by the turn of the millennium; owing to the positive development over the last decades, the accessible oil reserves are prospected to last until *ca.* 2060 [1]. The coverage of natural gas and coal is believed to be *ca.* 60 and 200 years, respectively [2]. The rapidly increasing demand for energy in concert with human-made climate change, which is strongly associated with the combustion of enormous amounts of fossil fuels to produce CO_2, requires fundamental rethinking. As one solution, CO_2-neutral, renewable energies should be used instead of fossil ones. In order to stop global warming, the European Union (EU) revised and sharpened their directive: It is targeted that at least 42.5% of the energy, which is consumed in the EU by 2023, stems from renewable sources (the previous goal was defined as 32% by 2030) [3]. A form of energy might be considered renewable if it is derived from natural sources whose rate of replenishment is faster than the rate of consumption. This holds true for a number of renewable technologies, which rely on wind, water, terrestrial heat, or sunlight; typically, biomass-derived fuels and electricity are also counted as renewable [4]. From these, photovoltaic (PV) technology offers the advantage of the direct conversion of sunlight into high-quality electric energy. The first practical solar cell, which exhibited an efficiency of *ca.* 6%, was demonstrated by the Bell Laboratories in 1958 [5, 6]; a few years later, in 1958, such cells were already utilized as the power source on the Vanguard-I satellite. In the 1990s, silicon-based cells were also eventually introduced for terrestrial use. The price for Si-based PVs has decreased remarkably from \$76.00 per Watt in 1977 to \$0.30 and, accordingly, the PV market is still dominated by this technology. The basic principle of such a cell, which is comprised of an *n*-doped and a *p*-doped layer of crystalline Si, is shown in Figure 6.1 (*n*- and *p*-type doping are achieved by P and B atoms, respectively, within the semiconductor material) [7]. Photocurrent is generated if a photon, whose energy is higher than the absorber's energy band gap (E_g), hits the *p*–*n* junction. As a result, charge carriers (*i.e.*, electrons and holes) are generated, which – due to their inherent electric field – migrate toward the electrodes, where one can grip electric energy.

Terpyridines: Multifaceted Ligands for Diverse Applications, First Edition.
Ulrich S. Schubert, Andreas Winter, and George R. Newkome.

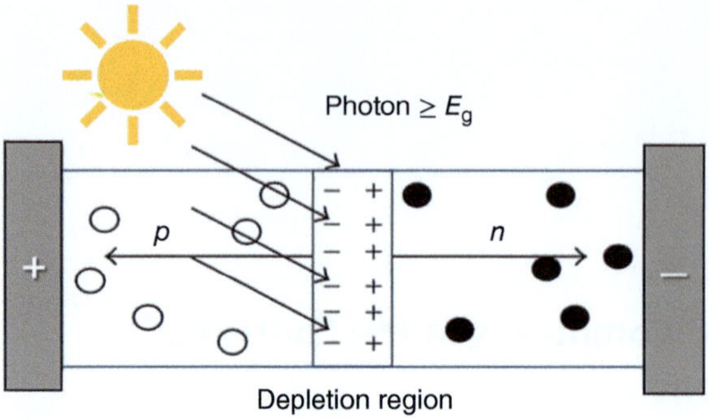

Figure 6.1 Schematic representation of a basic *p–n* junction solar cell including the relevant charge-transport processes [7] / with permission of John Wiley & Sons / CC BY 3.0.

Already in 1839, Becquerel observed the photoelectrochemical effect when AgX-coated Pt electrodes were illuminated in aqueous solution (X = Cl or Br) [8, 9]. Photoconductivity was observed for selenium, as an inorganic semiconductor, about 35 years later [10, 11]. Later, photoconductivity was also found first for organic compounds (*e.g.*, anthracene) [12, 13]. In his famous publication, Einstein provided the theoretical explanation of the PV effect [14], which was proven experimentally by Millikan [15]. In 1918, Czochralski developed a method, which paved the way toward the production of monocrystalline metals and metalloids, including silicon [16]. This, in turn, initiated intensified research on solar cells and the first Si-based cell was reported in 1941, followed by Ge- or CdS-containing ones a few years later [17]. This research led to the aforementioned discovery in the Bell Laboratories [6] and the first applications in spacecraft.

The common classification scheme for solar cells takes the utilized materials and the involved production technology into account. As shown in Figure 6.2, the 1st generation PV technology is based on the Si-wafer technology, which emerged with the manufacturing of integrated circuits and was later adapted for solar cells [18]. These traditional cells still dominate the market with a share of *ca.* 95% by 2020 [19] and offer record lab-cell efficiencies of 24.7% and 20.4% in the case of mono- and multi-crystalline Si wafers, respectively [20]. In order to reduce the costs on the one hand and to improve the device efficiency on the other, the 2nd generation technology came up. Regarding the costs, one has to be aware that the overall cost of a PV module is half of the cost of installation; the cost of the cells, in turn, is about half of the entire module cost [21]. To meet this concern, *i.e.*, to fabricate cheaper modules, thin-film devices were developed, whose absorber layer was only a few μm thick (instead of 100–200 μm in traditional cells). This resulted in a significant cost reduction with respect to materials, but initially at the disadvantage of efficiency. On the other hand, low-cost manufacturing techniques (*e.g.*, spray coating, vacuum deposition, *etc.*) could be used and lead to the development of flexible devices, which could *inter alia* be used for "everyday" devices, such as wearable electronics [22]. The 2nd generation technology is superior in terms of energy-payback time and greenhouse-gas emission and, thus, is generally considered the eco-friendlier solution [23]. Noteworthy, owing to improved production techniques, the efficiency has also geared up significantly: Thin-film PV devices now feature efficiency of

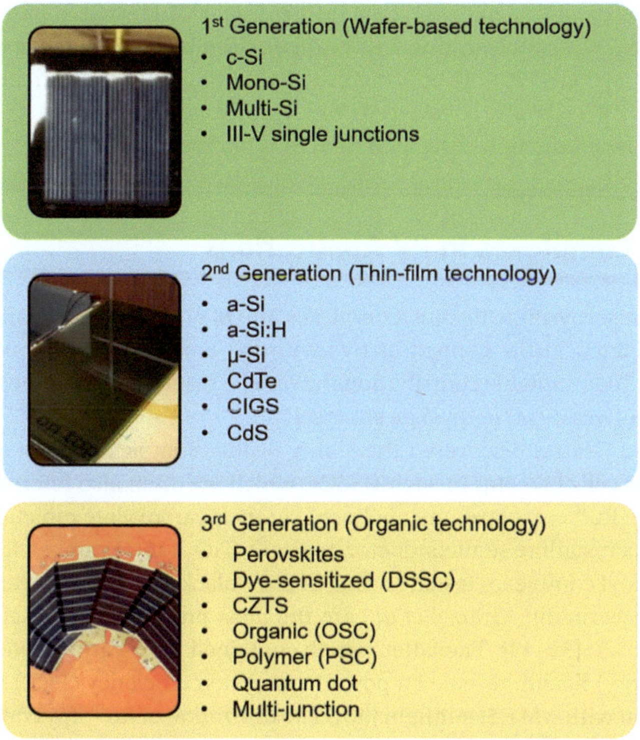

Figure 6.2 Classification scheme for PV technologies [17] / MDPI / CC BY 4.0.

>20% [a Cd-free CIGS cell afforded the record value of *ca.* 23.4%; CIGS: copper indium gallium (di)selenide] [20].

The 3rd generation technology is mainly based on the 2nd generation's thin-film technology, but does no longer rely on the conventional *p*–*n* junction (Figure 6.1) [17]. Firstly, the multi-junction cells are comprised of multiple *p*–*n* junctions made of different absorber materials. This approach allows for the tuning of E_g and absorption of a broader wavelength range; thus, the solar-cell efficiency (SCE) can be improved significantly [24]. Secondly, and with relevance to this chapter, different types of materials, such as organic compounds, polymers, or metal complexes, have moved into the focus of interest. New types of PV devices, namely dye-sensitized solar cells (DSSCs) [25–28] and organic or polymer PV cells (OPV and PPV, respectively), have emerged. Regarding the latter, the bulk-heterojunction (BHJ) cells represent one important cell type [29–31]. The so-called perovskite solar cells (PSCs) represent the latest innovation in the field of PV devices. In such cells, perovskite-type materials, such as hybrid organic–inorganic lead or tin halides, are utilized as the light-absorbing layer [32, 33]. In particular, methylammonium lead halides (MALHs) and the all-inorganic cesium-lead halides ($CsPbX_3$; X = Cl, Br, or I) offer the advantages of low-cost production and ease of manufacturing. The efficiency of simple single-junction cells showed a tremendous evolution from an initial performance of below 4% in 2009 [34] to almost 26% in 2021 [35]; in Si-based tandem cells, even 29.8% was reached, thus outcompeting the record efficiency of conventional *p–n*-junction cells [20]. Although considerable stability issues remain [36], perovskite-based PV cells are considered for commercialization since this technology combines low production costs with high SEC performances.

6.2 Terpyridine Complexes in PV Applications

But at which point came terpyridine ligands/complexes into play? Terpyridine complexes were first considered in the context of PVs with the emergence of the 3rd generation technology. Today notable contributions have been made to three of these technologies: DSSCs, polymer, and perovskite solar cells.

In 1991, O'Regan and Grätzel discovered the ability of Ru^{II} complexes to act as sensitizing dyes in solar cells [37, 38]. In such DSSCs, which are often also referred to as "Grätzel cells," the Ru^{II} complexes are anchored *via* their carboxylate moieties to the surface of a nanocrystalline semiconductor, typically TiO_2 [28]. From a wide range of Ru^{II}–oligopyridyl complexes tested so far, N3 and "black dye," which were both introduced by Nazeeruddin, Grätzel *et al.*, are the most prominent representatives (**1** and **2**, Figure 6.3) [39–42]. The latter, which contained a *tris*-carboxylated 2,2′:6′,2″-terpyridine (tpy) ligand, revealed a power-conversion efficiency (PCE) of 10.4% upon illumination with AM1.5[1] sunlight [40]. The utilization of Ru^{II}–tpy complexes in the context of DSSCs is discussed in Section 6.3.

The second application of tpy complexes in the context of PVs is related to polymer solar cells. The first example for an OPV cell, which was comprised of a donor-acceptor bilayer, was reported by Tang in 1986; this proof-of-concept device revealed a PCE of *ca.* 1% [44]. Similar to a conventional *p–n*-junction, the interface of the layers acted as the heterojunction, where holes and electrons

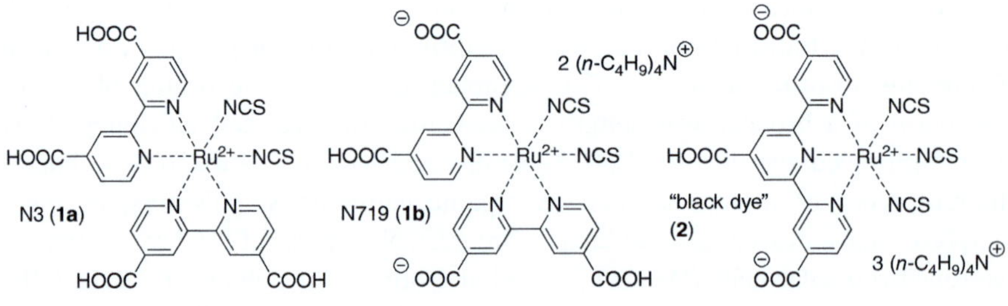

Figure 6.3 Schematic representation of the archetypical Ru^{II} complexes **1** and **2**, which are used as sensitizers, for DSSC applications.

1 The air mass (AM) coefficient, which refers to the optical path length through the earth's atmosphere, is commonly used to describe the solar spectrum after the sun's radiation has passed through the atmosphere [43] Conventionally, solar cells are characterized at AM1.5 sunlight. This mimics a 1.5 multiple of the atmosphere's thickness, *i.e.*, a solar zenith angle of 48.2°. Practically, standardized solar-cell tests are performed at 109,870 lx, which corresponds to an irradiance of 1000.4 W m^{-2}.

were generated and, subsequently, transported to the electrodes. Since then, numerous donor-acceptor combinations have been tested, and fullerene (C_{60}) and its derivatives were identified as the most efficient acceptors (*i.e.*, *n*-type components) for this application [45]. A big move toward PPV cells was taken by Heeger *et al.* [46] as well as Yoshino *et al.* [47], who independently demonstrated the remarkably fast electron transfer between conjugated polymers and C_{60} derivatives. The first PPV cell, comprising a planar heterojunction, was reported in 1993 [48]. The main limitations of the planar-junction cell configuration, which are the small surface area and the required long charge-carrier lifetime, were overcome by changing to a bulk heterojunction [49]. In a BHJ, the donor and acceptor components are mixed and co-processed to form the bulk active layer. Hiramoto *et al.* reported this concept first [50], whereas Heeger *et al.* [51] as well as Friend *et al.* [52] provided the functional BHJ-type PPVs, in which a polymer-fullerene and a polymer-polymer blend, respectively, represented the active layer. Since then, this concept has been applied to numerous materials and the efficiency has gradually been improved within 30 years of research [31]. Very recently, the record efficiency of PPV devices has been lifted to *ca.* 19.3% [53]. Besides the wide range of conjugated polymers [54–58], metal-containing polymers have also been utilized, as active materials, for PPV applications: Pt^{II}–polyynes, main- and side-chain polymers with Zn^{II}–porphyrin moieties, polymers with coordinated [$Re(CO)_3Cl$] complexes, polymers with cyclometalated Pt^{II} or Ir^{III} complexes as well as polymers with Ru^{II}–tpy complexes within the main chain or the side-chains [59]. The PV performance of these materials is detailed in Section 6.4.

As already mentioned in Section 6.1, perovskite-based solar cells have experienced a tremendous evolution since their first reports in 2009 [34]. The very high efficiencies in concert with the low production costs make them promising alternatives to conventional Si-based cells [60]. Detrimentally, perovskite-based cells are prone to degradation due to humidity, UV irradiation or temperature; thus, stability represents one limiting factor regarding large-scale commercialization [61]. It has been demonstrated that additives can be used for perovskite-crystal engineering, *i.e.*, modulation of the crystal's dimensionality and healing its structural defects. Moreover, one might use surface passivation to decrease the sensitivity to moisture. For these purposes, ionic liquids [62], polymers [63], self-assembled materials [64], and Lewis acids/bases [65] have *inter alia* been used. Regarding the latter, oligopyridines, such as tpy, have been shown to be applicable for both crystal engineering [66] and surface passivation [67, 68]. Furthermore, perovskite-type materials containing tpy-coordinated Ga^{III} or $In^{III/V}$ ions were reported [69, 70]. In Section 6.5, the utilization of tpy ligands in the context of perovskite-based devices is highlighted.

6.3 Terpyridine Complexes in Dye-Sensitized Solar Cells

In 1991, O'Regan and Grätzel published the first DSSC, which contained a Ru^{II} dye as the sensitizer [37]. The fundamental idea was to mimic the photosystem of plants, *i.e.*, the conversion of sunlight into energy. Unlike Nature, which utilizes chlorophyll, Grätzel *et al.* focused on chemical dyes, as (sun)light absorbers, which are anchored to a semiconductor of nanocrystalline morphology. The dye molecules will, after photoexcitation, transfer electrons into the semiconductor, thereby generating a potential difference between the electrodes and initiating the flow of an electric current. A redox couple, which is present in the organic electrolyte, completes the electrical circuit by transporting electrons back to the dye, which is reduced back to its ground state (GS). Regeneration of the dye by a redox system, most commonly the I^-/I_3^- couple, suppresses the electron recapture at the dye-superconductor interfaces. The redox system, in turn, is reactivated by electrons from the counter electrode. The voltage, which can be gripped from such DSSCs, represents the difference between the electron's Fermi level in the semiconductor and the electrolyte's redox potential. The basic working scheme of a DSSC, which generates electric power without suffering from any irreversible chemical transformations, is depicted in Figure 6.4a. The various processes, which account to the photoinduced generation of electric power in DSSCs, are listed in Figure 6.4b.

Most commonly, TiO_2 nanoparticles are used as the semiconductor in DSSCs. The particles, which have been sintered together to enable electronic conduction, are deposited onto glass slides (*e.g.*, by vacuum deposition, spin-coating, or printing techniques), which were pre-coated with a conductive, yet transparent substrate [*e.g.*, indium-tin oxide (ITO) or fluorine-doped tin oxide (FTO)]. A representative scanning-electron microscopy (SEM) image of such a TiO_2 ("anatase") film is shown

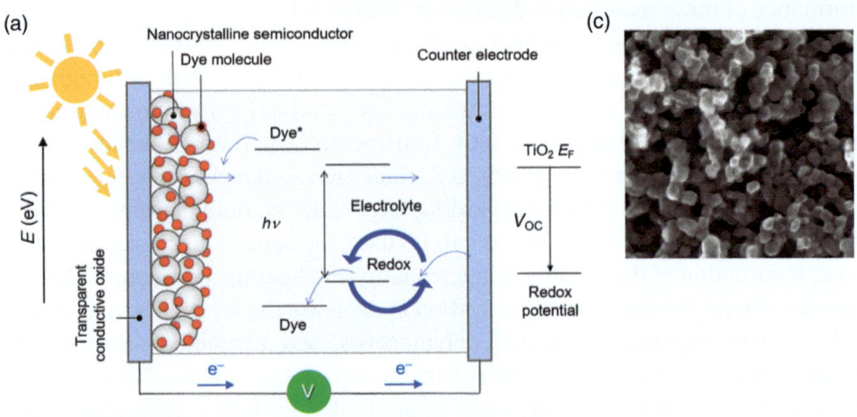

(b) Processes contributing to the generation of electric power in DSSCs:

(1) DS + $h\nu \rightarrow$ DS* (photoexcitation of the adsorbed dye)
(2) DS* $\rightarrow$ DS$^+$ + e$^-$ (electron transfer into the semiconductor)
(3) I_3^- + 2 e$^-$ $\rightarrow$ 3 I^- (reduction at the cathode)
(4) 2 DS$^+$ + 3 I^- $\rightarrow$ 2 DS + I_3^- (generation of the dye)

Performance-decreasing side reaction:

(1) DS$^+$ + e$^-$ $\rightarrow$ DS (recombination of charges at the semiconductor)
(2) I_3^- + 2 e$^-$ $\rightarrow$ 3 I^- (reduction at the anode)

Figure 6.4 (a) Schematic representation of the principle of a DSSC [71] / with permission of American Chemical Society. (b) Summary of the *in-operando* processes of a DSSC. Adapted from [72]. (c) Representative SEM image of a nanocrystalline TiO_2 film, which is typically used in DSSCs (DS denotes a dye sensitizer, which is anchored onto a semiconductor film) [28] / with permission of Elsevier.

in Figure 6.4c. The favorable film thickness is between 5 and 20 μm (corresponding to a TiO_2 mass of a few mg cm^{-2}); whereas the film porosity ranges from 50% to 60%, with an average pore size of *ca.* 15 nm [28]. TiO_2 is a low-cost, widely abundant material, which has found widespread use, as pigment, in daily-life applications: health-care products, paints, *etc.* However, its non-toxicity and biocompatibility have been questioned since TiO_2 has recently been listed as "potentially carcinogenic upon inhalation" [73]. Regarding its use in DSSCs some shortcomings of TiO_2 have also been identified: The nanoporous morphology features a large number of grain boundaries, at which charge-carrier recombination might lead to a loss of efficiency. Moreover, the efficiency in the near-infrared (NIR) regime is relatively poor [26]. Accordingly, other metal oxides, which possess suitable energy levels and high conductivities, have also been discussed as alternative photo-anode materials for DSSCs: ZnO [74], SnO_2 [75], and Nb_2O_5 [76], *etc.* Furthermore, hybrid semiconductors, comprised of TiO_2 and a carbonaceous material, have attracted significant attention [77]. In particular, carbon nanotubes (CNTs) have been incorporated into TiO_2 films to improve the DSSC performance, due to the nanotubes' high electron affinity and remarkable conductivity.

In either case, the dye (sensitizer) is chemisorbed onto the semiconductor's surface *via* carboxylate or phosphonate moieties, which enable a strong binding and facilitate the electron donation from the dye into semiconductor. The ideal sensitizer should exhibit a panchromatic, broadband absorptivity, which should ideally cover the full visible regime up to *ca.* 920 nm [28]. In such a case, the sensitizer should be capable of transforming standard AM1.5 sunlight into electricity in a highly efficient manner. In particular, polypyridyl complexes of Ru^{II} and Os^{II} ions were found to combine good PV performance with photochemical stability; accordingly, such complexes represent a highly favorable class of sensitizers. The "black dye" (BD or **2**, Figure 6.3), as one of the seminal Ru^{II} complexes for DSSC applications, afforded a PCE of >10% upon illumination with full sunlight (Figure 6.5a) [40]. The term "black dye" refers to its panchromatic absorptivity, which provided a high incident photon-to-current conversion efficiency (IPCE) of *ca.* 80% over a wide spectral range (Figure 6.5b). Notably, the PCE could be raised to *ca.* 11.1% after cautious structural

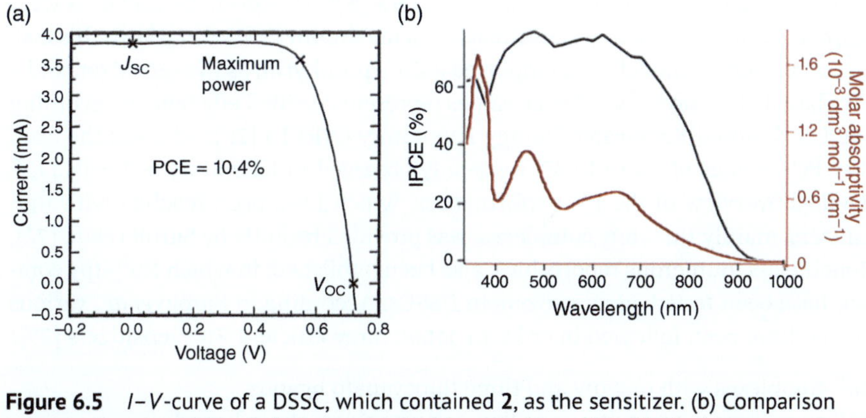

Figure 6.5 *I*–*V*-curve of a DSSC, which contained **2**, as the sensitizer. (b) Comparison between the absorption and photocurrent action spectrum of **2** attached to a nanocrystalline TiO_2 film. Adapted from [40].

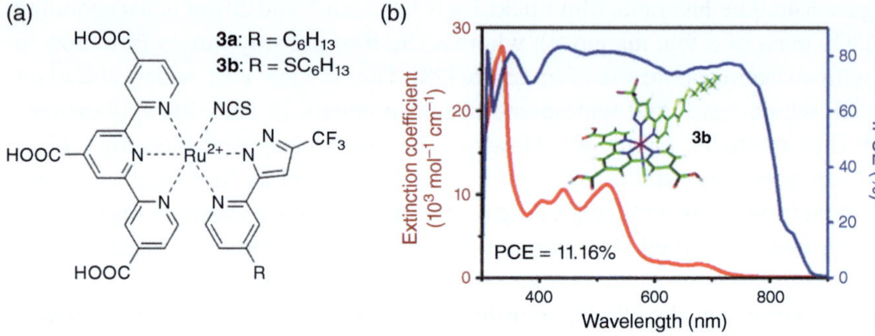

Figure 6.6 (a) Schematic representation of the "modified" Ru^{II} complexes **3** (counterions were omitted for clarity). (b) Comparison of the absorption and IPCE spectra of sensitizer **3b** [78] / with permission of Royal Society of Chemistry.

modification [78]. In Ru^{II} complexes **3**, two thiocyanato ligands were replaced by a bidentate 2-(pyrazol-3-yl)pyridine one (Figure 6.6). These sensitizers were superior when compared to **2**, not only regarding the improved PV performance but also the ease of synthesis and the operational stability (*i.e.*, PCE retention of 98% after illumination at 60 °C for 1000 h) represent notable improvements.

What is the substantial difference between Ru^{II} sensitizers containing bpy or tpy ligands? It has been shown that complexes of the latter type feature a lower band gap for the metal-to-ligand charge-transfer (MLCT) transition, which enables a better absorption behavior at longer wavelengths. This, in turn, facilitates a more efficient harvesting of sunlight [79]. However, a broader absorption does not necessarily imply an improved PV performance. In the case of N719 (**1b**, Figure 6.3), for example, the higher molar extinction coefficient accounted for the high PCE of 11.2% [80]. The better PV performance of N719 compared to that of N3 (**1a**, Figure 6.3) was, in turn, attributed to the stronger anchoring of the charged dye to the TiO_2 surface [81]. In general, Ru^{II}–bpy complexes are believed to afford a higher surface coverage due to the spatial orientation of the chemisorbed dye molecules [82]. The best performance with a Ru^{II}-based sensitizer was reported by Komiya *et al.*; these authors achieved a PCE of even 11.9% [83]. However, the Ru^{II} dyes were outcompeted in 2011, when a DSSC reached a maximum PCE of 12.3%; in this cell, Yeh, Zakeeruddin, Grätzel *et al.* employed a Zn^{II}–porphyrin, as the sensitizer, and a $Co^{II/III}$-based electrolyte [84]. These values represent certified efficiencies according to the US National Renewable Energy Laboratory (NREL) [20]; whereas the even higher PCE values of up to 14.3% were only observed in laboratory test cells [72]. A detailed overview of the PV performances, which have been reached with Ru^{II} sensitizers, mainly Ru^{II}–bpy complexes, was provided recently by Surolia *et al.* [72].

Nonetheless, numerous reports have also been published, in which Ru^{II}–tpy complexes have been tested, as sensitizers in DSSCs. According to Barolo *et al.*, various strategies have been followed in order to obtain more efficient Ru^{II} sensitizers [79]:

- Ru^{II} complexes with one tpy and three thiocyanato ligands
- Ru^{II} complexes with one tpy, one bpy, and one thiocyanato ligand

- Ru^{II} complexes with two tpy ligands
- Ru^{II} complexes with one tpy, one cyclometalating 2-phenylpyridine or 2-phenylpyrimidine, and one thiocyanato ligand
- Ru^{II} complexes with one tpy, one β-diketonate, and one thiocyanato ligand
- Ru^{II} complexes with one tpy and one pyrazole-containing ligand (the latter being bidentate or tridentate)
- Ru^{II} complexes with one tpy and one cyclometalating phenyl-bipyridine ligand
- Ru^{II} complexes with one tpy ligand and NHC-containing ligands (NHC: *N*-heterocyclic carbene)
- Ru^{II} complexes, as sensitizers for p-type semiconductors
- Combination of BD with organic co-sensitizers
- Beyond Ru^{II}: tpy complexes with other transition-metal ions

The PV characteristic of the Ru^{II} sensitizers, which are covered within the scope of this chapter, are summarized in Table 6.1. Specifically, the open-circuit voltage (V_{OC}), short-circuit current (J_{SC}), fill factor (FF), and PCE are listed. The V_{OC} and J_{SC} values can be directly extracted from the cell's I–V diagram (*i.e.*, intercept with the V- and I-axis, respectively) [96]. The fill factor (or "squareness"), in turn, represents the ratio of the cell's maximum power to the product of its V_{OC} and J_{SC} (these values are indicated exemplarily in the I–V-curve in Figure 6.5a). The aforementioned PCE is the ratio between the cell's power output and input (P_{out} and P_{in}, respectively); thereby, P_{out} is the product of V_{OC}, J_{SC}, and FF.

6.3.1 Ru^{II}–Tpy Sensitizers with Three Thiocyanato Ligands

Firstly, Ru^{II}–tpy complexes, which contained only a single COOH-moiety, as the anchoring group, were tested (**4** and **5**, Figure 6.7). However, these sensitizers mainly suffered from their relatively weak binding to the photoanode material; as a result, relatively low J_{SC} values and, thus, poor PV performances were obtained [85, 86]. The DSSCs, which contained the dyes **4b** and **4c** could be operated at conventional AM1.5 sunlight; whereas the **4a**-containing device was only used at the lower irradiance of 78 mW cm^{-2} [85]. The latter two complexes enabled a better charge injection, but were prone to aggregation due to their π-extended structure [86]. Falaras's carboxylated complex **5** revealed a stronger binding to the semiconductor (when compared to **4**) and, owing to the flexibility of the ligand, gave a higher surface coverage; the PV performance of the resulting DSSC was similar to those containing **4a** or **4c** [87]. None of these dyes was able to compete with N3 or BD in terms of PCE under comparable experimental conditions.

Kim *et al.* presented the Ru^{II} complexes **6** (Figure 6.7), which possessed similar LUMO levels as BD (**2**), but were superior in terms of absorption (*i.e.*, broader absorption range and higher extinction coefficients) as well as binding [88]. The improved photophysical properties were due to the interplay between the electron-accepting anchor group and the electron-donating thiophenyl spacer. The corresponding DSSCs afforded even slightly higher PCE values than analogous solar cells, which contained **1b** or **2** (*i.e.*, 6.5% and 7.5% *vs.* 5.8% and 6.1%, respectively). The increased PV performance of **6** was attributed to the improved IPCE

Table 6.1 Summary of the photovoltaic characteristics of DSSCs using Ru^{II} complexes as sensitizers.[a)]

Ru^{II} dye	V_{OC} (mV)	J_{SC} ($mA\,cm^{-2}$)	FF	PCE (%)	Relative PCE[b)]	Reference
N3 (**1a**)	720	18.20	0.730	10.00		[39]
Black dye (**2**)	720	20.53	0.704	10.40		[40]
N719 (**1b**)	846	17.73	0.720	11.18		[80]
3a	760	19.00	0.749	10.81	1.17[b)]	[78]
3b	740	20.40	0.739	11.16	1.21[b)]	[78]
4a	584	6.10	0.620	2.90	0.43[c)]	[85]
4b	630	12.80	0.710	5.70	0.69[b)]	[86]
4c	600	5.70	0.700	2.40	0.29[b)]	[86]
5	616	6.19	0.650	2.50	0.53[b)]	[87]
6a	709	12.90	0.714	6.50	1.07[b)]	[88]
6b	784	13.10	0.726	7.50	1.23[b)]	[88]
7	680	23.07	0.710	11.10	1.06[b)]	[89]
8a	750	17.00	0.715	9.10	1.07[b)]	[90]
8b	760	19.70	0.686	10.30	1.21[b)]	[90]
8c	720	17.80	0.690	8.86	1.04[b)]	[90]
8d	666	22.70	0.707	10.70	0.99[b)]	[91]
9a	680	18.10	0.710	8.70	0.80[b)]	[92]
9b	690	18.30	0.720	9.10	0.99[b)]	[92]
9c	670	18.70	0.700	8.70	0.95[b)]	[92]
9d	650	18.10	0.710	8.40	0.91[b)]	[92]
10a	605	13.05	0.708	5.60	0.64[b)]	[93]
10b	668	16.42	0.715	7.80	0.90[b)]	[93]
10c	712	17.73	0.711	9.00	1.03[b)]	[93]
12	630	18.54	0.700	8.18	0.69[b)]	[94]
13	610	17.19	0.710	7.47	0.76[b)]	[94]
14	670	21.13	0.715	10.18	0.92[b)]	[95]

a) Details on the electrolyte and the device configuration were omitted for clarity and can be found in the respective reference. In all DSSCs, an FTO/TiO_2 photoanode was used.
b) PCE of the respective dye divided by the PCE of the "black dye" (**2**) under the same experimental conditions.
c) PCE of the respective dye divided by the PCE of N3 (**1a**) under the same experimental conditions.

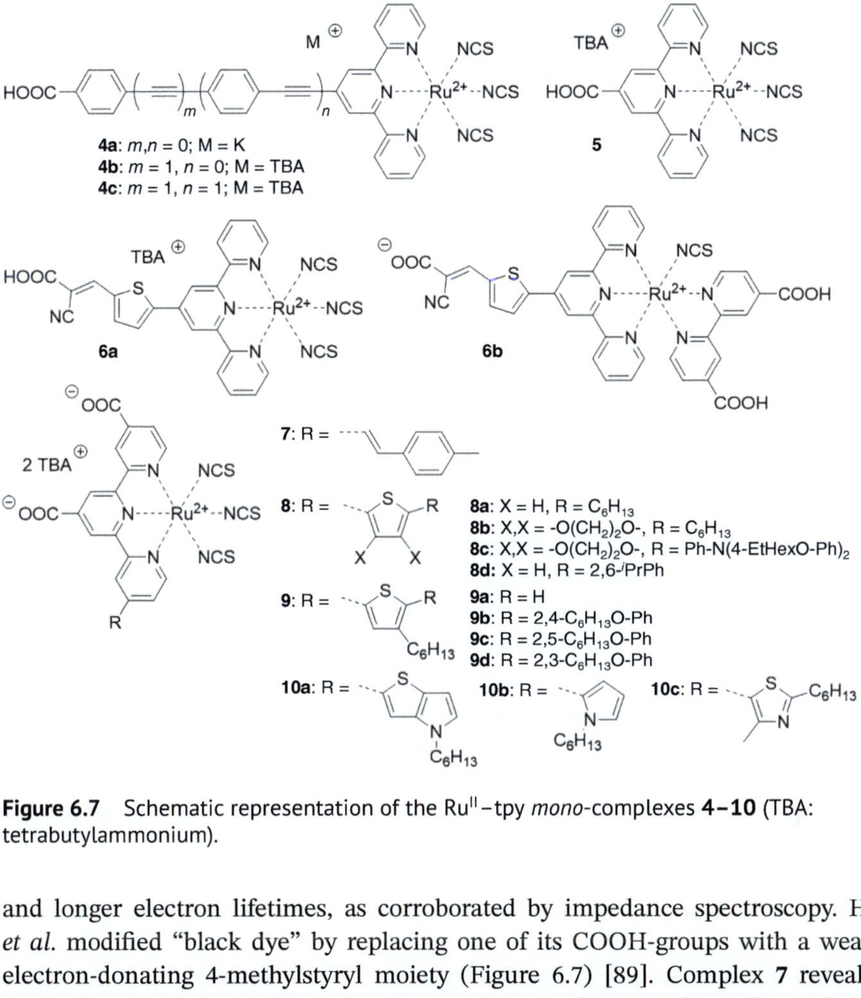

Figure 6.7 Schematic representation of the Ru^{II}–tpy *mono*-complexes **4–10** (TBA: tetrabutylammonium).

and longer electron lifetimes, as corroborated by impedance spectroscopy. Han *et al.* modified "black dye" by replacing one of its COOH-groups with a weakly electron-donating 4-methylstyryl moiety (Figure 6.7) [89]. Complex **7** revealed, due to its extended π-conjugated system, an increased MLCT absorption and more intense π–π* transitions; thus, the IPCE spectrum showed a better photoresponse over the entire spectral range. According to density functional theory (DFT) calculations, the electron injection into the TiO_2 conduction band was facilitated by the proper LUMO energy level of **7**. All things together accounted for the better PV performance of **7** relative to BD (the respective DSSCs provided PCE values of 11.1% and 10.5%).

In related work, Chi, Chou *et al.* [90] as well as Kimura, Mori *et al.* [92] introduced thiophene-based substituents instead of the aforementioned 4-methylstyryl one (**8** and **9**, Figure 6.7). In particular, complexes **8b** and **8c**, which contained an EDOT moiety, provided higher PCE values in initial studies than BD under the same experimental conditions (EDOT: 3,4-ethylenedioxythiophene; Table 6.1) [90]. The good PV performance of these complexes was attributed to their strong absorption in the visible regime (the bands at 390 and 475 nm, respectively, possessed very high molar extinction coefficients of >28,000 $L\,mol^{-1}\,cm^{-1}$). Even though a very low

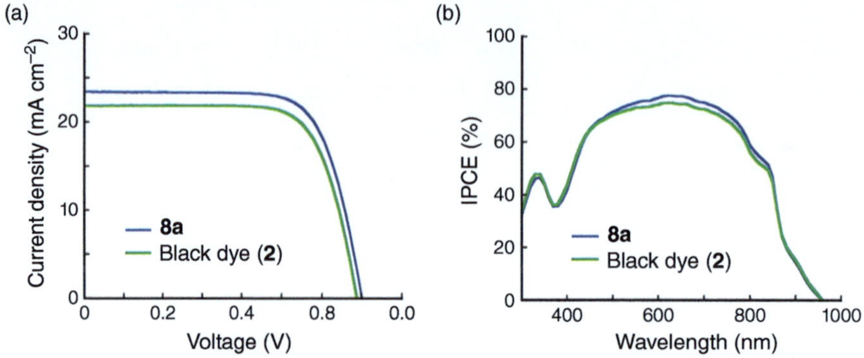

Figure 6.8 Comparison of (a) the *I*–*V*-curves and (b) IPCE spectra of DSSCs, which contained **8a** or **2**, as sensitizer [97] / with permission of Royal Society of Chemistry.

surface coverage was found for bulky complex **8d** (Figure 6.7), its PV performance was similar to those of **8a** and **2** under the same experimental conditions (Table 6.1) [91]. It was assumed that the bulky substituents acted as a protective layer to prevent the semiconductor from undergoing performance-reducing redox reactions with the electrolyte. In subsequent work, the performance of **8a** was improved significantly. Utilizing a low-temperature dye-adsorption method afforded a DSSC, whose PCE was 11.1%, thus higher than the corresponding DSSC based on BD (PCE of 10.9%; Figure 6.8) [97]. Remarkably, the **8a**-containing DSSC, which contained an ionic liquid [*i.e.*, 1-ethyl-3-methylimidazolium iodide, (emim)I], as the electrolyte, even delivered a maximum PCE of 11.9%. The complexes **9b–d** were equipped with a bulky dihexyloxyphenyl substituent, which also shielded the TiO_2 layer from the electrolyte [92]. The best results were obtained when **9b** was utilized as the sensitizer (the corresponding DSSC gave a similar PCE as the one which contained BD (**2**): 9.1% *vs.* 9.2%; Table 6.1).

Within the series of sensitizers **10** (Figure 6.7), the 4-methylthiazole derivative **10c** exhibited the best PV performance (Table 6.1) [93]. The superior PCE of this dye was attributed to the higher degree of conjugation in the tpy ligand, its improved light-harvesting capability, the higher TiO_2-surface coverage as well as the suppressed charge-carrier recombination at the photoanode-to-electrolyte interface.

Husson *et al.* [98] as well as Malapaka *et al.* [99] prepared 4,4″-dicarboxylated tpy ligands with electron-donating heteroaromatic substituents in 4′-position. However, the resulting Ru^{II} complexes **11** (Figure 6.9), as black-dye analogs, were not able to compete with **2** in terms of PV performance, *i.e.*, the PCE. Thus, the synthetic ease of the ligand synthesis *via* ring-assembly methods (see Section 1.5.3) came at the expense of practical applicability. It was proposed that substituents in the 4′-position hindered a proper anchoring of the complexes onto the TiO_2 and, thus, hampered the electron injection into the semiconductor.

Arakawa *et al.* compared the binding affinity of the dicarboxylated complexes **12** and **13** to TiO_2 to that of **2** [100]. Owing to its adjacent COOH moieties, **12** featured a superior absorptivity to the semiconductor surface than BD [100]; whereas steric

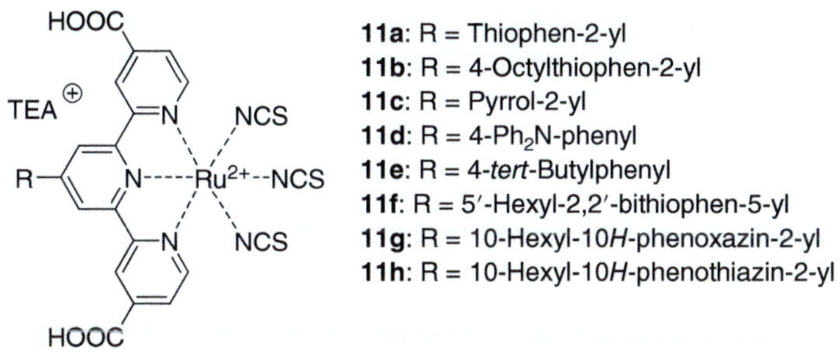

Figure 6.9 Schematic representation of the dicarboxylated complexes **11** (TEA: tetraethylammonium).

repulsion between **13** and the OH-groups on the TiO_2 surface resulted in a weaker binding and low surface coverage (Figure 6.10) [94]. On the other hand, **12** enabled a more effective electron injection than **12**, which gave rise to an overall better PV performance (PCE of 8.18% *vs.* 7.47%; Table 6.1). In subsequent work, these same authors also prepared the tetra*kis*-carboxylated complex **14**, which bound to the TiO_2 surface in a different way than **12** or **13** (Figure 6.10c). Moreover, dye

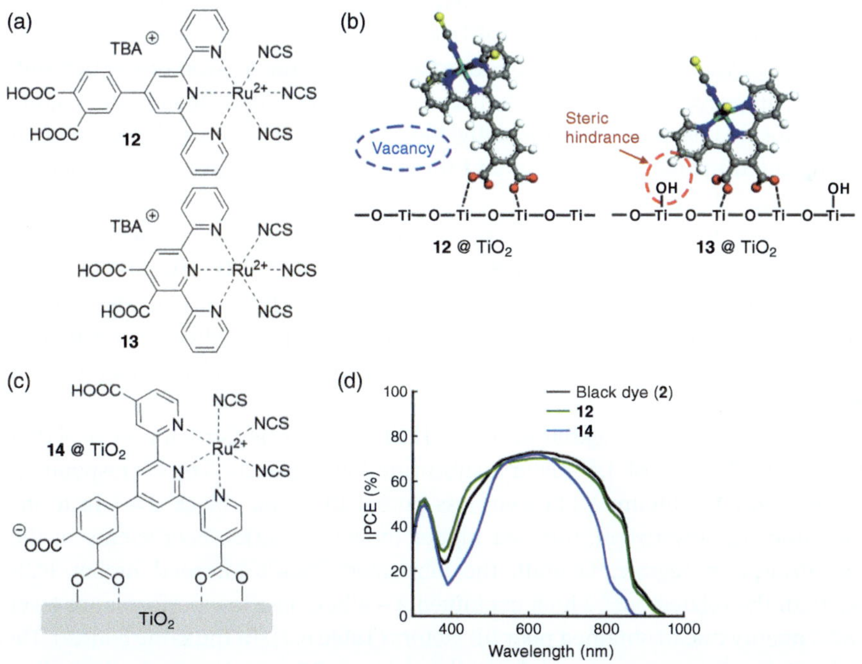

Figure 6.10 (a) Schematic representation of the Ru^{II} complexes **12** and **13**. (b) Schematic representation of the proposed model for binding of **12** (left) and **13** (right) to the TiO_2 surface *via* their COOH moieties [94] / with permission of Royal Society of Chemistry. (c) Schematic representation of the proposed binding of **14** to the TiO_2 surface. (d) IPCE spectra of DSSCs incorporating **2**, **12**, or **14** as sensitizer [95] / with permission of American Chemical Society.

14 showed an improved absorptivity and IPCE than **12** or **13** and provided, under optimized conditions, a very high PCE of 10.2% (for comparison, the analogous DSSC, which contained **2**, afforded a PCE of 10.8%).

6.3.2 Ru^{II}–Tpy Sensitizers with One Thiocyanato Ligand

Complexes of this type contain one each tpy and thiocyanato ligand as well as a bidentate one; thereby, the latter might be neutral (*e.g.*, bpy), anionic (*e.g.*, acac) or cyclometalating (*e.g.*, 2-phenylpyridinato) [79]. Such complexes are sometimes referred to as Ru^{II}–[3 + 2 + 1] sensitizers. The NCS^- ligand is of importance regarding the sensitizer's photophysical and electrochemical properties, mainly by a destabilization of the metal's t_{2g} orbital [101]. The replacement of two such ligands by a bidentate chelate was motivated by a number of reasons. Firstly, thiocyanate represents an ambident ligand, which is basically capable of binding *via* both of its nucleophilic sites. The undesired formation of complexes with *S*-coordinated ligands reduces the yield and requires tedious separation of isomers [102]. Moreover, the monodentate ligand is rather labile and might be displaced by nucleophiles, *e.g.*, from the electrolyte. Finally, the absence of an effective chromophore results in only a marginal contribution of the NCS^- ligand to the dye's absorptivity and in the high-energy regime. The use of bidentate ligands is beneficial to increase the stability, further manipulate the photophysical/electrochemical properties and, eventually, improve the IPCE. In particular, cyclometalated ligands are highly beneficial for this purpose. Detrimentally, the HOMOs of such complexes are often destabilized (compared to that of **2**) and the driving force for dye regeneration by the electrolyte is lowered.

One example of a mixed-ligand complex, which contained a tpy and a bpy ligand, was already mentioned in Section 6.3.1 (**6b** in Figure 6.7) [88]. The first example for such a complex, *i.e.*, phosphonate-containing **13** (Figure 6.11), was already reported in 1997 by Kalyanasundaram, Grätzel *et al.* [82]; whereas the related *tris*-carboxylated complex **14** (Figure 6.11) was only mentioned in a patent from 1998 [103].

More recently, Chandrasekharam *et al.* [104] as well as Giribabu *et al.* [105] explored the influence of 4,4′-distyryl-substituted bpy ligands. The corresponding Ru^{II} complexes **15** (Figure 6.11) were less potent than the "black dye" in terms of panchromatic absorption, but had higher molar extinction coefficients in the visible absorption regime. Overall, the fabricated DSSCs afforded higher PCE values than the related one, which contained **2** – albeit on a low performance level (<3.8%), mainly due to obtained poor fill factors (Table 6.2). In the same context, De Angelis, Malapaka *et al.* prepared the Ru^{II} complexes **16**, which contained pending dialkyl fluorene or *N*-alkyl carbazole moieties [106]. These complexes covered a broad absorption range with high extinction coefficients, but suffered from the inadequate location of the LUMO on the bpy ligand, thus too remote from the opposing COOH-anchoring sites and detrimental regarding the electron-injection step. Accordingly, these sensitizers afforded very low PCE values, which could

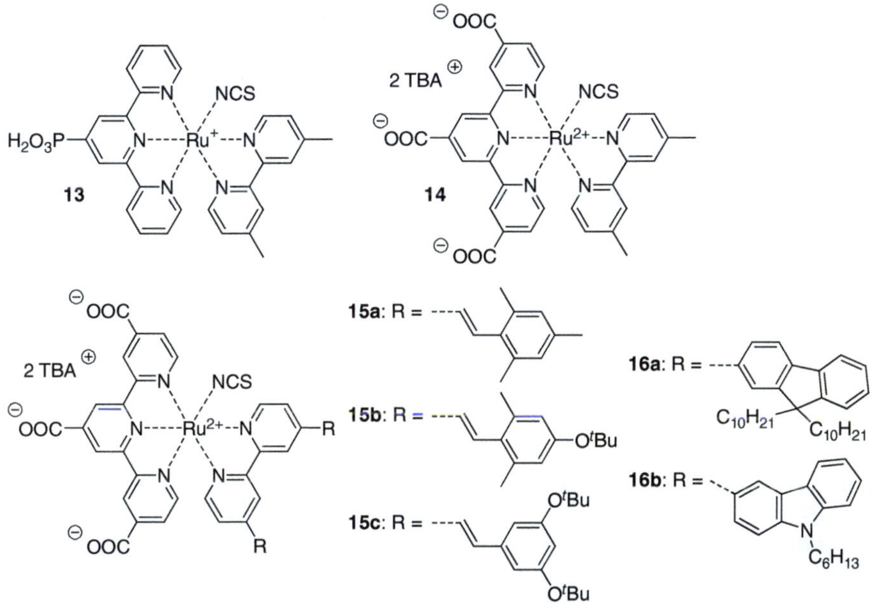

Figure 6.11 Schematic representation of Ru^{II}–tpy–bpy complexes **13–16**.

be increased only to 2.3% in the case of **16a**, if CDCA was used, as co-absorbent (CDCA: 3a,7a-dihydroxy-5b-cholic acid).

Furthermore, complex **7** (Figure 6.7) was modified by replacing two of its thiocyanato ligands by bpy derivatives (complexes **17**, Figure 6.12) [107]. Complex **17a**, which contained an unsymmetrical bpy ligand, showed an improved IPCE in the range from 300 to 700 nm when compared to "black dye." However, the overall PV performance of **17a** was lower than that of BD when using the otherwise same DSSC configuration (PCE of 5.19% *vs.* 6.44%, Table 6.2). A comparison of the PV activity of complexes **7** and **17a** was not provided. The tetra*kis*-carboxylated complex **17b** afforded, due to the absence of chromophores on the bpy ligand, an overall lower IPCE and PCE than **17a** or BD (Table 6.2).

The same bpy ligand as in **17b** was employed as the anchoring site in complexes **18** (Figure 6.12) [108]. Notably, these complexes missed the "conventional" COOH-moieties on the tpy ligand; instead, electron-donating, π-conjugated substituents were introduced in the 4,4′,4″-positions in order to increase the molar extinction coefficient. Independent from the used electrolyte (*i.e.*, liquid. *vs.* solid), complex **18a** was identified as the more potent sensitizer for DSSC applications – the maximum PCE was 5.4% (Table 6.2). The same concept, *i.e.*, electron-donating substituents on the tpy ligand and anchoring moieties on the bidentate chelate, was also applied in other complexes. For example, Falaras *et al.* tested complex **19** regarding its applicability to act as sensitizer in DSSCs, which was actually very poor (Figure 6.13a, Table 6.2) [109]. In related studies, also other carboxylated or sulfonated oligopyridyl ligands were used: 2-(pyridin-2-yl)-quinoline [116], phen [117], and others.

Table 6.2 Summary of the photovoltaic characteristics of DSSCs using Ru^{II} complexes as sensitizers.[a)]

Ru^{II} dye	V_{OC} (mV)	J_{SC} (mA cm^{-2})	FF	PCE (%)	Relative PCE[b)]	Reference
15a	613	11.9	0.43	3.15	1.17	[104]
15b	655	12.7	0.40	3.36	1.25	[104]
15c	610	8.6	0.72	3.76	0.84	[105]
16a	641	4.3	0.78	2.1	n/a	[106]
16b	590	1.5	0.75	0.7	n/a	[106]
17a	520	14.0	0.71	5.19	0.81	[107]
17b	490	13.3	0.70	4.63	0.72	[107]
18a	684	10.4	0.77	5.40	n/a	[108]
18b	632	10.0	0.78	4.90	n/a	[108]
19	584	4.3	0.43	1.64	n/a	[109]
20a	530	6.1	0.69	2.2	n/a	[110]
20b	580	9.1	0.70	3.7	n/a	[110]
21a	620	14.1	0.74	6.4	0.63	[111]
21b	670	18.7	0.73	9.1	0.90	[111]
21c	720	20.8	0.71	10.7	1.06	[111]
22a	530	15.9	0.63	5.31	n/a	[112]
22b	570	17.2	0.63	6.18	n/a	[112]
22c	640	18.0	0.66	7.6	n/a	[112]
22d	580	22.4	0.67	8.7	n/a	[113]
22e	590	22.0	0.68	8.8	n/a	[113]
22f	550	20.9	0.67	7.7	n/a	[113]
22g	520	19.0	0.64	6.3	n/a	[114]
23a	687	20.3	0.65	9.14	1.01	[115]
23b	668	21.7	0.64	9.33	1.03	[115]
23c	720	20.4	0.65	9.59	1.06	[115]
23d	714	21.6	0.65	10.05	1.11	[115]
24a	760	18.7	0.73	10.43	1.13	[78]
24b	730	20.1	0.72	10.51	1.14	[78]

a) Details on the electrolyte and the device configuration were omitted for clarity and can be found in the respective reference. In all DSSCs, an FTO/TiO_2 photoanode was used.

b) PCE of the respective dye-divided by the PCE of the "black dye" (**2**) under the same experimental conditions.

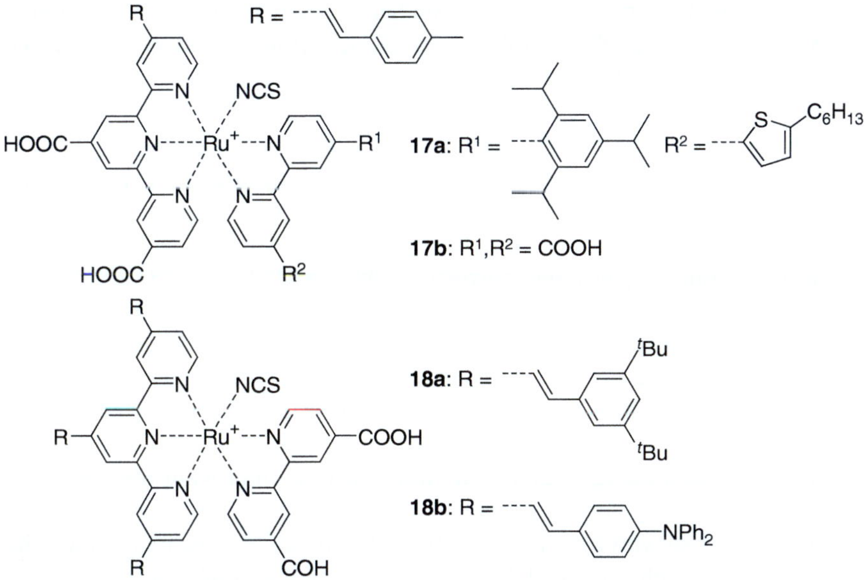

Figure 6.12 Schematic representation of Ru^{II} complexes **17** and **18** (counterions were omitted for clarity).

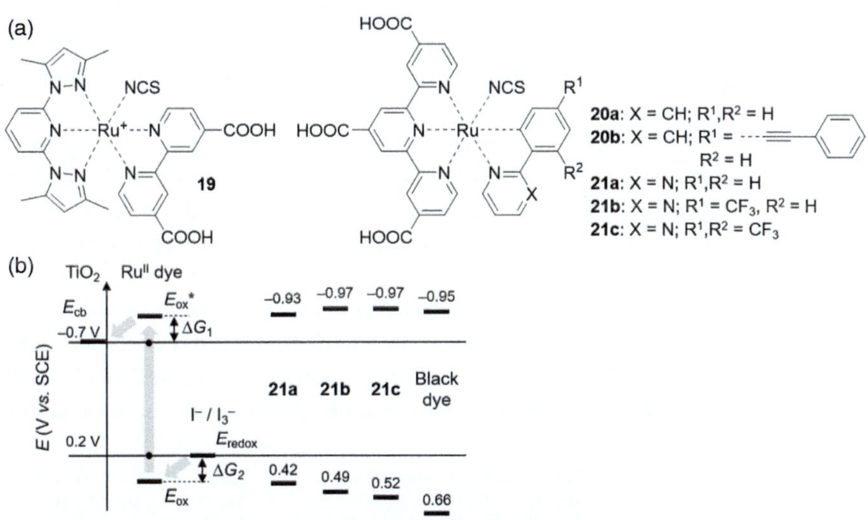

Figure 6.13 (a) Schematic representation of Ru^{II} complexes **19–21** (counterions were omitted for clarity). (b) Energy-levels diagrams of TiO_2, Ru^{II} sensitizers (**21** and **2**), and the I^-/I_3^- redox couple [111] / with permission of John Wiley & Sons.

Bidentate cyclometalating C^N ligands, such as 2-phenylpyridine, are commonly used in the context of Ir^{III} complexes for optoelectronic [118] or life-science applications [119]. Ligands of this type were also employed to substitute two of the thiocyanato ligands in Ru^{II} complexes of the black-dye family. It was intended to destabilize the metal's HOMO orbital *via* these strong σ-donor ligands and, thereby, lower the band gap and improve the absorption of light in the low-energy regime [79]. The first such complexes **20**, as analogs to **14**, were reported by Sugihara *et al.* (Figure 6.13a) [110]. These complexes efficiently absorbed light even in the NIR spectral regime; in particular, π-extended **20b** afforded an IPCE value of 10% at 900 nm. Detrimentally, the improved charge injection was cancelled by the sensitizers' low oxidation potentials, which hampered the dye's regeneration by the electrolyte (Table 6.2). In order to counteract this, these same authors modified the structure of the cyclometalating ligand by changing to a 2-phenylpyrimidine scaffold (**21**, Figure 6.13a) [111, 120]. Owing to the strongly electron-withdrawing CF_3-groups, the ΔG_2 energy gap was minimized, thus improving the capability for dye regeneration by the electrolyte (Figure 6.13b). In particular, the DSSC, which was sensitized with **21c** afforded a high PCE of 10.7%, thus higher than the analogous DSSC, which contained **2**, as the benchmark dye (Table 6.2).

In 2002, Sugihara, Arakawa *et al.* introduced β-diketonato chelates as strongly σ-donating ligands to destabilize the ground-state energy level of Ru^{II} dyes relative to that of BD (complexes **22** in Figure 6.14) [121]. DTF calculations revealed that the dyes' HOMO and LUMO were localized on the NCS^- and tpy ligand, respectively [113]. It was proposed that an appropriate sensitizer should possess a HOMO energy level at least 0.53 V *vs.* SCE (SCE: saturated calomel electrode); this requirement is met for those complexes, which contain a CF_3-group [122]. The first generation of complexes (**22a–c**) represented efficient panchromatic absorbers, which sensitized nanocrystalline TiO_2 with a maximum IPCE of 70% [121]. Moreover, the long alkyl chain of **22c** prevented the dye molecules from surface aggregation and shielded the photoanode from the electrolyte, thus suppressing undesired redox processes as reflected by the high V_{OC} [112]. Unlike for **22a/b**, **22c** required no co-adsorbent to show a reasonably good PV performance (*i.e.*, PCE of 7.6%; Table 6.2). In subsequent work, the substituents on the β-diketonato ligand were varied over a wide range [113, 122, 123]. The aryl-substituted derivatives **22d–f** absorbed light up to *ca.* 1000 nm and afforded maximum IPCE values of up to 80% [113, 114]. From these, complex **22e** featured the most notable PCE of 9.1%, at least under

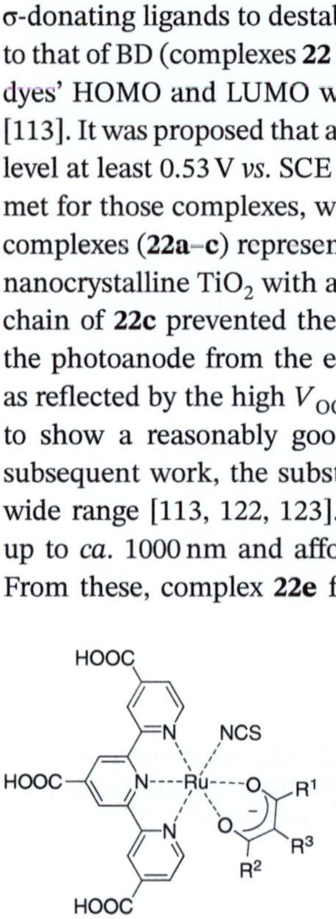

22a: $R^1 = CF_3$, $R^2 = Me$, $R^3 = H$
22b: $R^1 = CF_3$, $R^2 = C_4H_9$, $R^3 = H$
22c: $R^1 = CF_3$, $R^2 = C_{16}H_{33}$, $R^3 = H$
22d: $R^1 = CF_3$, $R^2 = C_6H_5$, $R^3 = H$
22e: $R^1 = CF_3$, $R^2 = 4\text{-F-}C_6H_4$, $R^3 = H$
22f: $R^1 = CF_3$, $R^2 = 4\text{-Cl-}C_6H_4$, $R^3 = H$
22g: $R^1 = CF_3$, $R^2 = 4\text{-}Ph_2N\text{-}C_6H_4$, $R^3 = H$
22h: R^1, $R^2 = C_6H_5$, $R^3 = H$
22i: R^1, $R^2 = Me$, $R^3 = H$
22j: R^1, $R^2 = C_5H_{11}$, $R^3 = H$
22k: R^1, R^2, $R^3 = Me$
22l: R^1, $R^2 = {}^tBu$, $R^3 = H$

Figure 6.14 Schematic representation of the β-diketonato-containing Ru^{II} complexes **22**.

optimized conditions. Unfortunately, the PV performance of the sensitizers was not compared to that of a reference dye within an analogous device configuration. Owing to the poor performance of the DSSCs, which contained complexes **22h–l**, their characteristics are not included in Table 6.2.

As mentioned beforehand, a remarkably high PV performance was observed for DSSCs, which contained complexes **3** as sensitizers (Figure 6.5, Table 6.1) [78]. This class of charge-neutral complexes, which contained a 2-(pyrazol-3-yl)pyridine ligand, was introduced by Chou, Chi *et al.* [115]. The first generation of complexes **23** was equipped with an electron-donating styryl substituent on the bidentate chelate (Figure 6.15a). These complexes represented superior sensitizers than the "black dye" – the higher molar extinction coefficients and the broader absorption range (up to 850 nm) were attributed to the HOMO destabilization by the coordinated anionic ligand. The highest PCE within the series was found for **23d**, which was higher than for BD (*i.e.*, 10.05% *vs.* 9.07%). Moreover, the DSSC, which was sensitized by **23d**, exhibited a remarkable photostability – the initial PCE was retained to *ca.* 91% after 1000 h of illumination with AM1.5 sunlight. Besides the utilization of 2,6-*bis*(pyrazol-3-yl)pyridine ligands to obtain *bis*-tridentate Ru^{II} complexes, Chou, Wei *et al.* also paid attention to the modification of the tpy ligand. Specifically, the substitution of an outer pyridine ring by a *tert*-butyl-substituted quinoline moiety afforded complexes **24** (Figure 6.15a) whose low-energy absorption bands

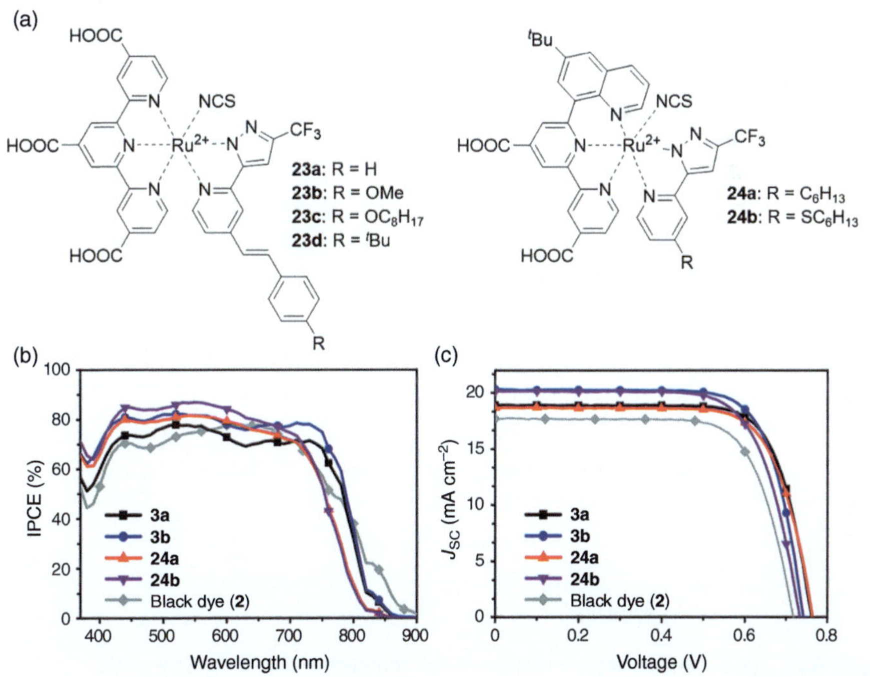

Figure 6.15 (a) Schematic representation of Ru^{II} complexes **23** and **24** (counterions were omitted for clarity). (b) IPCE spectra of DSSCs, which were sensitized by complexes **3**, **24**, or **2**. (c) *I*–*V* curves of these same DSSCs [78] / with permission of Royal Society of Chemistry.

were hypsochromically shifted; the increased molar extinction coefficients of these bands accounted for an overall improved IPCE and J_{SC} (Figure 6.15b,c) [78]. Even though a very good PV performance was observed for **24** – better than that of BD under identical experimental conditions – were reached, these complexes were outcompeted in terms of PCE by their aforementioned counterparts **3**, which contained the conventional *tris*-carboxylated tpy ligand (Tables 6.1 and 6.2).

6.3.3 Ru^{II}–Tpy Sensitizers Without Thiocyanato Ligands

Various Ru^{II} complexes, which did not contain NCS^- ligands, have also been studied regarding their capability to sensitize TiO_2. In this respect, one has to distinguish between black-dye analogs, in which all three NCS^- ligands were replaced by monodentate, non-heteroaromatic ones, and Ru^{II} *bis*-tridentate complexes.

Regarding the former, Segawa *et al.* presented the phosphine-coordinated sensitizer **25a** (Figure 6.16a), which – unlike most other Ru^{II}–polypyridyl complexes – provided an alternative path to the ^{3}MLCT state *via* a spin-forbidden absorption from the singlet GS [124]. Thereby, the considerable energy loss is typically the result of intersystem crossing (ISC) from an initially populated ^{1}MLCT state to the long-lived ^{3}MLCT one. The DSSC, which was sensitized by **25a** afforded a remarkably high J_{SC} of 26.8 mA cm^{-2}. The performance of the DSSC could be improved by adding 4-*tert*-butylpyridine (TBP) to the electrolyte (a maximum PCE

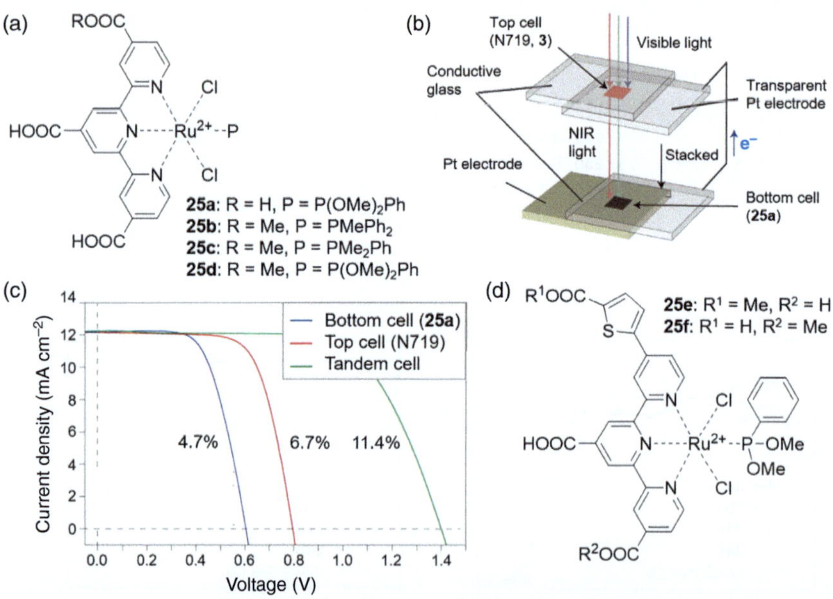

Figure 6.16 (a) Schematic representation of Ru^{II} complexes **25a–d**. (b) Schematic representation of the tandem DSSC, which was sensitized by **25a** and N719. (c) *I–V*-curves of the individual layers and the tandem DSSC upon illumination with AM1.5 sunlight. (d) Schematic representation of Ru^{II} complexes **25e/f** [124] / with permission of Springer Nature.

of 10.0% was obtained at a TBP concentration of 0.2 M; Table 6.3). Moreover, a tandem DSSC, which contained **25a** in combination with N719 (**3**), as the sensitizers, gave a PCE of 12.1% under illumination with low-intensity sunlight (*i.e.*, 35.5 mW cm^{-2}; Figure 6.16b,c). In subsequent work, complex **25c** was identified as a panchromatic absorber, which offered PV activity even up to 1100 nm [125]. The DSSC, which used **25c**, featured a high J_{SC} of 30.3 mA cm^{-2} as well as a PCE of 10.2% upon illumination with AM1.5 sunlight. The capability of **25c** to absorb light in the NIR regime was used to combine this sensitizer with a lead-bromide perovskite, which is known for its efficient harvesting of visible light. The thusly obtained tandem cell afforded a maximum PCE of *ca.* 22% when a spectral-splitting system was employed. This example highlights how the PV performance of the emerging class of perovskite solar cells, which are discussed in Section 6.5, can be improved by Ru^{II}–tpy complexes, as sensitizer, or tpy ligands, as additives.

More recently, Kinoshita *et al.* increased the intensity of the spin-forbidden single-to-triplet absorption in the NIR regime by *ca.* 20% by introducing a thiophene substituent on the tpy ligand (**25e/f**, Figure 6.16d) [126]. In particular, the DSSCs, which were sensitized by **25e** revealed a very good panchromatic light-to-energy conversion efficiency up to 1000 nm; this DSSC provided a maximum PCE of 9.16% (Table 6.3). Unlike for **25f**, dye aggregation was less pronounced due to a shielding effect of the methyl ester on the thiophene group. In the presence of deoxycholic acid (DCA), aggregation was suppressed further and the DSSC, which contained **25e**, gave a high PCE of even 10.2%.

The possibility to assemble heteroleptic *bis*-tridentate Ru^{II} complexes has been utilized in different fields of research. Such mixed-ligand complexes have, *e.g.*, been integrated into macromolecular architectures or used in the context of energy-/electron-transfer arrays. With respect to sensitizers for DSSCs, Falaras *et al.* tested dye **26**, which contained a phosphonic-acid moiety as the anchoring group, as well as a 4-iodophenyl substituent (Figure 6.17) [127]. The latter was introduced to enable interactions with the I^-/I_3^- redox couple and, thus, facilitate

Table 6.3 Summary of the photovoltaic characteristics of DSSCs using Ru^{II} complexes as sensitizers.[a)]

Ru^{II} dye	V_{OC} (mV)	J_{SC} (mA cm^{-2})	FF	PCE (%)	Relative PCE[b)]	Reference
25a	664	21.4	0.70	10.0	n/a	[124]
25c	552	30.3	0.60	10.2	n/a	[125]
25d	586	21.9	0.61	7.78	n/a	[126]
25e	598	25.0	0.61	9.16	n/a	[126]
25f	550	20.4	0.63	7.03	n/a	[126]

a) Details on the electrolyte and the device configuration were omitted for clarity and can be found in the respective reference. In all DSSCs, an FTO/TiO_2 photoanode was used.
b) PCE of the respective dye divided by the PCE of a reference dye under the same experimental conditions.

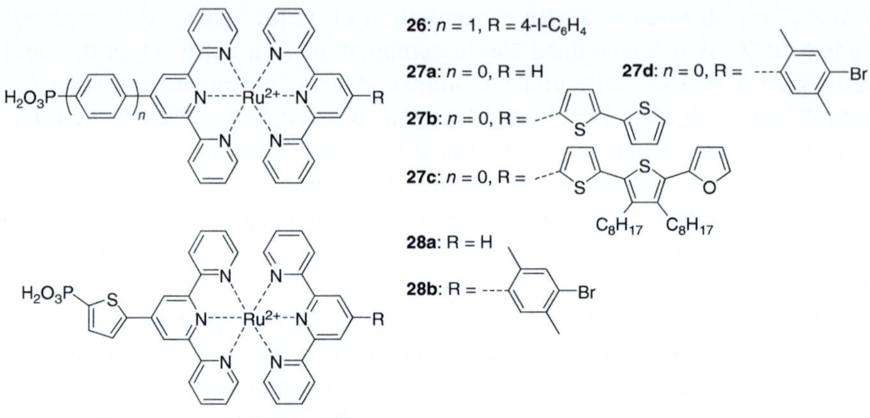

Figure 6.17 Schematic representation of the heteroleptic Ru^{II} *bis*-tpy complexes **26–28** (counterions were omitted for clarity).

the reductive dye regeneration. However, the solid-state device, in which a SnO_2 photoanode was sensitized by **26**, afforded a very low PCE of only 0.1% [solid electrolyte: poly(ethylene glycol) with LiI, I_2, and powdered TiO_2]. The related complexes **27** (Figure 6.17), which were aimed to improve the interaction with the hole-transport material (HTM) *via* their oligothiophenyl moieties, also showed an overall poor PV performance (PCE of ≤0.16%) [128]. The dyes' LUMOs were located on the remote substituents, which was detrimental for the electron injection into the semiconductor. Accordingly, the unsubstituted dye **27a** displayed the highest activity within this series of complexes. A slight improvement in PV performance was achieved by changing the position of the thiophenyl moiety (**28**, Figure 6.17) [129]. If placed between the anchoring group and the Ru^{II}–tpy complex, the electron-injection efficiency was increased. The DSSC, which was sensitized by **28b** afforded a maximum PCE of 0.5%; whereas the analogous device, which contained N3 (**1a**, Figure 6.3), revealed an eight-times higher efficiency. Similar results in terms of poor PV performance were made by Krebs *et al.*, who studied a library of homoleptic and heteroleptic Ru^{II} *bis*-tpy complexes [130]. None of these dyes was able to compete with the "black dye," which was the by far more potent sensitizer for the tested DSSC configuration.

As detailed in Section 6.4, Ru^{II}–tpy complexes have been integrated into polymers. These materials have been utilized in the context of polymer-based solar cells, specifically of the BHJ-type. Krebs *et al.* prepared the tpy-terminated π-conjugated macroligands **29** (Figure 6.18) and assembled their homoleptic [Ru(**29a**)$_2$]$^{2+}$ and [Ru(**29b**)]$^{2+}$ complexes as well as the heteroleptic [(**29a**)Ru(**29b**)]$^{2+}$ complex [131]. These metallopolymers, which did not possess any anchoring groups, were absorbed onto TiO_2. The highest DSSC performance was found for the cell, which was sensitized by the heteroleptic donor-acceptor assembly (PCE of *ca.* 0.1%). Notably, a Co$^{II/III}$ redox couple was used as the electrolyte to prevent polymer degradation by the conventional I^-/I_3^- system. However, dye regeneration by such Co^{II}–polypyridyl complexes is inefficient owing to the kinetically

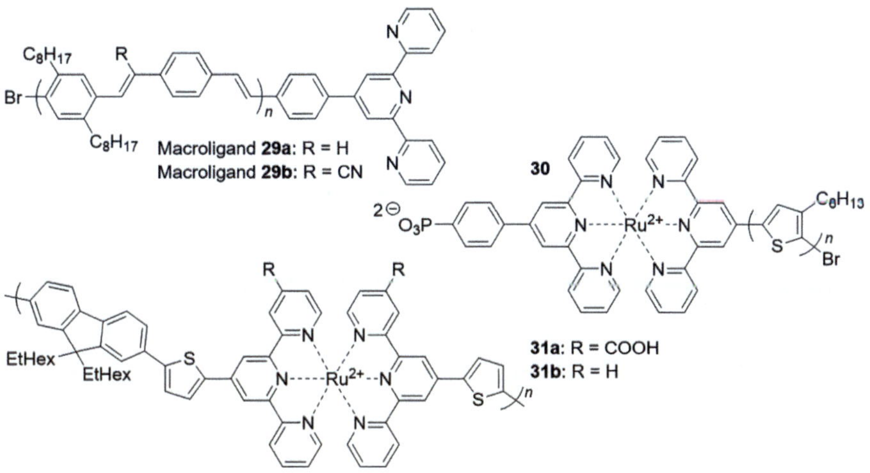

Figure 6.18 Schematic representation of the macroligands **29** and ruthenopolymers **30** and **31** (counterions were omitted for clarity).

slow electron-donor capability. Caramori, Gros *et al.* demonstrated that the process could be accelerated significantly by adding a Fe^{II} *tris*-bpy complexes, as a redox-comediator [132]. Sensitizer **30**, which contained a P3HT tail, featured a low photocatalytic activity, *i.e.*, PCE of *ca.* 0.05% [P3HT: poly(3-hexylthiophene), Figure 6.18] [133]. For comparison, the performance of the analogous DSSC, which was sensitized by BD, was *ca.* 6.5 times higher. In related work, Chan *et al.* prepared the conjugated metallopolymers **31** (Figure 6.18) [134]. Owing to its COOH moieties, **31a** efficiently anchored onto the TiO_2 surface and, thus, represented the more potent sensitizer than non-carboxylated **31b** (*i.e.*, PCE of 0.13% *vs.* 0.05%). In summary, the use of "ruthenopolymers" in the context of DSSCs resulted in disappointing PCE values and, thus, this concept was not pursued further.

In the context of their work on Ru^{II} complexes **16** (Figure 6.11), Malapaka *et al.* also investigated the heteroleptic Ru^{II} dye **32** (Figure 6.19) [106]. Although its electronic structure was unfavorable (*i.e.*, location of the LUMO on the remote tpy ligand), the corresponding DSSC afforded a relatively low PCE, which was, albeit, higher when compared to those based on the aforementioned Ru^{II} *bis*-tpy complexes (Table 6.7). Panchromatic sensitization up to 1000 nm was achieved with the pyrene-equipped complex **33** (Figure 6.19); however, the relatively low IPCE accounted for its inferior PV performance – higher PCE values were obtained for the DSSCs, which contained the dibromo-substituted complex **34** or **2** (Table 6.7) [135]. Mughal, Ahmed *et al.* prepared a library of homoleptic *bis*-tpy complexes with Fe^{II}, Co^{II}, Zn^{II}, and Ru^{II} ions [136]. In order to reduce the energy density at the metal centers, the ligands were equipped with electron-withdrawing groups – including chalcone or flavone moieties. However, neither of these complexes showed a reasonable PV activity (*i.e.*, the PCE remained ≤0.11%). Similar results in DSSC performance were also obtained when using the donor-acceptor-type complexes **35** as sensitizers (Figure 6.19) [137, 138]. The highest PCE of 0.71% was reported for

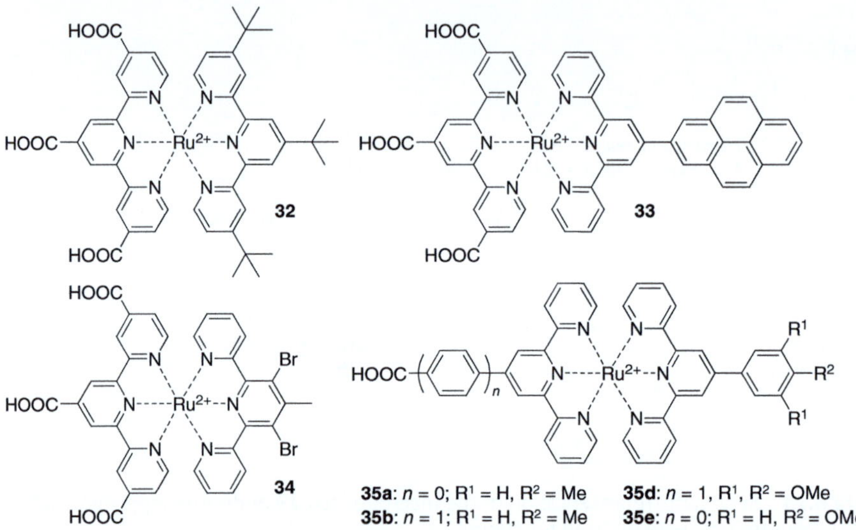

35a: n = 0; R¹ = H, R² = Me **35b**: n = 1; R¹ = H, R² = Me **35c**: n = 1; R¹ = H, R² = OMe **35d**: n = 1, R¹, R² = OMe **35e**: n = 0; R¹ = H, R² = OMe **35f**: n = 0; R¹, R² = OMe

Figure 6.19 Schematic representation of the Ru^{II} *bis*-tpy complexes **32–35** (counterions were omitted for clarity).

35b; for comparison, the "black dye" provided a PCE of 4.1% when incorporated into a DSSC of analogous configuration.

In total, the PV performance, which has been reached with Ru^{II} *bis*-tpy complexes, is short of meeting that of "black dye" or its aforementioned analogs. This finding can *inter alia* be ascribed to the unfavorable photophysics of such complexes (*i.e.*, short excited-state lifetimes) due to the distorted octahedral complex geometries [144]. Accordingly, *bis*-tridentate Ru^{II} complexes, in which one tpy ligand was replaced by a tridentate non-tpy one, were also considered for DSSC applications. For example, Islam *et al.* observed a relatively high PEC of 7.9% for their sensitizer **36**, which contained the tridentate DETA ligand (DETA: diethylenetriamine; Figure 6.20, Table 6.7) [139]. Furthermore, charge-neutral Ru^{II} complexes with a tridentate, dianionic 2,6-dipyrazolylpyridine ligand were prepared. In line with their studies on complexes **3** (Figure 6.6) [78], which represented high-performance sensitizers, complexes containing a tridentate 2,6-*bis*(pyrazol-5-yl)pyridine ligand were also utilized. Firstly, Chi *et al.* varied the substituent on the ligands' pyridine ring (**37**, Figure 6.20) [140]. The inferior sensitization capability of **37** in the NIR regime was levelled out by their superior IPCE performance in the visible range; overall, IPCEs – higher than that of **2** under same experimental conditions – were obtained. Accordingly, the DSSCs, which were sensitized by **37** afforded, at least in the case of the substituted derivatives (**37b–d**), a superior PV performance (Figure 6.21, Table 6.4). It was assumed that the "bulky ligand effect" allowed for a better dye packing on the TiO_2 surface and, thus, prevented the interfacial recombination of charges. Moreover, the high V_{OC} values were believed to result from the dye's dipole moment (μ), which was pointing more toward the semiconductor surface (for comparison, the negatively charged NCS^- ligands, as in BD, orientate μ in the opposite direction). In subsequent work, these same authors varied the structure of

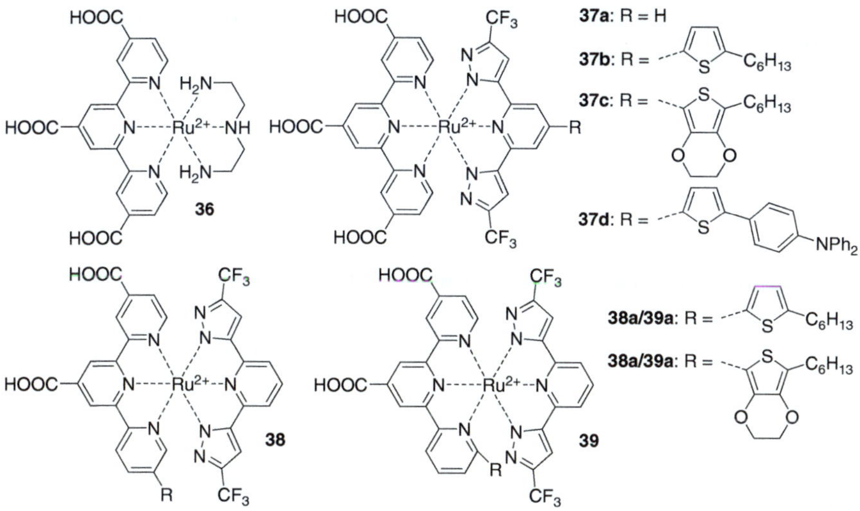

Figure 6.20 Schematic representation of the charge-neutral Ru^{II} complexes **36–39** (counterions were omitted for clarity).

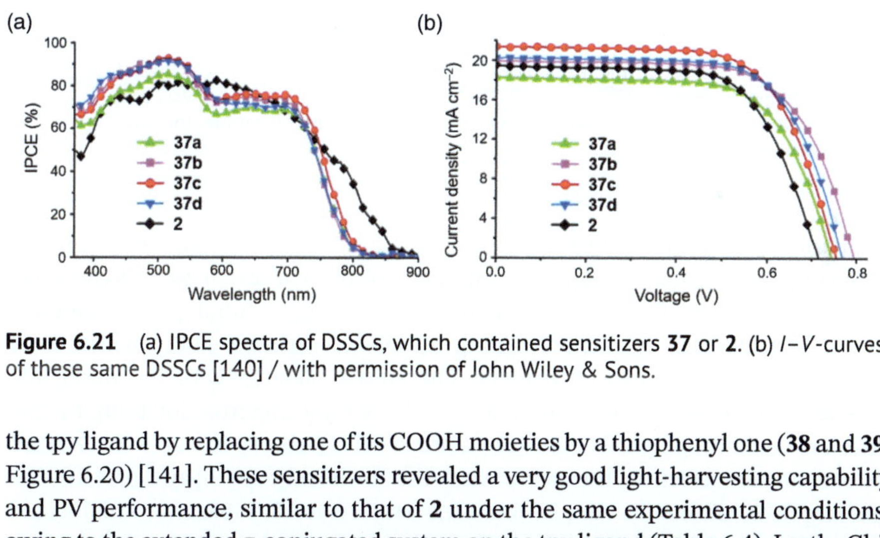

Figure 6.21 (a) IPCE spectra of DSSCs, which contained sensitizers **37** or **2**. (b) *I*–*V*-curves of these same DSSCs [140] / with permission of John Wiley & Sons.

the tpy ligand by replacing one of its COOH moieties by a thiophenyl one (**38** and **39**, Figure 6.20) [141]. These sensitizers revealed a very good light-harvesting capability and PV performance, similar to that of **2** under the same experimental conditions, owing to the extended π-conjugated system on the tpy ligand (Table 6.4). Lastly, Chi, Chou, Wei *et al.* prepared a series of related sensitizers, where one of the pyridine rings of the tpy ligand was replaced by a quinoline one (similar to complexes **24**, Figure 6.15a). These panchromatic dyes **40** efficiently absorbed light across the entire UV–vis–NIR regime. Due to the presence of the bulky *tert*-butyl-group in **40c/d**, dye aggregation on the TiO_2 surface was suppressed. This, in turn, resulted in an increase in J_{SC}, without affecting the V_{OC}. Under the tested DSSC configuration, complexes **40c/d** outcompeted BD in terms of PCE (Table 6.4).

Related to this, Dietzek, Berlinguette, Schubert *et al.* used a tridentate ligand in which the aforementioned pyrazolate moieties were replaced by 1,2,3-triazolates as a surrogate for the NCS^- ligands of BD (**40a**, Figure 6.22) [145]. This work

Table 6.4 Summary of the photovoltaic characteristics of DSSCs using Ru^{II} complexes as sensitizers.[a)]

Ru^{II} dye	V_{OC} (mV)	J_{SC} (mA cm^{-2})	FF	PCE (%)	Relative PCE[b)]	Reference
32	594	2.7	0.80	1.3	n/a	[106]
33	360	6.3	0.64	1.47	0.24	[135]
34	450	10.1	0.71	3.25	0.52	[135]
36	670	17.5	0.68	7.9	n/a	[139]
37a	740	18.2	0.68	9.11	0.98	[140]
37b	790	20.0	0.67	10.5	1.14	[140]
37c	760	21.4	0.66	10.7	1.16	[140]
37d	770	20.3	0.68	10.5	1.14	[140]
38a	710	17.7	0.70	8.80	1.05	[141]
38b	680	18.7	0.67	8.53	1.02	[141]
39a	710	19.0	0.68	9.21	1.10	[141]
39b	650	17.7	0.70	8.05	0.96	[141]
40a	610	8.9	0.70	4.00	0.66	[142]
40b	623	13.3	0.73	6.10	0.81	[143]

a) Details on the electrolyte and the device configuration were omitted for clarity and can be found in the respective reference. In all DSSCs, an FTO/TiO_2 photoanode was used.
b) PCE of the respective dye divided by the PCE of the "black dye" (**2**) under the same experimental conditions.

was performed in the context of using 1*H*-1,2,3-triazoles, as pyridine mimics, in oligopyridyl ligands [146]. The straightforward Cu^{I}-catalyzed alkyne-azide cycloaddition (CuAAC) was utilized to prepare charge-neutral [147], dianionic as well as tridentate ligands containing mesoionic carbenes [142], which were coordinated *inter alia* to Ru^{II} centers. The studies dealing with the mesoionic carbenes are discussed in Section 6.3.5. The strong electron-donation by the triazolate-containing ligand resulted in a broad MLCT absorption up to *ca.* 700 nm and an efficient electron transfer toward the anchoring tpy ligand of **40a**. The resulting DSSC afforded a high PCE of 4.0% (*vs.* 6.1% for **2**; Table 6.4), which could be increased to 4.9% in the presence of DCA as the co-absorbent. Islam, Otsuki *et al.* even reported a slightly improved performance for their dye **40b**, which contained tetrazolate moieties instead of the 1,2,3-triazolate ones (Figure 6.22, Table 6.4).

6.3.4 Ru^{II}–Tpy Sensitizers with a Tridentate, Cyclometalating Ligand

Berlinguette, Schubert *et al.* applied the same approach, *i.e.*, the CuAAC reaction, to prepare 1,3-*bis*(1,2,3-triazol-4-yl)benzene derivatives, as designated cyclometaling ligands [148]. Unlike for **40**, the resulting complexes **41** contained a ligand of the N^C^N-type, in which the pending alkyl chains were located on the triazoles'

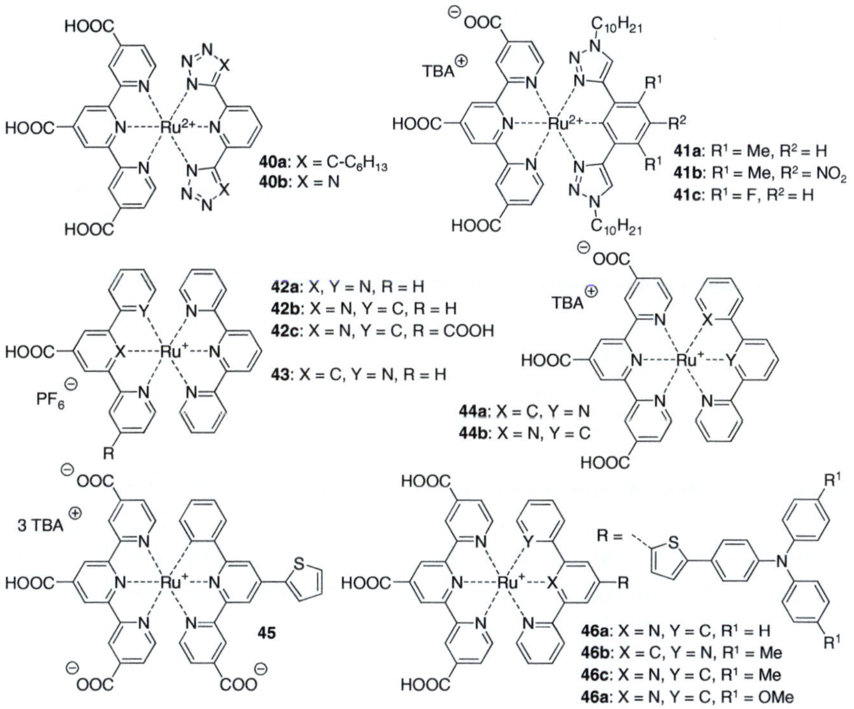

Figure 6.22 Schematic representation of Ru^{II} complexes **40–46** (counterions were omitted for clarity).

N^3-atoms (Figure 6.22). Notably, complex **41a**, which missed electron-withdrawing substituents and, thus, possessed the highest HOMO energy level within this series, provided the best DSSC performance of 4.0% (*vs.* 5.2% for **2** under comparable conditions; Table 6.5). The efficiency could be raised to 4.5% when **41a** was co-absorbed with CDA.

This work represents one example of the popular concept of using cyclometalating chelates to manipulate the photophysical and electrochemical properties of Ru^{II} complexes. This strategy was initially employed in the context of photocatalysis and later widely adapted for Ru^{II}-based sensitizers [154]. The first such study was performed by van Koten *et al.*, who compared the IPCE of complexes **42** (Figure 6.22) to that of N719 (**1b**, Figure 6.3), as the reference dye [155]. From these, **42c** and N719 showed similar IPCE spectra with an onset at *ca.* 800 nm and both gave photocurrents of *ca.* 12 mA cm^{-2} (electrolyte: γ-butyrolactone containing LiI and I_2). Whereas complex **43** (Figure 6.22), which had the anchoring group on the cyclometaling ring, provided only a very poor IPCE up to 700 nm [156]. This behavior was attributed to the unfavorable location of the LUMO, which hampered the injection of electrons into the TiO_2 conduction band. Accordingly, the N^N^C motif (as in **42b/c**) was identified to be more appropriate than the alternative N^C^N one (as in **43**). Overall, cyclometallation *via* a phenyl ring resulted in a red-shifted and broadened absorption profile when comparing the spectra of such Ru^{II} complexes to their non-cyclometalated counterparts.

Grätzel *et al.* inverted the molecular structure, *i.e.*, the cyclometaling moiety was part of the opposite ligand, but not of the anchoring one [149]. For both complexes **44a** and **44b** (Figure 6.22), a high PCE of *ca.* 8.3% was reported (N719 achieved 9.03% within an identical device setup). Kisserwan and Ghaddar modified the structure of **44a** by introducing a thienyl- and a COOH-substituent on the cyclometalating ligand [150]. These moieties were used to improve the dye's light-harvesting capability on the one hand and the binding to the TiO_2 on the other (**45**, Figure 6.22). Using an ionic liquid, which contained CuI, as the electrolyte, afforded a maximum PCE of 5.7%; whereas in the absence of the CuI additive, a PCE of 4.5% (*vs.* 6.2% for N719; Table 6.5) was obtained. The positive effect of CuI on the cell's PCE was attributed to an electrocatalytic process, which facilitated the dye's regeneration. In subsequent work, these same authors further varied the structure of the cyclometaling ligand *via* electron-withdrawing (*e.g.*, COOH or F) or electron-donating substituents (*e.g.*, OMe) [157]. However, all these dyes were less potent sensitizers than **45**, reaching a maximum PCE of 3.06%. Moreover, the DSSC, which was sensitized by **45**, revealed a good operational stability, *i.e.*, the initial PCE was retained to *ca.* 93% after continuous illumination with 1.5 a.m. sunlight at 60 °C for 24 h.

Table 6.5 Summary of the photovoltaic characteristics of DSSCs using cyclometalated Ru^{II} complexes as sensitizers.[a)]

Ru^{II} dye	V_{OC} (mV)	J_{SC} (mA cm^{-2})	FF	PCE (%)	Relative PCE[b)]	Reference
41a	660	8.1	0.70	4.0	0.78	[148]
41b	490	6.3	0.58	1.9	0.37	[148]
41c	570	5.8	0.70	2.5	0.49	[148]
44a	n/a	n/a	n/a	8.32	0.92	[149]
44b	n/a	n/a	n/a	8.32	0.92	[149]
45	590	12.2	0.63	4.5	0.73	[150]
46a	672	15.0	0.71	7.13	0.82	[151]
46b	608	8.3	0.77	3.90	0.45	[151]
46c	653	13.0	0.75	6.33	0.73	[151]
46d	682	16.7	0.71	8.02	0.93	[151]
47a	393	12.5	0.63	3.09	0.37	[152]
47b	424	15.6	0.63	4.16	0.49	[152]
47c	540	17.8	0.67	6.44	0.76	[152]
48	546	18.9	0.62	6.38	n/a	[153]

a) Details on the electrolyte and the device configuration were omitted for clarity and can be found in the respective reference. In all DSSCs, an FTO/TiO_2 photoanode was used.
b) PCE of the respective dye divided by the PCE of the "black dye" (**2**) under the same experimental conditions.

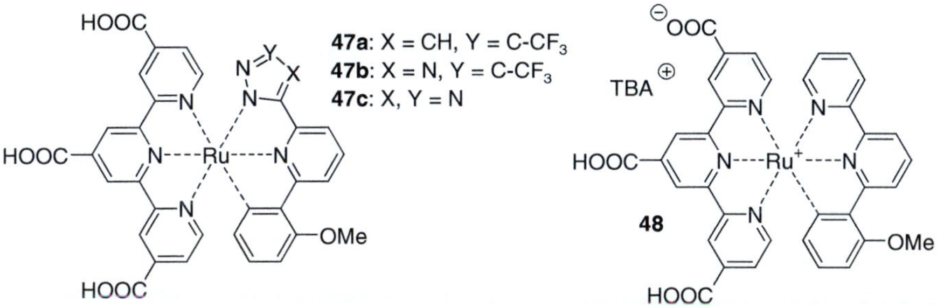

Figure 6.23 Schematic representation of the cyclometallated Ru^{II} complexes **47–48**.

Berlinguette *et al.* performed an extensive study, in which in-total 17 Ru^{II} complexes were screened regarding their light-harvesting capability [151]. Thereby, the nature of the tridentate ligands (*i.e.*, N^N^N, C^N^N, and N^C^N motives) as well as the substituents on the lateral triphenylamino moiety were varied. The lesson learned from this study was as follows: The complexes, which contained cyclometalating ligands (specifically of the C^N^N-type), absorbed light over a broad spectral range, owing to the electronic dissymmetry at the Ru^{II} center. Moreover, if the anchoring ligand was an N^C^N chelate, the LUMO was unfavorably located and the charge injection was insufficient. However, if the other ligand was cyclometalating, not only were the light-harvesting properties improved but the LUMO was also located on the anchoring ligand, thus accounting for an efficient injection of electrons. Accordingly, the four representatives **46** (Figure 6.22), which matched these criteria, were incorporated into DSSCs. The overall best PV performance was observed for dye **46d**, which afforded a PCE of 8.02% (Table 6.5).

Similar to the aforementioned azolate-containing dyes **40** (Figure 6.22), Islam, Otsuki *et al.* also studied the cyclometalated, charge-neutral complexes **47**, which varied in the number of *N*-atoms in the azolate ring (Figure 6.23) [152]. With increasing electron-withdrawing character of the azolate ligand, V_{OC} and J_{SC} gradually increased. Thus, the best overall performance was found for **47c**, which afforded a maximum PCE of 6.44% (Table 6.5). The related negatively charged complex **48** afforded a moderate PCE of 6.38% (*vs.* 7.85% for N719; Figure 6.23, Table 6.5), which resulted from the low FF value of 0.62; this, in turn, was due to the DSSC's high J_{SC} in combination with the low V_{OC}. It was further emphasized that the TBA cations were crucial for shielding the photoanode against unfavorable charge-recombination reactions.

6.3.5 Ru^{II}–Tpy Sensitizers with a Tridentate, NHC-Containing Ligands

The use of *N*-heterocyclic carbenes (NHCs), as strongly σ-donating ligands for metal-ion complexation, represents a highly active field of research [158], which arouse from the first isolation of an imidazol-2-ylidene derivative by Arduengo *et al.* [159]. Regarding Ru^{II} sensitizers for DSSC applications, the presumably first study was published by Kang, Chung *et al.* [160], who compared the PV performance

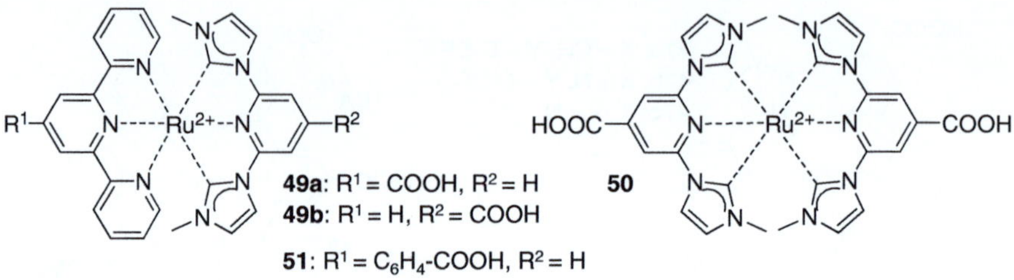

Figure 6.24 Schematic representation of the NHC-containing Ru^{II} complexes **49–51** (counterions were omitted for clarity).

of the heteroleptic complexes **49a/b** to that of homoleptic **50** (Figure 6.24). It was assumed that the interplay of a strongly σ-donating and a moderately π-donating ligand in **49a/b** would result in a small HOMO–LUMO energy gap, which would be beneficial for DSSC applications. However, these dyes exhibited a very poor PCE of ≤0.15% (*vs.* 6.34 for N719), which was mainly due to the low J_{SC}, *i.e.*, the insufficient electron injection (efficiency of below 5%). This behavior was attributed to the unfavorable orientation of the transition-dipole moment. The homoleptic dye **50** provided a slightly higher, yet very low, PCE of 0.48%. Later, various structural variants of **49** were presented, in which the imidazol-2-ylidene moieties were separated from the pyridine ring *via* CH_2 groups [161] or replaced by even stronger σ-donating benzoannulated NHCs [162]. However, PV studies on these dyes were not performed. Recently, Naziruddin *et al.* reported two further Ru^{II} *bis*-tridentate complexes, which contained two NHC moieties within one of the ligands (*e.g.*, **51**; Figure 6.24) [163]. Even though the charge injection could be improved, these sensitizers were far from being competitors for the standard dyes, such as **1** or **2** (Figure 6.3), regarding their operational PCE.

6.3.6 Beyond the Ru^{II}-Tpy System: Complexes with Other Transition-Metal Ions

The vast majority of metal-containing sensitizers for DSSC applications contain Ru^{II} ions, with complexes **1** and **2** representing the archetypal examples. Although high PCE values have been obtained with these dyes, they certainly have some shortcomings. Ruthenium is a low-abundant metal with a content in the Earth's crust of only 1 ppb. Thus, its use is always hampered by the material's relatively high costs and the industrial device production is problematic. Therefore, other transition-metal ions have been tested, as alternatives to Ru^{II} for the formation of light-harvesting tpy complexes, similar to BD.

Although osmium is exceedingly rare and even more expensive than Ru, as its lighter group-8 homolog, its tpy complexes were discussed as alternatives to the aforementioned Ru^{II}-containing ones. It was assumed that the so-called heavy-atom effect would enable the direct spin-forbidden transition from the GS to the low-energy ^{3}MLCT state upon light absorption [126]. As a consequence, an improved IPCE performance, in particular in the NIR spectral regime, was observed

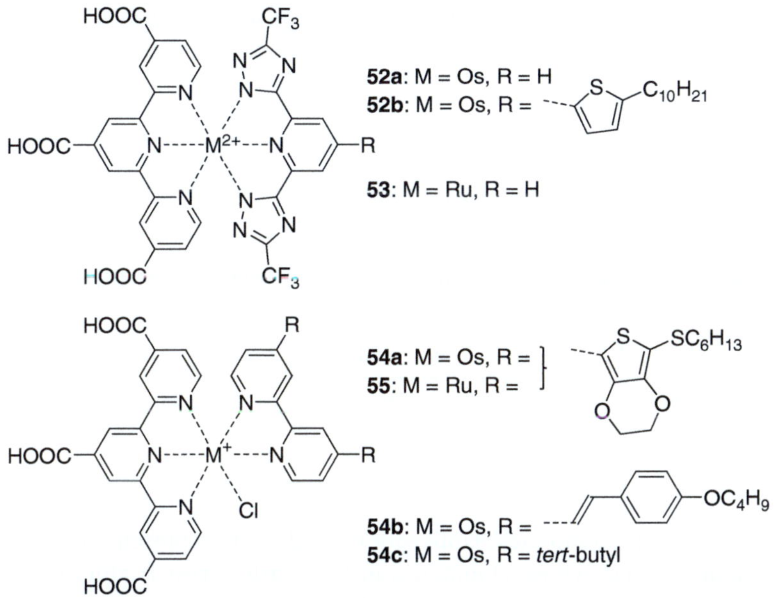

Figure 6.25 Schematic representation of the Ru^{II} and Os^{II}-containing sensitizers **52–55** (counterions were omitted for clarity).

(*e.g.*, IPCE of *ca.* 50% at 900 nm) [164]. The heteroleptic Os^{II} complexes **52** were tested as sensitizers in DSSCs (Figure 6.25) [165]. Panchromatic absorption up to 900 nm and good operational stability (*i.e.*, 1000 h of continuous illumination with AM1.5 sunlight at 60 °C) were found. The best PV performance (*i.e.*, a maximum PCE of 8.85%) was reported for **52a** when TBA deoxycholate was used as a co-absorbent. Under optimized conditions, **52b** even represented the more potent sensitizer when compared to its Ru^{II} analog **53** or N749 (Figure 6.25, Table 6.6). In a similar fashion, the sensitizing capability of complexes **54a** and **55** was compared (Figure 6.25) [166]. Owing to the spin-orbit coupling (SOC), the Os^{II} complex **54a** featured broadband absorption, which provided an IPCE beyond 1000 nm. The observed current density of *ca.* 19.4 mA cm^{-2} was significantly higher than that for the Ru^{II} dye **55**. However, in terms of practical PCE, the latter was slightly superior (Table 6.6). A similar effect, *i.e.*, panchromatic absorption beyond 1000 nm, was also observed for the Os^{II} sensitizer **54b**, which contained 4-alkoxystyrene moieties (Figure 6.25) [167]. The light-harvesting capability of this dye afforded a broad photocurrent response, which reached up to 1100 nm, and a PCE of 6.1% (Table 6.6). Whereas the related complex **54c** showed an inferior IPCE over the entire spectral range and, thus, overall poorer PV performance. Despite these promising results, the Ru^{II}-containing dyes are intrinsically more stable in their parent oxidation state, thus guaranteeing a better operational stability [79]. Overall, the Os^{II} sensitizers, which have been studied to date, are not superior to the state-of-the-art Ru^{II} sensitizers and will, also for cost reasons, not represent serious alternatives to those.

Table 6.6 Summary of the photovoltaic characteristics of DSSCs using Os^{II} and Ru^{II} complexes, as sensitizers.[a)]

Dye	V_{OC} (mV)	J_{SC} (mA cm^{-2})	FF	PCE (%)	Relative PCE[b)]	Reference
52a (Os)	560	20.1	0.66	7.47	0.91	[165]
52b (Os)	600	23.3	0.63	8.85	1.08	[165]
53 (Ru)	640	18.0	0.72	8.25	1.00	[165]
54a (Os)	474	19.4	0.62	5.7	0.98	[166]
54b (Os)	538	18.8	0.60	6.1	n/a	[167]
55 (Ru)	533	16.7	0.69	6.1	1.05	[166]

a) Details on the electrolyte and the device configuration were omitted for clarity and can be found in the respective reference. In all DSSCs, an FTO/TiO_2 photoanode was used.
b) PCE of the respective dye divided by the PCE of N749 under the same experimental conditions.

And what about Fe^{II}–tpy complexes? Unlike for Ru and Os, Fe represents a cheap and readily available metal, whose abundance in the Earth's crust is almost 6%. Accordingly, intensive research is devoted to the development of photoactive complexes, in which the "traditional" heavy metal ions are replaced by earth-abundant 3d transition-metal ions [168–170]. In this context, the Fe^{II} center, having a $3d^6$ configuration, is the most notable representative. A manifold of Fe^{II} complexes, which possess photoactive MLCT excited states, has already been reported [171]. However, the realization of sufficiently long-lived excited states remains problematic and requires a cautious ligand design [172]. As a first proof-of-concept, Ferrere and Gregg demonstrated the ability of Fe^{II}–polypyridyl complexes to photosensitize TiO_2 [173]. Specifically, $[Fe(dcbpy)_2(CN)_2]$, representing the Fe^{II} counterpart of the N3 dye, was able to sensitize nanocrystalline TiO_2 despite its very short-lived, non-emissive excited state (dcbpy: 2,2′-bipyridine-4,4′-dicarboxylic acid). However, when compared to their Ru^{II} counterparts, Fe^{II}–polypyridyl complexes are hardly suited for application in DSSCs because of their very poor PV performance of <1%. This restriction is mainly due to the unfavorable photophysics of such complexes, in particular the ultrafast decay of the lowest MLCT excited state within fs to ps. The $3d^6$-configured Fe^{II} ion establishes a ligand field of lower strength when compared to those formed by $4d^6$ or $5d^6$ centers (*i.e.*, Ru^{II} or Os^{II}, respectively). Due to the strong contraction of the 3d-orbitals and the resulting decreased ligand-field strength, Fe^{II} complexes possess metal-centered (MC) ^5T states of low energy. Photoexcitation of a Fe^{II}–polypyridyl complex results in the initial population of the MLCT states, which are subsequently deactivated to the ^5T state *via* ISC in an ultrafast fashion. These MC ^5T states are photochemically inactive and, thus, not capable of participating in the electron-injection step. In order to utilize Fe^{II}–polypyridyl complexes, as sensitizers, for DSSC applications, a substantial prolongation of the excited-state lifetime represents the major challenge. To reach this goal, Wärnmark *et al.* introduced NHC moieties, as strong σ-donors, to destabilize the MC states (*e.g.*, **56a**; Figure 6.26) [175]. This strategy resulted in a significant prolongation of the excited-state lifetime (*i.e.*, 9 ps for **56a** *vs.* 145 fs for $[Fe(tpy)_2]^{2+}$) [176].

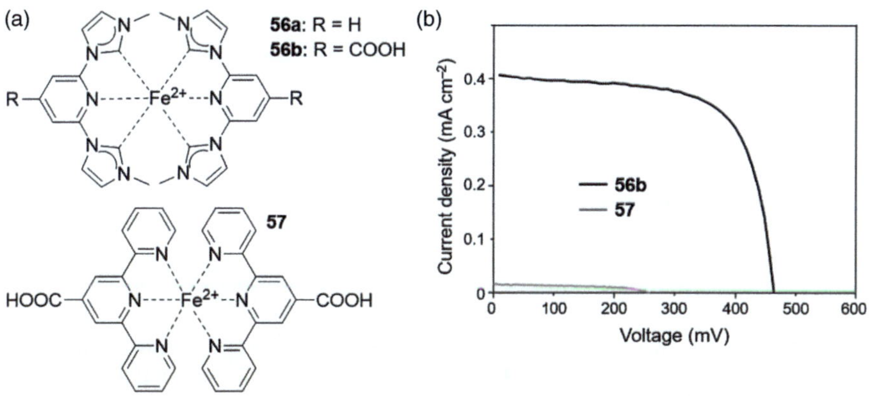

Figure 6.26 (a) Schematic representation of the Fe^{II} complexes **56** and **57** (counterions were omitted for clarity). (b) *I*–*V*-curves of DSSC, which were sensitized by **56b** and **57** [174] / with permission of John Wiley & Sons.

However, **56a** was only barely absorbing light in the spectral range where the solar spectral irradiance has its maximum (*i.e.*, 500–800 nm). The carboxylated derivative **56b** possessed an even longer MLCT lifetime of 16 ps and absorbed light up to *ca.* 600 nm (MLCT absorption band at 520 nm) [174]. Accordingly, this complex was tested regarding its capability to sensitize TiO_2. However, the IPCE of the DSSC was poor (*i.e.*, 2.5% at 500 nm). The very low J_{SC} of 0.41 mA cm^{-2} accounted for the overall low PV performance – the DSSC provided a PCE of 0.13%; for comparison, the DSSC, which contained the related Fe^{II}–tpy complex **57**, as the sensitizer, produced virtually no photocurrent (PCE of 0.01%; Figure 6.26b). Thus, Fe^{II} complexes, which contain NHC ligands, represent superior sensitizers when compared to the conventional complexes with polypyridyl ligands. However, their PV performance remains low when compared to that offered by the benchmark Ru^{II} sensitizers, such as N719 [175].

In alternative approaches, various authors have attempted to increase the MLCT lifetime by optimizing the coordination geometry and/or manipulating the MC/MLCT energy levels *via* strongly π-accepting ligands [168]. With respect to the latter, homoleptic and heteroleptic Fe^{II} complexes with the dcpp ligand were prepared [dcpp: 2,6-*bis*(2-carboxypyridyl)pyridine] [177, 178]. The homoleptic $[Fe(dpq)_2]^{2+}$ complex possessed an almost perfect octahedral geometry, which gave rise to a significant stabilization of the MLCT state [dqp: 2,6-di(quinoline-8-yl)pyridine] [179]. The MLCT lifetime of this complex was 450 fs, thus significantly longer than for the related $[Fe(tpy)_2]_{2+}$ complex (τ = 145 fs) [180]. However, none of these complexes has yet been studied in the context of DSSC applications.

Dixon *et al.* predicted that by using cyclometalating ligands, the order between the two relevant states might be reversed, *i.e.*, stabilization of the ^{3}MLCT state at the cost of the ^{3}MC one [181]. It was calculated that this behavior should be most pronounced for the twofold cyclometalated Fe^{II} complexes **58b–d** (Figure 6.27a). Indeed, the *mono*-cyclometalated complex **58a** verified the predicted ^{3}MLCT stabilization: This complex showed a 5.5 times increased ^{3}MLCT lifetime (relative to that of $[Fe(tpy)_2]^{2+}$) as well as a bathochromically shifted absorption profile,

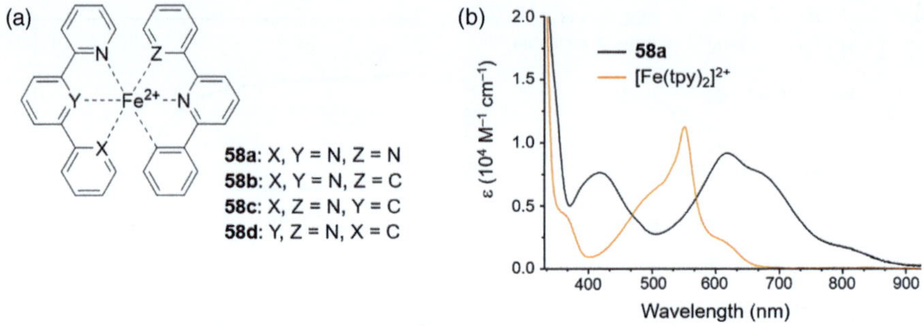

Figure 6.27 (a) Schematic representation of the cyclometalated Fe^{II} complexes **58**. (b) UV/vis/NIR absorption spectra of **58a** and $[Fe(tpy)_2]^{2+}$ (10^{-5} M in MeCN) [182] / with permission of John Wiley & Sons.

which even reached into the NIR regime (Figure 6.27a) [182]. These effects are highly beneficial regarding a potential use of such complexes in DSSCs; however, experimental studies in this direction are not yet at hand.

Aside from the plethora of tpy complexes of the divalent group-8 metal ions, complexes with other transition-metal ions have also been considered in the context of DSSC applications. This holds true for the complexes of Ir^{III} and Pt^{II}, which possess octahedral or square-planar geometries, due to their $5d^6$ and $5d^8$ electron configuration, respectively. Sakai *et al.* demonstrated that Pt^{II}–tpy complexes represent versatile photocatalysts for the hydrogen-evolution reaction (HER) in water [183]. Moreover, Pt^{II}–tpy complexes with alkynyl ligands might act as sensitizers in photocatalytic oxidation reactions (*e.g.*, olefin epoxidation, *etc.*) [184]. Such complexes have attracted a widespread interest due to their rich photophysical properties [185]. For example, their UV/vis absorption profile can be modulated *via* the electronic properties of the ligands – the tpy as well as the alkynyl derivative. Yam *et al.* showed that the absorption range could even be extended into the NIR regime when using appropriate donor-acceptor ligands [186]. In line with these studies, these same authors tested the donor-acceptor complexes **59** regarding their capability to sensitize nanocrystalline TiO_2 (Figure 6.28) [187]. In particular, complex **59d**, which contained the most extended π-extended chromophore, revealed a remarkably long CSS lifetime of 3.4 μs (CSS: charge-separated state). The CSS originated from the intramolecular electron transfer from the donor-type chromophore to the electron-poor tpy ligand. The directionality of this process facilitated the electron injection when the Pt^{II} dye was anchored onto the TiO_2 semiconductor. The resulting DSSC afforded a PCE of 3.6% (the other Pt^{II} complexes provided lower PCE values of ≤2.7%). For comparison, the N3 dye gave a significantly higher PCE of 10.4% under otherwise identical conditions. It was assumed that self-quenching and charge-recombination processes accounted for the only moderate PV performance of **59d**.

Cyclometalated Ir^{III} complexes play a prominent role in light-emitting devices, due to their remarkably luminescent properties [188]. However, their use for solar-cell applications has been considered only sparsely. This restriction is

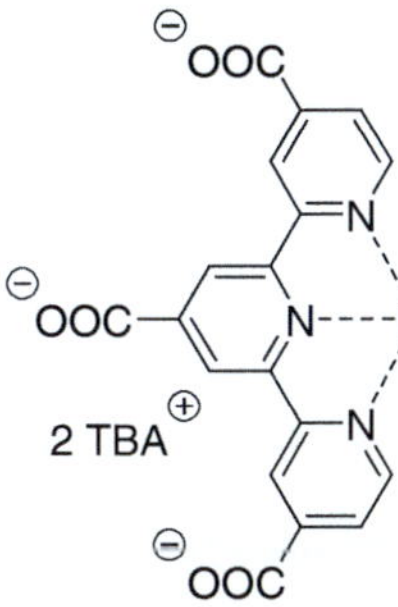

59a: R = thien-2-yl
59b: R = 2,2-bithien-5-yl
59c: R = C_6H_4-NMe_2
59d: R =

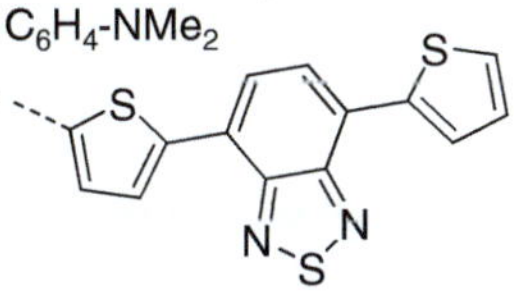

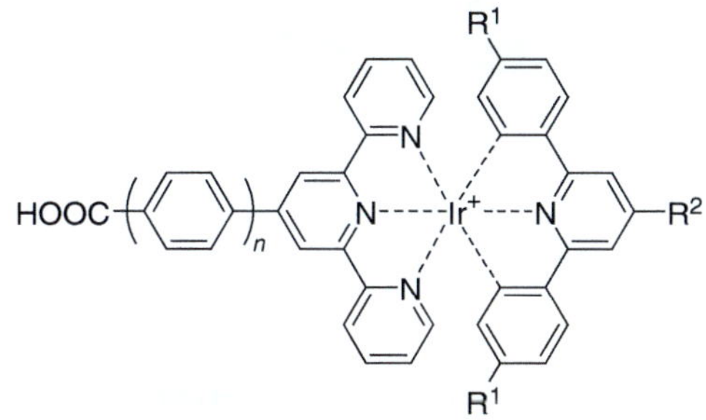

60a: n = 1, R^1, R^2 = H
60b: n = 1, R^1 = H, R^2 = Ph
60c: n = 0, R^1, R^2 = H
60d: n = 0, R^1 = Me, R^2 = H

Figure 6.28 Schematic representation of Pt^{II} and Ir^{III} complexes for DSSC application.

mainly due to the relatively narrow absorption profile, which – unlike those of Ru^{II}–polypyridyl complexes – already tails off by *ca.* 550 nm. Thus, low V_{OC} values are typically obtained and the overall PCE is typically low. Besides a broad range of *bis*-cyclometalated, *tris*-bidentate Ir^{III} complexes, a series of *bis*-tridentate systems has also been studied regarding their light-harvesting capability. Nakabayashi *et al.* reported the heteroleptic Ir^{III} complexes **60**, which contained a carboxylated tpy ligand as well as a *bis*-cyclometalating 2,6-diphenylpyridine derivative, as a strong ligand-field ligand (Figure 6.28) [189]. All four complexes possessed appropriate HOMO and LUMO energy levels, which facilitated the dye reduction by the electrolyte on the one hand and the electron injection into the TiO_2 on the other. However, the overall PV performance was relatively low – the maximum IPCE and PCE were 66% and 2.16%, respectively. This, in turn, was mainly attributed to the insufficient overlap of the dyes' absorption spectra with the solar radiation spectrum.

But let us turn back to the aspect of light harvesting by tpy complexes of earth-abundant d-block metal ions [190]. Schmuttenmaer, Crabtree, Brudvig, Batista *et al.* reported the sensitization of TiO_2 nanoparticles by a Mn^{II}–tpy *mono*-complex. The most notable aspect of this work was not the sensitization itself, but the use of a different anchoring group, *i.e.*, the water-compatible hydroxamate moiety [191]. This anchoring group represents a versatile alternative to the state-of-the-art COOH-group: It offers stability over the wide pH regime from 2 to 10.25, whilst being as efficient for the electron injection as the carboxylate and superior to phosphonate [192]. To the best of our knowledge, the hydroxamate anchor has not yet been combined with further tpy complexes, as designated sensitizers. Housecroft, Constable *et al.* focused on the sensitization of TiO_2 by Zn^{II}–tpy *bis*-complexes [193]. Owing to the metal's $3d^{10}$ electron configuration, such complexes do not exhibit MLCT transitions. In order to circumvent this limitation, the heteroleptic complexes **61** (Figure 6.30), bearing anchoring groups

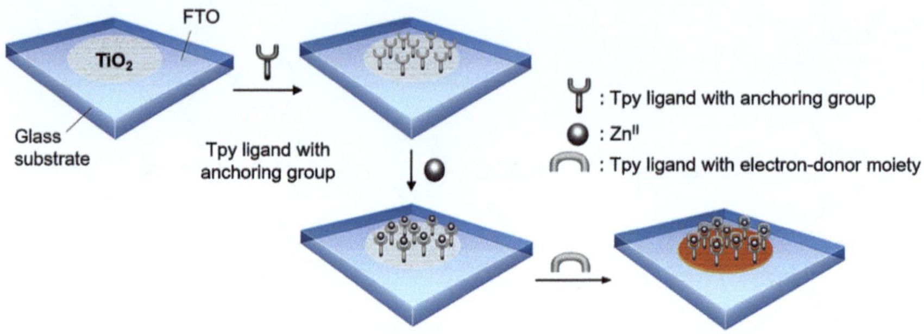

Figure 6.29 Schematic representation of the stepwise assembly of Zn^{II} sensitizers on a TiO_2 surface. Adapted from [194].

on one tpy ligand and a strong electron donor on the other, were assembled on the TiO_2-surface in a stepwise fashion (the intrinsic kinetic lability of Zn^{II}–tpy *bis*-complexes prevents the formation of heteroleptic, solution-stable complexes; Figure 6.29). Within the first series of Zn^{II} dyes, the best PV performance was reached for **61b**, when assembled from $ZnCl_2$ [193–195]. However, the maximum PCE of 0.71% corresponded to only 10% of the PCE, which was offered by N719 under identical experimental conditions. In subsequent work, these same authors replaced the methoxy groups on the donor-type ligand by longer alkoxy moieties [194]. Although the J_{SC} and V_{OC} of the solar cells were increased when using these modified sensitizers, the overall PCE remained low. A slight improvement in PCE was found when the dyes' absorption in the visible regime was increased [195]. For this purpose, dendritic triphenylamine moieties were installed on the outer tpy ligand. This resulted in a "hole stabilizing surface" and afforded higher J_{SC} and V_{OC} values, but the overall PCE remained far below that provided by N719, as the reference sensitizer (*i.e.*, to the best, only 3% of its PCE were reached) (Figure 6.30).

Finally, Co^{II}–tpy complexes are worth mentioning in this section. However, their use in the context of DSSC is not related to sensitization – such complexes have been utilized as redox mediators. The $Co^{II/III}$ redox couple represents one versatile candidate to replace the commonly used I^-/I_3^- system in DSSCs [196]. The advantageous use of transition-metal complexes (mainly of 3d metal ions), as redox mediators, results from typically reversible electrochemical behavior and the

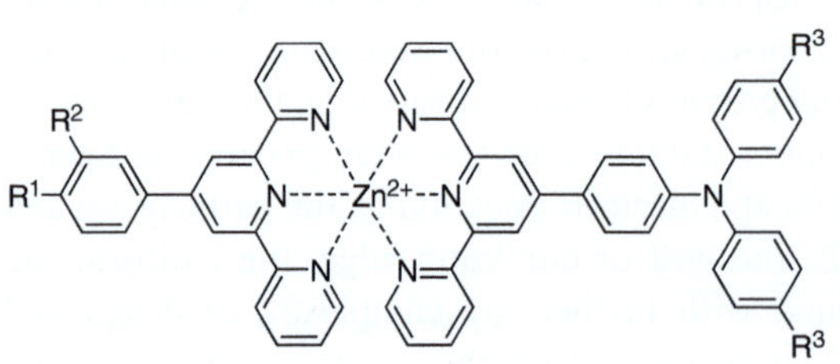

61a: R^1, R^2 = COOH, R^3 = H
61b: R^1, R^2 = COOH, R^3 = OMe
61c: R^1, R^2 = OH, R^3 = H
61d: R^1 = PO_3H_2, R^2 = H, R^3 = OMe

Figure 6.30 Schematic representation of the heteroleptic Zn^{II} complexes **61**, which were assembled onto a TiO_2 surface *via* the stepwise procedure, which is depicted in Figure 6.29 (counterions were omitted for clarity).

ability to modulate the redox potential *via* the choice of metal ions and organic ligands; moreover, such complexes possess more positive Fermi energy levels, absorb visible light less well and feature an increased stability when compared to the I^-/I_3^- system [197]. It has been stated that these metal complexes, which undergo outer-sphere one-electron-transfer processes (according to the Marcus theory), only require a thermodynamic driving force (ΔG^0) of *ca.* −0.2 eV to enable the efficient regeneration of the sensitizer. This, in turn, is highly beneficial to increase the DSSC's VOC [198, 199].

In 2010, Boschloo *et al.* reported a DSSC, which contained D35 (Figure 6.31) as an organic sensitizer, and $[Co(bpy)_3]^{2+}$ as the redox mediator [200]. This device, whose electrolyte was iodine-free, provided a remarkably high PCE of 6.7% upon illumination with AM1.5 sunlight (J_{SC} = 10.7 mA cm^{-2}; V_{OC} = 920 mV). For comparison, an analogous DSSC, which contained the conventional iodide/triiodide-based electrolyte, showed a lower PCE of 5.5% under identical experimental conditions. In subsequent work, these same authors compared the dye-regeneration efficiency of various Co^{II}–polypyridyl complexes [199]. For this purpose, the regeneration of the D35 dye in the presence of these redox mediators was analyzed. The Co^{II}–tpy complexes **62** (Figure 6.31) afforded similar values in terms of ΔG^0, IPCE, and regeneration efficiency as their counterparts from the $[Co(bpy)_3]^{2+}$ family, but the regeneration rates ($t_{1/2}$) of **62** were faster (6.9 μs for **62a** *vs.* 8.3 μs for $[Co(bpy)_3]^{2+}$). The acceleration of dye regeneration was attributed to the low-spin (LS) state of **62a**, which requires a lower reorganization energy compared to high-spin (HS) states, as typically present in $[Co(bpy)_3]^{2+}$ complexes [201]. In one further study, Lacroix *et al.* used a mixture of $[Co(phen)_3]^{2+}$ (phen: 1,10-phenanthroline) and **63**, as the redox mediator (Figure 6.31) [202]. The combination of these Co^{II} complexes established a redox cascade, which facilitated the electron transport from the poly(3,4-ethylenedioxythiophene) (PEDOT) counter electrode through the electrolyte to the photoanode, thus reducing charge recombination and improving the dye regeneration. Accordingly, the corresponding DSSC, which was sensitized by D35, provided a PCE of 5.1%; in the absence of **63**, as redox co-mediator in the electrolyte, the PCE dropped to 4.2%.

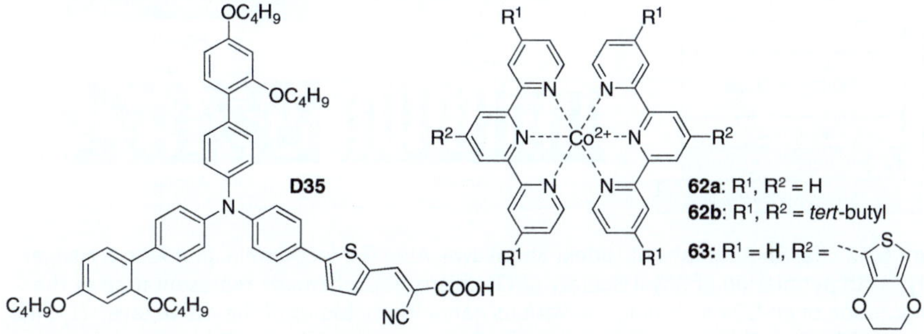

Figure 6.31 Schematic representation of sensitizer D35 and Co^{II} redox mediator **62** and **63**.

Despite the promising results of Co^{II}–tpy complexes, $[Co(bpy)_3]^{2+}$ and $[Co(phen)_3]^{2+}$ represent preferentially used redox mediators in DSSC electrolytes [197]. A maximum PCE of 24.5% for a DSSC, which contained a $[Co(bpy)_3]^{2+}$-based electrolyte and an organic dye, was reported by Lee *et al.* [203]. Detrimentally, the molecular size of such complexes is rather large and the mass-transport and diffusion processes are therefore rather slow (compared to the I^-/I_3^- redox system). Moreover, the large reorganization energy between the metal's + II and + III oxidation state results in considerably high energetic costs for the dye regeneration, which lowers the overall device efficiency.

6.4 Terpyridine Complexes in Polymer Photovoltaics

In 1977, Heeger, MacDiarmid, and Shirakawa reported the electronic conductivity of doped polyacetylene (Figure 6.32a) [205, 206]. This seminal work, for which these authors received the 2000 *Nobel Prize in Chemistry*, represents the birth of organic electronics and paved the way toward the application of conjugated polymers, as active materials, in the fields of OPVs, organic light-emitting diodes (OLEDs), organic field-effect transistors (OFETs), *etc.* [207]. As mentioned beforehand (Section 6.2), Heeger *et al.* were also the first who demonstrated a functional planar heterojunction cell, in which a layer of C_{60} was deposited from vacuum onto a spin-coated layer of MEH-PPV, as the conjugated polymer [MEH-PPV: poly(2-methoxy-5-(2-ethyl-hexyloxy)-*p*-phenylenevinylene)] [48]. Although such a

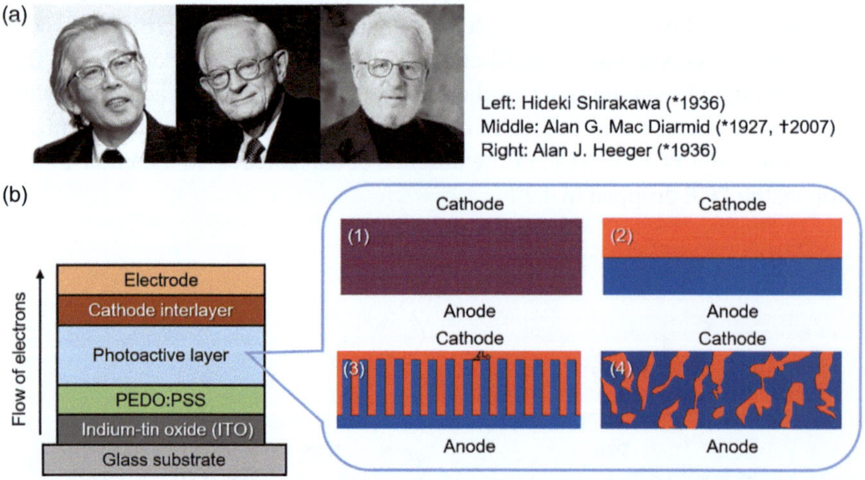

Figure 6.32 (a) Photographs of Hideki Shirakawa, Alan G. MacDiarmid, and Alan J. Heeger [204] / with permission of Royal Society of Chemistry. (b) Schematic representation of the configuration of an OPV including the various nanomorphologies of the active layer: (1) Fine mixture of donor and acceptor molecules, (2) donor-acceptor bilayer, (3) ideal morphology of a BHJ cell and (4) typical morphology obtained after solution processing (the donor and acceptor phase are depicted in blue and red, respectively) [31] / Elsevier / CC BY 4.0.

planar bilayer configuration offers the advantage of good charge transport (given that the lifetime of the charge carriers is so long that they can reach the electrodes), it has the downside of a small surface area at the donor-acceptor interface (Figure 6.32b) [31]. On the other hand, charge generation will be facilitated if the donor and acceptor molecules form a fine dispersion/mixture; however, such a morphology comes at the expense of the charge transport, which will be rather poor. According to Monte-Carlo modeling, a highly ordered, checkered layer morphology should enable an optimal performance in terms of charge separation and transport (the ideal domain width would be twice the exciton diffusion length) [208]. In practice, such a morphology can hardly be realized since phase separation of the donor and acceptor polymers in course of the processing will lead to other morphologies, according to the theories of polymer physics. In a typical BHJ, the layer morphology features connected domains (some tens of nm in size) as well as donor and acceptor molecules "dissolved" in the respective other domains [31]. In order to optimize the interfacial area in such a BHJ, post-processing treatment of the active layer, *e.g.*, by thermal annealing might be beneficial [209]. Moreover, the processing parameters (*e.g.*, choice of solvent and additives) were found to have a significant influence on the morphology [210].

Regarding the active material in BHJ, conjugated polymers offer the advantages of flexibility, comparably low preparation costs, and ease of processing. Numerous studies dealing with the synthesis and processing of conjugated polymers have been published over the last decades. The performance of BHJ devices depends on the efficiency of light absorption, diffusion of exciton to the donor-acceptor interface, light-induced charge separation, and the mobility of the thusly formed charge carriers (*i.e.*, electrons and holes) [59]. These aspects represent – along with the improvement of device stability and durability – the main aspects of research [211]. Metal-containing polymers, as the active material, came into play when Schanze *et al.* raised the question if triplet states could be utilized to improve charge separation within the BHJ [212]. In the following, a plethora of Pt^{II}-polyyne polymers have been prepared and tested for PV applications [59]. These active materials have, however, no relevance within the scope of this chapter and the reader is referred to the excellent review articles available in literature [59, 213, 214]. The integration of Ru^{II} complexes into polymers attracted interest to improve the spectral overlap between the sunlight and the absorption profile of the active material. As mentioned in Section 6.2, the MLCT transitions of Ru^{II}–tpy complexes at 500–600 nm can be utilized for light harvesting and contribute to the photosensitivity of the polymer, in which they are incorporated. In addition to this, such polymers can readily be processed from solution (*e.g.*, by spin coating or inkjet printing), typically feature a high thermal stability and take part in redox processes due to the easily accessible and highly reversible Ru^{II}/Ru^{III} redox couple.

Two different strategies were applied to combine Ru^{II}–tpy complexes with conjugated polymers: Firstly, metallopolymers were prepared, in which conjugated segments were linked *via* $[Ru(tpy)_2]^{2+}$ moieties. These main-chain polymers were synthesized by Pd^0-catalyzed cross-coupling procedures to overcome the limitations of the alternative strategy, *i.e.*, the polymerization-by-complexation method, in

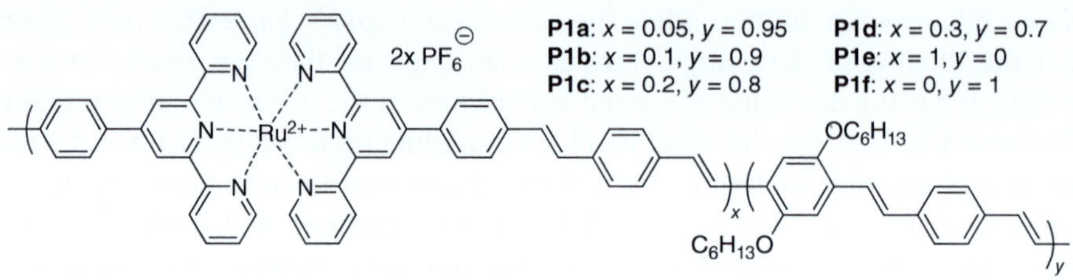

Figure 6.33 Schematic representation of the main-chain polymers **P1** [217] / John Wiley & Sons.

which conjugated *bis*-tpy ligands are combined with the respective metal ion. The latter represents a versatile approach to prepare such polymers based on the 1st row transition-metal ions (*e.g.*, Fe^{II}, Ni^{II}, Zn^{II}, *etc.*), but is less applicable to those metal ions that require harsh conditions to achieve complexation (*e.g.*, Ru^{II}, Os^{II}, Ir^{III}) [215, 216]. The second type of metallopolymers is comprised of a π-conjugated backbone to which Ru^{II}–tpy complexes (similar to Grätzel's "black dye," Figure 6.3) have been grafted.

The first main-chain polymers, which differed in the content of $[Ru(tpy)_2]^{2+}$ moieties within the backbone, were reported by Chan *et al.* (**P1**, Figure 6.33) [217]. Unlike for the all-organic polymer **P1f**, which showed a strong π–π* absorption band at *ca.* 450 nm, conjugation was interrupted in the metallopolymers due to the octahedral geometry of the complexes. Instead, the metallopolymers featured an intense MLCT absorption band at 507 nm, which is characteristic of $[Ru(tpy)_2]^{2+}$ complexes. The polymers' photoconductivity gradually increased with increasing content of complexes (at least up to 30%). Thus, the combination of electron-donating conjugated polymers with photosensitizing Ru^{II} dyes is beneficial for photocurrent generation. In a subsequent study, the beneficial influence of the metal sites on charge-carrier mobilities within the materials was demonstrated [218]. **P1e** and sulfonated polyaniline (SPAN), as the donor and acceptor polymers, respectively, were fabricated into a PV device *via* a layer-by-layer (LBL) process [219]. However, the performance of the PV device with the configuration ITO|(**P1e**/SPAN)$_{20}$|Al in terms of power-conversion efficiency (η_p) and PCE was rather poor (*i.e.*, *ca.* 10^{-3}% and *ca.* 2% at 510 nm, respectively, Table 6.7).

Even though the polymerization-by-complexation method is generally unfavorable in the case of Ru^{II} ions, Mikroyannidis, Hsu *et al.* prepared linear and hyperbranched polymers *via* this route (**P2–P5**, Figure 6.34) [220]. These authors utilized $RuCl_3 \cdot xH_2O$ as the precursor, which was reduced to Ru^{II} ions under the applied reaction conditions. The fabricated PV devices contained a bilayer heterojunction, which was comprised of the respective metallopolymer and a blend of methyl[6,6]-phenyl-C_{61}-butyrate (PCBM) and P3HT; the bilayer was processed on top of a PEDOT layer, as the hole-injection material. Within this series, the solar cell, which contained **P2** as the donor metallopolymer, provided the highest PCE value of 0.71% (Table 6.7).

Lin *et al.* employed the same polymerization-by-complexation method for the synthesis of the main-chain ruthenopolymers **P6** (Figure 6.35a) [221]. In this example,

Table 6.7 Summary of the photovoltaic characteristics of solar cells with Ru^{II}-containing polymers as the donor material.

Polymer	V_{OC} (V)	J_{SC} (mA cm^{-2})	FF	PCE (%)	Reference
P1e[a)]	0.68	0.04	0.21	0.02	[219]
P2[b)]	0.48	4.20	0.35	0.71	[220]
P3[b)]	0.22	5.23	0.29	0.33	[220]
P4[b)]	0.56	2.09	0.36	0.42	[220]
P5[b)]	0.30	4.15	0.28	0.35	[220]
P6a[c)]	0.61	2.18	0.34	0.45	[221]
P6b[c)]	0.42	1.14	0.30	0.14	[221]
P6c[c)]	0.58	1.14	0.33	0.22	[221]
P7[c)]	0.66	2.22	0.39	0.57	[222]
P8[c)]	0.70	3.50	3.71	0.90	[222]
P9[d)]	0.67	6.17	0.48	1.99	[223]
P10[d)]	0.73	7.12	0.51	2.66	[223]
P11[c)]	0.12	2.58	0.38	0.12	[224]
P12[c)]	0.18	1.60	0.29	0.08	[224]
P13a[d)]	0.58	4.59	0.43	1.19	[225]
P13b[d)]	0.53	4.31	0.33	0.82	[225]

a) Multilayer heterojunction with alternating layers of polymer and SPAN.
b) Bilayer heterojunction with polymer and PCBM:P3HT.
c) BHJ with polymer/PCBM blend.
d) BHJ with polymer/PC_{71}BM blend.

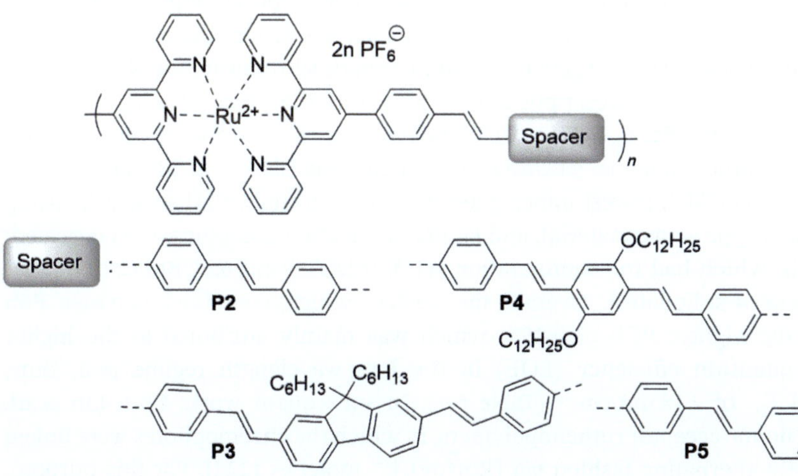

Figure 6.34 Schematic representation of the main-chain and hyperbranched metallopolymers (**P2–P4** and **P5**, respectively). Adapted from [220].

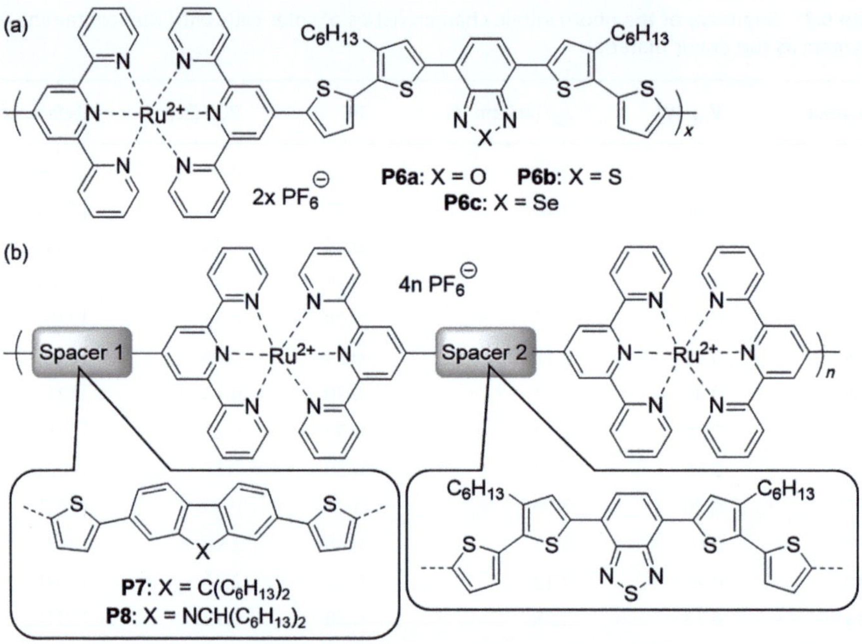

Figure 6.35 Schematic representation of (a) the main-chain polymers **P6**. Adapted from [221] and (b) the alternating copolymers **P7** and **P8**. Adapted from [222].

the Ru^{II}–tpy *bis*-complexes linked the conjugated segments, which possessed a donor-acceptor-type structure (*i.e.*, a bithiophene-flanked benzochalcodiazole moiety). Strong interaction of the planar-structured metallopolymer strands evoked strong intrachain charge-transfer (ICT) interactions, which – in concert with the complexes' MLCT transitions – resulted in a broad absorption ranging from 260 to 750 nm. This effect was most pronounced for **P6c**, whose chromophore contained the largest and least electronegative Se atom: The bathochromic tailing of the absorption band afforded the lowest optical bandgap within the series (**P6a**: E_g = 1.73 eV; **P6b**: E_g = 1.74 eV; **P6c**: E_g = 1.63 eV). The heteroatom's electronegativity (EN) also affected the ruthenopolymer's LUMO – with increasing EN, the LUMO energy level was lowered (LUMO: lowest unoccupied molecular orbital). BHJ solar cells using **P6**, as donor-type active material, and PCBM, as electron acceptor, were fabricated; these cells, which had the configuration [ITO|PEDOT:PSS|**P6**:PCBM|Ca|Al] [PSS: poly(styrene 4-sulfonate)]. Overall, the device, which contained polymer **P6b** featured the highest PCE of 0.45%, which was mainly attributed to the higher external quantum efficiency (EQE) in the long-wavelength regime and, thus, increased J_{SC} of 2.18 mA cm^{-2} (Table 6.7). In subsequent work, Chu, Lin *et al.* prepared donor-acceptor ruthenopolymers, in which the chromophores were linked in a strictly alternating fashion *via* $[Ru(tpy)_2]^{2+}$ moieties [222]. For this purpose, *bis*-terpyridine ligands, which contained either a dialkyl fluorene or an *N*-alkyl carbazole moiety, were combined with the ligand from **P6b** with Ru^{II} ions in a step-wise fashion to afford copolymers **P7** and **P8** (Figure 6.35b). The donor-acceptor concept provided the coverage of a broad absorption regime from *ca.* 250 to 750 nm; the strong absorption bands at *ca.* 560 nm were assigned to overlapping ICT and MLCT transitions. Variation of the donor component allowed for the modulation of E_g and of the PV performance of the copolymers. A lower E_g value was found for **P8**, which contained the strongly electron-donating carbazole moieties (*i.e.*, **P7**: 1.82 eV *vs.* **P8**: 1.76 eV); likewise, the HOMO energy level of **P8** was significantly lower than that of **P7**, which, in turn, provided a higher V_{OC} of 0.70 V for the fabricated BHJ solar cell (Table 6.7). The superior EQE of **P8** gave also rise to higher J_{SC} as well as PCE values when compared to the device fabricated from **P7**.

The best PV performance of a main-chain metallopolymer with incorporated $[Ru(tpy)_2]^{2+}$ moieties was reported by Li, Peng *et al.* in 2013 [223]. In these polymers, the π-conjugated segment was comprised of a donor-acceptor-donor triad, in which two 4*H*-cyclopenta[2,1-*b*;3,4-*b*]dithiophene moieties flanked a benzothiadiazole one (**P9** and **P10**, Figure 6.36a). These polymers were highly soluble in organic solvents and thermally very stable (*i.e.*, onset decomposition temperatures of *ca.* 415 °C). In particular, the fluorinated polymer **P10** displayed a high hole mobility, due to strong non-covalent interactions between the polymer chains, as well as a more appropriate match with the energy levels of $PC_{71}BM$, as the electron acceptor ($PC_{71}BM$: methyl [6,6]-phenyl-C_{71}-butyrate). The BHJ device, which contained a blend of **P10** and $PC_{71}BM$ in a 1:2 ratio, featured the remarkably high PCE of 2.66%; whereas the analogues solar cell, which contained **P9**, provided a lower, yet reasonably high PCE of 1.99% (Table 6.7). This PV performance is competitive compared with those reported for other types of metal-containing polymers (*e.g.*, the Pt^{II}–polyyne

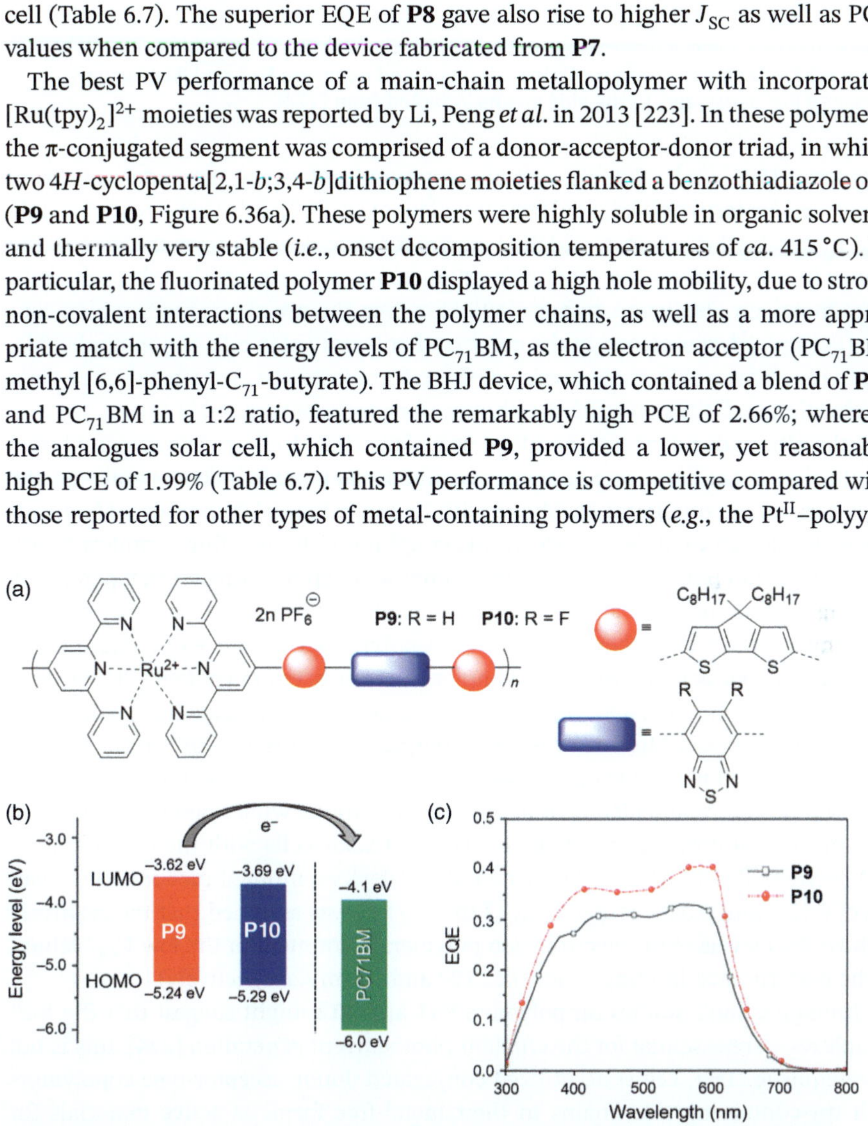

Figure 6.36 (a) Schematic representation of the main-chain polymers **P9** and **P10**. (b) Schematic representation of the energy levels of the polymers and $PC_{71}BM$. (c) EQE curves of the BHJ solar cells, which contained a polymer / PC71BM (1:2 ratio) blend, as the active layer [223] / with permission of Royal Society of Chemistry.

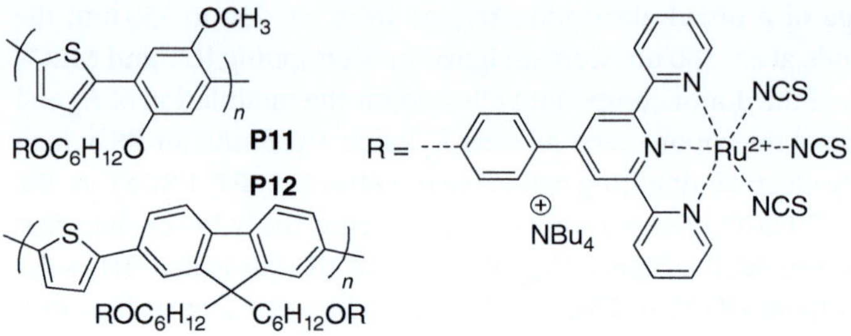

Figure 6.37 Schematic representation of conjugated polymers **P11** and **P12**, which contained Ru^{II} sensitizers in the side chains. Adapted from [224].

polymers) [59, 226]. Notably, metallo-supramolecular polymers derived from conjugated *bis*-terpyridine ligands and metal ions (*i.e.*, Fe^{II}, Ru^{II}, Co^{II}) are also widely used in one further context, which is electrochromic devices. In such devices, the change in color, contrast, or other properties results from reversible redox processes of the metal ion, the ligand, or both [59]. However, the use of polymers, which contain $[Fe(tpy)_2]^{2+}$ complexes, in the context of polymer solar cells is non-applicable, due to the unfavorable photophysics of these complexes (*i.e.*, very fast excited-state deactivation and, thus, insufficient photon-to-energy conversion) [175].

In the alternative concept, the metal complexes (*i.e.*, Ru^{II}–tpy mono-complexes) are attached to the conjugated polymer backbone *via* flexible linkers. Unlike for the former case, *i.e.*, conjugated backbone with integrated metal complexes, this metallopolymer design excludes the electronic coupling of the pending complexes with the conjugated chain; thus, both components will behave (almost) independently in terms of photophysical and electrochemical properties [227]. Chan *et al.* applied this strategy for the synthesis of side-chain polymers **P11** and **P12** (Figure 6.37), which contained pending Ru^{II}–tpy complexes, reminiscent of Grätzel's "black dye" (Section 6.3) [224]. First, the conjugated polymers with pending tpy side chains were prepared by a Suzuki-Miyaura cross-coupling; subsequently, the Ru^{II}-trithiocyanato complexes were installed *via* a two-step procedure. Due to the presence of the Ru^{II} dyes, the polymers provided a broad absorption profile, which ranged from 400 to 750 nm, thus almost into the NIR regime. BHJ solar cells with the configuration ITO|polymer/PCBM|Al were fabricated; these devices afforded moderate PCE values of 0.12% and 0.08%, respectively (Table 6.7). It was assumed that the relatively high HOMO levels of the electron-rich polymers accounted for the low V_{OC} values, as the performance-limiting factor (*i.e.*, 120 and 180 mV, respectively).

Although Chan's studies on polymers **P11** and **P12** might suggest that the Ru^{II} complexes were essential for the efficient photocurrent generation [224], this is not a prerequisite. Tsai, Lee *et al.* utilized conjugated donor-acceptor-type copolymers with tpy-containing side chains in their metal-free form, as active materials for BHJ solar cells [225]. For this purpose, the copolymers **P13** (Figure 6.38a) were prepared by a Stille cross-coupling and fabricated into devices with the overall configuration ITO|PEDOT:PSS|**P13**/$PC_{71}BM$|Ca|Ag. The compatibility of **P13** with the acceptor provided high-quality films, in which the $PC_{71}BM$ molecules were

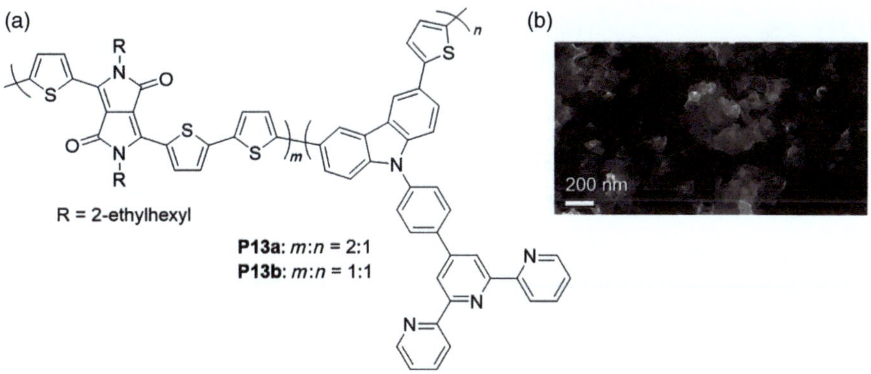

Figure 6.38 (a) Schematic representation of the tpy-containing copolymers **P13**. (b) Representative TEM image of a **P13a**/$PC_{71}BM$ film (1:2 ratio) after annealing at 100 °C for 10 min [225] / with permission of Royal Society of Chemistry / CC BY-NC 3.0.

homogeneously distributed in the polymer matrix, as evidenced by atomic-force and transmission-electron microscopy imaging (AFM and TEM, respectively; Figure 6.38b). The best PV performance, *i.e.*, a PCE of 1.19%, was found when the BHJ layer was comprised of a **P13a**/$PC_{71}BM$ 1:2 blend (Table 6.7). Notably, other conjugated polymers, which contain tpy ligands in the side chains, have also been synthesized. However, these materials were not considered in the context of PV applications; instead, their utilization for cation sensing or, after complexation with Ir^{III} or Ln^{III} ions, in polymer light-emitting diodes (PLEDs) has been envisioned [228].

6.5 Terpyridine Ligands and Complexes in the Context of Perovskite Solar Cells

Perovskites are a class of inorganic materials, which basically possess the same crystal structure as the parent mineral $CaTiO_3$, named perovskite. The mineral itself was discovered in the Ural Mountains by Rose in 1839 and named after the Russian mineralogist L.A. Perovski (Figure 6.39a) [231]. Perovskite's crystal structure was described first by Goldschmidt [232] and published later by Megaw [233]. The cubic crystal structure, which is depicted in Figure 6.39b, possesses the general formula ABX_3, in which A and B represent cations of different sizes. Taking Goldschmidt's tolerance factor (t) and the octahedral factor (μ) into account, one can conclude on the formability and stability of a perovskite-type material as well as on its crystallographic structure [234]. Based on an idealized model (*i.e.*, A, B, and X representing solid spheres), these values can be determined from the A–X and B–X distances, respectively. In the case of halide perovskites (X = halide anion), for example, t and μ will be in the range from 0.81 to 1.11 and 0.44 to 0.90, respectively [234]. It has been noted that the cubic structure can only be ascertained if $0.89 < t < 1.00$; for lower tolerance values, crystal structures of lower symmetry will

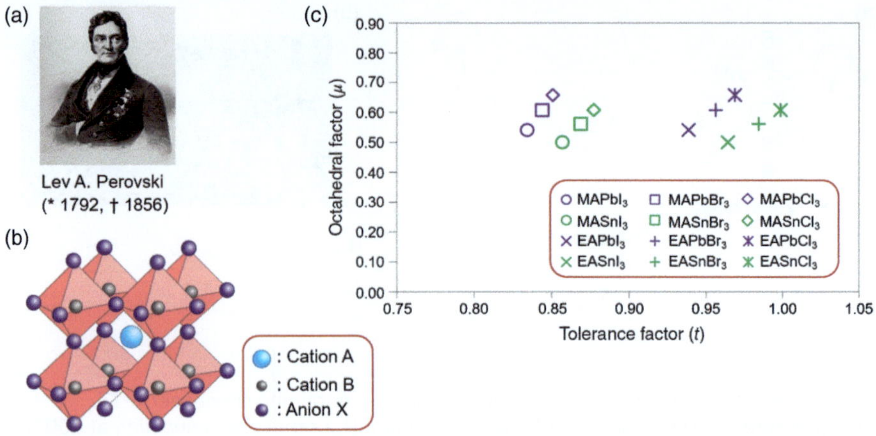

Figure 6.39 (a) Photograph of Lev A. Perovski [229] / with permission of John Wiley & Sons. (b) Schematic representation of the perovskite ABX_3 structure [230]. (c) Calculated t and μ factors for various hybrid halide perovskites [230] / with permission of Springer Nature.

be favored (*e.g.*, tetragonal or orthorhombic ones). One has to distinguish the purely inorganic perovskites from the hybrid organic–inorganic ones, in which A, being the larger cation, is organic [*e.g.*, methylammonium (MA), ethylammonium (EA), formamidinium (FA), *etc.*]. In hybrid halide perovskites, iodide is the preferentially used anion, but also mixed-halide materials are known (here, I^- is accompanied by Br^- or Cl^-) [230]. In photovoltaically relevant materials, Pb^{II} typically represents the smaller B-cation. Admittedly, Sn^{II} ions also afford perovskite-type structures with less distorted geometry and even more appropriate bandgaps [235]. However, tin(II)iodide-containing materials are prone to decomposition, due to the ease of oxidation of the Sn^{II} centers to Sn^{IV} ones. Owing to relativistic effects, which stabilize the +II oxidation state in the case of Pb, this issue is not relevant for such materials. By far, $MAPbI_3$ represents the most important hybrid perovskite material, with the mixed-halide analogs, *i.e.*, $MAPbI_{3-x}Cl_x$ and $MAPbI_{3-x}Br_x$ also having some practical relevance. In order to provide a structure–property relationship, the t and μ values for a library of hybrid halide perovskites are shown in Figure 6.39c [230].

Organic–inorganic halide perovskites were sporadically studied already in the 1880s [236], but their first practical use was proposed by Mitzi *et al.*, who were interest in "organic–inorganic electronics," *i.e.*, thin-film transistors and light-emitting diodes (LEDs) [237, 238]. These authors already observed the PV properties of these materials, but did not continue research in this direction owing to the toxicity of Pb on the one hand and the poor stability of tin halides on the other [230]. Eventually, the utilization of perovskite-type materials in the context of PV applications emerged with the first reports on functional devices by Miyasaka *et al.* [34]. In this groundbreaking work, perovskite nanocrystals were assembled onto TiO_2, as an *n*-type semiconductor (similar to the approach known from DSSCs, Figure 6.40a). The devices, which contained $MAPbI_3$, delivered a PCE of 3.8%;

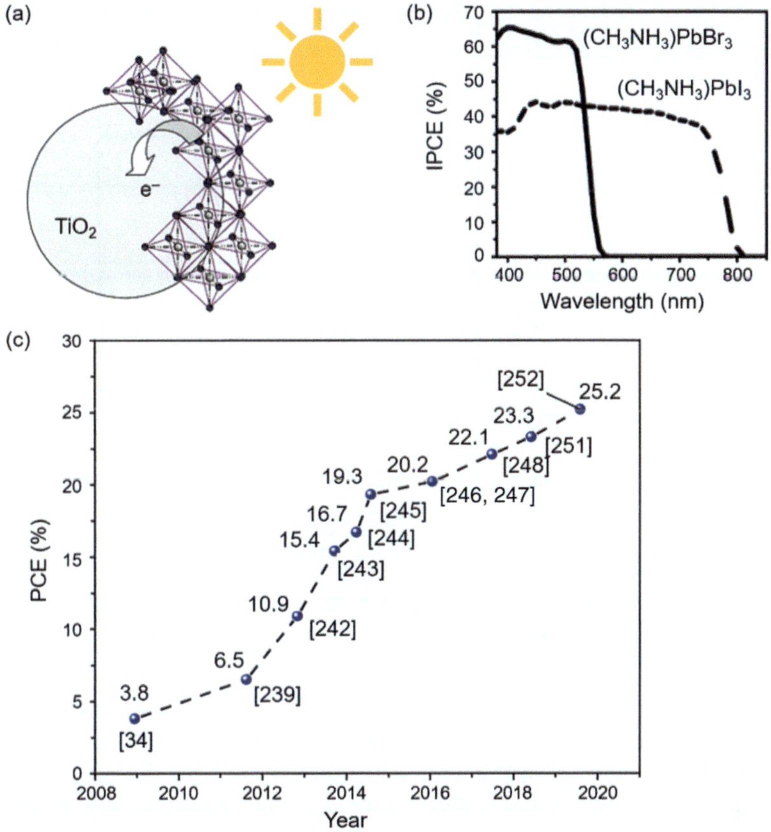

Figure 6.40 (a) Schematic representation of the function of perovskite nanocrystals as sensitizer for TiO_2 [34]. (b) Comparison of the IPCE action spectra of the photovoltaic cells, which contained $MAPbBr_3/TiO_2$ (solid line) or $MAPbI_3/TiO_2$ (dashed line) [34] / with permission of American Chemical Society. (c) Evolution of the PCE of perovskite solar cells from 2009 to 2020 [239] / with permission of John Wiley & Sons.

whereas a high photovoltage of 0.96 V was obtained when using $MAPbI_3$, as the sensitizer (Figure 6.40b). However, these cells, which used the I^-/I_3^- system as the redox shuttle in a liquid electrolyte, were only fairly stable. The development of the reached PCE values from this initial study up to 2020 is depicted in Figure 6.40c. The device performance was significantly increased by using surface-modified TiO_2 and an optimized protocol for the deposition of the perovskite absorber [240]. Park's solar cell afforded a maximum PCE of 6.5% upon illumination with AM1.5 sunlight, thus significantly higher than N719 (Figure 6.3), when processed into a DSSC with comparable device configuration (PCE of 3.9%); however, the limitation of cell degradation, due to dissolution of the active material in the liquid electrolyte, remained. As the next notable innovation, solid-state perovskite solar cells were developed independently by Snaith *et al.* as well as Grätzel *et al.* The latter authors deposited $MAPbI_3$, as the photoactive material, onto mesoporous TiO_2, followed by a layer of 2,2′,7,7′-tetra*kis*[*N*,*N*′-di(4-methoxyphenyl)amino]-9,9′-spirbiuorene

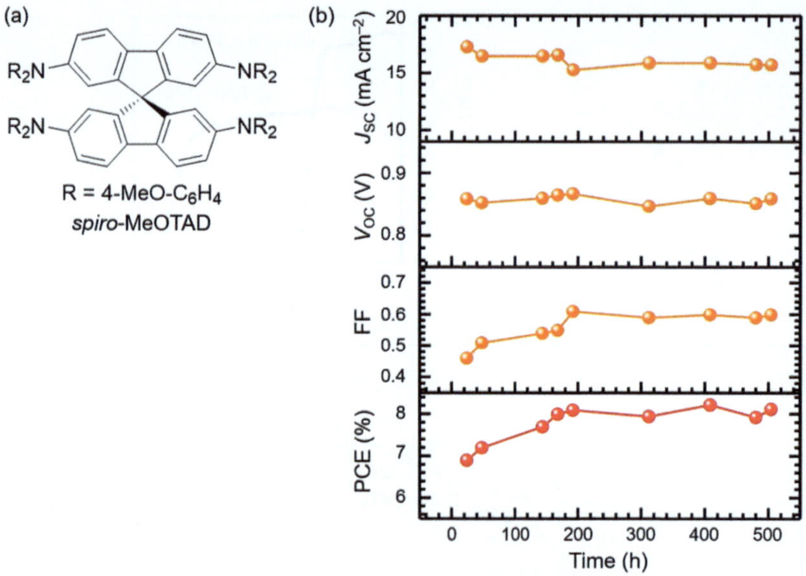

Figure 6.41 (a) Schematic representation of *spiro*-MeOTAD as an HTM. (b) Stability of the $MAPbI_3$-containing cell upon illumination with 1.0 a.m. sunlight (stored in air at room temperature without encapsulation) [241, 242] / Springer Nature / CC BY 4.0.

(*spiro*-MeOTAD, Figure 6.41a) [241]. This compound, as the HTM, facilitated the electron transfer to the semiconductor and, thus, a superior device performance (maximum PCE of 9.7%) and stability were observed (Figure 6.41b). Today, *spiro*-MeOTAD represents one of the most commonly used HTMs in perovskite solar cells [239]. Whereas Snaith *et al.* utilized insulating Al_2O_3 instead of the conventional TiO_2 semiconductor. Their low-cost, single-junction solar cell provided a PCE of 10.9% [243]. These authors also showed that nanostructuring represents no strict requirement to achieve high device efficiencies: A planar heterojunction was assembled by vacuum deposition of $MAPbI_{3-x}Cl_x$ onto a compact TiO_2 layer. The solar cell with the overall configuration FTO|TiO_2|$MAPb_{I-x}Cl_x$|*spiro*-MeOTAD|Ag revealed a PCE of 15.4% [244].

Yang *et al.* paid particular attention to the fabrication of the active layer and the influence of additives on the device performance [245]. In this context, doping of the TiO_2 layer with yttrium atoms was found to be highly beneficial for the withdrawal and transport of electrons. The thusly configured cells featured a maximum and average PCE of 19.3% and 16.6%, respectively. $FAPbI_3$ features a broader absorption range when compared to those of $MAPbI_3$ or $EAPbI_3$. In order to take advantage of this, Seok *et al.* applied an intramolecular exchange, in which the initial layer of PbI_2(DMSO) was treated with iodine [246]. This particular two-processing afforded an active layer of high quality in terms of crystallinity and grain size; thus, a maximum PCE of 20.2% was reached. A similar PV performance was also reported by Nazeeruddin *et al.* for their spiro-MeOTAD-containing solar cell [247]. It is well known that structural defects in the active layer affect the device performance due to charge-carrier recombination. Thus, dense and uniform layers will account for the good device performance. In one approach to prepare such a layer, layers of FA-lead iodide, which also contain small amounts of MA-lead bromide, Seok *et al.* processed $FAPBrI_3$, which contained some $FAPbBr_3$, in the presence of I_3^- ions [248]. The improved I^- management within the active layer helped raise the device's PCE to above 22%. Finally, a PCE of 23.7% was found for a solar cell, which contained the black α-phase of $FAPbI_3$ as the active material. This material features an appreciable low bandgap of *ca.* 1.5 V in thin films, but rapidly reorganized to the less suited δ-phase under ambient conditions. Stabilization of the α-phase was achieved by doping the perovskite material with methylenediammonium dichloride [249]. Remarkably, the expensive HTM *spiro*-MeOTAD could be replaced by the cheaper P3TH without affecting the device performance to a significant extent; a maximum PCE of 22.7% was reached in this case [250]. This aspcct of cost reduction becomes relevant regarding the upscaling of perovskite solar cells. Further increase in PCE was achieved by You *et al.*, who used phenethylammonium iodide for the passivation of surface defects [251]. The to-date highest certified PCE for a perovskite-type solar cell is 25.2%, thus reaching >80% of the thermodynamic efficiency limit [252].

Having briefly discussed the tremendous evolution of device performance since 2009 (Figure 6.40c), the shortcomings also need to be addressed at this point. Degradation still represents one limiting factor regarding perovskite commercialization. Recently, Duan, Yang *et al.* reviewed the various aspects of cell degradation and the possible pathways to counteract this issue [253]. Briefly, structural (as in the case of the aforementioned $FAPbI_3$ α-phase) and thermal instability, sensitivity to oxygen and moisture, sensitivity to UV and/or visible as well as ion migration represent the most relevant factors, which might decrease the device stability. Hasnain, Singh *et al.* pointed out that the lifetime of perovskite-based solar cells is by far shorter than <30 years, which is offered by the state-of-the-art silicon-based technology [254]. Regarding perovskite solar cells, the record stabilities are 10,000 h (*i.e.*, *ca.* 1 year) [255], 5200 h [256], and 4000 h [257]. However, device stabilization does not necessarily mean device improvement in terms of PCE [254]. Various strategies have been applied to resolve the degradation of solar cells. In particular, the field of additive engineering (active, electron-transport as well as hole-transport layer) represents one active field of current research.

Moreover, dimensionality engineering, the fabrication of low-dimensional perovskites, as alternatives to the conventional 3D ones, represents a versatile strategy to improve the device stability and manipulate the optoelectronic properties of the active material. In 2D perovskite films with layer structures, for example, the organic spacers, which act as insulators, suppress the out-of-plane transport of charges. The first use of such a 2D perovskite material, as the light absorber, in a solar-cell application was reported in 2013 by Kanatzidis *et al.* [258]. The improved stability of devices using 2D perovskites is mainly ascribed to increased hydrophobicity (*i.e.*, passivation against moisture) and the suppression of ion migration [259]. Besides 2D perovskites, 1D and 0D materials with connected octahedra or isolated octahedra, respectively, have also been reported [260]. Hybrid materials in which perovskites of different dimensionalities are combined have attracted enormous interest regarding applications in solar cells. Li, Fan *et al.* fabricated a 1D@3D perovskite, in which a 1D (bpy)PbI_2 material was deposited onto a 3D perovskite *via* a self-assembly process [261]. The solar cell, which contained this hybrid material, provided a PCE of 21.2% and long-term operational stability in the presence of water. The aforementioned device, which showed an operational lifetime of *ca.* 400 days, was comprised of a 2D–3D hybrid structure [255].

Against this background, Zhang, Li *et al.* prepared low-dimensional perovskites by the hydrothermal reaction of PbI_2 and tpy, followed by processing from solution [66]. Depending on the applied reaction conditions, either 0D $(tpy)_2PbI_6$, 1D $(tpy)_4Pb_5I_{18}$, or 1D $(tpy)_2Pb_3I_6$ single crystals were obtained (Figure 6.42a). Not only was the dimensionality of the materials modulated, but this engineering approach also enabled these authors to manipulate the electronic properties. Bandgaps in the range from 1.61 to 1.80 eV were calculated; these values were significantly lower than those reported for the 1D (bpy)PbI_2 perovskites (2.23–2.31 eV [261]). The conduction-band and valence-band minima (CBM and VBM, respectively), both the hybrid perovskites were determined by ultraviolet photoelectron spectroscopy (UPS): The VBM value of the 1D materials was *ca.* 5.40 eV, thus similar to that of $MAPbI_3$, as one of the standard 3D perovskites. In subsequent work, these same authors fabricated a solar cell, which contained a 1D-2D-3D hybrid perovskite, as the active material [262]. In this work, $CsPbI_2Br$, which represented a thermally stable, but moisture-sensitive 3D perovskite, was combined with two tpy-modified Pb^{II} iodides – namely 1D $(tpy)_2Pb_3I_6$ and 2D (tpy)Pb_3I_6. Owing to the hydrophobicity of the incorporated low-dimensional compounds, the hybrid material offered improved stability against water. Moreover, the interfacial heterostructure provided well-balanced HOMO/LUMO energy levels and charge-transfer channels, which facilitated the charge-carrier transport (Figure 6.42b). The multidimensional perovskite solar cells gave a high PCE of 16.1% (*i.e.*, >40% higher than a reference PCE with 3D $CsPbI_2Br$, as the active material) and showed an operational stability of >1000 h under ambient conditions (Figure 6.42c).

In related work, Li *et al.* prepared low-dimensional perovskites by the hydrothermal reaction of tpy, PbI_2, and $Ni(acac)_2$ (acac: acetylacetonate); thereby, the dimension and crystal structures were controlled by the reactants' stoichiometry [263]. The dimensionality of the materials increased with increasing concentration of PbI_2 in the reaction mixture: At the lowest concentration, 0D $[Ni(tpy)_2]PbI_4$ single crystals were obtained; whereas a high PbI_2 concentration favored the formation of 1D $[Ni(tpy)_2]Pb_2I_6$. A 0D–1D hybrid perovskite, $[Ni(tpy)_2]_2Pb_3I_{10}$, was formed at a medium PbI_2 concentration. In particular, the 1D perovskite revealed a high thermal stability as well as a good tolerance against water and air. Moreover, $[Ni(tpy)_2]Pb_2I_6$ absorbed visible light up to *ca.* 760 nm and possessed an E_g of 1.81 eV. According to UPS, $[Ni(tpy)_2]_2Pb_3I_{10}$ and $[Ni(tpy)_2]Pb_2I_6$ had CBM of −2.25 and −2.37 eV, respectively (Figure 6.43a). Li, Fan *et al.* also prepared a series of 0D lead-free perovskites, which contained $Ni^{II/III}$–tpy *bis*-complexes [264]. $[Ni(tpy)_2]BiBr_6$, $[Ni(tpy)_2]SbBr_6$, $[Ni(tpy)_2]In_2Br_8$, and $[Ni(tpy)_2]NiBr_4$ were prepared by a hydrothermal synthesis. X-ray diffraction (XRD) analysis of the single crystals revealed that the bulky $[Ni(tpy)_2]^{3+/2+}$ moieties occupied the A-sites within the perovskite lattice in a

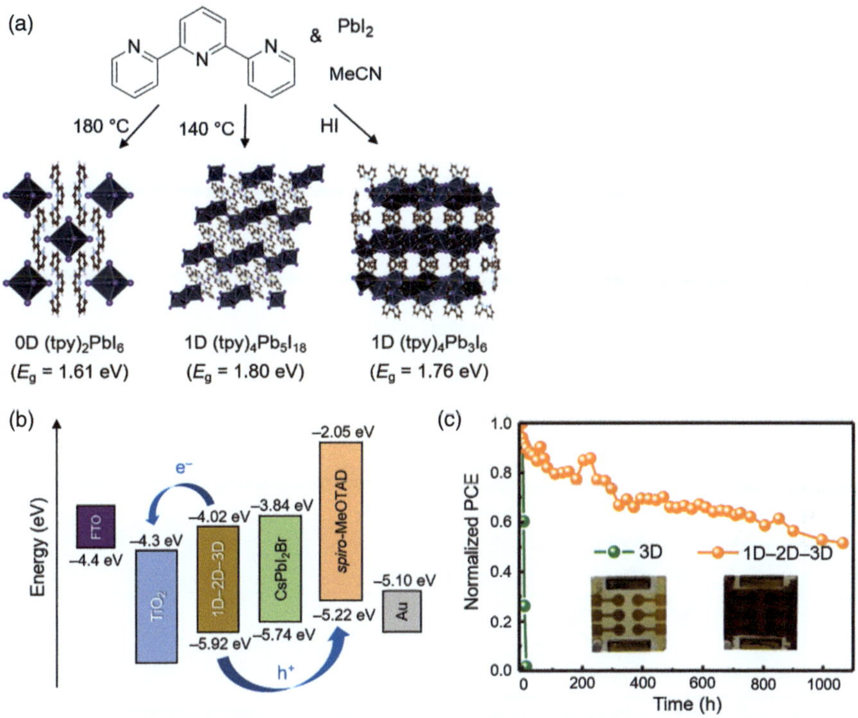

Figure 6.42 (a) Schematic representation of the hydrothermal synthesis of three different low-dimensional perovskites; the solid state structures and bandgaps of these materials are also shown [66] / with permission of Royal Society of Chemistry. (b) Schematic diagram of the band-energy alignment for the 1D-2D-3D perovskite solar cells [262]. (c) Comparison of the long-term stability of the 1D-2D-3D hybrid and pristine 3D perovskite solar cells upon illumination with white LED light (100 mW cm^{-2}) under ambient conditions (*i.e.*, 25 °C and 30% relative humidity in air) [262] / with permission of John Wiley & Sons.

highly regular fashion. This particular subgroup arrangement allowed the $Ni^{II/III}$ complexes on the one hand and the B-site metal halides (*i.e.*, $[B^{III}X_6]^{3-}$ octahedra or $[B^{II}X_4]^{2-}/[B^{III}X_4]^{-}$ tetrahedra) on the other to retain their photophysical properties. This resulted in a remarkable broadband absorption of light, which reached from 200 to beyond 1200 nm (Figure 6.43b); E_g values from 1.38 eV for $[Ni(tpy)_2]BiBr_6$, to 0.65 eV for $[Ni(tpy)_2]NiBr_4$ were determined from these spectra. DFT calculations and ligand-field considerations suggested that a plurality of transitions between the A-site complexes and the B-site metal-halide polyhedron accounted for this absorption behavior. The improved environmental stability of the materials was *inter alia* ascribed to the stabilization of the complexes due to π-back binding between the d^7-configured Ni^{III} center and the tpy ligand. In one further approach, a 0D $[Ga(tpy)_2]PbI_5$ perovskite was obtained [69]. Notably, in this work, the otherwise sparsely acknowledged $[Ga(tpy)_2]^{3+}$ complex was incorporated [265]. The calculated CBM and VBM energy levels of $[Ga(tpy)_2]PbI_5$ (*i.e.*, −3.66 and −5.46 eV, respectively) were similar to those of $MAPbI_3$, as one conventional 3D perovskite material (Figure 6.43a) and, thus, more appropriate for solar cell applications

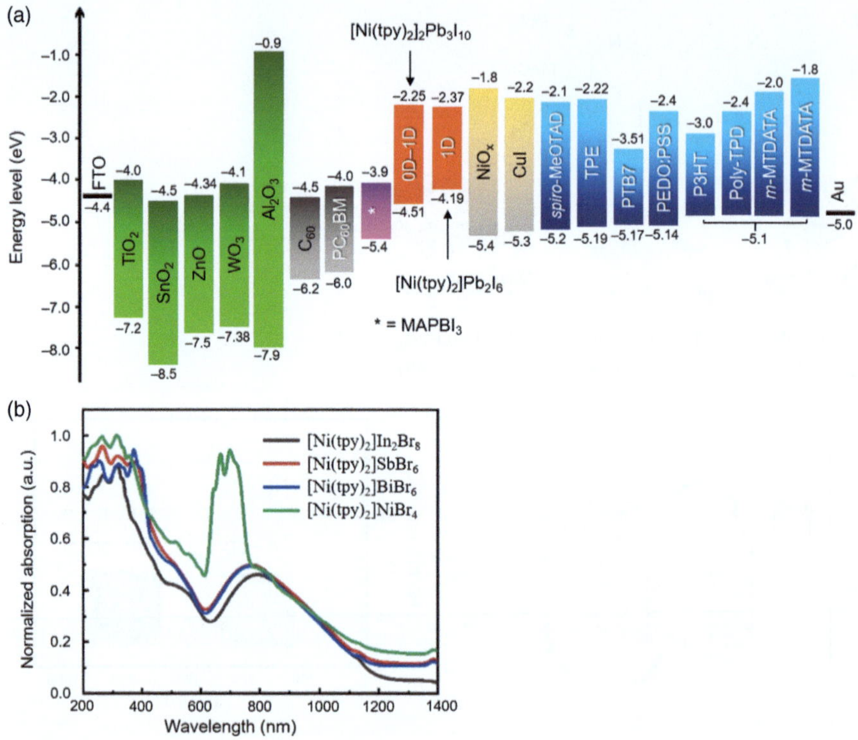

Figure 6.43 (a) Bandgap alignment of the $[Ni(tpy)_2]^{2+}$-containing perovskite single crystals with common electron- and hole-transport materials [263] / with permission of Royal Society of Chemistry. (b) UV/vis/NIR absorption spectra of the Pb-free perovskites [264] / with permission of American Chemical Society.

(in particular when compared to transition-metal-containing low-dimensional perovskites, such as the aforementioned Ni^{II}-containing ones [263]; Figure 6.43a). Remarkably, the bandgap of *ca.* 0.57 eV was even lower than those reported for the "broadband absorbers" with $Ni^{II/III}$–tpy *bis*-complexes [69]. DFT calculations gave evidence for a mixed coordination mode, in which both σ- and π-bonds between the Ga^{III} center and tpy accounted for the high degree of charge distribution on the ligands. It was proposed that this structural characteristic contributed to the facile electron-transfer processes within the perovskite.

Not only dimension engineering but also additive engineering represents a highly active field of research, which targets the improvement of perovskite solar cells in terms of stability and performance [266]. Numerous additives have already been tested to improve the interfaces between the perovskite and the charge-transport layers (*i.e.*, HTL and ETL). In this respect, the passivation of defect states on perovskite surfaces by Lewis acids or bases represents one promising strategy. Snaith *et al.* showed that simple Lewis bases, such as thiophene or pyridine, can be used for this purpose [267]. The treatment of $MAPBI_{3-x}Cl_x$ film with these additives efficiently reduced the rate of non-radiative recombination in the active material.

Upon annealing, MA cations and halide anions can be removed from the crystal lattice, leaving behind sub-coordinated Pb^{II} sites on the surface and the grain boundaries; these sites, in turn, represent trap states, which assist charge-carrier recombination and reduce the device efficiency [266]. It has been proposed that, *e.g.*, pyridine molecules might occupy the vacant coordination sites and, thus, enable the "electronic passivation" of the perovskite. In the case of a $MAPbI_{3-x}Cl_x$, the PCE of the device was raised from *ca.* 13 to 16.5% when electronic passivation with pyridine was conducted [267]. Likewise, poly(4-vinylpyridine) (P4VP) was used for defect passivation; here the PCE as well as the shelf-lifetime were significantly improved [63]. A different strategy was followed by Park *et al.*, who added the Lewis base (*i.e.*, pyridine, bpy, or tpy) to the perovskite precursor solution [268]. By doing so, the additive already influenced the crystal growth and enabled the *in situ* passivation of defects in the $FA_{0.88}Cs_{0.12}PbI_3$ structure. The bpy and tpy additives significantly improved the solar-cell performance (*i.e.*, the PCE when compared to device, which contained no additives); whereas pyridine (py), as the additive, had a detrimental effect on the PCE. Thus, the additive (*i.e.*, its denticity and size) plays a crucial role in the crystallization process and electronic passivation. The improved performance and stability of the solar cells were attributed to the ligands' ability to improve the active layers' morphology and, thereby, increase the charge-carrier lifetimes. The optimized additive concentrations were identified as 1% (bpy) and 0.2% (tpy); the stability of the thusly prepared solar cells under various experimental conditions is shown in Figure 6.44.

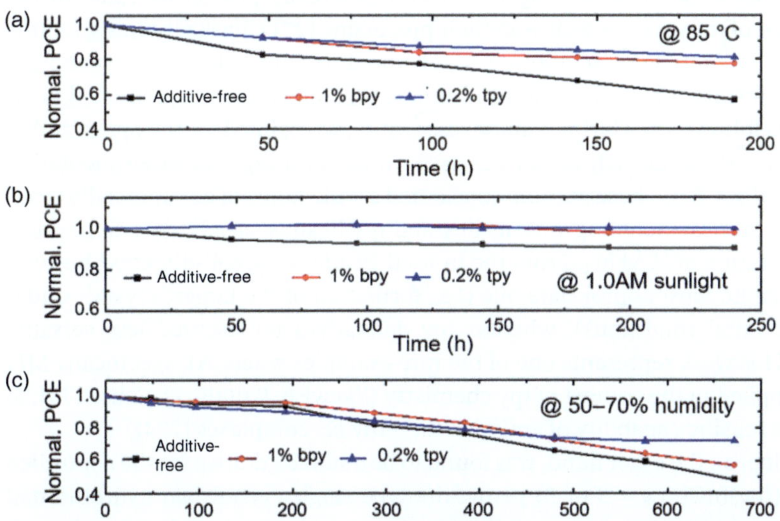

Figure 6.44 (a) Thermal stability of the unencapsulated devices upon heating to 85 °C under N_2 atmosphere. (b) Photostability of the unencapsulated devices at ambient temperature under N_2 atmosphere. (c) Moisture stability of the unencapsulated devices when exposed to a relative humidity of 50–70% at room temperature in the dark [268] / with permission of Royal Society of Chemistry.

Related to this, Duan, Tang *et al.* analyzed the electronic structure and steric demand of the same three additives [269]. The calculation of the electronic structure revealed that bpy enabled the strongest binding to Pb^{II} ions, followed by tpy and py. The superior electron donation of the bpy ligand, in concert with its steric demand, resulted in superior PV characteristics: The cell, which contained bpy-modified $CsPbIBr_2$, as the active material, afforded a maximum PCE of 11.4%; the operational and shelf lifetimes of such cells was 20 h and 60 days, respectively. For comparison, the PCE values of solar cells, which contained the pristine and tpy-modified perovskite, were somewhat lower (*i.e.*, 8.5% and 9.8%, respectively). DFT calculations and *ab initio* molecular-dynamics simulations (MDSs) on $MAPbI_3$ revealed that surface-coordinated bpy molecules were strongly bound (with a low degree of fluctuation), suppressed the charge-carrier recombination, and shielded the active layer against moisture [68]. Coordinated tpy ligands, on the other hand, offered a better charge-transfer capability, but a higher degree of fluctuation opened pathways for charge-carrier recombination and diffusion of water into the lattice. Thus, overall lower efficiency and stability are evidenced.

Recently, Tan, Xiang, Yan *et al.* summarized how the design and improvement of perovskite solar cells can be assisted by machine-learning (ML) workflows [270]. The use of ML in the field of materials science has emerged tremendously since it offers significant advantages over the aforementioned calculations or simulations: Firstly, the latter methods require huge computational resources and are still time-intensive. Moreover, it is difficult to apply them to complex systems, which are comprised of multiple components (*i.e.*, perovskite solar cells). Machine learning, being a subset of artificial intelligence (AI), has already proven its value in the development of materials as well as in their processing [271]. Basically, in a scientific ML approach, computers are allowed to take hold of human learning capabilities to support, inspire, or even substitute humans in their research [272, 273]. This strategy was followed by Hartono, Buonassisi, *et al.*, who aimed for the optimization of perovskite-crystal growth *via* an MS-assisted high-throughput experimentation (HTE) workflow [67]. These authors considered a wide range of commercially available oligopyridines, including various bpy and tpy ligands, as additives to improve the crystallization of $MAPbI_3$. From the in-total 29 additives, 6,6′-dimethyl-bpy was identified as the most appropriate one (*i.e.*, formation of the largest crystals under the experimental conditions); whereas tpy derivatives represented less versatile additives. This work represents one of the rare examples where AI, specifically ML, has been applied in the context of tpy chemistry (likewise, Baitalik *et al.* used ML to analyze the sensing capability of emissive Ru^{II}–tpy *bis*-complexes [274]).

Terpyridine, on the other hand, was found to be a suited additive for the controlled growth of (tpy)$InCl_x$ ($x = 3$ or 5) perovskite-type single crystals *via* hydrothermal synthesis in solution [70]. Although this example is not directly linked to solar-cell applications, this work is mentioned here for completeness. In the case of (tpy)$InCl_3$, the crystals had a layered structure, in which the layers were comprised of pseudo-octahedral complexes aligned along the crystals' *b*-axis (Figure 6.45a,b) [70]. Whereas a typical 0D perovskite crystal structure was observed for (tpy)$InCl_5$: The bulky $[In_2Cl_{10}]^{4-}$ anions were isolated from each other within a layered chain

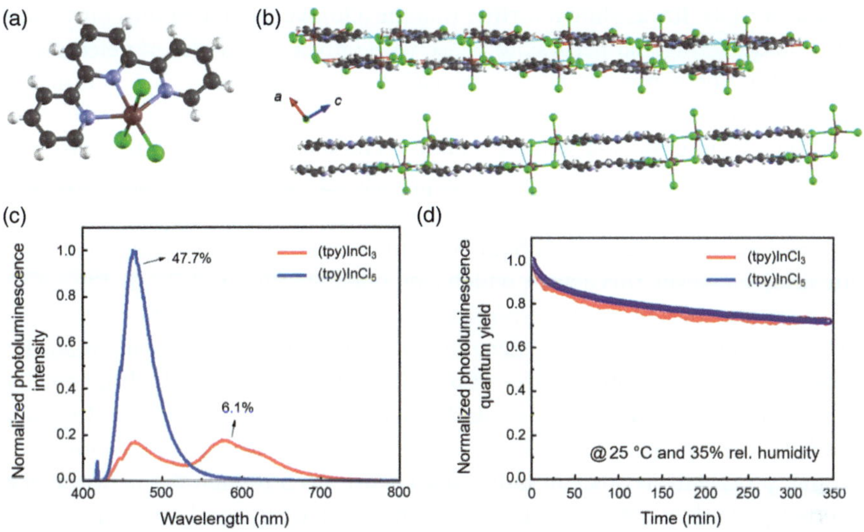

Figure 6.45 (a) Solid-state structure of (tpy)InCl_3. Adapted from [70]. (b) Solid-state structure of (tpy)InCl_3 (top) and (tpy)InCl_5 (bottom) as viewed along the (010) crystallographic orientation. Adapted from [70]. (c) Room-temperature emission spectra of (tpy)InCl_3 and (tpy)InCl_5. (d) Stability of these same materials upon irradiation with UV light for 350 min [70] / with permission of John Wiley & Sons.

of partially protonated tpy moieties (Figure 6.45b). Unlike for (tpy)InCl_5, which provided strong emission upon excitation with UV light (λ_{em} = 465 nm, *ca.* 48% quantum yield), (tpy)IrCl_3 was only a poor emitter (λ_{em} = 465 and 575 nm; *ca.* 6% quantum yield; Figure 6.45c). In either case, a good photostability under ambient conditions was observed – the initial emission efficiency was retained to 72% after 350 min of continuous irradiation with UV light (λ_{ex} = 365 nm; Figure 6.45d). Rather untypical for perovskite-type materials, the emission was short-lived (lifetime of *ca.* 2 ns). It was proposed that this behavior could still be beneficial for some optoelectronic applications (*e.g.*, lasers, quantum emitters, photodetectors, *etc.*). DFT calculations suggested that the superior photophysical properties of (tpy)InCl_5 were due to the material's crystal structure, which enabled a more efficient exciton quenching whilst suppressing pathways for the non-radiative decay.

A fundamentally different strategy to modulate perovskite crystallization was followed by Tsukagoshi, Chen, Nishihara *et al.*, who utilized a CONASH additive, as the nucleation seed (CONASH: coordination nanosheet) [275]. CONASHs, as 2D polymers based on metal-to-ligand coordination, have attracted considerable interest due to the photophysical and electrochemical properties; these properties can be utilized for various purposes: Energy storage, optoelectronics, *etc.* [276]. The CONASH, comprised of a rigid *tris*-terpyridine ligand and Zn^{II} ions, was prepared *via* a liquid/liquid interfacial synthesis and subsequently fragmented by sonication [275]. The thusly obtained fragments contained vacant tpy sites, which coordinated PbI_2, as the perovskite precursor. These species, in turn, acted as heterogeneous nucleation seeds for the high-conversion and uniform growth of perovskite crystals. It was shown further that the additive facilitated the conversion of PbI_2 into the tetragonal perovskite phase with only a low degree of defects (*i.e.*, residual PbI_2). The perovskite films were electronically passivated by coordinated CONASH fragments and, thus, afforded a prolonged emission lifetime (*ca.* 25 ns on an average) and improved thermal stability. The solar cell, which contained the modified perovskite as the active material, provided a PCE of almost 20%, whereas the PCE of non-modified perovskite was lower (the reference device gave only 18%).

The first perovskite solar cell with an inverted configuration was reported in 2013 (Figure 6.46). However, this device, which contained $MAPBI_3$ and C_{60}, as the active and electron-transport material, respectively, showed a poor PCE of only 3.9% [277]. A few years later, the remarkable advantages of the inverted design were realized: Improved operational stability under ambient conditions due to the hydrophobicity of the rear ETM layer, reduced hysteresis due to the fullerene's electron conductivity (which is faster than that of TiO_2) and ability to fabricate devices at low temperatures (and, prospectively, also on large scale) [278]. Even though the performance of inverted perovskite solar cells has been improved significantly within the last decade, the reported PCE values still stand behind those reachable with regular device configurations. This limitation is mainly due to the considerably large V_{OC} loss, which results from the non-radiative charge-carrier recombination at defect states in the active material or at the interfaces on the one hand and inappropriate energy levels of the component materials on the other. Accordingly, numerous attempts have already been made to counteract this by additive engineering, *i.e.*, interface passivation.

In one recent example, Chen, Chan, Chou *et al.* paid attention to the HTM-to-perovskite interface [279]. PTAA represents one of the most commonly used HTMs in inverted perovskite solar cells [PTAA: poly(*bis*(4-phenyl)(2,4,6-trimethylphenyl) amine)] [278]. Detrimentally, PTAA is very hydrophobic and, thus, the compatibility with the polar perovskite is rather poor. Moreover, PTAA possesses an energetically high HOMO level and its ability to contribute to defect passivation is limited. It was shown that the interfacial contact between the HTM and the perovskite and, thus, the entire device performance, could be significantly improved by introducing a supporting layer of a π-conjugated tpy derivative **64** (Figure 6.47a) [279]. By doing so, the

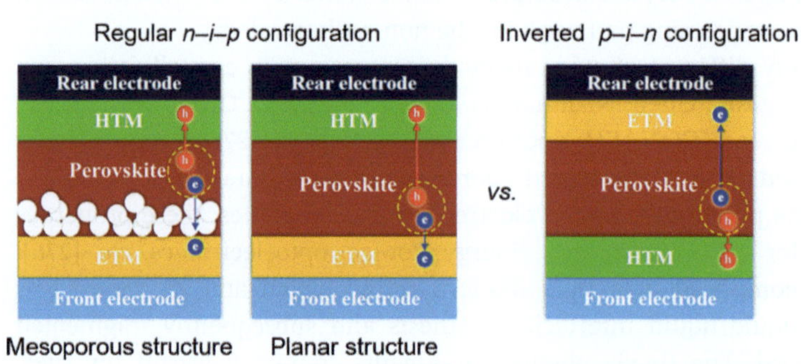

Figure 6.46 Schematic representation of the configuration of perovskite solar cells: Regular *vs.* inverted structure [278] / with permission of Royal Society of Chemistry.

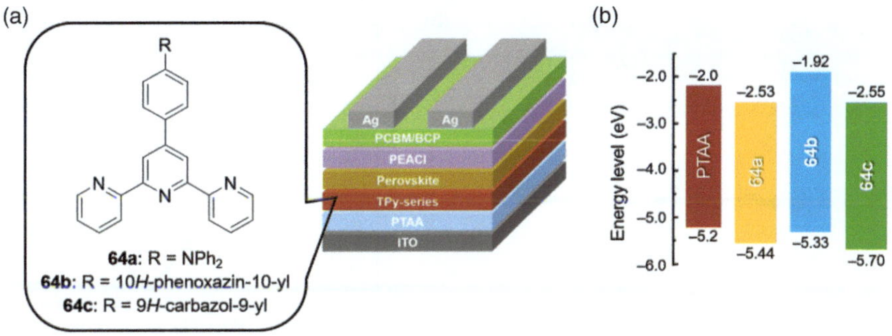

Figure 6.47 (a) Schematic representation of tpy derivatives **64**, which were incorporated, as an HTM-supporting layer, into inverted perovskite solar cells (PEACH: phenyl-ethylammonium chloride; BCP: bathocuproine). (b) Representation of the HOMO/LUMO energy levels of **64** in comparison to those of PTAA [279] / John Wiley & Sons / CC BY 4.0.

properties of the HTL were modulated in terms of energy level and hydrophilicity. In particular, phenoxazine-equipped **64c**, whose energy levels matched best with those of PTAA, enhanced the interfacial charge transport and accounted for defect passivation (Figure 6.47b). Thus, the inverted perovskite solar cell, which contained a combined PTAA/**64c** layer, afforded the best performance (*i.e.*, PCE of >22%; Table 6.8),

In a related study, Wang, Ma *et al.* also targeted the improvement of the HTM-to-perovskite interface. However, these authors studied an all-inorganic configuration, which was comprised of Cs-doped NiO_x (HTM), $CsPBiBr_2$ (perovskite), and PCBM/BCP (ETM) [280]. NiO_x represents a notable HTM for inverted perovskite cells due to its straightforward processability, stability against moisture, and favorable PV properties [281]. The used perovskite material, $CsPbIBr_2$, offers the advantage of a narrow band gap (*ca.* 2.05 eV) and a high thermal stability; accordingly, this perovskite has attracted interest regarding its application in all-inorganic assemblies [282]. In order to adjust the energy levels of these two important materials, two modifications were conducted [280]: Firstly, the NiO_x layer was doped with Cs ions, in order to modulate the material's Fermi level. Secondly, a layer of Grätzel's BD (**2**, Figure 6.3) was inserted between the HTL

Table 6.8 Summary of the photovoltaic characteristics of the inverted perovskite solar cells with PTAA/**64**, as the combined HTM layer.[a)]

Terpyridine	V_{OC} (V)	J_{SC} (mA cm^{-2})	FF	PCE (%)
None	1.12	23.46	0.78	20.45
64a	1.12	23.61	0.78	20.69
64b	1.15	23.96	0.81	22.36
64c	1.14	23.74	0.79	21.56

a) The solar cells had the overall configuration ITO|PTAA|**64**|perovskite|PEACl|PCBM/BCP|Ag.

Adapted from [279].

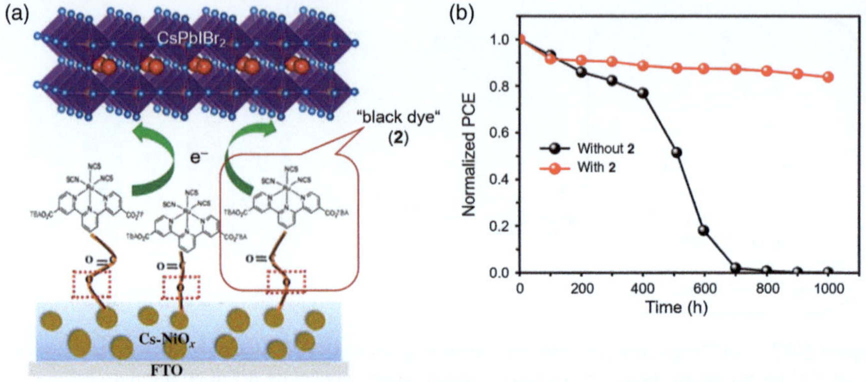

Figure 6.48 (a) Schematic representation of the inverted perovskite solar cell, which contained an additional layer of **2** for energy-level adjustment. (b) Stability of the inverted perovskite devices upon continuous illumination at room temperature and a relative humidity of 65% in air [280] / with permission of American Chemical Society.

and the active layer (Figure 6.48a). The incorporated Ru^{II} complexes not only supported the growth of the $CsPBiBr_2$ crystals, but they also suppressed the charge recombination at the interface and facilitated the injection of holes into the HTL, due to the eligible HOMO/LUMO energy levels of **2**. The thusly improved PV device provided a relatively high PCE of *ca.* 9.5% as well as a significant resistance against moisture when compared to a reference solar cell (Figure 6.48b).

Regarding the ETM in inverted perovskite solar cells, PCBM is commonly used since this C_{60} derivative possesses appropriate energy levels (matching those of perovskite) and enables an efficient acceptance/transport of electrons [283]. On the other hand, the fabrication of homogeneous, defect-free PCBM films is demanding; poor or incomplete coverage of the active layer will open pathways for low performance or even device degradation due to energy loss [284] or the diffusion of air/water into the active layer [285]. The ability of PCBM to contribute to defect passivation is low and, moreover, the molecules are prone to aggregation upon prolonged illumination or heating. It has been shown that buffer layers, which are capable of exciton blocking (*e.g.*, bathocuproine), can help level out these drawbacks. However, device fabrication becomes more complicated and might come along with reduced (thermal) stability [286]. In order to replace PCBM in the ETL, Tu, Zhou, Zhan *et al.* synthesized the fullerene dyads C_{60}-C_n-tpy, in which a tpy moiety was linked to C_{60} *via* an alkyl spacer [287]. In particular, the derivative with a C_8-spacer provided the lowest energy disorder and afforded films with improved morphological stability. An inverted perovskite solar cell with $Cs_{0.05}FA_{0.90}MA_{0.05}PbI_{2.85}Br_{0.15}$, as a hybrid perovskite, and a C_{60}-C_8-tpy/C_{60} blend, as the ETM, was fabricated and gave a PCE of 21.7%, which was retained to 96% after 1200 h of continuous operation (*i.e.*, illumination with white LED light under N_2 atmosphere). The analogous device, which contained $FAPbI_3$ as the light absorber, even afforded a PCE of >23% (Figure 6.49). In either case, the PCE of the cells, which utilized a conventional PCBM/C_{60} ETM blend, gave inferior results.

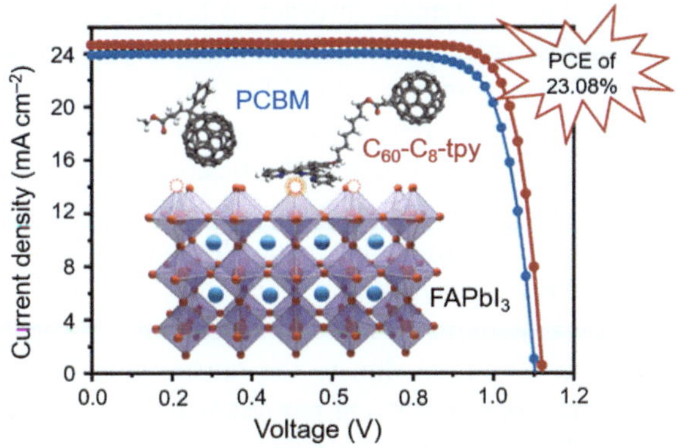

Figure 6.49 *J–V*-curves of $FAPbI_3$-based solar cells with PCBM/C_{60} or C_{60}-C_8-tpy, as the ETM [287] / with permission of American Chemical Society.

6.6 Concluding Remarks

DSSCs represent one of 3rd-generation technologies developed in the field of PVs (Figure 6.2). These thin-film devices do not require *p–n* junctions, as in the earlier generations, and can be fabricated into flexible and/or transparent forms. Thereby, the production costs are decreased significantly since mostly solution-based processes (*e.g.*, spin-coating, printing, *etc.*) can be utilized. DSSCs contain a light-absorbing dye, a sensitizer, which donates electrons into the electrode after its photoexcitation. The dye is eventually regenerated by the redox system, which is provided by the electrolyte. In order to prevent their dissolution into the electrolyte, the dye molecules are typically anchored to a semiconductor's surface *via* robust carboxylate or phosphonate moieties. Starting from Grätzel's groundbreaking research [37], Ru^{II}–polypyridyl complexes have found widespread use as sensitizers for DSSCs [72]. Among these, the complexes, which featured a panchromatic absorption profile, provided the highest PCE values of >11%, at least regarding lab-scale devices. Whereas higher certified and lab-scale performances have been reached with sensitizers not from the Ru^{II}–polypyridyl family (*i.e.*, 12.3% and 15.2%, respectively) [72]. Among the tpy-containing sensitizers, the "black dye" (**2**, N749; Figure 6.3) afforded a maximum PCE of 10.4% under optimized conditions [40]. Since then, numerous structural variants of this complex have been prepared and examined regarding their sensitizing capability in DSSCs [79]. Although decent PV performances have been found for lab-scale devices, the results could (not yet) compete with those obtained for DSSCs, which contained metalloporphyrin or record-holding all-organic sensitizers [288]. Moreover, the high material costs that are related to the use of Ru^{II} sensitizers are hardly compatible with the rapidly increasing demand for PV devices, in particular, not for large-scale and stationary ones. Whereas regarding portable and flexible solar cells, Ru^{II}-containing DSSCs might remain one alternative.

Organic and polymeric PV cells (OPVs and PPVs, respectively) represent one further type of thin, light-weight devices, which might even fully avoid the use of metal compounds (excluding the electrodes). However, transition-metal complexes, such as Ru^{II}–tpy systems, have been incorporated into the structure of π-conjugated polymers in order to broaden their absorption profile and increase the photosensitivity of the active material [211]. However, the PCE performance of these materials remains poor. The improvement of the PCE without affecting the device stability represents one major challenge when dealing with OPVs and PPVs. Recently, the record solar-cell efficiency of PPVs has been raised to >19% [53]. It is doubtful that metallopolymers, which contain Ru^{II}–polypyridyl complexes, will ever catch up to the tailor-made all-organic polymers, which have been presented as active materials in recent years.

Perovskite solar cells represent a relatively young PV technology, which has gained enormous attention due to the outstanding PCE values that have already been reached. Very recently, a Si-perovskite tandem cell achieved the record PCE of 34.6% [289]. Thus, performances beyond the Shockley–Queisser limit can be accessed by cautious selection and processing of the active material. Thereby, organic molecules can be utilized to control the crystal growth and provide organic–inorganic perovskites of low dimensionality. Such hybrid materials feature improved stability (compared to 3D perovskites) and very good photophysical properties. Moreover, such molecules can also be used for the passivation of electric defects. In these contexts, tpy ligands have successfully been employed and their large-scale use is not limited by costly transition metals, as in the case of DSSCs or PPVs.

References

1 *Erdöl – Resourcen zu Reserven machen*, https://www.bveg.de/die-branche/erdgas-und-erdoel-in-deutschland/erdoelreserven-in-deutschland/ (last accessed: 2025-04-01).
2 B. Li, L. Wang, B. Kang, P. Wang, Y. Qiu, *Sol. Energy Mater. Sol. Cells* 2006, 90, 549–573.
3 *Renewable Energy Directive*, https://energy.ec.europa.eu/topics/renewable-energy/renewable-energy-directive-targets-and-rules/renewable-energy-directive_en (last accessed: 2025-04-01).
4 *What is Renewable Energy?*, https://www.un.org/en/climatechange/what-is-renewable-energy (last accessed: 2025-04-01).
5 R. Stuewer, *APS News* 2009, 18(4), 2.
6 D. M. Chapin, C. S. Fuller, G. L. Pearson, *J. Appl. Phys.* 2004, 25, 676–677.
7 N. Gupta, G. F. Alapatt, R. Podila, R. Singh, K. F. Poole, *Int. J. Photoenergy* 2009, 2009, 154059.
8 A.-E. Becquerel, *Compt. Rend. Acad. Sci.* 1839, 9, 145–149.
9 A.-E. Becquerel, *Compt. Rend. Acad. Sci.* 1839, 9, 561–567.
10 W. Smith, *Nature* 1873, 7, 303–303.
11 W. G. Adams, R. E. Day, *Proc. R. Soc. London* 1876, 25, 113–117.
12 A. Pochettino, *Atti Reale Accad. Lincei, Rendiconti* 1906, 15, 355.
13 M. Volmer, *Ann. Phys.* 1913, 345, 775–796.
14 A. Einstein, *Ann. Phys.* 1905, 322, 132–148.
15 R. A. Millikan, *Phys. Rev.* 1916, 7, 355–388.

16 J. Czochralski, Z. Phys, *Chem.* 1918, 92U, 219–221.
17 R. A. Marques Lameirinhas, J. P. N. Torres, J. P. de Melo Cunha, *Energies* 2022, 15, 1823.
18 O. K. Simya, P. Radhakrishnan, A. Ashok, in *Handbook of Nanomaterials for Industrial Applications*, (Ed.: C. M. Hussain), Elsevier, Amsterdam, 2018, pp. 751–767.
19 *Photovoltaics Report*, https://www.ise.fraunhofer.de/content/dam/ise/de/documents/publications/studies/Photovoltaics-Report.pdf (last accessed: 2025-04-01).
20 M. A. Green, E. D. Dunlop, M. Yoshita, N. Kopidakis, K. Bothe, G. Siefer, X. Hao, *Prog. Photovolt.: Res. Appl.* 2023, 31, 651–663.
21 G. E. Eperon, M. T. Hörantner, H. J. Snaith, *Nat. Rev. Chem.* 2017, 1, 0095.
22 A. d. S. Ravasco, J. P. N. Torres, R. Lameirinhas, *Am. J. Eng. Appl. Sci.* 2021, 14, 337–350.
23 J. Peng, L. Lu, H. Yang, *Renew. Sustain. Energy Rev.* 2013, 19, 255–274.
24 A. Baiju, M. Yarema, *Front. Energy Res.* 2022, 10, 971918.
25 J. Gong, K. Sumathy, Q. Qiao, Z. Zhou, *Renew. Sustain. Energy Rev.* 2017, 68, 234–246.
26 V. Sugathan, E. John, K. Sudhakar, *Renew. Sustain. Energy Rev.* 2015, 52, 54–64.
27 S. A. P. Mahalingam, H. B. Abdullah, *Renew. Sustain. Energy Rev.* 2016, 63, 245–255.
28 M. Grätzel, *J. Photochem. Photobiol. C: Photochem. Rev.* 2003, 4, 145–153.
29 S. Rafique, S. M. Abdullah, K. Sulaiman, M. Iwamoto, *Renew. Sustain. Energ. Rev.* 2018, 84, 43–53.
30 L. Lu, T. Zheng, Q. Wu, A. M. Schneider, D. Zhao, L. Yu, *Chem. Rev.* 2015, 115, 12666–12731.
31 M. C. Scharber, N. S. Sariciftci, *Prog. Polym. Sci.* 2013, 38, 1929–1940.
32 J. S. Manser, J. A. Christians, P. V. Kamat, *Chem. Rev.* 2016, 116, 12956–13008.
33 L. Hamers, *ScienceNews* 2017, (1), 192.
34 A. Kojima, K. Teshima, Y. Shirai, T. Miyasaka, *J. Am. Chem. Soc.* 2009, 131, 6050–6051.
35 H. Min, D. Y. Lee, J. Kim, G. Kim, K. S. Lee, J. Kim, M. J. Paik, Y. K. Kim, K. S. Kim, M. G. Kim, T. J. Shin, S. Il Seok, *Nature* 2021, 598, 444–450.
36 T. A. Chowdhury, M. A. Bin Zafar, M. Sajjad-Ul Islam, M. Shahinuzzaman, M. A. Islam, M. U. Khandaker, *RSC Adv.* 2023, 13, 1787–1810.
37 B. O'Regan, M. Grätzel, *Nature* 1991, 353, 737–740.
38 M. Grätzel, *Nature* 2001, 414, 338–344.
39 M. K. Nazeeruddin, A. Kay, I. Rodicio, R. Humphry-Baker, E. Mueller, P. Liska, N. Vlachopoulos, M. Graetzel, *J. Am. Chem. Soc.* 1993, 115, 6382–6390.
40 M. K. Nazeeruddin, P. Péchy, T. Renouard, S. M. Zakeeruddin, R. Humphry-Baker, P. Comte, P. Liska, L. Cevey, E. Costa, V. Shklover, L. Spiccia, G. B. Deacon, C. A. Bignozzi, M. Grätzel, *J. Am. Chem. Soc.* 2001, 123, 1613–1624.

41 M. K. Nazeeruddin, P. Péchy, M. Grätzel, *Chem. Commun.* 1997, 1997, 1705–1706.
42 M. K. Nazeeruddin, S. M. Zakeeruddin, J.-J. Lagref, P. Liska, P. Comte, C. Barolo, G. Viscardi, K. Schenk, M. Graetzel, *Coord. Chem. Rev.* 2004, 248, 1317–1328.
43 *Air mass (solar energy)*, https://en.wikipedia.org/wiki/Air_mass_(solar_energy) (last accessed: 2025-04-01).
44 C. W. Tang, *Appl. Phys. Lett.* 1986, 48, 183–185.
45 J. C. Hummelen, B. W. Knight, F. LePeq, F. Wudl, J. Yao, C. L. Wilkins, *J. Org. Chem.* 1995, 60, 532–538.
46 N. S. Sariciftci, L. Smilowitz, A. J. Heeger, F. Wudl, *Science* 1992, 258, 1474–1476.
47 S. Morita, A. A. Zakhidov, K. Yoshino, *Solid State Commun.* 1992, 82, 249–252.
48 N. S. Sariciftci, L. Smilowitz, A. J. Heeger, F. Wudl, *Synth. Met.* 1993, 59, 333–352.
49 G. Li, R. Zhu, Y. Yang, *Nat. Photonics* 2012, 6, 153–161.
50 M. Hiramoto, H. Fujiwara, M. Yokoyama, *J. Appl. Phys.* 1992, 72, 3781–3787.
51 G. Yu, J. Gao, J. C. Hummelen, F. Wudl, A. J. Heeger, *Science* 1995, 270, 1789–1791.
52 J. J. M. Halls, C. A. Walsh, N. C. Greenham, E. A. Marseglia, R. H. Friend, S. C. Moratti, A. B. Holmes, *Nature* 1995, 376, 498–500.
53 J. Fu, P. W. K. Fong, H. Liu, C.-S. Huang, X. Lu, S. Lu, M. Abdelsamie, T. Kodalle, C. M. Sutter-Fella, Y. Yang, G. Li, *Nat. Commun.* 2023, 14, 1760.
54 S. Günes, H. Neugebauer, N. S. Sariciftci, *Chem. Rev.* 2007, 107, 1324–1338.
55 H. Zhou, L. Yang, W. You, *Macromolecules* 2012, 45, 607–632.
56 H. J. Son, F. He, B. Carsten, L. Yu, *J. Mater. Chem.* 2011, 21, 18934–18945.
57 S. Xiao, Q. Zhang, W. You, *Adv. Mater.* 2017, 29, 1601391.
58 A. R. Murad, A. Iraqi, S. B. Aziz, S. N. Abdullah, M. A. Brza, *Polymers* 2020, 12, 2627.
59 S. Götz, S. Zechel, M. D. Hager, G. R. Newkome, U. S. Schubert, *Prog. Polym. Sci.* 2021, 119, 101428.
60 J. Y. Kim, J.-W. Lee, H. S. Jung, H. Shin, N.-G. Park, *Chem. Rev.* 2020, 120, 7867–7918.
61 J. Chen, X. Cai, D. Yang, D. Song, J. Wang, J. Jiang, A. Ma, S. Lv, M. Z. Hu, C. Ni, *J. Power Sources* 2017, 355, 98–133.
62 T. Niu, L. Chao, W. Gao, C. Ran, L. Song, Y. Chen, L. Fu, W. Huang, *ACS Energy Lett.* 2021, 6, 1453–1479.
63 L. Zuo, H. Guo, D. W. de Quilettes, S. Jariwala, N. De Marco, S. Dong, R. DeBlock, D. S. Ginger, B. Dunn, M. Wang, Y. Yang, *Sci. Adv.* 2017, 3, e1700106.
64 H. Li, J. Shi, J. Deng, Z. Chen, Y. Li, W. Zhao, J. Wu, H. Wu, Y. Luo, D. Li, Q. Meng, *Adv. Mater.* 2020, 32, 1907396.
65 S. Wang, A. Wang, X. Deng, L. Xie, A. Xiao, C. Li, Y. Xiang, T. Li, L. Ding, F. Hao, *J. Mater. Chem. A* 2020, 8, 12201–12225.
66 Y. Yuan, Y. Xian, Y. Long, Y. Zhang, N. U. Rahman, Y. Zhang, J. Fan, W. Li, *RSC Adv.* 2021, 11, 24816–24821.

67 N. T. P. Hartono, M. Ani Najeeb, Z. Li, P. W. Nega, C. A. Fleming, X. Sun, E. M. Chan, A. Abate, A. J. Norquist, J. Schrier, T. Buonassisi, *Cryst. Growth Des.* 2022, 22, 5424–5431.

68 N. Chen, W. Zhang, Q.-S. Li, *Nanoscale* 2023, 15, 10730–10739.

69 Y. Xian, S. Yuan, Y. Zhang, Y. Yuan, Y. Zhang, J. Fan, W. Li, *J. Phys. Chem. Lett.* 2021, 12, 5571–5579.

70 Y. Zhang, S. Yuan, Y. Yuan, Y. Bao, Q. Ran, E. Liu, J. Fan, W. Li, *Adv. Opt. Mater.* 2022, 10, 2102041.

71 J. M. Cole, G. Pepe, O. K. Al Bahri, C. B. Cooper, *Chem. Rev.* 2019, 119, 7279–7327.

72 N. Tomar, A. Agrawal, V. S. Dhaka, P. K. Surolia, *Solar Energy* 2020, 207, 59–76.

73 European Chemical Agency, *Guide on the classification and labelling of titanium dioxide*, https://echa.europa.eu/-/new-guide-available-on-classifying-and-labelling-titanium-dioxide (last accessed: 2025-04-01).

74 D. Pourjafari, G. Oskam, in *Nanomaterials for Solar Cell Applications*, (Eds.: S. Thomas, E. H. M. Sakho, N. Kalarikkal, S. O. Oluwafemi, J. Wu), Elsevier, 2019, pp. 145–204.

75 Q. Wali, R. Jose, in *Nanomaterials for Solar Cell Applications*, (Eds.: S. Thomas, E. H. M. Sakho, N. Kalarikkal, S. O. Oluwafemi, J. Wu), Elsevier, 2019, pp. 205–285.

76 B. N. Nunes, L. A. Faustino, A. V. Muller, A. S. Polo, A. O. T. Patrocinio, in *Nanomaterials for Solar Cell Applications*, (Eds.: S. Thomas, E. H. M. Sakho, N. Kalarikkal, S. O. Oluwafemi, J. Wu), Elsevier, 2019, pp. 287–322.

77 M. Batmunkh, M. J. Biggs, J. G. Shapter, *Small* 2015, 11, 2963–2989.

78 S.-W. Wang, C.-C. Chou, F.-C. Hu, K.-L. Wu, Y. Chi, J. N. Clifford, E. Palomares, S.-H. Liu, P.-T. Chou, T.-C. Wei, T.-Y. Hsiao, *J. Mater. Chem. A* 2014, 2, 17618–17627.

79 D. Saccone, C. Magistris, N. Barbero, P. Quagliotto, C. Barolo, G. Viscardi, *Materials* 2016, 9, 137.

80 M. K. Nazeeruddin, F. De Angelis, S. Fantacci, A. Selloni, G. Viscardi, P. Liska, S. Ito, B. Takeru, M. Grätzel, *J. Am. Chem. Soc.* 2005, 127, 16835–16847.

81 A. S. Najm, S. A. Alwash, N. H. Sulaiman, M. S. Chowdhury, K. Techato, *Environ. Prog. Sustain. Energy* 2023, 42, e13955.

82 S. M. Zakeeruddin, M. K. Nazeeruddin, P. Pechy, F. P. Rotzinger, R. Humphry-Baker, K. Kalyanasundaram, M. Grätzel, V. Shklover, T. Haibach, *Inorg. Chem.* 1997, 36, 5937–5946.

83 R. Komiya, A. Fukui, N. Murofushi, N. Koide, R. Yamanaka, H. Katayama, in Technical Digest, 21st International Photovoltaic Science and Engineering Conference, *Vol.* 2, Fukuoka, 2011, pp. 2C-50-80.

84 A. Yella, H.-W. Lee, H. N. Tsao, C. Yi, A. K. Chandiran, M. K. Nazeeruddin, E. W.-G. Diau, C.-Y. Yeh, S. M. Zakeeruddin, M. Grätzel, *Science* 2011, 334, 629–634.

85 Z.-S. Wang, C.-H. Huang, Y.-Y. Huang, B.-W. Zhang, P.-H. Xie, Y.-J. Hou, K. Ibrahim, H.-J. Qian, F.-Q. Liu, *Sol. Energy Mater Sol. Cells* 2002, 71, 261–271.

86 T. Funaki, M. Yanagida, N. Onozawa-Komatsuzaki, Y. Kawanishi, K. Kasuga, H. Sugihara, *Sol. Energy Mater Sol. Cells* 2009, 93, 729–732.

87 G. C. Vougioukalakis, T. Stergiopoulos, G. Kantonis, A. G. Kontos, K. Papadopoulos, A. Stublla, P. G. Potvin, P. Falaras, *J. Photochem. Photobiol. A: Chem.* 2010, 214, 22–32.

88 K. L. V. Joseph, A. Anthonysamy, R. Easwaramoorthi, D. V. Shinde, V. Ganapathy, S. Karthikeyan, J. Lee, T. Park, S.-W. Rhee, K. S. Kim, J. K. Kim, *Dyes Pigm.* 2016, 126, 270–278.

89 Y. Numata, S. P. Singh, A. Islam, M. Iwamura, A. Imai, K. Nozaki, L. Han, *Adv. Funct. Mater.* 2013, 23, 1817–1823.

90 S.-H. Yang, K.-L. Wu, Y. Chi, Y.-M. Cheng, P.-T. Chou, *Angew. Chem. Int. Ed.* 2011, 50, 8270–8274.

91 H. Ozawa, K. Fukushima, A. Urayama, H. Arakawa, *Inorg. Chem.* 2015, 54, 8887–8889.

92 M. Kimura, J. Masuo, Y. Tohata, K. Obuchi, N. Masaki, T. N. Murakami, N. Koumura, K. Hara, A. Fukui, R. Yamanaka, S. Mori, *Chem. Eur. J.* 2013, 19, 1028–1034.

93 C.-Y. Chen, Y.-M. Feng, T.-Y. Wu, Y.-C. Liu, S.-Y. Chen, T.-Y. Lin, H.-H. G. Tsai, C.-G. Wu, *ACS Appl. Energy Mater.* 2021, 4, 13461–13470.

94 H. Ozawa, K. Fukushima, T. Sugiura, A. Urayama, H. Arakawa, *Dalton Trans.* 2014, 43, 13208–13218.

95 H. Ozawa, T. Sugiura, R. Shimizu, H. Arakawa, *Inorg. Chem.* 2014, 53, 9375–9384.

96 *Solar cells: A guide to theory and measurement*, https://www.ossila.com/pages/solar-cells-theory (last accessed: 2025-04-01).

97 H. Ozawa, T. Sugiura, T. Kuroda, K. Nozawa, H. Arakawa, *J. Mater. Chem. A* 2016, 4, 1762–1770.

98 J. Dehaudt, J. Husson, L. Guyard, F. Oswald, D. Martineau, *Renew. Energy* 2014, 66, 588–595.

99 G. Koyyada, V. Botla, S. Thogiti, G. Wu, J. Li, X. Fang, F. Kong, S. Dai, N. Surukonti, B. Kotamarthi, C. Malapaka, *Dalton Trans.* 2014, 43, 14992–15003.

100 H. Ozawa, S. Oura, R. Shimizu, H. Arakawa, *Chem. Lett.* 2012, 41, 1406–1408.

101 K. Kalyanasundaram, Ed, *Dye-Sensitized Solar Cells*, Taylor & Francis Inc., New York, 2010.

102 M. K. Nazeeruddin, M. Grätzel, *J. Photochem. Photobiol. A: Chem.* 2001, 145, 79–86.

103 M. Gratzel, M. K. Nazeeruddin, P. Pechy, Patent WO-9850393, 1998.

104 M. Chandrasekharam, G. Rajkumar, C. S. Rao, T. Suresh, Y. Soujanya, P. Y. Reddy, *Adv. OptoElectron.* 2011, 2011, 432803.

105 L. Giribabu, V. K. Singh, M. Srinivasu, C. V. Kumar, V. G. Reddy, Y. Soujanya, P. Y. Reddy, *J. Chem. Sci.* 2011, 123, 371–378.

106 G. Koyyada, P. Kumar Ch, P. Salvatori, G. Marotta, M. G. Lobello, O. Bizzarri, F. De Angelis, C. Malapaka, *Inorg. Chim. Acta* 2016, 442, 158–166.

107 C. H. S. S. Pavan Kumar, V. Anusha, K. Narayanaswamy, K. Bhanuprakash, A. Islam, L. Han, S. P. Singh, M. Chandrasekharam, *Inorg. Chim. Acta* 2015, 435, 46–52.

108 L. Giribabu, T. Bessho, M. Srinivasu, C. Vijaykumar, Y. Soujanya, V. G. Reddy, P. Y. Reddy, J.-H. Yum, M. Grätzel, M. K. Nazeeruddin, *Dalton Trans.* 2011, 40, 4497–4504.

109 T. Stergiopoulos, S. Karakostas, P. Falaras, *J. Photochem. Photobiol. A: Chem.* 2004, 163, 331–340.

110 T. Funaki, M. Yanagida, N. Onozawa-Komatsuzaki, K. Kasuga, Y. Kawanishi, M. Kurashige, K. Sayama, H. Sugihara, *Inorg. Chem. Commun.* 2009, 12, 842–845.

111 T. Funaki, H. Funakoshi, O. Kitao, N. Onozawa-Komatsuzaki, K. Kasuga, K. Sayama, H. Sugihara, *Angew. Chem. Int. Ed.* 2012, 51, 7528–7531.

112 A. Islam, S. P. Singh, M. Yanagida, M. R. Karim, L. Han, *Int. J. Photoenergy* 2011, 2011, 757421.

113 A. Islam, F. A. Chowdhury, Y. Chiba, R. Komiya, N. Fuke, N. Ikeda, K. Nozaki, L. Han, *Chem. Mater.* 2006, 18, 5178–5185.

114 S. Gao, A. Islam, Y. Numata, L. Han, *Appl. Phys. Express* 2010, 3, 062301.

115 B.-S. Chen, K. Chen, Y.-H. Hong, W.-H. Liu, T.-H. Li, C.-H. Lai, P.-T. Chou, Y. Chi, G.-H. Lee, *Chem. Commun.* 2009, 2009, 5844–5846.

116 B. N. Mongal, A. Pal, T. K. Mandal, J. Datta, S. Naskar, *Polyhedron* 2015, 102, 615–626.

117 S. Erten-Ela, S. Sogut, K. Ocakoglu, *Mater. Sci. Semicond. Process.* 2014, 23, 159–166.

118 C. Ulbricht, B. Beyer, C. Friebe, A. Winter, U. S. Schubert, *Adv. Mater.* 2009, 21, 4418–4441.

119 C. Caporale, M. Massi, *Coord. Chem. Rev.* 2018, 363, 71–91.

120 H. Kusama, T. Funaki, K. Sayama, *J. Photochem. Photobiol. A: Chem.* 2013, 272, 80–89.

121 A. Islam, H. Sugihara, M. Yanagida, K. Hara, G. Fujihashi, Y. Tachibana, R. Katoh, S. Murata, H. Arakawa, *New J. Chem.* 2002, 26, 966–968.

122 A. Islam, S. P. Singh, L. Han, *Int. J. Photoenergy* 2011, 2011, 204639.

123 A. Islam, F. A. Chowdhury, Y. Chiba, R. Komiya, N. Fuke, N. Ikeda, L. Han, *Chem. Lett.* 2005, 34, 344–345.

124 T. Kinoshita, J. T. Dy, S. Uchida, T. Kubo, H. Segawa, *Nat. Photonics* 2013, 7, 535–539.

125 T. Kinoshita, K. Nonomura, N. Joong Jeon, F. Giordano, A. Abate, S. Uchida, T. Kubo, S. I. Seok, M. K. Nazeeruddin, A. Hagfeldt, M. Grätzel, H. Segawa, *Nat. Commun.* 2015, 6, 8834.

126 T. Kinoshita, M. Otsubo, T. Ono, H. Segawa, A. C. S. Appl, *Energy Mater.* 2021, 4, 7052–7063.

127 T. Stergiopoulos, I. M. Arabatzis, M. Kalbac, I. Lukes, P. Falaras, *J. Mater. Process. Technol.* 2005, 161, 107–112.

128 C. Houarner, E. Blart, P. Buvat, F. Odobel, *Photochem. Photobiol. Sci.* 2005, 4, 200–204.

129 C. Houarner-Rassin, F. Chaignon, C. She, D. Stockwell, E. Blart, P. Buvat, T. Lian, F. Odobel, *J. Photochem. Photobiol. A: Chem.* 2007, 192, 56–65.

130 V. Duprez, M. Biancardo, F. C. Krebs, *Sol. Energy Mater. Sol. Cells* 2007, 91, 230–237.

131 V. Duprez, M. Biancardo, H. Spanggaard, F. C. Krebs, *Macromolecules* 2005, 38, 10436–10448.

132 S. Caramori, J. Husson, M. Beley, C. A. Bignozzi, R. Argazzi, P. C. Gros, *Chem. Eur. J.* 2010, 16, 2611–2618.

133 F. C. Krebs, M. Biancardo, *Sol. Energy Mater. Sol. Cells* 2006, 90, 142–165.

134 H. T. Chan, C. S. K. Mak, A. B. Djurišić, W. K. Chan, *Macromol. Chem. Phys.* 2011, 212, 774–784.

135 T. Swetha, P. J. S. Rana, A. Islam, S. P. Singh, *Sol. Energy* 2019, 188, 305–311.

136 E. U. Mughal, R. J. Obaid, A. Sadiq, M. A. Alsharif, N. Naeem, S. Kausar, A. A. Altaf, R. S. Jassas, S. Ahmed, R. I. Alsantali, S. A. Ahmed, *Dyes Pigm.* 2022, 201, 110248.

137 N. Jain, A. Mary, V. Manjunath, R. Sakla, R. S. Devan, D. A. Jose, A. R. Naziruddin, *Eur. J. Inorg. Chem.* 2021, 2021, 5014–5023.

138 N. Jain, A. Mary, T. Singh, S. Gadiyaram, J. Joshi, D. A. Jose, A. R. Naziruddin, *Eur. J. Inorg. Chem.* 2023, 26, e202300210.

139 A. Islam, S. Prakash Singh, L. Han, *Funct. Mater. Lett.* 2011, 04, 21–24.

140 C.-C. Chou, K.-L. Wu, Y. Chi, W.-P. Hu, S. J. Yu, G.-H. Lee, C.-L. Lin, P.-T. Chou, *Angew. Chem. Int. Ed.* 2011, 50, 2054–2058.

141 K.-L. Wu, C.-H. Li, Y. Chi, J. N. Clifford, L. Cabau, E. Palomares, Y.-M. Cheng, H.-A. Pan, P.-T. Chou, *J. Am. Chem. Soc.* 2012, 134, 7488–7496.

142 S. Sinn, B. Schulze, C. Friebe, D. G. Brown, M. Jäger, E. Altuntaş, J. Kübel, O. Guntner, C. P. Berlinguette, B. Dietzek, U. S. Schubert, *Inorg. Chem.* 2014, 53, 2083–2095.

143 G. Wu, R. Kaneko, Y. Zhang, Y. Shinozaki, K. Sugawa, A. Islam, L. Han, I. Bedja, R. K. Gupta, Q. Shen, J. Otsuki, *J. Power Sources* 2016, 307, 416–425.

144 E. A. Medlycott, G. S. Hanan, *Chem. Soc. Rev.* 2005, 34, 133–142.

145 S. Sinn, B. Schulze, C. Friebe, D. G. Brown, M. Jäger, J. Kübel, B. Dietzek, C. P. Berlinguette, U. S. Schubert, *Inorg. Chem.* 2014, 53, 1637–1645.

146 B. Schulze, U. S. Schubert, *Chem. Soc. Rev.* 2014, 43, 2522–2571.

147 B. Schulze, C. Friebe, M. D. Hager, A. Winter, R. Hoogenboom, H. Görls, U. S. Schubert, *Dalton Trans.* 2009, 2007, 787–794.

148 B. Schulze, D. G. Brown, K. C. D. Robson, C. Friebe, M. Jäger, E. Birckner, C. P. Berlinguette, U. S. Schubert, *Chem. Eur. J.* 2013, 19, 14171–14180.

149 E. Yoneda, M. Grätzel, M. K. Nazeeruddin, Patent EP-2036955, 2009.

150 H. Kisserwan, T. H. Ghaddar, *Dalton Trans.* 2011, 40, 3877–3884.

151 K. C. D. Robson, B. D. Koivisto, A. Yella, B. Sporinova, M. K. Nazeeruddin, T. Baumgartner, M. Grätzel, C. P. Berlinguette, *Inorg. Chem.* 2011, 50, 5494–5508.

152 G. Wu, R. Kaneko, A. Islam, Y. Zhang, K. Sugawa, L. Han, Q. Shen, I. Bedja, R. K. Gupta, J. Otsuki, *J. Power Sources* 2016, 331, 100–111.

153 R. Kaneko, G. Wu, K. Sugawa, J. Otsuki, A. Islam, L. Han, I. Bedja, R. K. Gupta, *J. Organomet. Chem.* 2017, 833, 61–70.

154 P. G. Bomben, K. C. D. Robson, B. D. Koivisto, C. P. Berlinguette, *Coord. Chem. Rev.* 2012, 256, 1438–1450.
155 S. H. Wadman, J. M. Kroon, K. Bakker, M. Lutz, A. L. Spek, G. P. M. van Klink, G. van Koten, *Chem. Commun.* 2007, 2007, 1907–1909.
156 S. H. Wadman, J. M. Kroon, K. Bakker, R. W. A. Havenith, G. P. M. van Klink, G. van Koten, *Organometallics* 2010, 29, 1569–1579.
157 H. Kisserwan, A. Kamar, T. Shoker, T. H. Ghaddar, *Dalton Trans.* 2012, 41, 10643–10651.
158 L. Mercs, M. Albrecht, *Chem. Soc. Rev.* 2010, 39, 1903–1912.
159 A. J. Arduengo III, R. L. Harlow, M. Kline, *J. Am. Chem. Soc.* 1991, 113, 361–363.
160 H.-J. Park, K. H. Kim, S. Y. Choi, H.-M. Kim, W. I. Lee, Y. K. Kang, Y. K. Chung, *Inorg. Chem.* 2010, 49, 7340–7352.
161 J. Dinda, S. Liatard, J. Chauvin, D. Jouvenot, F. Loiseau, *Dalton Trans.* 2011, 40, 3683–3688.
162 A. R. Naziruddin, C.-L. Kuo, W.-J. Lin, W.-H. Lo, C.-S. Lee, B.-J. Sun, A. H. H. Chang, W.-S. Hwang, *Organometallics* 2014, 33, 2575–2582.
163 N. Jain, A. Mary, V. Manjunath, R. Sakla, R. S. Devan, D. A. Jose, A. R. Naziruddin, *J. Organomet. Chem.* 2022, 959, 122203.
164 S. Altobello, R. Argazzi, S. Caramori, C. Contado, S. Da Fré, P. Rubino, C. Choné, G. Larramona, C. A. Bignozzi, *J. Am. Chem. Soc.* 2005, 127, 15342–15343.
165 K.-L. Wu, S.-T. Ho, C.-C. Chou, Y.-C. Chang, H.-A. Pan, Y. Chi, P.-T. Chou, *Angew. Chem. Int. Ed.* 2012, 51, 5642–5646.
166 R. Juwita, J.-Y. Lin, S.-J. Lin, Y.-C. Liu, T.-Y. Wu, Y.-M. Feng, C.-Y. Chen, H.-H. Gavin Tsai, C.-G. Wu, *J. Mater. Chem. A* 2020, 8, 12361–12369.
167 T. Kinoshita, J.-i. Fujisawa, J. Nakazaki, S. Uchida, T. Kubo, H. Segawa, *J. Phys. Chem. Lett.* 2012, 3, 394–398.
168 N. Sinha, O. S. Wenger, *J. Am. Chem. Soc.* 2023, 145, 4903–4920.
169 O. S. Wenger, *J. Am. Chem. Soc.* 2018, 140, 13522–13533.
170 C. Förster, K. Heinze, *Chem. Soc. Rev.* 2020, 49, 1057–1070.
171 P. Dierks, Y. Vukadinovic, M. Bauer, *Inorg. Chem. Front.* 2022, 9, 206–220.
172 Y. Liu, P. Persson, V. Sundström, K. Wärnmark, *Acc. Chem. Res.* 2016, 49, 1477–1485.
173 S. Ferrere, B. A. Gregg, *J. Am. Chem. Soc.* 1998, 120, 843–844.
174 T. Duchanois, T. Etienne, C. Cebrián, L. Liu, A. Monari, M. Beley, X. Assfeld, S. Haacke, P. C. Gros, *Eur. J. Inorg. Chem.* 2015, 2015, 2469–2477.
175 T. Duchanois, L. Liu, M. Pastore, A. Monari, C. Cebrián, Y. Trolez, M. Darari, K. Magra, A. Francés-Monerris, E. Domenichini, M. Beley, X. Assfeld, S. Haacke, P. C. Gros, *Inorganics* 2018, 6, 63.
176 Y. Liu, T. Harlang, S. E. Canton, P. Chábera, K. Suárez-Alcántara, A. Fleckhaus, D. A. Vithanage, E. Göransson, A. Corani, R. Lomoth, V. Sundström, K. Wärnmark, *Chem. Commun.* 2013, 49, 6412–6414.
177 L. L. Jamula, A. M. Brown, D. Guo, J. K. McCusker, *Inorg. Chem.* 2014, 53, 15–17.

178 A. K. C. Mengel, C. Förster, A. Breivogel, K. Mack, J. R. Ochsmann, F. Laquai, V. Ksenofontov, K. Heinze, *Chem. Eur. J.* 2015, 21, 704–714.

179 M. Darari, A. Francés-Monerris, B. Marekha, A. Doudouh, E. Wenger, A. Monari, S. Haacke, P. C. Gros, *Molecules* 2020, 25, 5991.

180 T. Reuter, A. Kruse, R. Schoch, S. Lochbrunner, M. Bauer, K. Heinze, *Chem. Commun.* 2021, 57, 7541–7544.

181 I. M. Dixon, F. Alary, M. Boggio-Pasqua, J.-L. Heully, *Dalton Trans.* 2015, 44, 13498–13503.

182 J. Steube, L. Burkhardt, A. Päpcke, J. Moll, P. Zimmer, R. Schoch, C. Wölper, K. Heinze, S. Lochbrunner, M. Bauer, *Chem. Eur. J.* 2019, 25, 11826–11830.

183 K. Sakai, H. Ozawa, *Coord. Chem. Rev.* 2007, 251, 2753–2766.

184 A. Winter, G. R. Newkome, U. S. Schubert, *ChemCatChem* 2011, 3, 1384–1406.

185 K. M.-C. Wong, V. W.-W. Yam, *Coord. Chem. Rev.* 2007, 251, 2477–2488.

186 K. M.-C. Wong, W.-S. Tang, X.-X. Lu, N. Zhu, V. W.-W. Yam, *Inorg. Chem.* 2005, 44, 1492–1498.

187 E. C.-H. Kwok, M.-Y. Chan, K. M.-C. Wong, W. H. Lam, V. W.-W. Yam, *Chem. Eur. J.* 2010, 16, 12244–12254.

188 E. Zysman-Coleman, Ed, *Iridium(III) in Optoelectronic and Photonics Applications*, John Wiley & Sons, Chichester, 2017.

189 Y. Shinpuku, F. Inui, M. Nakai, Y. Nakabayashi, *J. Photochem. Photobiol. A: Chem.* 2011, 222, 203–209.

190 B. Bozic-Weber, E. C. Constable, C. E. Housecroft, *Coord. Chem. Rev.* 2013, 257, 3089–3106.

191 W. R. McNamara, R. C. Snoeberger Iii, G. Li, C. Richter, L. J. Allen, R. L. Milot, C. A. Schmuttenmaer, R. H. Crabtree, G. W. Brudvig, V. S. Batista, *Energy Environ. Sci.* 2009, 2, 1173–1175.

192 K. L. Materna, R. H. Crabtree, G. W. Brudvig, *Chem. Soc. Rev.* 2017, 46, 6099–6110.

193 B. Bozic-Weber, E. C. Constable, N. Hostettler, C. E. Housecroft, R. Schmitt, E. Schönhofer, *Chem. Commun.* 2012, 48, 5727–5729.

194 N. Hostettler, S. O. Fürer, B. Bozic-Weber, E. C. Constable, C. E. Housecroft, *Dyes Pigm.* 2015, 116, 124–130.

195 N. Hostettler, I. A. Wright, B. Bozic-Weber, E. C. Constable, C. E. Housecroft, *RSC Adv.* 2015, 5, 37906–37915.

196 M. Freitag, G. Boschloo, *Curr. Opin. Electrochem.* 2017, 2, 111–119.

197 A. B. Muñoz-García, I. Benesperi, G. Boschloo, J. J. Concepcion, J. H. Delcamp, E. A. Gibson, G. J. Meyer, M. Pavone, H. Pettersson, A. Hagfeldt, M. Freitag, *Chem. Soc. Rev.* 2021, 50, 12450–12550.

198 Y. Saygili, M. Stojanovic, H. Michaels, J. Tiepelt, J. Teuscher, A. Massaro, M. Pavone, F. Giordano, S. M. Zakeeruddin, G. Boschloo, J.-E. Moser, M. Grätzel, A. B. Muñoz-García, A. Hagfeldt, M. Freitag, A. C. S. Appl, *Energy Mater.* 2018, 1, 4950–4962.

199 S. M. Feldt, P. W. Lohse, F. Kessler, M. K. Nazeeruddin, M. Grätzel, G. Boschloo, A. Hagfeldt, *Phys. Chem. Chem. Phys.* 2013, 15, 7087–7097.

200 S. M. Feldt, E. A. Gibson, E. Gabrielsson, L. Sun, G. Boschloo, A. Hagfeldt, *J. Am. Chem. Soc.* 2010, 132, 16714–16724.

201 E. Mosconi, J.-H. Yum, F. Kessler, C. J. Gómez García, C. Zuccaccia, A. Cinti, M. K. Nazeeruddin, M. Grätzel, F. De Angelis, *J. Am. Chem. Soc.* 2012, 134, 19438–19453.

202 S. Koussi-Daoud, D. Schaming, L. Fillaud, G. Trippé-Allard, F. Lafolet, E. Polanski, K. Nonomura, N. Vlachopoulos, A. Hagfeldt, J.-C. Lacroix, *Electrochim. Acta* 2015, 179, 237–240.

203 S. Venkatesan, W.-H. Lin, H. Teng, Y.-L. Lee, *ACS Appl. Mater. Interfaces* 2019, 11, 42780–42789.

204 N. Hall, *Chem. Commun.* 2003, 2003, 1–4.

205 C. K. Chiang, C. R. Fincher Jr., Y. W. Park, A. J. Heeger, H. Shirakawa, E. J. Louis, S. C. Gau, A. G. MacDiarmid, *Phys. Rev. Lett.* 1977, 39, 1098–1101.

206 H. Shirakawa, E. J. Louis, A. G. MacDiarmid, C. K. Chiang, A. J. Heeger, *J. Chem. Soc., Chem. Commun.* 1977, 1977, 578–580.

207 R. M. Pankow, B. C. Thompson, *Polymer* 2020, 207, 122874.

208 P. K. Watkins, A. B. Walker, G. L. B. Verschoor, *Nano Lett.* 2005, 5, 1814–1818.

209 F. Padinger, R. S. Rittberger, N. S. Sariciftci, *Adv. Funct. Mater.* 2003, 13, 85–88.

210 J. Peet, J. Y. Kim, N. E. Coates, W. L. Ma, D. Moses, A. J. Heeger, G. C. Bazan, *Nature Mater.* 2007, 6, 497–500.

211 P. Liu, G. Shi, X. Chen, *Front. Chem.* 2020, 8, 592055.

212 F. Guo, Y.-G. Kim, J. R. Reynolds, K. S. Schanze, *Chem. Commun.* 2006, 1887–1889.

213 J. M. Hamilton, M. J. Anhorn, K. A. Oscarson, J. H. Reibenspies, R. D. Hancock, *Inorg. Chem.* 2011, 50, 2764–2770.

214 W.-Y. Wong, X.-Z. Wang, Z. He, A. B. Djurišić, C.-T. Yip, K.-Y. Cheung, H. Wang, C. S. K. Mak, W.-K. Chan, *Nature Mater.* 2007, 6, 521–527.

215 A. Winter, U. S. Schubert, *Chem. Soc. Rev.* 2016, 45, 5311–5357.

216 U. S. Schubert, A. Winter, G. R. Newkome, *Ruthenium-Containing Polymers*, Springer Int. Publishing, Cham, 2021, pp. 15–185.

217 W. Y. Ng, W. K. Chan, *Adv. Mater.* 1997, 9, 716–719.

218 W. K. Chan, X. Gong, W. Y. Ng, *Appl. Phys. Lett.* 1997, 71, 2919–2921.

219 K. Y. K. Man, H. L. Wong, W. K. Chan, A. B. Djurišić, E. Beach, S. Rozeveld, *Langmuir* 2006, 22, 3368–3375.

220 P. D. Vellis, J. A. Mikroyannidis, C.-N. Lo, C.-S. Hsu, *J. Polym. Sci., Part A: Polym. Chem.* 2008, 46, 7702–7712.

221 H. Padhy, D. Sahu, I. H. Chiang, D. Patra, D. Kekuda, C.-W. Chu, H.-C. Lin, *J. Mater. Chem.* 2011, 21, 1196–1205.

222 H. Padhy, M. Ramesh, D. Patra, R. Satapathy, M. K. Pola, H.-C. Chu, C.-W. Chu, K.-H. Wei, H.-C. Lin, *Macromol. Rapid Commun.* 2012, 33, 528–533.

223 K. Feng, X. Shen, Y. Li, Y. He, D. Huang, Q. Peng, *Polym. Chem.* 2013, 4, 5701–5710.

224 K. W. Cheng, C. S. C. Mak, W. K. Chan, A. M. Ching Ng, A. B. Djurišić, *J. Polym. Sci., Part A: Polym. Chem.* 2008, 46, 1305–1317.

225 S.-H. Wang, T.-W. Wang, H.-C. Tsai, P.-C. Yang, C.-F. Huang, R.-H. Lee, *RSC Adv.* 2020, 10, 9525–9535.

226 Q. Lv, R. Yu, R. Shi, Z. a. Tan, *Mater. Chem. Front.* 2023, 7, 5063–5103.

227 C.-L. Ho, W.-Y. Wong, *Coord. Chem. Rev.* 2011, 255, 2469–2502.

228 A. Wild, A. Winter, F. Schlütter, U. S. Schubert, *Chem. Soc. Rev.* 2011, 40, 1459–1511.

229 E. A. Katz, *Helv. Chim. Acta* 2020, 103, e2000061.

230 M. A. Green, A. Ho-Baillie, H. J. Snaith, *Nat. Photon.* 2014, 8, 506–514.

231 *Perovskite Mineral Data*, http://webmineral.com/data/Perovskite.shtml (last accessed: 2025-04-01).

232 V. M. Goldschmidt, *Naturwissenschaften* 1926, 14, 477–485.

233 H. D. Megaw, *Nature* 1945, 155, 484–485.

234 C. Li, X. Lu, W. Ding, L. Feng, Y. Gao, Z. Guo, *Acta Crystallogr. B* 2008, 64, 702–707.

235 P. Umari, E. Mosconi, F. De Angelis, *Sci. Rep.* 2014, 4, 4467.

236 H. Topsöe, *Z. Kristallogr.* 1884, 8, 246–296.

237 D. B. Mitzi, S. Wang, C. A. Feild, C. A. Chess, A. M. Guloy, *Science* 1995, 267, 1473–1476.

238 D. B. Mitzi, K. Chondroudis, C. R. Kagan, *IBM J. Res. Dev.* 2001, 45, 29–45.

239 H. D. Pham, T. C.-J. Yang, S. M. Jain, G. J. Wilson, P. Sonar, *Adv. Energy Mater.* 2020, 10, 1903326.

240 J.-H. Im, C.-R. Lee, J.-W. Lee, S.-W. Park, N.-G. Park, *Nanoscale* 2011, 3, 4088–4093.

241 H.-S. Kim, C.-R. Lee, J.-H. Im, K.-B. Lee, T. Moehl, A. Marchioro, S.-J. Moon, R. Humphry-Baker, J.-H. Yum, J. E. Moser, M. Grätzel, N.-G. Park, *Sci. Rep.* 2012, 2, 591.

242 N. J. Jeon, H. G. Lee, Y. C. Kim, J. Seo, J. H. Noh, J. Lee, S. I. Seok, *J. Am. Chem. Soc.* 2014, 136, 7837–7840.

243 M. M. Lee, J. Teuscher, T. Miyasaka, T. N. Murakami, H. J. Snaith, *Science* 2012, 338, 643–647.

244 M. Liu, M. B. Johnston, H. J. Snaith, *Nature* 2013, 501, 395–398.

245 H. Zhou, Q. Chen, G. Li, S. Luo, T.-b. Song, H.-S. Duan, Z. Hong, J. You, Y. Liu, Y. Yang, *Science* 2014, 345, 542–546.

246 W. S. Yang, J. H. Noh, N. J. Jeon, Y. C. Kim, S. Ryu, J. Seo, S. I. Seok, *Science* 2015, 348, 1234–1237.

247 M. Saliba, S. Orlandi, T. Matsui, S. Aghazada, M. Cavazzini, J.-P. Correa-Baena, P. Gao, R. Scopelliti, E. Mosconi, K.-H. Dahmen, F. De Angelis, A. Abate, A. Hagfeldt, G. Pozzi, M. Graetzel, M. K. Nazeeruddin, *Nat. Energy* 2016, 1, 15017.

248 W. S. Yang, B.-W. Park, E. H. Jung, N. J. Jeon, Y. C. Kim, D. U. Lee, S. S. Shin, J. Seo, E. K. Kim, J. H. Noh, S. I. Seok, *Science* 2017, 356, 1376–1379.

249 H. Min, M. Kim, S.-U. Lee, H. Kim, G. Kim, K. Choi, J. H. Lee, S. I. Seok, *Science* 2019, 366, 749–753.

250 E. H. Jung, N. J. Jeon, E. Y. Park, C. S. Moon, T. J. Shin, T.-Y. Yang, J. H. Noh, J. Seo, *Nature* 2019, 567, 511–515.

251 Q. Jiang, Y. Zhao, X. Zhang, X. Yang, Y. Chen, Z. Chu, Q. Ye, X. Li, Z. Yin, J. You, *Nat. Photonics* 2019, 13, 460–466.

252 J. J. Yoo, G. Seo, M. R. Chua, T. G. Park, Y. Lu, F. Rotermund, Y.-K. Kim, C. S. Moon, N. J. Jeon, J.-P. Correa-Baena, V. Bulović, S. S. Shin, M. G. Bawendi, J. Seo, *Nature* 2021, 590, 587–593.

253 R. Wang, M. Mujahid, Y. Duan, Z.-K. Wang, J. Xue, Y. Yang, *Adv. Funct. Mater.* 2019, 29, 1808843.

254 S. Khatoon, S. Kumar Yadav, V. Chakravorty, J. Singh, R. Bahadur Singh, M. S. Hasnain, S. M. M. Hasnain, *Mater. Sci. Energy Technol.* 2023, 6, 437–459.

255 G. Grancini, C. Roldán-Carmona, I. Zimmermann, E. Mosconi, X. Lee, D. Martineau, S. Narbey, F. Oswald, F. De Angelis, M. Grätzel, M. K. Nazeeruddin, *Nat. Commun.* 2017, 8, 15684.

256 S. Bai, P. Da, C. Li, Z. Wang, Z. Yuan, F. Fu, M. Kawecki, X. Liu, N. Sakai, J. T.-W. Wang, S. Huettner, S. Buecheler, M. Fahlman, F. Gao, H. J. Snaith, *Nature* 2019, 571, 245–250.

257 Z. Wang, Q. Lin, F. P. Chmiel, N. Sakai, L. M. Herz, H. J. Snaith, *Nat. Energy* 2017, 2, 17135.

258 D. H. Cao, C. C. Stoumpos, O. K. Farha, J. T. Hupp, M. G. Kanatzidis, *J. Am. Chem. Soc.* 2015, 137, 7843–7850.

259 T. L. Leung, I. Ahmad, A. A. Syed, A. M. C. Ng, J. Popović, A. B. Djurišić, *Commun. Mater.* 2022, 3, 63.

260 T. Zhu, X. Gong, *InfoMat* 2021, 3, 1039–1069.

261 P. Liu, Y. Xian, W. Yuan, Y. Long, K. Liu, N. U. Rahman, W. Li, J. Fan, *Adv. Energy Mater.* 2020, 10, 1903654.

262 K. Liu, S. Yuan, Y. Xian, Y. Long, Q. Yao, N. U. Rahman, Y. Guo, M. Sun, Q. Xue, H.-L. Yip, A. Cabot, W. Li, J. Fan, *Small* 2021, 17, 2100888.

263 Y. Zhang, M. Zang, H. Yin, P. Liu, Y. Xian, W. Li, *CrystEngComm* 2020, 22, 1904–1908.

264 Y. Zhang, Y. Xian, S. Yuan, Y. Zhang, Y. Yuan, W. Li, J. Fan, *J. Phys. Chem. C* 2021, 125, 15579–15589.

265 A. Winter, G. R. Newkome, U. S. Schubert, *Inorg. Chem. Front.* 2024, 11, 342–399.

266 A. Mahapatra, D. Prochowicz, M. M. Tavakoli, S. Trivedi, P. Kumar, P. Yadav, *J. Mater. Chem. A* 2020, 8, 27–54.

267 N. K. Noel, A. Abate, S. D. Stranks, E. S. Parrott, V. M. Burlakov, A. Goriely, H. J. Snaith, *ACS Nano* 2014, 8, 9815–9821.

268 J. Chen, S.-G. Kim, X. Ren, H. S. Jung, N.-G. Park, *J. Mater. Chem. A* 2019, 7, 4977–4987.

269 J. Zhang, J. Duan, Q. Zhang, Q. Guo, F. Yan, X. Yang, Y. Duan, Q. Tang, *Chem. Eng. J.* 2022, 431, 134230.

270 Y. Liu, X. Tan, J. Liang, H. Han, P. Xiang, W. Yan, *Adv. Funct. Mater.* 2023, 33, 2214271.

271 A. Y.-T. Wang, R. J. Murdock, S. K. Kauwe, A. O. Oliynyk, A. Gurlo, J. Brgoch, K. A. Persson, T. D. Sparks, *Chem. Mater.* 2020, 32, 4954–4965.

272 M. I. Jordan, T. M. Mitchell, *Science* 2015, 349, 255–260.

273 A. L. Blum, P. Langley, *Artifi. Intell.* 1997, 97, 245–271.

274 A. Sahoo, S. Bhattacharya, S. Jana, S. Baitalik, *Dalton Trans.* 2023, 52, 97–108.

275 Y.-C. Wang, C.-H. Chiang, C.-J. Su, J.-W. Chang, C.-Y. Lin, C.-C. Wei, S.-K. Huang, H. Maeda, W.-B. Jian, U. S. Jeng, K. Tsukagoshi, C.-W. Chen, H. Nishihara, *J. Mater. Chem. A* 2023, 11, 7077–7084.

276 R. Sakamoto, K. Takada, X. Sun, T. Pal, T. Tsukamoto, E. J. H. Phua, A. Rapakousiou, K. Hoshiko, H. Nishihara, *Coord. Chem. Rev.* 2016, 320-321, 118–128.

277 J.-Y. Jeng, Y.-F. Chiang, M.-H. Lee, S.-R. Peng, T.-F. Guo, P. Chen, T.-C. Wen, *Adv. Mater.* 2013, 25, 3727–3732.

278 X. Lin, D. Cui, X. Luo, C. Zhang, Q. Han, Y. Wang, L. Han, *Energy Environ. Sci.* 2020, 13, 3823–3847.

279 C.-M. Hung, C.-C. Wu, P.-H. Tsao, C.-D. Lung, C.-H. Wang, I.-C. Ni, C.-C. Chu, C.-H. Cheng, W.-Y. Kuang, C.-I. Wu, H.-C. Chen, Y.-T. Chan, P.-T. Chou, *Adv. Energy Sustain. Res.* 2023, 4, 2300042.

280 S. Yang, L. Wang, L. Gao, J. Cao, Q. Han, F. Yu, Y. Kamata, C. Zhang, M. Fan, G. Wei, T. Ma, A. C. S. Appl, *Mater. Interfaces* 2020, 12, 13931–13940.

281 W. Chen, Y. Wu, J. Fan, A. B. Djurišić, F. Liu, H. W. Tam, A. Ng, C. Surya, W. K. Chan, D. Wang, Z.-B. He, *Adv. Energy Mater.* 2018, 8, 1703519.

282 B. Yang, M. Wang, X. Hu, T. Zhou, Z. Zang, *Nano Energy* 2019, 57, 718–727.

283 Y. Fang, C. Bi, D. Wang, J. Huang, *ACS Energy Lett.* 2017, 2, 782–794.

284 J. Lian, B. Lu, F. Niu, P. Zeng, X. Zhan, *Small Methods* 2018, 2, 1800082.

285 Q. Bao, X. Liu, S. Braun, M. Fahlman, *Adv. Energy Mater.* 2014, 4, 1301272.

286 R. Sharif, A. Khalid, S. W. Ahmad, A. Rehman, H. G. Qutab, H. H. Akhtar, K. Mahmood, S. Afzal, F. Saleem, *Nanoscale Adv.* 2023, 5, 3803–3833.

287 Y. Jiang, J. Wang, H. Zai, D. Ni, J. Wang, P. Xue, N. Li, B. Jia, H. Lu, Y. Zhang, F. Wang, Z. Guo, Z. Bi, H. Xie, Q. Wang, W. Ma, Y. Tu, H. Zhou, X. Zhan, *J. Am. Chem. Soc.* 2022, 144, 5400–5410.

288 Y. Ren, D. Zhang, J. Suo, Y. Cao, F. T. Eickemeyer, N. Vlachopoulos, S. M. Zakeeruddin, A. Hagfeldt, M. Grätzel, *Nature* 2023, 613, 60–65.

289 https://www.longi.com/en/news/2024-snec-silicon-perovskite-tandem-solar-cells-new-world-efficiency/ (last accessed: 2024-11-01).

7

The Bright Side of Terpyridine Chemistry: Light-Emitting Tpy Ligands and Complexes

7.1 Introduction

Owing to their remarkable photophysical properties, complexes of the Ru^{II} *tris*-bpy family have found widespread use in optoelectronic applications (bpy: 2,2′-bipyridine): Emitters in organic light-emitting diodes (OLEDs) or light-emitting electrochemical cells (LEECs), photo-redox catalysis (*e.g.*, water oxidation/reduction reaction, CO_2 reduction), photodynamic therapy (PDT), *etc.* [1]. In particular, the parent complex, $[Ru(bpy)_3]^{2+}$, which was first reported by Lehn and Sauvage in 1977 [2], has been extensively studied regarding its photophysical properties [3, 4]. Fortunately, the ^{3}MLCT and ^{3}MC states possess energy levels, which disfavor the population of the latter by internal conversion (IC, Figure 7.1a). As the most notable consequences, $[Ru(bpy)_3]^{2+}$ features an excited-state lifetime (τ) of *ca.* 1 μs and a high photoluminescence quantum yield (Φ_{PL}) of *ca.* 10% at room temperature in solution. Detrimentally, the use of unsymmetrically substituted bpy ligands results in stereoisomers, denoted as the Δ- and Λ-form, with absence of stereospecificity in the synthesis. Although the asymmetric synthesis of Ru^{II} *tris*-bidentate polypyridyl complexes has been reported [5, 6], this approach is not generally applicable and one has to opt for chromatography with chiral stationary phases [7], capillary electrophoresis [8], or the crystallization-based separation *via* enantiomerically pure anions [9].

As elaborated by Sauvage, Balzani, Barigelletti *et al.*, in their seminal review, the corresponding Ru^{II} *bis*-tridentate complexes are achiral. The complex $[Ru(tpy)_2]^{2+}$ represents the archetypal example (tpy: 2,2′,6′,2″-terpyridine). Such complexes can be prepared free of stereoisomers, even when dealing with unsymmetrically substituted ligands. Moreover, the incorporation of such complexes into linear arrays is enabled if 4′-functionalized tpy ligands are used. The wide range of such structures includes dyads and triads [10] as well as multinuclear assemblies and, eventually, 1D coordination polymers [11]. As one further notable aspect, the tridentate ligands should bring along increased (photo-)chemical stability, due to the so-called chelate effect (see Section 1.3) [12]. Thus, ligand dissociation/substitution is largely suppressed in complexes of the $[Ru(tpy)_2]^{2+}$ family. Despite their structural similarity, the photophysical properties of $[Ru(bpy)_3]^{2+}$ and $[Ru(tpy)_2]^{2+}$ are fundamentally different. The coordination geometry of the latter is strongly distorted (Figure 7.2a)

Terpyridines: Multifaceted Ligands for Diverse Applications, First Edition.
Ulrich S. Schubert, Andreas Winter, and George R. Newkome.

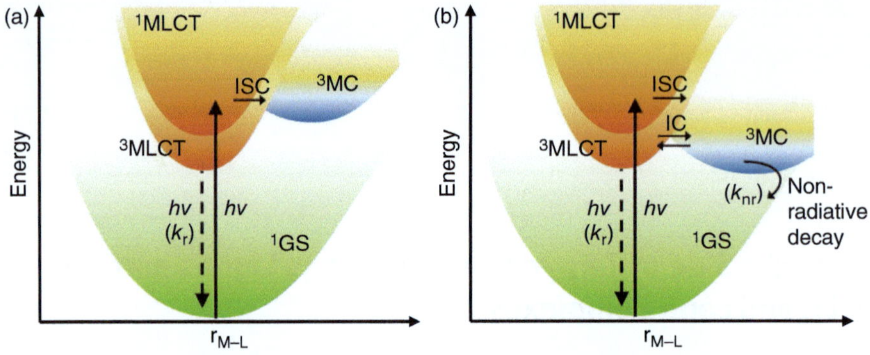

Figure 7.1 Comparison of the electronic-state diagrams of (a) $[Ru(bpy)_3]^{2+}$ and (b) $[Ru(tpy)_2]^{2+}$ showing the photoexcitation and the subsequent processes affecting their emission properties (GS: ground state, MLCT: metal-to-ligand charge transfer, MC: metal-centered, ISC: intersystem crossing, IC: internal conversion, k_r = radiative decay rate, k_{nr} = non-radiative decay rate). In either case, the ^{1}MLCT state is initially populated upon photoexcitation, followed by fast ISC to the ^{3}MLCT state; this state, in turn, radiatively decays to the GS (*i.e.*, the complex is emissive) or undergoes an IC to the ^{3}MC state, which decays to the GS in a non-radiative fashion (*i.e.*, the complex is non-emissive) [1] / with permission of Elsevier.

and, thus, the ligand-field strength is decreased. This strongly affects the positioning of the ^{3}MLCT and ^{3}MC energy levels: The lifetime of the former is decreased since thermal IC to the non-emissive ^{3}MC state is enabled (Figure 7.1b). Thus, $[Ru(tpy)_2]^{2+}$ complexes are hardly emissive at ambient temperature in solution (τ = 0.1–0.2 ns, $\Phi_{PL} \leq 0.0005\%$) [15]. Substantial effort has been made to improve the excited-state properties of such complexes *via* electron-donating/-withdrawing substituents on the tpy ligands, alternative heterocycles, or carbocycles as well as modulation of the chelate's bite angle [16–18]. In this context, the complex $[Ru(dqp)_2]^{2+}$ has to be highlighted [dqp: 2,6-*bis*(quinoline-8-yl)pyridine] [14]. Owing to the geometry-optimized ligand design, the complex features an ideal octahedral geometry (Figure 7.2b) and a remarkably stabilized ^{3}MLCT state, as expressed by the superior lifetime of *ca.* 3 μs. For comparison, the photophysical properties of the three aforementioned complexes are summarized in Table 7.1.

Table 7.1 Selected photophysical properties of archetypal Ru^{II}–polypyridyl complexes.

	λ_{abs} (nm)$^{a)}$	λ_{PL} (nm)$^{b)}$	τ (ns)	Φ_{PL} (%)
$[Ru(bpy)_3]^{2+}$	450	630	1,150.00	0.089
$[Ru(tpy)_2]^{2+}$	476	629	0,000.25	–
$[Ru(dqp)_2]^{2+}$	491	700	3,000.00	0.02

a) Measured in MeCN solution; only the low-energy MLCT band is listed.
b) Measured in deoxygenated EtOH/MeOH solution at room temperature.
Adapted from [17].

(a) (b)

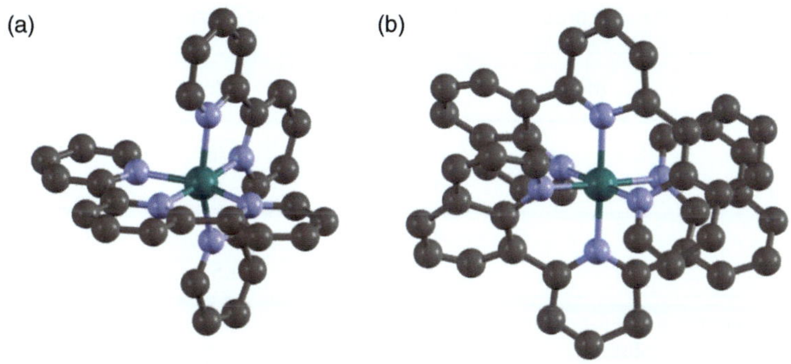

Figure 7.2 Solid-state structures of complexes (a) $[Ru(tpy)_2](PF_6)_2$. Adapted from [13] and (b) $[Ru(dqp)_2](PF_6)_2$. Adapted from [14]. The figures were generated from the original data sets using the Mercury 2022.3.0 software (hydrogen atoms, counterions and solvent molecules were omitted for clarity; CCDC deposition numbers 959787 and 628238, respectively).

However, the long quest to obtain Ru^{II} *bis*-tridentate polypyridyl complexes with long-lived excited states will not be elaborated at this stage. Instead, reference to the comprehensive reviews on this topic, which have been published elsewhere, is given [16, 17, 19]. The lessons learned from Ru^{II} have been adapted to other transition-metal ions, in particular, the earth-abundant first-row representatives (*e.g.*, Cr^{III} [20], V^{III} [21], *etc.*). More examples in this respect were summarized by Wenger *et al.* in their reviews on this topic [22, 23]. Despite their unfavorable photophysics, $[Ru(tpy)_2]^{2+}$ complexes have been tested for practical applications, *e.g.*, as emitters in LEECs (Section 7.4) [24, 25]. Luminescence is not limited to Ru^{II}–tpy complexes, tpy complexes of the 3rd-row transition-metal ions have also attracted interest for optoelectronic applications, specifically OLEDs: Re^{I}, Os^{II}, Ir^{III}, Pt^{II}, Au^{III} (Section 7.3) [26–28]. Moreover, the Zn^{II}–tpy and Bi^{III}–tpy systems are worth mentioning. Unlike for the complexes of heavy transition-metal ions, the origin of their emission is fundamentally different and will be dealt with in Sections 7.5 and 7.7. The same holds true for the complexes with lanthanide(III) ions, in which the "antenna effect" is employed, *i.e.*, the tpy ligand acts as a sensitizer to enable the 4f–4f emission of Ln^{III} ions in the visible to near-infrared (NIR) regime (Section 7.6) [29]. Last but not the least, tpy ligands themselves represent potent emitters if equipped with appropriate π-conjugated chromophores (Section 7.2) [30]. However, owing to the broadness of the topic, only selected aspects concerning light-emitting tpy complexes can be discussed within the scope of this chapter.

7.2 Luminescent π-Conjugated Terpyridines and Hybrid Materials

The photophysical properties of tpy ligands can be modulated *via* their substituents – for synthetic ease, preferentially in positions opposite to the ligand's *N*-atoms. Numerous studies have been reported, where this concept has been

Table 7.2 Selected photophysical properties of tpy in solution and in the solid state.

	λ_{abs} (nm)	λ_{PL} (nm)[a]	τ (ns)	Φ_{PL} (%)
Solution[a]	279	335	n/a	0.003[b]
Amorphous solid	305[c]	348	n/a	<0.01[d]
Needle crystals	305	353	0.8	<0.01[d]
Plate crystals	308	365	4.5	0.2[d]

The standard errors were not listed for clarity.
a) Measured in cyclohexane solution.
b) Determined using 2-aminopyridine, as the standard.
c) Determined from the diffraction reflection spectrum.
d) Determined using anthracene, as the standard.
Adapted from [32].

applied to functionalize tpy ligands. In particular, the so-called Kröhnke-type tpy ligands, *i.e.*, derivatives containing (het)aryl-substituents in their 4′-position, have predominantly been studied [30]. The parent tpy ligand is barely emissive in solution or in the solid state (*i.e.*, as amorphous powder or needle crystals) [31]; nonetheless, when converted into plate crystals, an intense blue emission was observed [32]. Thus, luminescence of tpy depends on the molecular packing in the crystal lattice. Compared to the needle crystals, the emission maximum was bathochromically shifted by 12 nm and the excited state was stabilized as expressed by the significantly prolonged lifetime (*i.e.*, 4.5 ns *vs.* 9.8 ns; Table 7.2). Remarkably, the heat-mode interconversion of the two polymorphs enabled the on–off switching of the luminescence with high reversibility.

However, when aryl-substituents are attached in the ligand's 4′-position, Φ_{PL} is increased dramatically [33]. For example, Williams *et al.* reported Φ_{PL} values up to 60% for their arylated derivatives [34]. In one application, the tpy ligand **1** (Figure 7.3a) was used as a stabilizing agent for Ag nanocrystals (NCs) [35]. The

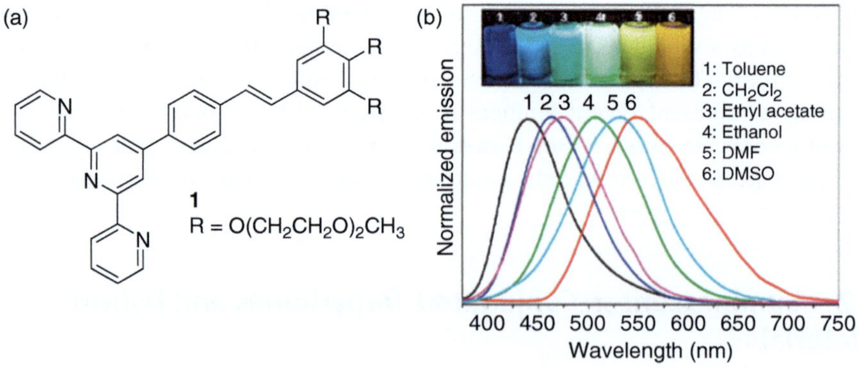

Figure 7.3 (a) Schematic representation of tpy ligand **1**. (b) Solvent-dependency of the emission of **1**-stabilized Ag NC. Reproduced with permission from [35] / The Royal Society of Chemistry.

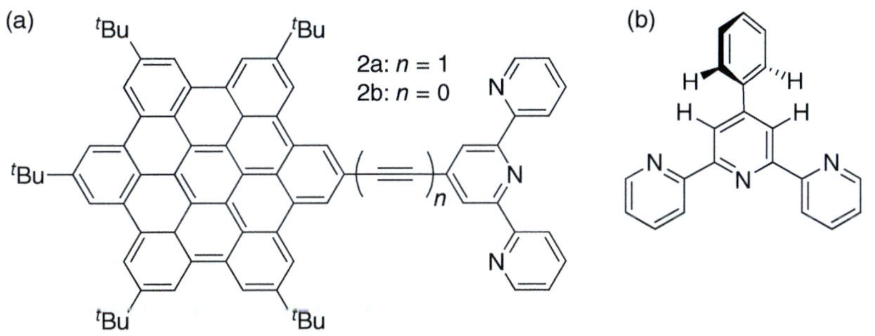

Figure 7.4 (a) Schematic representation of the HBC-equipped tpy ligands **2**. (b) Schematic representation of the non-planarity in Kröhnke-type ligands due to steric repulsion of adjacent hydrogen atoms.

NCs were prepared by the *in situ* reduction of $AgPF_6$ by $NaBH_4$ in the presence of **1**. These NCs, whose diameter was in the range from 8 to 14 nm, had ligand molecules attached to their surfaces *via* Ag-to-N interactions. Thereby, the emission of **1** was significantly enhanced. Moreover, the hybrid materials exhibited a pronounced polarity dependency of the emission, *i.e.*, the emission maximum was bathochromically shifted with increasing solvent polarity (Figure 7.3b). Remarkably, these assemblies were stable in solution for several months.

Murphy and Draper attached the "superaromatic" hexa-*peri*-hexabenzocoronene (HBC) moiety to tpy *via* its 4′-position (**2**, Figure 7.4a) [36]. The UV/vis absorption and emission properties of **2a** and **2b** were essentially a combination of those of their subunits, *i.e.*, the HBC and tpy parts. The excited-state lifetimes were similar for both compounds (*ca.* 20 ns), but the emission intensity of **2a** was lower, presumably due to the more efficient electronic coupling of the subunits *via* the ethynyl bridge. It is well-known that the excited-state planarization in Kröhnke-type terpyridines is hampered due to steric repulsion of adjacent protons [37]. For this reason, the phenyl and central pyridine rings are not coplanar with each other, but twisted by *ca.* 25°; this, in turn, results in a mismatch between the GS and excited-state geometries and, ultimately, lowers the degree of conjugation (Figure 7.4b) [38]. Similar observations were also made by Gao *et al.*, who attached a "superaromatic" corannulene moiety to a tpy fragment, either *via* a C—C single bond or an ethynyl moiety [39].

The aspect of excited-state planarization was analyzed further by Dietzek *et al.*, who selected the π-conjugated tpy **3** as a model substrate (Figure 7.5a) [40, 41]. Detailed photophysical measurements revealed that an emissive molecular geometry was obtained within ps, which decayed on the ns-timescale into a long-lived non-emissive triplet state [40]. Subsequent studies corroborated that the excited-state planarization in solution was dependent on the environment, *i.e.*, the solvent's polarity, viscosity, and temperature [41]. Specifically, the rate for planarization increased with decreasing solvent viscosity, whereas the excited-state lifetime decreased with increasing temperature. It was elaborated that the processes represented barrier-free motions. Ligand **3** as well as its ditopic counterpart were further utilized for the preparation of luminescent complexes and coordination polymers

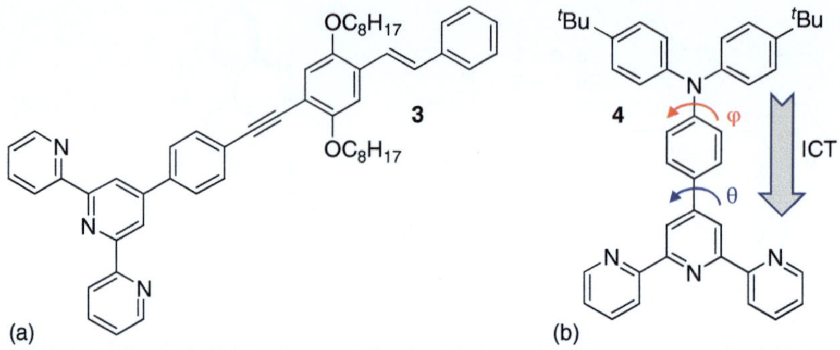

Figure 7.5 Schematic representation of the π-conjugated tpy ligands **3** and **4**. The latter represents a push-pull chromophore capable to under ICT process (θ and φ denote the dihedral angles of the molecule's two torsions).

with Zn^{II} ions (see Section 7.5). Related to this, Maroń, Cannelli *et al.* showed that the photophysics of the triarylamine-substituted tpy ligand **4** (Figure 7.5b) were also highly sensitive to the solvent's properties (*i.e.*, polarity, viscosity, and pH value). Owing to the molecule's push-pull character, intramolecular charge transfer (ICT) processes were enabled. Depending on the environment, Φ_{PL} values in the range from 0.48 to 0.83 were observed, whereas the excited-state lifetime was 1.53–6.56 ns [42]. Temperature- and viscosity-dependent studies by the means of ultrafast spectroscopy indicated the existence of two emitting excited states with different structural conformations, due to the molecule's twofold torsion [43]. In detail, ISC, rapidly populating the triplet manifold, represented the dominant process if polarity and viscosity effects could be neglected. Whereas the time scale of light-induced conformational changes could be tuned over two orders of magnitude (*i.e.*, fast in highly polar solvents and slow in highly viscous solvents.

Zych *et al.* prepared a library of thermally stable, highly luminescent tpy derivatives, in which an electron-rich chromophore was linked to a 4′-phenyl-tpy fragment *via* an ethynyl bridge (**5**, Figure 7.6a) [44, 45]. The presence of the ethynyl bridge had a distinct impact on the compounds' photophysical properties when compared to those of analog tpy derivatives missing the ethynyl moiety: the UV/vis absorption and emission maxima of **5** were shifted to lower energies and their Φ_{PL} values were significantly higher (up to 84%). In the case of derivatives **5c–f**, homogeneous films were prepared by spin-coating them from $CHCl_3$ solutions [44]. These films, which represented molecular glasses, were also highly emissive. This holds particularly true when **5** was incorporated into the binary PVK:PBD matrix (50:50 in wt%), which is known to facilitate the transport of charge carriers in OLED devices [PVK: poly(*N*-vinylcarbazole); PBD: 2-*tert*-butylphenyl-5-biphenyl-1,3,4-oxadiazole]. Furthermore, OLEDs, which contained these host-guest blends as the active layers, were fabricated and the electroluminescence (EL) of these devices was analyzed (see Section 7.3). Overall, the ethynyl linkage is superior to other linkages (*e.g.*, C—C single bond, vinyl or 1*H*-1,2,3-triazol-1,4-diyl, *etc.*) since it enables the most efficient electronic coupling between the tpy site and the attached chromophore [46].

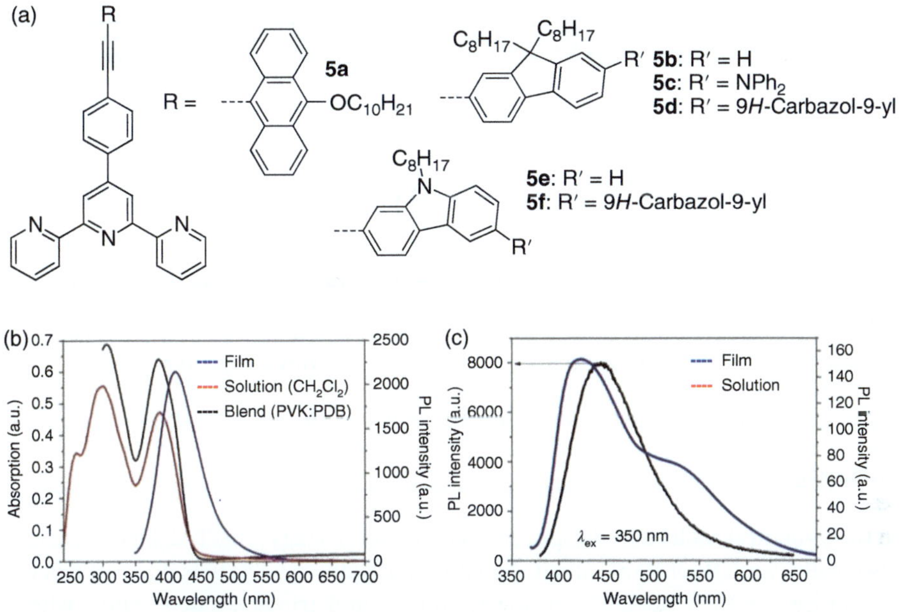

Figure 7.6 (a) Schematic representation of the π-conjugated tpy ligands **5**. (b) UV/vis absorption spectra of **5c** in solution and as a thin-solid film; the PL spectrum of the **5c**/PVK blend is also shown. (c) Comparison of the PL spectra of **5d** in solution and as a thin-solid film [44] / with permission of John Wiley & Sons.

Numerous ditopic ligands, in which the tpy moieties are linked *via* extended π-conjugated systems, have also been studied regarding their photophysical properties. A systematic overview of this type of compound was already provided by Schubert *et al.* [30] and a detailed discussion goes far beyond the scope of this section. Many of these ditopic tpy ligands were employed for the synthesis of coordination polymers with divalent first-row transition-metal ions. In this context, the Zn^{II}-containing polymers were of particular interest regarding optoelectronic applications (see Section 7.5). Whereas the metallopolymers, which contained Fe^{II}, Co^{II}, Ru^{II}, or Os^{II} ions, were used as active materials for electrochromic [47] or, sporadically, energy-storage [48] as well as photovoltaic devices (see Sections 5.4 and 6.3, respectively) [49].

As the last notable example, Sasade, Kido *et al.* reported on a series of tpy derivatives, which exhibited thermally activated delayed fluorescence (TADF) [50]. These TADF dyes represent the 3rd generation OLED emitters. Although the first evidence for TADF of eosin was already reported in 1961 [51], the practical application of the concept for light-emitting devices was recognized by Adachi *et al.* almost 40 years later [52–54]. These materials are the successors of the phosphorescent emitters (2nd generation), which harvest singlet and triplet excitons due to the efficient spin–orbit coupling (SOC) of their heavy transition-metal ions (*e.g.*, Ir^{III}, Pt^{II}, *etc.*). As a result, the internal quantum efficiency (IQE) of 100% can be reached with such molecules (see Section 7.3) resulting in 100% IQE [55]. Detrimentally, such emitters suffer from some considerable drawbacks: strong exciton annihilation, limitations

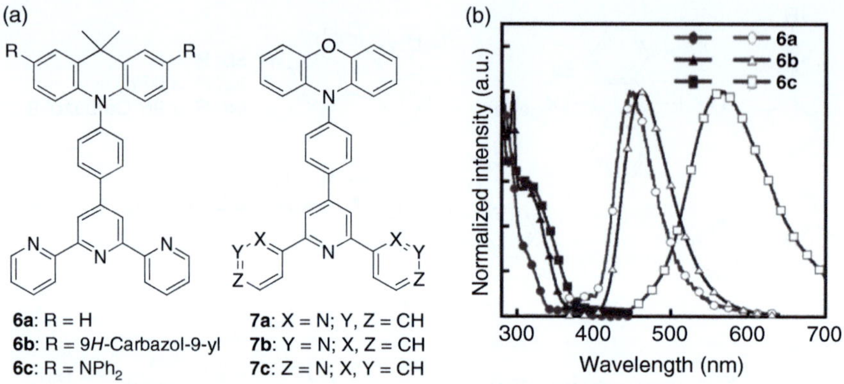

Figure 7.7 (a) Schematic representation of the TADF emitters **6** and **7** [50, 59]. (b) UV/vis absorption and emission spectra of tpy derivatives 6 (10^{-5} M in toluene) [50] / with permission of John Wiley & Sons.

in the structural molecular design, and high material costs [56]. TADF emitters can also realize an IQE of 100%, albeit *via* a different pathway. TADF is due to a reverse ISC process, which involves the lowest singlet and triplet excited states, whose energy gap can thermally be overleaped. Thus, singlet and triplet excitons can be harvested under electrical excitation. Moreover, TADF emitters can be synthesized at lower costs and higher flexibility regarding the molecular design. The recent progress made in the development of potent TADF emitters for OLED applications has been subject of various review articles [57, 58]. Against this background, the tpy derivatives **6** were studied (Figure 7.7a) [50]. All three dyes were thermally stable and displayed intense blue to green emissions in dilute solutions (Figure 7.7b). The emitters were processed into solid films using DPEPO, as the host material [DPEPO: *bis*(2-(diphenylphosphino)phenyl)ether oxide]. These films featured luminescence with $\Phi_{PL} > 46\%$ and delayed-fluorescence lifetimes of 15–320 μs. As detailed in Section 7.3, compounds were further utilized, as the active materials, in OLEDs.

More recently, Adachi, Zhang *et al.* utilized the isomeric tpy derivatives **7**, as TADF emitters (Figure 7.7a) [59]. These authors paid particular attention to one shortcoming of non-doped OLEDs, which is emission quenching due to intermolecular triplet interaction. It was shown that compounds **7** were able to establish 3D frameworks *via* intermolecular hydrogen-bonding (*H*-bonding) interactions. In these robust environments, non-radiative decay processes and triplet-exciton quenching *via* the $\pi\pi$-stacking of triplets was restricted. Moreover, the strictly horizontal dipole orientation was beneficial for the emitters' overall OLED performance. In particular, the TADF emitter **7b** afforded a high maximum external quantum efficiency (EQE) of 23.6% with the low roll-off of 7.2% at 1000 cd m^{-2} (see Section 7.3.1).

7.3 OLEDs

OLEDs represent (often colored) light sources, which are applied in displays (*e.g.*, smartphones, laptops, TV screens, *etc.*) or in lighting panels [60]. Owing to the increasing interest in display and lighting technologies, research in OLEDs has been intensified over the last 30 years – this holds true for academia and industry. In particular, OLEDs are – for a number of reasons – superior to other technologies regarding display applications. Unlike liquid-crystal displays (LCDs), OLED displays emit their own visible light and, thus, a backlight becomes obsolete. They can even display deep-black levels and consume power and produce light only when active, thus they are more energy efficient than LCDs. When operated under low ambient-light conditions (*i.e.*, in crepuscular light or dark rooms), OLED screens are superior to LCDs in terms of contrast ratio (regardless of the LCD backlight technology). As further benefits, the improved resolution, wider viewing angles, high contrast colors, and faster response times have to be named [61]. OLEDs can not only be fabricated thinner and lighter than LCDs but transparent as well as flexible displays have also been realized. The latter is highly useful for roll-up or foldaway devices, such as stand-up displays or smartphones [62].

However, OLED technology has considerable shortcomings which still need to be overcome. These are related to the reduction of materials and fabrication costs on the one hand and device optimization on the other (*i.e.*, efficiency, lifetime, light output, driving voltage, *etc.*) [60, 63]. One particular challenge is related to the lifetime of blue OLEDs, which are relatively short-lived when compared to the red and green ones. This holds particularly true for the phosphorescent emitters of the 2nd generation (*vide infra*) and the recently developed emitters, which display TADF, represent one promising strategy to overcome the rapid degradation of blue emitters and, thus, extend the device lifetime [64].

The working principle of OLEDs is fundamentally different from that of inorganic LEDs, which utilize a *p-n* diode structure [65]. Emission of light occurs when current flows through the semiconductor material. The recombination of electrons and holes in the semiconductor releases energy in the form of photons. Doping is commonly used to create the *p*-type and *n*-type regions in the semiconductor host, thus modulating its conductivity. Basically, OLEDs can be described as organic materials sandwiched between two electrodes, *i.e.*, the anode and cathode; the entire assembly is deposited on a conductive and transparent substrate [66]. Thereby, the organic material features an electric conductivity due to the delocalization of π-electrons. The conductivity level is in between those of insulators and conductors; thus, such materials are commonly referred to as organic semiconductors. Their highest occupied and lowest unoccupied molecular orbitals (HOMO and LUMO, respectively) correspond, in fair approximation, to inorganic semiconductors' valence and conduction bands. In the simplest design, the OLED contains only a single organic layer. For example, Friend *et al.* used such a straightforward configuration in their seminal work in this field; these authors used poly(*p*-phenylene vinylene) (PPV) as the active material [67]. Whereas Tang and VanSlyke opted for a bilayer design, which was prepared by successive vapor deposition of the organic materials [68]. Today's OLEDs are mostly multilayer architectures, which are comprised of various layers of organic semiconductors (Figure 7.8a) [69]. Typically, a transparent and conductive substrate (TCS), such as indium-tin oxide (ITO), serves as the anode and a cathode is comprised of a low work-function metal (*e.g.*, Ca or Ba), coupled to an electron-injection

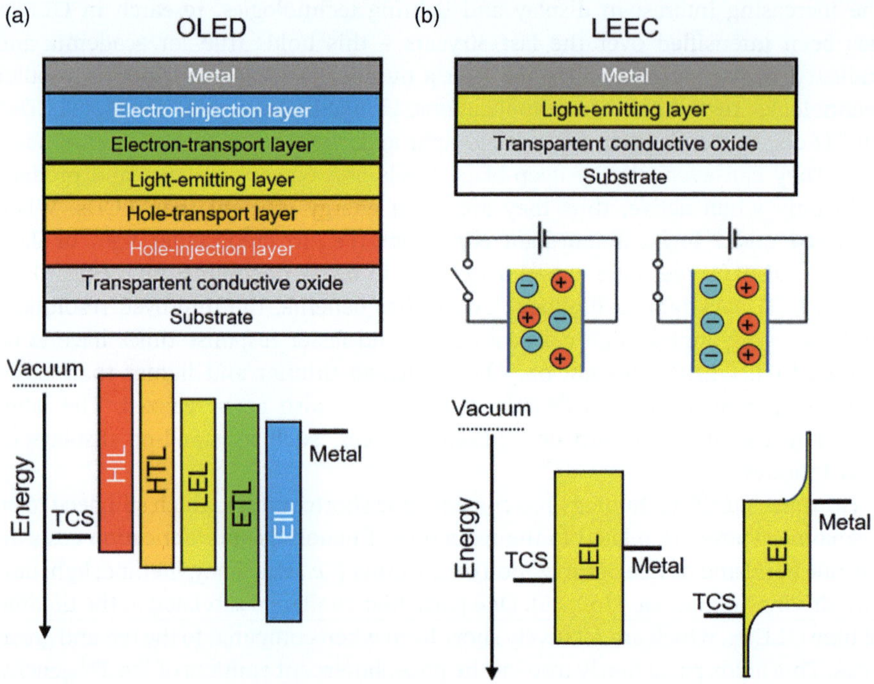

Figure 7.8 (a) Schematic representation of the configurations of a typical (a) OLED and (b) LEEC [25] / with permission of John Wiley & Sons. A multilayer OLED is comprised of several layers of materials, which are processed stepwise from solution or by thermal vacuum evaporation. Electrons are injected from a low work-function metal or a chemically n-doped electron-injection layer; in either case, rigorous encapsulation is required to counteract their instability in air. In contrast, a LEEC consists of only one optoelectronically active layer whose negative counterions are displaced when an external bias is applied. This setup enables an efficient hole and electron injection from air-stable metals.

layer (EIL). Owing to the intrinsic instability of these components against moisture and air, rigorous encapsulation is required. The sophisticated multilayer configuration enables a balanced injection of charge carriers, *i.e.*, electrons and holes, at the electrons upon applying an external voltage and their migration through the respective electron- and hole-transport layers (ETLs and HTLs, respectively) to the light-emitting layer, where they eventually recombine and release energy as photons upon relaxation to the ground state [70]. Moreover, the balanced configuration of layers also accounts for the operational stability of the OLED over hundreds to thousands of hours [71]. If the energy of the emitted photons is in the visible range, one can use them as a lighting source or in displays [69]. Despite the commercial success of the OLED technology, which began with products from Eastman Kodak Co. in 1987 [72], some shortcomings still remain. These are mainly related to the required multilayer design, whose cost-intensive production is hardly compatible with the needs of large-scale applications of the lighting/display market. For a detailed view on the history of OLED technology, reference to a review on this topic by Bräse *et al.* is given at this stage [60].

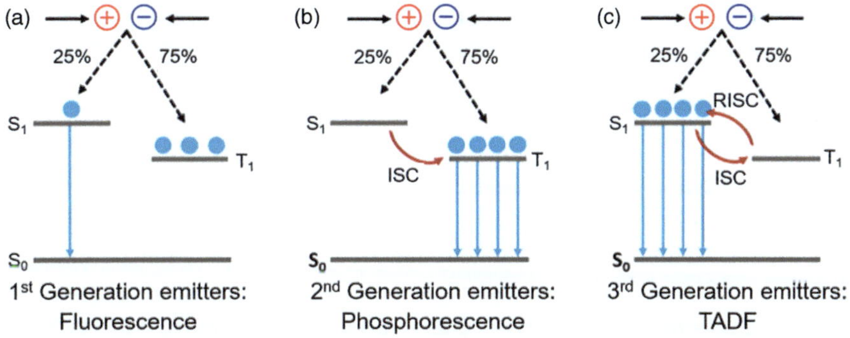

Figure 7.9 Schematic representation of the mechanism for emission in (a) 1st, (b) 2nd, (c) 3rd generation luminophores [74] / with permission of Edinburgh Instruments.

Choudhary *et al.* listed the key requirements for each of the layers [69]. Within the scope of this section, the focus is mainly on the light-emitting layer, which is responsible for some of the most relevant device characteristics: emission color, contrast ratio, external efficiency, *etc.* An unmanageable amount of literature dealing with the OLED concept has been published over the past decades. Thereby, the development of emitter materials represents one core aspect of interest ever since. Following the conventional classification, three classes of emitters can be distinguished when taking the mechanisms leading to light emission into account [73]. Fluorescent organic dyes represent the 1st generation OLED emitters. For example, Alq_3 was used by Tang and VanSlyke in their OLED from 1987 [68]. However, such dyes are intrinsically limited in performance, since emission is only possible from the transition from one singlet state to another (*i.e.*, the $S_1 \rightarrow S_0$ transition), whereas, if states of different multiplicity are populated *via* ISC, fluorescence is spin-forbidden and the decay to the GS will proceed in a non-radiative fashion. For this reason, the IQE of such emitters is only *ca.* 25%, whilst the major share of energy is lost as heat energy or crystal vibration (Figure 7.9a). The maximum EQE of OLEDs, which use 1st generation emitters, is 5% (thereby, effects of optical out-coupling are ignored [75]). Some representative examples for fluorescent emitters of the 1st generation are shown in Figure 7.10.

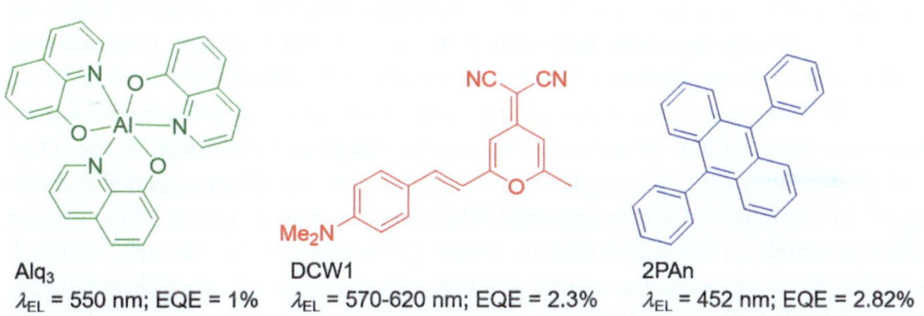

Figure 7.10 Schematic representation of three 1st generation fluorescent emitters.

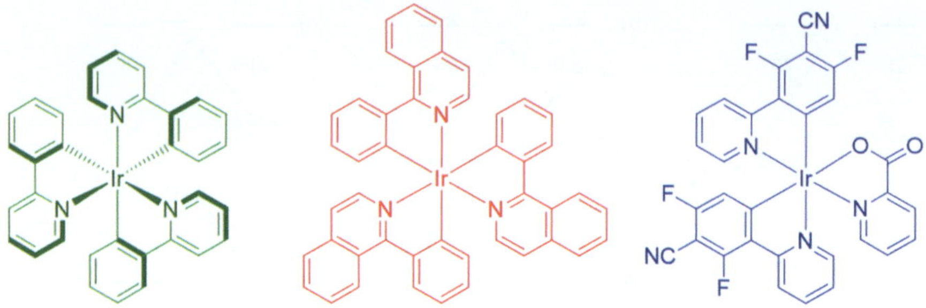

fac-Ir(ppy)$_3$
λ_{EL} = 510 nm; EQE = 8%

Ir(pic)$_3$
λ_{EL} = 620 nm; EQE = 12.3%

FCNIrpic
λ_{EL} = 466 nm; EQE = 22.1%

Figure 7.11 Schematic representation of three representative phosphorescent Ir^{III} complexes.

Secondly, phosphorescent emitters were introduced to overcome the IQE limitation of fluorescent dyes. In particular, complexes of heavy transition-metal ions have gained interest, due to the SOC, which allows for the harvesting of both singlet and triplet excitons. Unlike for fluorescent emitters, ISC enables the $S_1 \rightarrow T_1$ transition (Figure 7.9b). The thusly populated T_1 state decays to the GS in a radiative fashion, *i.e.*, by phosphorescence. This approach allows the 2nd generation emitters to reach an IQE of up to 100%, which is crucial for increasing the EQE values of OLEDs. This concept was introduced by Forrest *et al.* in 2000 [76] and, since then, numerous transition-metal complexes have been studied, as phosphorescent emitters for OLEDs [77]. Among these, cyclometalated Ir^{III} complexes have gained the most notable attention [78]. This is due to the ability to adjust the emission color over a very broad spectral range *via* the cyclometalating and/or ancillary ligands; Ir^{III} complexes showing green, orange-red, or blue emission have been reported (Figure 7.11). The complex *fac*-Ir(ppy)$_3$, which displays green phosphorescence, represents the most prominent derivative from the family of *tris*-cyclometalated complexes (*fac*: facial; ppy: 2-phenylpyridinato) [79]. In a similar fashion, phosphorescent complexes have also been prepared with other third-row transition-metal ions: Re^{I}, Os^{II}, Pt^{II}, and Au^{III} [77].

Since the heavy-metal complexes combined high stability with significantly improved photophysical performance (*i.e.*, increased IQE), the 2nd generation emitters were successfully incorporated into commercial OLED devices [60]. However, these heavy metals also represent some of the least earth-abundant elements and their practical use is always limited by environmental concerns and high material costs [63]. Thus, the focus of interest has been shifted toward emitters, which are efficient in terms of IQE, whilst avoiding the use of high-cost materials. To reach this goal, phosphorescent Cu^{I} complexes and organic dyes, which are capable of showing TADF, have *inter alia* been explored. The latter are considered the 3rd generation emitters, which, in the meantime, have gained interest also in industrial research [60]. For TADF emitters, the energy gap between the S_1 and T_1 states is small (*i.e.*, $\Delta E_{ST} < 0.1$ eV) [69]. If the lifetime of the T_1 excitons is sufficiently long, they can be up-converted to the S_1 state *via* a reverse intersystem crossing (RISC) process;

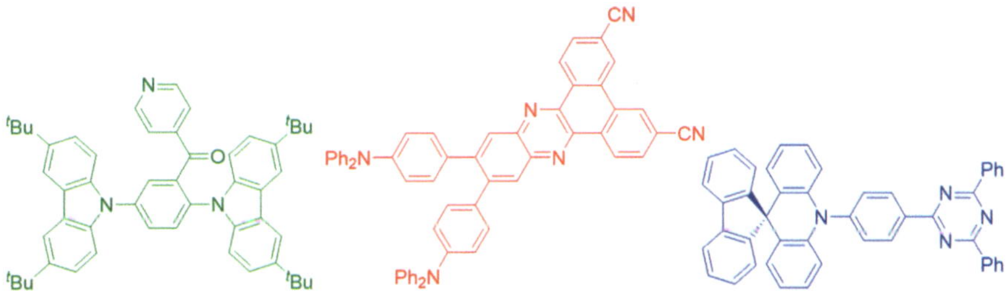

Figure 7.12 Schematic representation of three representative TADF emitters.

this formally spin-forbidden process can readily occur upon thermal activation (Figure 7.9c) [80]. The thusly repopulated S_1 can radiatively relax to the GS by fluorescence; since formally no excitons are lost *via* non-radiative relaxation, the theoretical IQE is 100%, similar to that of a phosphorescent emitter. Reprehensively, some TADF emitters, which provided EQEs of >25%, are depicted in Figure 7.12.

The use of tpy ligands in the context of OLED emitters has been discussed on two counts. Firstly, heavy-metal complexes with tpy ligands might display phosphorescence and, hence, can be classified as 2nd generation emitters. Moreover, tpy ligands in their metal-free state might show fluorescence or TADF and, thus, represent OLED emitters themselves.

7.3.1 Metal-Free Terpyridines, as OLED Emitters

Interest in metal-free tpy ligands for OLED applications arises from the quest to obtain luminophores which display intense blue emission. According to the CIE-1931 color space, which was created by the Commission internationale de l'éclairage (CIE) in 1931 [81], the *xy*-coordinates for deep-blue emission are defined as $y < 0.15$ and $(x + y) < 0.30$ (Figure 7.13a). However, to reach this goal, phosphorescent emitters are of limited applicability, owing to their limited stability [60]. Whereas stable and blue-emitting fluorophores are more accessible [64]. In addition to this, such dyes could also be used as host materials for red-/orange-emitting phosphorescent emitters [84]. Against this background, Zhen, Yang, Zhang *et al.* prepared the donor–acceptor dyes **8**, in which the electron-rich diphenylamine and electron-poor tpy moieties were linked in a linear fashion *via* a biphenyl bridge (Figure 7.13b) [85]. Variation of the position of the *N*-atoms enabled the modulation of the acceptor strength of the tpy subunit. This had a profound influence on the emission color, LUMO energy levels and charge-carrier properties; whereas the triplet energy levels of **8a–c** were almost identical. Multilayer OLEDs using either of the tpy derivatives **8**, as the non-doped emissive layer, afforded blue EL with EQEs of 2.7–4.9% (Table 7.3). Taking the aforementioned theoretical EQE limit of 5% into account, the tpy derivatives **8** represented remarkably efficient blue-fluorescent emitters. Moreover, the compounds served as practical hosts for a red-emitting Ir^{III} complex. The best performance in terms of EQE was obtained when the Ir^{III}

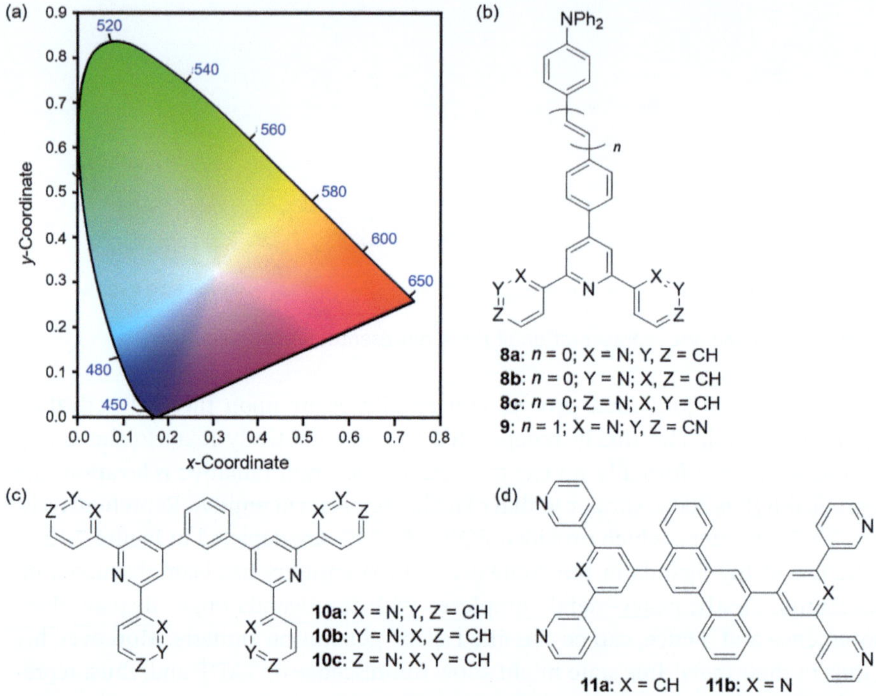

Figure 7.13 (a) Representation of the CIE-1931 color space showing all colors visible to the human eye based on the 2° standard observer (the numbers in blue correspond to the wavelengths of the monochromatic light in the spectral range) [82, 83] / with permission of Wikimedia Commons / Public Domain. (b) Schematic representation of the tpy derivatives **8** and **9**. (c) Schematic representation of the bis-tpy derivatives **10**. (d) Schematic representation of compounds **11**.

phosphor was blended into a matrix of **8b** (EQE of *ca.* 20%). At the optimized dopant concentration of 0.1%, the dual function of the matrix compound **8a** was capitalized to generate a warm white light with the CIE coordinates (0.43; 0.34) at 1000 cd m^{-2}. Thus, a WOLED was fabricated, which contained only one single-doped emissive layer (WOLED: the white OLED).

Wang *et al.* explored the influence of a vinyl moiety within the chromophore by comparing the performance of **8a** and **9**, as fluorescent OLED emitters (Figure 7.13c) [86]. The thermally stable dyes (*i.e.*, decomposition temperatures >330 °C) displayed distinct differences in their photophysical properties: The excited-state lifetime of **9** was significantly shorter and its LUMO level mismatched with that of BCP, as the selected ETL material (BCP: 2,9-dimethyl-4,7-diphenyl-1,10-phenanthroline). Whereas the energy levels of **8a** were more balanced, this, in turn, enabled its use as an OLED emitter. The fabricated multilayer device provided blue EL (λ_{EL} = 460 nm) with maximum CE and PE values of 3.6 cd A^{-1} and 0.55 lm W^{-1}, respectively (Table 7.3). With the optimized device configuration, a brightness of 3,000 cd m^{-2} was reached, thus outcompeting the performance of analogous OLED, which contained the emitter **9**, by a factor of *ca.* 28.

Table 7.3 Summary of the electroluminescent properties of the OLED devices.

Emitter	V_{on} (V)	CE (cd A^{-1})	PE (lm W^{-1})	EQE	CIE (*x*; *y*)	References
Fluorescent emitters (1st generation)						
8aa)	3.0	5.6	4.4	4.9	(0.15; 0.11)	[85]
8ab)	6.0	3.6	0.55	n/a	(0.19; 0.22)	[86]
8ba)	2.6	5.3	5.6	3.8	(0.15; 0.17)	[85]
8ca)	2.7	5.0	5.2	2.7	(0.17; 0.26)	[85]
9b)	6.0	n/a	n/a	n/a	n/a	[86]
TADF fluorescent (3rd generation)						
6ac)	7.76	1.29	0.52	1.08	(0.16; 0.18)	[50]
6bc)	5.94	5.40	2.86	3.02	(0.18; 0.28)	[50]
6cc)	4.83	47.3	30.8	14.8	(0.36; 0.56)	[50]
6cd)	3.40	71.4	65.8	21.9	(0.35; 0.57)	[50]
7aa)		24.7	20.3	8.2	(0.33; 0.56)	[59]
7ae)		43.6	44.2	15.9	(0.24; 0.48)	[59]
7ba)		79.8	83.8	23.6	(0.34; 0.56)	[59]
7be)		46.2	50.0	16.3	(0.27; 0.50)	[59]
7ca)		50.4	52.8	16.4	(0.38; 0.55)	[59]
7ce)		55.3	62.0	18.0	(0.28; 0.54)	[59]

a) Device configuration: ITO|TAPC|TCTA|**X**|TmPyPB|LiF|Al.
b) Device configuration: ITO|NPB| TAPC|**X**|BCP|LiF|Al.
c) Device configuration: ITO|TAPC|**X**@mCP/DPEPO| B3PyPB|LiF|Al.
d) Device configuration: ITO|TAPC|**X**@ TCTA/CBP| B3PyPB|LiF|Al.
e) Device configuration: ITO|TAPC|TCTA|**X**@mCP|TmPyPB|LiF|Al.

Before turning attention to tpy derivatives, as 3rd generation OLED emitters, one further notable application, which is the use of terpyridines, as ETL materials, has to be mentioned. Interest in this topic arouse from the prominent role of pyridine derivatives used in this context [87, 88]. Compared to other *N*-heteroaromatic systems, pyridine is favored due to its superior electron-accepting nature and the ability to modulate the electron-transporting ability *via* molecular modifications. For example, Sasabe, Kito *et al.* varied the position of the peripheral *N*-atoms in compounds **10** (Figure 7.13c) [89]. Depending on the substitution pattern, either intramolecular (**10a**) or intermolecular *H*-bonding (**10b** and **10c**) was observed in the solid state. In particular, **10b** showed the highest degree of alignment and the most pronounced horizontal molecular orientation, which is beneficial for both the electron injection at the cathode and the electron mobility. As a proof of concept, these authors used **10b** as the ETL in a multilayer OLED, which contained *fac*-Ir(ppy)$_3$, as the phosphorescent emitter. This OLED provided PE and EQE values of 104 lm W^{-1} and 23.5%, respectively, at a luminance of 100 cd m^{-2}. For

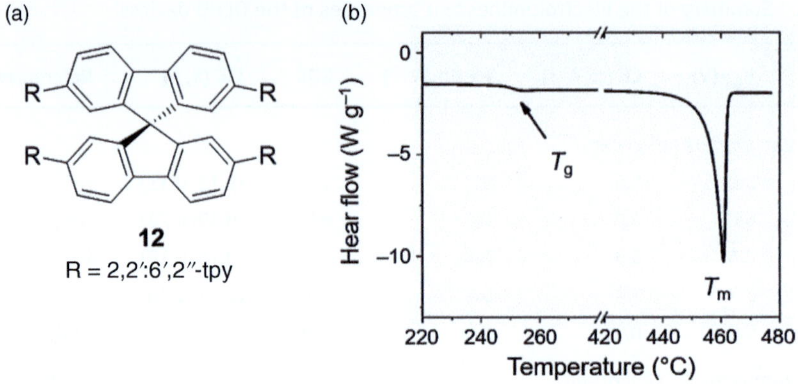

Figure 7.14 (a) Schematic representation of the spiro-compound **12**. (b) DSC diagram of **12** [91] / with permission of Elsevier.

comparison, when using **10a** as the ETL material, an inferior OLED performance was observed, presumably due to the molecules' lower degree of ordering in the ETL (*i.e.*, PE = 77.8 lm W^{-1}; EQE = 21.2). In subsequent work, these same authors utilized the chrysene-containing compounds **10** as ETL materials for green-emitting OLEDs (Figure 7.13d) [90]. In particular, a highly stable phosphorescent OLED was obtained with **10b**. This device afforded a low turn-on voltage of 2.4 V at 1 cd m^{-2} and an EQE of 17.5% at 1,000 cd m^{-2}. Moreover, the operational lifetime was remarkably long: The LT_{50} of 19,000 h at 1,000 cd m^{-2} was derived from log-term measurements at the current density of 25 mA cm^{-2} (LT_{50}: 50% of the initial luminance). The LT_{50} of this OLED was *ca.* 1.5 times longer when using **10a** as the ETL material. The difference in performance was attributed to the lower electron affinity of **10b**, which facilitated the electron uptake from the electrode; the improved horizontal orientation of molecules due to *H*-bonding increased the carrier mobility and, thus, enhanced the device stability. Although intermolecular *H*-bonding was disabled, tetra*kis*-tpy **12** (Figure 7.14a) represented a notable ETL material [91]. The thermal properties of **12** were analyzed by differential scanning calorimetry (DSC; Figure 7.14b) and thermogravimetric analysis (TGA). The very high glass-transition temperature (T_g) of *ca.* 250 °C suggested a good morphological stability of the material's amorphous phase. This behavior – along with its electron-transport properties – accounted for the long LT_{50} times of OLEDs, which contained **12**, as the ETL material. For example, the OLED with $Ir(mppy)_3$, as the phosphorescent emitter, had a low turn-on voltage of 2.23 V and a maximum PE of 97.8 lm W^{-1}; the LT_{50} was 101 h at an initial luminance of 10,000 cd m^{-2}.

But let us come back to the use of tpy derivatives, as OLED emitters – specifically in the context of TADF. As mentioned beforehand, the tpy derivatives **6** featured blue to green emissions in dilute toluene solutions (Figure 7.7) [50]. Multilayer OLEDs were fabricated, in which **6** was combined mCP/DPEPO mixture, as the host material [mCP: *N,N*-dicarbazoyl-3,5-benzene; DPEPO: *bis*(2-(diphenylphosphino)phenyl)ether oxide]. Remarkably, the blue emitters **6a** and **6b** provided EQE values >9% at the low brightness of 1 cd m^{-2}, thus even

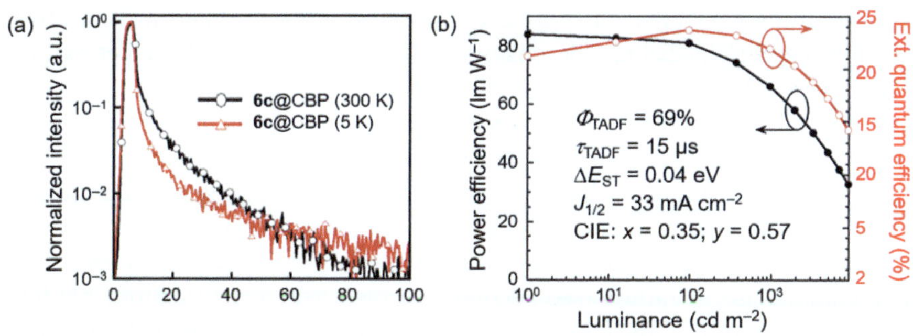

Figure 7.15 (a) Transient PL decay curves of a **6c**@CBP film at two different temperatures. (b) PE-L-EQE characteristics of a **6c**-containing OLED (host material: TCTA/CBP) [50] / with permission of John Wiley & Sons.

exceeding the theoretical EQE limit of fluorescent dyes. Owing to the small ΔE_{ST} value of **6c**, intense and long-lived green TADF was observed, when this dye was blended into the host matrix ($\Phi = 69\%$, $\tau_{TADF} = 15\,\mu s$; Figure 7.15a). The OLED, which contained this blend, as the emissive layer, afforded a maximum EQE of 20.3% at the brightness of $1\,cd\,m^{-2}$. The performance was improved further by using a double-emission layer structure, which was comprised of **6c**/TCTA and **6c**/CBP blends (the hosts possessed more appropriate energy levels) [TCTA: *tris*(4-9*H*-carbazol-9-ylphenyl)amine; CBP: 4,4′-*bis*(9*H*-carbazol-9-yl)biphenyl]. Under these optimized conditions, the OLED gave high PE and EQE values at the luminance of $1\,cd\,m^{-2}$ (*i.e.*, $80\,lm\,W^{-1}$ and 23.7%, respectively). The efficiency roll-off was very low, as expressed by the EQE of 21% at the high brightness of $1{,}000\,cd\,m^{-2}$. This work represented, to the best of our knowledge, the first study on tpy-based TADF emitters.

In one further study, the phenoxazine-equipped terpyridine derivatives **7** were tested as TADF emitters (Figure 7.7) [59]. As pointed out before, intermolecular *H*-bonding established a 3D framework which prevented triplet excitons from the non-radiative quenching *via* $\pi\pi$-stacking and, at the same time, enabled a horizontal molecular orientation. This held particularly true for derivative **7b**, which represented an efficient TADF emitter – even in non-doped OLEDs. In the absence of a host material, a maximum EQE of 23.6% at a low luminance; the EQE roll-off was as low as 7.2% when going up to $1{,}000\,cd\,m^{-2}$. Remarkably, the performance of the OLED with the non-doped emissive layer was better than that of the corresponding mCP-doped OLED [mCP: 1,3-di(9*H*-carbazol-9-yl)benzene; Table 7.3]. This unusual trend was rationalized by the advantageous features of the 3D framework, which could only be established in the non-doped environment. The positive effect of the doped layer could not compensate for the disturbance of the supramolecular structure.

7.3.2 Ir^{III}–Tpy Complexes: Phosphorescent OLED Emitters?

Phosphorescent Ir^{III} complexes represent one of the most prominent classes within the OLED emitters from the 2nd generation [78]. Aside their application in light-emitting devices, such complexes have also attracted interest from other fields of research: biolabeling, biosensing and bioimaging [92, 93], singlet oxygen ($^1O_2$) sensitizers [94], PDT [95], photoredox catalysis [96], *etc.* The outstanding role of Ir^{III} phosphors is due to their (photo)chemical stability, the readily accessible luminescent excited states, the high Φ_{PL} values and, last but not least, the ability for emission-color tuning *via* ligand-structure design [78, 97]. Regarding the use in OLEDs, the field is dominated by charge-neutral *tris*-bidentate complexes and the full range of emission colors has been realized. Cationic Ir^{III} complexes find a predominant use in LEECs (see Section 7.4) [98]. Whereas comparatively few papers dealing with Ir^{III}–tpy complexes have been published so far (161 by March 2025, according to a SciFinder^n survey). Taking the binding mode of the tpy ligand into account, one has to distinguish between complexes in which the ligand is coordinated in a hypodentate (*i.e.*, mono- or bidentate) or the "conventional" tridentate fashion (see Section 1.3).

In the case of bidentate coordination, the tpy ligand acts as a bpy one, without using its third pyridine ring. Typically, such Ir^{III} complexes carry two *mono*-cyclometaling, bidentate ligands; in this respect, the well-known complex $[Ir(ppy)_2(bpy)]^+$ represents the chemical lead. Complexes of this kind have predominantly been employed as photocatalysts for the hydrogen-evolution reaction (HER), but also luminescent devices using the complexes, as emissive materials, have been fabricated. Bernhardt *et al.* emphasized that the use of cationic Ir^{III} complexes in this field is advantageous regarding the simplicity of the device configuration [99]. Unlike for the neutral Ir^{III} phosphors, which are commonly incorporated into an organic matrix within a multilayer architecture comprising of various charge-transport/-blocking materials (Figure 7.8a), their cationic counterparts allow for a simpler design. Thc devices are comprised of a single-layer of the neat complex, which is sandwiched between two air-stable electrodes (*e.g.*, ITO and Au; Figure 7.8b). This is due to the presence of mobile ions, which redistribute once an electric field is applied, and facilitate the injection of charges at the electrodes. Moreover, the charge-transport mechanism involves charge-carrier hopping; thus, device operation from an alternating current (AC) power source and at low turn-on voltages (<3 V) is enabled. These devices, which fundamentally differ from OLEDs, are commonly referred to as LEECs. We will come back to this topic in Section 7.4, since various complexes from the $[Ir(ppy)_2(tpy)]^+$ family have also been utilized in this context.

Dumur *et al.* compared the applicability of copolymers **12**, as the emitters in OLEDs and LEECs (Figure 7.16a) [100]. These material represent one example for the use of polymers, bearing luminescent complexes in the pending side chains, in the context of OLED applications [49]. The random copolymers **12**, which contained a polystyrene (PS) backbone, were prepared by nitroxide-mediated polymerization (NMP) and characterized by size-exclusion chromatography (SEC). These measurements suggested the presence of an average of three (**12a**) and ten (**12b**) Ir^{III} complexes per polymer chain. The polymers could be processed from solution and revealed good film-forming properties. However, when combined with PVK, as the host material, phase segregation occurred due to the incompatibility of polymers.

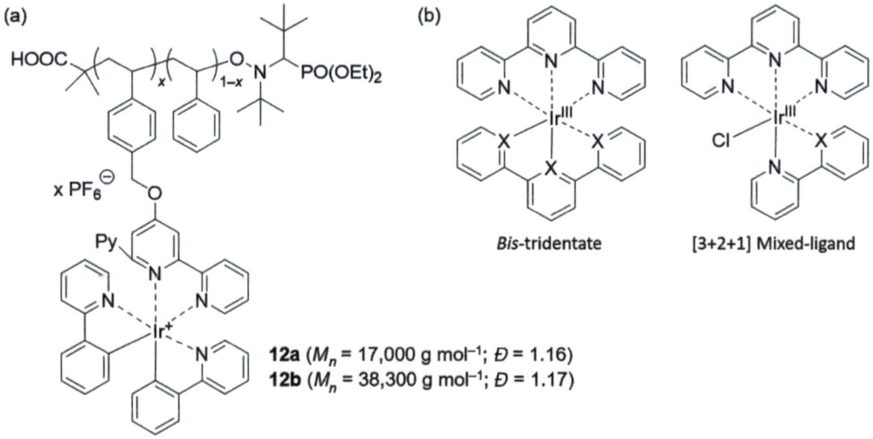

12a (M_n = 17,000 g mol^{-1}; $Đ$ = 1.16)
12b (M_n = 38,300 g mol^{-1}; $Đ$ = 1.17)

Figure 7.16 (a) Schematic representation of copolymers **12**. (b) Schematic representation of the two different coordination modes, which are conceivable with tridentate ligands (X = C or N).

Moreover, the insulating nature of the PS backbone and the absence of charge carriers hampered the injection of holes. Thus, the multilayer OLED provided only a very low brightness and poor EQE. Whereas the multilayer LEEC, in which **12** was spin-coated on-top of a PVK layer, gave a significant light-emission performance. In a similar fashion, further polymers bearing $[Ir(ppy)_2(tpy)]^+$ complexes were tested as emitters for LEECs (see Section 7.4).

Concerning the tridentate coordination of tpy, two different complex architectures are conceivable – *bis*-tridentate and mixed-ligand ones, in which a tpy, a bidentate (*e.g.*, ppy$^-$ or bpy) and a monodentate ligand (*e.g.*, Cl^-) are present (Figure 7.16b). However, the latter are rare and have mainly been used for photocatalytic solar-fuel generation or CO_2 reduction [101].

When dealing with *bis*-tridentate Ir^{III} complexes one has to be aware of the implications on the photophysical properties that this architecture brings along. This issue was analyzed thoroughly by Williams *et al.* already in 2008 [28]. These authors considered the *bis*-tridentate Ir^{III} complexes of tpy, as a N^N^N-type ligand, as well as those of their *N*-deficient, cyclometalating counterparts 2,6-di(pyridin-2-yl)benzene (dpb) and 2,6-diphenylpyridine (dpp); the latter represent N^C^N- and C^N^C-type ligands, respectively. According to these authors, the degree of cyclometallation played a fundamental role regarding the luminescence properties of these isostructural complexes. As summarized in Figure 7.17, the luminescence of $[IrN_6]^{3+}$ complexes is mainly of ligand-centered (LC) character; whereas, MLCT states dominate the emission of *tris*-cyclometalated $[IrN_3C_3]$ species. In between, the emissive states of *mono*-cyclometallated $[IrN_5C]^{2+}$ complexes are largely of ligand-to-ligand charge transfer (LLCT) character. Regarding the excited states of the *bis*-cyclometallated complexes, one has to take the position of the cyclometalating sites into account (*i.e.*, whether they are in the same or different ligands). In either scenario – the excited state is of mixed LLCT/MLCT character.

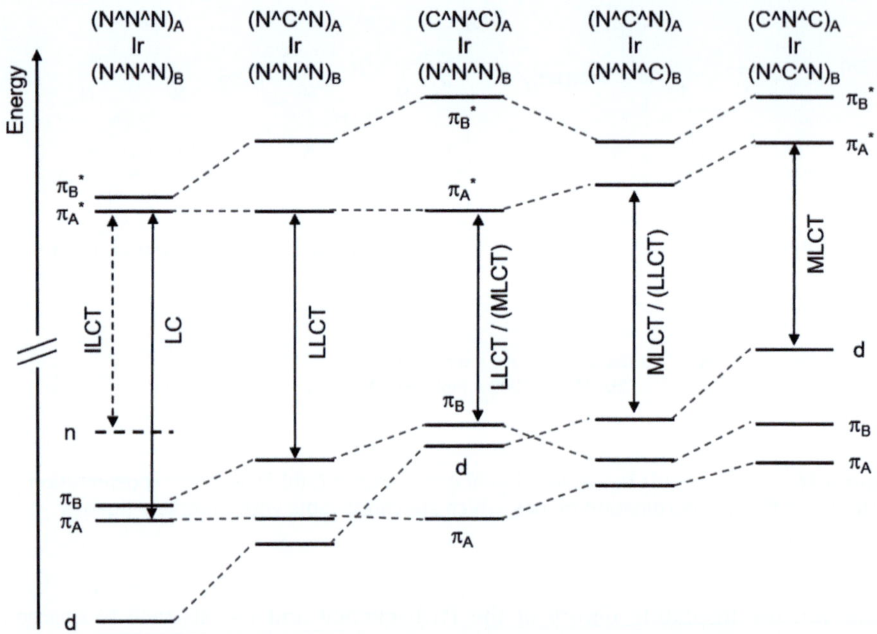

Figure 7.17 Simplified representation of the energy-level diagram showing the influence of cyclometallation on the frontier orbitals of the respective *bis*-tridentate Ir^{III} complexes, which results in excited states of different character (the dashed "n" orbital represents an electron-rich n-/π-orbital, which is localized on a remote substituent/group) [28] / with permission of Royal Society of Chemistry.

This trend can be rationalized by taking a closer look at the effect of cyclometallation on the metal's and ligands' relevant orbitals (Figure 7.17). Firstly, in the case of $[Ir(tpy)_2]^{3+}$ complexes, the metal's d-orbitals are very low in energy and the HOMO is ligand-based; thus, emission will arise from a LC π–π* state. However, if strongly electron-donating substituents are introduced in the ligand's 4′-position, the HOMO might be relocated away from the tpy moiety into the substituent; in this case, intraligand charge transfer (ILCT) states will come into play (*i.e.*, n → π* or π → π*). If one replaces one tpy ligand by a *mono*-cyclometalating dpb one, the d-orbitals are destabilized and, at the same time, the π-/π*-orbitals of the relevant ligand are raised. Since the orbitals of the tpy ligand are hardly affected by this, the localization of the HOMO and LUMO might be changed (*i.e.*, HOMO on the dpb and LUMO on tpy). This, in turn, entails the population of emissive LLCT states with low contributions from MC ones. Such an unfavorable scenario is typically characterized by low radiation constants (k_r) and very weak emission [102]. What happens if one introduces one further Ir—C bond? If the second Ir—C bond of complexes $[IrN_4C_2]^+$ is installed in the same ligand as the first one, the shift in energy levels of the d- and π_B-/π_B*-orbitals is basically repeated [28]. The energy level of the d-orbital is approximately the same as that of the π_B-orbital and the emission will be of a MLCT/LLCT character. The considerable contributions from the metal's d-orbitals are beneficial for the emission lifetime and quantum

Table 7.4 Summary of the photophysical properties of archetypal *bis*-tridentate Ir^{III} complexes.[a)]

Complex	λ_{PL} (×10²)	Φ_{PL} (×10²)	τ (ns)	k_r (×10⁴ s⁻¹)	References
$[Ir(tpy)]^{3+}$	458, 491, 523	3.0	1,200	2.5	[103]
$[Ir(tpy)(dpx)]^{2+}$[b)]	502	$<10^{-2}$	n/a	n/a	[104]
$[Ir(dpx)(pbpy)]^{+}$[b)]	632	2.3	120	19	[105]
[Ir(dpx)(dpp)][b)]	585	321	3,800	5.5	[104]

a) Measurements were performed in MeCN solution at room temperature; cationic complexes were used as their PF_6^- salts.
b) dpx: 3,5-dimethyl-2,6-di(pyridine-2-yl)benzene; pbpy: 6-phenyl-2,2-bipyridine.
Adapted from [102].

yield. In *bis*-cyclometalated complexes, where each ligand forms one Ir—C bond, the metal's d-orbitals are similarly affected, but the destabilization effect is now distributed over both ligands' orbitals and no longer concentrated on a single one. This brings along strong contributions from the metal's orbitals, thus increasing the MLCT character of the emission. Such complexes feature k_r values, which are one order-of-magnitude higher as for the former also *bis*-cyclometalated complexes [102]. Finally, in the case of *tris*-cyclometallated $[IrN_3C_3]$ complexes, the d-orbitals are raised further in energy, thus effectuating a mere MLCT emissive state with long excited-state lifetime and high Φ_{PL}. For clarity, the photophysical properties of some archetypal Ir^{III} complexes are summarized in Table 7.4. This short discussion gives a general overview on the photophysical characteristics of the various *bis*-tridentate complexes. However, one has to be aware, that the picture becomes more complicated when taking specific ligand effects into account. Nonetheless, the presented idealized view reflects the trends observed for numerous Ir^{III} complexes.

Owing to these shortcomings and the sometimes-intricate synthesis, *bis*-tridentate Ir^{III} complexes have not been utilized for OLED applications, at least not when carrying tpy ligands. Whereas complexes with dpb-type ligands have gained considerably more attention – this holds even more true for complexes with Pt^{II} ions [102].

7.3.3 Phosphorescent OLED Emitters Based on Other Third-Row Transition-Metal Ions

7.3.3.1 Pt^{II}–Tpy Complexes

Similar to Ir^{III} complexes, Pt^{II} complexes have also attracted enormous interest regarding their use as phosphorescent emitters for OLED applications [106]. As mentioned beforehand, the heavy-atom effect enables a strong SOC, which facilitates both the ISC and the formally spin-forbidden radiative decay of triplet states. Notably, the SOC constant of Pt is actually the second largest of all. The excited-state properties of Pt^{II} complexes can be tuned over a wide range and the emission can even be shifted to the NIR spectral regime. Indeed, Pt^{II} complexes represent one of the most efficient NIR OLED emitters known so far, with one representative being the record holder for the longest wavelength NIR OLED emission with reasonable EQE [106]. The Pt^{II} complexes parallel their Ir^{III} counterparts regarding the ability to modulate the emission characteristics *via* the ligand structure and the degree of mixing between metal's d-orbitals and the ligands' molecular orbitals. However, the most striking difference to the d^6-configured Ir^{III} complexes, which possess an octahedral geometry, is the square-planar structure, which results from the metal's d^8-electron configuration. The vacant axial positions enable the self-assembly *via* metallophilic Pt···Pt and/or ligand-to-ligand interactions (*e.g.*, ππ-stacking). This behavior can be utilized, for example, to assemble superstructures, whose luminescence is responsive to external stimuli (*e.g.*, solvent polarity, pH value, temperature, presence of salts/additives, *etc.*) [107].

As summarized by Williams *et al.* elsewhere, cyclometalated Pt^{II} complexes are the most interesting candidates for use as triplet OLED emitters [108]. Unlike pyridyl ligands, which represent π-acceptor ligands, the cyclometalating phenyl moieties are very strong σ-donors and strong-field ligands. Their coordination has a strong influence on the excited-state properties due to the destabilization of the metal's d-orbitals. The acetylide ligands, which are also widely used in the field of Pt^{II} coordination chemistry have basically the same influence. Numerous examples with bidentate, tridentate, and even tetradentate ligands have already been reported.

Regarding tridentate ligands, various *N*-deficient tpy analogs have been employed: 6-phenyl-2,2′-bipyridine (pbpy), dpb, and dpp [108, 109]. However, the discussion of these potential OLED emitters goes far beyond the scope of this section. EQE values >20% have been reported for these OLEDs, which displayed green, red, or orange emission; moreover, the unique excimer emission, which resulted from metallophilic interactions, allowed for the fabrication of efficient single-dopant WOLEDs [106]. A comprehensive review focusing on the properties and applications of Pt^{II}–tpy complexes was provided by Newkome *et al.* in 2008 [26]. These authors presented the biomedical and photocatalytic applications of these complexes (see Chapters 4 and 8, respectively). However, owing to the overall inferior luminescence characteristics, Pt^{II}–tpy complexes have not yet been used as OLED emitters.

7.3.3.2 Au^{III}–Tpy Complexes

Similar to their isoelectronic Pt^{II} counterparts, Au^{III}–tpy complexes have a substantial medicinal relevance [26, 110]. However, Au^{III} complexes are prone to reduction to afford Au^{I} and/or Au^{0} species, if a suitable oxidant is present. This behavior paved the way for applications in catalysis, specifically the oxidative coupling to give C—C bonds [111]. Although stabilization of the +III state can be achieved by chelating polypyridyl ligands, such as tpy, luminescence is almost absent in such complexes. Owing to the electrophilicity of the Au^{III} center, the vacant d_{x2-y2} orbital is low-lying; thus, the energy of the ligand-to-metal charge transfer (LMCT) excited states is close to that of the emissive intraligand (IL) ones. The former are thermally accessible and, therefore, the presumptive luminescence is quenched [112]. As for the related Pt^{II} complexes, strong-field ligands have been employed to improve the emission properties (*e.g.*, cyclometalating, acetylide, or NHC ligands).

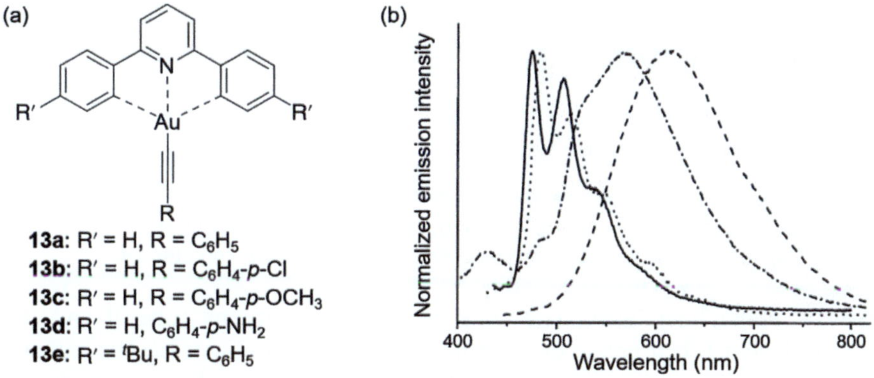

13a: R′ = H, R = C_6H_5
13b: R′ = H, R = C_6H_4-*p*-Cl
13c: R′ = H, R = C_6H_4-*p*-OCH_3
13d: R′ = H, C_6H_4-*p*-NH_2
13e: R′ = ^{t}Bu, R = C_6H_5

Figure 7.18 (a) Schematic representation of the charge-neutral Au^{III} complexes **13**. (b) Normalized room-temperature emission spectra of **13a** (—), **13d** (- - -), and **13e** (••••) in degassed CH_2Cl_2 solution as well as of **13a** as thin-solid film (•–•–) [115] / with permission of John Wiley & Sons.

The aim of this approach is essentially the increase of the metal's d–d* energy gap [113]. In particular, tridentate cyclometalating ligands have widely been used in this context [112–114]. For example, Yam *et al.* reported the emissive Au^{III} complexes **13**, which contained both the *bis*-cyclometalating dpp as well as a strongly σ-donating acetylide ligand (Figure 7.18a) [115]. The archetypal complex **13a** revealed an emission maximum at 476 nm in CH_2Cl_2 solution; whereas, the thin-solid film exhibited orange emission (λ_{PL} = 568 nm; Figure 7.18b). The redshift of the emission maximum was rationalized by ππ-stacking and/or aurophilic interactions. Although complexes **13** were weak emitters at room temperature in solution (Φ_{PL} < 0.01), the acetylide ligand had a strong contribution to their emission, since the related [Au(dpp)Cl] complex was only emissive in the solid state and at low temperature.

In the following, the basic structure of **13a** has been modified to provide complexes, whose Φ_{PL} exceeded 75%, even in the solid state [116]. Push-pull complexes, which contained electron-rich diarylamine moieties on the acetylide ligand, were used as phosphorescent OLED emitters. The devices, which contained these well-designed emitters, afforded high EQEs of up to 11%, thus being competitive to OLEDs based on complexes from the $Ir(ppy)_3$ family [117, 118]. Despite this notable progress in the OLED performance of Au^{III} emitters, their future is presumably not so bright: Au also represents a sparsely abundant element in Earth's crust and, thus, can hardly be regarded as an alternative to the beforementioned Pt and Ir.

7.3.3.3 Re^I–Tpy Complexes

Charge-neutral Re^I–tricarbonyl complexes represent one further class of phosphorescent complexes, which have – owing to their remarkable photophysical properties – been discussed *inter alia* in the context of dye-sensitized solar cells (DSSCs) [119], photocatalytic CO_2 reduction [120], supramolecular constructs [121] and OLEDs [122]. Among others, complexes of the general formula *fac*-[Re(N^N)$(CO)_3$L] (Figure 7.19a) have been studied to a significant extent – both experimentally as well as theoretically. In these complexes, N^N represents a

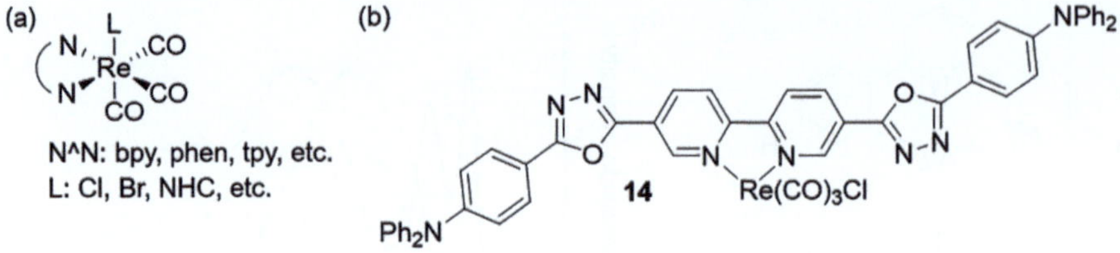

Figure 7.19 (a) Schematic representation of the general formula of neutral [Re(N^N)(CO)$_3$L] complexes. (b) Schematic representation of the Re^I complex **14**.

bidentate polypyridyl ligand [*e.g.*, bpy or 1,0-phenanthroline (phen)] and the auxiliary ligand L is a halide or a strong σ-donor [*e.g.*, a *N*-heterocyclic carbene (NHC)] [123]. Wrighton *et al.* were the first to report on the phosphorescence of *fac*-[Re(phen)(CO$_3$)Cl] and related complexes; these authors already pointed to the long excited-state lifetimes and large Stokes shifts [124, 125]. The particular interest in using such complexes, as emitters, for OLED applications arouse from Chan's observation of EL from the multifunctional complex **14**, which offered both emission and charge-carrier capability (Figure 7.19b) [126, 127]. However, the OLED performance suffered from the rough morphology of the **14**/polycarbonate emissive layer (L_{max} = 730 cd m^{-2}, EQE = 0.1%). In the following various other Re^I complexes, including the parent *fac*-[Re(bpy)(CO)$_3$Cl] complexes, were tested, as OLED emitters [122].

These studies revealed that the neutral Re^I complexes were not able to compete with the Ir^{III}- or Pt^{II}-based emitters in terms Φ_{PL}, which is at least one order-of-magnitude lower, and EQE, which hardly exceeds 10%. Moreover, one is limited in the range of available emission colors, *i.e.*, a genuine/deep blue emission cannot be realized [122]. On the other hand, the Re^I–tricarbonyl complexes also offer some notable benefits: ease-of-synthesis (*i.e.*, mild conditions, straightforward purification, and high yields), relatively short excited-state lifetimes of <1 μs and very good luminescent properties.

The excited-state properties of *fac*-[Re(N^N)(CO)$_3$L] have thoroughly been analyzed [128]. In general, emission arises from the lowest triplet excited states, which are populated by rapid vibrational relaxation and ISC from higher vibrational energy levels. The nature of the involved transitions (*i.e.*, MLCT, LLCT, and IL) and, thus, the photophysical/-chemical properties strongly depend on the nature of the ligands. For example, the relative energy of the π*-orbitals of the N^N ligand can be varied by electron-donating or -withdrawing substituents; whereas the axial ligand exerts influence *via* its ligand-field strength. Due to the strong SOC of the Re^I ion, most of these complexes are luminescent. However, this does not hold true for the complex *fac*-[Re(tpy)(CO)$_3$Cl], in which the tpy ligand is coordinated in the $\kappa^2 N^1,N^2$ mode (**15a**, Figure 7.19) [129]. The very weak luminescence in solution at room temperature was attributed to a competition between emission from the thermally coupled MLCT and LC states. Whereas owing to the increased energy gap between these two states in the Re^I complexes with bpy or phen ligands, this deactivating competition is diminished [130]. Moreover, the triplet-triplet annihilation (TTA) represents a general issue in *fac*-[Re(N^N)(CO)$_3$L] complexes, which limits its

OLED efficiency at high current density. A significant effort has been made to counteract this and, thereby, two strategies have mainly been followed. Firstly, bidentate ligands with different ligand-field strengths have been developed in order to manipulate the orbitals' energy levels in the complex [122]. Alternatively, TTA can be suppressed by steric hindrance; for this purpose, electron-donating or -withdrawing substituents have been introduced on the chelating ligand. Against this background, Zheng, Accorsi *et al.* reported on a series of *fac*-[Re(tpy)$(CO)_3$Cl] complexes (**15b–f**, Figure 7.19), in which the electron-donating carbazole or triarylamine moieties were introduced in the ligands 4′-position [131]. Although the luminescence in solution at room temperature remained low (*i.e.*, $\Phi_{PL} < 1.5\%$), the emission spectra corroborated the anticipated decrease of the triplet-energy level E_{T1} due to the presence of the electron-donating substituents. The emission of these complexes ($\lambda_{PL} \sim 600$ nm) was of $d_{Re}\pi \rightarrow \pi_{tpy}{}^*$ MLCT character and red-shifted compared to that of **15a**. In the solid state, the luminescence intensity increased and appeared blue-shifted ($\lambda_{PL} \sim 550$ nm) relative to those observed in solution. A subsequent theoretical evaluation by the means of density-functional theory (DFT) and time-dependent DFT (TD-DFT) suggested the potential applicability of these complexes, as phosphorescent OLED emitters [132].

In the following, various other 4′-modified tpy ligands were utilized for the synthesis of luminescent Re^I complexes [133–143]. However, only a handful of these have been tested, as emitters, for OLED applications. For example, OLEDs were fabricated, which contained complex **15g**, as the emitter – either in the neat form or within a blend with the PVK:PBD host [137]. These devices displayed red or yellowish-orange emission, respectively. A similar behavior was also observed for the related complex **15h** [135]. In a further study, Machura, Schab-Balcerzak *et al.* reported the use of complexes **15i–k**, as phosphorescent OLED emitters (Figure 7.20) [136]. These complexes possessed suitable HOMO/LUMO levels, which matched with those of PVK, as the host material (Figure 7.21a). Efficient energy transfer from the polymer host to the complex accounted for the observed emission. The OLEDs, which contained a doped **15i**@PVK emissive layer, displayed red or orange light under the applied voltage (Figure 7.21b,c).

In 1993, Orrell *et al.* observed the transformation of complex *fac*-[Re(tpy)$(CO)_3$Br] into *mer,cis*-[Re(tpy)$(CO)_2$Br]; thereby, the tpy changed its coordination from the bidentate $\kappa^2 N^1,N^2$ mode to the tridentate $\kappa^3 N$ one (Figure 7.22) [144]. This strategy was taken up and optimized further by Hightower *et al.* [145, 146] as well as Hanan *et al.* [147]. It was *inter alia* shown in this context, that the auxiliary ligand

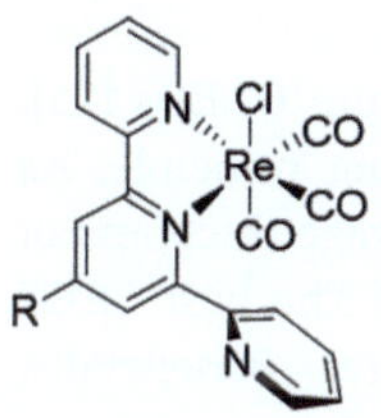

15a: R = H
15b: R = 9*H*-Carbazol-9-yl
15c: R = NPh_2
15d: R = N(4-^{t}Bu-Ph)$_2$
15e: R = NPh(naphthalen-1-yl)
15f: R = NPh(naphthalen-2-yl)
15g: R = C_6H_4-(pyrrolidin-1-yl)
15h: R = C_6H_4-(9*H*-carbazol-9-yl)
15i: R = 1-Methylpyrrol-2-yl
15j: R = 2,2′-Bitiophen-5-yl
15k: R = 3,4-Ethylenedioxythiophen-1-yl

Figure 7.20 Schematic representation of the Re^I complexes **15**.

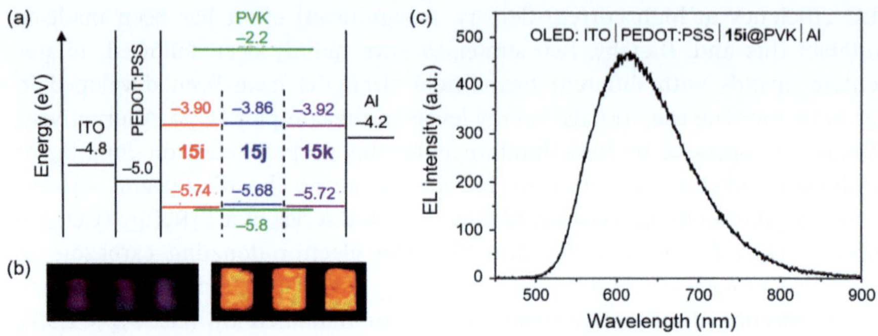

Figure 7.21 (a) Energy-level diagram of the materials incorporated in the OLED device. (b) Photographs of the OLEDs under applied voltages [the emissive layer contained neat PVK (left) or a **15i**@PVK blend (right)]. (c) EL spectrum of the OLED, which contained a **15i**@PVK blend, as the emissive layer. Reproduced with permission from [136] / The Royal Society of Chemistry.

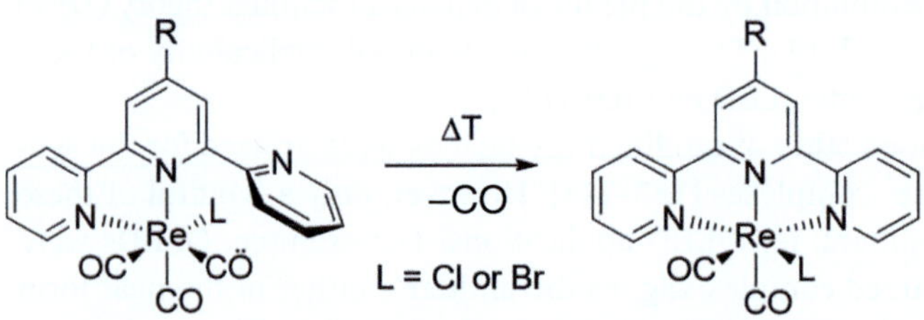

Figure 7.22 Schematic representation of the formation of neutral Re^{I} κ^3N-complexes.

L could be replaced by *N*-heteroaromatics or tertiary phosphines to afford cationic Re^{I} complexes. The change in coordination mode had a profound influence on the complexes' photophysical and electrochemical properties, due to the destabilization of the HOMO energy level [147]. As a result, the absorption profiles were broadened over the full vis spectral regime (*i.e.*, low-energy maxima at *ca.* 700 nm, tailing out to 800 nm). The complexes with bromide, as the axial ligand, were non-emissive; whereas the cationic complexes, which contained neutral pyridine or triphenylphosphine ligands, represented panchromatic absorbers and NIR emitters with emission maxima in the range from 840 to 950 nm and 780 to 800 nm, respectively.

7.4 Light-Emitting Electrochemical Cells

As mentioned beforehand, LEECs are fundamentally different from OLEDs [25]. This holds true for both the architecture as well as the working principle. As shown in Figure 7.8b, the light-emitting layer, which has an average thickness of 100–200 nm, is typically sandwiched in between two electrodes. The best LEEC performances have been reached with such simple architectures. Beneficially, the active layer can be prepared from solution (*i.e.*, by spin-coating or printing technologies). LEEC devices are insensitive to the electrodes' work functions and, thus, air-stable metals can be used as the anode and cathode. Moreover, air-sensitive charge-injection layers are also not required. Therefore, elaborate device encapsulation becomes obsolete and the fabrication of large-area LEECs is enabled, even at affordable costs.

The first functional LEEC was presented in 1995 by Pei *et al.* [148, 149]. In this seminal work, the active layer was comprised of a mixture of an inorganic salt (*i.e.*, lithium triflate) and a two-component polymer blend. The latter was formed from MEH-PPV, a π-conjugated luminescent polymer, and PEG, as an ion-conductive polymer [MEH-PPV: poly[5-(2′-ethylhexyloxy)-2-methoxy-1,4-phenylene viny lene); PEG: poly(ethylene glycol)]. Shortly after this seminal demonstration of a polymer-based LEC (PLEEC), the alternative concept, which is a LEEC based on ionic transition-metal complexes (iTMC-LEECs) was also reported. Murray, Wightman *et al.* employed a metallopolymer film, as the single light-emitting material [150]. The ionic polymer film was prepared by the *in situ* electropolymerization of $[Ru(vbpy)_3](PF_6)_2$. (vbpy: 4-vinyl-4′-methyl-2,2′ridine). In this archetypal iTMC-LEEC, the cationic Ru^{II} complexes and their PF_6^- anions are of importance for a number of reasons [25]. These species promote the injection of charges at the electrodes as well as their transport through the device; moreover, the *tris*-bidentate Ru^{II} complexes are responsible for the orange luminescence. In today's iTMC-LEECs, cationic complexes from the $[Ir(ppy)_2(bpy)]^+$ family are preferentially used [98, 151]. Since their discovery, numerous studies have been devoted to the improvement of materials, device architectures, driving schemes, *etc.* As summarized by Bolink, Armaroli *et al.*, notable achievements regarding emission color as well as device efficiency, turn-on time, and stability have been made in last *ca.* 30 years [25].

Before turning attention to the use of tpy complexes in the field of (mainly) iTMC-LEECs, the working principle of LEECs is briefly addressed. Without going too much into detail, two alternative rationales have been provided to explain the operational mechanism of LEECs – the electrodynamical and electrochemical doping (ED and ECD, respectively) [25]. These models share the idea that the charge-injection barriers at the electrodes are reduced by the migration of ions in the active layer as soon as a bias is applied. In the ED model (Figure 7.23a), the formation of an electric double layer at the electrodes is assumed, due to the accumulation of ions. This results in a sharp decrease of the electric potential near the electrodes' interfaces and facilitates the respective injection of charges (*i.e.*, electrons and holes). Emission of light originates from the so-called field-free region in the active layer's bulk, where the charge carriers are paired.

The second model is also based on the accumulation of ions at the electrodes [25]. However, the formation of highly conductive *p*- and *n*-doped regions is considered here (Figure 7.23b). The addition of electrons is referred to negative (*n*-type) doping, whereas the generation of holes corresponds to the positive (*p*-type) doping. Over time, the doped regions widen, until eventually an intrinsic (*i*, undoped) junction is obtained in between. Across this *i*-region, the applied potential decreases sharply; this, in turn, promotes charge recombination and the emission of light.

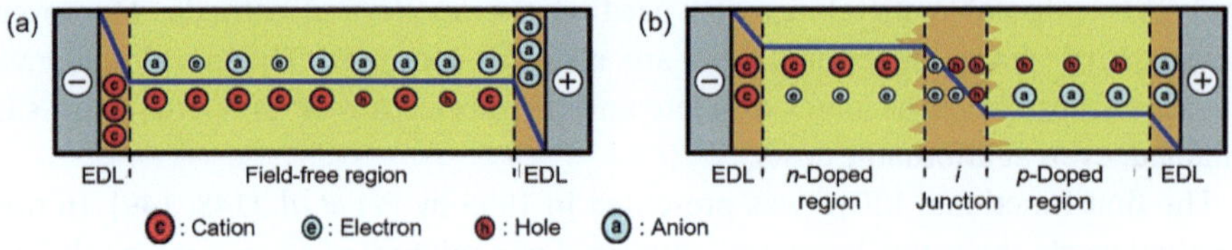

Figure 7.23 Schematic representation of the potential profile as well as the electronic/ionic charge distribution in a LEC during steady-state operation, according to (a) the ED and (b) ECD model [25] / with permission of John Wiley & Sons. The blue line sketches the respective potential profile; whereas the electronic/ionic charge carriers are illustrated by cyan (negatively charged) and red (positively charged) symbols, respectively. The high- and low-field regions in the bulk are highlighted in orange and yellow, respectively. In the low-field regions, negative and positive charges are mutually compensated.

7.4.1 Terpyridine Complexes in LEECs

The presumably first mention of a tpy complex in the context of LEECs was provided by Bolink *et al.*, who used the heteroleptic Ru^{II} complex **16a**, as the light-emitting material (Figure 7.24a) [152]. The LEEC had the configuration ITO|**16**/PMMA|Au and afforded light of deep-red color (CIE coordinates: $x = 0.717$, $y = 0.282$) when operated at low voltage [PMMA: poly(methyl methacrylate)]. PMMA, which is actually an insulator, is commonly used in iTMC-LEECs to improve the film quality and increases the distance between complexes to suppress self-quenching; thereby, the device's EQE is increased and lifetime prolonged. [153] However, the obtained current efficiency (CE) was very low (*ca.* 0.01 cd A^{-1}) when compared to that of an analogous LEEC, which contained $[Ru(bpy)_3](PF_6)_2$/PMMA, as the light-emitting layer (in this case, the CE was *ca.* 100 times higher) [152]. On the other hand, the latter device was significantly less stable than the LEEC, which contained complex **16a** (*i.e.*, the LT_{50} of <1 min *vs. ca.* 30 min). In subsequent work, Bolink *et al.* also tested the heteroleptic complexes **17**, which contained an 1,3,5-triazine-based ligand (Figure 7.24b) [154]. Such complexes possess superior excited-state properties (*i.e.*, increased Φ_{PL} and τ) than complexes from the $[Ru(tpy)_2]^{2+}$ family, due to the increased degree of planarity (see Section 1.3) [155]. For example, the excited-state lifetime of **17a** at room temperature in solution is *ca.* 50 times longer as that of parent

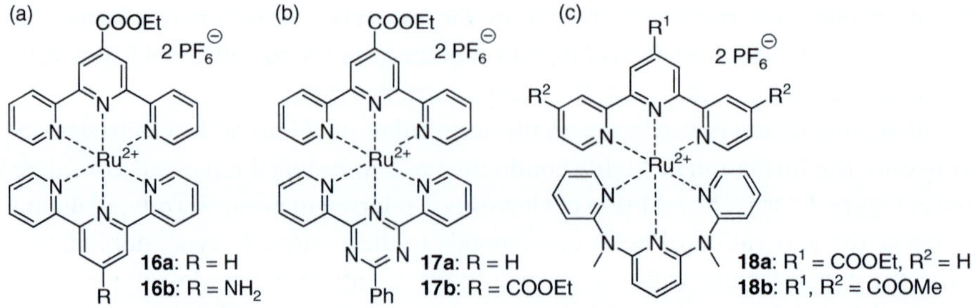

Figure 7.24 Schematic representation of the Ru^{II} complexes **16–18**, as emitters for LECs.

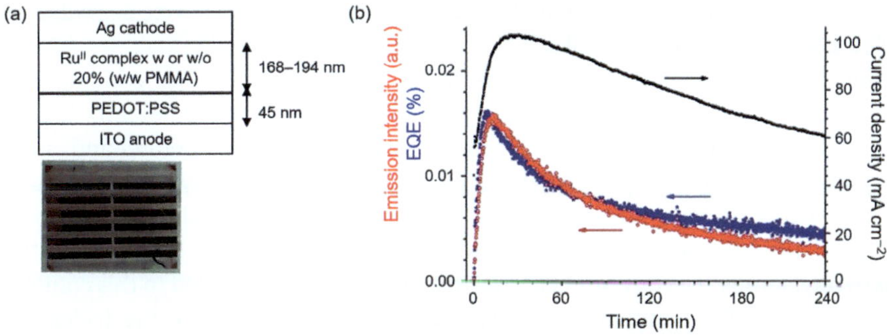

Figure 7.25 (a) Schematic representation of the LEEC configuration, which was used by Heinze *et al.*; a photograph of such a LEC is also shown. (b) Performance of the LLEC with **18a** (without PMMA) at the applied voltage of 3 V: Evolution of the current density (black), emission intensity (red), and EQE (blue) over time. Reproduced with permission from [24] / John Wiley & Sons.

$[Ru(tpy)_2](PF_6)_2$ [154]. The LEECs afforded a deep-red light emission with slightly improved EQE of 0.005%. Owing to the decreased mobility of the PF_6^- counterions in the **17b**/PMMA film, the turn-on time of this LEEC was longer; this LEEC was also more stable, as expressed by the longer LT_{50} of 360 h (*vs.* 120 h for the device with **17a**/PMMA). Since both the luminance and current density showed a similar dependence *vs.* time, loss of charge injection/transport rather than dye degradation was assumed to be the reason for the decline of performance over time.

Heinze *et al.* followed two further strategies to increase Φ_{PL} and τ of Ru^{II} *bis*-tridentate complexes [24]. Firstly, a push-pull-type substitution, as in complex **16b** (Figure 7.24) is suitable to increase the energy difference between the ^{3}MLCT and ^{3}MC states by lowering the former *via* the ester moiety, whilst simultaneously raising the latter due to the presence of the electron-donating NH_2-group. For complex **16b**, a Φ_{PL} of 0.18% and an excited-state lifetime of 34 ns were reported. LEECs with a two-layer configuration were fabricated, in which either the neat complex or a blend with 20 wt% PMMA was used, as the light-emitting layer (Figure 7.25a). In the case of **16b**, luminescence was only observed in the presence of PMMA; the very low EQE of 0.001% was obtained at the turn-on voltage of 4 V.

In the second approach, Heinze *et al.* aimed for the improvement of the excited-state characteristics by optimizing the octahedral complex geometry [24]. For this purpose, the ddpd ligand, which offered a larger bite angle, was employed to prepare complexes **18** [ddpd: *N*,*N*′-dimethyl-*N*,*N*′-di(pyridin-2-yl)pyridine-2,6-diamine; Figure 7.24] Due to their improved luminescence lifetimes and quantum yields, (*e.g.*, $\Phi_{PL} = 0.45$ and 1.1% for **18a** and **18b**, respectively), these complexes were tested as red-to-NIR emitters in LEECs. In the case of **18b**, a remarkably low-energy emission at 755 nm was observed (CIE coordinates: $x = 0.731$, $y = 0.269$). Whereas the highest luminence (3.36 cd m^{-2}) and EQE were reached with **18a** in the presence of PMMA and at the applied voltage of 6 V. Thus, dilution of the complexes with PMMA had a positive effect on the EQE but required higher applied voltages to overcome the polymer's insulating nature. In either case, the maximum EQE was rapidly reached, but the high applied voltages accounted for the device degradation within a few hours (Figure 7.25b).

Unlike Ru^{II} complexes, which often suffer from their low Φ_{PL}, cationic Ir^{III} complexes are better suited for LEEC applications due to their inherently higher Φ_{PL} values and, thus, theoretically higher EQEs. As mentioned beforehand, such complexes have intensively been studied [98, 151]. Most commonly, three bidentate ligands – two *mono*-cyclometalating and a neutral *N*-heteroaromatic one – are coordinated to the Ir^{III} center. Besides the widespread use of bidentate ligands from the polypyridyl family (*i.e.*, bpy or phen), tpy ligands have also been used in this context. Similar to the aforementioned Re^{I} complexes (see Section 7.3.3), the tridentate ligand is coordinated in hypodentate $\kappa^2 N^1,N^2$ mode. The lifetime of iTMC-LEECs can be increased, *e.g.*, by steric shielding of the metal centers. Bolink, Constable *et al.* paid particular attention to this aspect. These authors introduced complexes $[Ir(ppy)_2(pbpy)]^+$ in which the pending phenyl moiety established intramolecular $\pi\pi$-stacking interactions with one of the cyclometalating ligands in a face-to-face fashion [156]. The resulting cage-like structure retained its conformation in the GS as well as in the excited state and was highly stable against reactions with water. The LEEC, which contained the parent complex $[Ir(ppy)_2(pbpy)](PF_6)$ in combination with a small amount of the ionic liquid $(bmim)(PF_6)$ afforded, under optimized conditions, an average luminance of 200 cd m^{-2} at the driving voltage of 3 V; the LT_{50} of this device was >3,000 h (bmim: 1-butyl-3-methylimidazolium). The stabilization of iTMC-LEEC emitters by intramolecular $\pi\pi$-stacking has become one of standard design features [25]. Against this background, Housecroft, Bolink *et al.* also prepared the complexes **19** and **20**, which also exhibited the rigid cage-like conformation (Figure 7.26) [157]. LEECs were fabricated, whose light-emitting layers were comprised of a mixture of the respective complex with $(bmim)(PF_6)$. These devices featured rapid turn-on times but suffered from the low EQE of <0.5%. The highest luminance of up to 60 cd m^{-2} was reached with the LEECs containing one of the complexes **19b–d**.

Notably, attempts have also been made to incorporate $[Ir(ppy)_2(tpy)]^+$ complexes into polymers in order to improve the quality of the light-emitting layer and to "dilute" the complexes. As mentioned earlier, polymer **12**, which contained

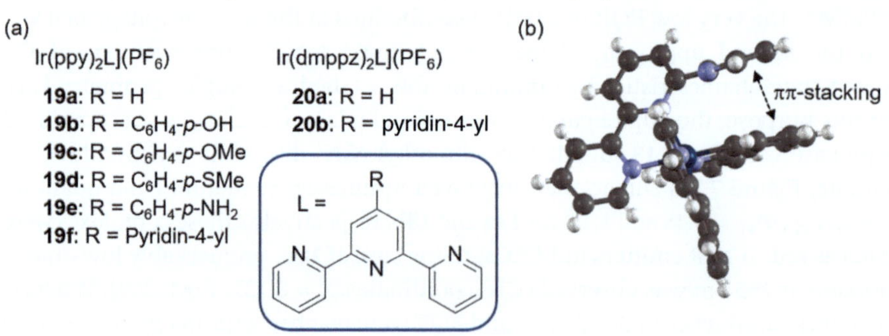

Figure 7.26 (a) Schematic representation of the cationic Ir^{III} complexes **19** and **20**. (b) Solid-state structure of $[Ir(ppy)(tpy)](PF_6)$ (counterions and solvent molecules were omitted for clarity). Adapted from [157].

such complexes in the pending side chains, were tested as emitters in OLEDs and LEECs (Figure 7.16a) [100]. Only the LEEC, which contained polymer **12b**, provided a current of 2.5 mA m^{-2} and a brightness of *ca.* 70 cd m^{-2} under the high-voltage conditions of 24 V. The poor LEEC performance was attributed to the absence of mobile ions (despite the PF_6^- anions) and the unlikeliness of charge hopping between remote Ir^{III} complexes. Since EL was absent in the case of **12a**, it was assumed that the critical content of Ir^{III} centers per polymer chain was *ca.* 5 – charge hopping could not occur between the too remote complex sites at low "concentration." A similar strategy was followed by Schubert *et al.* who attached complexes from the $[Ir(ppy)_2(tpy)]^+$ family to the termini of preformed polymers, *i.e.*, PS or PEG [158–160]. The polymers were processed into thin-solid films by solution-based methods, *i.e.*, spin-coating or inkjet printing. In particular, the latter technique allowed for the material-efficient fabrication of films with excellent control over thickness and morphology. Tew *et al.* opted for a different approach, in which tpy moieties were introduced in the side chains of a diblock copolymer, which was comprised of a PS and a polyacrylate segment (the latter carrying the tpy moieties) [161]. Subsequently, the polymer was treated with $Ir(tpy)Cl_3$ in order to graft the Ir^{III} complexes onto the polymer; however, under the unfavorably harsh conditions, only a degree of grafting of *ca.* 35% was reached. Functional LEECs containing either of the aforementioned polymers were not reported.

Finally, one further approach to prepare metallopolymers for LEEC applications was reported by Chan *et al.* [162]. These authors prepared the main-chain polymers **21** by the polycondensation of the homoleptic amino-functionalized $[Ru(tpy)_2]^{2+}$ complex and aromatic *bis*-anhydrides (Figure 7.27). The polymers exhibited charge-carrier mobilities in the range of 10^{-4} cm^2 V^{-1} s^{-1}, it was thus assumed that the electron-poor *bis*-imide moieties played a substantial role in the

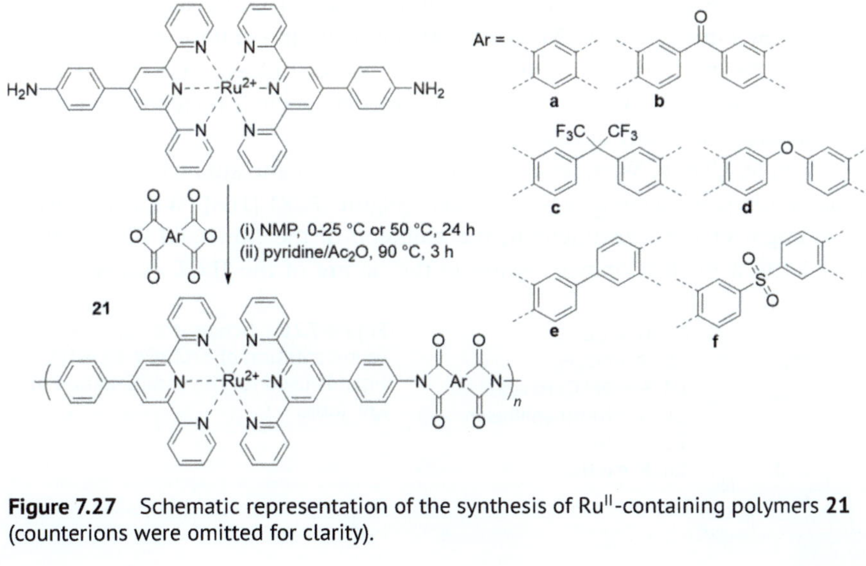

Figure 7.27 Schematic representation of the synthesis of Ru^{II}-containing polymers **21** (counterions were omitted for clarity).

charge-transport process. When used as the light-emitting material in single-layer LEECs, an EQE and luminance of up to 0.1% and 120 cd m^{-2} were observed.

Although a plethora of polymers, which contain tpy complexes within the main chain, have been reported [11, 30, 49], further studies in the context of LEEC applications have not been provided. However, a wide range of Zn^{II}–tpy coordination polymers with remarkable photophysical properties has been presented; this topic will be illuminated in the subsequent Section 7.5.

7.5 Light-Emitting Zn^{II}–Tpy Complexes and Coordination Polymers

Against the background of using earth-abundant, cheap and non-harmful metal complexes for optoelectronic applications, Zn^{II} complexes have gained significant importance. However, Zn represents a first-row transition metal and, thus, Zn^{II} complexes are seldom phosphorescent but fluorescent (see Section 7.3). Note that phosphorescence will only occur for elements with an atomic number of >40 (as a rule of thumb) [163]. As an early exception from this, Wang's complex $[Zn(pmbp)_2](ClO_4)_2$ is worth highlighting; this complex possessed an excited-state lifetime of 6 μs at room temperature in the solid state or at 77 K in MeOH solution (pmbp: 4,4′-diphenyl-6,6′-dimethyl-2,2′-bipyrimidine) [164]. Although the emitters, which were discussed in Sections 7.3 and 7.4, are superior in terms of Φ_{PL} and τ, their high costs will always hamper large-scale applications. This brings Zn^{II} complexes back into the game, since the common precursor salts are *ca.* 1,400 times cheaper than the Ir^{III} precursors [163]. Moreover, Zn^{II} complexes can be prepared under comparatively mild conditions and in high yields; whereas the synthesis of *bis*- or *tris*-cyclometalated Ir^{III} complexes requires overall more forcing reaction conditions. Regarding OLED applications, the first Zn^{II} complexes were tested already in 2000 [165]. In particular, Znq_2-containing OLED featured a highly intense yellow emission of 16,200 cd m^{-2} at the operating voltage of 20 V [Znq_2: zinc(II) *bis*(8-hydroxyquinoline)]. Although Zn^{II}–tpy complexes were not included in Dumur's extensive review on this topic, various such complexes have been employed in light-emitting devices.

For example, Cheng, Wang *et al.* performed a systematic study, in which various complexes $[ZnL_2](PF_6)_2$ were screened (Figure 7.28) [166]. With increasing donor-strength of the substituent R, the emission of the thin-solid films shifted from violet to a reddish orange, owing to the nature of the ILCT excited states.

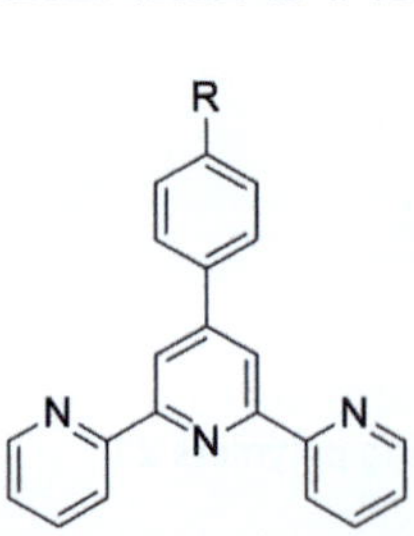

L1: R = ^{t}Bu
L2: R = OC_6H_{12}
L3: R = 9*H*-Carbazol-9-yl
L4: R = NPh(naphthalen-1-yl)
L5: R = NPh_2
L6: R = S^{t}Bu

Figure 7.28 Schematic representation of Kröhnke-type tpy ligands used for the complexation of Zn^{II} ions.

Multilayer OLEDs were fabricated, using these complexes as the emitters. From these, $[Zn(L3)_2](PF_6)_2$ afforded the highest brightness and CE at the turn-on voltage of *ca.* 10 V (*i.e.*, 2,250 cd m^{-2} and 5.28 cd A^{-1}). This study revealed that the emission properties of Zn^{II}–tpy complexes mainly depend on the nature of the π-conjugated chromophore. The Zn^{II} ion possesses a closed-shell d^{10}-electron configuration and participation in the ligand-based $\pi \rightarrow \pi^*$ transitions is disabled. Thus, the emission of the complexes is solely of ILCT character [167]. A wide range of Zn^{II}–tpy complexes has been studied regarding their photophysical properties [168]. Sindhu *et al.* compared the emission behavior of $[Zn(tpy)_2]^{2+}$ complexes in the solid state to that observed in solution. As a common trend, the solid-state emission maxima appeared bathochromically shifted, due to non-covalent interactions, *i.e.*, ππ-stacking of phenyl moieties and/or excimer formation [169].

Schubert, Popp *et al.* demonstrated that aggregation and crystallization could largely be overcome by blending Zn^{II}–tpy complexes into a polymer matrix. Owing to its solubilizing alkyl chains, complex **22** possessed a high compatibility with PMMA. Homogeneous films, which contained *ca.* 5 wt% of **22** in PMMA were prepared by inkjet printing. These films showed only a weak redshift of the emission band (*i.e.*, *ca.* 1,000 cm^{-1} relative to that in solution) and a remarkably high Φ_{PL} of 27%. It was proposed that the undesirable interactions could be restrained further, by covalently linking the complexes and the polymer matrix [170, 171] (Figure 7.29).

Moreover, Zn^{II}–tpy complexes have also attracted attention due to their nonlinear optical (NLO) properties [173, 174]. For example, two-photon absorption (2PA) of light by the complex $[Zn(L6)_2](PF_6)_2$ (Figure 7.28) was reported by Sun, Tian *et al.* [175]. This complex possessed a moderate 2PA cross section in the NIR regime and exhibited superior excited-state properties, *i.e.*, prolonged τ and increased Φ_{PL}. The improved emission behavior of this complex was ascribed to its two-photon excited fluorescence. Moreover, the complex was employed as a 2PA probe for the *in vitro* and *in vivo* bioimaging of HepG2 cells and zebrafish larva, respectively, using a two-photon scanning microscope. Related to this, Li, Tian *et al.* designed the Zn^{II} complex **23**, which even exhibited three-photon activity triggered by the complex's behavior [176]. The AIE concept, as introduced in 2001

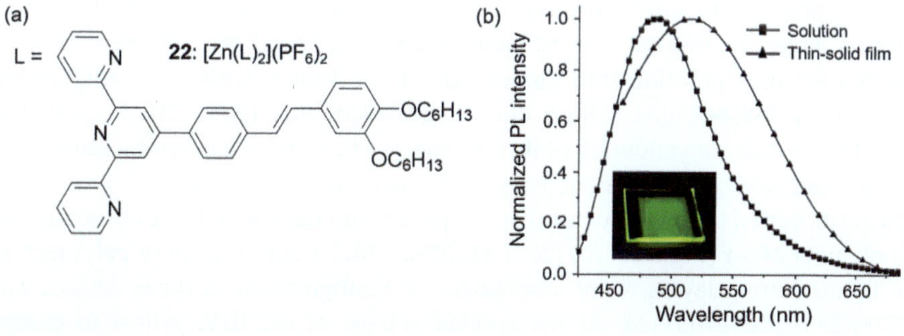

Figure 7.29 (a) Schematic representation of Zn^{II} complex **22**. (b) Comparison of the emission spectra of **22** in a dilute MeCN solution and PMMA matrix. Reproduced with permission from [172] / John Wiley & Sons.

by Tang *et al.* [177], has inspired widespread research aiming for the development of AIEgens, as luminophores for applications in optoelectronics or biomedicine [178]. Aggregation-induced quenching (AIQ) represents one limiting factor in the efficiency of OLED or LEEC devices, due to the diminishing of emission by aggregation/interaction of emitter molecules at high concentrations. This behavior is also detrimental regarding applications of hydrophobic luminophores under physiological conditions or, to be more general, in water. Whereas in the case of AIE, barely or even non-emissive molecules aggregate (typically in the presence of water) and become highly emissive, basically due to the restriction of intramolecular motion, which suppresses non-radiative deactivation pathways. Among many other systems, Zn^{II} complexes, including Zn^{II}–tpy ones, have been discussed as potential AIEgens [179].

Fluorescence quenching (or sometimes enhancement) of Zn^{II}–tpy complexes has also been used in the context of sensing applications. The wide range of analytes, which can be detected by such complexes *inter alia* includes sulfonamide antibiotics [180], nerve gases or nerve-gas mimics [181, 182], phosphates [183–185], *etc.*; likewise, terpyridines can also be utilized for the detection of Zn^{II} ions in solution due to "turn-on" fluorescence [186]. However, the discussion of the broad field of sensing applications cannot be provided within the scope of this section.

Besides the aforementioned mononuclear complexes, also coordination polymers, in which π-conjugated segments have been linked *via* $[Zn(tpy)_2]^{2+}$ moieties, have been the topic of substantial research [11, 30, 49, 187]. Some selected examples, highlighting the diversity of the incorporated chromophores, are shown in Figure 7.30, and reference is given to many other contributions in this field, that are not covered herein for the lack of space [188–200]. One of the first series of such polymers was reported by Che *et al.* [201]. These authors prepared the polymers by the self-assembly of the respective *bis*-tpy ligands with $Zn(OAc)_2 \times 2H_2O$ in NMP, as a high-boiling, aprotic solvent. This protocol represents the most applicable procedure for preparing such polymers and has been adapted by many groups of researchers. Che's polymers featured high thermal stability and intense violet to yellow ILCT emission in solution as well as in thin-solid films (the color of the emitted light was dependent on the nature of the ligands' chromophores). In most cases, the Φ_{PL} values of the films were even higher than those of the respective ligands in solution. From the in-total nine different polymers, **24a** and **24b** (Figure 7.30) were selected for incorporation into monolayer OLED devices with the configuration ITO|PEDOT:PSS|**24**|Ca|Al. These devices provided blue (**24a**; CIE coordinates: $x = 0.15, y = 0.21$) or yellow EL (**24b**; $x = 0.46, y = 0.50$) with a maximal luminance of 1,700 and 2,380 cd m^{-2}, respectively, and a turn-on voltage of *ca.* 6 V.

Lin *et al.* varied the substituent(s) on the phenyl moieties within the chromophore of polymers **24c–f** (Figure 7.30) [202]. OLEDs, which contained these polymers, as the light-emitting layers, were fabricated; the configuration of these devices was ITO|PEDOT:PSS|**24**|LiF|Al. At the applied voltage of *ca.* 10 V, yellow to orange EL was obtained with polymers **24d–f**. Within this set of OLED devices, the **24e**-containing one revealed the best performance in terms of PE (0.33 cd·A^{-1} at 14 V), EQE (1.02%), and brightness (931 cd·m^{-2} at 14 V; Figure 7.31). In related

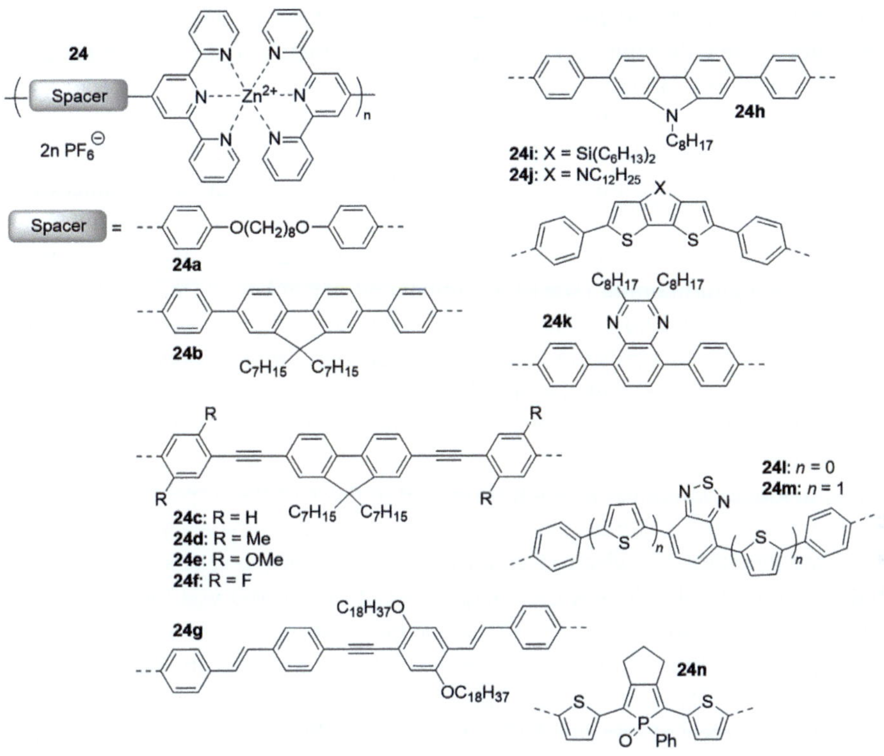

Figure 7.30 Schematic representation of the Zn^{II}-containing coordination polymers **24**.

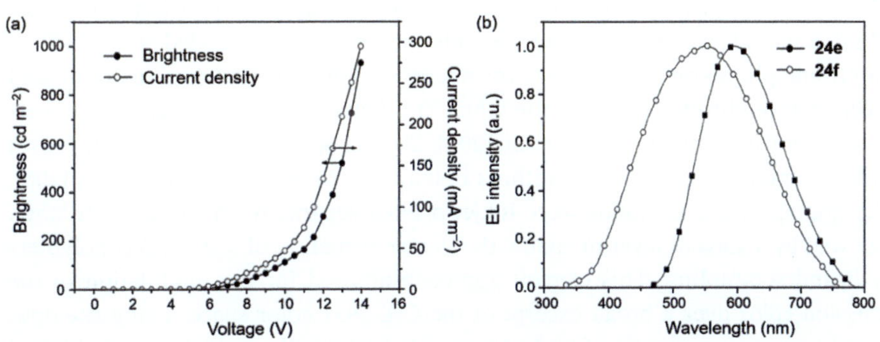

Figure 7.31 (a) Current–voltage-brightness characteristics of the OLED device, which contained **24e**, as the light-emitting layer. (b) Normalized EL spectra of the OLED devices, which contained **24e** or **24f**, as the light-emitting layer [202] / with permission of John Wiley & Sons.

work, Schubert *et al.* studied the thermal, photophysical, and electrochemical properties of ten different Zn^{II} coordination polymers [203]. Despite its poor solubility, polymer **24g** (Figure 7.30) was processed into homogeneous thin-solid films by spin coating and inkjet printing. These films exhibited yellow fluorescence and, when incorporated into a single-layer OLED, orange (albeit very weak) EL. In subsequent work, Schubert *et al.* paid particular attention to the processing of such polymers [204]. It was shown that homogeneous films with a thickness of 150–200 nm could generally be prepared by inkjet printing at 40–50 °C from a *N*,*N*-dimethylformamide (DMF)/acetophenone solvent mixture. Furthermore, the processability and film-formation behavior of Zn^{II} polymers were significantly improved, by attaching polymer chains to the ligand's chromophores instead of the conventionally used short alkyl groups [170, 171].

In one further study, Chen, Li *et al.* studied the Zn^{II} polymers **24h–j** (Figure 7.30) [205]. The emission maxima of these materials in the solid state were significantly shifted to longer wavelengths when compared to those of the respective ligands. It was assumed that the complexation by Zn^{II} ions increased the electron-deficiency of the tpy moieties: The electron-donor chromophore of the spacer increased the HOMO energy level, whereas the energy level of the LUMO was lowered due to metal-ion-complexation. As a result, the polymers **24h–j** featured very low optical bandgaps of 2.07, 1.97, and 1.56 eV, respectively. A dark red, but very weak, emission was observed for **24j**, as a thin-solid film (λ_{PL} = 586 nm, Φ_{PL} = 0.06%).

Wong, Schubert *et al.* presented a further approach for the tuning of the emission color [206]. Firstly, these authors prepared a library of 10 different Zn^{II} coordination polymers, from which the representatives **24k–m** (Figure 7.30) were selected for subsequent use, due to their distinct blue (**24k**; λ_{PL} = 443 nm), green (**24l**; λ_{PL} = 503 nm), and red emission colors (**24m**; λ_{PL} = 606 nm). Owing to the kinetic lability of $[Zn(tpy)_2]^{2+}$ complexes in solution, the mixing of the three homopolymers afforded statistical copolymers, whose emission properties depended on the ratio of the chromophores on the one hand and energy-transfer process between donor- and acceptor-type chromophores on the other. This behavior was analyzed by Schubert *et al.* already in a previous publication [207]. In order to prepare thin-solid films with various emission colors, these authors opted for a combinatorial approach [206]: The homopolymer solutions were inkjet-printed separately on top of each other, followed by a solvent layer to enable the *in situ* formation of statistical copolymers by ligand scrambling. This elegant approach allowed for the modulation of the emission color over a broad excerpt of the CIE-1931 color space. Using the optimized homopolymer ratio of 80:10:10 provided a thin film, which revealed broad emission over whole vis spectral region with PL maxima at 452 and 608 nm. Thus, the emission color was perceived as white (CIE coordinates: $x = 0.311$, $y = 0.307$).

In a similar fashion, also 2D constructs, *e.g.*, metallomacrocycles [208] or rosettes, have been prepared (see Chapter 10) [209]. Regarding the latter, the emissive properties in solution could be tuned *via* the number of rosette generations and the solvent composition. Newkome *et al.* reported a hexagonal macrocycle, which was comprised of each six Zn^{II} ions and angular *bis*-tpy ligands [208]. This assembly provided green EL with a maximum luminance of 39 cd m^{-2} and CE of 0.16 cd A^{-1} when fabricated into an OLED with multilayer configuration.

7.6 Lanthanide(III)–Tpy Complexes

The lanthanides, which represent the major fraction of the rare-earth elements, feature remarkable photophysical and magnetic properties, which can be ascribed to their unique electronic configuration, *i.e.*, the shielding of their 4f orbitals by 5s and 5p ones [210–215]. This gives, for example, rise to intense emission or high magnetic moments for Ln^{III} complexes. Except for La^{III} and Lu^{III}, all other Ln^{III} ions are basically emissive in the visible or NIR regime (*i.e.*, Yb^{III}, Nd^{III}, and Er^{III}) [215]. In practice, Ln^{III} centers are barely emissive since their direct excitation is precluded due to the ions' small absorption cross sections, low molar absorption coefficients (below 10 L mol^{-1} cm^{-1}), and Laporte-forbidden 4f → 4f electronic transitions. It is well-known that this obstacle can be overcome by the so-called "antenna effect" [216, 217]. As shown in Figure 7.32, the indirect excitation of a Ln^{III} ion occurs *via* the chromophore of the coordinated ligands (the "antenna"). The S_1 excited state is initially populated, followed by ISC to the T_1 excited state, which is capable of transferring energy to the Ln^{III} ion. Alternatively, energy can be directly transferred from the ligand-based S_1 state to the ion's 4f* orbitals. The thusly excited Ln^{III} decays to the GS in a radiative fashion. However, sensitization will only be efficient if the T_1 energy level of the antenna systems matches with the energy level of the Ln^{III} ion [218]. One has to keep this aspect in mind when aiming for light-emitting Ln^{III} complexes.

Among other potential ligands for Ln^{III} ions, those from the tpy family represent versatile antenna ligands due to their chelating properties and π-conjugated system. The former is beneficial to control the nuclearity, *i.e.*, the formation of discrete assemblies, whereas the relatively large π-system can act as an antenna for light absorption and, thus, sensitize the Ln^{III} luminescence *via* the aforementioned mechanism. However, unlike for tpy complexes of transition-metal ions, the coordination chemistry of Ln^{III} ions is less straightforward: High and variable

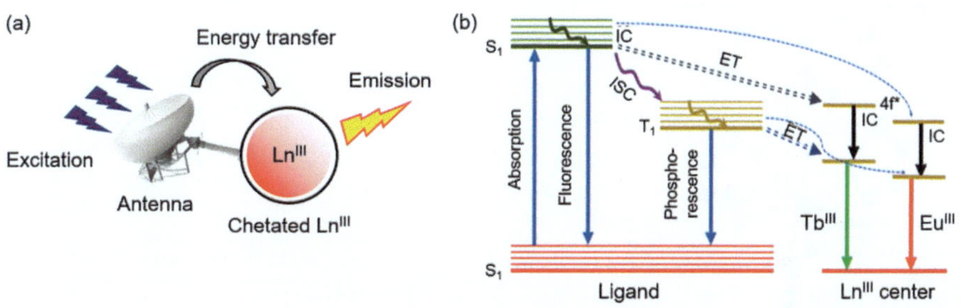

Figure 7.32 (a) Schematic representation of the antenna effect, *i.e.*, the indirect excitation of a Ln^{III} ion. (b) Simplified schematic representation of the mechanism of the Ln^{III} 4f emission due to the antenna effect [217] / with permission of John Wiley & Sons.

coordination numbers (CNs) as well as the notoriously poor control over the coordination geometry (*i.e.*, the stereoselective binding) represent complicating factors when dealing with Ln^{III} complexes in general. Recently, Li *et al.* provided a comprehensive review on discrete Ln^{III}–tpy complexes. These authors looked back on the *ca.* 60-year-long history of this topic and presented the recent achievements regarding structural diversity and complexity on the one hand as well as the range of applications on the other [29]. According to them, the formation of Ln^{III}–tpy complexes can be improved in different ways. Firstly, the self-assembly can be influenced by the choice of the counterions, which might act as capping ligands (*e.g.*, ClO_4^-, halides, AcO^-, NO_3^-, SCN^-, *etc.*). Moreover, auxiliary or secondary ligands can play a vital role in coordination chemistry, in particular, when attached to the tpy ligand in 6,6″-positions (*e.g.*, COOH, pyridine, tetrazole, *etc.*). In the following, some examples of light-emitting Ln^{III}–tpy complexes will be presented.

In one approach, Ziessel, Bünzli *et al.* used the tpy ligand **25**, which carried a boradiazaindacenyl moiety in 4′-position [219]. Due to the high molar absorption coefficient and Φ_{PL} in concert with its (photo)chemical robustness, this **25** represented a promising antenna ligand for the complexation of Ln^{III} ions (Figure 7.33a). The obtained complexes [Ln(**25**)]$(NO_3)_3$ possessed a nona-coordinate geometry (Figure 7.33b) and featured – in the case of Yb^{III}, Nd^{III}, and Er^{III} – NIR luminescence in the range from 900 to 1,600 nm (λ_{ex} = 514 nm). In particular, complex [Yb(**25**)][$(NO_3)_3$] had an excited-state lifetime of 9.7 μs in CH_2Cl_2 solution at room temperature; the Φ_{PL} of 0.31% corresponded to the ligand's sensitization efficiency of *ca.* 63%.

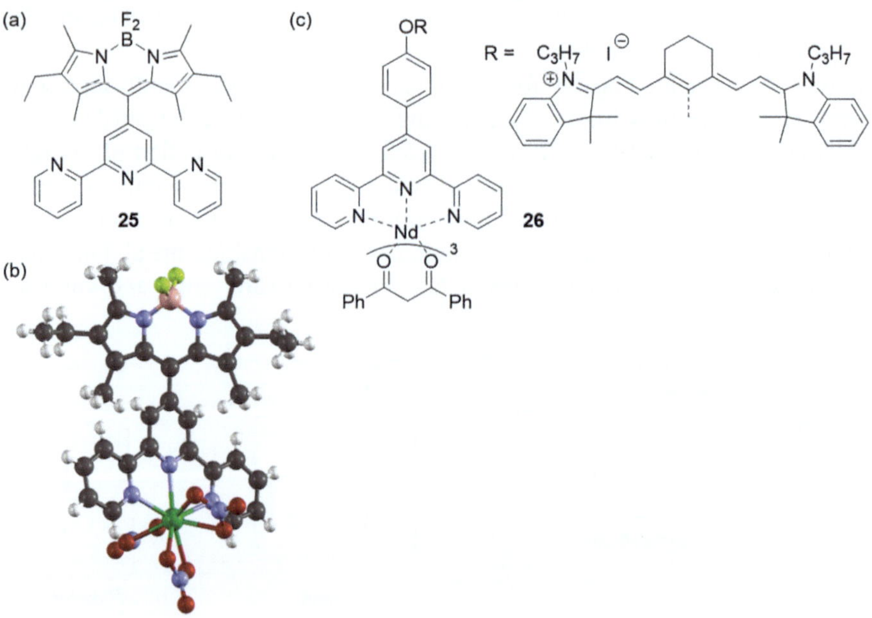

Figure 7.33 (a) Schematic representation of ligand **25**. (b) Solid-state structure of [Yb(**25**)]$(NO_3)_3$ (solvent molecules were omitted for clarity). Adapted from [219]. (c) Schematic representation of the Nd^{III} complex **26**.

A similar strategy was also followed, *e.g.*, by Li, Müller *et al.*, who reported solid-state luminescence for the mononuclear $[Ln(ptpy)](NO_3)_3$ complexes (Ln = Eu, Gd, Tb, or Dy; ptpy: 4′-phenyl-tpy) [220]. Most notably, $[Eu(ptpy)](NO_3)_3$ exhibited red emission (CIE coordinates: $x = 0.655$, $y = 0.345$) with high Φ_{PL} and remarkably long τ_{PL} (*i.e.*, 16.8% and 0.55 ms, respectively). The same tpy ligand was also combined with $LnCl_3$ salts to afford two series of coordination polymers: $[Ln(ptpy)]Cl_3$ and $[Ln(ptpy)(py)]Cl_3$ (py: pyridine) [221]. However, the former was only obtained with Ln = La, Ce, Pr, or Nd; whereas La^{III} and Ce^{III} did not give the polymer of the latter type. Coordination polymers $[Ln(ptpy)(py)]Cl_3$ (Ln = Pr, Ho, Er, and Tm) featured – in addition to their characteristic NIR luminescence – also emission in the visible regime.

Recently, Peón *et al.* prepared the Nd^{III} complex **26**, which contained a cyanine dye, as the antenna system (Figure 7.33c) [222]. Such chromophores are of particular interest due to their NLO properties, which result from their symmetry and extended conjugation [223]. The luminescence of this complex, which also contained three β-diketonate ligands, was due to the 2PA by the ligand in the 800–900 nm region and subsequent energy transfer to the Nd^{III} center.

Mazzanti *et al.* reported on the synthesis of pentadentate ligands H_2**27** (Figure 7.34) [224]. The extension of denticity was realized by thermal cycloaddition of the corresponding 6,6′-dicyano-substituted tpy precursors. The complexes $(NHEt_3)[Ln(\textbf{27})_2]$ were comprised of the $[Ln(\textbf{27})_2]^-$ anion, in which the Ln^{III} centers featured the high CN of 10 (Ln =. Nd, Eu, and Tb). Likewise, deca-coordinate Ln^{III} ions were also observed when tpy ligands with carboxylic-acid moieties in 6,6′-positions were used, as the pentadentate ligands. The pentadentate ligands efficiently sensitized the luminescence of Ln^{III} ions in the visible to NIR regions with high Φ_{PL} values. The complex $(NHEt_3)[Eu(\textbf{27a})_2]$ displayed intense red luminescence (*i.e.* $\Phi_{PL} = 35\%$, $\tau = 2.5$ ms); whereas for the NIR-emitting Nd^{III} complexes, the best emission performance was noted for $(NHEt_3)[Nd(\textbf{27c})_2]$ (*i.e.*, $\Phi_{PL} = 0.29\%$).

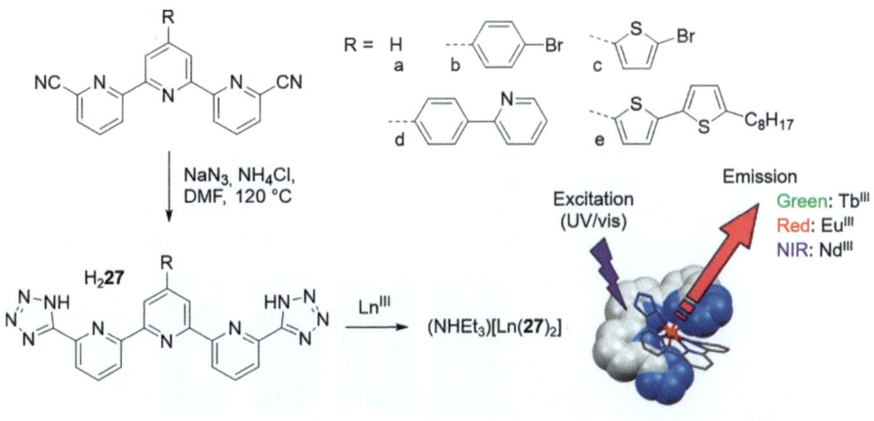

Figure 7.34 Schematic representation of the synthesis of ligands H_2**27** and their *bis*-pentadentate complexes $(NHEt_3)[Ln(\textbf{27})]$. A representation of the sensitized emission is also shown [224] / with permission of John Wiley & Sons.

Besides these examples for mononuclear Ln^{III}–tpy emitters, studies are also available where such complexes have been incorporated into multinuclear, *i.e.*, polymeric ensembles. In this context, conventional coordination frameworks of various topologies [225] as well as organic polymers carrying luminescent Ln^{III}–tpy entities have been reported. Concerning the latter, one has to distinguish between polymers, which carry tpy moieties for Ln^{III} complexation at the chain ends, and those that possess such ligand sites, as the pending side chains. The former have been utilized for the assembly of stimuli-responsive and self-healing metallopolymers (see Chapter 9) [226–229]. Polymers with tpy-containing side chains have been extensively studied, as macroligands for the binding of transition-metal ions (*e.g.*, Ru^{II} or Os^{II}). In particular, Schubert *et al.* heavily contributed to this field of research [18, 230]. However, the binding of Ln^{III} ions to preformed polymers has been studied to a minor extent.

Tew *et al.* were the first authors who prepared the metallopolymers **28** (Figure 7.35) by coordinating Ln^{III} ions to the binding sites of a tpy-equipped diblock copolymer, which was synthesized *via* atom-transfer radical polymerization (ATRP). The coordination of $Ln(OTf)_3$ salts was studied in solution as well as in the solid state and the resulting polymers showed the expected luminescence behavior according to the respective [Ln(tpy)]$(OTf)_3$ complexes] (Ln: Dy, Eu, Tb, and Sm) [232]. In subsequent work, these same authors also prepared the "alloy" polymer **28e**, which contained Eu^{III} and Tb^{III} centers in a 1:1 ratio [233]. Unlike for the 1:1 mixture of **28b** and **28c**, yellow solid-state luminescence was observed for **28e**. It was assumed that the presence of "metastable" Eu^{II} species accounted for this behavior. The formation of such species is possible in the case of low-energy CT states; it was proposed that the presence of Tb^{III} centers in close proximity to the Eu^{III} ones facilitated the reduction of the latter. Moreover, a wide range of emission colors between blue, red, and green was realized by coordinating Dy^{III}, Eu^{III}, and Tb^{III} ions in various ratios [231]. Polymer **28f**, which contained the three Ln^{III} ions in the optimized 1.8:1:1 ratio, exhibited the emission of "true white" light when processed into a thin-solid film. The CIE coordinates of the emission (*i.e.*, $x = 0.311$ and $y = 0.328$)

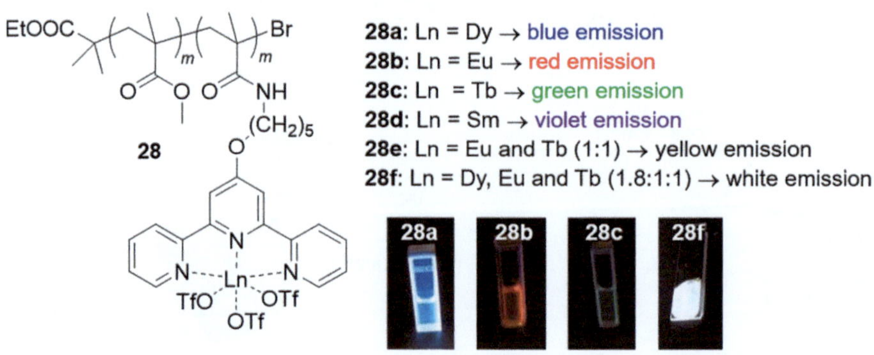

Figure 7.35 Schematic representation of Tew's metallopolymers **28**. The photographs show the luminescence of the respective thin-solid films (prepared by drop-casting onto glass slides) [231] / with permission of John Wiley & Sons.

were close to the theoretical values (*i.e.*, $x = 0.313$ and $y = 0.329$). The blue-emitting polymer **28a** was combined with $[Ru(bpy)_3]Cl_2$ in a 2:1 ratio to afford a hybrid material, which displayed "near-white" emission upon excitation at 350 nm [234]. It was shown that energy transfer from the excited Dy^{III}–tpy complexes to the Ru^{II} centers accounted for the observed emission behavior (the efficiency of the ET was *ca.* 76%).

Related to Tew's work, Turchanin *et al.* fabricated tpy-functionalized carbon nanomembranes (CNMs) on Au surfaces [235]. The cross-linking of preassembled monolayers on Au substrates upon irradiation with low-energy electrons represents a versatile method to prepare ultrathin (*ca.* 1 nm), large (*ca.* 10 cm^2), free-standing, and mechanically stable nanosheets [236]. The fabrication process was tolerant to the presence of lateral functionalities, such as tpy moieties, which remained mostly intact under the applied conditions [235]. The tpy moieties on the CNMs' surfaces were used for the reversible binding of Eu^{III} or Tb^{III} ions. The formation of the surface-anchored Ln^{III}–tpy complexes was proven by high-resolution X-ray photoelectron spectroscopy (XPS) and luminescence measurements. The latter revealed the characteristic emission bands of the Eu^{III}–tpy and Tb^{III}–tpy complexes, respectively. Before this, Wang *et al.* prepared luminescent hybrid materials, which were comprised of a silica network with pending Eu^{III}–tpy and/or Tb^{III}–tpy complexes [237].

A different type of polymer, namely the tpy-functionalized poly(ionic liquid) (PIL, **29**, Figure 7.36a), was reported by Li *et al.* [238]. Upon addition of $EuCl_3$, a metallo-supramolecular polymer network was formed by inter- and intrachain complexation; the resulting material was studied regarding its solid-state luminescence. Owing to the confined environment, both the quantum yield and the lifetime were increased when compared to those of mononuclear Eu^{III}–tpy complexes [240]. A similar behavior in terms of improved photostability was also observed by Martinez, Rocha *et al.*, who focused on elastin-like polymers with pending tpy moieties (Figure 7.36b) [239]. Elastin-like polymers are of general interest due to their durable, highly tunable, biodegradable, and stimuli-responsive structure [241]. The biopolymers were expressed by genetically modified *Escherichia coli* and subsequently modified by coupling tpy ligands to the polypeptides' pending COOH moieties [239].

Two further side-chain polymers were discussed in the context of Eu^{III}–tpy complexes. Firstly, the water-soluble copolymers **30** featured different luminescence properties, which depended on the nature of the secondary ligand [242]. In the case of **30a**, an AIE-like behavior was found, whereas AIQ was observed when the Eu^{III} centers possessed an alternative antenna ligand, such as the β-diketonate one in **30b** (Figure 7.37a). Finally, free-radical polymerization (FRP) was used for the synthesis of the copolymers **31**, which differed in the ratio of the incorporated monomers (Figure 7.37b). These polymers were highly emissive in dilute solution as well as in the solid state. In particular, the carbazole-based emission was suppressed in the solid state and energy was transferred to the Eu^{III} complexes with an efficiency of up to 40%. Double-layer OLEDs, which displayed pure-red EL, were fabricated. When using **31a**, as the light-emitting material, a maximum brightness of 68.2 cd m^{-2} was

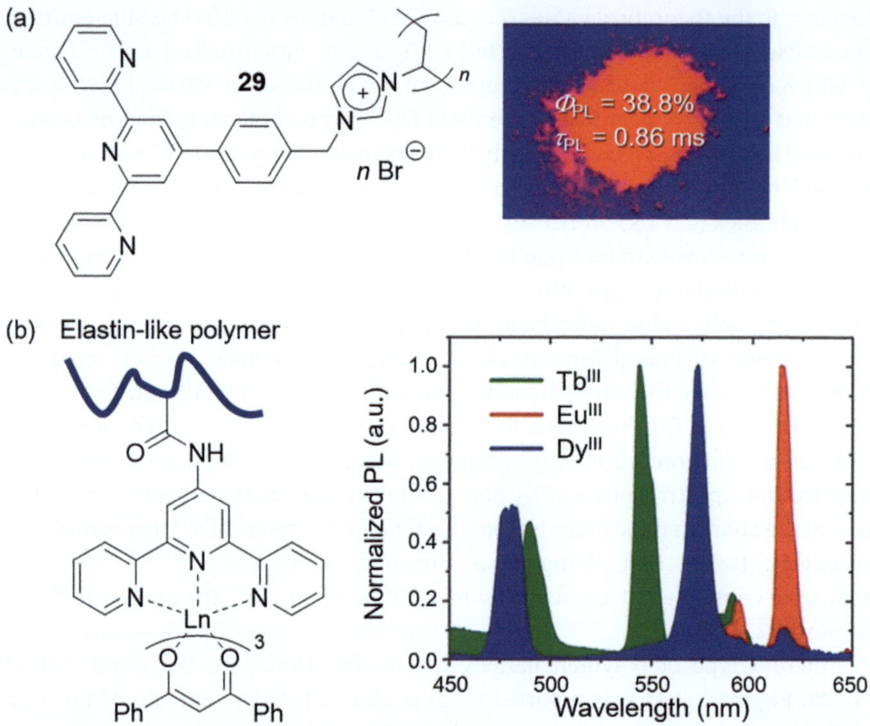

Figure 7.36 (a) Schematic representation of polymer **29**. The photograph shows the intense red solid-state luminescence of the $EuCl_3$/**29** polymer [238] / with permission of John Wiley & Sons. (b) Schematic representation of the tpy-equipped elastin-like polymer; the luminescence spectra of the polymers obtained by Ln^{III} complexation are also shown [239] / with permission of John Wiley & Sons.

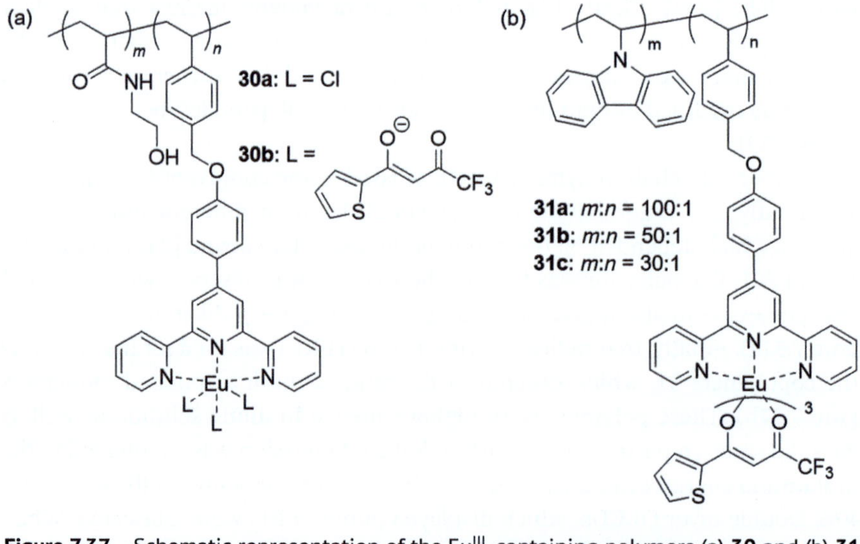

Figure 7.37 Schematic representation of the Eu^{III}-containing polymers (a) **30** and (b) **31**.

obtained at the turn-on voltage of 17 V; the EQE was 0.37% at the luminance of 9.5 cd m^{-2}.

Various aromatic carboxylates have been used to prepare coordination compounds with Ln^{III} salts *via* hydrothermal reactions: 4-chlorobenzoic acid [243], 2,4-dichlorobenzoic acid [244, 245], 2,4-dimethylbenzoic acid [246, 247], thiophene-2-carboxylic acid (TCA) [248], *etc.* The latter ligand also represented a versatile auxiliary for the formation of coordination compounds with $BiCl_3$ and tpy [249]. These solid-state emitters are rare examples for the use of main-group elements in the context of light-emitting tpy complexes (see Section 7.7).

7.7 Miscellaneous Light-Emitting Tpy Complexes

To the end of this chapter, we would like to prescind from the tpy complexes of transition-metal or rare-earth-metal ions, as emitters, and pay attention to light-emitting materials that have been derived from main-group elements and tpy ligands. This represents an often overlooked field of research; nonetheless, notable achievements have been made in recent years [250].

7.7.1 Perovskite-Based Light-Emitting Diodes

As already addressed in Section 5.5, metal-halide perovskites represent an emerging class of materials for solar-cell applications. They entered the competition for power-conversion efficiency (PCE) in 2009 [251]. Since then, the PCE of such devices has been improved dramatically from initially 3.8–25.7% in 2021 (considering only single-junction cells) [252]. Likewise, the use of such materials in light-emitting devices, specifically perovskite LEDs (PeLEDs), is also a rapidly developing field of research. The first such device was reported in 2014 by Friend *et al.*, who achieved a luminance of 364 cd m^{-2} at the current density of 123 mA cm^{-2} when using $MAPbBr_3$, as a green emitter, in multilayer PeLED (EQE of 0.1%, MA = methylammonium) [253]. In the last years, considerable progress has been made in boosting the EQE from <1% to >30% for green and red emitters; whereas an EQE of *ca.* 24% has recently been reached for NIR emitters [254–258]. However, unlike for the "record" PCEs of solar cells (see Chapter 6), the EQEs do not represent certified values and a comparison of data from different labs is constrained. Until now, a certification infrastructure analogous to that of the solar-cell community has not been installed [259]. The widespread use of $MAPbX_3$ or $CsPbX_3$ (X = Cl, Br and/or I) in PeLEDs is due to the ease of processing from solution, high color purity with narrow spectral width (fwhm of *ca.* 20 nm), ability to adjust the bandgap over a wide range, and large charge-carrier mobility. The related perovskite LEECs (PeLEECs), whose function basically relies on the mobility of ions, are still not competitive to PeLEDs in terms of device performance [153, 260].

The photophysical properties of perovskite emitters can *inter alia* be modulated *via* the choice of the ammonium cation and organic additives/ligands. As important, the operational PeLED lifetime can be improved by dimension control on

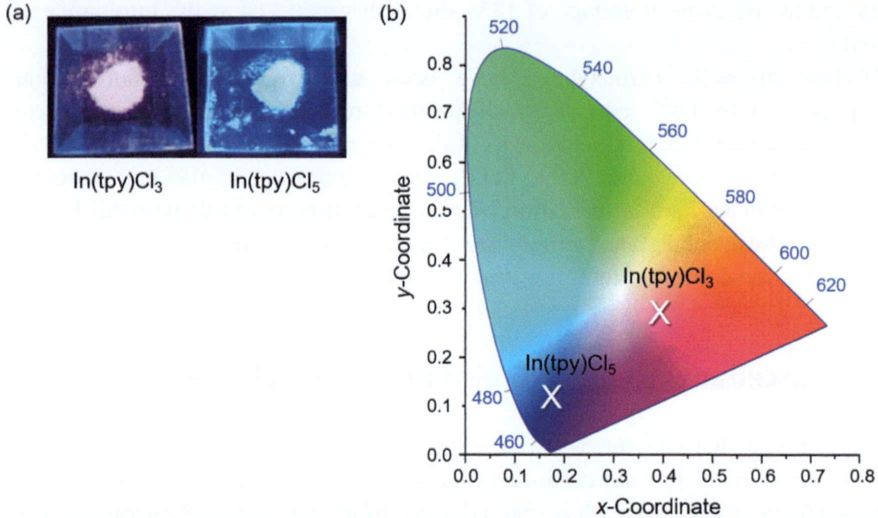

Figure 7.38 (a) Photographs of $In(tpy)Cl_3$ (left) and $In(tpy)Cl_5$ (right) showing their luminescence upon excitation with UV light ($\lambda_{ex} = 365$ nm). (b) Ranking of the emission of $In(tpy)Cl_3$ and $In(tpy)Cl_5$ in the CIE-1931 color space [263] / with permission of John Wiley & Sons.

the one hand and ligand engineering on the other [261, 262]. Although being more prominently used in the context of perovskite solar cells (*i.e.*, for passivation and dimension control), polypyridyl ligands (*e.g.*, bpy, phen, and tpy) have also been employed for the stabilization of PeLED emitters. Fan, Li *et al.* prepared $In(tpy)Cl_3$ and $In(tpy)Cl_5$ by hydrothermal synthesis [263]. Both compounds revealed solid-state luminescence under ambient conditions (Figure 7.38); in particular, a high Φ_{PL} of 47.6% was observed for $In(tpy)Cl_5$, whereas $In(tpy)Cl_3$ was a rather poor emitter ($\Phi_{PL} \approx 6.1\%$). For consistency, these materials have been discussed further in the context of tpy ligands and perovskite materials mainly for solar-cell applications (see Section 5.5).

7.7.2 Bi^{III}–Tpy Complexes as Solid-State Emitters

As already mentioned briefly in Section 3.6, Bi^{III}–tpy compounds might display interesting luminescence properties [249, 264]. Recently, bismuth-organic compounds, such as coordination polymers or metal–organic frameworks (MOFs), have attracted considerable interest due to their remarkable photophysical properties. The growing interest in such materials has also to be seen in the context of abundance as well as eco-friendliness and low toxicity – in this respect, bismuth is considered superior to many of the heavy transition metals, that are needed for optoelectronic applications (*e.g.*, Ru, Ir, Pt, *etc.*) [265]. Moreover, Bi^{III} offers a highly flexible coordination chemistry and, thus, offers plenty of room for the modulation of the photophysical properties *via* structural regulation. The Bi^{III} center possesses a unique $[Xe]4f^{14}5d^{10}6s^2$ electron configuration, which combines the closed-shell characteristics of d^{10} metal ions (*e.g.*, Zn^{II}, Ag^{I}, Au^{I}, *etc.*) with the features associated with the $6s^2$ lone pair of electrons [266, 267]. Most luminescent Bi^{III}-containing coordination compounds have been assembled from aromatic carboxylates and polypyridyl ligands (*e.g.*, bpy, phen, or tpy); a broad range of emission colors, including "white" light, has already been realized with this ligand toolbox. However, the mechanism for emission is usually far from being trivial, owing to the numerous contributing ET pathways, which include CT and/or ligand-based processes. Specifically, LMCT, MLCT, intraligand transitions (*i.e.*, $\pi^* \rightarrow n$ or $\pi^* \rightarrow \pi$) as well as intra-configurational $^1P^1 \rightarrow {}^1S_0$ and $^3P_1 \rightarrow {}^1S_0$ transitions, which originate from the $6s^2$-$5d^{10}$ Bi^{III} center, might be involved [249]. To complicate things further, intervalence CT has also been observed for some Bi^{III} compounds [268]. Moreover, the emission of some halobismuthates has been assigned to combinations of metal-centered sp excited states and halogen-to-metal charge transfer (XMCT) [269]. These excited-state features are the unique selling point of Bi^{III} emitters, whose emission properties (*i.e.*, color and lifetime) can be modulated *via* the coordinated donor ligands as well as the preparation method [270, 271].

The luminescent complex $[Bi_2Cl_6(tpy)_2]$ (**32**) was prepared *via* the solvothermal reaction of $BiCl_3$ and tpy in MeCN (dipicolinic acid was used as an additive) [249]. Steady-state and time-resolved photophysical studies, reinforced by DFT calculations, were performed to clarify the origin of the compound's yellow-green solid-state emission, which decayed biexponentially (λ_{em} = 537 nm, τ_1 = 10 µs, τ_2 = 196 µs; Figure 7.39a). The experiments suggested that mainly LLCT and LMCT transitions contributed to the long excited-state lifetime of this complex. In subsequent work, Réal, Knope *et al.* prepared a series of Bi^{III} coordination compounds by the solvothermal reaction of BiX_3 (X = Cl or Br), tpy, and/or TCA in aqueous MeOH (**33–38**, Table 7.5) [264]. Moreover, a series with heterobimetallic compounds of the general formula $[Bi_{1-x}Eu_xCl(tpy)(\kappa^2\text{-TC})_2]$ was also synthesized by doping **34a** with small amounts of Eu^{III} ions (**39**; x = 0.001, 0.005, 0.01, or 0.05; TC: thiophene-2-carboxylate). All these materials were studied by steady-state and time-resolved spectroscopy regarding their solid-state luminescence properties. Green to yellow luminescence was observed for **33–38**, whereas **39** featured an emission with contribution from both organobismuth- and Eu^{III}-centered luminescence (Figure 7.39b,c). The emission color of **39** could be varied from yellow *via* orange to red by increasing the content of the Eu^{III} dopant (Figure 7.39c). According to DFT calculations, halide-metal-to-ligand charge transfer (XMLCT) processes contributed significantly to the materials' photophysical properties. The occurrence of this seldomly observed pathway for visible-light absorption highlights the unique characteristics of the luminescence of organobismuth halides. Likewise, intense solid-state luminescence was also reported for the organobismuth phases $[BiX_y(TTA)_{3-y}(tpy)]$, which were doped with up to 3 mol% of Eu^{III} ions (X = NO_3, Cl, Br or Br_2; TTA: 2-thenoyltrifluoroacetone) [272].

Despite this notable progress in the field of organobismuth chemistry, it remains questionable if Bi^{III}-based luminophores will ever be competitive with other prominent emitters, which are traditionally based on heavy transition-metal ions (*e.g.*, Ir^{III} or Ru^{II}). Although Bi is more abundant in Earth's upper continental crust than

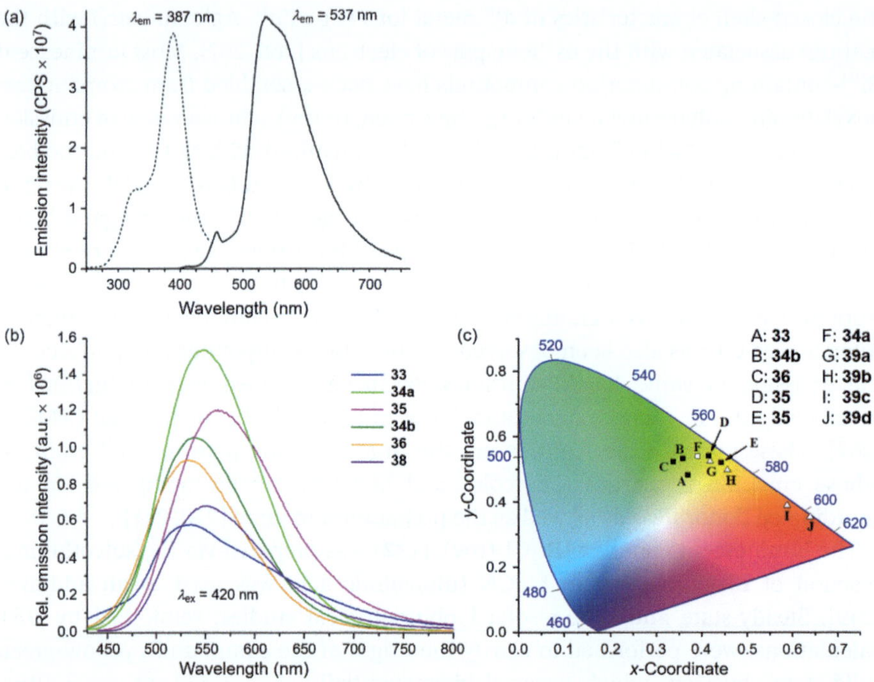

Figure 7.39 (a) Emission (solid line) and excitation spectra (dashed line) of a solid sample of **32** [249] / with permission of Royal Society of Chemistry. (b) Solid-state luminescence spectra of various samples at room temperature. (c) CIE chromaticity diagram for the compounds listed in Table 7.5 [264] / with permission of American Chemical Society.

the highly siderophile metals (*e.g.*, Pt, Ru, Ir, Au, *etc.*), it is still considerably rare (Figure 7.40) and, thus, the synthesis of organobismuth compounds on a large scale will also be cost-intensive.

7.8 Concluding Remarks

Luminescent tpy complexes have been the subject of intense research already for several decades. The enormous number of reported complexes can be handled by applying two classification schemes, *i.e.*, the nature of the incorporated metal ion and the practical use. Regarding the latter, the light-emitting devices – namely OLEDs and LEECs – play the most relevant role. Besides these technologies, which basically translate electricity into light, also bioimaging (*i.e.*, the detection of emission in cells) [110], and sensing (*i.e.*, the enhancement or depletion of emission in the presence of certain analytes) [275] has attracted widespread interest. In the field of OLEDs, tpy complexes of heavy transition-metal ions (*i.e.*, Ru^{II}, Ir^{III}, Pt^{II}, or Re^{I}) have been employed. However, the performance of these phosphorescent emitters could not compete with those of the state-of-the-art emitters, *e.g.*, charge-neutral *bis*- or *tris*-cyclometalated Ir^{III} complexes [78]. Likewise, the LEEC technology, in particular the iTMC-LEEC variant, is dominated by the cationic *bis*-cyclometalated Ir^{III} complexes and, to a minor extent, complexes from the $[Ru(bpy)_3]^{2+}$ family [25, 151].

Table 7.5 Summary of the Bi^{III} coordination compounds and the color of their solid-state emission.

Complex	Composition	CIE coordinates
32	$[Bi_2(tpy)_2Cl_6]$	n/a
33	$[Bi_2Cl_4(tpy)_2(\kappa^2\text{-TC})_2]$	$x = 0.37; y = 0.48$
34a	$[BiCl(tpy)(\kappa^2\text{-TC})_2]^{a),b)}$	$x = 0.39; y = 0.54$
34b	$[BiCl(tpy)(\kappa^2\text{-TC})_2]^{a)}$	$x = 0.36; y = 0.53$
35	$[BiBr(tpy)(\kappa^2\text{-TC})_2]^{b)}$	$x = 0.44; y = 0.52$
36	$[BiCl(tpy)(\kappa^2\text{-TC})_2][Bi(tpy)(\kappa^2\text{-TC})_3]\cdot 0.55(TCA)$	$x = 0.35; y = 0.52$
37	$[BiBr_3(tpy)(MeOH)]$	n/a
38	$[BiBr_2(tpy)(\kappa^2\text{-TC})][BiBr_{1.16}(tpy)(\kappa^2\text{-TC})_{1.84}]$	$x = 0.41; y = 0.54$
39a	$[Bi_{0.999}Eu_{0.001}Cl(tpy)(\kappa^2\text{-TC})_2]$	$x = 0.42; y = 0.52$
39b	$[Bi_{0.995}Eu_{0.005}Cl(tpy)(\kappa^2\text{-TC})_2]$	$x = 0.46; y = 0.50$
39c	$[Bi_{0.99}Eu_{0.01}Cl(tpy)(\kappa^2\text{-TC})_2]$	$x = 0.59; y = 0.39$
39d	$[Bi_{0.95}Eu_{0.05}Cl(tpy)(\kappa^2\text{-TC})_2]$	$x = 0.64; y = 0.35$

a) **34a** and **34b** were polymorphs.
b) **34a** and **35** were isomorphs.
Adapted from [264].

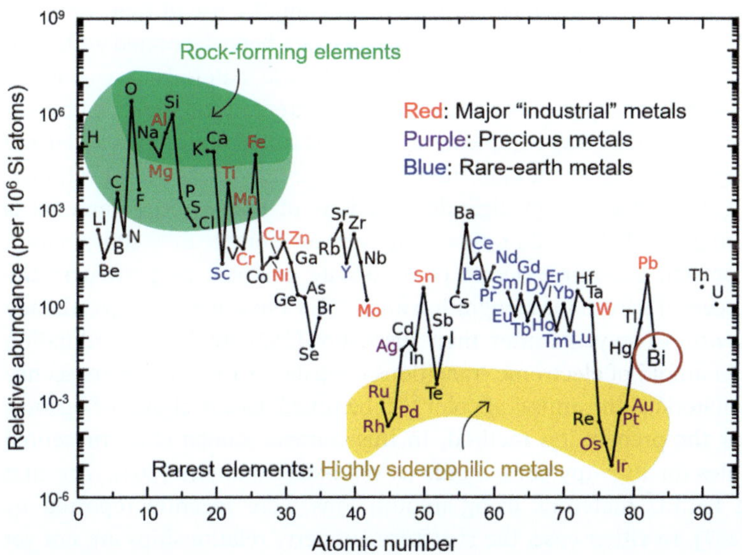

Figure 7.40 Relative abundance of the chemical elements in Earth's upper continental crust. The relative abundance is defined as the number of the respective element's atoms per 10^6 atoms of Si [273, 274] / with permission of U.S. Geological Survey.

The Zn^{II}–tpy complexes represent a unique class of emitters, whose fluorescence is – owing to the absence of MLCT transitions – mainly of LC character. Thus, the π-conjugated chromophore, which is attached to the tpy ligand, dictates the photophysical properties. Moreover, such complexes can typically be prepared in a straightforward fashion under milder conditions as required, *e.g.*, for the synthesis of Ru^{II}–tpy or Ir^{III}–tpy *bis*-complexes. Such complexes have not only been employed as emitters in OLEDs but also in the context of other applications: NLO materials [276], bioimaging [110], (chemo)sensing [277], *etc.* Moreover, the self-assembly of Zn^{II} ions with π-conjugated *bis*-tpy ligands affords organo-soluble metallopolymers, which can be processed into thin-solid films. A wide range of such materials has been prepared and (in some cases) incorporated into light-emitting devices, whose performances were only moderate to low [49]. The reversibility of the Zn^{II}–tpy interaction represents one further notable aspect, which could be used for emission-color tuning. Schubert, Wong *et al.* demonstrated that a broad range of emission colors, including white light, could be obtained by mixing two or three different Zn^{II}–tpy metallopolymers [206]. The combination of properties (*i.e.*, ease-of-synthesis, low toxicity and costs, intense fluorescence, and reversibility) makes Zn^{II}–tpy complexes/polymers highly interesting for use in various applications and, thus, one can expect that research on these materials will be continued.

In Ln^{III}–tpy complexes, the metal-centered emission is stimulated by the ligands *via* the so-called "antenna effect." Thus, the origin of emission is fundamentally different from the fluorescence or phosphorescence of complexes with transition-metal ions [29]. Discrete Ln^{III}–tpy complexes have been utilized for chemosensing (due to emission quenching) and bioimaging, thus paralleling the use of their Zn^{II}–tpy counterparts. Moreover, thin polymer films or CNMs have been decorated with such emitters [235, 278]; however, practical applications of these materials have (not yet) been deduced. Detrimentally, the use of rare-earth metals entails relatively high costs and, thus, luminescent Ln^{III}–tpy complexes will only be suited for highly specialized applications (*e.g.*, in biological studies), but not for large-scale device applications.

Finally, the combination of tpy ligands with ions of main-group elements is worth mentioning [250]. Perovskite-type materials as well as Bi^{III} coordination compounds represent emerging classes of solid-state emitters. In particular, the PeLEDs have received enormous recognition and EQE values of $>20\%$ have already been reached within ten years after their discovery [252]. In Bi^{III} coordination compounds, a manifold of electronic transitions contributes to their luminescence [249]. In such materials, the emission color can be tuned *via* the choice of ligands, dopants, and/or the preparation method. In this context, Knope *et al.* presented the first examples for Bi^{III}–tpy compounds as solid-state emitters [264]. The first tpy-containing PeLED materials using indium ions were recently reported by Fan, Li *et al.* [257]. In either case, the structure–property relationships are not yet fully understood, but both types of materials certainly possess potential for use in practical devices and deserve intensified research.

References

1 M. T. Rupp, N. Shevchenko, G. S. Hanan, D. G. Kurth, *Coord. Chem. Rev.* 2021, 446, 214127.
2 J.-M. Lehn, J.-P. Sauvage, *Nouv. J. Chim.* 1977, 1, 449–451.
3 S. Campagna, F. Puntoriero, F. Nastasi, G. Bergamini, V. Balzani, in *Photochemistry and Photophysics of Coordination Compounds, Part I*, (Eds.: V. Balzani, S. Campagna), Springer, Berlin & Heidelberg, 2007, pp. 117–214.
4 A. Juris, V. C. Balzani, F. Barigelletti, S. Campagna, P. Belser, A. von Zelewsky, *Coord. Chem. Rev.* 1988, 84, 85–277.
5 C. Fu, M. Wenzel, E. Treutlein, K. Harms, E. Meggers, *Inorg. Chem.* 2012, 51, 10004–10011.
6 E. Meggers, *Chem. Eur. J.* 2010, 16, 752–758.
7 P. Sun, A. Krishnan, A. Yadav, F. M. MacDonnell, D. W. Armstrong, *J. Mol. Struct.* 2008, 890, 75–80.
8 C. Jiang, M.-Y. Tong, D. W. Armstrong, S. Perera, Y. Bao, F. M. MacDonnell, *Chirality* 2009, 21, 208–217.
9 O. Hamelin, J. Pécaut, M. Fontecave, *Chem. Eur. J.* 2004, 10, 2548–2554.
10 E. Baranoff, J.-P. Collin, L. Flamigni, J.-P. Sauvage, *Chem. Soc. Rev.* 2004, 33, 147–155.
11 A. Winter, U. S. Schubert, *Chem. Soc. Rev.* 2016, 45, 5311–5357.
12 A. E. Martell, R. D. Hancock, R. J. Motekaitis, *Coord. Chem. Rev.* 1994, 133, 39–65.
13 S. Mallakpour, M. Dinari, H. Hadadzadeh, M. Daryanavard, F. Roudi, *J. Fluoresc.* 2014, 24, 1841–1848.
14 M. Abrahamsson, M. Jäger, T. Österman, L. Eriksson, P. Persson, H.-C. Becker, O. Johansson, L. Hammarström, *J. Am. Chem. Soc.* 2006, 128, 12616–12617.
15 J.-P. Sauvage, J.-P. Collin, J.-C. Chambron, S. Guillerez, C. Coudret, V. Balzani, F. Barigelletti, L. De Cola, L. Flamigni, *Chem. Rev.* 1994, 94, 993–1019.
16 E. A. Medlycott, G. S. Hanan, *Coord. Chem. Rev.* 2006, 250, 1763–1782.
17 L. Hammarström, O. Johansson, *Coord. Chem. Rev.* 2010, 254, 2546–2559.
18 U. S. Schubert, A. Winter, G. R. Newkome, *Terpyridine-Based Materials*, Wiley-VCH, Weinheim, 2011.
19 E. A. Medlycott, G. S. Hanan, *Chem. Soc. Rev.* 2005, 34, 133–142.
20 S. Otto, M. Grabolle, C. Förster, C. Kreitner, U. Resch-Genger, K. Heinze, *Angew. Chem. Int. Ed.* 2015, 54, 11572–11576.
21 M. Dorn, J. Kalmbach, P. Boden, A. Päpcke, S. Gómez, C. Förster, F. Kuczelinis, L. M. Carrella, L. A. Büldt, N. H. Bings, E. Rentschler, S. Lochbrunner, L. González, M. Gerhards, M. Seitz, K. Heinze, *J. Am. Chem. Soc.* 2020, 142, 7947–7955.
22 O. S. Wenger, *J. Am. Chem. Soc.* 2018, 140, 13522–13533.
23 C. Wegeberg, O. S. Wenger, *JACS Au* 2021, 1, 1860–1876.

24 A. Breivogel, M. Park, D. Lee, S. Klassen, A. Kühnle, C. Lee, K. Char, K. Heinze, *Eur. J. Inorg. Chem.* 2014, 2014, 288–295.
25 R. D. Costa, E. Ortí, H. J. Bolink, F. Monti, G. Accorsi, N. Armaroli, *Angew. Chem. Int. Ed.* 2012, 51, 8178–8211.
26 I. Eryazici, C. N. Moorefield, G. R. Newkome, *Chem. Rev.* 2008, 108, 1834–1895.
27 B. Happ, A. Winter, M. D. Hager, U. S. Schubert, *Chem. Soc. Rev.* 2012, 41, 2222–2255.
28 J. A. G. Williams, A. J. Wilkinson, V. L. Whittle, *Dalton Trans.* 2008, 16, 2081–2099.
29 X. Yu, Y. Hu, C. Guo, Z. Chen, H. Wang, X. Li, *Supramol. Mater.* 2022, 1, 100017.
30 A. Wild, A. Winter, F. Schlütter, U. S. Schubert, *Chem. Soc. Rev.* 2011, 40, 1459–1511.
31 G. Albano, V. Balzani, E. C. Constable, M. Maestri, D. R. Smith, *Inorg. Chim. Acta* 1998, 277, 225–231.
32 T. Mutai, H. Satou, K. Araki, *Nat. Mater.* 2005, 4, 685–687.
33 T. Mutai, K. Araki, *Curr. Org. Chem.* 2007, 11, 195–211.
34 W. Goodall, K. Wild, K. J. Arm, J. A. G. Williams, *J. Chem. Soc. Perkin Trans.* 2002, 2, 1669–1681.
35 Y. Gao, J. Wu, Q. Zhao, L. Zheng, H. Zhou, S. Zhang, J. Yang, Y. Tian, *New J. Chem.* 2009, 33, 607–611.
36 F. A. Murphy, S. M. Draper, *J. Org. Chem.* 2010, 75, 1862–1870.
37 C. P. Brock, R. P. Minton, *J. Am. Chem. Soc.* 1989, 111, 4586–4593.
38 Y.-Q. Fang, N. J. Taylor, F. Laverdière, G. S. Hanan, F. Loiseau, F. Nastasi, S. Campagna, H. Nierengarten, E. Leize-Wagner, A. Van Dorsselaer, *Inorg. Chem.* 2007, 46, 2854–2863.
39 D. Wu, T. Shao, J. Men, X. Chen, G. Gao, *Dalton Trans.* 2014, 43, 1753–1761.
40 R. Siebert, D. Akimov, M. Schmitt, A. Winter, U. S. Schubert, B. Dietzek, J. Popp, *ChemPhysChem* 2009, 10, 910–919.
41 R. Siebert, A. Winter, U. S. Schubert, B. Dietzek, J. Popp, *J. Phys. Chem. C* 2010, 114, 6841–6848.
42 A. M. Maroń, O. Cannelli, E. C. Socie, P. Lodowski, B. Machura, *Molecules* 2022, 27, 7071.
43 A. M. Maroń, O. Cannelli, E. C. Socie, P. Lodowski, M. Oppermann, B. Machura, M. Chergui, *Phys. Chem. Chem. Phys.* 2024, 26, 6265–6276.
44 D. Zych, A. Slodek, M. Matussek, M. Filapek, G. Szafraniec-Gorol, S. Krompiec, S. Kotowicz, M. Siwy, E. Schab-Balcerzak, K. Bednarczyk, M. Libera, K. Smolarek, S. Maćkowski, W. Danikiewicz, *ChemistrySelect* 2017, 2, 8221–8233.
45 D. Zych, A. Slodek, M. Matussek, M. Filapek, G. Szafraniec-Gorol, S. Maślanka, S. Krompiec, S. Kotowicz, E. Schab-Balcerzak, K. Smolarek, S. Maćkowski, M. Olejnik, W. Danikiewicz, *Dyes Pigments* 2017, 146, 331–343.
46 A. Wild, C. Friebe, A. Winter, M. D. Hager, U.-W. Grummt, U. S. Schubert, *Eur. J. Org. Chem.* 2010, 2010, 1859–1868.
47 M. Higuchi, *Bull. Jpn. Soc. Coord. Chem.* 2022, 79, 68–77.

48 S. Mondal, T. Yoshida, S. Maji, K. Ariga, M. Higuchi, A. C. S. Appl, *Mater. Interfaces* 2020, 12, 16342–16349.
49 P. Liu, G. Shi, X. Chen, *Front. Chem.* 2020, 8, 592055.
50 H. Sasabe, Y. Hayasaka, R. Komatsu, K. Nakao, J. Kido, *Chem. Eur. J.* 2017, 23, 114–119.
51 C. A. Parker, C. G. Hatchard, *Trans. Faraday Soc.* 1961, 57, 1894–1904.
52 A. Endo, M. Ogasawara, A. Takahashi, D. Yokoyama, Y. Kato, C. Adachi, *Adv. Mater.* 2009, 21, 4802–4806.
53 A. Endo, K. Sato, K. Yoshimura, T. Kai, A. Kawada, H. Miyazaki, C. Adachi, *Appl. Phys. Lett.* 2011, 98, 083302.
54 H. Uoyama, K. Goushi, K. Shizu, H. Nomura, C. Adachi, *Nature* 2012, 492, 234–238.
55 R. Ieuji, K. Goushi, C. Adachi, *Nat. Commun.* 2019, 10, 5283.
56 H.-T. Feng, J. Zeng, P.-A. Yin, X.-D. Wang, Q. Peng, Z. Zhao, J. W. Y. Lam, B. Z. Tang, *Nat. Commun.* 2020, 11, 2617.
57 Y.-Z. Shi, H. Wu, K. Wang, J. Yu, X.-M. Ou, X.-H. Zhang, *Chem. Sci.* 2022, 13, 3625–3651.
58 H.-Z. Li, F.-M. Xie, Y.-Q. Li, J.-X. Tang, *J. Mater. Chem. C* 2023, 11, 6471–6511.
59 Y. Shi, K. Wang, Y. Tsuchiya, W. Liu, T. Komino, X. Fan, D. Sun, G. Dai, J. Chen, M. Zhang, C. Zheng, S. Xiong, X. Ou, J. Yu, J. Jie, C.-S. Lee, C. Adachi, X. Zhang, *Mater. Horiz.* 2020, 7, 2734–2740.
60 G. Hong, X. Gan, C. Leonhardt, Z. Zhang, J. Seibert, J. M. Busch, S. Bräse, *Adv. Mater.* 2021, 33, 2005630.
61 H.-W. Chen, J.-H. Lee, B.-Y. Lin, S. Chen, S.-T. Wu, *Light Sci. Appl.* 2018, 7, 17168–17168.
62 S.-M. Lee, J. H. Kwon, S. Kwon, K. C. Choi, *IEEE Trans. Electron Devices* 2017, 64, 1922–1931.
63 D. Volz, M. Wallesch, C. Fléchon, M. Danz, A. Verma, J. M. Navarro, D. M. Zink, S. Bräse, T. Baumann, *Green Chem.* 2015, 17, 1988–2011.
64 I. Siddiqui, S. Kumar, Y.-F. Tsai, P. Gautam, K. Shahnawaz, J.-T. Kesavan, L. Lin, K.-H. Khai, A. Chou, S. Choudhury, J.-H. J. Grigalevicius, *Nanomaterials* 2023, 13, 2521.
65 N. Zheludev, *Nat. Photonics* 2007, 1, 189–192.
66 A. Islam, M. Rabbani, M. H. Bappy, M. A. R. Miah, N. Sakib, *International Conference on Informatics, Electronics and Vision (ICIEV)*, Dhaka, Bangladesh, IEEEE, 2013, pp. 1–5.
67 J. H. Burroughes, D. D. C. Bradley, A. R. Brown, R. N. Marks, K. Mackay, R. H. Friend, P. L. Burns, A. B. Holmes, *Nature* 1990, 347, 539–541.
68 C. W. Tang, S. A. VanSlyke, *Appl. Phys. Lett.* 1987, 51, 913–915.
69 J. Bauri, R. B. Choudhary, G. Mandal, *J. Mater. Sci.* 2021, 56, 18837–18866.
70 K. Müllen, U. Scherf, Eds, *Organic Light Emitting Devices*, Wiley-VCH, Weinheim, 2005.
71 Z. Li, Z. R. Li, H. Meng, Eds, *Organic Light-Emitting Materials and Devices*, CRC Press, Boca Raton, 2006.
72 S. A. VanSlyke, C. W. Tang, L. C. Roberts, Patent US-4720432A, 1987.

73 M. Koden, *OLED Displays and Lighting*, John Wiley & Sons Ltd., Chichester, 2017.

74 *TADF: What is Thermally Activated Delayed Fluorescence?*, https://www.edinst.com/blog/tadf-thermally-activated-delayed-fluorescence/ (last accessed: 2025-04-01).

75 A. Salehi, X. Fu, D.-H. Shin, F. So, *Adv. Funct. Mater.* 2019, 29, 1808803.

76 M. A. Baldo, M. E. Thompson, S. R. Forrest, *Nature* 2000, 403, 750–753.

77 H. Xu, R. Chen, Q. Sun, W. Lai, Q. Su, W. Huang, X. Liu, *Chem. Soc. Rev.* 2014, 43, 3259–3302.

78 E. Longhi, L. De Cola, in *Iridium(III) in Optoelectronic and Photonics Applications*, (Ed.: E. Zysman-Coleman), John Wiley & Sons, Chichester, 2017, pp. 205–274.

79 M. A. Baldo, S. Lamansky, P. E. Burrows, M. E. Thompson, S. R. Forrest, *Appl. Phys. Lett.* 1999, 75, 4–6.

80 Z. Yang, Z. Mao, Z. Xie, Y. Zhang, S. Liu, J. Zhao, J. Xu, Z. Chi, M. P. Aldred, *Chem. Soc. Rev.* 2017, 46, 915–1016.

81 T. Smith, J. Guild, *Trans. Opt. Soc.* 1931, 33, 73.

82 *What is the CIE Color Space?*, https://luminusdevices.zendesk.com/hc/en-us/articles/4414846186253-What-is-the-CIE-Color-Space-What-s-the-difference-between-CIE-1931-and-CIE-1976 (last accessed: 2025-04-01).

83 *CIE 1931 Color Space*, https://en.wikipedia.org/wiki/CIE_1931_color_space (last accessed: 2025-04-01).

84 D. H. Kim, N. S. Cho, H.-Y. Oh, J. H. Yang, W. S. Jeon, J. S. Park, M. C. Suh, J. H. Kwon, *Adv. Mater.* 2011, 23, 2721–2726.

85 C.-L. Liu, C.-J. Zheng, X.-K. Liu, Z. Chen, J.-P. Yang, F. Li, X.-M. Ou, X.-H. Zhang, *J. Mater. Chem. C* 2015, 3, 1068–1076.

86 C. Fan, X. Wang, J. Luo, *Opt. Mater.* 2017, 64, 489–495.

87 D. Chen, S.-J. Su, Y. Cao, *J. Mater. Chem. C* 2014, 2, 9565–9578.

88 A. P. Kulkarni, C. J. Tonzola, A. Babel, S. A. Jenekhe, *Chem. Mater.* 2004, 16, 4556–4573.

89 Y. Watanabe, R. Yoshioka, H. Sasabe, T. Kamata, H. Katagiri, D. Yokoyama, J. Kido, *J. Mater. Chem. C* 2016, 4, 8980–8988.

90 T. Owada, H. Sasabe, Y. Sukegawa, T. Watanabe, T. Maruyama, Y. Watanabe, D. Yokoyama, J. Kido, *J. Mater. Chem. C* 2020, 8, 3200–3205.

91 X. Guo, Z. Tang, W. Yu, Y. Wang, Z. Zhao, J. Gu, Z. Liu, B. Qu, L. Xiao, Z. Chen, *Org. Electron.* 2021, 89, 106048.

92 B. H. Jhun, D. Song, S. Y. Park, Y. You, *Top. Curr. Chem.* 2022, 380, 35.

93 K. K.-W. Lo, K. Y. Zhang, *RSC Adv.* 2012, 2, 12069–12083.

94 R. Gao, D. G. Ho, B. Hernandez, M. Selke, D. Murphy, P. I. Djurovich, M. E. Thompson, *J. Am. Chem. Soc.* 2002, 124, 14828–14829.

95 H. Huang, S. Banerjee, P. J. Sadler, *ChemBioChem* 2018, 19, 1574–1589.

96 T. M. Monos, C. R. J. Stephenson, in *Iridium(III) in Optoelectronic and Photonics Applications*, (Ed.: E. Zysman-Coleman), John Wiley & Sons Ltd., Chichester, 2017, pp. 541–581.

97 J. C. Deaton, F. N. Castellano, in *Iridium(III) in Optoelectronic and Photonics Applications*, (Ed.: E. Zysman-Coleman), John Wiley & Sons Ltd., Chichester, 2017, pp. 1–69.

98 A. F. Henwood, E. Zysman-Colman, *Top. Curr. Chem.* 2016, 374, 36.

99 M. S. Lowry, S. Bernhard, *Chem. Eur. J.* 2006, 12, 7970–7977.

100 F. Dumur, Y. Guillaneuf, A. Guerlin, G. Wantz, D. Bertin, F. Miomandre, G. Clavier, D. Gigmes, C. R. Mayer, *Macromol. Chem. Phys.* 2011, 212, 1616–1628.

101 J. A. Porras, I. N. Mills, W. J. Transue, S. Bernhard, *J. Am. Chem. Soc.* 2016, 138, 9460–9472.

102 J. A. G. Williams, *Chem. Soc. Rev.* 2009, 38, 1783–1801.

103 J.-P. Collin, I. M. Dixon, J.-P. Sauvage, J. A. G. Williams, F. Barigelletti, L. Flamigni, *J. Am. Chem. Soc.* 1999, 121, 5009–5016.

104 A. J. Wilkinson, A. E. Goeta, C. E. Foster, J. A. G. Williams, *Inorg. Chem.* 2004, 43, 6513–6515.

105 V. L. Whittle, J. A. G. Williams, *Inorg. Chem.* 2008, 47, 6596–6607.

106 K. Li, G. S. Ming Tong, Q. Wan, G. Cheng, W.-Y. Tong, W.-H. Ang, W.-L. Kwong, C.-M. Che, *Chem. Sci.* 2016, 7, 1653–1673.

107 B. Z. Momeni, A. S. Abd-El-Aziz, *Coord. Chem. Rev.* 2023, 486, 215113.

108 J. A. G. Williams, S. Develay, D. L. Rochester, L. Murphy, *Coord. Chem. Rev.* 2008, 252, 2596–2611.

109 J. Kalinowski, V. Fattori, M. Cocchi, J. A. G. Williams, *Coord. Chem. Rev.* 2011, 255, 2401–2425.

110 R. Abhijnakrishna, K. Magesh, A. Ayushi, S. Velmathi, *Coord. Chem. Rev.* 2023, 496, 215380.

111 M. N. Hopkinson, A. D. Gee, V. Gouverneur, *Chem. Eur. J.* 2011, 17, 8248–8262.

112 C. Bronner, O. S. Wenger, *Dalton Trans.* 2011, 40, 12409–12420.

113 C. Fan, C. Yang, *Chem. Soc. Rev.* 2014, 43, 6439–6469.

114 R. Kumar, C. Nevado, *Angew. Chem. Int. Ed.* 2017, 56, 1994–2015.

115 V. W.-W. Yam, K. M.-C. Wong, L.-L. Hung, N. Zhu, *Angew. Chem. Int. Ed.* 2005, 44, 3107–3110.

116 M.-C. Tang, D. P.-K. Tsang, Y.-C. Wong, M.-Y. Chan, K. M.-C. Wong, V. W.-W. Yam, *J. Am. Chem. Soc.* 2014, 136, 17861–17868.

117 K. M.-C. Wong, X. Zhu, L.-L. Hung, N. Zhu, V. W.-W. Yam, H.-S. Kwok, *Chem. Commun.* 2005, 23, 2906–2908.

118 V. K.-M. Au, K. M.-C. Wong, D. P.-K. Tsang, M.-Y. Chan, N. Zhu, V. W.-W. Yam, *J. Am. Chem. Soc.* 2010, 132, 14273–14278.

119 A. S. Polo, M. K. Itokazu, N. Y. Murakami Iha, *Coord. Chem. Rev.* 2004, 248, 1343–1361.

120 Y. Kuramochi, O. Ishitani, H. Ishida, *Coord. Chem. Rev.* 2018, 373, 333–356.

121 S.-S. Sun, A. J. Lees, *J. Am. Chem. Soc.* 2000, 122, 8956–8967.

122 G.-W. Zhao, J.-H. Zhao, Y.-X. Hu, D.-Y. Zhang, X. Li, *Synth. Met.* 2016, 212, 131–141.

123 W.-M. Xue, M. C.-W. Chan, Z.-M. Su, K.-K. Cheung, S.-T. Liu, C.-M. Che, *Organometallics* 1998, 17, 1622–1630.

124 M. Wrighton, D. L. Morse, *J. Am. Chem. Soc.* 1974, 96, 998–1003.

125 P. J. Giordano, S. M. Fredericks, M. S. Wrighton, D. L. Morse, *J. Am. Chem. Soc.* 1978, 100, 2257–2259.

126 X. Gong, P. K. Ng, W. K. Chan, *Adv. Mater.* 1998, 10, 1337–1340.

127 W. K. Chan, P. K. Ng, X. Gong, S. Hou, *Appl. Phys. Lett.* 1999, 75, 3920–3922.

128 A. Vlček, M. Busby, *Coord. Chem. Rev.* 2006, 250, 1755–1762.

129 A. Juris, S. Campagna, I. Bidd, J.-M. Lehn, R. Ziessel, *Inorg. Chem.* 1988, 27, 4007–4011.

130 L. A. Worl, R. Duesing, P. Chen, L. D. Ciana, T. J. Meyer, *J. Chem. Soc. Dalton Trans.* 1991, 1991, 849–858.

131 D. Wang, Q.-L. Xu, S. Zhang, H.-Y. Li, C.-C. Wang, T.-Y. Li, Y.-M. Jing, W. Huang, Y.-X. Zheng, G. Accorsi, *Dalton Trans.* 2013, 42, 2716–2723.

132 G. Velmurugan, P. Venuvanalingam, *Dalton Trans.* 2015, 44, 8529–8542.

133 B. N. Ghosh, F. Topić, P. K. Sahoo, P. Mal, J. Linnera, E. Kalenius, H. M. Tuononen, K. Rissanen, *Dalton Trans.* 2015, 44, 254–267.

134 T. Klemens, A. Świtlicka-Olszewska, B. Machura, M. Grucela, H. Janeczek, E. Schab-Balcerzak, A. Szlapa, S. Kula, S. Krompiec, K. Smolarek, D. Kowalska, S. Mackowski, K. Erfurt, P. Lodowski, *RSC Adv.* 2016, 6, 56335–56352.

135 K. Choroba, A. Maroń, A. Świtlicka, A. Szłapa-Kula, M. Siwy, J. Grzelak, S. Maćkowski, T. Pedzinski, E. Schab-Balcerzak, B. Machura, *Dalton Trans.* 2021, 50, 3943–3958.

136 T. Klemens, A. Świtlicka-Olszewska, B. Machura, M. Grucela, E. Schab-Balcerzak, K. Smolarek, S. Mackowski, A. Szlapa, S. Kula, S. Krompiec, P. Lodowski, A. Chrobok, *Dalton Trans.* 2016, 45, 1746–1762.

137 T. Klemens, A. Świtlicka, A. Szlapa-Kula, S. Krompiec, P. Lodowski, A. Chrobok, M. Godlewska, S. Kotowicz, M. Siwy, K. Bednarczyk, M. Libera, S. Maćkowski, T. Pędziński, E. Schab-Balcerzak, B. Machura, *Appl. Organomet. Chem.* 2018, 32, e4611.

138 T. Klemens, A. Świtlicka, S. Kula, M. Siwy, K. Łaba, J. Grzelak, M. Szalkowski, S. Maćkowski, E. Schab-Balcerzak, B. Machura, *J. Lumin.* 2019, 209, 346–356.

139 M. Małecka, B. Machura, A. Świtlicka, S. Kotowicz, G. Szafraniec-Gorol, M. Siwy, M. Szalkowski, S. Maćkowski, E. Schab-Balcerzak, *Spectrochim. Acta Part A* 2020, 231, 118124.

140 K. Choroba, S. Kotowicz, A. Maroń, A. Świtlicka, A. Szłapa-Kula, M. Siwy, J. Grzelak, K. Sulowska, S. Maćkowski, E. Schab-Balcerzak, B. Machura, *Dyes Pigments* 2021, 192, 109472.

141 A. Szlapa-Kula, M. Małecka, A. M. Maroń, H. Janeczek, M. Siwy, E. Schab-Balcerzak, M. Szalkowski, S. Maćkowski, T. Pedzinski, K. Erfurt, B. Machura, *Inorg. Chem.* 2021, 60, 18726–18738.

142 J. Palion-Gazda, A. Szłapa-Kula, M. Penkala, K. Erfurt, B. Machura, *Molecules* 2022, 27, 7147.

143 J. Palion-Gazda, K. Choroba, A. M. Maroń, E. Malicka, B. Machura, *Molecules* 2024, 29, 1631.

144 E. W. Abel, V. S. Dimitrov, N. J. Long, K. G. Orrell, A. G. Osborne, H. M. Pain, V. Šik, M. B. Hursthouse, M. A. Mazid, *J. Chem. Soc. Dalton Trans.* 1993, 4, 597–603.

145 D. R. Black, S. E. Hightower, *Inorg. Chem. Commun.* 2012, 24, 16–19.
146 B. A. Frenzel, J. E. Schumaker, D. R. Black, S. E. Hightower, *Dalton Trans.* 2013, 42, 12440–12451.
147 T. Auvray, B. Del Secco, A. Dubreuil, N. Zaccheroni, G. S. Hanan, *Inorg. Chem.* 2021, 60, 70–79.
148 Q. Pei, Y. Yang, G. Yu, C. Zhang, A. J. Heeger, *J. Am. Chem. Soc.* 1996, 118, 3922–3929.
149 Q. Pei, G. Yu, C. Zhang, Y. Yang, A. J. Heeger, *Science* 1995, 269, 1086–1088.
150 K. M. Maness, R. H. Terrill, T. J. Meyer, R. W. Murray, R. M. Wightman, *J. Am. Chem. Soc.* 1996, 118, 10609–10616.
151 A. F. Henwood, E. Zysman-Colman, in *Iridium(III) in Optoelectronic and Photonics Applications*, (Ed.: E. Zysman-Coleman), John Wiley & Sons Ltd., Chichester, 2017, pp. 275–357.
152 H. J. Bolink, L. Cappelli, E. Coronado, P. Gaviña, *Inorg. Chem.* 2005, 44, 5966–5968.
153 K. Youssef, Y. Li, S. O'Keeffe, L. Li, Q. Pei, *Adv. Funct. Mater.* 2020, 30, 1909102.
154 H. J. Bolink, E. Coronado, R. D. Costa, P. Gaviña, E. Ortí, S. Tatay, *Inorg. Chem.* 2009, 48, 3907–3909.
155 M. I. J. Polson, E. A. Medlycott, G. S. Hanan, L. Mikelsons, N. J. Taylor, M. Watanabe, Y. Tanaka, F. Loiseau, R. Passalacqua, S. Campagna, *Chem. Eur. J.* 2004, 10, 3640–3648.
156 H. J. Bolink, E. Coronado, R. D. Costa, E. Ortí, M. Sessolo, S. Graber, K. Doyle, M. Neuburger, C. E. Housecroft, E. C. Constable, *Adv. Mater.* 2008, 20, 3910–3913.
157 E. C. Constable, C. E. Housecroft, G. E. Schneider, J. A. Zampese, H. J. Bolink, A. Pertegás, C. Roldan-Carmona, *Dalton Trans.* 2014, 43, 4653–4667.
158 E. Holder, V. Marin, D. Kozodaev, M. A. R. Meier, B. G. G. Lohmeijer, U. S. Schubert, *Macromol. Chem. Phys.* 2005, 206, 989–997.
159 E. Tekin, E. Holder, V. Marin, B.-J. de Gans, U. S. Schubert, *Macromol. Rapid Commun.* 2005, 26, 293–297.
160 E. Holder, V. Marin, M. A. R. Meier, U. S. Schubert, *Macromol. Rapid Commun.* 2004, 25, 1491–1496.
161 K. A. Aamer, G. N. Tew, *J. Polym. Sci. Part A Polym. Chem.* 2007, 45, 1109–1121.
162 W. Y. Ng, X. Gong, W. K. Chan, *Chem. Mater.* 1999, 11, 1165–1170.
163 F. Dumur, *Synth. Met.* 2014, 195, 241–251.
164 Q.-D. Liu, R. Wang, S. Wang, *Dalton Trans.* 2004, 14, 2073–2079.
165 T. Sano, Y. Nishio, Y. Hamada, H. Takahashi, T. Usuki, K. Shibata, *J. Mater. Chem.* 2000, 10, 157–161.
166 X. Chen, Q. Zhou, Y. Cheng, Y. Geng, D. Ma, Z. Xie, L. Wang, *J. Lumin.* 2007, 126, 81–90.
167 P.-T. Chou, Y. Chi, *Chem. Eur. J.* 2007, 13, 380–395.
168 R. Diana, B. Panunzi, *Molecules* 2020, 25, 4984.

169 R. Lakshmanan, N. C. Shivaprakash, S. N. Sindhu, *J. Lumin.* 2018, 196, 136–145.
170 A. Winter, C. Friebe, M. D. Hager, U. S. Schubert, *Macromol. Rapid Commun.* 2008, 29, 1679–1686.
171 A. Wild, A. Teichler, C. von der Ehe, A. Winter, M. D. Hager, B. Yao, B. Zhang, Z. Xie, W.-Y. Wong, U. S. Schubert, *Macromol. Chem. Phys.* 2013, 214, 1072–1080.
172 A. Winter, C. Friebe, M. Chiper, U. S. Schubert, M. Presselt, B. Dietzek, M. Schmitt, J. Popp, *ChemPhysChem* 2009, 10, 787–798.
173 F. Tessore, D. Roberto, R. Ugo, M. Pizzotti, S. Quici, M. Cavazzini, S. Bruni, F. De Angelis, *Inorg. Chem.* 2005, 44, 8967–8978.
174 P. Shi, Q. Jiang, X. Zhao, Q. Zhang, Y. Tian, *Dalton Trans.* 2015, 44, 8041–8048.
175 Q. Zhang, X. Tian, Z. Hu, C. Brommesson, J. Wu, H. Zhou, S. Li, J. Yang, Z. Sun, Y. Tian, K. Uvdal, *J. Mater. Chem. B* 2015, 3, 7213–7221.
176 Z. Feng, D. Li, M. Zhang, T. Shao, Y. Shen, X. Tian, Q. Zhang, S. Li, J. Wu, Y. Tian, *Chem. Sci.* 2019, 10, 7228–7232.
177 J. Luo, Z. Xie, J. W. Y. Lam, L. Cheng, H. Chen, C. Qiu, H. S. Kwok, X. Zhan, Y. Liu, D. Zhu, B. Z. Tang, *Chem. Commun.* 2001, 18, 1740–1741.
178 Y. Hong, J. W. Y. Lam, B. Z. Tang, *Chem. Soc. Rev.* 2011, 40, 5361–5388.
179 R. Diana, B. Panunzi, *Molecules* 2021, 26, 4176.
180 Q. Zhao, J. Wang, *Materi. Adv.* 2022, 3, 7569–7578.
181 A. Wild, A. Winter, M. D. Hager, U. S. Schubert, *Chem. Commun.* 2012, 48, 964–966.
182 C.-J. Li, J.-G. Guo, S.-L. Cai, S.-R. Zheng, W.-G. Zhang, *Inorg. Chem. Commun.* 2016, 73, 16–20.
183 A. Wild, A. Winter, M. D. Hager, U. S. Schubert, *Analyst* 2012, 137, 2333–2337.
184 J. Harathi, R. S. Kumar, S. K. A. Kumar, D. Saravanakumar, S. Senthilkumar, K. Thenmozhi, *Sensors Actuators B Chem.* 2023, 390, 133967.
185 P. Das, A. Ghosh, M. K. Kesharwani, V. Ramu, B. Ganguly, A. Das, *Eur. J. Inorg. Chem.* 2011, 2011, 3050–3058.
186 T. Mandal, A. Hossain, A. Dhara, A. Al Masum, S. Konar, S. K. Manna, S. K. Seth, S. Pathak, S. Mukhopadhyay, *Photochem. Photobiol. Sci.* 2018, 17, 1068–1074.
187 C.-L. Ho, W.-Y. Wong, *Coord. Chem. Rev.* 2011, 255, 2469–2502.
188 R. Dobrawa, F. Würthner, *Chem. Commun.* 2002, 17, 1878–1879.
189 X.-y. Wang, A. Del Guerzo, R. H. Schmehl, *Chem. Commun.* 2002, 2002, 2344–2345.
190 P. D. Vellis, J. A. Mikroyannidis, C.-N. Lo, C.-S. Hsu, *J. Polym. Sci. Part A Polym. Chem.* 2008, 46, 7702–7712.
191 Y.-Y. Chen, Y.-T. Tao, H.-C. Lin, *Macromolecules* 2006, 39, 8559–8566.
192 R. Dobrawa, M. Lysetska, P. Ballester, M. Grüne, F. Würthner, *Macromolecules* 2005, 38, 1315–1325.
193 P. Bláhová, J. Zedník, I. Šloufová, J. Vohlídal, J. Svoboda, *Soft Mater.* 2014, 12, 214–229.

194 P. Štenclová, K. Šichová, I. Šloufová, J. Zedník, J. Vohlídal, J. Svoboda, *Dalton Trans.* 2016, 45, 1208–1224.

195 A. Wild, F. Schlütter, G. M. Pavlov, C. Friebe, G. Festag, A. Winter, M. D. Hager, V. Cimrová, U. S. Schubert, *Macromol. Rapid Commun.* 2010, 31, 868–874.

196 T. Vitvarová, J. Svoboda, M. Hissler, J. Vohlídal, *Organometallics* 2017, 36, 777–786.

197 A. Langlois, P. B. J. St, P.-L. Onge, A. Karsenti, S. R.-G. Younus, *J. Phys. Chem. B* 2021, 125, 9579–9587.

198 I. Šloufová, T. Urválková, M. Hissler, J. Vohlídal, *Polymers* 2022, 14, 5207.

199 M. Hrma, K. Šichová, J. Svoboda, J. Vohlídal, *Polymer* 2017, 122, 22–33.

200 M. Yuan, F. Wang, Y.-K. Tian, *RSC Adv.* 2018, 8, 40794–40797.

201 S.-C. Yu, C.-C. Kwok, W.-K. Chan, C.-M. Che, *Adv. Mater.* 2003, 15, 1643–1647.

202 Y.-Y. Chen, H.-C. Lin, *J. Polym. Sci. Part A Polym. Chem.* 2007, 45, 3243–3255.

203 A. Winter, C. Friebe, M. Chiper, M. D. Hager, U. S. Schubert, *J. Polym. Sci. Part A Polym. Chem.* 2009, 47, 4083–4098.

204 C. Friebe, A. Wild, J. Perelaer, U. S. Schubert, *Macromol. Rapid Commun.* 2012, 33, 503–509.

205 F. Li, Z. Hu, H. Qiao, L. Liu, J. Hu, X. Chen, J. Li, *Dyes Pigments* 2016, 132, 142–150.

206 A. Wild, A. Teichler, C.-L. Ho, X.-Z. Wang, H. Zhan, F. Schlütter, A. Winter, M. D. Hager, W.-Y. Wong, U. S. Schubert, *J. Mater. Chem. C* 2013, 1, 1812–1822.

207 F. Schlütter, A. Wild, A. Winter, M. D. Hager, A. Baumgaertel, C. Friebe, U. S. Schubert, *Macromolecules* 2010, 43, 2759–2771.

208 S.-H. Hwang, P. Wang, C. N. Moorefield, J.-C. Jung, J.-Y. Kim, S.-W. Lee, G. R. Newkome, *Macromol. Rapid Commun.* 2006, 27, 1809–1813.

209 G.-Q. Yin, H. Wang, X.-Q. Wang, B. Song, L.-J. Chen, L. Wang, X.-Q. Hao, H.-B. Yang, X. Li, *Nat. Commun.* 2018, 9, 567.

210 J.-C. G. Bünzli, *Acc. Chem. Res.* 2006, 39, 53–61.

211 J.-C. G. Bünzli, C. Piguet, *Chem. Soc. Rev.* 2005, 34, 1048–1077.

212 H. B. Kagan, *Chem. Rev.* 2002, 102, 1805–1806.

213 Z. Zhu, M. Guo, X.-L. Li, J. Tang, *Coord. Chem. Rev.* 2019, 378, 350–364.

214 A. J. Amoroso, S. J. A. Pope, *Chem. Soc. Rev.* 2015, 44, 4723–4742.

215 S. V. Eliseeva, J.-C. G. Bünzli, *Chem. Soc. Rev.* 2010, 39, 189–227.

216 M. Hasegawa, H. Ohmagari, H. Tanaka, K. Machida, *J. Photochem. Photobiol. C Photochem. Rev.* 2022, 50, 100484.

217 S. Afzal, U. Maitra, *Helv. Chim. Acta* 2022, 105, e202100194.

218 M. J. Beltrán-Leiva, E. Solis-Céspedes, D. Páez-Hernández, *Dalton Trans.* 2020, 49, 7444–7450.

219 R. F. Ziessel, G. Ulrich, L. Charbonnière, D. Imbert, R. Scopelliti, J.-C. G. Bünzli, *Chem. Eur. J.* 2006, 12, 5060–5067.

220 L. L. Cai, Y. T. Hu, Y. Li, K. Wang, X. Q. Zhang, G. Muller, X. M. Li, G. X. Wang, *Inorg. Chim. Acta* 2019, 489, 85–92.

221 A. E. Sedykh, D. G. Kurth, K. Müller-Buschbaum, *Eur. J. Inorg. Chem.* 2019, 2019, 4564–4571.

222 J. Durán-Hernández, L. Muñoz-Rugeles, Ó. Guzmán-Méndez, M. M. Reza, A. Cadena-Caicedo, V. García-Montalvo, J. Peón, *J. Phys. Chem. A* 2022, 126, 2498–2510.
223 G. S. Gopika, P. M. H. Prasad, A. G. Lekshmi, S. Lekshmypriya, S. Sreesaila, C. Arunima, M. S. Kumar, A. Anil, A. Sreekumar, Z. S. Pillai, *Mater. Today Proc.* 2021, 46, 3102–3108.
224 E. S. Andreiadis, R. Demadrille, D. Imbert, J. Pécaut, M. Mazzanti, *Chem. Eur. J.* 2009, 15, 9458–9476.
225 S.-R. Zheng, R.-L. Chen, Z.-M. Liu, X.-L. Wen, T. Xie, J. Fan, W.-G. Zhang, *CrystEngComm* 2014, 16, 2898–2909.
226 P. Chen, Q. Li, S. Grindy, N. Holten-Andersen, *J. Am. Chem. Soc.* 2015, 137, 11590–11593.
227 D. Zhao, J. Yang, X. Tian, J. Wei, Q. Li, Y. Wang, *Chem. Eng. J.* 2022, 434, 134806.
228 O. Kotova, S. J. Bradberry, A. J. Savyasachi, T. Gunnlaugsson, *Dalton Trans.* 2018, 47, 16377–16387.
229 T. Sato, M. Higuchi, *Chem. Commun.* 2012, 48, 4947–4949.
230 U. S. Schubert, H. Hofmeier, G. R. Newkome, *Modern Terpyridine Chemistry*, Wiley-VCH, Weinheim, 2006.
231 R. Shunmugam, G. N. Tew, *Polym. Adv. Technol.* 2007, 18, 940–945.
232 R. Shunmugam, G. N. Tew, *J. Polym. Sci. Part A Polym. Chem.* 2005, 43, 5831–5843.
233 R. Shunmugam, G. N. Tew, *J. Am. Chem. Soc.* 2005, 127, 13567–13572.
234 R. Shunmugam, G. N. Tew, *Polym. Adv. Technol.* 2008, 19, 596–601.
235 Z. Tang, E. Chulanova, M. Küllmer, A. Winter, J. Picker, C. Neumann, K. Schreyer, F. Herrmann-Westendorf, A. Arnlind, B. Dietzek, U. S. Schubert, A. Turchanin, *Nanoscale* 2021, 13, 20583–20591.
236 A. Turchanin, *Chimia* 2019, 73, 473–479.
237 B.-H. Tong, S.-J. Wang, J. Jiao, F.-R. Ling, Y.-Z. Meng, B. Wang, *J. Photochem. Photobiol. A Chem.* 2007, 191, 74–79.
238 Q. Ru, Z. Xue, Y. Wang, Y. Liu, H. Li, *Eur. J. Inorg. Chem.* 2014, 2014, 469–474.
239 K. Ghosh, E. R. M. Balog, J. L. Kahn, D. P. Shepherd, J. S. Martinez, R. C. Rocha, *Macromol. Chem. Phys.* 2015, 216, 1856–1861.
240 C. Yang, J. Xu, Y. Zhang, Y. Li, J. Zheng, L. Liang, M. Lu, *J. Mater. Chem. C* 2013, 1, 4885–4901.
241 T. Kowalczyk, K. Hnatuszko-Konka, A. Gerszberg, A. K. Kononowicz, *World J. Microbiol. Biotechnol.* 2014, 30, 2141–2152.
242 Z. Zhang, H. Chang, Y. Kang, X. Li, H. Jiang, B. Xue, Y. Wang, X. Lü, X. Zhu, *Sensors Actuators B Chem.* 2019, 282, 999–1007.
243 K. P. Carter, S. J. A. Pope, C. L. Cahill, *CrystEngComm* 2014, 16, 1873–1884.
244 Y. Wang, Q.-Q. Zhao, N. Ren, J.-J. Zhang, L.-N. Geng, S.-P. Wang, *J. Therm. Anal. Calorim.* 2016, 126, 1703–1712.
245 Y. Wang, P.-P. Shen, N. Ren, J.-J. Zhang, L.-N. Geng, S.-P. Wang, S.-K. Shi, *RSC Adv.* 2016, 6, 70770–70780.

246 L. Ying-Ying, R. Ning, W. Shu-Ping, Z. Jian-Jun, *Inorg. Chim. Acta* 2020, 510, 119755.

247 J.-Y. Zhao, N. Ren, J.-J. Zhang, *Polyhedron* 2021, 194, 114892.

248 R. J. Batrice, J. A. Ridenour, R. L. Ayscue Iii, J. A. Bertke, K. E. Knope, *CrystEngComm* 2017, 19, 5300–5312.

249 A. K. Adcock, R. L. Ayscue, L. M. Breuer, C. P. Verwiel, A. C. Marwitz, J. A. Bertke, V. Vallet, F. Réal, K. E. Knope, *Dalton Trans.* 2020, 49, 11756–11771.

250 A. Winter, G. R. Newkome, U. S. Schubert, *Inorg. Chem. Front.* 2024, 11, 342–399.

251 A. Kojima, K. Teshima, Y. Shirai, T. Miyasaka, *J. Am. Chem. Soc.* 2009, 131, 6050–6051.

252 H. Min, D. Y. Lee, J. Kim, G. Kim, K. S. Lee, J. Kim, M. J. Paik, Y. K. Kim, K. S. Kim, M. G. Kim, T. J. Shin, S. Il Seok, *Nature* 2021, 598, 444–450.

253 Z.-K. Tan, R. S. Moghaddam, M. L. Lai, P. Docampo, R. Higler, F. Deschler, M. Price, A. Sadhanala, L. M. Pazos, D. Credgington, F. Hanusch, T. Bein, H. J. Snaith, R. H. Friend, *Nat. Nanotechnol.* 2014, 9, 687–692.

254 A. Fakharuddin, M. K. Gangishetty, M. Abdi-Jalebi, S.-H. Chin, A. R. bin Mohd Yusoff, D. N. Congreve, W. Tress, F. Deschler, M. Vasilopoulou, H. J. Bolink, *Nat. Electron.* 2022, 5, 203–216.

255 Y. Yu, Y. Tang, B. Wang, K. Zhang, J.-X. Tang, Y.-Q. Li, *Laser Photonics Rev.* 2023, 17, 2200608.

256 R. Yu, C. Li, B. Zhao, Z. a. Tan, *Laser Photonics Rev.* 2024, 18, 2300780.

257 Y. Liu, F. Di Stasio, C. Bi, J. Zhang, Z. Xia, Z. Shi, L. Manna, *Adv. Mater.* 2024, 36, 2312482.

258 C. Yi, G. Zhang, R. Lu, Z. Wang, Y. Wang, X. Cao, K. Ni, N. Chen, G. Zheng, W. Huang, J. Wang, *Nano Lett.* 2024, 24, 10196–10201.

259 S.-H. Jeong, J. Park, T.-H. Han, F. Zhang, K. Zhu, J. S. Kim, M.-H. Park, M. O. Reese, S. Yoo, T.-W. Lee, *Joule* 2020, 4, 1206–1235.

260 Y.-D. Lin, C.-W. Lu, H.-C. Su, *Chem. Eur. J.* 2023, 29, e202202985.

261 Q. Zhang, D. Zhang, B. Cao, S. Poddar, X. Mo, Z. Fan, *ACS Nano* 2024, 18, 8557–8570.

262 Y. Liu, L. K. Ono, Y. Qi, *InfoMat* 2020, 2, 1095–1108.

263 Y. Zhang, S. Yuan, Y. Yuan, Y. Bao, Q. Ran, E. Liu, J. Fan, W. Li, *Adv. Opt. Mater.* 2022, 10, 2102041.

264 R. L. Ayscue III, V. Vallet, J. A. Bertke, F. Réal, K. E. Knope, *Inorg. Chem.* 2021, 60, 9727–9744.

265 R. Mohan, *Nat. Chem.* 2010, 2, 336–336.

266 P. C. Ford, A. Vogler, *Acc. Chem. Res.* 1993, 26, 220–226.

267 H. Nikol, A. Vogler, *J. Am. Chem. Soc.* 1991, 113, 8988–8990.

268 J. R. Sorg, T. Schneider, L. Wohlfarth, T. C. Schäfer, A. Sedykh, K. Müller-Buschbaum, *Dalton Trans.* 2020, 49, 4904–4913.

269 A. Vogler, H. Nikol, *Pure Appl. Chem.* 1992, 64, 1311–1317.

270 L. A. Maurer, O. M. Pearce, F. D. R. Maharaj, N. L. Brown, C. K. Amador, N. H. Damrauer, M. P. Marshak, *Inorg. Chem.* 2021, 60, 10137–10146.

271 A. C. Marwitz, A. D. Nicholas, L. M. Breuer, J. A. Bertke, K. E. Knope, *Inorg. Chem.* 2021, 60, 16840–16851.

272 A. K. Adcock, A. C. Marwitz, L. A. Sanz, R. Lee Ayscue, J. A. Bertke, K. E. Knope, *CrystEngComm* 2021, 23, 8183–8197.

273 G. B. Haxel, J. B. Hedrick, G. J. Orris, U.S. Geological Socitey, 2002, Fact Sheet 087-002.

274 *Abundance of the chemical elements*, https://en.wikipedia.org/wiki/Abundance_of_the_chemical_elements (last accessed: 2025-04-01).

275 R. R. Panicker, A. Sivaramakrishna, *Coord. Chem. Rev.* 2022, 459, 214426.

276 S. Di Bella, A. Colombo, C. Dragonetti, S. Righetto, D. Roberto, *Inorganics* 2018, 6, 133.

277 S. F. Kainat, M. B. Hawsawi, E. U. Mughal, N. Naeem, A. M. Almohyawi, H. M. Altass, E. M. Hussein, A. Sadiq, Z. Moussa, A. S. Abd-El-Aziz, S. A. Ahmed, *RSC Adv.* 2024, 14, 21464–21537.

278 R. Shunmugam, G. J. Gabriel, K. A. Aamer, G. N. Tew, *Macromol. Rapid Commun.* 2010, 31, 784–793.

8

Terpyridine Complexes for Electrocatalytic Applications – CO_2 Reduction, Water Splitting, and Beyond

8.1 Introduction

2,2′:6′,2″-Terpyridines (tpy) and their complexes play a significant role in the field of catalysis. In this context, the catalysis of organic reactions represents a broad field of research, which has been extensively reviewed by us [1, 2] as well as other groups of authors [3–7]. Although this research area is dominated by complexes of transition-metal ions, notable achievements have also been made in using tpy complexes of main-group elements for catalytic purposes (see Section 3.8) [8]. Among others, cross-coupling, hydroboration, dehalohydrogenation, and alcohol-/hydrocarbon-oxidation reactions have been successfully performed in the presence of tpy complexes as catalysts. When talking about transformations of functional groups, one has to distinguish between reactions, which involve solely transition-metal catalysis, and those that rely on photocatalysis, *i.e.*, the catalytic cycle starts with the excitation of a sensitizer (*e.g.*, the complex itself or a pending/external chromophore) with UV and/or visible light [1].

One further aspect, which is of significant importance, is the usage of tpy complexes in the context of energy-related applications. This holds true for the topics which have been covered in Chapters 5–7: energy storage (*i.e.*, battery systems), light-to-electricity interconversion (*i.e.*, photovoltaics), and the reversal of the former (*i.e.*, light-emitting devices). However, the exit from nuclear and fossil-fuel energy, as one of the urgent challenges to face climate change without producing enormous amounts of harmful waste, has promoted research on two prominent reactions: CO_2 reduction and the splitting of water. The former is related to the conversion of CO_2 into compounds of higher chemical energy (*e.g.*, MeOH or HCOOH) [9–11], whereas the latter concerns the oxygen-evolution [12–14] and/or hydrogen-evolution reaction (OER and HER, respectively) [15–17]. In order to reach these goals, numerous photo-, electro-, and photoelectrochemical approaches, which often employ transition-metal complexes, as the catalysts, have already been published. Within the scope of this chapter, solely the electrocatalytic approaches, which use metal–tpy complexes, as the catalysts, will be in the focus of interest. Before going into detail, the reader is reminded that the presented work is still restricted to fundamental research. The reported turnover numbers (TONs) and catalytic performances are way too low to allow large-scale or even industrial

Terpyridines: Multifaceted Ligands for Diverse Applications, First Edition.
Ulrich S. Schubert, Andreas Winter, and George R. Newkome.

application. Regarding the electrochemical CO_2 reduction (ERC), for example, a broad range of bulk-metal electrodes has been screened [18]. It has been shown that Cu provides a high faradaic efficiency for the formation of hydrocarbons, whereas the noble metals (*e.g.*, Ag or Au) are more suited when aiming for CO, as the reduction product. Similar considerations also hold true for the electrocatalytic HER, in which metal chalcogenides, carbides, nitrides, or phosphides as electrocatalysts are employed [19].

8.2 Electrochemical CO_2 Reduction, Catalyzed by Tpy Complexes

Carbon dioxide is one of the greenhouse gases, which enormously contributes to global warming and powers the climate crisis. Thus, the reduction of worldwide CO_2 emissions represents one of the key challenges for which efficient and sustainable solutions are urgently required. In this context, stringent decarbonization of industry/transportation, CO_2 capture, and long-term storage as well as chemical CO_2 recycling are in the focus of public discussion [20, 21]. The latter refers to the reintroduction of CO_2 into the "carbon cycle" by converting it into useful high-energy compounds [22]. Methanol represents the most relevant recycling product, which is, at once, a primary feedstock and versatile intermediate for the production of chemical compounds [9]. Furthermore, methanol can be employed in direct methanol fuel cells (DMFCs), which produce electrical energy from chemical energy [23]. Beneficially, methanol can be stored and transported in an easy and relatively safe fashion [24].

Besides the conventional catalytic hydrogenation of CO_2, which suffers from the extensive use of molecular hydrogen, various other approaches have been followed: photocatalysis, electrocatalysis, and photoelectrocatalysis [9–11]. In the following, the fundamental aspects of the electrocatalytic process will be discussed briefly.

8.2.1 Fundamental Aspects of the Electrochemical CO_2 Reduction

Research on the ERC can be traced back to the 19th century, when the formation of CO or formic acid from CO_2 was first documented [9]. Impressed by the consequences of the oil embargoes in the 1970s, research on the ERC was intensified and, in the mid-1980s, the formation of MeOH from CO_2 was eventually reported. In these studies, which used metal (*e.g.*, Ru or Mo) or semiconductor (*e.g.*, GaAs or InP) electrodes, poor current densities were reached [25–27]. As one notable milestone, the highly efficient formation of hydrocarbons at a Cu electrode was realized by Hori *et al.* [28]. Based on these findings, the development of product-selective ERC catalysts was prospected. Since then, numerous studies dealing with this topic have been published [29].

The basic principle of an ERC device is depicted in Figure 8.1. If electricity from whatever source is applied, the OER from water, as the anodic process, generates electrons, which are transported through the external circuit toward the cathode. Simultaneously, protons diffuse through the semipermeable membrane toward the cathode. There, electrons and protons are used to reduce CO_2 to form various

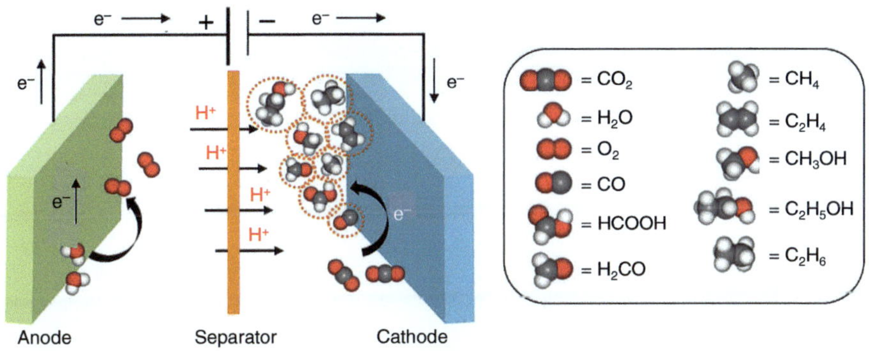

Figure 8.1 Schematic representation of an electrochemical CO_2 reduction system [30] / John Wiley & Sons / CC BY 4.0.

organic products. The product formation depends on the number of consumed electron equivalents, which ranges from two- to eight-electron reductions, and the applied reaction conditions (*e.g.*, pH value, CO_2 concentration, temperature, electrode material, *etc.*). Carbon monoxide, formic and oxalic acid, formaldehyde, MeOH, CH_4, ethylene, and EtOH might *inter alia* be formed [31]. Beneficially, all these low-carbon compounds are of particular virtue for further industrial use. For comparison, their market prices, as a function of their energetic costs, are summarized in Figure 8.2 [31].

Various proton-coupled CO_2 reductions, which afford these products in an aqueous environment, occur at moderately negative potentials (E^0) in the range of −0.25 to −0.60 V *vs.* the normal hydrogen electrode (NHE, Figure 8.3) [9]. However, one has to be aware that the reported potentials strongly depend on the applied measurement conditions (*e.g.*, solvent, pH value, temperature, reference electrode, *etc.*) and, thus, varying values for these processes can be found in the available literature. The direct one-electron reduction to afford the $CO_2^{\bullet-}$ radical anion requires significantly more energy than all other reduction processes. This is mainly due to the enormous reorganization energy that is needed to convert the linear CO_2 molecule into the angular geometry of the radical anion [32].

One has to take one further limiting factor into account when talking about the ERC, which is the intrinsically poor solubility of CO_2 in aqueous media. Under idealized conditions, the CO_2 solubility is as low as *ca.* 34 mM [33]; however, the solubility under experimental conditions is even lower due to the high ionic strength of aqueous solutions. Arpornwichanop *et al.* noted that this is detrimental regarding the CO_2 mass transport and, as a consequence, also for the reaction rates and current densities [9]. Various strategies have been applied to counteract this issue [34–36]. These include the use of gas-diffusion electrodes (with or without dispersed catalyst) and/or organic electrolytes, in which CO_2 is better soluble; moreover, flow-cell designs have been used in this context.

As the presumably most relevant limitation, electrons and protons might directly combine at the cathode; thus, the HER represents a considerable alternative reaction pathway. The competition between H_2 evolution and CO_2 reduction is due to the similar redox potentials, as listed in Figure 8.3. If both reactions occur

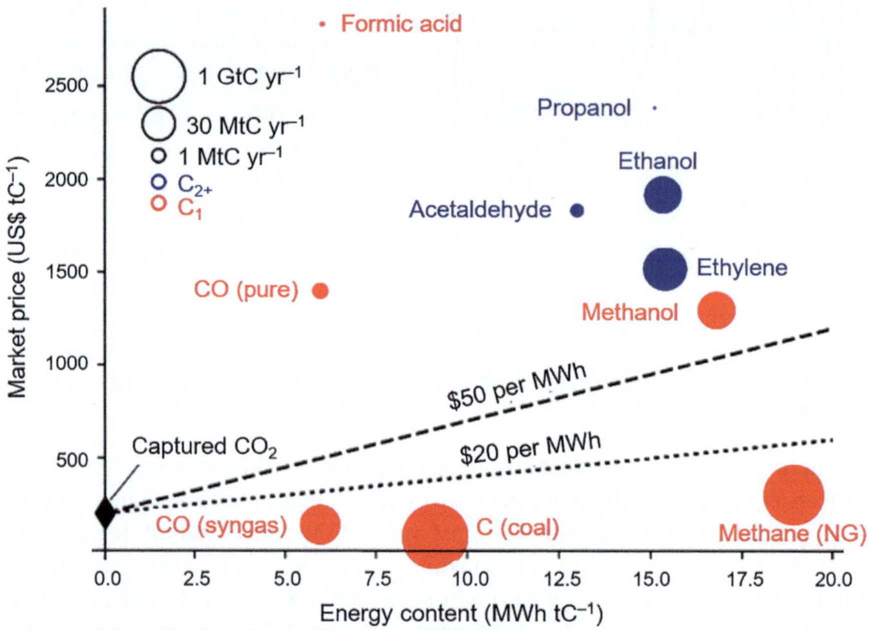

Figure 8.2 Market prices of selected CO_2 reduction products as a function of energy content. The lines represent minimum energy and CO_2 costs; capital costs are not considered [31] / with permission of American Chemical Society.

$$
\begin{array}{lll}
CO_2 + e^- \rightarrow CO_2^{\bullet -} & & E^0 = -1.90\ V \\
CO_2 + 2\,H^+ + 2\,e^- \rightarrow HCOOH & & E^0 = -0.61\ V \\
CO_2 + 2\,H^+ + 2\,e^- \rightarrow CO + H_2O & & E^0 = -0.52\ V \\
CO_2 + 4\,H^+ + 4\,e^- \rightarrow H_2CO + H_2O & & E^0 = -0.51\ V \\
CO_2 + 6\,H^+ + 6\,e^- \rightarrow CH_3OH + H_2O & & E^0 = -0.38\ V \\
CO_2 + 8\,H^+ + 8\,e^- \rightarrow CH_4 + 2\,H_2O & & E^0 = -0.24\ V \\
\hline
2\,H^+ + 2\,e^- \rightarrow H_2 & & E^0 = -0.42\ V
\end{array}
$$

Figure 8.3 Summary of the possible pathways for CO_2 reduction along with the corresponding reduction potentials (E^0 *vs.* the NHE at 25 °C and pH-neutral conditions) [9] / with permission of Elsevier / CC BY-NC-ND 4.0.

simultaneously, the product yield is decreased and the applied energy is (at least in parts) consumed to produce undesired H_2 without reducing CO_2. The ratio between formed H_2 and consumed CO_2 mainly depends on the activation barrier of the respective reactions, which, in turn, is determined by the choice of electrode material and catalyst. If the overpotential for H_2 formation is sufficiently high, the ERC is favored, and the product selectivity can be high [29].

In order to provide product selectivity, various types of electrodes and catalysts have been explored over the last decades. The main challenge is the formation of the desired reduction product, whilst suppressing the generation of the others as side products. Thereby, a high electrocatalytic efficiency in concert with electrode/catalyst robustness is desirable to make the ERC process as sustainable as possible. One has to distinguish between the actual ERC, which occurs at chemically inert electrodes, and the electrocatalytic variants, in which the catalyst is dissolved in the electrolyte or immobilized on the electrode [29]. In the general concept of catalysis, the latter corresponds to homogeneous and heterogeneous systems, respectively. Transition-metal catalysts have been widely used in these two scenarios. From the wide range of reported catalysts, metalloporphyrins and polypyridyl complexes have gained the most notable attention [37]. The broad field of polypyridyl complexes, as catalysts for CO_2 reduction, has recently been reviewed by Fontecave *et al.* [38]. These complexes mainly contain bidentate ligands from the bpy or phen families (bpy: 2,2′-bipyridine, phen: 1,10-phenanthroline) but also tridentate tpy or even tetradentate qtpy derivatives have commonly been employed (qtpy: 2,2′:6′,2″:6″,2‴-quaterpyridine). Traditionally, complexes from the 2nd- and 3rd-row transition-metal ions have been employed; recently, the focus of interest has been shifted toward their more abundant 1st-row counterparts, to account for improved sustainability and cost reduction.

8.2.2 Electrocatalytic CO_2 Reduction Using Tpy Complexes as Homogeneous Catalysts

The pioneering work on polypyridyl complexes, as catalysts for CO_2 reduction, was performed in the 1980s. Ziessel *et al.* reported that *fac*-[Re(bpy)$(CO)_3$Cl] mediated the ERC at low potential to provide CO, as the main product [39], whereas Tanaka *et al.* initially utilized $[Ru(bpy)_2(CO)_2]^{2+}$ and $[Ru(bpy)_2(CO)Cl]^{+}$ to catalyze the ERC in aqueous DMF [40, 41]. These studies revealed a pronounced influence of the pH value on the product distribution of the bulk CO_2 electrolysis: In a slightly acidic milieu (*i.e.*, pH ≈ 6) syngas, *i.e.*, a mixture of H_2 and CO, was produced, whereas formate was the favored product, if the process was run under more basic conditions (*i.e.*, pH ≈ 9.5). In the following, numerous variants of the archetypal catalysts were screened [38]. In this context, Tanaka *et al.* also studied the ERC in the presence of the mixed-ligand complex $[Ru(tpy)(bpy)(CO)]^{2+}$ (**1a**, Figure 8.4) [42, 43]. Due to its low operational stability at room temperature, the electrolysis in aqueous EtOH was performed at −20 °C. As a result of multi-electron reductions, a broad product distribution was obtained: Besides CO and HCOOH, also formaldehyde, MeOH, glycolic acid, and hydroxyacetic acid were formed, whereas $[Ru(bpy)_2(CO)_2]^{2+}$ exclusively afforded CO/HCOOH as the product. The different reactivity was attributed to the initial formation of the $[Ru(tpy)(bpy)(CHO)]^{+}$ species, which paved the way for the subsequent reduction and C–C coupling steps. Overall, as is typical for most of the complexes reported in literature, $[Ru(tpy)(bpy)(CO)]^{2+}$ performed at a relative negative potential and the CO_2 conversion rate was slow.

Moreover, Deronzier, Ziessel *et al.* reported that the complex [Ru(tpy)(CO)Cl_2] represented a catalyst for the ERC (independent from the complex's *cis-/trans*-stereochemistry) [44]. Moreover, the complex was electropolymerized to afford the $[Ru(tpy)(CO)]_n$ metallopolymer, which contained Ru^0—Ru^0 bonds. This material

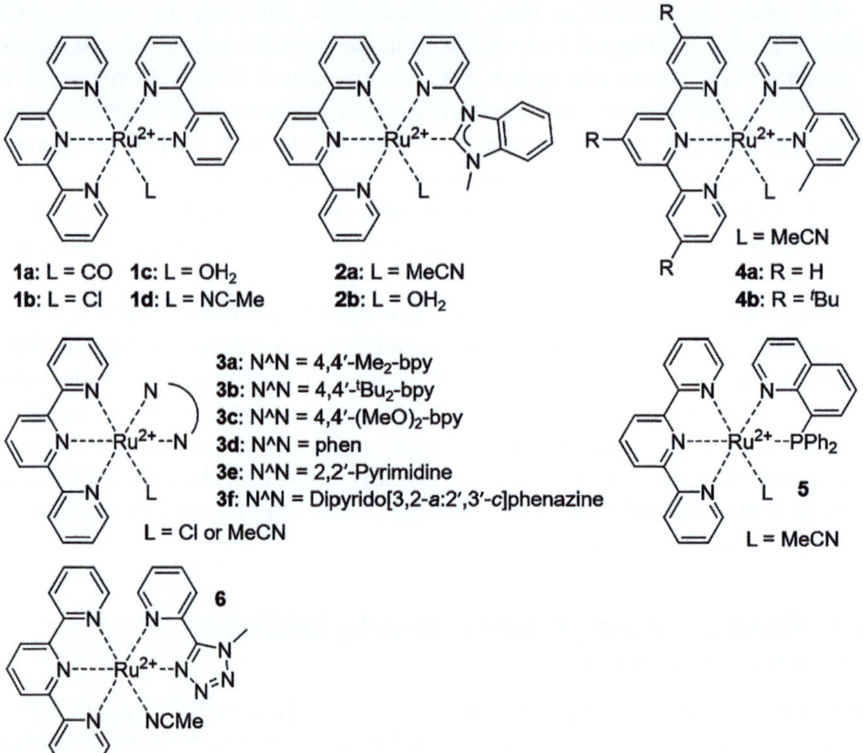

Figure 8.4 Schematic representation of Ru^{II} complexes **1–6** (counterions were omitted for clarity).

was also electrocatalytically active, either dissolved in DMSO or deposited on carbon-felt electrodes. Unlike the homogeneous electrocatalysis, the heterogeneous variant performed in various solvents, even including water (see Section 8.2.3).

Likewise, various variants of Tanaka's archetypal complex **1a** have been discussed in the context of electrocatalytic CO_2 reduction (Figure 8.4). The solvato complex **1d** was studied in detail by Meyer *et al.* [45]. At the potential of −2.21 V, the ERC afforded CO from CO_2-saturated MeCN with a faradaic efficiency of *ca.* 60%. The fundamental catalytic cycle, which was suggested by these authors, is presented in Figure 8.5. The solvato complex, which was formed from an appropriate precursor complex (*e.g.*, **1b/c**), experienced two successive one-electron reduction steps and entered the catalytic cycle as a charge-neutral species (the tpy and bpy ligands were believed to possess radical-anion character). Follow-up studies revealed that the ligand structure had a tremendous effect on the complex's electrocatalytic activity. For example, the Ru^{II} complex **2** (Figure 8.4), which contained an NHC-ligand, was identified to be the more efficient ERC catalyst than **1d** (NHC: *N*-heterocyclic carbene) [46]. The superior performance was expressed by the catalytic rate constants (k_{cat}), which were 5.5 and 19.1 s^{-1}, respectively. Meyer *et al.* attributed the

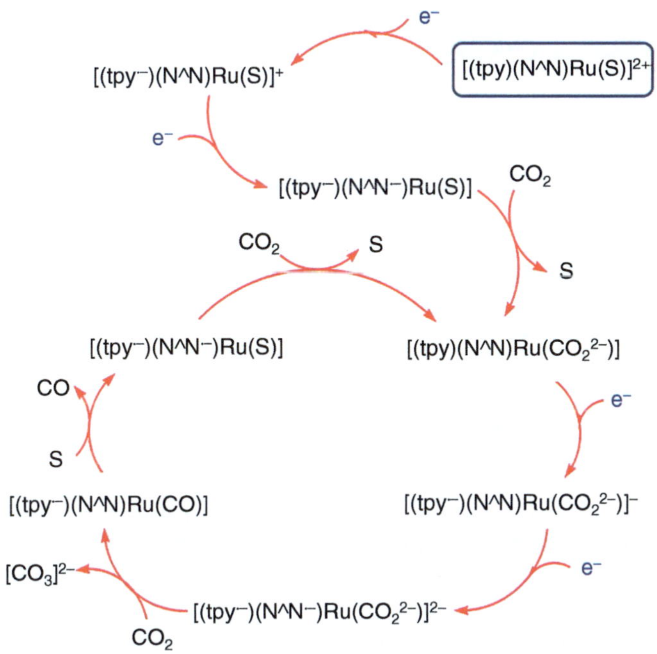

Figure 8.5 Schematic representation of the proposed mechanism of the ERC according to Meyer *et al.* (N^N denotes a bidentate ligand, such as bpy; S represents a coordinated solvent molecule) [45] / with permission of Royal Society of Chemistry.

high catalytic activity of **2** to an increased lability of the solvato ligand in the doubly reduced state, in particular when the strongly σ-donating NHC moiety was in *trans*-position to the solvato ligand. This structural requirement was elaborated by Miller *et al.* in their more recent mechanistic studies [47]. Since **2b** represented a likewise robust single-site catalyst for the electrochemical water oxidation [48], a two-component electrochemical cell was assembled, which allowed the formal splitting of CO_2 into CO and O_2. Thus, a single complex was used to achieve both the CO_2 reduction and OER [46]. Moreover, a tandem cell was presented, which combined the ERC with the two-electron alcohol oxidation [49]. The concomitant formation of syngas and benzaldehyde from CO_2 and benzylic alcohol, respectively, proceeded with an overall energy efficiency of *ca.* 18%. The anode half-cell operated under homogeneous conditions using **2b** as the electrocatalyst, whereas the immobilized $[Ru(Mebip)(bpy)(OH_2)]^{2+}$ complex served as a heterogeneous catalyst for the electro-oxidation [Mebip 2,6-*bis*(1′-methylbenzimidazol-2-yl)pyridine].

In line with Meyer's studies in this field, Ott *et al.* varied the properties of the bidentate ligand *via* its electron density [50]. This study revealed that the complexes, which contained electron-rich bidentate ligands, were more susceptible to ligand substitution and, thus, increased reactivity in the ERC was observed. Beyond these considerations, these authors proposed the coexistence of two mechanistic pathways: In the EEC case, the chemical reaction follows two sequential electron-transfer processes (Figure 8.6), whereas the sequence of electron transfer,

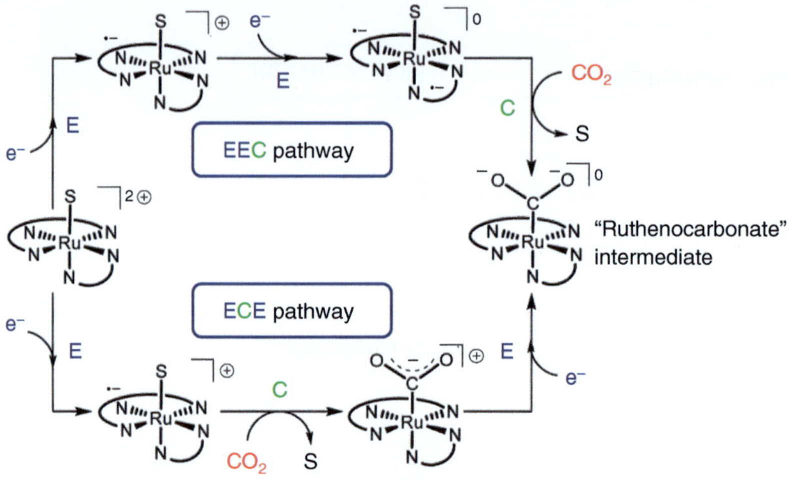

Figure 8.6 Schematic representation of the competing EEC and ECE pathways involved in the ERC process (N^N^N and N^N denote a tpy ligand and a bpy ligand, respectively). Adapted from [51].

chemical reaction, and electron transfer is referred to as the ECE mechanism. The former pathway is prevalent in the case of electron-rich bidentate ligands, whereas the ECE becomes dominant with increasing steric hindrance in the N^N ligand. In either case, the presence of singly reduced ligand moieties in the "resting state" of the catalytic cycle was assumed. These species were thought to facilitate the observed multi-electron reduction reactions, which proceeded with high turnover frequencies (TOFs) at increased overpotential. Examples for ERC catalysts, which provoked the ECE pathway, were also provided by Ott *et al.* [51, 52]. If the CH_3-group of complexes **4** was located in *trans*-position of the labile ligand, the ECE pathway (Figure 8.6) was activated by steric hindrance and the singly reduced species was involved in the CO_2 binding. In this case, the first reduction step of **4** was utilized for the ERC, which proceeded at low overpotential.

A similar effect, *i.e.*, increased catalytic activity due to increased electron density in the bidentate ligand, was assumed by Muckerman, Fujita *et al.*, who introduced electron-donating OH-groups in bpy's 4,4′- or 6,6′-positions [53]. Detrimentally, these ligands experienced reductive protonation under the ERC conditions. The thusly formed quinone-type structures withdrew electron density from the Ru^{II} center and, thus, decreased the catalytic activity far below that of the parent complex **1d**. This work reveals the necessity of cautious ligand design: Energy donation *via* +I substituents is beneficial to increase the electrocatalytic activity, whereas electron-donation *via* +M effects might render the ligands no longer non-innocent and catalytic activity might be diminished.

A different type of bidentate ligand, namely 8-diphenylphosphino-quinoline, was used by Masaoka *et al.* [54]. Similar to Ott's methyl-substituted complex **4**, complex **5** (Figure 8.4) enabled the CO_2 reduction *via* an ECE pathway at low overpotential. In this case, the bulky *P*-center was axially coordinated to the Ru^{II}

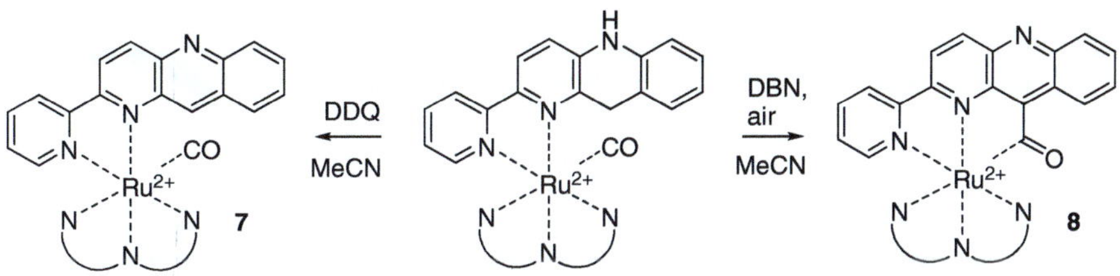

Figure 8.7 Schematic representation of the synthesis of the Ru^{II} complexes **7** and **8** from a common precursor (counterions were omitted for clarity).

center, thus labializing the opposite solvato ligand by its *trans*-effect and facilitating the CO_2 binding. Worth mentioning in this context, a mixed-ligand Ru^{II} complex, which contained tpy and a bidentate phosphine ligand, was reported by Meyer *et al.* already far beforehand [55]. The [Ru(tpy)(dppene)Cl] complex generated CO at the applied potential of −1.4 V *vs.* NHE [dppep: *cis*-1,2-*bis*(diphenylphosphino)ethene]. The NBu_4^+ cations, which were present in the nonaqueous electrolyte, served as sacrificial H^+ donors *via* Hofmann degradation.

Not only steric effects but also electronic ones might provoke the ERC to proceed *via* an ECE mechanism. This was demonstrated by the example of complex **6**, which contained a tetrazole moiety (Figure 8.4) [56]. Binding of CO_2 to afford the ruthenocarboxylate intermediate proceeded after the first one-electron reduction; thus, the ERC could be performed at the very low overpotential of 0.37 V – independent from the stereochemistry of **6** (*i.e.*, *cis-/trans*-isomerism). The faradaic efficiency for CO formation was *ca.* 86% (TOF = $0.032\,s^{-1}$). As the final notable example in the context of Ru^{II}–tpy complexes, as ERC catalysts, Tanaka *et al.* reported the related complexes **7** and **8**, which mimicked the natural NAD/NADH system. Using one common precursor complex, **7** and **8** were prepared by oxidation with DDQ and aerobic oxidation in the presence of DBN, respectively [DDQ: 2,3-dichloro-5,6-dicyano-1,4-benzoquinone; DBN: 1,5-diazabicyclo(4.3.0)non-5-ene; Figure 8.7] [57]. The *in situ* twofold reductive protonation of the coordinated benzo[*b*][1,5]naphthyridine moiety afforded the catalytically active species, which reduced CO_2 *via* a hydride-transfer mechanism [58]. Thereby, the selectivity of **8** to generate formate rather than CO in a bulk electrolysis at −1.53 V (*vs.* Fc/Fc^+) was higher than that of **7** (*i.e.*, 7 *vs.* 2.5); for comparison, the non-NAD-type complex [Ru(tpy)(bpy)(CO)] favored the evolution of CO (formate-to-CO selectivity of 0.125).

Beyond Ru^{II} complexes, a wide range of ERC catalysts, which are based on 4d/5d transition-metal ions, have been reported over the last decades: Re, Os, Rh, Ir, Pd, *etc.* [1, 38, 59]. However, very few of them contained tpy ligands. Firstly, $Rh(tpy)Cl_3$ and the heteroleptic complex **9**, which contained an unusual imidate ligand, catalyzed the electroreduction of CO_2 in aqueous DMF at the applied potential of −1.96 and −2.00 V, respectively (Figure 8.8a) [60]. The formation of formate proceeded with a good faradaic efficiency of *ca.* 80%, but the turnover remained overall low. Moreover, a small library of *mono*-cyclometalated Ir^{III}–tpy complexes was employed in the context of CO_2 reduction (Figure 8.8b); however, these complexes acted as

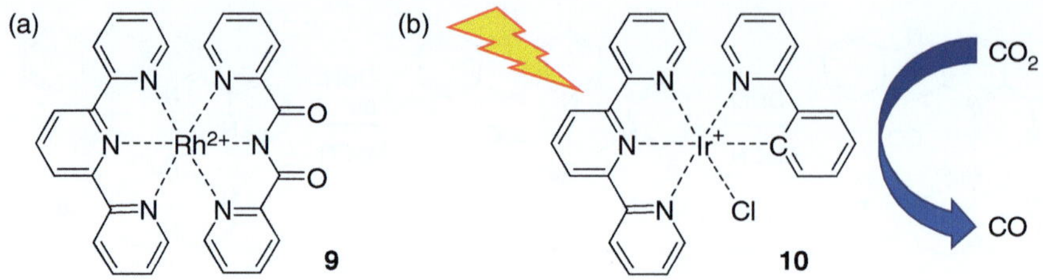

Figure 8.8 (a) Schematic representation of the Rh^{III} complex **9**. (b) Schematic representation of the Ir^{III} complexes **10**, which were used for the photocatalytic CO_2 reduction. (In both panels, the counterions were omitted for clarity).

photocatalysts only and are, thus, not considered further within the scope of this chapter.

In recent years, the focus of interest has shifted toward the use of complexes based on more abundant metals. This holds true for biological applications as well as those that are related to energy and catalysis. Regarding the use of complexes earth-abundant metals, as molecular catalysts for the ERC, Machan *et al.* emphasized that one has to be aware that the activity, electronic structure of catalytic intermediates, or product selectivity might be very different when comparing structurally related complexes of the metal ions of one group (*e.g.*, those of Fe, Ru, and Os) or row of the periodic table of elements (*e.g.*, Cr → Ni) [61]. The periodic trends in electrocatalytic performance (and deviations thereof) have been elaborated by those authors. Recently, Younus, Y. Zhang, Verpoort, S. Zhang *et al.* presented a detailed overview of the various earth-abundant molecular ERC catalysts [59]. Reference to this article is given for a broader view on this topic. Regarding the use of tpy complexes, Fontecave *et al.* provided a systematic study on the electrocatalytic activity of the homoleptic $[M(tpy)_2]^{2+}$ complexes (M = Mn, Fe, Co, Ni, Cu, and Zn) [62]. From these, only the Ni^{II} and Co^{II} complexes exhibited electrocatalytic activity. Notably, this result contradicts the outcome of earlier studies by Abruña *et al.*, who showed that the electrodeposited films of complexes $[M(tpy)_2]^{2+}$ (M = Fe, Co, Ni) represented heterogeneous catalysts for the ERC (see Section 8.2.3) [63, 64]. The homogeneous catalysis revealed differences in their product selectivity. The $[Ni(tpy)_2]^{2+}$ complex selectively generated CO, while the concurrent HER was fully suppressed, whereas $[Co(tpy)_2]^{2+}$ produced a mixture of CO and H_2. The different reactivity was attributed to the structure of the complexes in their "resting states" (Figure 8.9). A low-valent Ni^I species, as postulated for the corresponding Co complex, was absent and, thus, the formation of Ni–hydride intermediates, representing the prerequisite for H_2 evolution, was excluded. On the contrary, the proposed $[Co^I(tpy)]^+$ intermediate enabled both CO and H_2 formation. It was shown further that the CO/H_2 ratio could be adjusted *via* the applied potential: CO formation was favored in the low-potential regime, whereas with increasing potential, the H_2 fraction increased (at the applied potential of −2.1 V, a CO-to-H_2 ratio of 1 was reached). It is evident that the initial Co^{II} *bis*-complex represented a pre-catalyst, which – due to its saturated coordination sphere – was unable to catalyze the ERC itself. Thus, dissociation of one tpy ligand in the course of the

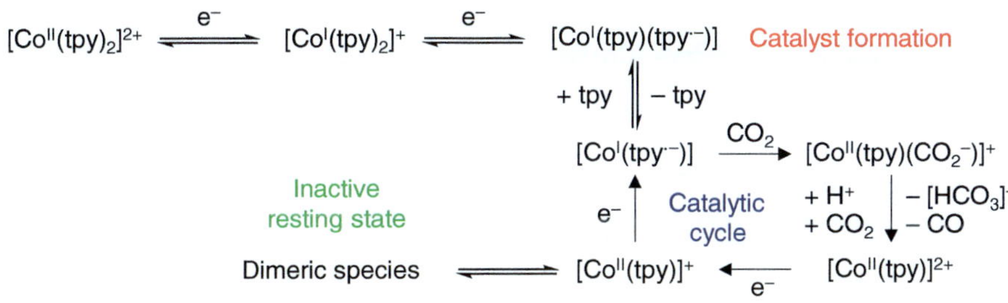

Figure 8.9 Schematic representation of the proposed mechanism for the ERC using $[Co(tpy)_2]^{2+}$, as the catalyst [62] / with permission of Royal Society of Chemistry.

initial two-electron reduction of $[Co(tpy)_2]^{2+}$ was assumed (Figure 8.9). In subsequent work, Fontecave *et al.* also demonstrated that the product selectivity could be modulated *via* substituents on the tpy ligand [65]. The presence of electron-donating groups favored the HER – the highest faradaic efficiency for the HER (23%) was observed when a MeO-substituent was present in the 4′-position. However, when electron donation to the ligand was absent, the CO formation became favored and the faradaic yield for H_2 formation dropped below 5%.

Cyanoacetic acid represents a commodity chemical, which is widely used as a CH-acidic compound in condensation reactions. Moreover, it has found widespread use in the industrial production *inter alia* of adhesives, pharmaceuticals, agrochemicals, dyes, and UV absorbers. From the available synthetic large-scale protocols, the electrocarboxylation of chloroacetonitrile is of particular interest due to its benignity and eco-friendliness [66]. Harmful reagents, such as inorganic cyanides, are avoided; synthesis falls back to chloroacetonitrile, as a comparatively cheap starting material, and the widely abundant CO_2. Thus, the synthesis of carboxylic acids *via* electrocarboxylation represents one method for chemical CO_2 fixation. The reaction is typically performed in a CO_2-saturated aprotic solvent and in the presence of a transition-metal complex, as the electrocatalyst. In this context, the homoleptic $[M(tpy)_2]^{2+}$ complexes were also employed (M = Co or Ni) [67, 68]. Reynes *et al.* performed extensive electrochemical experiments to elucidate the mechanism of the cyanoacetic-acid synthesis using $[Co(tpy)_2]^{2+}$ as the electrocatalyst. It was suggested that a Co^{I}–tpy *mono*-complex was initially formed by a one-electron reduction (Figure 8.10). To this species, chloroacetonitrile was added oxidatively. In the absence of CO_2, the one-electron reduction of the resultant Co^{III} intermediate afforded the Co^{II}–tpy *mono*-complex on one hand and an alkyl anion on the other. However, in the presence of CO_2, the latter was inserted into the Co^{III}–alkyl bond. After one further reduction step, the release of cyanoacetate eventually regenerated the catalyst. A similar mechanistic pathway was also proposed for the electrocarboxylation in the presence of the analogous $Ni[(tpy)_2]^{2+}$ complex [68]. Both electrocatalysts performed at lower overpotentials than the [Ni(salen)] complex (by *ca.* 0.4 V), as the state-of-the-art electrocatalyst for this reaction [salen: *N,N′-bis*(salicylidene)ethylenediamine] [69]. However, the latter afforded chloroacetic acid with a higher faradaic yield (*i.e.*, 71% *vs.* 60%) [67, 68].

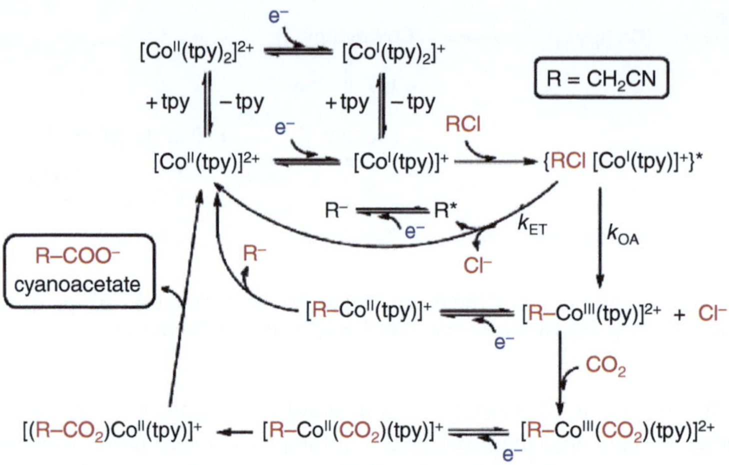

Figure 8.10 Schematic representation of the proposed mechanism for the electrocarboxylation of chloroacetonitrile [67] / with permission of Elsevier.

Interest in using Mn^{I} complexes, as ERC catalysts, arouse from the high chemical similarity between Mn and its homolog Re, whilst being by far more abundant in the Earth's crust [70]. The first such complex, [Mn(bpy)$(CO)_3$Br], used as an electrocatalyst for the CO_2 reduction, was reported by Chardon-Noblat, Deronzier *et al.* in 2011 [71]. This complex (and its derivatives) closely resembled their counterparts from the [Re(bpy)$(CO)_3$X] family (X = Cl or Br) and featured a high efficiency and stability in the catalytic ERC in the presence of a Brønsted acid. The CO_2 reduction was performed at a lower overpotential (*ca.* 0.4 V less than when using Re^{I} complexes) to afford CO with very high product selectivity. Against the background to screen other polypyridyl ligands in the coordination sphere of Mn^{I}–carbonyl complexes, Machan and Kubiak prepared the Mn^{I} complexes **11**, which differed in their coordination mode (Figure 8.11a) [72]. Both **11a** and **11b** acted as electrocatalysts in the ERC to generate CO, albeit with low efficiencies (phenol was used as the proton source). It was assumed that a dimeric species, which contained a Mn^{0}—Mn^{0} bond, was initially formed upon one-electron reduction; the subsequent reduction step afforded $[Mn(tpy)(CO)_2]^-$, as the catalytically active species. Remarkably, the structure of the latter was verified by low-temperature

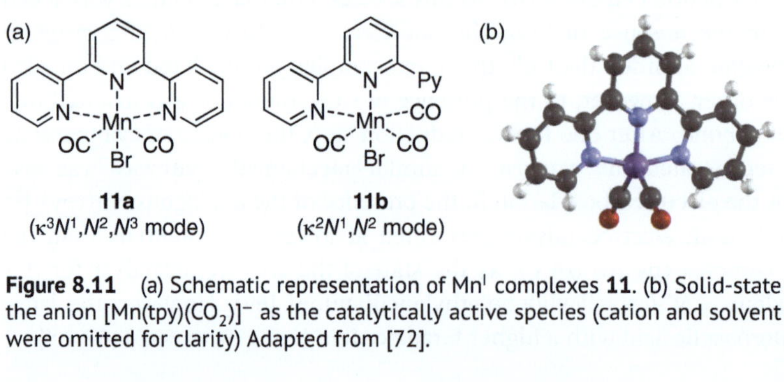

Figure 8.11 (a) Schematic representation of Mn^{I} complexes **11**. (b) Solid-state structure of the anion $[Mn(tpy)(CO_2)]^-$ as the catalytically active species (cation and solvent molecules were omitted for clarity) Adapted from [72].

X-ray diffraction (XRD) analysis (Figure 8.11b). The lower catalytic activity of **11** than that of the analogous Mn^I–bpy systems was presumably due to the improper tpy ligand structure, *i.e.*, catalyst decomposition could be counteracted by using tridentate ligands with increased donor strength and bulky substituents (to prevent dimerization). This proposal was confirmed experimentally at the example of Mn^I complexes bearing tridentate NHC-containing ligands [59].

8.2.3 Electrocatalytic CO_2 Reduction Using Tpy Complexes as Heterogeneous Catalysts

Besides the various soluble complexes, which were used as homogeneous catalysts for the ERC process to afford higher-energy organic compounds, their immobilization has also been studied to a significant extent. The thusly derived heterogeneous systems offer numerous advantages over their homogeneous counterparts: more efficient use of catalytic sites, improved structural integrity, minimization of deactivation processes, better control over the chemical environment around the catalytic center, high variability in solvent choice, accelerated electron-transfer kinetics between the electrode's surface and the catalyst, *etc.* [59]. Accordingly, this concept has also been widely applied to electrocatalytic CO_2 reduction. The immobilization of a molecular catalyst onto an electrode or conductive substrate might occur *via* physisorption, chemical bonding (covalent or non-covalent) or by integration into 2D/3D frameworks, *i.e.*, coordination nanosheets (CONASHs) or metal–organic frameworks (MOFs), respectively. In order to immobilize a complex, appropriate functional groups are required and, thus, the synthesis might become more complicated. This represents one shortcoming when compared to the homogeneous variant.

In one of the pioneering contributions in this field, Abruña *et al.* modified glassy-carbon electrodes (GCE) by coating them with insoluble metallopolymer films [63]. These films were generated by the *in situ* elctropolymerization of $[Co(vtpy)_2]^{2+}$ (vtpy: 4′-vinyl-tpy). As one notable finding, the electrocatalytic formation of formate in CO_2-saturated DMF required a *ca.* 1.0 V lower overpotential than when using a pristine GC electrode. Furthermore, the thickness of the metallopolymeric film had an influence on the electrocatalytic activity. If the surface coverage (Γ) was too high, the diffusion of CO_2 was hampered and the catalytic activity decreased. The nature of the counterion also played a considerable role: In this initial work, a balanced surface coverage of *ca.* 10^{-8} mol cm^{-2} was used. The reduction was even fully retarded, if HPO_4^{2-}anions were present. In subsequent work, metallopolymer films were fabricated by electropolymerization of the respective $[M(vtpy)_2]^{2+}$ complexes (M = Cr, Ni, Co, Fe, Ru, Os) [64, 73, 74]. The films, which contained 3d transition-metal ions, were electrocatalytically more active than those with Ru^{II} or Os^{II} ions. The highest current efficiency was reached with the films, which contained $[Cr(tpy)_2]^{2+}$ complexes, whereas the highest TON of >15,000 was obtained for the Fe^{II}-containing films [73]. Note, that the $[Fe(tpy)_2]^{2+}$ complex was electrocatalytically inactive when used under homogeneous conditions [62].

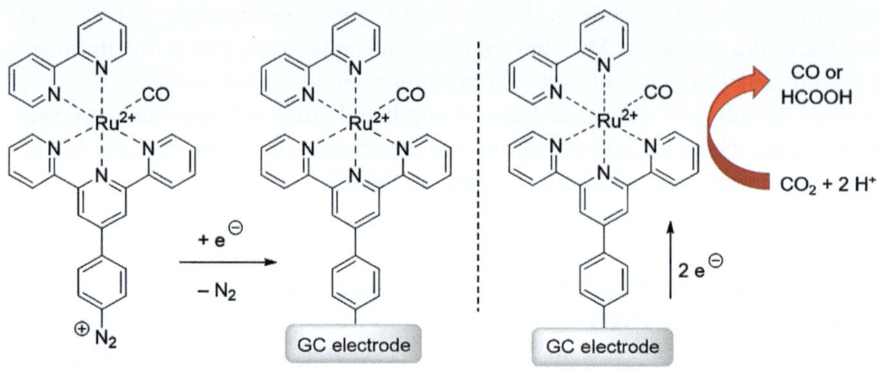

Figure 8.12 Schematic representation of the electrochemical functionalization of GC electrodes and their use as heterogeneous ERC catalyst (counterions were omitted for clarity). Adapted from [75].

A different strategy for the modification of GCEs was followed by Tanaka *et al.*, who grafted a multilayer of $[Ru(tpy)(bpy)(CO)]^{2+}$ complexes onto the electrode surface by *in situ* electroreduction of their diazonium moieties (Figure 8.12) [75]. The thusly obtained hybrid electrode produced CO and HCOOH in aqueous MeCN at the applied potential of −1.70 V. Related to this, Fontecave *et al.* prepared a tpy-functionalized GCE *via* the same strategy [76]. The subsequent treatment with Co^{II} ions afforded surface-anchored *mono*-complexes. Unlike for the aforementioned $[Co(tpy)_2]^{2+}$ complexes in solution, the CO_2 reduction proceeded inefficiently, presumably due to the low *in-operando* stability of the catalysts. Thus, the results obtained for the homogeneous electrocatalysis were not reproduced with this heterogeneous variant.

The immobilization of molecular ERC catalysts *via* ππ-stacking interactions between the catalyst's π-conjugated chromophore and the electrode's surface represents a commonly applied strategy, in particular when dealing with CNT-coated electrodes (CNT: carbon nanotube) [59]. This was, *e.g.*, demonstrated by McQueen and Goldsmith, who bound electrocatalytically active $[Co(tpy)_2]^{2+}$ complexes to single-walled carbon nanotubes (SWCNTs) *via* the complexes' pending pyren-1-yl substituents [77].

However, when dealing with metal-oxide or metal-chalcogenide semiconductors, the complexes have to be equipped with *O*- or *S*-containing anchoring groups, which enable a robust immobilization (*e.g.*, –COOH, $–PO_3H_2$, –SH, *etc.*). Numerous studies are available where self-assembled monolayers have been formed on electrode surfaces *via* such functionalities [78]. For example, Meyer *et al.* fabricated a hybrid electrode, in which Ott's catalyst **4a** was grafted onto a nanocrystalline TiO_2 film, followed by the deposition of a protective NiO overlayer [79]. The electrochemical cell, as shown in Figure 8.13, selectively generated CO from CO_2-saturated MeCN with a low overpotential. The bulk CO_2 electrolysis over 4 h afforded a maximum TON of 237. *Post-operando* XPS analysis of the hybrid electrode revealed its integrity, *i.e.*, absence of catalysis degradation during the ERC process (XPS: X-ray photoelectron spectroscopy). In a control experiment, the ERC was also performed under

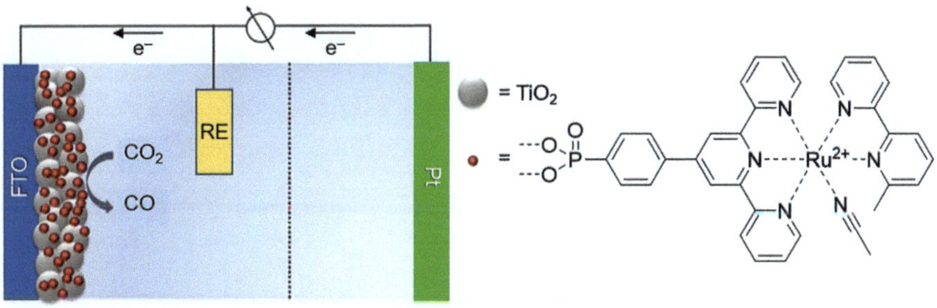

Figure 8.13 Schematic representation of an electrochemical cell for CO_2 reduction (FTO: fluorine-doped tin oxide, RE: reference electrode) [79] / with permission of John Wiley & Sons.

homogeneous conditions. Thereby, a significantly worse performance was observed (*i.e.* faradaic efficiency of 62.5%, TON = 3). This discrepancy was ascribed to the insufficient interaction of the catalyst with the electrode within a certain timeframe due to its diffusion limitation, whereas the immobilization allowed for a good interaction with the electrode and, thus, good electrocatalytic activity was achieved.

Furthermore, a hybrid electrode, in which the NHC-containing complex **2b** was incorporated, was reported by Meyer *et al.* [80]. A multilayer heterostructure was prepared by atomic layer deposition of TiO_2 onto an FTO electrode, followed by a thin layer of each CNT as well as TiO_2. The latter guaranteed surface stabilization and enabled the anchoring of the catalyst *via* its $-PO_3H_2$ moiety (thus, similar to the previous example; see Figure 8.13). After the loading of the electrode with the complex, an overlayer of TiO_2 was finally deposited. This heterostructured electrode supported the ERC to afford syngas with variable CO/H_2 ratios (depending on the applied voltage). However, after 15 min of electrocatalysis, the electrode changed its character due to an irreversible transformation of the immobilized complex; thus, CO generation was depleted and only the HER was observed thereafter. In one further study, Meyer *et al.* showed that the catalyst stability could be increased by using a sophisticated electrode design [81]. In this work, complex **2b** was combined with a nitrogen-doped porous carbon (NPC) electrode, which is well-known to facilitate C—C bond formation in the course of the ERC process [82]. The NPC material was fabricated by a two-step process, which involved the three-component assembly of a resin and its subsequent pyrolysis at 750 °C (Figure 8.14a) [81]. The material's high surface area and the porous graphene-like structure allowed the robust binding of complex molecules *via* their pending pyrene moieties. Unlike the embedment of **2b** into a TiO_2 matrix (*vide supra*), this hybrid electrode offered more exposed reactive sites and better access to the catalytic interfaces. The ERC in an aqueous environment favored the formation of C_2 compounds over the generation of syngas or MeOH. In particular, EtOH and acetate were obtained with faradaic efficiencies of 31% and 38%, respectively (potential range from −0.87 to −1.07 V *vs.* NHE), whereas the yield of syngas was lower and its CO-to-H_2 ratio depended on the applied potential. It was proposed that EtOH formation was due to C–C coupling between CO intermediates, which were coordinated to the Ru^{II} center or

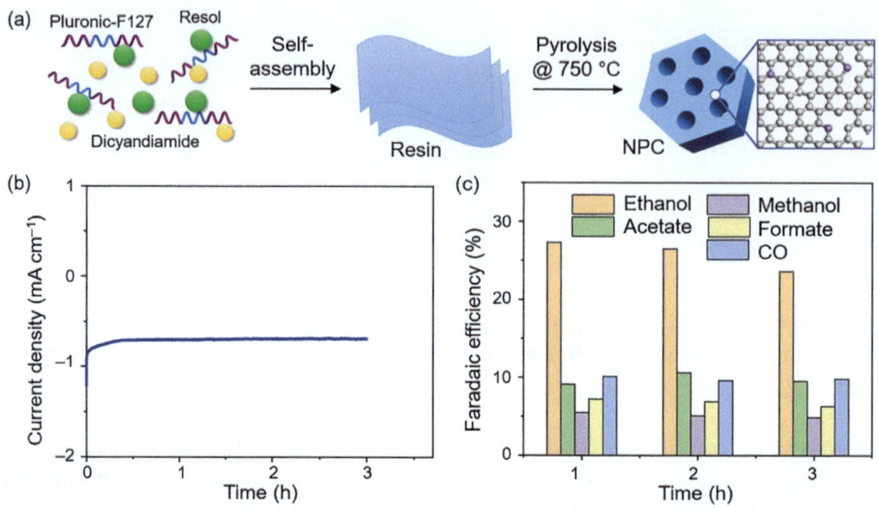

Figure 8.14 (a) Schematic representation of the two-step fabrication of *N*-doped porous carbon (NPC). (b) Evolution of the current density for the ERC on (Upper) and faradaic efficiencies (Lower) for CO_2 reduction on catalyst-loaded NPC over time (0.5 M aqueous $KHCO_3$ solutions; −0.97 V *vs.* SHE), (c) Faradaic efficiencies of various ERC products using the same experimental setup [81] / with permission of National Academy of Sciences.

resulted from protonation after detachment. Thus, the confinement of CO_2 (or its reduction intermediates) at the metal center in close proximity to the NPC surface was highly beneficial to obtain C_2 products (or, prospectively, even C_{2+} compounds) [83]. The operational stability of the hybrid electrode was demonstrated by CO_2 bulk electrolysis, which could be run for at least 3 h without significant loss of electrocatalytic activity (Figure 8.14b,c) [81].

Before closing this section, attention has to be turned briefly to the photoelectrochemical (PEC) CO_2 reduction, which has been realized with various heterogeneous catalysts. In a PEC cell, sunlight is harvested by a sensitizer and the collected energy is used to perform a chemical reaction and not to deliver electricity. This contrasts a PEC cell from a solar cell and is often referred to as "artificial photosynthesis." Although this concept is mostly applied to water splitting (*i.e.*, the light-driven electrolysis of water), it has also been employed in the context of CO_2 reduction [11]. The thereby used sensitizers are reminiscent of those that are used, *e.g.*, in the so-called dye-sensitized solar cells (DSSCs), which have been discussed beforehand in the context of solar cell applications (Chapter 6).

For example, in a straightforward approach, the $[Ru(tpy)(bpy)_2]^{2+}$ complex was anchored onto a nanostructured TiO_2 substrate *via* a COOH-substituent on the tpy ligand [84]. The thusly obtained hybrid material acted as a photocathode for CO_2 reduction. Upon irradiation with sunlight, CO_2 was reduced to MeOH at a potential of −0.4 *vs.* SHE. The PEC CO_2 reduction was run over 8 h and provided MeOH with a faradaic efficiency of *ca.* 64% and a TON of 62.6; the product selectivity was 45%, whereas if the sensitizer was dissolved in the electrolyte but not anchored to the TiO_2, the PEC performance dropped by *ca.* 50%; thus, the immobilization of the sensitizers facilitates the charge-separation step, *i.e.*, the electron injection into the semiconductor, from which CO_2 reduction proceeds.

In similar fashion, Chauvin *et al.* opted for a bimetallic photocathode design, in which both the sensitizer and the CO_2-reduction catalyst (*i.e.*, $[Ru(bpy)_3]^{2+}$ and $[Mn(tpy)(CO)_3Br]$, respectively) were immobilized onto TiO_2 *via* their lateral phosphonate moieties [85]. Upon irradiation with visible light, this photocathode reduced CO_2 to formic acid with high selectivity and a quantum yield (φ) of *ca.* 0.2%. Electrochemical experiments revealed that the electron transfer from the photoinduced $Ru^{II} \rightarrow Mn^{I}$ electron transfer proceeded through the semiconductor's conduction band and by hopping from one metal center to the other. Moreover, the one-electron-reduced Mn^0 complex represented the catalytically active species. Unlike for homogeneous scenario [72], where a dimeric Mn^0–Mn^0 complex, as the actual pre-catalyst, was formed, the heterogeneous PEC used the monomeric Mn^0 species [85]. The more efficient pathway, which was offered by the heterogeneous array of complexes, allowed the PEC CO_2 reduction to proceed with a TON of 27.

The use of heavy metal-based complexes (as sensitizer or catalyst) can be avoided, *e.g.*, by combining the complex $[Co(tpy)_2]^{2+}$ with an Si electrode. The PEC reduction of CO_2 to afford CO was realized under homogeneous as well as heterogeneous conditions. Kang *et al.* reported the CO generation with a faradaic efficiency of up to 83% from nonaqueous electrolytes, in which $[Co(tpy)_2]^{2+}$ was dissolved [86], whereas Reisner *et al.* used the corresponding phosphonate-equipped complex, which was anchored onto a *p*-type Si electrode, which was interfaced with a layer of mesoporous TiO_2 [87]. Notably, compared to the aforementioned photocathode, which was comprised of a Ru^{II} complex and a TiO_2 substrate, here the role of the components was inverted: the Si electrode acted as a light harvester and the Co^{II} complex represented the (pre-)catalyst. This hybrid photocathode enabled the CO_2-to-CO reduction even from a purely aqueous environment; thereby, a TON of *ca.* 330 was reached and the PEC process was operable for at least 24 h.

Finally, a different type of composite photocathode for the PCE CO_2 reduction was also presented by Reisner *et al.* [88]. The free-radical polymerization (FRP) afforded random copolymers, which contained tpy moieties for metal-ion complexation, phosphonate moieties for surface anchoring as well as hydrophobic alkyl chains (Figure 8.15a). The subsequent treatment with Co^{II} ions gave a partially cross-linked network, which was deposited onto a mesoporous inverse opal TiO_2 (IO-TiO_2). This heterostructured composite provided a hydrophobic environment around the Co^{II} complexes, where CO_2 reduction occurred selectively, even in the presence of water (Figure 8.15b,c). The so-called "outer-sphere effect" is known to promote the CO_2 reduction, whilst suppressing the concurrent HER [89]. Reisner *et al.* screened various electrode materials as supports for the composite to enable PEC CO_2 reduction from CO_2-purged aqueous MeCN [88]. From these, the Si electrode provided the best results in terms of syngas production over 6 h; thereby, CO fraction in the syngas was as high as 80%. For comparison, the PEC reaction, which used $[Co(tpy)_2]^{2+}$ as the homogeneous catalyst, only reached a CO fraction of 66%.

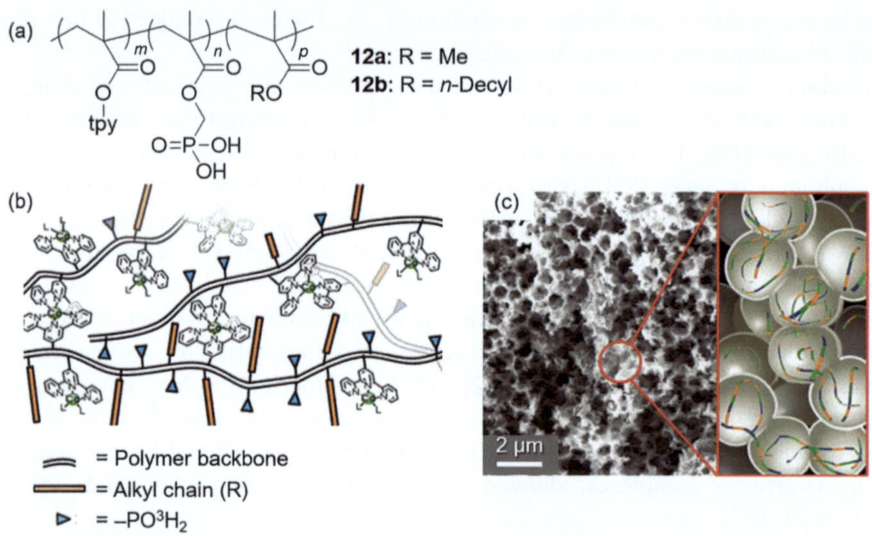

Figure 8.15 Schematic representation of the random copolymers **12**. (b) Schematic representation of such a copolymer after complexation with Co^{II} ions. (c) Representative SEM image and schematic representation of the metallopolymer-coated, mesoporous IO-TiO_2 scaffold [88] / with permission of John Wiley & Sons.

8.3 Electrochemical HER, Catalyzed by Tpy Complexes

Similar to the CO_2-reduction reactions, which have been discussed beforehand, the formation of molecular hydrogen (H_2), as a potential sustainable fuel source, has attracted enormous attention. In the context of keeping up the energy supply in a time of global warming and decline of fossil resources, the production of H_2 on a large scale becomes increasingly important [1]. Fuel cells represent one alternative energy system to the conventional combustion of fossil fuels. Fuel cells are attractive for numerous reasons: continuous and high energy density, low operational temperature, ease of fabrication, transportability, *etc.* [90]. In this context, H_2, MeOH, EtOH, $NaBH_4$, carbonate, and NH_3 are commonly employed as fuels. From these, H_2 is highly advantageous, due to its green combustion (water is the exclusive product) in concert with high energy density and output. However, the conventional technologies for H_2 production suffer from the high energetic costs and/or poor yield [91]. Moreover, the methods that utilize fossil fuels or biomass simultaneously produce CO_2 and CO, as undesirable greenhouse gases, whereas the splitting of water – either by electrolysis or photolysis – is an eco-friendly process, at least at first glance (Figure 8.16).

However, the simple transformation of H^+ to H_2, as one of the half-cell reactions of the water-splitting process, is not as simple as one might think. Although this reaction represents one of the references in electrochemistry (*i.e.*, the NHE with $E^0 = 0\,V$), its implementation into the emerging "hydrogen economy" is hampered by the enormous costs and challenges related to efficiency in terms of production

Steam reforming:	$CH_4 + H_2O \rightarrow CO + 3\,H_2$	@ 700–1000 °C	48% share
	$CO + H_2O \rightarrow CO_2 + H_2$	@ 360 °C	
Methane pyrolysis:	$CH_4 \rightarrow C + 2\,H_2$	@ 1065 °C	
Partial oil oxidation:	$C_{12}H_{24} + 6\,O_2 \rightarrow 12\,CO + 12\,H_2$		30% share
Partial coal oxidation:	$C_{24}H_{12} + 12\,O_2 \rightarrow 24\,CO + 6\,H_2$		18% share
Water splitting:	$2\,H_2O \rightarrow 2\,H_2 + O_2$	@ 50–80 °C	4% share

Figure 8.16 Summary of various methods for H_2 production, including the respective reaction temperatures and percentage share in worldwide production. Adapted from [91].

and storage [91]. In order to improve the efficiency of proton/water reduction, a plethora of catalytic systems have already been explored. In this context, the photocatalytic and electrochemical approaches are particularly worth mentioning [16, 91–94]. Before turning attention to the use of metal complexes in the context of the electrocatalytic HER, specifically of those containing tpy ligands, some general aspects of the electrocatalytic HER have to be recapped.

8.3.1 Fundamental Aspects of the Electrochemical HER

The HER represents a versatile approach to store energy in the chemical H—H bond. A plethora of strategies have been developed to overcome the most critical limitation, which is the high kinetic barrier for the reductive transformation of protons into molecular hydrogen. In particular, metallic Pt features a remarkably high electrocatalytic activity for the HER, but its use is restricted by the metal's low natural abundance and high costs [95]. Thus, intense research is devoted to the development of efficient and less cost-intensive catalytic systems, which employ more abundant metals. In this context, both heterogeneous materials as well as homogeneous transition-metal complexes are in the focus of interest. The latter allow the modulation of reactivity not only by the choice of the metal center but also by structural modification of the ligands.

Before turning attention to the use of transition-metal complexes as electrocatalysts for the HER, specifically those with tpy ligands, the mechanism for the HER itself needs to be briefly recapped. The electrochemical water-splitting reaction can be deconvoluted into two half-cell reactions, which are the anodic OER and the cathodic HER. The standard potential and Gibbs free energy for the electrolysis are 1.23 V (*vs.* NHE) and 273 kJ mol^{-1} (Figure 8.17) [97]. One has to be aware that the HER mechanism strongly depends on the electrolyte's pH value: the more basic the electrolyte is, the poorer and more sluggish the HER kinetics become. There are two fundamentally different devices available for the electrochemical splitting of water: the polymer-electrolyte electrolyzers and the alkaline ones [91]. The former suffer from the high costs since only noble-metal-based catalysts are compatible with the electrolytes used, whereas the latter have been commercialized since catalysts based on non-noble metals can deal with the alkaline environment [98].

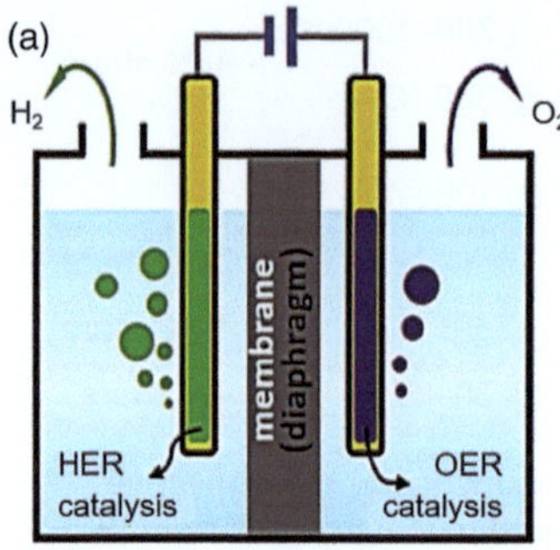

(b)

Acidic environment

$2\,H_2O$	$\rightarrow$	$O_2 + 4\,H^+ + 4\,e^-$	$E^0 = -1.23$ V
$4\,H^+ + 4\,e^-$	$\rightarrow$	$2\,H_2$	$E^0 = -0.00$ V
$2\,H_2O$	$\rightarrow$	$O_2 + 2\,H_2O$	$E^0 = -1.23$ V

Basic environment

$4\,OH^-$	$\rightarrow$	$O_2 + 2\,H_2O + 4\,e^-$	$E^0 = -0.40$ V
$4\,H_2O + 4\,e^-$	$\rightarrow$	$2\,H_2 + 4\,OH^-$	$E^0 = -0.83$ V
$2\,H_2O$	$\rightarrow$	$O_2 + 2\,H_2O$	$E^0 = -1.23$ V

Figure 8.17 (a) Schematic representation of an electrochemical cell for the splitting of water [96] / with permission of American Chemical Society. (b) Hall-cell reactions (*i.e.*, the HER and OER, respectively), in an acidic or basic environment.

The elemental steps of the HER in acidic and basic media are summarized in Figure 8.18. Firstly, under acidic conditions, the so-called Volmer step generates hydrogen atoms, which are adsorbed at the catalytic sites, by the combination of protons and electrons [97]. Subsequently, molecular hydrogen is formed by the reaction of an adsorbed hydrogen atom with a proton and an electron; thereby, a vacant catalytic site in proximity to an occupied one is utilized (Heyrovsky reaction). Alternatively, the Tafel step also yields H_2 by the combination of two adsorbed *H*-atoms. These three elemental steps are likewise involved in the HER under basic conditions. The Volmer and Heyrovsky reactions consume water molecules for the formation of absorbed hydrogen atoms and the evolution of H_2, respectively. In both steps, hydroxide is liberated. The Tafel step in a basic environment is essentially the same as under acidic conditions.

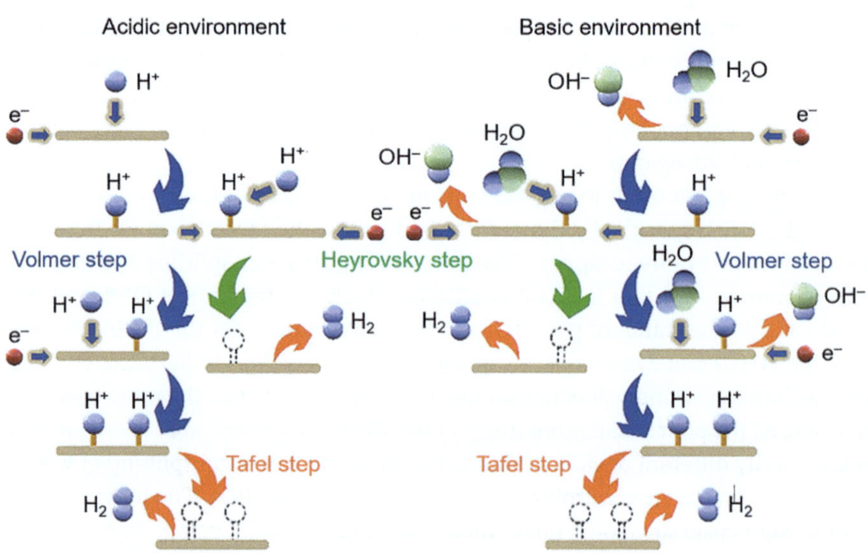

Figure 8.18 Schematic representation of the key steps of the HER in an acidic and basic environment [99] / John Wiley & Sons / CC BY 4.0.

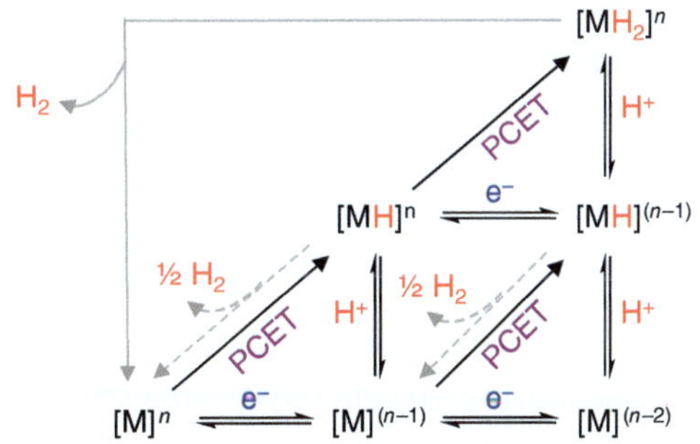

Figure 8.19 Schematic representation of the potential pathways for HER, if catalyzed by a transition-metal complex, which is denoted as [M] with the overall charge *n*. The dashed and solid gray arrows refer to the heterolytic and homolytic pathways, respectively [94] / with permission of Elsevier.

When using transition-metal complexes, as molecular catalysts for the electrochemical HER, one has to consider the various protonation and reduction, which contribute to the H_2 evolution. The metal center needs to be reduced to a low-valent state and then protonated to afford a metal–hydride species (Figure 8.19) [94]. As shown in various examples in Section 8.3.2, these species represent important intermediates in the catalytic cycle of proton reduction. Starting from these intermediates, different pathways for H_2 evolution have been proposed: Firstly, redox/protonation reactions between metal–hydrides might provide species of various valence states. The so-called homolytic pathway involves the oxidative coupling of two hydrides to afford H_2, whereas the H—H bond can also be formed *via* a heterolytic pathway, which is the protonation of the hydride by the present Brønsted acid. Transition-metal hybrids have been intensively studied in the context of organometallic chemistry [100], but the occurrence of such species with the catalytic cycle of the HER has been sparsely verified experimentally [101–103]. This, of course, can be rationalized by the fact that thermodynamically stable intermediates, which can be characterized spectroscopically or even be isolated, will hardly play any active role in catalysis. As emphasized by Thummel *et al.*, intermediates of too "low" energy are unlikely to occur in efficient catalytic cycles; this also holds true for the electrocatalytic HER [94]. Notably, the reactions of the metal–hydride species are not necessarily the rate-determining ones and, thus, the observed/apparent rate constant (k_{obs}) for proton catalysis might not always be related to the actual H_2-evolution step. This peculiarity makes it difficult for one to distinguish between the homolytic and heterolytic pathways for H_2 formation.

It has to be emphasized that, for practical reasons, experimental studies on the electrocatalytic water splitting typically focus on one of the half-cell reactions, *i.e.*, the HER or OER. One reason for doing so is to avoid the exothermic recombination of H_2 and O_2 to afford water in the so-called "Knallgas" reaction. Moreover, the anodic OER is thermodynamically less favored than the cathodic HER (*i.e.*, ΔG of 475 kJ per mol O_2) [104]. For these reasons, research is either focused on the electrochemical HER or OER.

8.3.2 Tpy Complexes as Catalysts for the HER

Although Ru^{II}–tpy complexes have successfully been employed as catalysts for the ERC, and the concomitant HER has been thoroughly studied, such complexes have not been utilized in the context of the electrocatalytic HER. Quite on the contrary, suppression of the HER in favor of the CO_2 reduction was the focus of interest (see Section 8.2). Overall, Ru^{II} complexes have only sparsely received recognition as actual HER electrocatalysts [105]. On the other hand, Ru^{II}-tpy *bis*-complexes have mainly been employed as a sensitizer, in the photocatalytic variant [1].

Unlike the complexes with heavy transition-metal ions, their counterparts with 3d metal ions are subject of intense research for electrocatalytic HER. This also holds true for the molecular complexes with ligands from the polypyridyl family [94]. In particular, Co^{II}–polypyridyl complexes have experienced widespread attention. Such complexes are capable of storing multiple reduction equivalents since the ligands not only stabilize the reduced metal centers but also play a non-innocent role by accommodating electrons in their π-conjugated system (*i.e.*, formation of radical-anion ligands).

As mentioned beforehand, Fontecave *et al.* recognized the ability of electrode-anchored Co^{II}–tpy *mono*-complexes to electrocatalyze the HER using DMF as the solvent and acetic acid as the proton source [76]. The bulk electrolysis at −1.50 V *vs.* Ag/AgCl provided a current density of 300 μA cm^{-2}, which was sustained for at least 3 h. Throughout that time, *ca.* 6 μmol h^{-1} of H_2 were produced, which corresponded to an overall TON of 24,000 after 3 h (assuming the coverage of the electrode with a monolayer of Co^{II} complexes). A faradaic yield of >90% was obtained, indicating that the HER represented the only relevant electrolysis process. Notably, the electrocatalytic HER using immobilized Co^{II}–tpy complexes was already reported by Abe and Kaneka beforehand [106]. These authors dispersed $[Co(tpy)_2]^{2+}$ complexes in a Nafion™ membrane, which was coated onto a pyrolytic graphite electrode. Thus, immobilization was due to the electrostatic interaction between the cationic complexes and the negatively charged membrane (Nafion™ possesses $-SO_3H$ moieties). The HER was performed at various pH values and two different mechanisms for the H_2 evolution were suggested. Firstly, in an acidic environment (*i.e.*, pH ≤ 4), a Co^{I} complex was identified as the catalytically active species, whereas at lower pH values, Co^{0} species were apparently involved. In the latter case, the HER was more efficient (*i.e.*, 20.3 *vs.* 7.4 μL h^{-1}).

Fontecave *et al.* performed a systematic study to elaborate the influence of the ligand structure on the electrocatalytic activity of the corresponding Co^{II} *bis*-complexes (**13**, Figure 8.20a) [107]. The parent $[Co(tpy)_2]^{2+}$ complex (**13a**) was prone to decomposition under the applied HER conditions (*i.e.*, acetic acid and MeCN, as the Lewis acid and solvent, respectively), whereas the amine-functionalized complexes (**13b–e**) revealed the highest electrocatalytic activity and enabled the H_2 evolution at low overpotentials. This held particularly true for **13b** and **13e**, which generated H_2 with faradaic efficiencies of 70% and 90%, respectively (Figure 8.20b). Density-functional theory (DFT) calculations were conducted to elucidate the reaction's mechanism and two different pathways were identified: Firstly, in the case of complexes **13a–b** and **13f–h**, the HER proceeded *via* the temporary reduction of the central pyridine ring of one of the ligands, thus resembling the biological NAD/NADH system. Thereby, the metal center retained its coordination sphere throughout the catalytic cycle, independent from the actual oxidation state

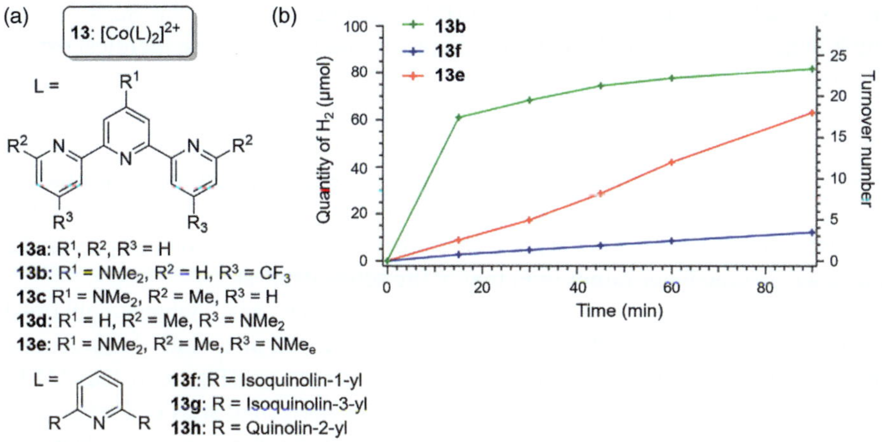

Figure 8.20 (a) Schematic representation of the Co^{II} complexes **13** (counterions were omitted for clarity). (b) Electrocatalytic HER performance of selected Co^{II} complexes **13** [107] / with permission of John Wiley & Sons.

(*i.e.*, Co^{II} or Co^{I}). However, the NMe_2 substituents of complexes **13c–e** acted as a proton relay: One outer pyridine ring was decoordinated and *N*-protonated; this enabled the subsequent formation of a Co^{II}–hydride intermediate, which facilitated the H_2-evolution step. According to the experimental data, the latter catalytic cycle provided the overall more efficient pathway for the HER.

A similar behavior in terms of high HER efficiency at low overpotential was also reported for quinolinyl-substituted Co^{II} complexes **14a/b** (Figure 8.21) [108, 109]. However, unlike complexes **13c–e**, the formation of a Co^{II}–hydride intermediate subsequent to partial ligand dechelation was excluded on the basis of electrochemical analyses. Instead, the presence of solvato complexes and the "non-innocent" role of the ligands was presumed (*i.e.*, protonated radical-anion species were presumably formed due to a proton-coupled two-electron reduction of a detached pyridine ring). The 2-chloroquinolin-3-yl-substituted complex (**14b**) featured a *ca.* 300 mV lower overpotential for proton reduction than the parent complex $[Co(tpy)_2]^{2+}$ [109]. Moreover, Padhi *et al.* showed that the electrocatalytic HER in the presence of Co^{II}–tpy complexes can be improved by introducing redox-mediating moieties on the tpy ligand [110]. The Fc-equipped complex **14c** (Figure 8.21) represented superior HER electrocatalyst to parent $[Co(tpy)_2]^{2+}$ complex in both organic and aqueous environments (Fc: ferrocene). For example, the HER in aq. DMF, using acetic acid as the proton source proceeded with a faradaic efficiency of *ca.* 90% in

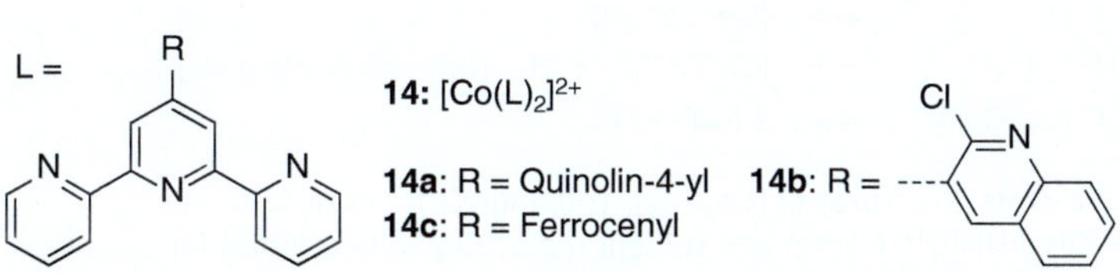

Figure 8.21 Schematic representation of Co^{II} complexes **14**.

the overpotential range from 655 to 955 mV. The observed TOF was $825\,s^{-1}$, thus much higher than that of the parent $[Co(tpy)_2]^{2+}$ under otherwise identical experimental conditions (TOF = $366\,s^{-1}$). Due to the presence of the electron-donating Fc moieties of **14c**, the overpotential for the HER was *ca.* 200 mV lower than when using $[Co(tpy)_2]^{2+}$ as the electrocatalyst. Similar experiments in terms of superior electrocatalytic activity of **14c** were also observed when the HER was run in aqueous media. It has to be noted that Fc and its derivatives represent versatile electrocatalysts themselves [111], but in Padhi's study, the Fc moieties played the role of a redox mediator rather than that of the actual catalyst.

In a different approach, the dinuclear complex **15a**, which contained 3,5-di(pyridin-2-yl)pyrazolate (dpp), as the *mono*-anionic and *bis*-bidentate bridge, was prepared by Fukuzumi, Nam, Llobet *et al.* (Figure 8.22a) [112]. These authors referred to the binuclear nature of the active sites of [FeFe]- and [NiFe]-hydrogenase, in which the close proximity of the metal centers is believed to have a synergistic effect regarding electron-/proton-transfer processes or the activation of certain chemical bonds [94]. Besides the catalytic four-electron reduction of O_2 (see Section 8.5) [112], the HER using **15b**, as the catalyst, was also studied [113]. Thereby, **15b** was synthesized from **15a** by oxidation with a Ce^{IV} salt in aqueous MeCN. The HER, which used trifluoroacetic acid (TFA) as the proton source, involved various mixed-valence species, from which the $[Co^{II}Co^{III}–H]^{2+}$ intermediate produced H_2 upon protonation (Figure 8.22b). However, this HER was not (yet) performed electrocatalytically, but rather used cobaltocene as a sacrificial electron donor.

The previous examples clearly show that – at least amongst the tpy complexes – the Co^{II}-containing complexes represent the most intensively studied HER electrocatalysts. Besides these, scattered studies have been published, where other 1st-row transition-metal ions have been utilized. For example, a Ni^{II}–tpy *bis*-complex, which

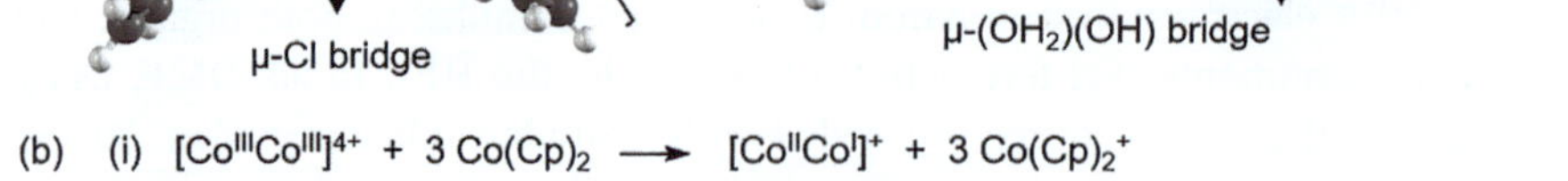

(b)

(i) $[Co^{III}Co^{III}]^{4+} + 3\,Co(Cp)_2 \longrightarrow [Co^{II}Co^{I}]^{+} + 3\,Co(Cp)_2^{+}$

(ii) $[Co^{II}Co^{I}]^{+} + H^{+} \rightleftharpoons [Co^{II}Co^{III}–H]^{+}$

(iii) $[Co^{II}Co^{III}–H]^{2+} + H^{+} \longrightarrow [Co^{II}Co^{III}]^{3+} + H_2$ (rate-determining step)

(iv) $[Co^{II}Co^{I}]^{+} + [Co^{II}Co^{III}]^{3+} \longrightarrow 2\,[Co^{II}Co^{II}]^{2+}$

Figure 8.22 (a) Solid-state structures of $[Co_2(tpy)_2(\mu\text{-dpp})(\mu\text{-Cl})]^{2+}$ (**15a**, left) and $[Co_2(tpy)_2(\mu\text{-dpp})(\mu\text{-(OH}_2\text{)(OH)})]^{4+}$ (cation and solvent molecules were omitted for clarity). Adapted from [112]. (b) Simplified representation of the reaction mechanism of the HER, which was catalyzed by **15b**. Adapted from [113].

contained the same tpy ligand as the Co^{II} complex **14b**, was reported by Padhi *et al.* [114]. This complex enabled the electrocatalytic HER in aqueous DMF using TFA as the proton source. The efficiency for H_2 generation was higher than that observed for the parent complex $[Ni(tpy)_2]^{2+}$ and required a *ca.* 180 mV lower overpotential. Similar to the behavior of complexes **14**, the pending-base effect (*i.e.*, the partial dechelation and subsequent protonation) represented one key factor for the catalytic activity; the pK_a value for this process was 4.6 (a matching acid-base equilibrium was not reported for **14b**). To the best of our knowledge, follow-up studies on Ni^{II}–tpy complexes, as HER electrocatalysts, have not yet been published.

Recently, Cu^{II}–polypyridyl complexes have also attracted some interest for application, as electrocatalysts in the HER [94]. Undoubtedly, copper plays a relevant role in biological redox processes, but the use of Cu^{II} complexes as catalysts for the HER is less explored, in particular when compared to the related Co^{II} systems. One reason for this is the relatively high reduction potential of copper's low-valent intermediates, which might occur in the catalytic cycle of the proton-reduction process [115]. So far, only two studies have dealt with the use of Cu^{II}–tpy complexes in this context. Firstly, Raj and Padhi reported the Cu^{II} complexes **16a-c**, which were assembled from Cu^{II} acetate and the respective tpy ligand in a MeOH/MeCN solvent mixture [116]. Depending on the nature of the ligand's substituent in 4′-position, the resulting complexes differed in the nuclearity and, thus, their molecular structure: According to single-crystal XRD analyses, **16a** and **16c** were mono- and binuclear complexes, respectively, whereas **16b** represented a 1D coordination polymer (Figure 8.23a). Independent of this, all three compounds catalyzed the electrochemical HER using acetic acid as the proton source. The highest catalytic activity in aqueous DMF was observed for **16a**, which featured a TOF of 1,473 s^{-1}. Independent from the original structure, the formation of $[Cu(L)(OAc)(OH_2)]^+$ complexes under the reaction conditions was proposed. These aquo complexes, whose pK_a values were *ca.* 4.5, were subsequently deprotonated to afford the corresponding hydroxo species [Cu(L)(OAc)(OH)]. However, when changing to acidic water, as the solvent, **16c** was identified as the most potent electrocatalyst for water reduction (*i.e.*, TOF value of 810 mol H_2 per h). In either case, proton-coupled electron transfer (PCET) processes generated Cu^{II}–H intermediates, which, in the presence of protons, produced H_2 and reconverted themselves to $[Cu(L)(OAc)(OH_2)]^+$. A similar behavior in terms of electrocatalytic HER from aqueous DMF or water, in the presence of acetic acid, was also observed for complexes **16d–e** (Figure 8.23), which contained the same quinolinyl substituents as the aforementioned Co^{II} complexes **14a–b** (Figure 8.21).

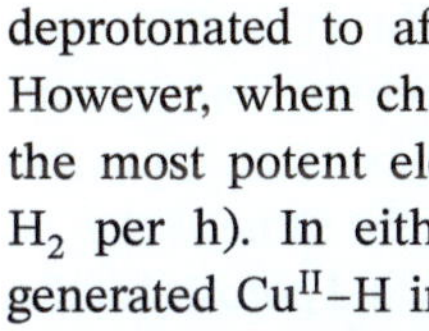

L1: R = Pyridin-3-yl
L2: R = Pyridin-4-yl
L3: R = Phenyl
L4: R = Quinolin-4-yl
L5: R = 2-Chloroquinolin-3-yl

16a: $[\{(OAc)_2Cu(L1)\}_2Cu(OAc)_2(H_2O)_2]$
16b: $\{[Cu(L2)(OAc)]Cl\}_n$
16c: $Cu(L3)(OAc)_2$
16d: $[Cu(L4)(OAc)(OH_2]Cl$
16e: $[Cu(L5)(OAc)(OH_2]Cl$

Figure 8.23 Schematic representation of the Cu^{II} complexes **16**.

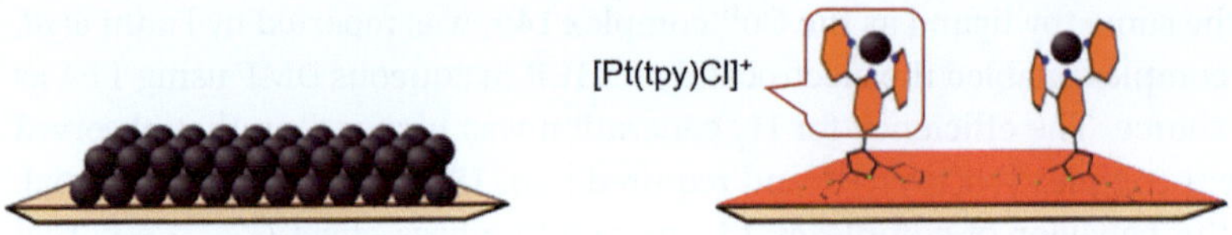

Figure 8.24 Schematic representation of the structure of a conventional bulk Pt electrode (left) and a metallopolymer-coated electrode (right) [118] / with permission of American Chemical Society.

Similar to the mechanism of the water reduction in the presence of **16a–c**, the deprotonated [Cu^{II}(L)(OAc)(OH)] complex afforded [Cu^0(L)(OAc)(OH_2)]$^-$ *via* a PCET pathway; this species, in turn, afforded the reactive Cu^{II}–H intermediate, from which the actual HER occurred and the catalytic cycle was closed. Unlike for **16a–c**, **16d–e** operated under more benign conditions (*i.e.*, at the pH value of 7 in phosphate-buffered water) and significantly lower overpotential when compared to the blank solution. Under these conditions, complex **16e** achieved a respectable TOF value of 7,174 mmol of H_2 per h.

As mentioned before, transition-metal catalysts, in particular from the Pt group, are highly versatile due to their high activities, which enable the HER with relatively low overpotentials [94]. However, the high cost and limited natural abundance of Pt hampers the large-scale application of the benchmark catalysts. It has been shown that the required amount of Pt can be reduced significantly by the deposition of Pt or Pt-alloy nanoparticles onto carbon-based electrodes. Such hybrid electrodes have been used as high-performance electrocatalysts for various applications, including the HER [117]. The emerging field of using nanoparticles for the electrocatalytic HER has been reviewed by Hu, Zhang, Wong *et al.* a few years ago [93]. In order to further reduce the amount of Pt, Nann *et al.* coated an Au electrode with a thin metallopolymer layer, which was comprised of a conductive polythiophene backbone and Pt^{II}–tpy complexes, which were exposed to the solvent (Figure 8.24) [118]. This spatial arrangement enabled a highly efficient utilization of metal centers since all the single sites could (at least in theory) participate in the electrocatalytic HER. Thus, the electrocatalytic performance of this hybrid electrode was almost the same as that of a conventional Pt-mesh electrode, despite the higher surface area and porosity of the latter. The strategy of using immobilized [Pt(tpy)Cl]$^+$ complexes for the electrocatalytic HER stands in line with Abruña's earlier work, in which electropolymerized metallopolymer films were used for the ERC (see Section 8.2.3).

8.4 Electrochemical OER Catalyzed by Tpy Complexes

As mentioned beforehand, the OER represents the cathodic half-cell reaction of water electrolysis, which – in the ideal case – simultaneously generates H_2 and O_2 (see Section 8.3.1). Not only this, but the water-oxidation reaction (WOR) to afford molecular oxygen, which naturally occurs in many life forms (*i.e.*, algae, cyanobacteria, and higher plants), is also one essential process that enables the existence of life on Earth. The natural photosystem II (PS-II) in green plants contains a μ-oxido Mn_4Ca cluster within its O_2-evolving complex. The WOR at this active site proceeds *via* the stepwise extraction of in-total four electrons and four protons from two water molecules to afford one O_2 molecule [119]. From the perspective of thermodynamics, this process is highly unfavorable (E^0 = 1.23 V *vs.* SHC at pH 0); the OER is an endothermic reaction, whose ΔG value is remarkably high (*i.e.*, +475 kJ per mol of formed O_2) [104]. Moreover, one has to take the kinetic barrier of the half-cell reaction into account, which is overpotential. This value, which also plays an important role in the electrocatalytic CO_2 reduction and HER, is, simply speaking, defined as "the difference between the equilibrium potential for a given reaction (also called the thermodynamic potential) and the potential at which the catalyst operates at a specific current under specific conditions" [120]. Thirdly, the OER is also more demanding than the HER when considering the kinetics. When the OER is performed electrochemically, each of four electrons and protons must be transferred at the anode; simultaneously, the intrinsically weak O—O single bond must be established within the first intermediate steps of the reaction cascade [104]. For these reasons, the overpotential of the electrochemical water oxidation is high – typically >400 mV. In order to overcome these shortcomings, catalysts, specifically transition-metal complexes, have been introduced, which enable the OER to proceed with high rates and at low overpotentials [121].

These catalysts are playing an essential role in the O—O bond formation, thereby decreasing the energetic costs of this process. According to Kandar and Grotjahn, one has to distinguish between two mechanistic pathways toward the O—O bond formation, *i.e.*, nucleophilic attack of water and bimolecular radical oxo-coupling (WNA and I2M, respectively; Figure 8.25) [104]. Initially, successive PCET reactions afford a high-valent metal-oxo intermediate, whose reactivity depends on both the metal center's number of d-electrons and ligand field. For the metal-oxo intermediate, two resonance structures can be assumed, from which the WNA and I2P pathways branch off. Firstly, water, as a nucleophile, might attack the highly electrophilic oxo ligand. The electrophilicity of the oxo ligand is dictated by the metal's oxidation state on the one hand and the molecular environment on the other [the latter is mainly due to the geometry of the ligand(s)]. When considering the involved orbitals, an overlap of water's σ-bonding orbital with the antibonding π*-orbital of the metal-oxo group might be assumed. This interaction results in the

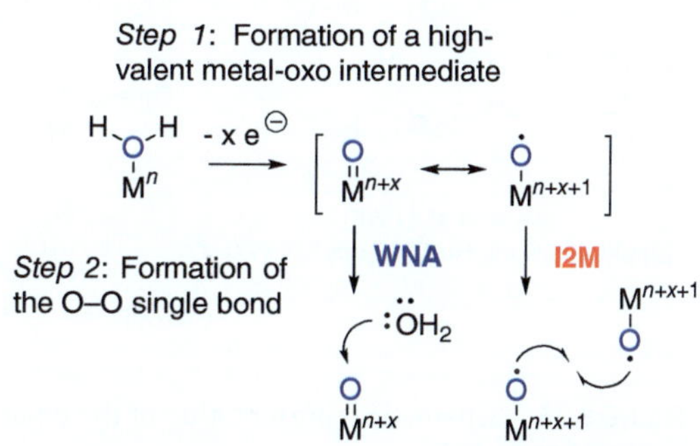

Figure 8.25 Schematic representation of the mechanistic pathways for the metal-catalyzed O–O bond formation. Adapted from [104].

formation of the O—O single bond to the disadvantage of the M—O double bond; this step involves the formal two-electron reduction of the metal center, whereas, in the alternative I2M pathway, the dimerization of two metal-oxo intermediates *via* radical-radical coupling establishes the O—O single bond. In this case, the singly occupied antibonding π*-orbitals of the metal-oxo fragments are involved [122].

Once the formation O—O single bond has been formed, the subsequent steps must proceed in a controlled fashion to afford O_2 as the product rather than oxidizing side products (*e.g.*, $HO^{\cdot}$ or H_2O_2), which could lead to catalyst degradation due to irreversible ligand and/or metal-center oxidation. Thus, electrocatalysts for the WOR need to be stable enough to withstand the forcing conditions of the OER for a longer period of time. However, molecular catalysts, *i.e.*, transition-metal complexes, which operate under homogeneous conditions, are highly susceptible to oxidative degradation. The complete loss of ligands due to oxidative degradation eventually affords metal-oxide species (*e.g.*, nanoparticles). These secondary materials might be (electro)catalytically inactive or represent (electro)catalysts themselves. In this case, the OER is not driven by the initial transition-metal complex (or its precursor), but by its decomposition product. One has to remember that metal oxides are often used as heterogeneous catalysts for the WOR [121].

A wide range of water-oxidation catalysts (WOCs), which contain Mn, Ir, Fe, Cu, Cu, or Ru ions, has already been reported [123–125]. From these, the Ru-containing ones have received the most recognition so far. Research on the WOCs can at least be traced back to 1959, when Gortsema and Cobble reported the OER from water, which was catalyzed by a Ru^{IV} species, which was generated *in situ* by reduction of RuO_4 by H_2O_2 or Fe^{II} ions; in this primal study, ceric ammonium nitrate (CAN) was used, as the sacrificial oxidant [126]. Later, some stable Ru^{IV}–oxo complexes with polypyridyl ligands were prepared by PCET reactions [127–129]. The first structurally defined WOC was then presented by Meyer *et al.* [130]. Their dinuclear μ-O-bridged complex $[(bpy)_2(H_2O)Ru^{III}\text{-O-}Ru^{III}(OH_2)(bpy)_2]^{4+}$, which is commonly known as the "blue dimer," has been extensively studied since its discovery in 1982 [121]. In the following, various dinuclear complexes, which contained tpy ligands, have been reported. Some selected examples (**17–19**), which belong to the 1st generation WOCs, are shown in Figure 8.26 [131–134].

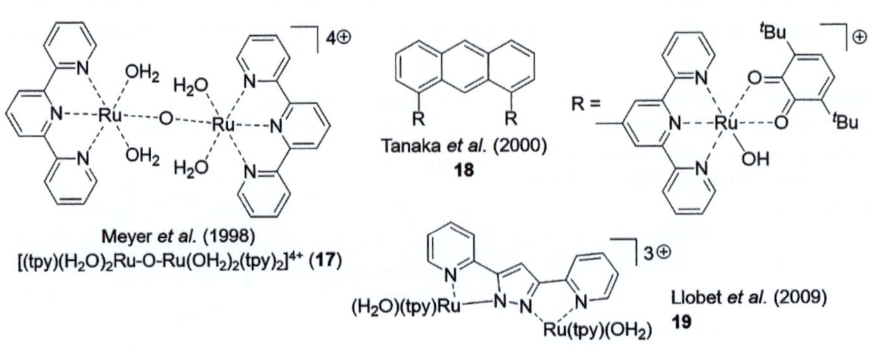

Figure 8.26 Schematic representation of the dinuclear WOCs **17–19** (counterions were omitted for clarity).

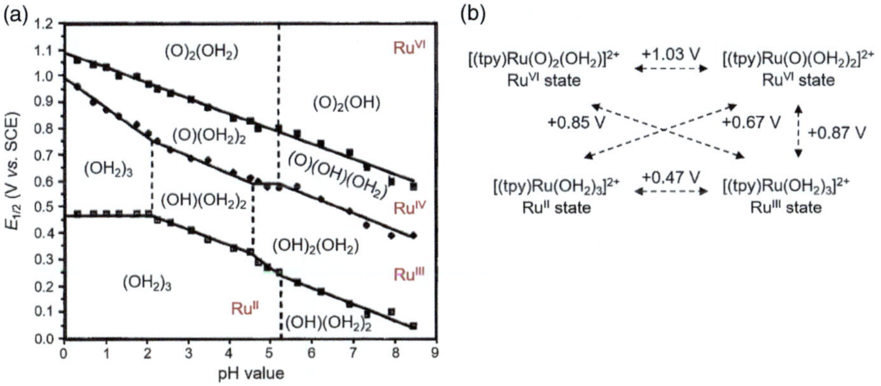

Figure 8.27 (a) Pourbaix or (potential-pH) diagram of $[Ru(tpy)(OH_2)_3]^{2+}$ showing the regions of stability for the various Ru oxidation states along with the corresponding protonation states; the respective pK_a values are indicated by the vertical dashed lines [135]. (b) Summary of the redox systems $Ru^{VI/IV}$, $Ru^{IV/III}$, and $Ru^{III/III}$ in aqueous environment (pH 1, potentials *vs.* SCE) [135] / with permission of American Chemical Society.

Unlike for the "blue dimer," the +IV oxidation state of **17** was not stable and both anation and oxidative cleavage resulted in rapid catalyst deactivation in the presence of CAN, as the oxidant [131]. One has to be aware that the potentials of the commonly involved Ru-based redox couples are highly pH-dependent and that the individual redox states possess defined regions of stability (Figure 8.27) [135]. These restrictions have to be considered when dealing with Ru–polypyridyl complexes, as WOCs. Tanaka's complex (**18**)$(SbF_6)_2$ represented the first electrocatalyst for the WOR, which functioned in absence of any sacrificial oxidant [132, 133]. Under homogeneous conditions, the electrochemical OER proceeded at +1.70 V in aqueous 2,2,2-trifluoroethanol with a faradaic efficiency of 91% and a TON of 21. However, when the catalyst was immobilized onto an ITO electrode, the TON for the water electrolysis at a pH value of 4 was raised to 33,500 (ITO: indium-tin oxide) [132]. Mechanistic studies revealed that both the $Ru^{II/III}$ redox couple as well as the quinone/semiquinone system contributed to the four-electron oxidation of water.

Thummel's discovery that mononuclear Ru^{II} complexes could also catalyze the OER [136] eventually resulted in a paradigm change, which was expressed by the striking slogan "One site is enough!" [137]. Thus, the commonly accepted thinking that the multi-electron nature of the OER could not be realized by a single-site complex was overcome. This way of thinking was based on the knowledge of the structure of the multinuclear O_2-evolving cluster in the PS-II on the one hand and the nonavailability of active mononuclear WOCs on the other [104]. Almost simultaneously, the groups of Thummel, Sakai, and Meyer reported a plethora of mononuclear Ru^{II}–polypyridyl complexes, which constituted the 2nd generation WOCs [1, 104]. However, all these complexes, which possessed the general formula $[Ru(tpy)(N^\wedge N)(OH_2)]^{2+}$ or $[Ru(tpy)(N^\wedge N)X]^{+}$, were only used as WOCs in the presence of CAN, as the sacrificial oxidant, but not under electrochemical conditions (N^N: bidentate polypyridyl ligand, X = Cl or Br). Nonetheless, a controversy arose

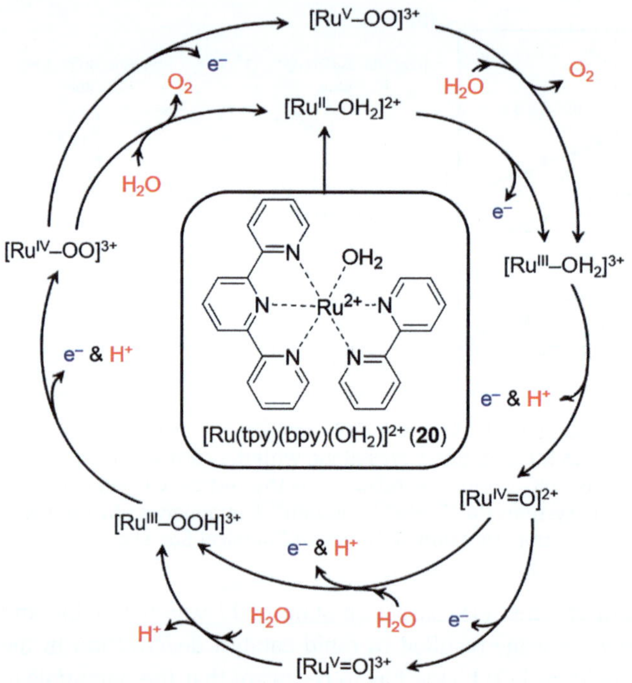

Figure 8.28 Schematic representation of the mechanism for the catalytic WOR, as suggested by Pushkar *et al.* (counterions were omitted for clarity). [138] / with permission of American Chemical Society.

regarding the mechanism of the WOC when using these catalysts. The initially proposed Ru^{V}–oxo intermediates [137] were ruled out by Pushkar *et al.*, who identified $[Ru^{IV}(tpy)(bpy)(O)]^{2+}$ to be the predominant steady-state species within the catalytic cycle – at least when using **20**, as the WOC (Figure 8.28) [138]. As the only paramagnetic species, Ru^{III}–peroxido intermediates were detected by electron paramagnetic resonance (EPR) spectroscopy. It was proposed that the formation of intermediate Ru^{V}–oxo complexes under the experimental conditions was disfavored for thermodynamic reasons.

Ott *et al.* as well as Morris *et al.* aimed for the stabilization of WOCs by incorporating them into a robust MOF structure, specifically from the UiO-67 type (UiO: University of Oslo) [139, 140]. In such MOFs, the $Zr_6(O)_4(OH)_4$ clusters are linked *via* rigid (typically aromatic) dicarboxylate ligands. For this purpose, the derivative **21** was co-assembled with the other molecular components to afford the hybrid electrode, which is shown in Figure 8.29a. Morris *et al.* performed the WOR at the potential of +1.71 V under pH-neutral conditions with a TOF of *ca.* $0.2\,s^{-1}$ (Figure 8.29b) [140], whereas Ott *et al.* applied a voltage of +1.5 V to generate O_2 continuously over several hours with the faradaic efficiency of *ca.* 82% [139]. In either case, the rigid-robust MOF environment ensured a good catalyst durability and prevented the metal centers of the WOCs from agglomerating into clusters or forming metal-oxide nanoparticles, as the degradation products. Notably, the

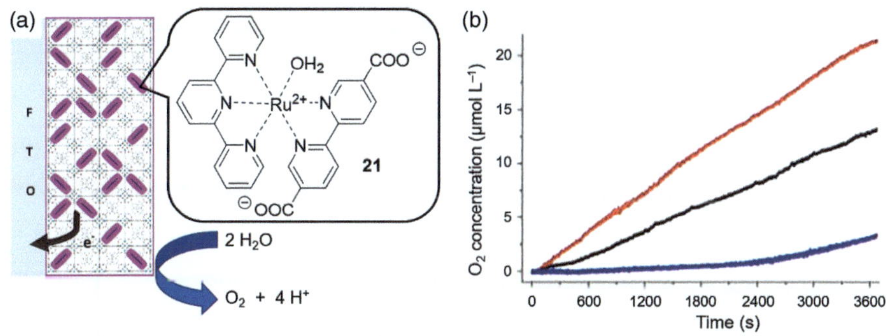

Figure 8.29 (a) Schematic representation of the UiO-67−**21**@FTO assembly, which was utilized as a heterogeneous OER electrocatalyst (counterions were omitted for clarity). (b) O_2 evolution provided by this same electrode at 1.71 V *vs.* SHE at pH 6.8 [140] / with permission of John Wiley & Sons.

strategy to immobilize single-site catalysts in MOF structures has also been widely employed for other purposes beyond the electrocatalytic OER [141].

However, with the emergence of the 3rd generation WOCs, which contain tetra- or even pentadentate polypyridyl-based ligands, the focus of interest shifted to these catalytically more potent systems. Exemplarily, the complex $[Ru(dpp)(pic)_2]^{2+}$ [dpp: 2,9-di(pyridine-2-yl)-1,10-phenanthroline; pic: 4-picoline] featured a superior TON of 416 and the O_2-evolution rate was as high as 0.330 μmol min^{-1} (TOF = 27.5×10^{-6} s^{-1}) [142]. After this, interest in molecular WOCs from the Ru^{II}–tpy family rapidly declined and only a handful of studies dealing with Ru^{II}-based WOCs have appeared in recent years.

In one example, Yagi *et al.* reported that the dinuclear complex **22**, whose structure was stabilized by intramolecular *H*-bonding interactions, acted as an electrocatalyst for the OER under homogeneous conditions (Figure 8.30a) [144]. Electrochemical studies revealed that the catalytic cycle involved a Ru^{III}–OH/Ru^{III}–OH species as the resting state. Furthermore, the formation of the O—O single bond proceeded not *via* the WNA pathway, as in the case of most mononuclear Ru^{II} WOCs, but involved the coupling of adjacent Ru–oxo species (corresponding to an intramolecular I2M pathway, see Figure 8.25). In subsequent work, complex **22** was anchored onto a TiO_2-coated ITO substrate *via* lateral carboxy moieties on the tpy ligands [143]. This hybrid electrode enabled the water oxidation under pH-neutral conditions with an overpotential of 530 mV and a high catalytic current of 5.1 mA cm^{-2} at 1.6 V *vs.* SCE (Figure 8.30b). Moreover, the bulk electrolysis at 1.16 V (*vs.* SCE) afforded O_2 with a faradaic efficiency of *ca.* 87%. *Post-operando* analysis revealed that *ca.* 75% of the complex molecules remained anchored to the electrode, thus showing the robustness of the electrode design.

Molecular WOCs have also been formed with the abundant 1st row transition-metal ions, thus repelling the use of costly Ru [145]. Regarding tpy complexes, the dinuclear μ-O-bridged mixed-valence Mn^{III}/Mn^{IV} complex **23** has received most recognition (Figure 8.31) [146, 147]. This complex, which is reminiscent of the O_2-evolving active cluster in the natural PS-II, catalyzed the WOR in the presence of

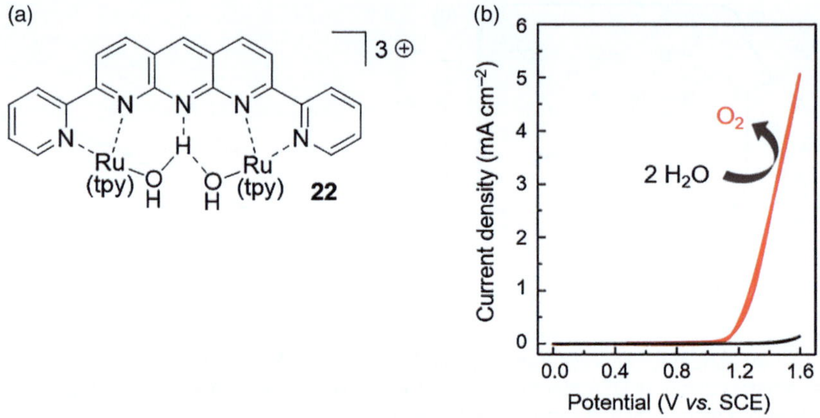

Figure 8.30 (a) Schematic representation of the dinuclear WOC **22** (counterions were omitted for clarity). (b) CV curve of the hybrid electrode in pH-neutral water upon scanning from 0 to +1.6 V (scan rate: 100 mV s^{-1}); O_2 evolution was observed above +1.1 V [143] / with permission of American Chemical Society.

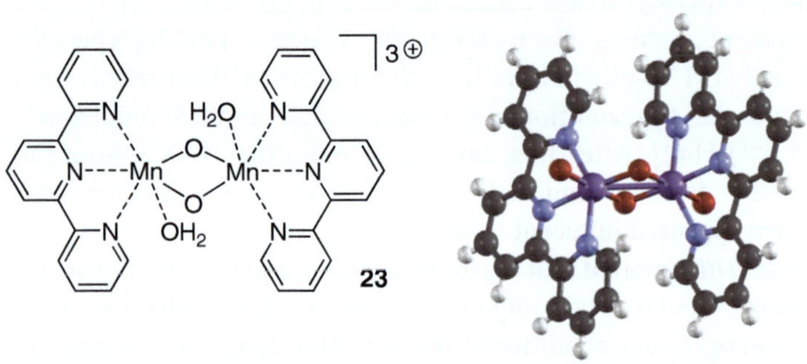

Figure 8.31 Schematic representation and solid-state structure of the mixed-valence dinuclear WOC **23** (anions and solvent molecules were omitted for clarity). Adapted from [146].

NaOCl or $KHSO_5$ ("oxone"), as the chemical oxidants. However, after several hours of operation, **23** decomposed into MnO_4^-, presumably due to disproportionation of the Mn^V/Mn^V intermediate. Moreover, **23** was prone to decomposition into insoluble Mn-oxo oligomers upon heating to 60 °C. It was speculated that this process, which *inter alia* involved a change in the valency, might also contribute to the loss of OER activity of the natural Mn_4CaO_5 cluster upon thermal treatment. Unexpectedly, the precipitate with the proposed composition $[Mn_2O_2(tpy)_x(H_2O)_y]_n$ itself represented a stable and heterogeneous WOC, which drove the chemical OER with higher efficiency than **23**.

More recently, a homoleptic Co^{II} complex, whose tpy ligand was equipped with a short tripeptide residue, acted as WOC in the photocatalytic OER using $[Ru(bpy)_3]^{2+}$ and $Na_2S_2O_8$, as the sensitizer and oxidant, respectively [148]. Unlike the parent complex $[Co(tpy)_2]^{2+}$, the "metallopeptoid" was stable under the experimental conditions and enabled the O_2 evolution with a maximum TON of 28.1 in 45 min upon irradiation with blue LED light (LED: light-emitting diode).

Aside from these selected examples for tpy complexes of the 1st-row transition-metal ions, the use of such complexes, as WOCs, is exceedingly rare and has, to the best of our knowledge, not been explored in the context of the electrochemical OER [149].

8.5 Miscellaneous Electrochemical Applications Catalyzed by Tpy Complexes

Beyond the three presumably most notable electrocatalytic processes (*i.e.*, ERC, HER, and OER), a variety of further reactions have been studied in a similar fashion and tpy complexes have also been utilized in this context. The miscellaneous electrocatalytic reactions, which are summarized in this section, can be distinguished between energy-related ones and those that target the interconversion of functional groups in organic chemistry.

8.5.1 Electrochemical Oxygen Reduction Catalyzed by Tpy Complexes

The electrochemical OER has been discussed beforehand (see Section 8.4). But why is the oxygen-reduction reaction (ORR), as the reversal process, also of technological interest? The ORR represents one of the key processes in fuel cells, specifically proton-exchange membrane fuel cells (PEMFCs) [150, 151]. Such devices transform, simply speaking, the chemical energy, which is released in the electrochemical reaction of H_2 and O_2, into electrical energy. In contrast to this, the combustion of H_2 exclusively produces thermal energy. In PEMFC devices, the four-electron reduction of molecular oxygen to afford water represents the cathode reaction. This process and its anodic counterpart, *i.e.*, the concomitant hydrogen-oxidation reaction (HOR), are commonly catalyzed by noble-metal nanoparticles, which have been deposited onto low-cost carbon-based electrodes (*e.g.*, carbon black or graphite) [152]. The reaction kinetics of the anodic HOR are typically fast, and this reaction can efficiently be run even at low catalyst loadings [153], whereas the kinetics of the ORR are intrinsically slow and high catalyst loadings are required to reach an "ideal" fuel-cell performance [154]. In order to improve the kinetics of the ORR on the one hand and reduce the costs on the other, research has been devoted to replacing the commonly used noble-metal-based catalysts (*e.g.*, Pt and Pd) with less expensive materials. Nitrides, carbides, and phosphides of earth-abundant metals represent some of the next-generation ORR catalysts [154–156]. In particular, the latter are of significant importance for various other applications (*e.g.*, photovoltaics, energy storage, and water splitting); however, their production is hampered by the required stringent reaction conditions (*i.e.*, inert atmosphere, high temperatures, *etc.*). In particular, the often-pyrophoric nature of the used phosphidating agent has to be considered [157].

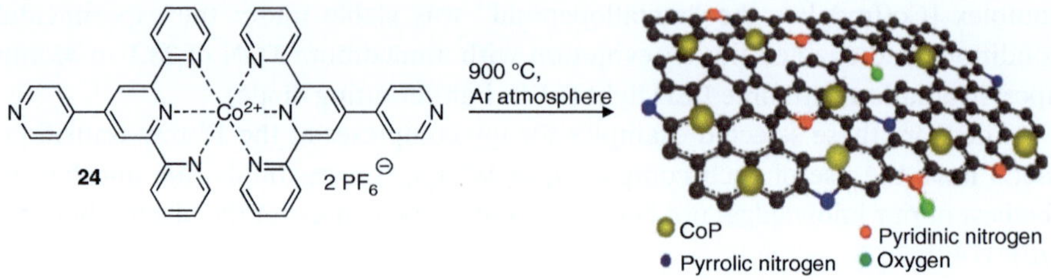

Figure 8.32 Schematic representation of the fabrication of NC-CoP nanoparticles by carbothermal reduction of **24** [158] / with permission of American Chemical Society.

In order to make the production process more applicable, Kumar and Raj utilized the Co^{II} complex **24** as the single-source precursor for the carbothermal synthesis of mesoporous nitrogen-doped carbon-embedded cobalt phosphide (NC-CoP) nanoparticles (Figure 8.32) [158]. Thereby, the phosphidation of Co was achieved by the complex's PF_6^- counterion; thus, the use of an external source for P was obsolete. Since neither $[Co(bpy)_3](PF_6)_2$ nor $[Co(tpy)_2](PF_6)_2$ afforded well-defined NC-CoP nanoparticles under otherwise identical preparation conditions, a proper N-to-C ratio, as present in **24**, was proposed to be crucial for the success of the carbothermal reduction. Nitrogen adsorption–desorption measurements revealed a large surface area of *ca.* 120 $m^2\ g^{-1}$. When used as a heterogeneous catalyst for the electrochemical ORR, a high catalytic activity and product selectivity were observed, *i.e.*, the four-electron reduction of O_2 to afford H_2O was highly favored over the competing peroxide formation, which is due to a two-electron reduction. Moreover, NC-CoP nanoparticles were electrocatalytically more active than pristine NC. It was proclaimed that the improved O_2 absorption and electron-transfer kinetics of the ORR were due to a synergy between the NC support and the CoP nanoparticles.

A similar strategy, *i.e.*, the pyrolysis of a precursor material, was followed by Dey *et al.* [159]. Unlike the previous example, in which a molecular precursor was used, these authors pyrolyzed the metallopolymer, which was formed from Fe^{II} ions and the *bis*-tridentate ligand **25** (Figure 8.33). According to XPS and X-ray absorption spectroscopy (XAS) measurements, the obtained material contained $Fe\text{-}N_4/C$ active sites and was rich in pyridinic and graphitic *N*-atoms. These structural features

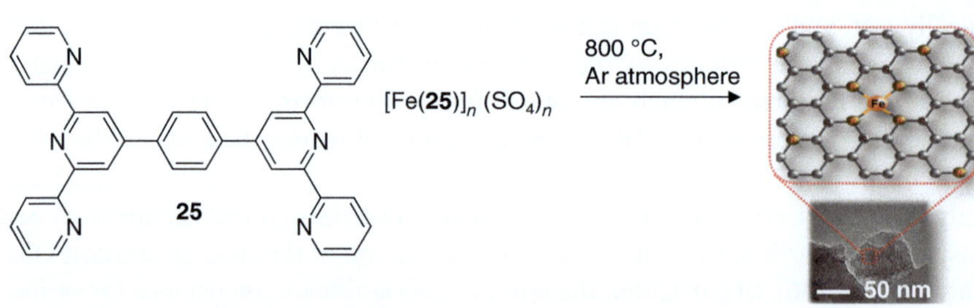

Figure 8.33 Schematic representation of the preparation of a Fe-N4/C electrocatalyst by pyrolysis of a precursor metallopolymer. A representative transmission-electron miscroscopy (TEM) image of the material is also shown [159] / with permission of American Chemical Society.

accounted for the high electrocatalytic activity and product selectivity on the one hand and superior stability, as expressed by the good tolerance against MeOH, on the other (compared to that of Pt/C). The Fe-N_4/C electrocatalyst featured higher onset (E_{onset}) and half-wave potentials ($E_{1/2}$) than Pt/C and outperformed this commercial catalyst at least in alkaline environments. In one application, the as-prepared Fe-N_4/C was utilized as the cathode material in a rechargeable Zn-air battery. This cell further incorporated RuO_2, as the OER catalyst for charging, and Zn foil, as the anode. This cell delivered a significantly higher power density compared to the corresponding cell, which contained a Pt/C cathode (*i.e.*, 225 *vs.* 177 mW cm^{-2} at current densities of 240 and 260 mA cm^{-2}). However, the performance of Fe-N_4/C in PEMFCs was not explored.

The aforementioned metallopolymer-coated electrode with exposed $[Pt(tpy)Cl]^+$ complexes (Figure 8.24) was also tested regarding its applicability in the electrocatalytic ORR [118]. However, this hybrid electrode could not compete with conventional Pt-based ones, due to the insufficient kinetic current densities and mass activities, which were related to the material's low electrochemically active surface area (*i.e.*, too small amount of Pt centers per cm^2).

Besides these examples for the ORR, in which tpy-based or tpy-derived materials were used, as the actual electrocatalysts, also the use of polymers containing metal–tpy complexes, as redox mediators in ORR enzyme electrodes has been explored. Such hybrid electrodes are of particular interest for enzymatic biofuel cells [160]. In such devices, the O_2-reducing enzyme is brought into contact with the electrode by a redox-polymer interlayer, which shuttles electrons between the electrode and the enzymes' reactive centers. Among others, metalloenzymes from the laccase family have been utilized; the active site of these enzymes is comprised of an array of four Cu^{II} centers (Figure 8.34a). Laccase-containing biocathodes for the ORR have already been prepared with different architectures and compositions [163]. In one approach, Heller *et al.* utilized the Os^{II}-containing

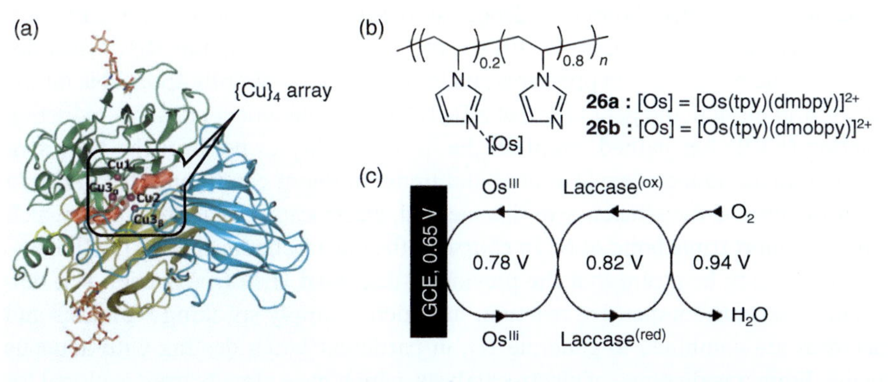

Figure 8.34 (a) Representation of the overall structure of *Coriolopsis caperata* laccase, as a representative example [161] / MDPI / CC BY 4.0. (b) Schematic representation of the Os^{II}-containing redox polymers **26** (counterions were omitted for clarity). (c) Schematic representation of the redox cascade of an ORR enzyme electrode (configuration: GCE|**26a**|laccase). Adapted from [162].

polymer **26a**, as the redox mediator, in which $[Os(tpy)(dmbpy)]^{2+}$ moieties were coordinated to one-fifth of the polymer's side chains (dmbpy: 4,4′-dimethyl-bpy; Figure 8.34b). Such medium density of Os^{II} centers per polymer chain provided sufficient charge-transport and mediation kinetics [164]. The enzyme electrode with the configuration GCE|**26a**|laccase operated in a citrate-buffered solution (pH value of 5) at 37.5 °C [163]. Under these almost physiological conditions, the O_2 reduction occurred at 0.7 V *vs.* NHE (Figure 8.34c). In subsequent work, the GCE was replaced by a carbon cloth, which was comprised of hydrophilic carbon fibers with an average diameter of 10 μm; thereby, the shear-stress stability of the composite electrode was improved significantly [165].

Moreover, the redox potential of the redox mediator was varied *via* the electron-donor strength of the substituents on the bpy ligands of the polymer-bound complexes from the $[Os(tpy)(bpy)]^{2+}$ and $[Os(bpy)_2Cl]^{+}$ families [166]. In view of a potential use of GCE-based enzyme electrodes in biofuel cells with a reversible H_2 anode, the optimum redox-mediator potential should be 0.66 V *vs.* NHE (at least when using laccase of *T. versicolor*, whose standard redox potential is $E^0 = 0.82$ V). Under these conditions, a maximum power output was predicted. This requirement was met by polymer **26b** (Figure 8.34b), which featured a redox potential of 0.69 V *vs.* NHE. However, proof-of-concept biofuel cells, which incorporated this redox mediator, were not assembled. In order to overcome the intrinsic limitation of laccases, *i.e.*, redox-inactivity at the physiological pH of 7, was overcome by using bilirubin oxidase from *Myrothecium verrucaria*, as the redox anion [167]. However, the redox mediators only contained coordinated complexes from the $[Os(bpy)_2Cl]^{+}$ family and research on those bearing $[Os(tpy)(bpy)]^{2+}$ moieties was discontinued.

8.5.2 Electrochemical Nitrogen Reduction Catalyzed by Tpy Complexes

Ammonia (NH_3) represents one of the most important raw materials, which has ever since been required *inter alia* for the production of resins and fertilizers. Moreover, its use in NH_3 fuel cells [168] or hydrogen storage [169] is a topic of significant interest. The Haber-Bosch process is still the technology for industrial NH_3 synthesis. However, the required high pressures and temperatures are hardly applicable for the on-site and on-demand generation of NH_3 [170]. The electrochemical N_2-reduction reaction (NRR) has gained attention due to the ability to perform the NH_3 synthesis from an aqueous environment and under ambient conditions [171, 172]. In order to improve the efficiency of this method, electrocatalysts are required, which break the inert triple bond of N_2 in course of the electrochemical NRR [173]. However, one has to be aware that the previously discussed HER (see Section 8.3) represents a serious competitive reaction, in which – simply speaking – protons and electrons are combined to generate H_2, in particular when dealing with aqueous media. From a wide range of electrocatalysts, which have already been explored for the NRR, single-site metal centers, immobilized in a heterogeneous environment, represent one promising variant. Covalent organic frameworks (COFs) are structurally defined 2D polymers that can be prepared in numerous topologies depending on the chemical structure of the used monomers. The presence of heteroatoms

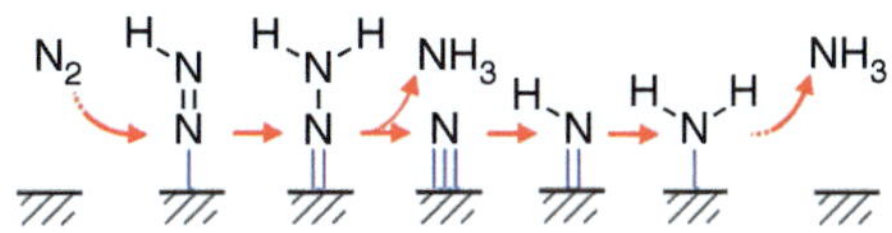

Figure 8.35 Schematic representation of the proposed elemental steps of the NRR, which occur at single-site electrocatalysts. Adapted from [175].

within the COF structure allows one to immobilize metal centers *via* coordinative bonds to the heteroatoms' lone electron pairs. COFs, which have been loaded with 3d transition-metal ions, have been utilized as electrocatalysts for the reduction of various molecules: O_2, NO_3^-, CO_2, *etc.* [174].

A recent computational study revealed that the loading of COFs, which provide tpy-like binding sites, with transition-metal ions should afford versatile heterogeneous NRR electrocatalysts [175]. In particular, the incorporation of Fe ions seems promising due to the theoretical onset potential of −0.49 eV (*vs.* the SHE) on the one hand and the ability to bind all relevant intermediates, which might be involved in the NRR, on the other (Figure 8.35). The binding of these intermediates should be facilitated by π-back-donation and σ-donation between the metal's 3d orbitals and the π* orbital of the N—N bonds. Regrettably, the experimental use of such metal-loaded COFs in the electrocatalytic NRR has not yet been demonstrated.

8.5.3 Examples for Electrochemical Functional Group Transformations Catalyzed by Tpy Complexes

Beyond the use of metal-terpyridine complexes as catalysts for the electrochemical reduction of protons/water, CO_2, N_2, or O_2 on the one hand and the oxidation of water on the other, other reactions have also been explored in the same context [1–3, 7]. It was shown, for example, that the Ru^{II}–solvato complexes, such as **1c/d** (Figure 8.4), can not only be used for the electrocatalytic CO_2 reduction or HER under homogeneous conditions but also the electrochemical carbonyl-to-alcohol reduction can be supported by such complexes. Meyer *et al.* reported that the hydride complex $[Ru(tpy)(bpy)H]^+$, which was generated electrochemically from **1b**, reduced acetone to 2-propanol, albeit the faradaic efficiency was rather low (*ca.* 28%). For the hydride species, a moderate ligand-acidity constant (pK_a^{LAC}) of 10 and a relatively high Ru^{II}—H bond-dissociation energy of 71 kcal mol^{-1} were determined [176]. These values might be favorable for the CO_2 reduction or HER (see Section 8.3), but less applicable for the carbonyl reduction (as revealed by rather poor faradaic efficiency for this reaction).

Ni^{II}–tpy complexes have been employed, as electrocatalysts, for the formation of C—C bonds, to a significant extent. Firstly, Budnikova *et al.* provided evidence for the active role of Ni^{I}–tpy species in C–C cross-coupling reactions, specifically the electrochemical fluoroalkylation of olefins [177, 178]. The catalytically active species was formed by the *in situ* one-electron reduction of the $[Ni(tpy)]Br_2$ pre-catalyst. Budnikova, Vicic, and Klein studied the mechanism of this reaction and the reader is referred to this review for an in-depth view on this topic [179]. As summarized in

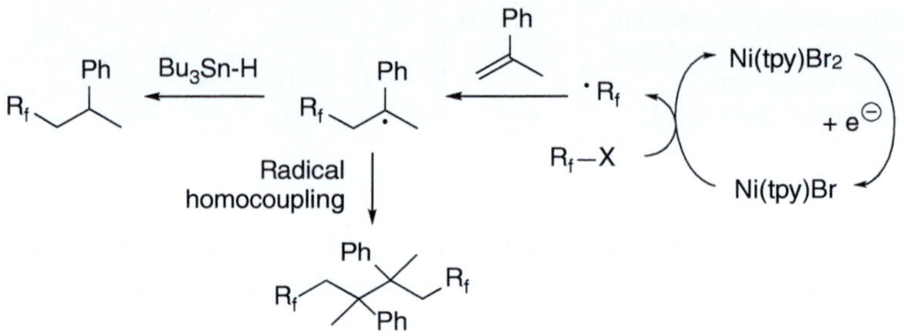

Figure 8.36 Schematic representation of the electrocatalytic fluoroalkylation of alkenes according to Budnikova *et al.* Adapted from [177].

Figure 8.36, reoxidation of the [Ni(tpy)]Br intermediate by a perfluoroalkyl halide afforded perfluoroalkyl radicals, which, in turn, added to the C—C bond of α-methyl styrene [177]. The thusly formed benzylic radicals were either trapped by treatment with tributyltin hydride or dimerized by radical homocoupling.

Moreover, Budnikova *et al.* also performed the electrocatalytic perfluoroalkylation of aromatic compounds [180]. Detrimentally, the original protocol for the cross-coupling between an aryl bromide or iodide and a perfluoroalkyl iodide required a non-noble metal, as the sacrificial anode material (Figure 8.37a). Later, a more comfortable protocol was established, which enabled the single-stage cross-coupling between the aforementioned substrates [181]. The Cu^{II} ions, which were released from the anode, played a substantial role in the catalytic cycle and were involved in the product formation step. These authors tested various combinations

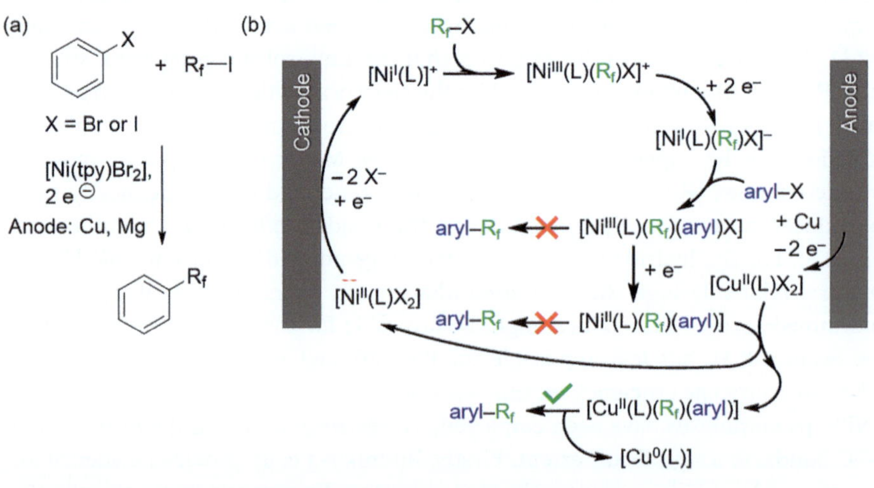

Figure 8.37 (a) Schematic representation of the electrocatalytic fluoroalkylation of bromo- or iodobenzene. Adapted from [180]. (b) Schematic representation of the proposed mechanism of the one-step electrocatalytic cross coupling between (hetero)aryl halides and perfluoroalkyl halides [179] / MDPI / CC BY 4.0.

of polypyridyl ligands and transition-metal ions to find the most potent electrocatalyst for the cross coupling. Presumably, due to its low reduction potential of −1.30 V *vs.* SHE, $[Ni(tpy)]Br_2$ offered the best performance (*i.e.*, yield of cross-coupling products), whereas lower yields were obtained when using a catalyst, such as $[Ni(bpy)]Br_2$, whose reduction potential was only −1.12 V. The proposed mechanism for the one-step electrocatalytic coupling is summarized in Figure 8.37b. The method was of wide scope: A broad range of heteroarenes, excluding only very electron-rich ones and thiophene derivatives, reacted with high conversion and good yields.

Finally, Zhou *et al.* reported a series of organic–inorganic hybrid materials, in which polyoxometalate (POM) clusters were assembled with Cu^{II}–tpy *mono*-complexes (Figure 8.38) [182]. The latter bridged the individual POM moieties and, thereby, compensated for the charges of the polyanionic clusters within the supramolecular structures. The aim of this study was to achieve a synergistic effect by combining two different electrocatalytically active species. Without going further into detail, POMs are well-known for their ability to serve as interim electron reservoirs and this behavior has been widely capitalized for electrocatalytic and energy-storage applications [183]. The obtained materials, which contained the bi-capped (PMo_8V_6) Keggin, (PW_{12}) Keggin, or (P_2W_{18}) Wells–Dawson motif, were fabricated into bulk-modified carbon-paste electrodes. These hybrid electrodes exhibited a dual electrocatalytic activity, *i.e.*, they catalyzed both the reduction of nitrite and the oxidation of ascorbic acid. In particular, the best dual-functional performance was observed for the hybrid, which contained the (PW_{12}) Keggin-type POM moieties. In subsequent work, further such hybrids were prepared, which contained structurally simple anions, *i.e.*, $[V_2O_6]^{2-}$ or $(Mo_3O_{10})^{2-}$, or polyanions, *i.e.*, (SiW_{12}), (Mo_3O_{10}), and (Mo_8O_{26}) [184–186]. Likewise, the hybrid electrodes, which contained either of these materials, featured dual-functional electrocatalytic activity toward nitrite reduction and ascorbic acid oxidation.

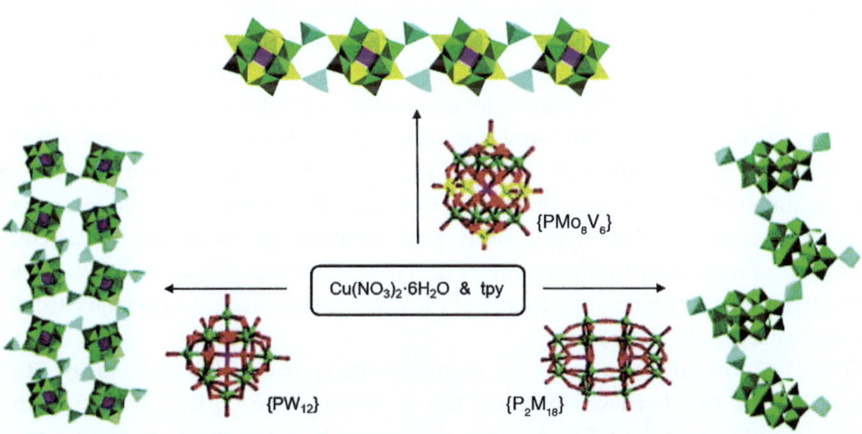

Figure 8.38 Schematic representation of the synthesis of hybrid POMs from Cu^{II} ions, tpy, and the respective POM precursors [182] / with permission of Royal Society of Chemistry.

8.6 Concluding Remarks

Terpyridine (tpy) ligands and their transition-metal complexes have found widespread use as catalysts in numerous applications. This also holds true for the broad field of biomedicine, where cancer-cell death (apoptosis) is induced/catalyzed by such complexes (see Section 8.4). The catalysis of functional-group transformations in organic chemistry represents an equally important field of research. Oxidations and reductions as well as cross-coupling reactions (*i.e.*, the formation of C—C bonds) have to be named exemplarily in this context. As reviewed elsewhere, tpy complexes have *inter alia* been evaluated in this context [1–3]. The majority of these catalytic reactions can be rationalized by the established concepts of transition-metal catalysis. However, electrocatalytic, photocatalytic, and photoelectrocatalytic variants play an important role and can be expected to come to forefront within the next few years. This holds particularly true for the energy-related applications. The decarbonation of energy supply and reduction of CO_2 emissions represent strong driving forces for intensified research in CO_2 reduction and water splitting.

The ERC aims for the conversion of CO_2 into higher-energy compounds, such as MeOH, formaldehyde, formic acid, or CO. In this respect, Ru^{II}–tpy complexes are discussed as potential electrocatalysts already since the 1980s [1]. More recently, Co^{II} complexes have gained considerable attention in order to replace cost-intensive Ru^{II} by cheaper, more abundant alternatives. This aspect plays a key role in current research on the photo- and electrochemistry of transition-metal complexes, which targets the use of 3d transition-metal complexes as next-generation emitters, photo(electro)catalysts, *etc.* [187–189]. Notable performance in the ERC under homogeneous conditions has been reported for the Ru^{II} electrocatalyst **2b** (Figure 8.4), which contained an NHC-type ligand [46]. This complex was also employed in tandem cells, which combined the ERC, as the anodic process, with the cathodic OER [46] or alcohol-to-aldehyde oxidation [49].

Since the HER represents a considerable side-reaction of the electrocatalytic CO_2 reduction, syngas, *i.e.*, a mixture of CO and H_2, is often obtained as the product. In the absence of CO_2, the same electrocatalysts can also be basically employed for the HER. However, Ru^{II}–tpy complexes have hardly received recognition in this context; instead, such complexes are better suited as sensitizers for the photocatalytic HER. On the other hand, Co^{II}–tpy complexes have been intensively screened regarding their applicability and structure–performance relationships have been established by Fontecave *et al.* [107]. However, these complexes could not compete with cobaloxime, as one of the most potent HER electrocatalysts [190] or the hydrogenases [191].

The OER formally represents the complementary half-cell process to the HER. However, for the discussed reasons, the electrocatalytic OER is mostly studied independently from the HER. Water oxidation can be catalyzed by transition-metal complexes and numerous Ru^{II}–polypyridyl complexes have already been tested, as WOCs [104]. In particular, the mononuclear Ru–tpy complexes have been studied to a significant extent. However, these WOCs have been outcompeted by the 3rd generation WOCs, which contained polypyridyl ligands of higher denticity. In particular, Llobet's WOC, which contained the flexible tpy-2,6″-dicarboxylate ligand, featured a remarkably high catalytic performance after electrochemical activation [192]. Due to the "internal base" effect, which was provided by one of the carboxylate moieties, the TOF for O_2 generation was as high as 50,000 s^{-1} at the pH value of 10.0. However, in most cases, the WOCs functioned as chemical catalysts only, *i.e.*, the OER was performed in the presence of sacrificial oxidants (*e.g.*, CAN, NaOCl, or oxone), whereas heterogeneous WOCs, in which the molecular catalysts were immobilized onto electrodes, displayed respectable electrocatalytic performances. As in the case of the HER and CO_2 reduction reaction, the replacement of Ru represents one major challenge to render the OER more sustainable. In this respect, tpy complexes are not competitive to other complexes, which are based on abundant transition-metal ions [193].

The ORR and NRR are further important processes, which can – at least in theory – be electrocatalyzed by tpy complexes. In the case of the ORR, which is of importance, as the anodic half-cell reaction, for fuel cells and metal-air batteries, tpy complexes have already been utilized. However, they were not used themselves as electrocatalysts, but served as precursors for the fabrication of metallized carbon-nitrogen materials by pyrolysis. These materials, in turn, represented promising anode materials for zinc-air batteries [159].

A wide range of reactions, including the highly relevant ERC, HER, OER, and ORR, can be electrocatalyzed by tpy complexes (or the materials fabricated thereof). However, these catalysts are still far from being close to large-scale or commercial application. This holds particularly true for the Ru^{II} complexes, which will always suffer from the metal's limited abundance and high materials costs; thus, intense research is required to improve the existing (or introduce new) tpy-containing catalyst systems [194].

References

1 A. Winter, U. S. Schubert, *ChemCatChem* 2020, 12, 2890–2941.

2 A. Winter, G. R. Newkome, U. S. Schubert, *ChemCatChem* 2011, 3, 1384–1406.

3 C. Wei, Y. He, X. Shi, Z. Song, *Coord. Chem. Rev.* 2019, 385, 1–19.

4 S. F. Kainat, M. B. Hawsawi, E. U. Mughal, N. Naeem, A. M. Almohyawi, H. M. Altass, E. M. Hussein, A. Sadiq, Z. Moussa, A. S. Abd-El-Aziz, S. A. Ahmed, *RSC Adv.* 2024, 14, 21464–21537.

5 S. Ranjan Jena, T. Mandal, J. Choudhury, *Chem. Rec.* 2022, 22, e202200165.

6 G. Chelucci, R. P. Thummel, *Chem. Rev.* 2002, 102, 3129–3170.

7 R. R. Panicker, A. Sivaramakrishna, *Coord. Chem. Rev.* 2022, 459, 214426.

8 A. Winter, G. R. Newkome, U. S. Schubert, *Inorg. Chem. Front.* 2024, 11, 342–399.

9 K. Wiranarongkorn, K. Eamsiri, Y.-S. Chen, A. Arpornwichanop, *J. CO2 Util.* 2023, 71, 102477.

10 S. Fang, M. Rahaman, J. Bharti, E. Reisner, M. Robert, G. A. Ozin, Y. H. Hu, *Nat. Rev. Methods Primers* 2023, 3, 61.

11 N. Nandal, S. L. Jain, *Coord. Chem. Rev.* 2022, 451, 214271.

12 R. Sen, S. Das, A. Nath, P. Maharana, P. Kar, F. Verpoort, P. Liang, S. Roy, *Front. Chem.* 2022, 10, 861604.

13 S. Lin, H. Huang, T. Ma, Y. Zhang, *Adv. Sci.* 2021, 8, 2002458.

14 Q. Shi, H. Duan, *Chem Catal.* 2022, 2, 3471–3496.

15 Y. Li, X. Wei, L. Chen, J. Shi, *Angew. Chem. Int. Ed.* 2021, 60, 19550–19571.

16 J. Corredor, M. J. Rivero, C. M. Rangel, F. Gloaguen, I. Ortiz, *J. Chem. Tech. Biotechnol.* 2019, 94, 3049–3063.

17 M. F. R. Samsudin, *Energies* 2023, 16, 5743.

18 D. Raciti, C. Wang, *ACS Energy Lett.* 2018, 3, 1545–1556.

19 X. Zou, Y. Zhang, *Chem. Soc. Rev.* 2015, 44, 5148–5180.

20 H. McLaughlin, A. A. Littlefield, M. Menefee, A. Kinzer, T. Hull, B. K. Sovacool, M. D. Bazilian, J. Kim, S. Griffiths, *Renew. Sustain. Energy Rev.* 2023, 177, 113215.

21 N. Bahman, M. Al-Khalifa, S. Al Baharna, Z. Abdulmohsen, E. Khan, *Rev. Environ. Sci. Biotechnol.* 2023, 22, 451–470.

22 A. D. N. Kamkeng, M. Wang, J. Hu, W. Du, F. Qian, *Chem. Eng. J.* 2021, 409, 128138.

23 X. Zhao, M. Yin, L. Ma, L. Liang, C. Liu, J. Liao, T. Lu, W. Xing, *Energy Environ. Sci.* 2011, 4, 2736–2753.

24 Z. Tian, Y. Wang, X. Zhen, Z. Liu, *Fuel* 2022, 320, 123902.

25 K. W. Frese Jr., S. C. Leach, *J. Electrochem. Soc.* 1985, 132, 259.

26 D. P. Summers, S. Leach, K. W. Frese, *J. Electroanal. Chem. Interfac. Electrochem.* 1986, 205, 219–232.

27 D. Canfield, K. W. Frese, *J. Electrochem. Soc.* 1983, 130, 1772.

28 Y. Hori, K. Kikuchi, S. Suzuki, *Chem. Lett.* 1985, 14, 1695–1698.

29 C. Costentin, M. Robert, J.-M. Savéant, *Chem. Soc. Rev.* 2013, 42, 2423–2436.

30 W. Zhang, Y. Hu, L. Ma, G. Zhu, Y. Wang, X. Xue, R. Chen, S. Yang, Z. Jin, *Adv. Sci.* 2018, 5, 1700275.

31 S. Nitopi, E. Bertheussen, S. B. Scott, X. Liu, A. K. Engstfeld, S. Horch, B. Seger, I. E. L. Stephens, K. Chan, C. Hahn, J. K. Nørskov, T. F. Jaramillo, I. Chorkendorff, *Chem. Rev.* 2019, 119, 7610–7672.

32 I. Janik, G. N. R. Tripathi, *J. Chem. Phys.* 2016, 144, 154307.

33 E. R. Irtem, T. Andreu, A. Parra, M. D. Hernández-Alonso, S. García-Rodríguez, J. M. Riesco-García, G. Penelas-Pérez, J. R. Morante, *J. Mater. Chem. A* 2016, 4, 13582–13588.

34 S. Liang, N. Altaf, L. Huang, Y. Gao, Q. Wang, *J. CO2 Util.* 2020, 35, 90–105.

35 D. Wakerley, S. Lamaison, J. Wicks, A. Clemens, J. Feaster, D. Corral, S. A. Jaffer, A. Sarkar, M. Fontecave, E. B. Duoss, S. Baker, E. H. Sargent, T. F. Jaramillo, C. Hahn, *Nat. Energy* 2022, 7, 130–143.

36 D. Ma, T. Jin, K. Xie, H. Huang, *J. Mater. Chem. A* 2021, 9, 20897–20918.

37 F. Franco, S. Fernández, J. Lloret-Fillol, *Curr. Opin. Electrochem.* 2019, 15, 109–117.

38 N. Elgrishi, M. B. Chambers, X. Wang, M. Fontecave, *Chem. Soc. Rev.* 2017, 46, 761–796.

39 J. Hawecker, J.-M. Lehn, R. Ziessel, *J. Chem. Soc., Chem. Commun.* 1984, 1984, 328–330.

40 H. Ishida, H. Tanaka, K. Tanaka, T. Tanaka, *J. Chem. Soc., Chem. Commun.* 1987, 131–132.

41 H. Ishida, K. Tanaka, T. Tanaka, *Organometallics* 1987, 6, 181–186.

42 H. Nagao, T. Mizukawa, K. Tanaka, *Inorg. Chem.* 1994, 33, 3415–3420.

43 H. Nagao, T. Mizukawa, K. Tanaka, *Chem. Lett.* 2006, 22, 955–958.

44 S. Chardon-Noblat, P. Da Costa, A. Deronzier, S. Maniguet, R. Ziessel, *J. Electroanal. Chem.* 2002, 529, 135–144.

45 Z. Chen, C. Chen, D. R. Weinberg, P. Kang, J. J. Concepcion, D. P. Harrison, M. S. Brookhart, T. J. Meyer, *Chem. Commun.* 2011, 47, 12607–12609.

46 Z. Chen, J. J. Concepcion, M. K. Brennaman, P. Kang, M. R. Norris, P. G. Hoertz, T. J. Meyer, *Proc. Natl. Acad. Sci. U.S.A.* 2012, 109, 15606–15611.

47 S. Gonell, M. D. Massey, I. P. Moseley, C. K. Schauer, J. T. Muckerman, A. J. M. Miller, *J. Am. Chem. Soc.* 2019, 141, 6658–6671.

48 Z. Chen, J. J. Concepcion, T. J. Meyer, *Dalton Trans.* 2011, 40, 3789–3792.

49 Y. Wang, S. Gonell, U. R. Mathiyazhagan, Y. Liu, D. Wang, A. J. M. Miller, T. J. Meyer, A. C. S. Appl, *Energy Mater.* 2019, 2, 97–101.

50 T. A. White, S. Maji, S. Ott, *Dalton Trans.* 2014, 43, 15028–15037.

51 B. A. Johnson, S. Maji, H. Agarwala, T. A. White, E. Mijangos, S. Ott, *Angew. Chem. Int. Ed.* 2016, 55, 1825–1829.

52 B. A. Johnson, H. Agarwala, T. A. White, E. Mijangos, S. Maji, S. Ott, *Chem. Eur. J.* 2016, 22, 14870–14880.

53 L. Duan, G. F. Manbeck, M. Kowalczyk, D. J. Szalda, J. T. Muckerman, Y. Himeda, E. Fujita, *Inorg. Chem.* 2016, 55, 4582–4594.

54 S. K. Lee, M. Kondo, G. Nakamura, M. Okamura, S. Masaoka, *Chem. Commun.* 2018, 54, 6915–6918.

55 C. M. Bolinger, B. P. Sullivan, D. Conrad, J. A. Gilbert, N. Story, T. J. Meyer, *J. Chem. Soc., Chem. Commun.* 1985, 1985, 796–797.

56 B. Giri, A. Mahata, T. Kella, D. Shee, F. De Angelis, S. Maji, *J. Catal.* 2022, 405, 15–23.

57 D. Ghosh, T. Fukushima, K. Kobayashi, S. Sen, S. Kitagawa, T. Kato, K. Tanaka, *Dalton Trans.* 2017, 46, 4373–4381.

58 D. Ghosh, K. Kobayashi, T. Kajiwara, S. Kitagawa, K. Tanaka, *Inorg. Chem.* 2017, 56, 11066–11073.

59 H. A. Younus, N. Ahmad, W. Ni, X. Wang, M. Al-Abri, Y. Zhang, F. Verpoort, S. Zhang, *Coord. Chem. Rev.* 2023, 493, 215318.

60 P. Paul, B. Tyagi, A. K. Bilakhiya, M. M. Bhadbhade, E. Suresh, G. Ramachandraiah, *Inorg. Chem.* 1998, 37, 5733–5742.

61 C. Jiang, A. W. Nichols, C. W. Machan, *Dalton Trans.* 2019, 48, 9454–9468.

62 N. Elgrishi, M. B. Chambers, V. Artero, M. Fontecave, *Phys. Chem. Chem. Phya.* 2014, 16, 13635–13644.

63 H. C. Hurrell, A.-L. Mogstad, D. A. Usifer, K. T. Potts, H. D. Abruña, *Inorg. Chem.* 1989, 28, 1080–1084.

64 C. R. Arana, S. Yan, M. Keshavarz-K, K. T. Potts, H. D. Abruña, *Inorg. Chem.* 1992, 31, 3680–3682.

65 N. Elgrishi, M. B. Chambers, M. Fontecave, *Chem. Sci.* 2015, 6, 2522–2531.

66 R. Matthessen, J. Fransaer, K. Binnemans, D. E. De Vos, *Beilstein J. Org. Chem.* 2014, 10, 2484–2500.

67 D. Chen, P.-L. Fabre, O. Reynes, *Electrochim. Acta* 2011, 56, 8603–8610.

68 D. Chen, P.-L. Fabre, O. Reynes, N. Chouini-Lalanne, V. Sartor, *Inorg. Chim. Acta* 2021, 524, 120442.

69 A. A. Isse, A. Gennaro, *J. Electrochem. Soc.* 2002, 149, D113.

70 A. Sinopoli, N. T. La Porte, J. F. Martinez, M. R. Wasielewski, M. Sohail, *Coord. Chem. Rev.* 2018, 365, 60–74.

71 M. Bourrez, F. Molton, S. Chardon-Noblat, A. Deronzier, *Angew. Chem. Int. Ed.* 2011, 50, 9903–9906.

72 C. W. Machan, C. P. Kubiak, *Dalton Trans.* 2016, 45, 17179–17186.

73 J. A. Ramos Sende, C. R. Arana, L. Hernandez, K. T. Potts, M. Keshevarz-K, H. D. Abruña, *Inorg. Chem.* 1995, 34, 3339–3348.

74 C. R. Arana, M. Keshavarz, K. T. Potts, H. D. Abruña, *Inorg. Chim. Acta* 1994, 225, 285–295.

75 Y. Tsukahara, T. Wada, K. Tanaka, *Chem. Lett.* 2010, 39, 1134–1135.

76 N. Elgrishi, S. Griveau, M. B. Chambers, F. Bedioui, M. Fontecave, *Chem. Commun.* 2015, 51, 2995–2998.

77 E. W. McQueen, J. I. Goldsmith, *J. Am. Chem. Soc.* 2009, 131, 17554–17556.

78 A. Tverdokhlebova, I. Sterin, O. Smutok, E. Katz, *J. Solid State Electrochem.* 2024, 28, 711–755.

79 T.-T. Li, B. Shan, W. Xu, T. J. Meyer, *ChemSusChem* 2019, 12, 2402–2408.

80 Y. Wang, S. L. Marquard, D. Wang, C. Dares, T. J. Meyer, *ACS Energy Lett.* 2017, 2, 1395–1399.

81 Y. Liu, X. Fan, A. Nayak, Y. Wang, B. Shan, X. Quan, T. J. Meyer, *Proc. Natl. Acad. Sci. U.S.A.* 2019, 116, 26353–26358.

82 T. Li, X. An, D. Fu, *Energy Fuels* 2023, 37, 8160–8179.

83 T. Yan, X. Chen, L. Kumari, J. Lin, M. Li, Q. Fan, H. Chi, T. J. Meyer, S. Zhang, X. Ma, *Chem. Rev.* 2023, 123, 10530–10583.

84 J. Liu, H. Shi, Q. Shen, C. Guo, G. Zhao, *Appl. Catal. B: Environ.* 2017, 210, 368–378.

85 L. Le-Quang, M. Stanbury, S. Chardon-Noblat, J.-M. Mouesca, V. Maurel, J. Chauvin, *Chem. Commun.* 2019, 55, 13598–13601.

86 L. Chen, Z. Wang, P. Kang, *Chin. J. Catal.* 2018, 39, 413–420.

87 J. J. Leung, J. Warnan, K. H. Ly, N. Heidary, D. H. Nam, M. F. Kuehnel, E. Reisner, *Nat. Catal.* 2019, 2, 354–365.

88 J. J. Leung, J. A. Vigil, J. Warnan, E. Edwardes Moore, E. Reisner, *Angew. Chem. Int. Ed.* 2019, 58, 7697–7701.

89 W. W. Kramer, C. C. L. McCrory, *Chem. Sci.* 2016, 7, 2506–2515.

90 L. Fan, Z. Tu, S. H. Chan, *Energy Rep.* 2021, 7, 8421–8446.

91 S. Aralekallu, L. K. Sannegowda, V. Singh, *Int. J. Hydrogen Energy* 2023, 48, 16569–16592.
92 A. Gupta, B. Likozar, R. Jana, W. C. Chanu, M. K. Singh, *Int. J. Hydrogen Energy* 2022, 47, 33282–33307.
93 J. Zhu, L. Hu, P. Zhao, L. Y. S. Lee, K.-Y. Wong, *Chem. Rev.* 2020, 120, 851–918.
94 L. Tong, L. Duan, A. Zhou, R. P. Thummel, *Coord. Chem. Rev.* 2020, 402, 213079.
95 A. Kazemi, F. Manteghi, Z. Tehrani, *ACS Omega* 2024, 9, 7310–7335.
96 B. You, Y. Sun, *Acc. Chem. Res.* 2018, 51, 1571–1580.
97 D. K. Tiwari, A. Ray, A. Roy, S. Das, in *Water Remediation*, (Eds.: S. Bhattacharya, A. B. Gupta, A. Gupta, A. Pandey), Springer Singapore, Singapore, 2018, pp. 135–153.
98 N. Dubouis, A. Grimaud, *Chem. Sci.* 2019, 10, 9165–9181.
99 Y. Wang, M. Zhang, Y. Liu, Z. Zheng, B. Liu, M. Chen, G. Guan, K. Yan, *Adv. Sci.* 2023, 10, 2207519.
100 N. V. Belkova, L. M. Epstein, O. A. Filippov, E. S. Shubina, *Chem. Rev.* 2016, 116, 8545–8587.
101 N. Wang, H. Lei, Z. Zhang, J. Li, W. Zhang, R. Cao, *Chem. Sci.* 2019, 10, 2308–2314.
102 D. C. Lacy, G. M. Roberts, J. C. Peters, *J. Am. Chem. Soc.* 2015, 137, 4860–4864.
103 S. J. C. Robinson, D. M. Heinekey, *Chem. Commun.* 2017, 53, 669–676.
104 J. M. Kamdar, D. B. Grotjahn, *Molecules* 2019, 24, 494.
105 V. Kaim, M. Joshi, M. Stein, S. Kaur-Ghumaan, *Int. J. Hydrogen Energy* 2023, 48, 30718–30731.
106 T. Abe, M. Kaneko, *J. Mol. Catal. A: Chem.* 2001, 169, 177–183.
107 S. Aroua, T. K. Todorova, V. Mougel, P. Hommes, H.-U. Reissig, M. Fontecave, *ChemCatChem* 2017, 9, 2099–2105.
108 K. Majee, S. K. Padhi, *New J. Chem.* 2019, 43, 3856–3865.
109 E. Ahmad, K. Majee, J. Patel, B. Das, S. K. Padhi, *Eur. J. Inorg. Chem.* 2017, 2017, 3409–3418.
110 S. Kumar Padhi, E. Ahmad, S. Rai, B. Panda, *Polyhedron* 2020, 187, 114677.
111 A. V. Sariga, *Top. Curr. Chem.* 2023, 381, 32.
112 S. Fukuzumi, S. Mandal, K. Mase, K. Ohkubo, H. Park, J. Benet-Buchholz, W. Nam, A. Llobet, *J. Am. Chem. Soc.* 2012, 134, 9906–9909.
113 S. Mandal, S. Shikano, Y. Yamada, Y.-M. Lee, W. Nam, A. Llobet, S. Fukuzumi, *J. Am. Chem. Soc.* 2013, 135, 15294–15297.
114 E. Ahmad, S. Rai, S. K. Padhi, *Int. J. Hydrogen Energy* 2019, 44, 16467–16477.
115 E. A. Lewis, W. B. Tolman, *Chem. Rev.* 2004, 104, 1047–1076.
116 S. Rai, S. K. Padhi, *Electrochim. Acta* 2020, 364, 137277.
117 M. Li, K. Duanmu, C. Wan, T. Cheng, L. Zhang, S. Dai, W. Chen, Z. Zhao, P. Li, H. Fei, Y. Zhu, R. Yu, J. Luo, K. Zang, Z. Lin, M. Ding, J. Huang, H. Sun, J. Guo, X. Pan, W. A. Goddard, P. Sautet, Y. Huang, X. Duan, *Nat. Catal.* 2019, 2, 495–503.
118 S. Elmas, W. Beelders, S. J. Bradley, R. Kroon, G. Laufersky, M. Andersson, T. Nann, A. C. S. Sustain, *Chem. Eng.* 2017, 5, 10206–10214.

119 M. D. Kärkäs, O. Verho, E. V. Johnston, B. Åkermark, *Chem. Rev.* 2014, 114, 11863–12001.

120 A. M. Appel, M. L. Helm, *ACS Catal.* 2014, 4, 630–633.

121 J. D. Blakemore, R. H. Crabtree, G. W. Brudvig, *Chem. Rev.* 2015, 115, 12974–13005.

122 T. A. Betley, Q. Wu, T. Van Voorhis, D. G. Nocera, *Inorg. Chem.* 2008, 47, 1849–1861.

123 D. G. H. Hetterscheid, J. N. H. Reek, *Angew. Chem. Int. Ed.* 2012, 51, 9740–9747.

124 W. Rüttinger, G. C. Dismukes, *Chem. Rev.* 1997, 97, 1–24.

125 B. M. Hunter, H. B. Gray, A. M. Müller, *Chem. Rev.* 2016, 116, 14120–14136.

126 F. P. Gortsema, J. W. Cobble, *J. Am. Chem. Soc.* 1959, 81, 5516–5516.

127 B. A. Moyer, T. J. Meyer, *J. Am. Chem. Soc.* 1978, 100, 3601–3603.

128 B. A. Moyer, T. J. Meyer, *Inorg. Chem.* 1981, 20, 436–444.

129 R. A. Binstead, B. A. Moyer, G. J. Samuels, T. J. Meyer, *J. Am. Chem. Soc.* 1981, 103, 2897–2899.

130 S. W. Gersten, G. J. Samuels, T. J. Meyer, *J. Am. Chem. Soc.* 1982, 104, 4029–4030.

131 E. L. Lebeau, S. A. Adeyemi, T. J. Meyer, *Inorg. Chem.* 1998, 37, 6476–6484.

132 T. Wada, K. Tsuge, K. Tanaka, *Inorg. Chem.* 2001, 40, 329–337.

133 T. Wada, K. Tsuge, K. Tanaka, *Angew. Chem. Int. Ed.* 2000, 39, 1479–1482.

134 C. Sens, I. Romero, M. Rodríguez, A. Llobet, T. Parella, J. Benet-Buchholz, *J. Am. Chem. Soc.* 2004, 126, 7798–7799.

135 S. A. Adeyemi, A. Dovletoglou, A. R. Guadalupe, T. J. Meyer, *Inorg. Chem.* 1992, 31, 1375–1383.

136 R. Zong, R. P. Thummel, *J. Am. Chem. Soc.* 2005, 127, 12802–12803.

137 J. J. Concepcion, J. W. Jurss, J. L. Templeton, T. J. Meyer, *J. Am. Chem. Soc.* 2008, 130, 16462–16463.

138 Y. Pushkar, D. Moonshiram, V. Purohit, L. Yan, I. Alperovich, *J. Am. Chem. Soc.* 2014, 136, 11938–11945.

139 B. A. Johnson, A. Bhunia, S. Ott, *Dalton Trans.* 2017, 46, 1382–1388.

140 S. Lin, Y. Pineda-Galvan, W. A. Maza, C. C. Epley, J. Zhu, M. C. Kessinger, Y. Pushkar, A. J. Morris, *ChemSusChem* 2017, 10, 514–522.

141 Y.-S. Wei, M. Zhang, R. Zou, Q. Xu, *Chem. Rev.* 2020, 120, 12089–12174.

142 H.-W. Tseng, R. Zong, J. T. Muckerman, R. Thummel, *Inorg. Chem.* 2008, 47, 11763–11773.

143 Y. Tanahashi, S. Nagai, Y. Tsubonouchi, M. Hirahara, T. Sato, E. A. Mohamed, Z. N. Zahran, K. Saito, T. Yui, M. Yagi, A. C. S. Appl, *Energy Mater.* 2020, 3, 12172–12184.

144 M. Hirahara, S. Nagai, K. Takahashi, K. Saito, T. Yui, M. Yagi, *Inorg. Chem.* 2015, 54, 7627–7635.

145 A. Singh, L. Spiccia, *Coord. Chem. Rev.* 2013, 257, 2607–2622.

146 J. Limburg, J. S. Vrettos, L. M. Liable-Sands, A. L. Rheingold, R. H. Crabtree, G. W. Brudvig, *Science* 1999, 283, 1524–1527.

147 J. Limburg, G. W. Brudvig, R. H. Crabtree, *J. Am. Chem. Soc.* 1997, 119, 2761–2762.

148 S. Pahar, K. Majee, G. Maayan, *Eur. J. Inorg. Chem.* 2024, 27, e202300501.

149 L.-H. Zhang, S. Mathew, J. Hessels, J. N. H. Reek, F. Yu, *ChemSusChem* 2021, 14, 234–250.

150 S. J. Peighambardoust, S. Rowshanzamir, M. Amjadi, *Int. J. Hydrogen Energy* 2010, 35, 9349–9384.

151 M. Shao, Q. Chang, J.-P. Dodelet, R. Chenitz, *Chem. Rev.* 2016, 116, 3594–3657.

152 D. Banham, S. Ye, *ACS Energy Lett.* 2017, 2, 629–638.

153 M. Shao, *J. Power Sources* 2011, 196, 2433–2444.

154 Y. Nie, L. Li, Z. Wei, *Chem. Soc. Rev.* 2015, 44, 2168–2201.

155 C. R. Raj, A. Samanta, S. H. Noh, S. Mondal, T. Okajima, T. Ohsaka, *J. Mater. Chem. A* 2016, 4, 11156–11178.

156 M. Sun, H. Liu, J. Qu, J. Li, *Adv. Energy Mater.* 2016, 6, 1600087.

157 R. Prins, M. E. Bussell, *Catal. Lett.* 2012, 142, 1413–1436.

158 M. M. Kumar, C. R. Raj, A. C. S. Appl, *Nano Mater.* 2019, 2, 643–648.

159 S. Sarkar, A. Biswas, N. Kamboj, R. S. Dey, *Inorg. Chem.* 2020, 59, 13453–13464.

160 N. Mano, A. de Poulpiquet, *Chem. Rev.* 2018, 118, 2392–2468.

161 M. Loi, O. Glazunova, T. Fedorova, A. F. Logrieco, G. Mulè, *J. Fungi* 2021, 7, 1048.

162 S. C. Barton, H.-H. Kim, G. Binyamin, Y. Zhang, A. Heller, *J. Am. Chem. Soc.* 2001, 123, 5802–5803.

163 A. Le Goff, M. Holzinger, S. Cosnier, *Cell Mol. Life Sci.* 2015, 72, 941–952.

164 J. W. Gallaway, S. A. Calabrese Barton, *J. Electroanal. Chem.* 2009, 626, 149–155.

165 S. C. Barton, H.-H. Kim, G. Binyamin, Y. Zhang, A. Heller, *J. Phys. Chem. B* 2001, 105, 11917–11921.

166 J. W. Gallaway, S. A. Calabrese Barton, *J. Am. Chem. Soc.* 2008, 130, 8527–8536.

167 N. Mano, H.-H. Kim, Y. Zhang, A. Heller, *J. Am. Chem. Soc.* 2002, 124, 6480–6486.

168 G. Jeerh, M. Zhang, S. Tao, *J. Mater. Chem. A* 2021, 9, 727–752.

169 M. Aziz, A. T. Wijayanta, A. B. D. Nandiyanto, *Energies* 2020, 13, 3062.

170 A. Vojvodic, A. J. Medford, F. Studt, F. Abild-Pedersen, T. S. Khan, T. Bligaard, J. K. Nørskov, *Chem. Phys. Lett.* 2014, 598, 108–112.

171 C. Guo, J. Ran, A. Vasileff, S.-Z. Qiao, *Energy Environ. Sci.* 2018, 11, 45–56.

172 M. Wang, S. Liu, T. Qian, J. Liu, J. Zhou, H. Ji, J. Xiong, J. Zhong, C. Yan, *Nat. Commun.* 2019, 10, 341.

173 A. R. Singh, B. A. Rohr, M. J. Statt, J. A. Schwalbe, M. Cargnello, J. K. Nørskov, *ACS Catal.* 2019, 9, 8316–8324.

174 K. Kamiya, *Chem. Sci.* 2020, 11, 8339–8349.

175 K. Ohashi, K. Iwase, T. Harada, S. Nakanishi, K. Kamiya, *J. Phys. Chem. C* 2021, 125, 10983–10990.

176 B. E. Rennie, R. G. Eleftheriades, R. H. Morris, *J. Am. Chem. Soc.* 2020, 142, 17607–17629.

177 D. Mikhaylov, T. Gryaznova, Y. Dudkina, M. Khrizanphorov, S. Latypov, O. Kataeva, D. A. Vicic, O. G. Sinyashin, Y. Budnikova, *Dalton Trans.* 2012, 41, 165–172.

178 Y. B. Dudkina, M. N. Khrizanforov, T. V. Gryaznova, Y. H. Budnikova, *J. Organomet. Chem.* 2014, 751, 301–305.

179 Y. H. Budnikova, D. A. Vicic, A. Klein, *Inorganics* 2018, 6, 18.

180 M. Khrizanforov, T. Gryaznova, O. Sinyashin, Y. Budnikova, *J. Organomet. Chem.* 2012, 718, 101–104.

181 M. Khrizanforov, V. Khrizanforova, V. Mamedov, N. Zhukova, S. Strekalova, V. Grinenko, T. Gryaznova, O. Sinyashin, Y. Budnikova, *J. Organomet. Chem.* 2016, 820, 82–88.

182 W. Zhou, Y. Zheng, G. Yuan, J. Peng, *Dalton Trans.* 2019, 48, 2598–2605.

183 Y. Zhang, Y. Li, H. Guo, Y. Guo, R. Song, *Mater. Chem. Front.* 2024, 8, 732–768.

184 Y. Zheng, Y. Tan, W. Zhou, X. Hao, X. Liu, J. Peng, *Inorg. Chem.* 2021, 60, 12323–12330.

185 W. Zhou, Y. Guan, Y. Zheng, X. Chen, J. Li, X. Liu, J. Peng, *Cryst. Growth Des.* 2022, 22, 6588–6597.

186 W. Zhou, P. Liu, Y. Zheng, X. Liu, Y. Zhang, G. Yuan, J. Peng, *Dalton Trans.* 2019, 48, 16350–16357.

187 O. S. Wenger, *J. Am. Chem. Soc.* 2018, 140, 13522–13533.

188 P. Dierks, Y. Vukadinovic, M. Bauer, *Inorg. Chem. Front.* 2022, 9, 206–220.

189 C. Förster, K. Heinze, *Chem. Soc. Rev.* 2020, 49, 1057–1070.

190 V. Artero, M. Chavarot-Kerlidou, M. Fontecave, *Angew. Chem. Int. Ed.* 2011, 50, 7238–7266.

191 H. Ji, L. Wan, Y. Gao, P. Du, W. Li, H. Luo, J. Ning, Y. Zhao, H. Wang, L. Zhang, L. Zhang, *J. Energy Chem.* 2023, 85, 348–362.

192 R. Matheu, M. Z. Ertem, J. Benet-Buchholz, E. Coronado, V. S. Batista, X. Sala, A. Llobet, *J. Am. Chem. Soc.* 2015, 137, 10786–10795.

193 M. Kondo, H. Tatewaki, S. Masaoka, *Chem. Soc. Rev.* 2021, 50, 6790–6831.

194 Y. Gong, J. Yao, P. Wang, Z. Li, H. Zhou, C. Xu, *Chin. J. Chem. Eng.* 2022, 43, 282–296.

9

Staying in Good Shape? Tpy-Containing Polymers for Self-Healing and Shape-Memory Applications

9.1 Introduction

Many polymers, including metallopolymers, are able to alter their properties if their environmental conditions are changed [1]. This behavior, which is referred to as stimuli-responsiveness [2], is related to the reversibility of certain structural elements [3]. Such reversibility can be installed *via* dynamic covalent bonds as provided *inter alia* by hydrazones [4], acylhydrazones [5, 6], disulfides [7, 8], or boronates [9, 10]. Further examples for functional groups, which can be utilized for dynamic covalent chemistry (DCvC), can be found in Zhang's review on this particular topic [11]. The combination of different DCvC moieties might afford functional materials with multiple, ideally orthogonal stimuli-responsiveness. In this respect, Matile *et al.* demonstrated the orthogonality of disulfide, hydrazone, and boronate moieties (*i.e.*, their selective cleavage and reformation) [12]. Dynamics can also be introduced *via* supramolecular, non-covalent interactions [13] and, in particular, hydrogen-bonding (*H*-bonding) interactions have widely been employed for this purpose [14–16]. Besides this, other types of non-covalent interactions have also been discussed in this context: ππ-stacking [17], host-guest complexation [18], halogen bonding (*X*-bonding) [19] as well as metal-to-ligand coordination [1, 20, 21]. Regarding the latter, the reversibility of coordination represents the origin of stimuli-responsiveness.

Unlike most of these supramolecular forces (excluding *H*-bonding), metal-to-ligand coordination is particularly sensitive to the stoichiometry, *i.e.*, the desired property and, thus, the ability to respond to a certain stimulus will only be given at a proper ratio between metal ions and present ligand sites [22, 23]. Bonding/debonding within a metallopolymer upon applying an external stimulus is typically accompanied by a drastic change in properties: coloration, luminescence, viscosity, *etc.* [24]. A wide range of external stimuli has already been applied to induce the switching of metallopolymers' properties: temperature, light, pH value, mechanical stress, exposure to chemicals, *etc.* (Figure 9.1). Moreover, combinations of these have afforded multi-responsive metallopolymers. These various types of stimuli-responsive metallopolymers were highlighted by Schubert *et al.* in a recent review article [21]. One unique feature of metallopolymers is their novel ability to respond to redox chemistry. The oxidation of a metal center increases its

Terpyridines: Multifaceted Ligands for Diverse Applications, First Edition.
Ulrich S. Schubert, Andreas Winter, and George R. Newkome.

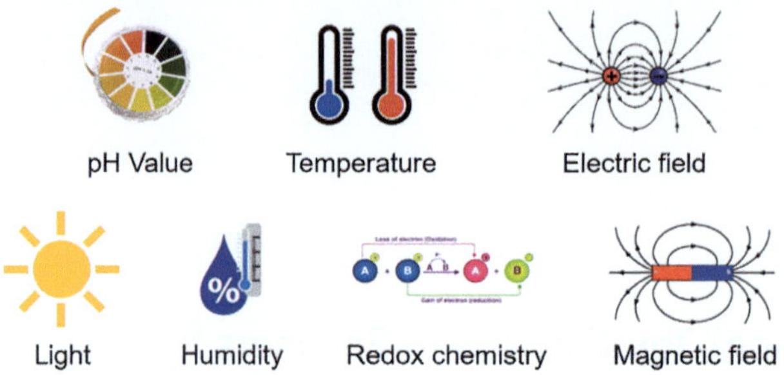

Figure 9.1 Representation of various external stimuli for addressing polymers [25] / MDPI / CC BY 4.0.

Lewis-hardness, thus leading to different coordination properties (*i.e.*, changes in coordination geometry and/or variation of bond strengths). In the most extreme case, even depolymerization might occur if the oxidized metal center is no longer compatible with the ligand moieties.

The ability of metallopolymers to (reversibly) change properties upon an external stimulus has paved the way to notable applications in the fields of catalysis, sensing, and drug delivery, to name just a few [21, 26, 27]. In addition to these, smart materials have been designed that display self-healing and/or shape-memory behavior [28–30]. These aspects represent comparably new topics discussed in the context of metallopolymers, but their use for practical applications has already been demonstrated [31]. However, research is still in its infancy and the involved mechanism(s), *i.e.*, the molecular process(es), which occur in the solid state, are far from being fully understood.

9.2 Some Fundamental Aspects of Self-Healing in Polymers

Today's society faces a rapid increase in waste, which pollutes the environment (including the oceans) and is harmful to animate beings. Synthetic polymers, which are mostly discarded by the end of their life cycle, represent a major share of this waste. The end of the life cycle is defined by the material's fatigue, weakening, and damage after its use under often-forcing operational conditions (*e.g.*, mechanical stress). In order to counteract this and to increase the sustainability of polymers, which are still mainly based on fossil fuels, polymers with prolonged service lifetimes have been developed. Unlike the traditional strategy to improve the lifetime by strengthening the material, recovery and restoration of the operational performance have moved into the focus of interest. Self-healing materials are characterized by the ability to recover at least partially from local mechanical damage (*i.e.*, scratches or cracks; Figure 9.2). The healing efficiency refers to what percentage the original properties (*i.e.*, mechanical, thermal, optoelectronic, *etc.*) have been restored after the healing process; thereby, a healing efficiency of 100% states that the performance of the undamaged polymers has been fully regained.

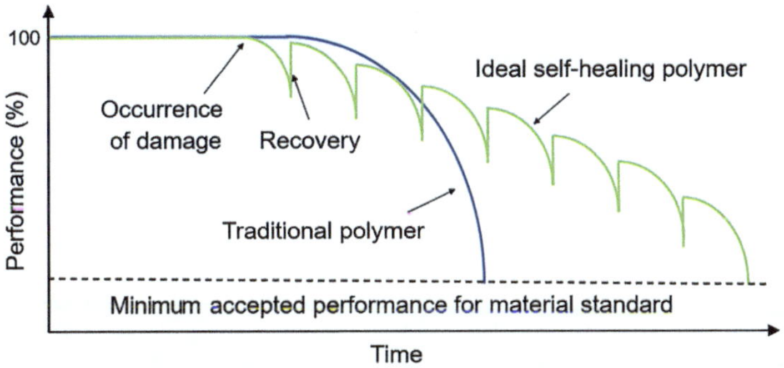

Figure 9.2 Comparison of the performance of a traditional polymer and a self-healing polymer over time [32] / with permission of John Wiley & Sons / CC BY-NC-ND 4.0.

In order to realize self-healing, two different strategies have been proposed – the extrinsic and intrinsic self-healing approaches [33]. Firstly, extrinsic self-healing relies on the embedding of microcapsules within the polymer matrix, from which a healing agent is released when the capsules break upon the material's damage (Figure 9.3a). This strategy was introduced by White *et al.* [35] and has, since then, been adapted for the self-healing of different types of polymers [36–38]. For example, the healing agent can be comprised of a monomer, which polymerizes in the presence of a catalyst, which is either dispersed within the polymeric matrix or also released from the capsules [39]. The crack closures and the material's properties are largely restored. Since this process is initiated as soon as damage occurs, this kind of self-healing is often referred to as "autonomous" healing [38]. Unlike earlier attempts to achieve self-healing in polymers, White's strategy allowed for the healing of many sites owing to the dispersed microcapsules. Before this, polymer composites were prepared, which contained a single hollow fiber, which was filled with the healing agent (*e.g.*, a non-reacting adhesive or an air-curing agent) [40]. If a microcrack damages the surface of the fiber within the composite, the healing agent will leak out and close the crack. Later on, Dry *et al.* developed improved internal delivery systems, such as a three-part MAA adhesive (MMA: methyl methacrylate) [41–43]. However, White's microcapsules were more efficient in healing scratches/cracks than Dry's hollow fibers. Detrimentally, extrinsic self-healing is always limited in the number of healing cycles. Once the healing agent is exhausted, another crack cannot be repaired (at least not, if it occurs at the same position). In order to overcome this restriction, a biomimetic approach was developed, in which a microvascular network was used [44]. This material mimicked the capillary network of human skin and allowed for the repeated healing of the polymer matrix. Alternatively, the healing agent can be pumped through the microvascular network and, thus, the healing agent is continuously available for scratch healing, even at the same position [45].

The alternative strategy, which is the intrinsic self-healing of polymers, does not require any embedded healing agent, but is based on the presence of reversible groups within the polymer structure (Figure 9.3b) [46]. These reversible sites can

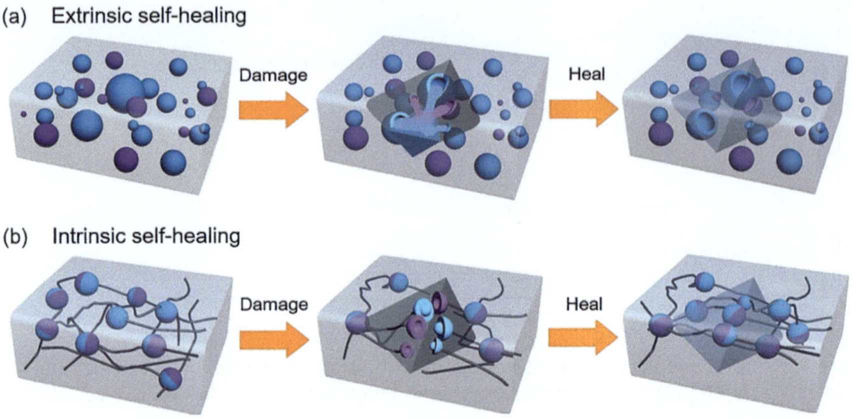

Figure 9.3 Schematic representation of the (a) extrinsic and (b) intrinsic healing of polymers [34] / with permission of John Wiley & Sons.

be cleaved after damage to the material by applying an external stimulus (*e.g.*, light or heat). This, in turn, results in an increased flexibility/mobility of the polymer by which the crack can be healed. Subsequently, reconnection of the reversible groups reinstalls the initial polymer structure and, thus, the original properties are largely retrieved. The design of practicably useful self-healing materials is nontrivial and numerous approaches have already been followed [47]. Among these, the Diels–Alder reaction represents the presumably best studied one [48]. First used by Wudl *et al.* at the example of furan and maleimide moieties within a polymer [49, 50], the thermally induced Diels–Alder chemistry is a highly versatile tool to realize self-healing in polymer networks [51]. Moreover, intrinsic self-healing has been realized due to the reversibility of binding in polymer networks containing disulfide, acylhydrazone, imine, or urea moieties as well as by transesterification reactions [21, 52].

Besides these examples that utilize the concept of DCvC, non-covalent interactions have also been employed to introduce self-healing behavior: ππ-stacking, *H*-bonding, ionic interactions, host-guest chemistry, metal-to-ligand coordination, *etc.* [53]. These supramolecular interactions feature a high degree of reversibility, *i.e.*, multiple switching between the bonded and debonded state is enabled. However, they differ in their dissociation energies, which – as a first approximation – controls the time for bond rearrangement (Figure 9.4a) [52]. Saito, Sokolov *et al.* pointed out that this time represented the longest relaxation time in a dynamic network and, thus, can be equalized with the terminal relaxation time, which marks the beginning of the material's viscoelastic flow [54]. It was further shown that the kinetics of a self-healing process correlated with the latter and, thus, were highly dependent on the dissociation energy; nonlinear rheology measurements were performed to corroborate this proposal [55]. Thus, the time for bond rearrangement represents one critical parameter for a material's self-healing performance.

Saito and Sokolov also paid particular attention to the mechanisms which control the rearrangement of the dynamic bonds and analyzed their implications on

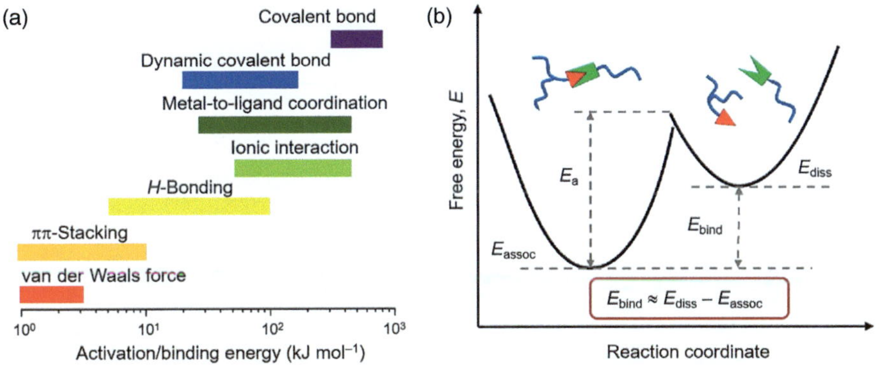

Figure 9.4 (a) Overview of the typical dissociation energy for the various types of dynamic/reversible bonds. (b) Simplified illustration of the critical energy parameters of bond rearrangement [52] / with permission of American Chemical Society.

the macroscopic scale, *i.e.*, the self-healing process [52]. Owing to restrictions in the available space, only the important aspects will be recapitalized within the scope of this section. The dynamic bond rearrangement is governed by two critical parameters, which are E_{bind} on the one hand and E_a on the other (Figure 9.4b). The former represents the difference in Gibbs free energy between the bonded/associated state and the debonded/dissociated state (*i.e.*, E_{assoc} and E_{diss}, respectively). Albeit being often confused in literature, the physical meaning of E_{bind} and E_a is fundamentally different. The former refers to the degree of dissociation at the point of equilibrium and determines whether bond rearrangement at a given temperature will proceed *via* a dissociative or an associative mechanistic pathway. In the case of polymers, which possess associative dynamic bonds, the E_{bind}/RT ratio will be high (*i.e.*, ≫1–3); R and T denote the gas constant (8.314 J K^{-1} mol^{-1}) and temperature (in K), respectively. Such polymers are commonly referred to as vitrimers (*vide infra*), whereas the rate of bond dissociation at a specific condition (*e.g.*, viscoelasticity) is expressed by the E_a value. Noteworthy, E_a can be even higher if E_{bind} is relatively low (as for some dissociative dynamic bonds). Beyond these fundamental considerations, the dimensions of these energies depend on numerous parameters, such as the presence of certain catalysts or the close vicinity of other reversible bonds [52].

The term vitrimer was coined in 2011 by Leibler *et al.*, who pointed to the similarity of the viscosity-temperature relationship of "covalent adaptive networks" (CANs) and vitreous silica, which both displayed an Arrhenius-like dependence [56]. This behavior has attracted tremendous interest from materials scientists and has been the subject of various comprehensive reviews [57–59]. As mentioned before, vitrimers are characterized by an associative bond rearrangement, *i.e.*, their cross-linking density of the polymer network is fixed since a cross-link will only be broken when a new one to another position has been formed (Figure 9.5a), whereas the dissociative exchange mechanism implies the breakage of cross-linking bonds before new ones are established (Figure 9.5b).

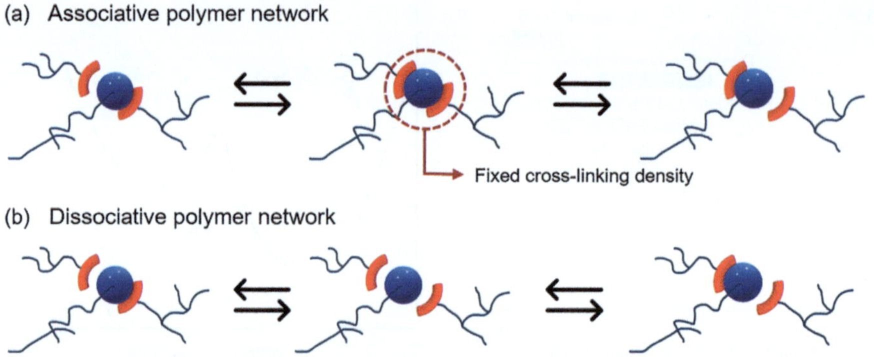

Figure 9.5 Schematic representation of the dynamic bond rearrangement in an (a) associative and (b) dissociative polymer network [59] / with permission of John Wiley & Sons.

However, the differentiation between the two distinct mechanisms is idealized. In practice, one will not be able to decide on either of these just by looking at the physical properties. Indeed, the polymers, which exchange bonds by a dissociative mechanism, might feature almost identical rheological properties as vitrimers [60, 61]. This will be if the bond formation is significantly faster than the dissociation of bonds. In such a case, the share of dissociated bonds will be small and the material's viscosity-temperature dependence still obeys Arrhenius' law. These polymers are classified as vitrimeric or vitrimer-like [62].

Let us now come back to polymer self-healing, which is based on non-covalent interactions. Presumably, *H*-bonding represents the most famous example [14]. Owing to the relatively weak nature of this supramolecular force, a high degree of reversibility is provided and, thus, high self-healing vitality can be obtained. However, this brings along one grave shortcoming – the often too soft materials possess insufficient mechanical properties [63]. Nonetheless, *H*-bonding has widely been applied in the field of self-healing materials [64, 65] and, notably, even the materials' softness could be overcome, at least to some extent [66, 67]. This concept has also been extended toward *X*-bonding, as the reversible interaction, within self-sealing polymer networks [68]. Regarding $\pi\pi$-stacking, the supramolecular interaction between "orthogonal" π-conjugated systems, *i.e.*, a donor- and an acceptor-type one, is of high relevance [69, 70]. This rather weak attraction between typically aromatic systems has also been utilized for the synthesis of self-healing polymers [71–74]. Finally, the interaction of metal ions with organic ligands has been explored to a significant extent; in this case, the reversibility of binding is described as complexation/decomplexation and clustering [21, 31, 75–79]. Unlike most of the other non-covalent interactions, metal-to-ligand complexation introduced further properties, which cannot (or hardly) be realized by all-organic polymers: redox activity, luminescence, dielectrics, magnetism, catalysis, *etc.* [59]. The field of self-healing metallopolymers was pioneered by Rowan, Weder *et al.*, who demonstrated the optical self-healing of a Zn^{II}-containing metallopolymer [80]. At about the same time, Holten-Andersen *et al.* followed by biomimetic strategy, in which self-healing was

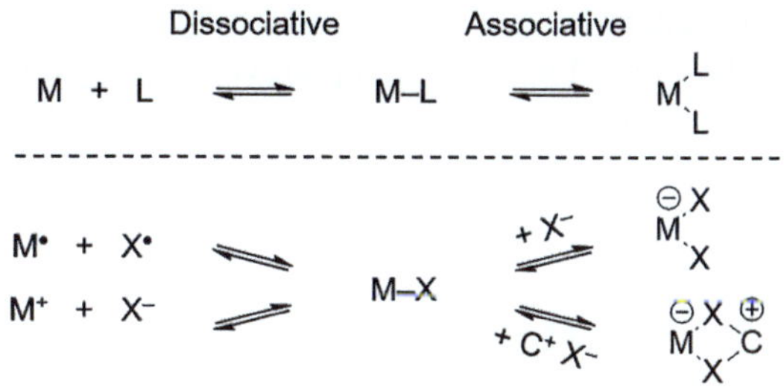

Figure 9.6 Schematic representation of the possible exchange mechanisms of metal-to-ligand bonds (L and X denote neutral and anionic ligands, respectively; Cat^+ represents an auxiliary cation). Adapted from [59].

due to the reversible complexation of Fe^{III} ions by polymer-bound catechol moieties [81]. Uflyand *et al.* classified self-healing polymers by the type of the organic ligand, which is attached to the polymer backbone [77], whereas Schubert *et al.* distinguished between three major classes: ionomers, biological and bioinspired polymers as well as synthetic metallopolymers [21]. To extend this list further, self-healing has also been reported for metal–organic frameworks (MOFs) [82]. Among others, tridentate chelates, such as 2,6-*bis*(1′-methylbenzimidazol-2-yl)pyridine (Mebip) and 2,2′:6′,2″-terpyridine (tpy), have been commonly used, as ligand moieties, in synthetic self-healing polymers (either as pending side chains or as the terminal units in telechelic polymers).

Before turning attention to the use of metal–tpy complexes in the context of self-healing polymers, the healing mechanism *via* metal–ligand exchange will briefly be evaluated. Can self-healing metallopolymers be classified as vitrimers? This question was raised recently by Poli *et al.* [59]. According to these authors, one has to distinguish between metal-to-ligand interactions using neutral (L-type) and anionic (X-type) ligands (Figure 9.6). This discrimination is reasonable since M—L bonds are able to dissociate and form without the separation of charges (this holds particularly true for the class of weak-field ligands), whereas the cleavage of an M—X bond is energetically unfavorable for both the homolytic as well as heterolytic scenarios. In the former case, an unlikely reduction of the metal center and the formation of $X^{\bullet}$ radicals (or $X^{\bullet -}$ radical anions) is premised. The homolytic cleavage to afford M^+ and X^- species is hampered in organic polymer matrices, which typically possesses low permittivity. However, this process might readily occur in an aqueous environment, *e.g.*, in hydrogels. The associative exchange process, in turn, might be supported by an auxiliary cation (a proton in the simplest case), thus providing charge neutrality during the exchange. In the case of tpy-containing polymers, a dependency on the cross-linking metal salt was found [83]. Strongly coordinating anions (*e.g.*, chloride, acetate, *etc.*) allowed for a faster healing than weakly coordinating ones. This trend is indicative of a dynamic ligand-exchange processes, *i.e.*, partial decomplexation represented the main contributor for the crosslink migration, in which the anion played a non-innocent role [84]. Thus, in special cases, where the exchange process is applicable, such polymers can be referred to as “coordination vitrimers” [59].

Beyond self-healing, *i.e.*, the restoration of the material’s properties during its period of application, Schubert *et al.* pointed to one further potential use of the reversibility of binding, which targets the end of the lifetime, *i.e.*, the recycling [21].

The material could be dissembled by applying an external trigger (*e.g.*, removal of the metal ions by a competitive ligand, protonation and/or precipitation). For further reading on this rather new concept, reference to recent literature is given at this stage [77, 85, 86].

9.3 The Metal-to-Tpy Linkage in Polymer Chemistry

Much has already been said on this prominent application of metal–tpy chemistry and numerous reviews on this topic are already available [87–93]. Therefore, only some fundamental aspects are included here for completeness and to reach the reader who is less familiar with the field "supramacromolecular" chemistry.

Tpy ligands can be introduced at the termini or in the side chains of polymers *via* different strategies [87]. For example, homo- and heterotelechelic macroligands can be prepared by end-group modification of appropriate α- and/or ω-functionalized precursor polymers (*e.g.*, homopolymers, block/statistical copolymers, *etc.*; Figure 9.7a–c). Alternatively, such materials are also accessible when utilizing tpy-equipped initiators (Figure 9.7d). The polymerization of tpy-containing polymers affords polymers which carry tpy moieties in the side chains (this approach is also adaptable to the synthesis of copolymers; Figure 9.7e).

These macroligands can be combined with transition-metal ions and further (macro)ligands to obtain metallo-supramolecular block copolymers, chain-extended polymers, graft co-polymers, polymer networks, *etc.* Lohmeijer and Schubert compared this approach, which is modular and highly versatile, to playing LEGO™ with macromolecules [87]. In particular, the directional and robust coordination chemistry of Ru^{III}/Ru^{II} ions (see Section 1.4) has been utilized for the synthesis of various types of homopolymer and copolymer architectures [94]. Representatively, the tetrablock copolymer **1** is shown in Figure 9.8. The left building block was prepared by a sequential nitroxide-mediated polymerization (NMP) using a tpy-functionalized initiator, whereas the second tpy macroligand was prepared by the chain-end functionalization of poly(ethylene glycol) monomethylether (MPEG). The two macroligands were linked by a Ru^{II} ion *via* a two-step assembly

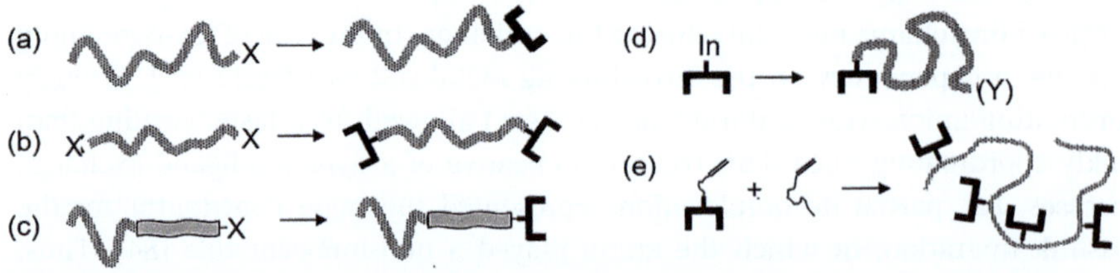

Figure 9.7 Schematic representation of the possible strategies to prepare tpy-equipped polymers (macroligands): (a) heterotelechelic homopolymer, (b) homotelechelic homopolymer, (c) heterotelechelic block copolymer, (d) heterotelechelic polymer *via* a tpy-functionalized initiator approach and (e) (co)polymerization of tpy-functionalized monomers [87] / with permission of John Wiley & Sons.

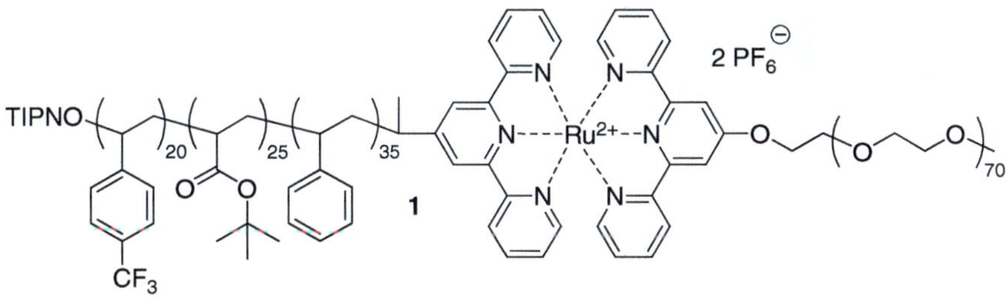

Figure 9.8 Schematic representation of metallo-supramolecular tetrablock copolymer **1** (TIPNO: 2,2,5-trimethyl-4-phenyl-3-azahexane-3-nitroxide).

procedure. Owing to the different solubility of the material's constituent blocks in selective solvents, micelles of various morphologies were obtained in alcohol solutions [95]. Micellization in solution as well as phase-separation in the melt or solid state represent the most notable applications of metallo-supramolecular block copolymers. Numerous studies on this topic have been published, mainly by the Schubert and Gohy groups [90, 91]. The self-assembly behavior of these materials can be understood by taking the fundamentals of polymer physics into account. However, this assumption overlooks the "non-innocent" role of the linking cationic transition-metal ion complex. Besides the ability to disassemble the architecture by cleaving the coordinative bond, the aforementioned optoelectronic or magneto-chemical properties, which are introduced by the complexes, have to be highlighted. Moreover, the charged nature of the metallo-supramolecular linkage can alter the physical properties of the overall block copolymer when compared to those of a related covalent one. This might hold true in solution, the melt, and the solid state. Based on these considerations, the metal–tpy linkage in such assemblies might even be regarded as a very short block or segment on its own, whose properties are distinctly different from those of the flanking blocks/segments. This aspect has been discussed in more detail against the background of block copolymers with "element blocks" [92].

In the overwhelming majority of metallo-supramolecular block copolymers, Ru^{II}–tpy *bis*-complexes were used. However, Fustin, Gohy *et al.* demonstrated that Co^{II} and Ni^{II} also provided sufficient kinetic stability to connect two macroligands in a heteroleptic fashion [96]. More recently, Chan *et al.* realized the preparation of solution-stable diblock copolymers, in which the constituent blocks were linked by a $[Zn(tpy)_2]^{2+}$ moiety [97–99]. Thereby, the intrinsic kinetic lability of the linkages was counterbalanced by the stabilizing $\pi\pi$-stacking interactions between the orthogonal tpy ligands (see Figure 1.18). The cumulated stability constant (β_{21}) for this complex, as determined by isothermal titration calorimetry (ITC), was significantly higher than that of the parent $[Zn(tpy)_2]^{2+}$ complex (*i.e.*, β_{21} of 10.15 *vs. ca.* 6.5) [100]. Thus, first-row transition-metal ions are available for the assembly of solution-stable and processable block copolymers and one is no longer restricted to the intricate and more expensive Ru^{II} ion, although the latter is superior in terms of kinetic stability.

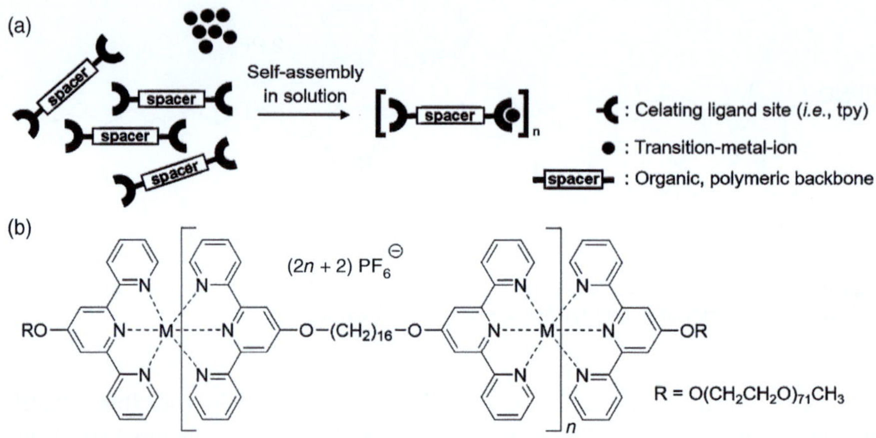

Figure 9.9 (a) Schematic representation of the synthesis of linear metallo-supramolecular polymers by the self-assembly of *bis*-tpy ligands [88] / with permission of John Wiley & Sons. (b) Schematic representation of a metallo-supramolecular AB_nA triblock copolymer.

Chain-extended polymers are comprised of macromolecular building blocks, which are connected with each other *via* their terminal functionalities. Such polymers are referred to as homo- and heterotelechelic ones, if they are equipped with two equal or different linking moieties, respectively. In the case of interest, tpy ligands represent the terminal functionalities and the self-assembly of these *bis*-tpy ligands into linear metallo-supramolecular polymers occurs *via* a polycondensation-like mechanism (Figure 9.9a) [88]. This concept was applied to various types of α,ω-tpy-functionalized polymers: poly(ethylene glycol) (PEG) [101–108], poly(tetrahydrofuran) (PTHF) [109] and Pluronics™ [109, 110]. In most of these studies, Ru^{II} ions were used for the metallo-supramolecular polymerization but also transition-metal ions, which afforded kinetically less stable assemblies, were employed (*e.g.*, Fe^{II}, Co^{II}, Ni^{II}, *etc.*) [105–108]. Furthermore, Schubert *et al.* established a protocol in which such chain-extended polymers were end-capped by complexation with a tpy-functionalized macroligand. Solution-stable metallo-supramolecular AB_nA triblock copolymers were obtained *via* this procedure (Figure 9.9b) [111].

In a similar fashion, *also bis*-tpy ligands, which contained a rigid, mostly π-conjugated spacer, have been extensively employed to obtain linear coordination polymers [88, 89]. The Zn^{II}-containing polymers have already been discussed in the context of light-emitting materials (see Section 7.5) [112]. Owing to the redox activity of the incorporated Ru^{II} and/or Fe^{II} center, these coordination polymers were studied regarding electrochromic [113–115] and/or energy-storage applications [116]. With respect to the latter, reference to Section 5.4 is given at this stage.

In 1988, Potts and Usifer reported the free-radical polymerization (FRP) of 4- and 4′-vinyl-functionalized tpy as well as copolymerization of the latter with styrene [117]. In this primal study, the formation of insoluble cross-linked networks upon addition of transition-metal ions was also shown. The metal-free polymers could

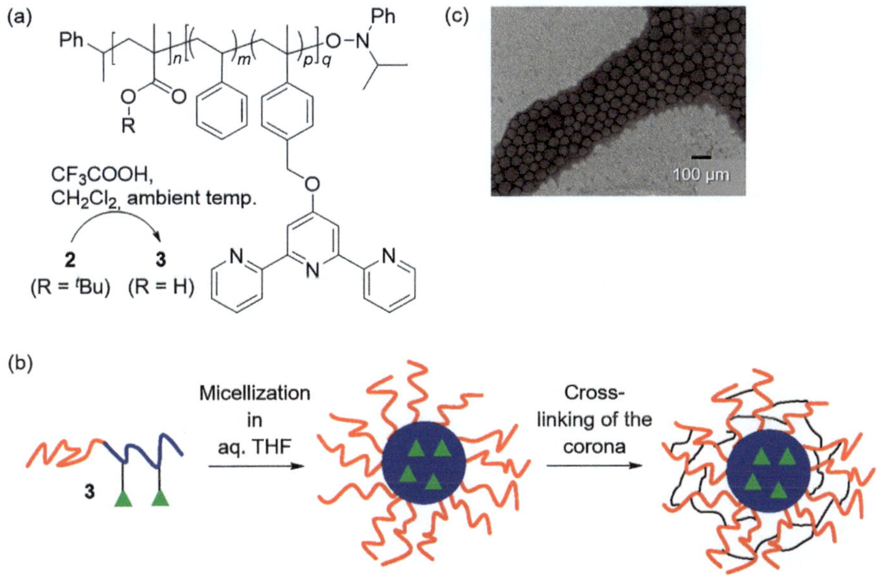

Figure 9.10 (a) Schematic representation of the block-random copolymers **2** and **3**. (b) Schematic representation of the nanoparticle synthesis *via* micellization of **3** in aq. THF and subsequent cross-linking in the corona. (c) Representative TEM image of these nanoparticles after deposition onto a carbon-coated Cu grid and staining with phosphotungstic acid [118] / with permission of American Chemical Society.

be regenerated by treatment with hot concentrated HCl. Thus, the reversibility of metallo-supramolecular cross-linking was already fundamentally proven by these authors. In subsequent years, research on side-chain polymers was continued, but the focus of interest shifted more and more toward precise polymer architectures, which can hardly be realized by FRP [*i.e.*, defined (block co)polymers, low dispersities (Đ), *etc.*]. For this purpose, the various controlled radical polymerization (CRP) methods have gained significant attention. Among these, atom-transfer radical polymerization (ATRP), NMP, and reversible addition-fragmentation chain-transfer (RAFT) polymerization represent the most notable techniques, which have been applied to prepare polymers with tpy-equipped side chains [93].

In one representative example, O'Reilly *et al.* prepared the block-random copolymer **2** by NMP; the subsequent saponification of the ester moieties afforded the copolymer **3** (Figure 9.10a) [118]. Owing to the amphiphilic nature of **3**, micellization occurred in aqueous tetrahydrofuran (THF). Finally, the cross-linking of the hydrophilic core by amidation with an aliphatic diamine, as the cross-linking agent, afforded stable nanoparticles (Figure 9.10b). Dynamic light scattering (DLS) and transmission-electron microscopy (TEM) corroborated the formation of micelles and nanoparticles, respectively, with narrow size distributions (Figure 9.10c). The presence of tpy moieties within the nanoparticles' cores allowed for their loading with Ru^{II} or Fe^{II} ions. This process, which could be monitored by UV/vis absorption spectroscopy [*i.e.*, appearance of the characteristic metal-to-ligand charge-transfer

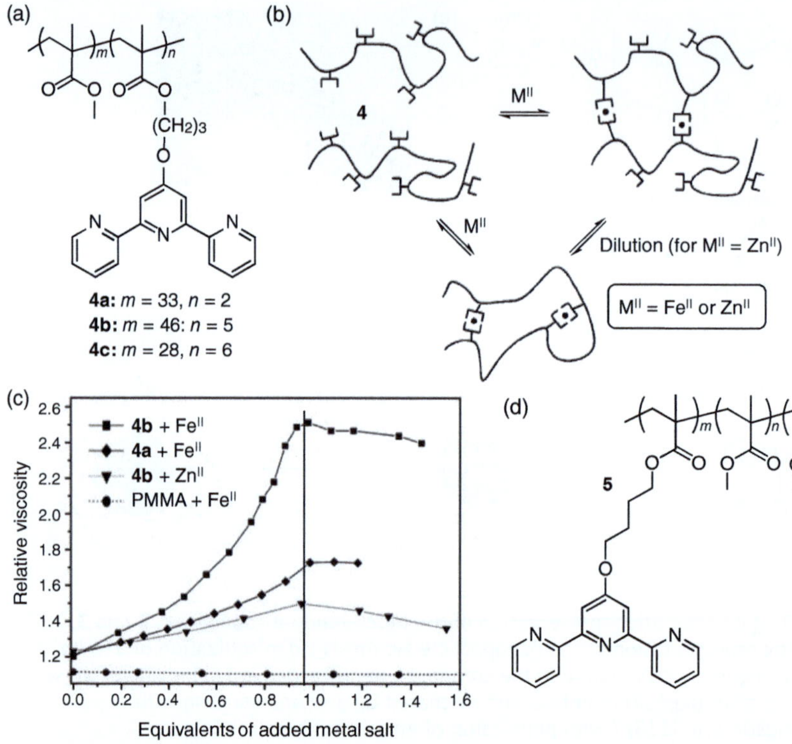

Figure 9.11 (a) Schematic representation of copolymers **4**. (b) Schematic representation of the metallo-supramolecular cross-linking of **4**. (c) Changes of the relative viscosity of **4** as a function of added M^{II} ions (M = Fe or Zn) [119] / with permission of John Wiley & Sons. (d) Schematic representation of the random terpolymer **5**. Adapted from [120].

(MLCT) bands at 390 nm (for Ru^{II}) and 550 nm (for Fe^{II})], had no substantial impact on the particle size and shape.

Cross-linking of side-chain polymers with transition-metal ions to afford "metallo-supramolecular gels" was explored by Hofmeier and Schubert [119]. These authors studied the changes in viscosity which arouse upon the stepwise addition of Fe^{II} or Zn^{II} ions to poly(methyl methacrylate) (PMMA)-type polymers **4**, which carried tpy moieties in the side chains (Figure 9.11a/b). The size of the thusly formed aggregates was dependent on the polymers' tpy content, amount of added metal ions, and concentration (Figure 9.11c). As important, the choice of the metal salt had an influence on the material's properties, *i.e.*, the polymer networks, which were formed with Zn^{II} ions, revealed lower viscosities than the Fe^{II}-containing ones. Beyond the point of equivalence, a gradual decrease of the viscosity was observed due to the formation of polymer-bound tpy *mono*-complexes at the cost of the cross-linking *bis*-complexes. Owing to the lower kinetic stability of the $[Zn(tpy)_2]^{2+}$ cross-links, this behavior was more expressed in the case of Zn^{II} ions.

Likewise, Tew *et al.* reported the formation of highly cross-linked metallo-supramolecular networks when side-chain polymers – similar to those of

Schubert *et al.* (Figure 9.11a) – were combined with Cu^{II} ions [121, 122]. As noted by Tew *et al.*, such supramolecular gels are of particular interest regarding potential applications *inter alia* as rheology modifiers, stimuli-responsive materials (*e.g.*, sensors or actuators), self-healing systems, and drug carriers [123]. Thus, the aforementioned studies can be regarded as the conceptual groundwork for the work on self-healing metallopolymers, which emerged a few years later (see Section 9.4).

Before eventually turning attention to the self-healing properties of tpy-containing polymers, one further polymer, which was capable of both metallo-supramolecular as well as covalent cross-linking, has to be mentioned. The random terpolymers **5**, which contained different ratios of tpy sites and oxirane functionalities, were studied by Schubert *et al.* (Figure 9.11d) [120]. Dual cross-linking was achieved by metal-to-ligand coordination on the one hand and ring-opening polymerization, catalyzed by the Lewis acid $AlCl_3$, on the other. In particular, a very high cross-linking density (Q) of almost 1.00 was obtained for the dual cross-linked material, whereas significantly lower Q values were found for the polymer networks based only on metallo-supramolecular or covalent cross-links (*i.e.*, *ca.* 0.06 *vs.* *ca.* 0.30). In addition to this, further approaches for the combination of covalent and metallo-supramolecular cross-linking have also been reported. For this purpose, terpolymers, which contained other reactive groups than the oxirane moieties of **5**, were prepared. Covalent cross-linking was realized by the photochemical cross-linking of terminal olefins, thermal curing of OH-groups with *bis*-isocyanates as well as the ring-opening of oxetanes, which was mediated by Lewis acids [124, 125].

9.4 Self-Healing Polymers Based on Metal–Tpy Complexes

The reversibility of metal-to-ligand interactions has been recognized as a versatile strategy to enable self-healing in polymer materials [21]. Besides the aforementioned fundamental studies on the metallo-supramolecular crosslinking of side-chain polymers, further work has been done by taking advantage of the reversibility of metal–tpy complexation. Firstly, Fustin, Bailly, Gohy *et al.* reported on the formation of block copolymer micelles, which contained tpy ligands, as the external moieties of their coronas [126, 127]. In one of their studies, these micelles were interconnected to afford a supramolecular network, which possessed stimuli-responsive properties [127]. Rheology measurements showed that deformations introduced by mechanical stress were relaxed instantaneously, as soon as the stress was released. As one further part of their extensive work on the self-assembly of tpy-equipped polymers in the presence of metal ions, Fustin, Gohy *et al.* also studied the hydrogel formation of the side-chain copolymers **6**, which were prepared by RAFT polymerization (Figure 9.12a) [128]. Copolymer **6b** was comprised of a water-compatible block, in which the tpy moieties were incorporated, and a hydrophobic polystyrene (PS) segment. This amphiphilic polymer assembled into spherical micelles in water and, subsequently, metallo-supramolecular cross-linking to provide hydrogels was achieved by adding Ni^{II} ions (Figure 9.12b). Thereby, the critical concentration for

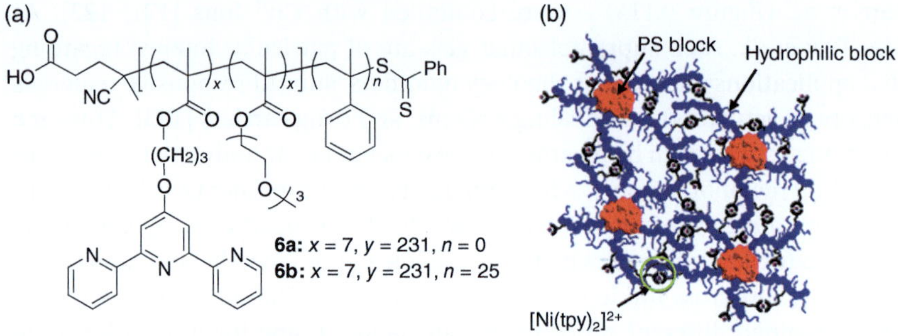

Figure 9.12 (a) Schematic representation of the tpy-equipped side-chain polymers **6**. (b) Schematic representation of the supramolecular hydrogel formed from **6b** and Ni^{II} ions [128] / with permission of Royal Society of Chemistry.

gel formation was *ca.* 4 wt%, thus lower than for the hydrophilic polymer **6a**. The gels were analyzed further by rheology. It was assumed that the micellization of **6b**, due to the presence of the PS block, provided additional cross-linking sites within the network structure. Accordingly, the hydrogel obtained from **6b** and Ni^{II} ions was mechanically more robust than the one formed from **6a**.

A different strategy to form a metallo-supramolecular network was followed by Seiffert *et al.*, who prepared the four-arm star-shaped macroligand **7** (Figure 9.13), which formed metallo-supramolecular gels upon addition of transition metal ions (*e.g.*, Mn^{II}, Co^{II}, Zn^{II}) [129]. It was proposed that **7**, whose arms featured a very low dispersity, should provide a highly regular polymer network with almost monodisperse meshes (Figure 9.14a). However, this idealized topology is only present in covalent organic frameworks (COFs) [130] or coordination nanosheets (CONASHs) [131], which are formed from monodisperse organic compounds and, in the case of the latter, transition-metal ions, whereas a lower degree of homogeneity is typically obtained for macromolecular components, which is due to structural imperfections, *i.e.*, entanglement of chains, loops, and dangling chain ends (Figure 9.14b) [129]. Furthermore, irregularity might also occur on the larger length scale. For example, the cross-linking density of polymer networks/gels will be spatially inhomogeneous (Figure 9.14c) if the copolymerization of the monomers and cross-linkers has been performed in an uncontrolled fashion (*e.g.*, by FRP). Regarding covalently cross-linked materials, this inhomogeneity results in weaker properties compared to those of networks/gels with idealized structure (at the

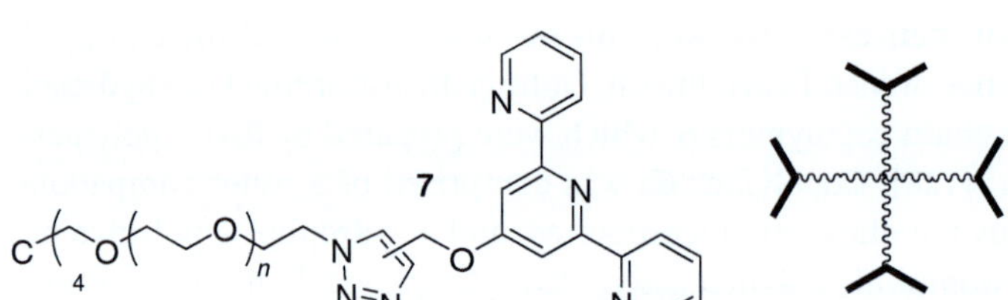

Figure 9.13 Schematic representation of Seiffert's four-arm tetra*kis*-tpy ligand. Adapted from [129].

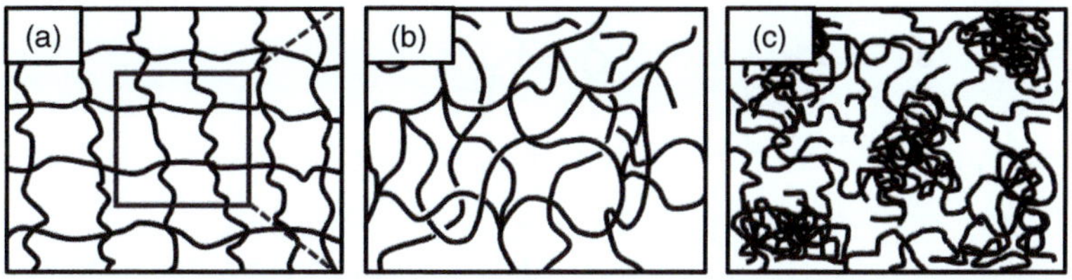

Figure 9.14 Simplified representation of the different polymer-network topologies: (a) Homogeneous polymer network with monodisperse mesh sizes. (b) Irregular polymer network with polydisperse mesh-size distribution and structural imperfections (*i.e.*, chain entanglement, loops, dangling chain ends). (c) Heterogeneous polymer network due to the clustering of cross-linking moieties [129] / with permission of Royal Society of Chemistry.

same overall cross-linking density). It is assumed that the dense agglomerates of cross-links simply represent "supernodes," which are less suited to store elastic energy than homogeneously distributed cross-linking sites. The same basically also holds true for their supramolecular counterparts, in which clustering or $\pi\pi$-stacking of the non-covalent cross-linking moieties gives rise to an inhomogeneous topology. Unlike the clustering of covalent cross-linking sites, the structural inhomogeneity of supramolecular networks/gels has even been identified as beneficial regarding the mechanical properties, in particular, in the case of those formed by *H*-bonding [132–137]. The latter behavior was also observed by Seiffert *et al.* for some of their metallo-supramolecular gels [129, 138]. The materials assembled from **7** and Co^{II} ions were structurally less homogeneous than their Zn^{II}-containing counterparts. One could assume that the kinetic lability of the $[Zn(tpy)_2]^{2+}$ cross-links allowed for better self-sorting and -repair in the course of the network formation.

All of the aforementioned studies on the metallo-supramolecular polymer-network formation were performed in solution or in the gel state. However, when aiming for self-healing applications, one has to deal with the solid state and the structure–property relationships established for the materials in solution (or as gels) might not be the same in the solid state. Nevertheless, some fundamental assumptions hold true: The strength of the coordinative bond will have a profound impact on the self-healing ability of a metallo-supramolecular polymer network. Weak metal-to-ligand interaction will be beneficial for the self-healing process, but too weak binding will be detrimental regarding the mechanical properties [139].

The first tangible study on the self-healing behavior of a tpy-equipped polymer was performed by Schubert, Hager *et al.* in 2013 [140]. Unlike the pioneering work on this topic by Rowan, Weder *et al.* [80], these authors opted for the metallo-supramolecular cross-linking of side-chain polymers instead of the formation of linear coordination polymers since improved mechanical properties were assumed for the resulting metallo-supramolecular networks, even in the absence of stabilizing additives (*e.g.*, cellulose nanocrystals) [140]. The PMMA-type polymers **8** were prepared by RAFT copolymerization and metallo-supramolecular cross-linking was achieved by adding Fe^{II} ions (Figure 9.15). The cross-linked polymers were thermally stable at least up to 200 °C; thus, self-healing at elevated temperatures was prospected. Indeed, the healing of mechanical damages, *i.e.*, scratches, was possible at *ca.* 100 °C (Figure 9.16). This study revealed further

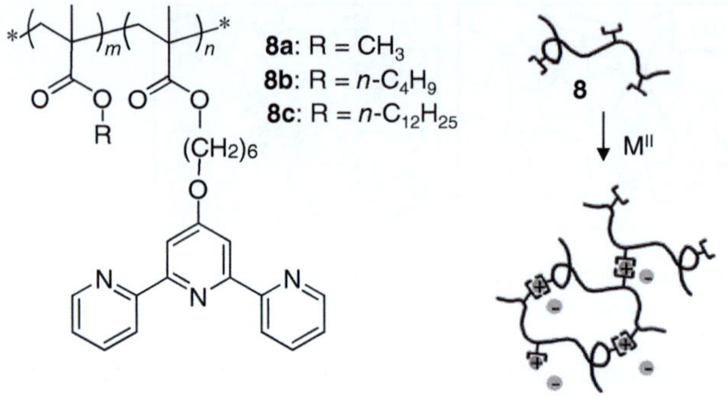

Figure 9.15 Schematic representation of copolymers **8** (the chain-end groups were omitted for clarity), from which metallo-supramolecular networks were formed upon addition of transition-metal ions (*e.g.*, Fe^{II} or Cd^{II}) [140] / with permission of John Wiley & Sons.

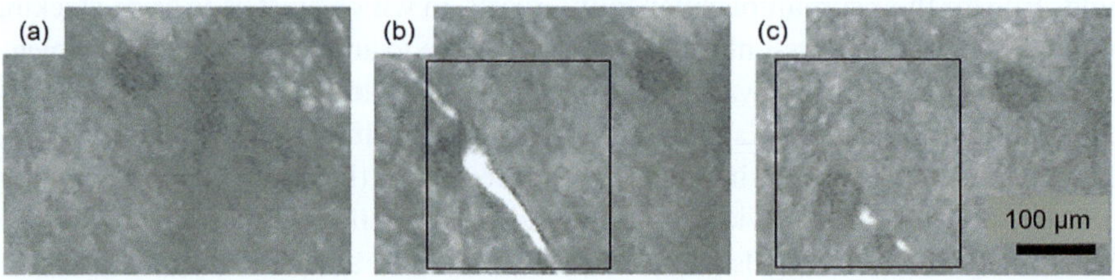

Figure 9.16 Representative scanning-electron microscopy (SEM) images showing the self-healing of the cross-linked polymer prepared from **8c** and $FeSO_4$: (a) Original polymer film. (b) Scratched polymer film. (c) Scratch healing after 82 h at 100 °C [140] / with permission of John Wiley & Sons.

that structural flexibility of the polymer backbone was crucial for the self-healing process. Unlike the relatively stiff network formed from **8a**, the lauryl residues of **8c** accounted for sufficient flexibility in the cross-linked state and, thus, a good crack-healing ability was provided.

In subsequent work, these authors also extended their work to Cd^{II} ions, as the cross-linking species [141]. It was assumed that the lower binding strength of the $[Cd(tpy)_2]^{2+}$ linkage could be beneficial for the self-healing process since the increased reversibility should enable self-healing at even lower temperatures. The combination of polymer **8c** with $Cd(OAc)_2$ provided a polymer network, which featured an improved self-healing performance (*i.e.*, cracks were healed at 80° within a few minutes). The self-healing mechanism was studied by solid-state NMR spectroscopy as well as small-angle X-ray scattering (SAXS) and differential scanning calorimetry (DSC) measurements. It was concluded that the superior self-healing ability of this material was due to the presence of the AcO^- anions, which favored a different complexation mode. Unlike the cross-linking with the Cd^{II} halides, where $[Cd(tpy)_2]^{2+}$ complexes are typically formed, μ-OAc-bridged clusters, such as $[(tpy)Cd(\mu\text{-}OAc)_2Cd(tpy)]^{2+}$, were assumed in this case. Such clusters are often observed in Cd^{II}–tpy coordination polymers [142]; this aspect has already been

Table 9.1 Summary of the self-healing behavior of polymer networks formed from **8b** and transition-metal salts.

Ion	NO_3^-	AcO^-	Cl^-
Zn^{II}	☹	☹	☹
Ni^{II}	☹	☹	☹
Co^{II}	☺ (@ 120 °C)	☹	☺ (@ 90 °C)
Mn^{II}	☹	☺ (@ 100 °C)	☺ (@ 60 °C)

☹: no healing below 150 °C; ☹: healing of only very small scratches; ☺: healing above the given temperature. For all measurements, the same healing time of 17 h was applied.
Adapted from [83].

addressed in Chapter 2, when dealing with coordination compounds of the "wrong" tpy ligands, *i.e.*, the various constitutional isomers of the 2,2′:6′,2″-motif. Combined quantum mechanical/molecular mechanical (QM/MM) simulations were performed to gain further insight into the self-healing mechanism [84]. According to these theoretical studies, which focused on the cross-linking with $FeSO_4$ or $CdBr_2$, partial decomplexation represented the main driving force for the thermal self-healing of the metallopolymers. Despite the promising self-healing ability of the Cd^{II}-containing polymers, practical applications of these systems are unlikely due to the harmful nature of Cd^{II} salts. In order to overcome this obstacle, Hager, Schubert *et al.* screened various salts of first-row transition-metal ions, as a cheaper and eco-friendly alternative to Cd^{II} ions [83]. In particular, self-healing studies were performed using in-total 12 transition-metal salts as the cross-linking agents; the results are summarized in Table 9.1. When using $MnCl_2$ for the cross-linking of **8b** (Figure 9.15), the best self-healing behavior was found, *i.e.*, scratch healing occurred at the healing temperature of 60 °C. Not only were mechanically induced scratches efficiently healed, but also the damage, which was caused by ablation with a high-intensity fs-laser beam, was repaired, although after longer healing times only [143]. The efficiency of self-healing of the polymer network **8b**/$MnCl_2$ was attributed to the presence of rather labile [(tpy)Mn[μ-Cl)$_2$Mn(tpy)] moieties, as the actual cross-linking moieties.

Interestingly, the rheology behavior of the metallopolymers could be correlated to the scratch-healing ability. The crossover of the storage and the loss modulus (G' and G'', respectively) gives evidence for the material's self-healing behavior: If such a crossover was present, the metallo-supramolecular network showed self-healing at a specific temperature [83].

Self-healing metallopolymers can not only be formed from transition-metal ions, but lanthanide(III) ions can also be used for this purpose (*i.e.*, $Ln^{III} = Eu^{III}$, Tb^{III}, or Dy^{III}) [144]. Hager, Schubert *et al.* combined polymer **8b** (Figure 9.15) with various $Ln(NO_3)_3$ salts to obtain the corresponding metallo-supramolecular networks, in which each Ln^{III} center coordinated three tpy ligands. This characteristic of the Ln^{III} coordination chemistry has also been mentioned in the context of Ln^{III}–tpy complexes, as emitters (see Section 7.6) [145]. In such complexes, the metal centers are luminescent due to the ligands' antenna effect. This behavior was also found for the cross-linked metallopolymers, but was not studied further. The materials' properties were examined by microscopy imaging and indentation measurements [144]. The stoichiometry of complexation as well as its intrinsically labile nature were verified by ITC measurements. The Vickers-hardness test revealed the highest hardness for the Eu^{III}-containing polymer network; thus, the poorest self-healing performance was observed for this metallopolymer, whereas the Tb^{III}- and Dy^{III}-containing materials were significantly softer and, thus, self-healing occurred more readily (*i.e.*, at 100 °C within 20 h). For the sake of completeness, the attempt to prepare a self-healing Pt^{II}-containing polymer has also to be mentioned [146]. Unlike the previous metallo-polymers, for which cross-linking was achieved by the addition of a metal salt, a different strategy was followed in this case. First, polymer-bound $[(tpy)PtCl]^+$ complexes were formed by reacting **8b** with the $Pt(cod)Cl_2$ precursor complexes (cod: cycloocta-1,5-diene); subsequently, cross-linking was achieved by adding a tetravalent pyridine linker. Despite the lability of the metallo-supramolecular cross-links, the self-healing performance was inferior to that obtained with first-row transition-metal or lanthanide ions.

The limited self-healing performance of (metallo-)supramolecular polymers is often due to their mechanical properties, *i.e.*, the insufficient mobility of the cross-linked network. As a solution to overcome this shortcoming, Hager, Schubert *et al.* opted for block-copolymer architectures, in which a hard block accounted for the mechanical performance and a soft one enabled mobility and, thus, allowed for self-healing [147]. PS was chosen as the hard segment and the soft segment was formed from *n*-butyl acrylate and a tpy-equipped monomer (*e.g.*, **9**; Figure 9.17a). This particular polymer design was inspired by the self-healing ability of mussel byssus threads, whose protein, the so-called precools, possesses a block copolymer-like structure, *i.e.*, a collagen-rich central domain is flanked on both sides with flexible histidine-rich parts, in which cross-linking occurs by complexation of Fe^{III} ions [21]. The bioinspired block copolymers were prepared in a highly controlled fashion by sequential RAFT polymerization [147]. Different copolymer architectures were prepared and subsequently cross-linked by transition-metal ions. In particular, the polymer network, which was derived from polymer **9** and $MnCl_2$ exhibited self-healing at 100 °C on a remarkably short time scale of only a few minutes (Figure 9.17b). The quantitative analysis of the self-healing kinetics revealed that self-healing of the Mn^{II}-containing polymers occurred *via* a two-step mechanism: Firstly, the low glass-transition temperature (T_g) of the hard PS block brought along increased mobility. The elastic recovery of this segment was followed by a second step, which was attributed to the reversibility of the metal-to-ligand interactions. Rheology measurements revealed a lifetime of the metallo-supramolecular bonds of *ca.* 40 s, thus within the desirable regime of 10–100 s [148]. In the case of other transition-metal ions, the metallo-supramolecular cross-links were too stable and, thus, self-healing proceeded slower or, for Ni^{II} ions, did not occur at all [149].

More recently, Hager *et al.* introduced a new approach for the quantification of scratch-healing of polymers [150]. This method analyzed the damage mechanically and optically using the MST^3 scratch tester and laser-scanning microscopy (LSM), respectively. The data of the latter were processed by a MATLAB™ script, which

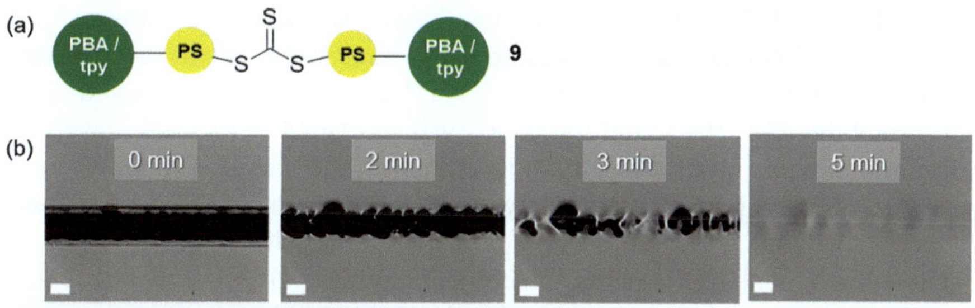

Figure 9.17 (a) Simplifier schematic representation of diblock copolymer, such as **9**. (b) Microscopy images showing the self-healing progress of the polymer network formed from **9** and $MnCl_2$ (heating @ 100 °C; the scale bars are 100 μm) [147] / with permission of American Chemical Society.

generated high-quality 3D images of the scratches and allowed for the quantification of the self-healing process. In one application, the self-healing kinetics *inter alia* of **8b** and **9**, cross-linked by Mn^{II} ions, were re-evaluated [151]. This analysis suggested that the self-healing of such metallopolymers occurs even *via* a three-step mechanism. An initial time (induction period) is required to allow crack closure to occur. Thereafter, a first healing period follows, which is characterized by a high crack-healing efficiency. However, this phase typically lasts less than half of the overall healing time. In the final phase, healing occurs with much slower kinetics. Thus, the time required for complete scratch healing closure is governed by the first and the last period.

A different type of polymer structure, namely poly(2-isopropenyl-2-oxazoline) (PiPOx), was used by Hoogenboom *et al.* [152]. This water-soluble polymer has attracted considerable interest due to the ease of functionalization by post-polymerization chemistry. This strategy has *inter alia* been applied to synthesize hydrogels, thermoresponsive copolymers with tunable solubility, molecular brushes, or materials for photonics [153]. A covalently cross-linked polymer was prepared *via* a two-step procedure: First, tpy ligands were grafted onto the polymer by ring-opening of 2-oxazoline moieties (**10**, Figure 9.18); subsequently, cross-linking was achieved by reaction with azelaic acid [152]. The properties of the resulting hydrogels could be modulated by the added transition-metal ions (*i.e.*,

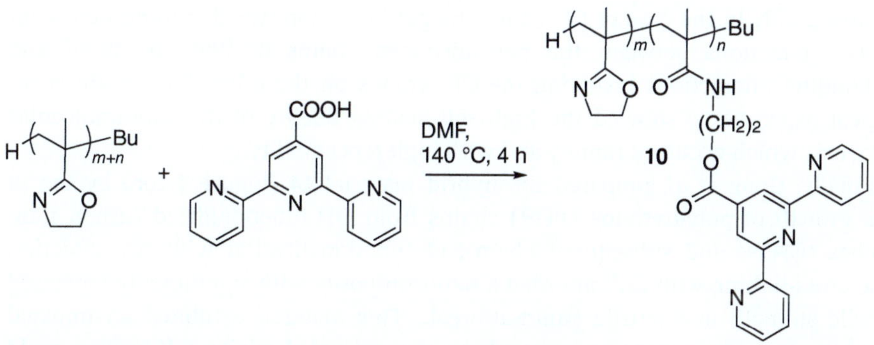

Figure 9.18 Schematic representation of the tpy-equipped PiPox **10**.

Fe^{II}, Ni^{II}, Co^{II}, and Zn^{II}). Owing to the low kinetic stability of the $[Zn(tpy)_2]^{2+}$ linkage, the Zn^{II}-containing hydrogel featured remarkably high compression strength and toughness (*i.e.*, 9 MPa and 99 MJ m^{-3}, respectively) as well as efficient self-recoverability (*i.e.*, the initial toughness was recovered to 98% after swelling for 60 min in absence of external stimuli). Moreover, the metallo-supramolecular cross-linking of **10** afforded hydrogels, which – at least in the case of Zn^{II} and Co^{II} ions – featured gel–sol transition properties and self-healing behavior at ambient temperature [154]. Notably, the Zn^{II}-containing hydrogel could be molded into various geometrical shapes; at the hands of these objects, self-healing was demonstrated even on the macroscopic scale.

Besides these examples, where the repair of mechanical damage was the focus of interest, the reversibility of metal-to-ligand interactions could also be used to "repair" catalysts, *i.e.*, restore catalytic activity after deactivation. For example, Song *et al.* prepared the cross-linked polymer **11** by ring-opening metathesis polymerization (ROMP; Figure 9.19) [155]. This strategy has been widely applied in the past to prepare side-chain metallopolymers using appropriately functionalized norbornene monomers [156]. The combination of **11** with the Ru^{II} complex **12** and Mg^{II} ions afforded a metallo-supramolecular gel, which was loaded with catalytically active $[Ru(bpy)_3]^{2+}$ moieties [155]. As the test reaction, the oxidative coupling of benzylamine to *N*-benzylbenzaldimine, which was photocatalyzed by the gel, was explored. After catalyst deactivation, the inactive Ru^{II} complexes were cleaved from the polymer by adding F^- ions; subsequently, the photoactive polymer could be regenerated by adding fresh **12** and Mg^{II} ions (Figure 9.19). Thus, the high reversibility of Mg^{II}–tpy complexation was utilized for the reactivation of a photocatalytically active material.

The non-cross-linked version of **11** (*i.e.*, $z = 0$ in Figure 9.19) was also prepared and combined with various transition-metal ions to obtain metallogels [157]. From these, the Zn^{II}-containing gel displayed self-healing in the absence of an external stimulus. The gel was cut into pieces and these pieces were put together again; within 5–10 min, the initial mechanical properties were fully restored.

Moreover, several non-polymeric tpy derivatives have also been used to obtain adaptive and self-healing coordination compounds [158–160]. In one example, Rissanen *et al.* reported the spontaneous gel formation of **13** in the presence of Fe^{II}, Co^{II} or Ni^{II} ions (Figure 9.20a) [160]. However, the gelation was anion-dependent and occurred only in the case of Cl^- salts. The gel formation was due to non-covalent F⋯H interactions between the perfluorinated chains on the one hand and *H*-bonding interactions involving the Cl^- anions on the other. Step-strain rheological experiments showed the high self-healing fidelity of the supramolecular network, which occurred rapidly and with high repeatability.

Finally, Gong *et al.* prepared the hybrid material **14** (Figure 9.20b) by the *in situ* growth of polyurethane (PUR) chains from OH-functionalized carbon nanotubes (CNTs) and subsequent chain-end functionalization with tpy moieties. The cross-linking with Zn^{II} afforded a nanocomposite with significantly increased tensile strength and tensile strain-at-break. This material exhibited an unusual combination of properties, *i.e.*, strength, toughness, and elasticity. Self-healing could

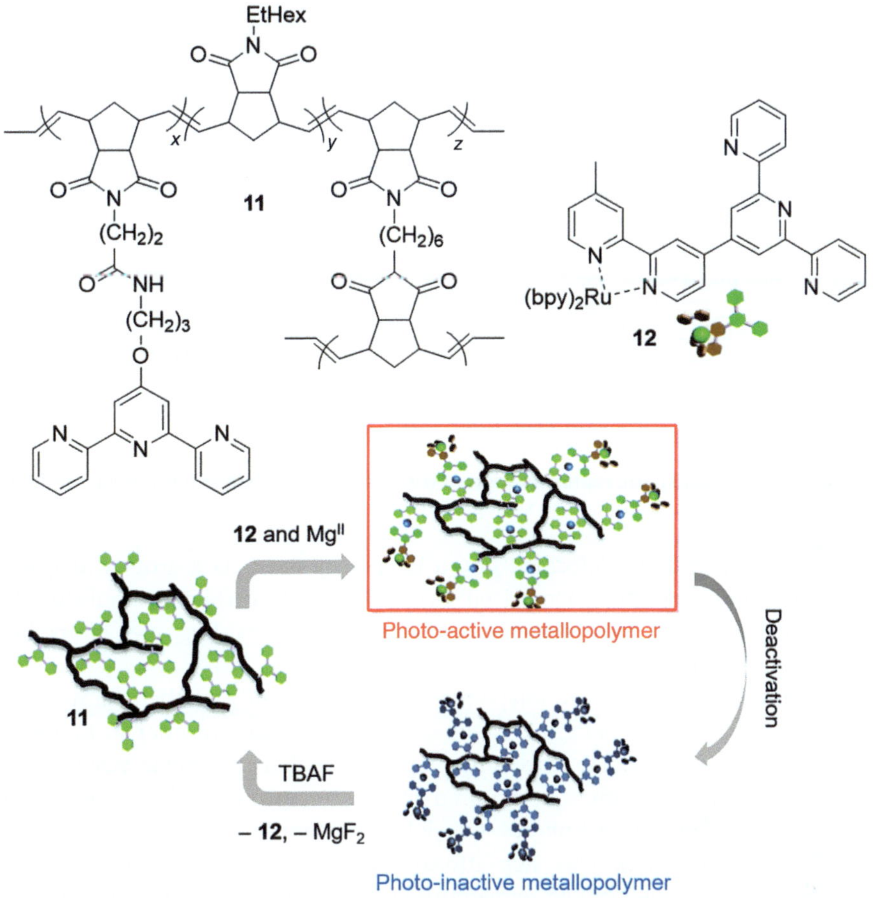

Figure 9.19 Schematic representation of the cross-linked polymer **11**, which gave a photocatalytically active gel upon addition of **12** and Mg^{II} ions [155] / with permission of Royal Society of Chemistry. After the loss of photocatalytic activity, the grafted Ru^{II} complex could be cleaved by F^- ions. The thusly re-obtained **12** could again be transformed into a photocatalytically active material by adding **12** and Mg^{II} ions.

be induced *via* various external stimuli including irradiation with near-infrared (NIR) light, temperature (*i.e.*, 90 °C for 1 h) or solvent annealing. In either case, the material recovered from the mechanical damage on relatively short timescales and with high efficiency (*i.e.*, >98%).

9.5 Shape-Memory Polymers – A Brief Introduction and Examples for Tpy-Containing Materials

The reversibility of metal-to-ligand interaction can not only be utilized for the self-healing of polymers, but one further application has recently gained significant attention, which are the shape-memory polymers (SMPs). Such materials

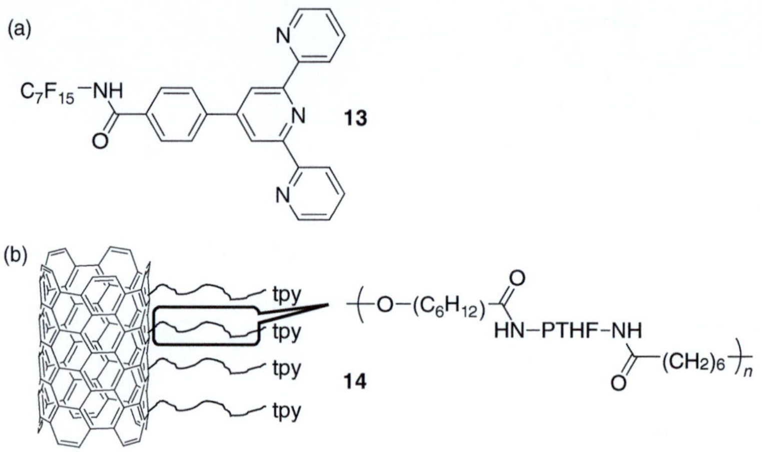

Figure 9.20 (a) Schematic representation of the perfluoroalkyl-substituted tpy ligand **13**. (b) Schematic representation of the CNT-PUR-tpy hybrid material **14**.

exhibit the ability to be deformed into a temporary fixed state, from which they can recover and adopt the "memorized" original state, if a certain external stimulus is applied [161–168]. This behavior is of importance regarding, *e.g.*, biomedical applications [169].

The basic principle for the shape-memory effect (SME), at the example of a temperature-responsive material, is shown in Figure 9.21. In order to make use of this effect, both a special processing method as well as a suitable design of the polymer-network structure are required [170]. Processing might be performed by one of the typical techniques for polymer manufacturing (*e.g.*, extrusion, spinning, pressing, *etc.*). This step affords the material in its permanent shape well above the so-called switch temperature (T_{sw}), which might correspond to the glass-transition, melting, or reaction temperature of the secondary phase (T_g, T_m, or T_{react}, respectively). Thus, the secondary phase can be expressed by a certain trigger (*e.g.*, temperature), whereas the primary/stable phase is comprised of an interpenetrating and/or cross-linked network. This network structure not only provides stability of the SMP but also accounts for the shaping back to its original form. Ideally, the cycle of deformation and restoration is repeatable many times. As said, the second phase becomes reversibly flexible if the external trigger is applied, and is able to fix its shape *via*, *e.g.*, crystallization or reversible bonding. The latter can be either covalent or non-covalent (*vide infra*).

However, the reversibility and repeatability of the SME still represent the limiting factor in most real-life applications of SMPs [168]. Regrettably, one can only recover the temporary shape by a new programming step, once the permanent shape has formed (Figure 9.22a). However, this one-way SME requires the customer/user of the SMP device to act for himself and hampers the commercialization of otherwise interesting and useful materials. Desirably, a shape-memory device is capable of reversibly changing its shape many times without presupposing the user to induce the re-shaping process. This two-way SME simply relies on the reformation of the temporary shape as soon as the stimulus, which triggers the shaping back to

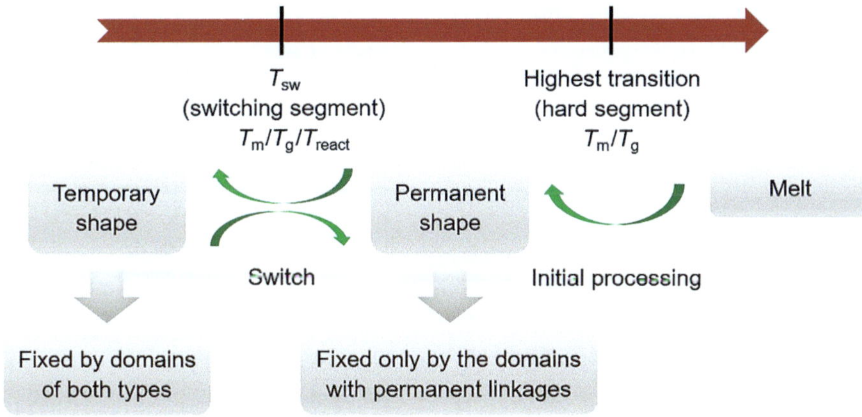

Figure 9.21 Schematic representation of the basic principle of the shape-memory effect in polymers [168] / with permission of Elsevier.

Figure 9.22 Schematic representation of the transition between different phases due to the SME. (a) One-way SME. (b) Two-way SME. (c) Dual and triple SME; thereby, each transition can be a one-way or two-way SME [168] / with permission of Elsevier.

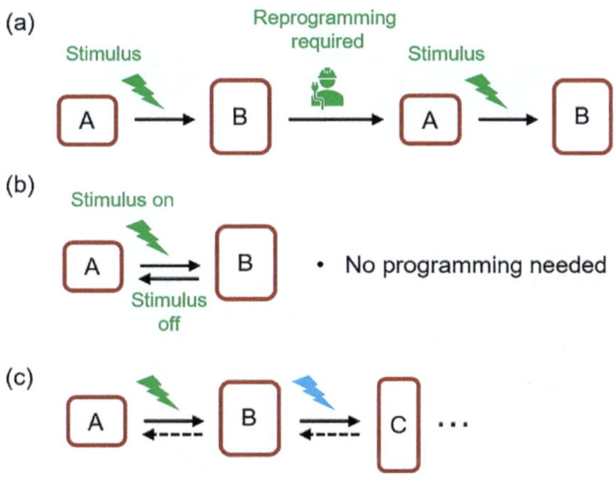

the permanent phase, is switched off (Figure 9.22b). Although significant progress has been made in designing SMPs, examples which feature the two-way SME remain rare. In addition to this dual SME, in which a temporary phase is transferred into the permanent one (named A and B, respectively), one might also envision more sophisticated systems that, *e.g.*, provide two temporary phases and a permanent one (Figure 9.22c) [171]. In such triple-shaped polymers, the temporary shape B needs to be programmed first, followed by the setting of the second one (*i.e.*, temporary shape A). Using a proper stimulus, one will be able to perform the A → B transition, whereas the permanent phase C is formed if the second trigger is applied. Thereby, either transformation might use the one-way or two-way SME. This concept can be developed further toward multiple SMPs which possess one permanent phase and $(n-1)$ temporary ones [164].

For example, the SME can rely on the reversibility of certain cycloaddition reactions, such as the thermal Diels–Alder or photoinduced [2 + 2]-addition reactions [172–175]. The reversible thiol–ene reaction represents one further possibility to design SMPs [176]. Furthermore, non-covalent interactions can be utilized to

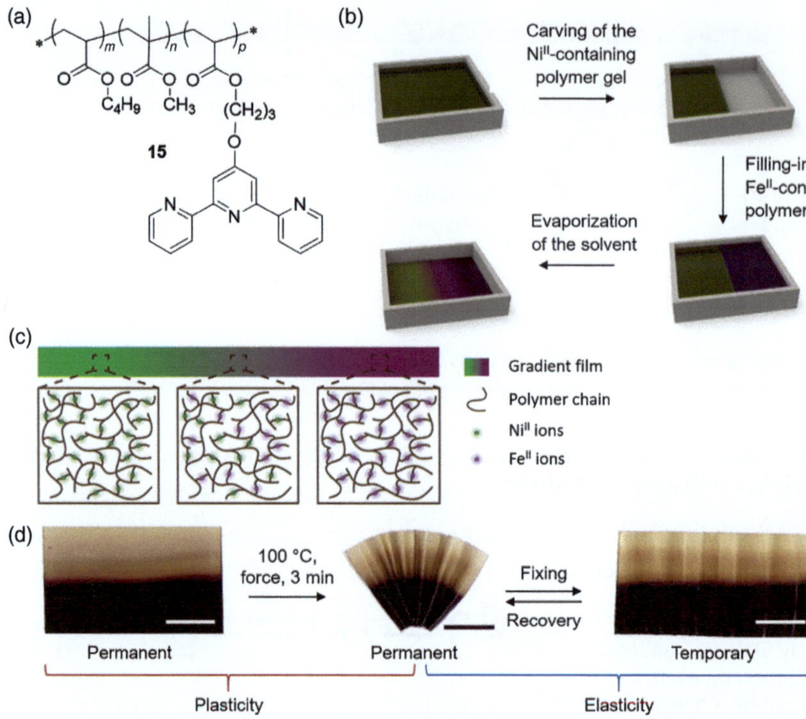

Figure 9.23 (a) Schematic representation of the terpolymer **15**. (b) Representation of the fabrication process of the gradient SMP. (c) Schematic representation of the spatial distribution of the metal complexes within the gradient network. (d) Visualization of the sample's shape-shifting versatility due to its elasticity and gradient plasticity (the scale bars represent 10 mm) [182] / with permission from John Wiley & Sons.

obtain an SME. Owing to the high degree of reversibility, *H*-bonding is frequently used in this context [177–179]. Besides this, ionic interactions and metal-to-ligand coordination, as the origin of the SME, have also attracted considerable attention [21]. The first tangible example of a shape-memory metallopolymer was reported by Kumpfer and Rowan [180]. In line with their previous work on self-healing polymers [181], these authors opted for the tridentate mebip ligand, which was attached to the chain ends of polybutadiene. A supramolecular network was formed by complexation of this telechelic polymer with Eu^{III} ions; subsequently, the permanent phase was established by covalent crosslinking *via* thiol–ene reaction [180].

Regarding tpy-containing polymers, Zhao, Xie *et al.* presented a metal-supramolecular SMP, which featured gradient thermal plasticity [182]. The random terpolymer **15** (Figure 9.23a), which was prepared by RAFT polymerization, was cross-linked by the coordination of Ni^{II} and Fe^{II} ions, respectively. The two metallopolymers were brought into contact *via* a fabrication process, which is outlined in Figure 9.23b. This process resulted in a partial mixing of the metal ions at the interfaces (Figure 9.23c). These complexes possessed different stability constants, which allowed them to access both the permanent as well as the temporary shapes.

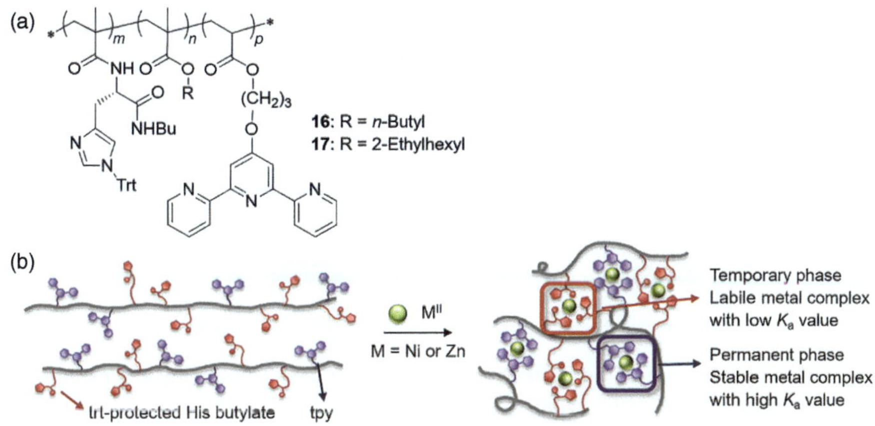

Figure 9.24 (a) Schematic representation of copolymers **16** and **17**. (b) Schematic representation of the formation of SMPs by the metallo-supramolecular cross-linking of **16** with Zn^{II} or Ni^{II} ions [183] / with permission of John Wiley & Sons.

The fabricated metallo-supramolecular polymer network exhibited solid-state plasticity and elasticity-based shape-memory properties (Figure 9.23d).

In continuation of the aforementioned work on self-healing metallopolymers (Section 9.4), Schubert *et al.* prepared copolymers **16**, which contained tpy and trt-protected histidine (His) moieties, as the binding sites for metal–ion complexation (trt: triphenylmethyl), in various ratios [183]. The complexation of **16** (Figure 9.24a) with either Zn^{II} or Ni^{II} ions afforded polymer networks, which displayed shape-memory characteristics. In these SMPs, the permanent and temporary phases were comprised of metal complexes, which differed significantly in their K_a values (Figure 9.24b). It was shown that the polymer composition (*i.e.*, the ratio of binding sites) and the choice of the metal salt had an influence on the switching temperature, which ranged from 85 to 115 °C. Particularly high switching temperatures were observed when polymers with a high overall ligand content and high tpy fraction were combined with Zn^{II} ions [*i.e.*, $(m+p) \gg n$ and $p \gg m$]. Cyclo-mechanic tests revealed very high strain-fixity and strain-recovery ratios of up to 99.8% and 95.4%, respectively. In subsequent work, these same authors prepared cross-linked metallopolymers from **16** and **17** by using combinations of metal salts (*i.e.*, Fe^{II}/Zn^{II}, Zn^{II}/Ni^{II}, and Fe^{II}/Ni^{II}). The use of two different, ligand-selective metal salts separated the activation temperatures, and removed interferences, thus improving the materials' shape-memory performance when compared to the aforementioned polymer networks, which contained only a single type of metal ions. The thusly prepared SMPs featured fixity and recovery rates of up to 99%.

Furthermore, Schubert *et al.* also prepared copolymers, which contained either tpy or trt-His moieties in the side chains (*i.e.*, **16** with $m = 0$ or $p = 0$) [184]. The mixing of these polymers in the presence of transition-metal ions provided interpenetrated polymer networks with tpy and trt-His complexes, as the strong and weak cross-linking moieties, respectively. These materials displayed very good shape-memory abilities, as expressed by the high strain-fixity/-recovery rates of

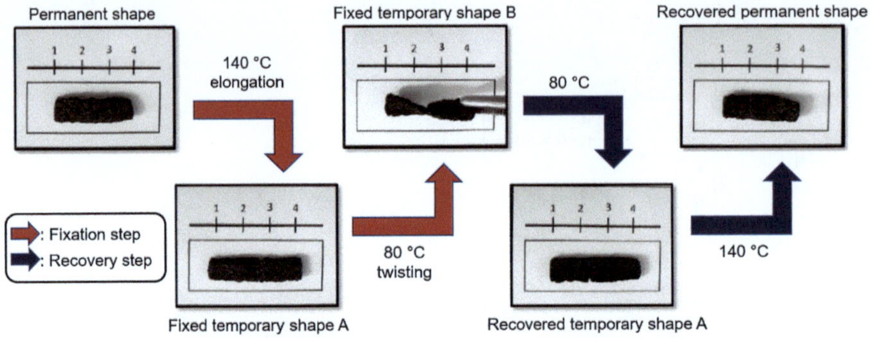

Figure 9.25 Visualization of the triple shape-memory test of the covalently and metallo-supramolecularly cross-linked polymer network [187] / MDPI / CC BY 4.0.

≥99%. Moreover, it was shown that a new permanent shape could be programmed without losing the shape-memory properties. This study revealed the remarkable self-healing ability of the materials: For example, in the case of the interpenetrated network, which was formed with Zn^{II} ions, self-healing occurred within 24 h at 140 °C with an efficiency of 98%.

Schubert *et al.* also synthesized polymers with pending 2-(1*H*-1,2,3-triazol-4-yl) pyridine (tripy) moieties, as bidentate ligands for metal–ion complexation. The metallo-supramolecular polymer networks formed thereof also showed shape-memory behavior [185, 186]. Finally, a covalently cross-linked polymer, which contained tridentate tpy and bidentate tripy moieties, was reported [187]. The addition of metal salts resulted in the formation of the respective homoleptic complexes, thus affording a dual cross-linked network. The complexes within this material possessed different K_a values and, thus, different activation temperatures. The presence of two individual switching units allowed for the fixation and recovery of two temporary shapes (Figure 9.25).

Very recently, Zechel, Schubert *et al.* demonstrated the processing of shape-memory metallopolymers by DLP-based 3D printing (DLP: digital light processing) (Figure 9.26a) [188, 189]. This specific technique relies on the *in situ* layer-by-layer hardening of the printed structures by photopolymerization [190]. For this purpose, photopolymerizable monomers, *i.e.*, Zn^{II}–tpy and Zn^{II}–His *bis*-complexes, were prepared and 3D printed in a 1:1 ratio and in the presence of a cross-linking agent (5 wt%) [188]. The DLP-based printing enabled the fabrication of optional motifs, which included rectangular rods, snails as well as hollow dodecahedrons (Figure 9.26b). The uniform distribution of both types of metallo-supramolecular linkages over the 3D-printed structures was confirmed by inductively coupled plasma optical emission spectroscopy and Fourier-transform Raman spectroscopy. The 3D-printed specimens exhibited remarkable shape-memory characteristics, *i.e.*, their thermomechanical analyses revealed quantitative fixity rates after mechanical deformation and recovery rates up to 99%. This holds true for both the snake motif and the delicate dodecahedron (Figure 9.26c).

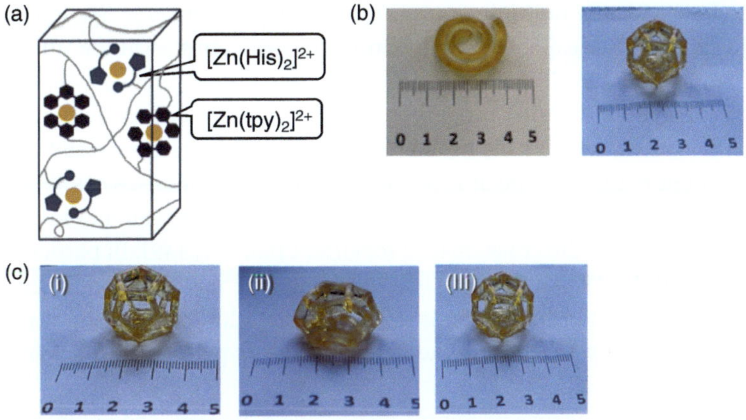

Figure 9.26 (a) Schematic representation of the 3D-printed cross-linked metallopolymer. (b) Representative photographs of the 3D-printed specimens – a spiral and a dodecahedron (the scale is in cm). (c) Series of photographs showing the shape-memory behavior of the metallopolymer: (i) specimen in its permanent shape; (ii) specimen in the temporary shape adopted upon deformation at 100 °C; (ii) specimen in its permanent shape after recovery at 100 °C in an oven (the scale is in cm) [188] / with permission of Taylor & Francis Group.

9.6 Concluding Remarks

Polymers represent an essential part of today's daily life. This is mainly due to the ease of processing and the relatively low production/manufacturing costs since fossil oil and gas sources are still the main feedstock for polymer synthesis. Advantageously, polymers can be on a large scale and with tailor-made properties for a plethora of possible applications. Owing to the outstanding role of this class of material, it has been suggested that the current period of human history might be referred to as the "plastics age," thus putting polymers in line with famous epoch-making materials: stone, bronze, and iron [191]. However, the widespread and often checkless use of polymer-based items has entailed serious environmental problems. Today, plastic waste is found in terrestrial and marine environments – ranging from the bottom of the oceans to the top of the highest mountains. Representing synthetic materials, their degradation in nature proceeds very slowly. One approach to counteract this is the use of bio-based or bio-derived polymers, which can be prepared from renewable sources or even CO_2 [192] and be more readily degradable, *e.g.*, by bacteria or fungi [193]. The consequent recycling (or upcycling) of plastic waste represents one further, although still cost-intensive and demanding alternative [194]. The traditional thermal recycling, simply speaking, the combustion to produce energy, becomes more and more unalluring due to the enormous CO_2 production.

One further approach to reduce plastic waste is to improve the durability of plastic items, *i.e.*, increase their lifetime. In this context, smart materials, which are able to adapt to changes in environmental conditions, have gained significant importance [195]. Such "intelligent" materials are not only of interest for applications as sensors or for controlled drug delivery but also enable the healing of a material's damage (*i.e.*, cracks or scratches) – ideally multiple times (Figure 9.2). As detailed in Section 9.2, reversibility in bonding – covalent as well as non-covalent ones – can be utilized to enable self-healing [29, 46, 53, 196]. Among others, metal-to-ligand coordination can be used to achieve self-healing properties [77]. As summarized in Section 9.4, tpy-containing polymers, which have been cross-linked with transition-metal ions, represent one notable class of self-healing polymers. Substantial progress has already been made to improve the self-healing performance by cautious design of the polymer structure and choice of the cross-linking metal salt [83, 147, 149]. In particular, metal complexes with low K_a values are beneficial since the reversible breakage of the cross-links will occur at lower temperatures and, thus, allow self-healing to proceed under mild conditions. Up to now, the best self-healing performance has been observed for the Mn^{II}–ty system.

SMPs represent one further type of stimuli-responsive polymers. The notable feature of these materials is their ability to "remember" the predetermined permanent shape [161, 162, 168, 197, 198]. As mentioned beforehand, SMPs are able to adopt a temporary shape if a certain stimulus is applied. This new shape is retained until one further external stimulus induces the re-establishment of the "memorized" permanent shape. The potential use of SMPs in various fields of research has already been the subject of various review articles: biomedicine [199], aerospace [200], flexible electronics [201], *etc.* The use of tpy-containing polymers in the context of SMPs represents a very young aspect of interest, with the first studies being published within the last few years. Nonetheless, such materials have already proven their shape-memory ability, which is – similar to the self-healing behavior – due to the reversible breakage of metal-to-ligand coordinative bonds.

In either case – self-healing or SMPs – the tpy-containing systems are still far from commercialization and the practical use in "real" applications still needs to be demonstrated. This holds particularly true, *e.g.*, for organic electronics, where the self-healing process should not affect its function. Still, the use of shape-memory metallopolymers in biomedical applications (*e.g.*, as vascular stent implants) requires the use of non-harmful metal salts. Thus, shape-memory performance and biocompatibility need to be compromised. The same also holds true for the coloration, which might be due to the presence of colored metal–tpy cross-links. Special attention has also been paid to the materials' mechanical properties, which require substantial improvement (most metal-containing polymers are still too soft).

To conclude, the use of tpy-containing polymers to obtain self-healing and/or shape-memory materials by the cross-linking with transition-metal salts represents a versatile, but not yet fully explored, strategy. Given that the self-healing/shape-memory performance as well as the mechanical properties are improved further, commercialization might be prospected for the future. In this respect, the recently demonstrated ability to process SMPs also by easy-to-handle 3D printing represents a notable achievement and milestone on the long road toward commercial use [188].

References

1 K. Y. Zhang, S. Liu, Q. Zhao, W. Huang, *Coord. Chem. Rev.* 2016, 319, 180–195.
2 K. M. Herbert, S. Schrettl, S. J. Rowan, C. Weder, *Macromolecules* 2017, 50, 8845–8870.
3 R. J. Wojtecki, M. A. Meador, S. J. Rowan, *Nat. Mater.* 2011, 10, 14–27.
4 T. Ono, S. Fujii, T. Nobori, J.-M. Lehn, *Chem. Commun.* 2007, 42, 4360–4362.
5 W. G. Skene, J.-M. Lehn, *Proc. Natl. Acad. Sci. USA* 2004, 101, 8270–8275.
6 J.-M. Lehn, *Prog. Polym. Sci.* 2005, 30, 814–831.
7 N. Morimoto, X.-P. Qiu, F. M. Winnik, K. Akiyoshi, *Macromolecules* 2008, 41, 5985–5987.
8 D. Roy, J. N. Cambre, B. S. Sumerlin, *Prog. Polym. Sci.* 2010, 35, 278–301.
9 A. P. Bapat, B. S. Sumerlin, A. Sutti, *Mater. Horiz.* 2020, 7, 694–714.
10 M. Gosecki, M. Gosecka, *Polymers* 2022, 14, 842.
11 Y. Jin, C. Yu, R. J. Denman, W. Zhang, *Chem. Soc. Rev.* 2013, 42, 6634–6654.
12 S. Lascano, K.-D. Zhang, R. Wehlauch, K. Gademann, N. Sakai, S. Matile, *Chem. Sci.* 2016, 7, 4720–4724.
13 C. R. South, C. Burd, M. Weck, *Acc. Chem. Res.* 2007, 40, 63–74.
14 L. Brunsveld, B. J. B. Folmer, E. W. Meijer, R. P. Sijbesma, *Chem. Rev.* 2001, 101, 4071–4098.
15 F. Liu, M. W. Urban, *Prog. Polym. Sci.* 2010, 35, 3–23.
16 R. Yerushalmi, A. Scherz, M. E. van der Boom, H.-B. Kraatz, *J. Mater. Chem.* 2005, 15, 4480–4487.
17 D. González-Rodríguez, A. P. H. J. Schenning, *Chem. Mater.* 2011, 23, 310–325.
18 O. Kretschmann, S. W. Choi, M. Miyauchi, I. Tomatsu, A. Harada, H. Ritter, *Angew. Chem. Int. Ed.* 2006, 45, 4361–4365.
19 R. Kampes, S. Zechel, M. D. Hager, U. S. Schubert, *Chem. Sci.* 2021, 12, 9275–9286.
20 H. Zhou, M. Chen, Y. Liu, S. Wu, *Macromol. Rapid Commun.* 2018, 39, 1800372.
21 S. Götz, S. Zechel, M. D. Hager, G. R. Newkome, U. S. Schubert, *Prog. Polym. Sci.* 2021, 119, 101428.
22 W. Weng, Z. Li, A. M. Jamieson, S. J. Rowan, *Soft Matter* 2009, 5, 4647–4657.
23 H. Hofmeier, U. S. Schubert, *Chem. Commun.* 2005, 19, 2423–2432.
24 A. J. McConnell, C. S. Wood, P. P. Neelakandan, J. R. Nitschke, *Chem. Rev.* 2015, 115, 7729–7793.
25 A. V. Maksimkin, T. Dayyoub, D. V. Telyshev, A. Y. Gerasimenko, *Nanomaterials* 2022, 12, 2272.
26 M. A. C. Stuart, W. T. S. Huck, J. Genzer, M. Müller, C. Ober, M. Stamm, G. B. Sukhorukov, I. Szleifer, V. V. Tsukruk, M. Urban, F. Winnik, S. Zauscher, I. Luzinov, S. Minko, *Nat. Mater.* 2010, 9, 101–113.
27 G. I. Dzhardimalieva, L. N. Rabinskiy, K. A. Kydralieva, I. E. Uflyand, *RSC Adv.* 2019, 9, 37009–37051.

28 M. D. Hager, P. Greil, C. Leyens, S. van der Zwaag, U. S. Schubert, *Adv. Mater.* 2010, 22, 5424–5430.

29 S. J. Garcia, *Eur. Polym. J.* 2014, 53, 118–125.

30 S. Billiet, X. K. D. Hillewaere, R. F. A. Teixeira, F. E. Du Prez, *Macromol. Rapid Commun.* 2013, 34, 290–309.

31 M. Enke, D. Döhler, S. Bode, W. H. Binder, M. D. Hager, U. S. Schubert, in *Self-healing Materials*, (Eds.: M. D. Hager, S. van der Zwaag, U. S. Schubert), Springer Int. Publ, Cham, 2016, pp. 59–112.

32 N. El Choufi, S. Mustapha, B. A. Tehrani, B. P. Grady, *Macromol. Rapid Commun.* 2022, 43, 2200164.

33 S. H. Pezzin, in *Multifunctional Epoxy Resins: Self-Healing, Thermally and Electrically Conductive Resins*, (Eds.: N. Hameed, J. C. Capricho, N. Salim, S. Thomas), Springer Nature, Singapore, 2023, pp. 107–138.

34 D. Chen, D. Wang, Y. Yang, Q. Huang, S. Zhu, Z. Zheng, *Adv. Energy Mater.* 2017, 7, 1700890.

35 S. R. White, N. R. Sottos, P. H. Geubelle, J. S. Moore, M. R. Kessler, S. R. Sriram, E. N. Brown, S. Viswanathan, *Nature* 2001, 409, 794–797.

36 M. Gragert, M. Schunack, W. H. Binder, *Macromol. Rapid Commun.* 2011, 32, 419–425.

37 S. H. Cho, H. M. Andersson, S. R. White, N. R. Sottos, P. V. Braun, *Adv. Mater.* 2006, 18, 997–1000.

38 X. K. D. Hillewaere, F. E. Du Prez, *Prog. Polym. Sci.* 2015, 49–50, 121–153.

39 J. M. Kamphaus, J. D. Rule, J. S. Moore, N. R. Sottos, S. R. White, *J. R. Soc. Interface* 2008, 5, 95–103.

40 C. Dry, N. Sottos, *1993 North American Conference on Smart Structures and Materials*, Vol. 1916, SPIE, Bellingham, 1993.

41 C. Dry, *Compos. Struct.* 1996, 35, 263–269.

42 C. Dry, W. McMillan, *1996 Symposium on Smart Structures and Materials*, Vol. 2718, SPIE, Bellingham, 1996.

43 C. Dry, W. McMillan, *Smart Mater. Struct.* 1996, 5, 297.

44 K. S. Toohey, N. R. Sottos, J. A. Lewis, J. S. Moore, S. R. White, *Nat. Mater.* 2007, 6, 581–585.

45 H. R. Williams, R. S. Trask, A. C. Knights, E. R. Williams, I. P. Bond, *J. R. Soc. Interface* 2008, 5, 735–747.

46 Y. Yang, M. W. Urban, *Chem. Soc. Rev.* 2013, 42, 7446–7467.

47 J. Dahlke, S. Zechel, M. D. Hager, U. S. Schubert, *Adv. Mater. Interfaces* 2018, 5, 1800051.

48 Y.-L. Liu, T.-W. Chuo, *Polym. Chem.* 2013, 4, 2194–2205.

49 X. Chen, M. A. Dam, K. Ono, A. Mal, H. Shen, S. R. Nutt, K. Sheran, F. Wudl, *Science* 2002, 295, 1698–1702.

50 X. Chen, F. Wudl, A. K. Mal, H. Shen, S. R. Nutt, *Macromolecules* 2003, 36, 1802–1807.

51 C. R. Ratwani, A. R. Kamali, A. M. Abdelkader, *Prog. Mater. Sci.* 2023, 131, 101001.

52 B. Li, P.-F. Cao, T. Saito, A. P. Sokolov, *Chem. Rev.* 2023, 123, 701–735.

53 F. Herbst, D. Döhler, P. Michael, W. H. Binder, *Macromol. Rapid Commun.* 2013, 34, 203–220.

54 S. Ge, M. Tress, K. Xing, P.-F. Cao, T. Saito, A. P. Sokolov, *Soft Matter* 2020, 16, 390–401.

55 T. Yan, K. Schröter, F. Herbst, W. H. Binder, T. Thurn-Albrecht, *Sci. Rep.* 2016, 6, 32356.

56 D. Montarnal, M. Capelot, F. Tournilhac, L. Leibler, *Science* 2011, 334, 965–968.

57 W. Denissen, J. M. Winne, F. E. Du Prez, *Chem. Sci.* 2016, 7, 30–38.

58 J. Zheng, Z. M. Png, S. H. Ng, G. X. Tham, E. Ye, S. S. Goh, X. J. Loh, Z. Li, *Mater. Today* 2021, 51, 586–625.

59 M. Murali, E. Manoury, R. Poli, *Eur. J. Inorg. Chem.* 2023, 26, e202300574.

60 A. Jourdain, R. Asbai, O. Anaya, M. M. Chehimi, E. Drockenmuller, D. Montarnal, *Macromolecules* 2020, 53, 1884–1900.

61 M. M. Obadia, A. Jourdain, P. Cassagnau, D. Montarnal, E. Drockenmuller, *Adv. Funct. Mater.* 2017, 27, 1703258.

62 B. R. Elling, W. R. Dichtel, *ACS Cent. Sci.* 2020, 6, 1488–1496.

63 P. Cordier, F. Tournilhac, C. Soulié-Ziakovic, L. Leibler, *Nature* 2008, 451, 977–980.

64 I. Gadwal, *Macromol* 2021, 1, 18–36.

65 Z. Xie, B.-L. Hu, R.-W. Li, Q. Zhang, *ACS Omega* 2021, 6, 9319–9333.

66 Y. Chen, A. M. Kushner, G. A. Williams, Z. Guan, *Nat. Chem.* 2012, 4, 467–472.

67 S. Chen, N. Mahmood, M. Beiner, W. H. Binder, *Angew. Chem. Int. Ed.* 2015, 54, 10188–10192.

68 R. Tepper, S. Bode, R. Geitner, M. Jäger, H. Görls, J. Vitz, B. Dietzek, M. Schmitt, J. Popp, M. D. Hager, U. S. Schubert, *Angew. Chem. Int. Ed.* 2017, 56, 4047–4051.

69 T. Aida, E. W. Meijer, S. I. Stupp, *Science* 2012, 335, 813–817.

70 F. J. M. Hoeben, P. Jonkheijm, E. W. Meijer, A. P. H. J. Schenning, *Chem. Rev.* 2005, 105, 1491–1546.

71 S. Burattini, H. M. Colquhoun, J. D. Fox, D. Friedmann, B. W. Greenland, P. J. F. Harris, W. Hayes, M. E. Mackay, S. J. Rowan, *Chem. Commun.* 2009, 44, 6717–6719.

72 S. Burattini, H. M. Colquhoun, B. W. Greenland, W. Hayes, *Faraday Discuss.* 2009, 143, 251–264.

73 L. R. Hart, J. H. Hunter, N. A. Nguyen, J. L. Harries, B. W. Greenland, M. E. Mackay, H. M. Colquhoun, W. Hayes, *Polym. Chem.* 2014, 5, 3680–3688.

74 L. R. Hart, N. A. Nguyen, J. L. Harries, M. E. Mackay, H. M. Colquhoun, W. Hayes, *Polymer* 2015, 69, 293–300.

75 B. Sandmann, S. Bode, M. D. Hager, U. S. Schubert, in *Hierarchical Macromolecular Structures: 60 Years after the Staudinger Nobel Prize II*, (Ed.: V. Percec), Springer Int. Pub., Cham, 2013, pp. 239–257.

76 H. Li, P. Yang, P. Pageni, C. Tang, *Macromol. Rapid Commun.* 2017, 38, 1700109.

77 G. I. Dzhardimalieva, B. C. Yadav, S. Singh, I. E. Uflyand, *Dalton Trans.* 2020, 49, 3042–3087.

78 C.-H. Li, J.-L. Zuo, *Adv. Mater.* 2020, 32, 1903762.

79 S. Basak, A. Bandyopadhyay, *J. Organomet. Chem.* 2021, 956, 122129.

80 M. Burnworth, L. Tang, J. R. Kumpfer, A. J. Duncan, F. L. Beyer, G. L. Fiore, S. J. Rowan, C. Weder, *Nature* 2011, 472, 334–337.

81 N. Holten-Andersen, M. J. Harrington, H. Birkedal, B. P. Lee, P. B. Messersmith, K. Y. C. Lee, J. H. Waite, *Proc. Natl. Acad. Sci. USA* 2011, 108, 2651–2655.

82 J. L. Obeso, M. T. Huxley, C. Leyva, J. Gabriel Flores, N. Martín-Guaregua, M. Viniegra, J. Aguilar-Pliego, J. Antonio de los Reyes, I. A. Ibarra, R. A. Peralta, *Coord. Chem. Rev.* 2023, 496, 215403.

83 S. Bode, M. Enke, R. K. Bose, F. H. Schacher, S. J. Garcia, S. van der Zwaag, M. D. Hager, U. S. Schubert, *J. Mater. Chem. A* 2015, 3, 22145–22153.

84 S. Kupfer, L. Zedler, J. Guthmuller, S. Bode, M. D. Hager, U. S. Schubert, J. Popp, S. Gräfe, B. Dietzek, *Phys. Chem. Chem. Phys.* 2014, 16, 12422–12432.

85 Z. Cheng, M. Yan, L. Cao, J. Huang, X. Cao, D. Yuan, Y. Chen, *Ind. Eng. Chem. Res.* 2019, 58, 3912–3920.

86 Z. F. Zhang, X. T. Liu, K. Yang, S. G. Zhao, *Macromol. Res.* 2019, 27, 803–810.

87 B. G. G. Lohmeijer, U. S. Schubert, *J. Polym. Sci. Part A Polym. Chem.* 2003, 41, 1413–1427.

88 M. Chiper, R. Hoogenboom, U. S. Schubert, *Macromol. Rapid Commun.* 2009, 30, 565–578.

89 A. Winter, U. S. Schubert, *Chem. Soc. Rev.* 2016, 45, 5311–5357.

90 C.-A. Fustin, P. Guillet, U. S. Schubert, J.-F. Gohy, *Adv. Mater.* 2007, 19, 1665–1673.

91 J.-F. Gohy, *Coord. Chem. Rev.* 2009, 253, 2214–2225.

92 A. Winter, U. S. Schubert, in *New Polymeric Materials Based on Element-Blocks*, (Ed.: Y. Chujo), Springer, Singapore, 2019, pp. 307–346.

93 U. S. Schubert, A. Winter, G. R. Newkome, *Terpyridine-Based Materials*, Wiley-VCH, Weinheim, 2011.

94 U. S. Schubert, A. Winter, G. R. Newkome, *Ruthenium-Containing Polymers*, Springer Int. Publishing, Cham, 2021, pp. 15–185.

95 J.-F. Gohy, C. Ott, S. Hoeppener, U. S. Schubert, *Chem. Commun.* 2009, 40, 6038–6040.

96 C. Mugemana, P. Guillet, S. Hoeppener, U. S. Schubert, C.-A. Fustin, J.-F. Gohy, *Chem. Commun.* 2010, 46, 1296–1298.

97 Y.-J. He, T.-H. Tu, M.-K. Su, C.-W. Yang, K. V. Kong, Y.-T. Chan, *J. Am. Chem. Soc.* 2017, 139, 4218–4224.

98 T.-H. Tu, T. Sakurai, S. Seki, Y. Ishida, Y.-T. Chan, *Angew. Chem. Int. Ed.* 2021, 60, 1923–1928.

99 W.-N. Wu, T.-H. Tu, C.-H. Pai, K.-H. Cheng, S.-H. Tung, Y.-T. Chan, C.-L. Liu, *Macromolecules* 2022, 55, 10670–10681.

100 T. Bätz, I. Anufriev, C.-H. Pai, T. Schuett, I. Nischang, S. Zechel, M. D. Hager, Y.-T. Chan, U. S. Schubert, A. C. S. Appl, *Polym. Mater.* 2023, 5, 1354–1363.

101 H. Hofmeier, S. Schmatloch, D. Wouters, U. S. Schubert, *Macromol. Chem. Phys.* 2003, 204, 2197–2203.

102 M. A. R. Meier, H. Hofmeier, C. H. Abeln, C. Tziatzios, M. Rasa, D. Schubert, U. S. Schubert, *e-Polymers* 2006, 6, 016.

103 M. Raşa, B. G. G. Lohmeijer, H. Hofmeier, H. M. L. Thijs, D. Schubert, U. S. Schubert, C. Tziatzios, *Macromol. Chem. Phys.* 2006, 207, 2029–2041.

104 U. S. Schubert, O. Hien, C. Eschbaumer, *Macromol. Rapid Commun.* 2000, 21, 1156–1161.

105 M. Chiper, M. A. R. Meier, J. M. Kranenburg, U. S. Schubert, *Macromol. Chem. Phys.* 2007, 208, 679–689.

106 S. Schmatloch, A. M. J. van den Berg, A. S. Alexeev, H. Hofmeier, U. S. Schubert, *Macromolecules* 2003, 36, 9943–9949.

107 S. Schmatloch, M. F. González, U. S. Schubert, *Macromol. Rapid Commun.* 2002, 23, 957–961.

108 S. Schmatloch, A. M. J. van den Berg, M. W. M. Fijten, U. S. Schubert, *Macromol. Rapid Commun.* 2004, 25, 321–325.

109 M. Chiper, R. Hoogenboom, U. S. Schubert, *Eur. Polym. J.* 2010, 46, 260–269.

110 M. Chiper, S. Hoeppener, U. S. Schubert, C.-A. Fustin, J.-F. Gohy, *Macromol. Chem. Phys.* 2010, 211, 2323–2330.

111 M. A. R. Meier, D. Wouters, C. Ott, P. Guillet, C.-A. Fustin, J.-F. Gohy, U. S. Schubert, *Macromolecules* 2006, 39, 1569–1576.

112 P. Liu, G. Shi, X. Chen, *Front. Chem.* 2020, 8, 592055.

113 M. Higuchi, *Polym. J.* 2009, 41, 511–520.

114 M. Higuchi, *Bull. Jpn. Soc. Coord. Chem.* 2022, 79, 68–77.

115 F. S. Han, M. Higuchi, D. G. Kurth, *Adv. Mater.* 2007, 19, 3928–3931.

116 H. Chen, J. Xing, W. Wang, X. Li, M. Shu, P. Gao, Y. Pan, J. Liu, *Sol. Energy Mater. Sol. Cells* 2022, 248, 111967.

117 K. T. Potts, D. A. Usifer, *Macromolecules* 1988, 21, 1985–1991.

118 A. D. Ievins, X. Wang, A. O. Moughton, J. Skey, R. K. O'Reilly, *Macromolecules* 2008, 41, 2998–3006.

119 H. Hofmeier, U. S. Schubert, *Macromol. Chem. Phys.* 2003, 204, 1391–1397.

120 H. Hofmeier, A. El-ghayoury, U. S. Schubert, *e-Polymers* 2003, 3, 053.

121 K. J. Calzia, G. N. Tew, *Macromolecules* 2002, 35, 6090–6093.

122 G. N. Tew, K. A. Aamer, R. Shunmugam, *Polymer* 2005, 46, 8440–8447.

123 R. Shunmugam, G. J. Gabriel, K. A. Aamer, G. N. Tew, *Macromol. Rapid Commun.* 2010, 31, 784–793.

124 H. Hofmeier, A. El-Ghayoury, U. S. Schubert, *J. Polym. Sci. Part A Polym. Chem.* 2004, 42, 4028–4035.

125 A. El-ghayoury, H. Hofmeier, B. de Ruiter, U. S. Schubert, *Macromolecules* 2003, 36, 3955–3959.

126 P. Guillet, C.-A. Fustin, C. Mugemana, C. Ott, U. S. Schubert, J.-F. Gohy, *Soft Matter* 2008, 4, 2278–2282.

127 P. Guillet, C. Mugemana, F. J. Stadler, U. S. Schubert, C.-A. Fustin, C. Bailly, J.-F. Gohy, *Soft Matter* 2009, 5, 3409–3411.

128 F. D. Jochum, J. Brassinne, C.-A. Fustin, J.-F. Gohy, *Soft Matter* 2013, 9, 2314–2320.

129 T. Rossow, S. Seiffert, *Polym. Chem.* 2014, 5, 3018–3029.

130 K. Geng, T. He, R. Liu, S. Dalapati, K. T. Tan, Z. Li, S. Tao, Y. Gong, Q. Jiang, D. Jiang, *Chem. Rev.* 2020, 120, 8814–8933.

131 R. Sakamoto, K. Takada, X. Sun, T. Pal, T. Tsukamoto, E. J. H. Phua, A. Rapakousiou, K. Hoshiko, H. Nishihara, *Coord. Chem. Rev.* 2016, 320–321, 118–128.

132 R. F. M. Lange, M. Van Gurp, E. W. Meijer, *J. Polym. Sci. Part A Polym. Chem.* 1999, 37, 3657–3670.

133 J.-L. Wietor, D. J. M. van Beek, G. W. Peters, E. Mendes, R. P. Sijbesma, *Macromolecules* 2011, 44, 1211–1219.

134 S. H. M. Söntjens, R. A. E. Renken, G. M. L. van Gemert, T. A. P. Engels, A. W. Bosman, H. M. Janssen, L. E. Govaert, F. P. T. Baaijens, *Macromolecules* 2008, 41, 5703–5708.

135 F. Herbst, K. Schröter, I. Gunkel, S. Gröger, T. Thurn-Albrecht, J. Balbach, W. H. Binder, *Macromolecules* 2010, 43, 10006–10016.

136 S. Sivakova, D. A. Bohnsack, M. E. Mackay, P. Suwanmala, S. J. Rowan, *J. Am. Chem. Soc.* 2005, 127, 18202–18211.

137 O. Colombani, C. Barioz, L. Bouteiller, C. Chanéac, L. Fompérie, F. Lortie, H. Montès, *Macromolecules* 2005, 38, 1752–1759.

138 T. Rossow, A. Habicht, S. Seiffert, *Macromolecules* 2014, 47, 6473–6482.

139 A. C. Jackson, F. L. Beyer, S. C. Price, B. C. Rinderspacher, R. H. Lambeth, *Macromolecules* 2013, 46, 5416–5422.

140 S. Bode, L. Zedler, F. H. Schacher, B. Dietzek, M. Schmitt, J. Popp, M. D. Hager, U. S. Schubert, *Adv. Mater.* 2013, 25, 1634–1638.

141 S. Bode, R. K. Bose, S. Matthes, M. Ehrhardt, A. Seifert, F. H. Schacher, R. M. Paulus, S. Stumpf, B. Sandmann, J. Vitz, A. Winter, S. Hoeppener, S. J. Garcia, S. Spange, S. van der Zwaag, M. D. Hager, U. S. Schubert, *Polym. Chem.* 2013, 4, 4966–4973.

142 C. E. Housecroft, E. C. Constable, *Molecules* 2021, 26, 3110.

143 M. Abend, C. Kunz, S. Stumpf, S. Gräf, S. Zechel, F. A. Müller, M. D. Hager, U. S. Schubert, *J. Mater. Chem. A* 2019, 7, 2148–2155.

144 S. Götz, S. Zechel, M. D. Hager, U. S. Schubert, *Polymers* 2020, 12, 838.

145 X. Yu, Y. Hu, C. Guo, Z. Chen, H. Wang, X. Li, *Supramol. Mater.* 2022, 1, 100017.

146 S. Götz, M. Abend, S. Zechel, M. D. Hager, U. S. Schubert, *J. Appl. Polym. Sci.* 2019, 136, 47064.

147 M. Enke, R. K. Bose, S. Bode, J. Vitz, F. H. Schacher, S. J. Garcia, S. van der Zwaag, M. D. Hager, U. S. Schubert, *Macromolecules* 2016, 49, 8418–8429.

148 R. K. Bose, N. Hohlbein, S. J. Garcia, A. M. Schmidt, S. van der Zwaag, *Phys. Chem. Chem. Phys.* 2015, 17, 1697–1704.

149 R. K. Bose, M. Enke, A. M. Grande, S. Zechel, F. H. Schacher, M. D. Hager, S. J. Garcia, U. S. Schubert, S. van der Zwaag, *Eur. Polym. J.* 2017, 93, 417–427.

150 M. Abend, L. Tianis, C. Kunz, S. Zechel, S. Gräf, F. A. Müller, U. S. Schubert, M. D. Hager, *Polym. Test.* 2020, 90, 106699.

151 M. Abend, S. Zechel, L. Tianis, C. Kunz, M. Enke, J. Dahlke, S. Gräf, F. A. Müller, U. S. Schubert, M. D. Hager, *Polym. Test.* 2021, 100, 107264.

152 X. Xu, F. A. Jerca, V. V. Jerca, R. Hoogenboom, *Adv. Funct. Mater.* 2019, 29, 1904886.

153 S. Nemati Mahand, S. Aliakbarzadeh, A. Moghaddam, A. Salehi Moghaddam, B. Kruppke, M. Nasrollahzadeh, H. A. Khonakdar, *Eur. Polym. J.* 2022, 178, 111484.

154 X. Xu, F. A. Jerca, V. V. Jerca, R. Hoogenboom, *Macromolecules* 2020, 53, 6566–6575.

155 D.-C. Jeong, J. Lee, Y. H. Ro, C. Song, *Polym. Chem.* 2017, 8, 1923–1931.

156 C. G. Hardy, J. Zhang, Y. Yan, L. Ren, C. Tang, *Prog. Polym. Sci.* 2014, 39, 1742–1796.

157 J. Lee, H. H. Moon, K. Paeng, C. Song, *Polymers* 2018, 10, 1173.

158 Y. Li, J. Guo, B. Dai, L. Geng, F. Shen, Y. Zhang, X. Yu, *J. Colloid Interface Sci.* 2018, 521, 190–196.

159 H. Yang, A. Wang, L. Zhang, X. Zhou, G. Yang, Y. Li, Y. Zhang, B. Zhang, J. Song, Y. Feng, *New J. Chem.* 2017, 41, 15173–15179.

160 L. Arnedo-Sánchez, S. Nonappa, S. Bhowmik, R. Hietala, M. Puttreddy, L. De Lahtinen, K. R. Cola, *Dalton Trans.* 2017, 46, 7309–7316.

161 C. Liu, H. Qin, P. T. Mather, *J. Mater. Chem.* 2007, 17, 1543–1558.

162 P. T. Mather, X. Luo, I. A. Rousseau, *Annu. Rev. Mater. Res.* 2009, 39, 445–471.

163 T. Xie, *Polymer* 2011, 52, 4985–5000.

164 J. Hu, Y. Zhu, H. Huang, J. Lu, *Prog. Polym. Sci.* 2012, 37, 1720–1763.

165 W. M. Huang, Y. Zhao, C. C. Wang, Z. Ding, H. Purnawali, C. Tang, J. L. Zhang, *J. Polym. Res.* 2012, 19, 9952.

166 L. Sun, W. M. Huang, Z. Ding, Y. Zhao, C. Wang, H. Purnawali, C. Tang, *Mater. Des.* 2012, 33, 577–640.

167 G. J. Berg, M. K. McBride, C. Wang, C. N. Bowman, *Polymer* 2014, 55, 5849–5872.

168 M. D. Hager, S. Bode, C. Weber, U. S. Schubert, *Prog. Polym. Sci.* 2015, 49–50, 3–33.

169 I. V. W. Small, P. Singhal, T. S. Wilson, D. J. Maitland, *J. Mater. Chem.* 2010, 20, 3356–3366.

170 A. Lendlein, O. E. C. Gould, *Nat. Rev. Mater.* 2019, 4, 116–133.

171 M. Behl, A. Lendlein, *J. Mater. Chem.* 2010, 20, 3335–3345.

172 J.-M. Raquez, S. Vanderstappen, F. Meyer, P. Verge, M. Alexandre, J.-M. Thomassin, C. Jérôme, P. Dubois, *Chem. Eur. J.* 2011, 17, 10135–10143.

173 C. Zeng, H. Seino, J. Ren, K. Hatanaka, N. Yoshie, *Macromolecules* 2013, 46, 1794–1802.

174 H. Jiang, S. Kelch, A. Lendlein, *Adv. Mater.* 2006, 18, 1471–1475.

175 A. Lendlein, H. Jiang, O. Jünger, R. Langer, *Nature* 2005, 434, 879–882.

176 F. Shohraty, J. Hniopek, J. Meurer, S. Zechel, M. Schmitt, J. Popp, M. D. Hager, *Macromol. Mater. Eng.* 2023, 308, 2300002.

177 S. Chen, H. Yuan, S. Chen, H. Yang, Z. Ge, H. Zhuo, J. Liu, *J. Mater. Chem. A* 2014, 2, 10169–10181.

178 T. Ware, K. Hearon, A. Lonnecker, K. L. Wooley, D. J. Maitland, W. Voit, *Macromolecules* 2012, 45, 1062–1069.

179 M. Guo, L. M. Pitet, H. M. Wyss, M. Vos, P. Y. W. Dankers, E. W. Meijer, *J. Am. Chem. Soc.* 2014, 136, 6969–6977.

180 J. R. Kumpfer, S. J. Rowan, *J. Am. Chem. Soc.* 2011, 133, 12866–12874.

181 G. L. Fiore, S. J. Rowan, C. Weder, *Chem. Soc. Rev.* 2013, 42, 7278–7288.

182 L. Yang, G. Zhang, N. Zheng, Q. Zhao, T. Xie, *Angew. Chem. Int. Ed.* 2017, 56, 12599–12602.

183 J. Meurer, J. Hniopek, T. Bätz, S. Zechel, M. Enke, J. Vitz, M. Schmitt, J. Popp, M. D. Hager, U. S. Schubert, *Adv. Mater.* 2021, 33, 2006655.

184 J. Meurer, T. Bätz, J. Hniopek, C. Bernt, S. Zechel, M. Schmitt, J. Popp, M. D. Hager, U. S. Schubert, *J. Mater. Chem. A* 2022, 10, 25106–25117.

185 J. Meurer, J. Hniopek, S. Zechel, M. Enke, J. Vitz, M. Schmitt, J. Popp, M. D. Hager, U. S. Schubert, *Polymers* 1889, 2019, 11.

186 J. Meurer, J. A. Rodriguez Agudo, S. Zechel, M. D. Hager, U. S. Schubert, *Macromol. Chem. Phys.* 2021, 222, 2000462.

187 J. Meurer, T. Bätz, J. Hniopek, M. Jäger, S. Zechel, M. Schmitt, J. Popp, M. D. Hager, U. S. Schubert, *Polymers* 1833, 2022, 14.

188 M. Klein, M. F. Agyemang, O. Müschke, M. Schmitt, J. Popp, M. D. Hager, S. Zechel, U. S. Schubert, *Virtual Phys. Prototyp.* 2025, 20, e2499470.

189 M. Klein, P. Fesser, S. Zechel, M. D. Hager, U. S. Schubert, *Chem. Eur. J.* 2025, 31, e202404267.

190 S. Swetha, T. J. Sahiti, G. S. Priya, K. Harshitha, A. Anil, *Interactions* 2024, 245, 178.

191 R. Porta, *FEBS Open Bio* 2021, 11, 948–953.

192 Y. Zhu, C. Romain, C. K. Williams, *Nature* 2016, 540, 354–362.

193 R. R. A. Silva, C. S. Marques, T. R. Arruda, S. C. Teixeira, T. V. de Oliveira, *Macromol* 2023, 3, 371–399.

194 H. Jung, G. Shin, H. Kwak, L. T. Hao, J. Jegal, H. J. Kim, H. Jeon, J. Park, D. X. Oh, *Chemosphere* 2023, 320, 138089.

195 M. Wei, Y. Gao, X. Li, M. J. Serpe, *Polym. Chem.* 2017, 8, 127–143.

196 S. Wang, M. W. Urban, *Nat. Rev. Mater.* 2020, 5, 562–583.

197 T. Dayyoub, A. V. Maksimkin, O. V. Filippova, V. V. Tcherdyntsev, D. V. Telyshev, *Polymers* 2022, 14, 3511.

198 A. Lendlein, S. Kelch, *Angew. Chem. Int. Ed.* 2002, 41, 2034–2057.

199 D. Rokaya, H. E. Skallevold, V. Srimaneepong, A. Marya, P. K. Shah, Z. Khurshid, M. S. Zafar, J. Sapkota, *J. Compos. Sci.* 2023, 7, 24.

200 Y. Liu, H. Du, L. Liu, J. Leng, *Smart Mater. Struct.* 2014, 23, 023001.

201 H. Gao, J. Li, F. Zhang, Y. Liu, J. Leng, *Mater. Horiz.* 2019, 6, 931–944.

10

Metallo-Supramolecular Nanosized Constructs Based on Metal–Terpyridine Complexes

10.1 Introduction

In all the previous chapters of this book, discrete (mainly mononuclear) complexes as well as 1D/2D coordination polymers have been the focus of interest. However, the combination of 2,2′:6′,2″-terpyridine (tpy) ligands with (mostly) transition-metal ions has also been utilized to create nanosized objects with highly sophisticated 2D/3D architectures. The preparation of such defined species *via* metallo-supramolecular self-assembly of molecular components will only occur, if the formation of ill-defined assemblies is avoided and the "order from chaos" pathway is preferred [1, 2]. In order to reach this goal, the coordination-driven self-assembly (CDSA), as a bottom-up approach, is commonly used. Thereby, oligotopic ligands and transition-metal ions are combined to afford "supramolecules" with precise control over size and shape. The ligands, which possess defined molecular structures, as well as the metal ions, which prefer defined coordination geometries, dictate the supramolecular structure, which can often be predicted. However, this approach will only be productive when taking the directionality and reversibility of the underlying metal-to-ligand coordinative bonds into account [3]. Within the last three decades, this concept has been exploited by numerous research groups: Lehn *et al.* [4, 5], Sauvage *et al.* [6], Stoddart *et al.* [7–9], Stang *et al.* [10, 11], Fujita *et al.* [12, 13], Newkome *et al.* [14, 15], Schmittel *et al.* [16, 17], Raymond *et al.* [18], as well as others [19–28]. The plethora of metallo-supramolecular constructs have been summarized in various recent reviews and the reader is referred to these articles for an in-depth reading on this topic. The range of potential applications is similarly broad and the use of metallo-supramolecular constructs in the context of chemical topology [29–31], molecular machines [32, 33], molecular magnets [34, 35], molecular weaving [36–39], host-guest chemistry (*i.e.*, molecular recognition and separation) [40–42], supramolecular catalysis [18, 23, 43, 44], luminescent materials [45], optical bio- or chemosensors [45–48], biomedicine (*i.e.*, drug delivery and theragnostic) [49, 50] has *inter alia* been discussed.

The assemblies contain metal complexes, as the connecting points. Among the wide range of potential ligands, the tpy system is of particular virtue due to its ability to form pseudo-octahedral *bis*-complexes with most of the (divalent) transition-metal ions (see Chapter 1). This coordination behavior combines

Terpyridines: Multifaceted Ligands for Diverse Applications, First Edition.
Ulrich S. Schubert, Andreas Winter, and George R. Newkome.

high directionality with the absence of stereoisomerism, which occurs, *e.g.*, in *tris*-bidentate complexes of unsymmetrical bpy ligands (bpy: 2,2′-bipyridine). Moreover, $[M(tpy)_2]^{2+}$ complexes feature subtle differences in their kinetic stability in solution, which ranges from highly labile (*e.g.*, Zn^{II} and Cd^{II}) to inert (*e.g.*, Ru^{II} or Os^{II}). One has to take this behavior into account when aiming for a defined molecular architecture. As exemplarily shown by Wesdemiotis, Newkome *et al.* at the example of the divalent group-8 ions, control over the self-assembly process is essentially lost when kinetically stable complexes are irreversibly formed, whereas the anticipated structure, as the equilibrium species, was obtained selectively only in the case of kinetically less stable complexes (*i.e.*, Fe^{II} ions) [51]. Thus, the thermodynamically stable species will preferentially be obtained if kinetic trapping of undesired species is disabled and metallo-supramolecular repair is enabled due to the reversibility of complexation. This concept has been widely employed, *e.g.*, to assemble discrete 2D/3D constructs [52–56], linear polymers [57–60], coordination nanosheets (CONASHs) [61, 62] and metal–organic frameworks [63].

Less commonly, the combination between tpy and bidentate (*e.g.*, bpy or phen) or monodentate ligands (*e.g.*, pyridine) has also been used to assemble structurally defined nanostructures [52]. These assemblies contained non-octahedral complexes of, *e.g.*, Cu^{I}, Cu^{II}, Pt^{II}, Pd^{II}, or Re^{I} ions. In particular, Schmittel *et al.* developed the so-called HETTAP concept, in which the coordination of each one tpy and sterically hindered phen ligand to Zn^{II} ions selectively afforded heteroleptic penta-coordinate complexes (HETTAP: heteroleptic terpyridine and phenanthroline complexation) [17]. Likewise, Lehn *et al.* capitalized on the preference of Cu^{II} ions to form penta-coordinate complexes of trigonal-bipyramidal geometry; these authors assembled various nanostructures based on heteroleptic $[Cu(tpy)(bpy)]^{2+}$ complexes [64, 65]. The assemblies, which were formed *via* these strategies, were already summarized in the preceding edition of this book and reference is given to the corresponding chapter there for further reading [66].

10.2 Octahedral Tpy Complexes – From Metallomacrocycles to 2D Fractal Architectures and Beyond

In 1999, Newkome *et al.* reported the synthesis of a metallo-supramolecular hexagon, which was assembled from a rigid *bis*-tpy ligand and Ru^{II} ions (Figure 10.1a) [68]. Due to the ligand's 120° angle, the hexagonal structure represented the favored geometric motif and the shape-persistency of the assembly was ensured by the inert nature of the $[Ru(tpy)_2]^{2+}$ moieties. In subsequent work, numerous variants of this metallomacrocycle were prepared by varying the structure of the rigid *bis*-tpy ligand and/or the transition-metal ion [14]. Depending on the ligand/metal combination, not only hexameric species were formed – also dimers, trimers, tetramers, or pentamers were identified. The success of the self-assembly strategy is mainly due to the pre-organized ligand structure, which facilitates the formation of discrete species, which are intrinsically favored for entropic reasons,

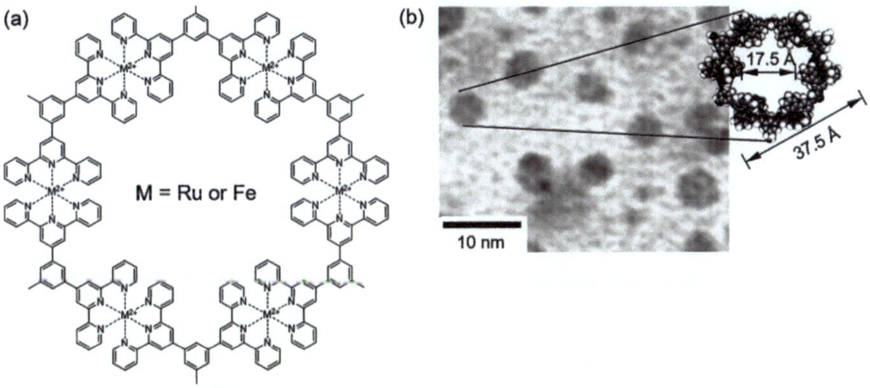

Figure 10.1 (a) Schematic representation of the hexagonal macrocycles. (b) Representative high-resolution TEM image of individual Fe^{II}-containing hexagons along with a computer-generated model for comparison [67] / with permission of John Wiley & Sons.

at the cost of ill-defined assemblies, such as linear oligomers or polymers [69]. The latter issue, *i.e.*, the competition between metallo-supramolecular polymerization and macrocyclization when using flexible *bis*-tpy ligands, was thoroughly analyzed by Schubert *et al.* [70, 71].

Wang *et al.* also paid attention to the competition between metallo-supramolecular polymerization and macrocyclization [72]. These authors elaborated that the self-assembly of a pre-organized *bis*-tpy ligand with Fe^{II} ions at ambient temperature afforded the metallopolymer, which represented the kinetic product (Figure 10.2), whereas, at elevated temperatures, the former was converted into a hexameric macrocycle, as the thermodynamically favored product. Likewise, the dynamics of the metallo-supramolecular self-assembly process were also probed by Li *et al.* [73]. These authors confirmed Wang's observation and further demonstrated that the exchange of coordinated metal ions might occur. Following the order of kinetic stability, a Zn^{II}-containing macrocycle was converted into an Fe^{II}-containing one without affecting the overall size and shape. These examples highlight the possibility of interconverting metallo-supramolecular constructs *via* certain parameters, *e.g.*, temperature. Furthermore, the structures of such assemblies might also be concentration-dependent. This not only holds true for metallomacrocycles [74], but also for 3D architectures, where the reversible

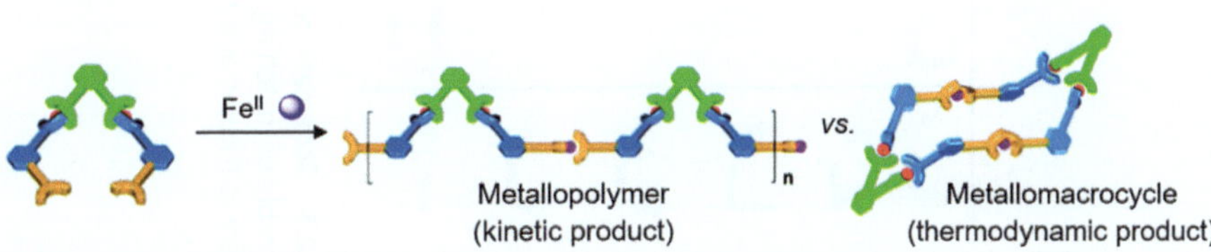

Figure 10.2 Schematic representation of the temperature-dependent self-assembly of a pre-organized *bis*-tpy ligand [72] / with permission of Royal Society of Chemistry.

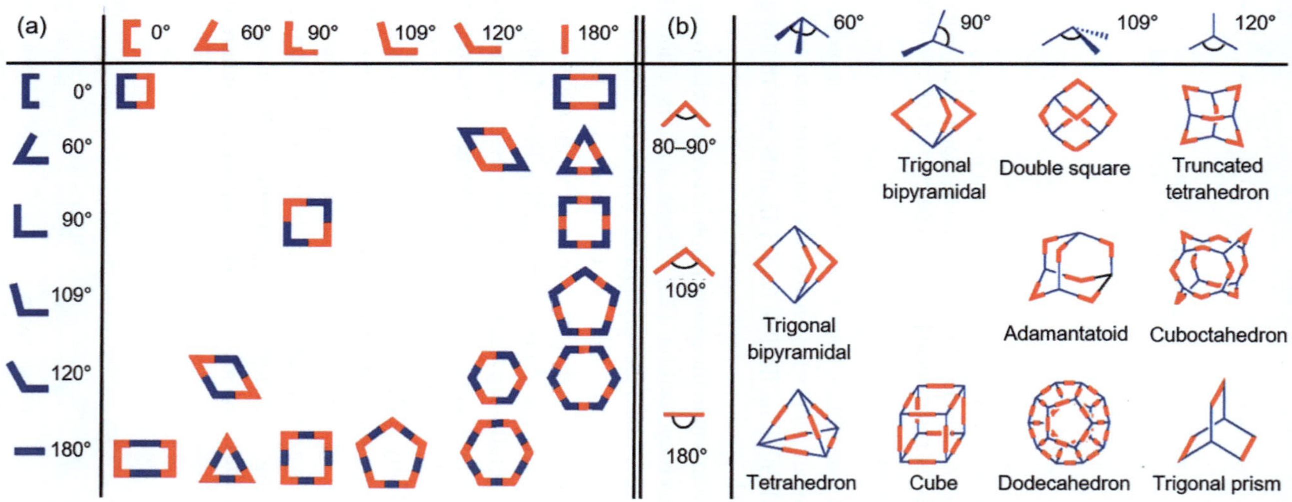

Figure 10.3 Schematic representation of the various combinations of polygons (a) and polyhedral (b), which result from the metallo-supramolecular self-assembly of matching components with predefined geometries. In panel (a), both components represent ditopic ligands, whereas, in panel (b) one component is tritopic and the other ditopic [10] / with permission of American Chemical Society.

interconversion of nanostructures has been observed for various examples (*vide infra*) [52].

Newkome *et al.* emphasized the importance of structural matching and stoichiometric control for obtaining metallo-supramolecular assemblies in a predictable and controlled fashion [10]. The molecular building blocks, which are linked by metal-ion complexation under a precise stoichiometric ratio, must be structurally rigid and offer appropriate angles between their binding sites. The proper combination of organic ligands, whose angular orientation might range from 0° to 180°, provides convex polygons (in the case of ditopic ligands, such as *bis*-tpy systems). The available polygons (or macrocycles), which can be formed by this strategy are displayed in Figure 10.3a (for clarity, only the geometrically most favored structures are shown). However, in practice, one will often deal with mixtures of macrocycles, which are simultaneously formed under the reaction conditions (*vide supra*) [10, 66]. The rings are typically planar, due to their rigid structures; however, steric constraints might result in a deviation from planarity and metallomacrocycles with folded conformations have also been reported [75–77].

The same considerations regarding the choice of components and their stoichiometry also apply when going from 2D to 3D assemblies. However, one has to use proper combinations of angular and linear components, which contain more than two (tpy) binding sites. In this respect, the metallo-supramolecular self-assembly of a tritopic component with a ditopic one represents the most common strategy (Figure 10.3b). The 3D constructs, which have been assembled *inter alia* by this approach, are dealt with in the subsequent Section 10.3.

The characterization of the early-days metallomacrocycles was mainly limited to NMR spectroscopy, conventional electrospray-ionization mass spectrometry (ESI MS) and, if suitable single crystals could be grown, X-ray diffraction (XRD) analysis. Moreover, high-resolution microscopy imaging [*e.g.*, transmission-electron microscopy (TEM)] can be utilized to visualize such assemblies; by doing so, their size and geometry can be determined as well as correlated to the corresponding computer-generated data (Figure 10.1b) [14]. A significant breakthrough in the characterization of metallo-supramolecular assemblies was achieved when Newkome, Wesdemiotis *et al.* combined ESI with travelling-wave ion-mobility (TWIM) MS [78]. This approach relies on the soft and nondestructive ionization of an assembly in the ESI chamber and the subsequent separation of (supramolecular) ions by charge and collision cross-section (CCS), which is – simply speaking – a function of size and shape, in the ion-mobility chamber. The resulting gas-phase separated ions can then be analyzed regarding their m/z ratios and isotopic patterns. This method allows one to deconvolute overlapping ions of the same m/z value according to their size/shape and net charge. For example, cyclic and linear assemblies or macrocycles of different sizes can be readily differentiated [79, 80]. Moreover, the coupling to energy-resolved (gradient) tandem mass spectrometry (gMS^2) enables one to study the supramolecular ions further regarding their gas-phase stability and/or fragmentation pathways. Today, this sophisticated technique is considered the state-of-the-art method for analyzing such metallo-supramolecular assemblies by mass-spectrometric means. Not only this, multidimensional MS can even be utilized to develop metallo-supramolecular chemistry further, *e.g.*, by revealing (unexpected) self-assembly modes, identifying isomers or conformers, distinguishing between isobaric ions and mapping the outcome of CDSA processes [81].

As the next level of complexity, Newkome *et al.* targeted the metallo-supramolecular self-assembly of fractals, specifically Sierpiński *n*-gons. These systems are named after the Polish mathematician W.F. Sierpiński, whose research was devoted to the theories of sets, numbers, functions, and topology (Figure 10.4a) [85], whereas the term "fractal" was coined by the French-American mathematician B.B. Mandelbrot to describe geometric shapes, which are self-similar across different scales or generations [86]. As the first example, the Sierpiński hexagonal gasket, which essentially represented the hexaflake motif but was missing the central hexagon, was reported in 2006 (Figure 10.4b) [83]. This gasket was assembled in a three-step procedure from a combination of V_2- and Y-type ligands (R = Me; Figure 10.5). The outer, smaller hexagons were formed first *via* the semi-self-assembly of these (in a 4:2 ratio) with Ru^{II} ions; subsequently, the self-assembly of the small hexagons with Fe^{II} ions gave the large hexagon in *ca.* 35% yield. The entire assembly, whose molar mass was *ca.* 38,700 g mol^{-1}, contained 36 Ru^{II} and 6 Fe^{II} ions. This work represents one elegant example where the stepwise self-assembly methodology was superior to one which is directed by the ligand geometry [55].

The stoichiometry-controlled self-assembly of the C6 component and a W-type derivative (R = OMe) with either Zn^{II} or Cd^{II} ions provided metallo-supramolecular assemblies, which were reminiscent of a six-spoked wheel (Figure 10.6a) [87]. The isolated yield was remarkably high (*i.e.*, up to 94%), presumably due to the efficient interplay of geometric and thermodynamic control. Thereby, the C6 component not only acted as a structure-controlling template during the self-assembly process but also represented the vertex for the assembly's triangular subunits. The spoked-wheel motif can also be considered as an arrangement of six equilateral triangles, which are joined facially around one center [52]. Taking this topology into account, one might deduce three different subsets of two of these triangles. The 1,2-, 1,3-, and 1,4-subsets have been referred to as centrally fused rhomboid,

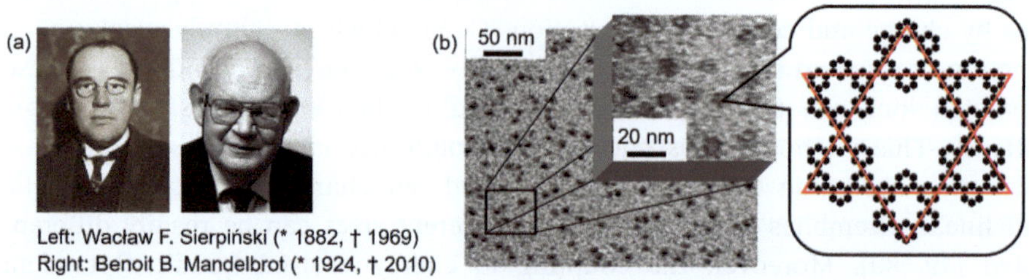

Figure 10.4 (a) Photographs of Wacław F. Sierpiński and Benoit B. Mandelbrot [82] / Wikimedia Commons / Public Domain; Rama / Wikimedia Commons / CC BY-SA 2.0 FR. (b) Representative TEM images of the Sierpiński hexagonal gasket [83, 84] / with permission of American Association for the Advancement of Science – AAAS.

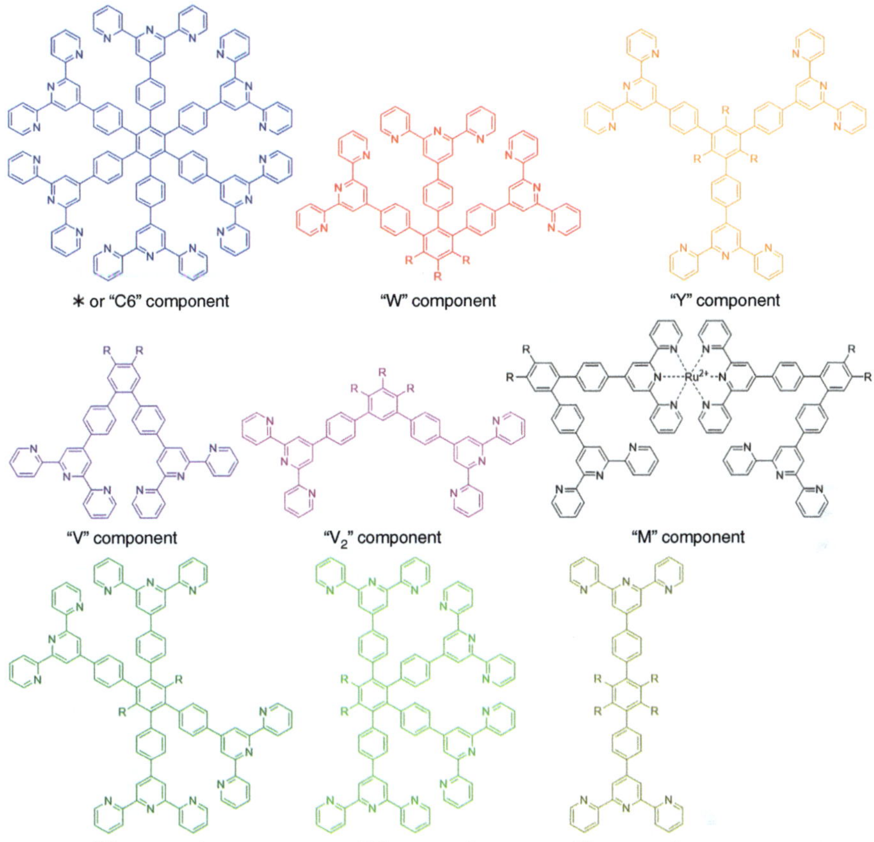

Figure 10.5 Schematic representation of molecular building blocks, which are commonly used as ligand monomers, for the self-assembly of shape-persistent nanosized constructs.

butterfly, and bowtie motif, respectively (Figure 10.6b). The former was prepared by a one-pot self-assembly of the aforementioned W-type building block and a complementary V-type component (R = OBn) [88]. Both Zn^{II} as well as Cd^{II} ions quantitatively afforded the centrally fused rhomboid assembly. The analogous "open" rhomboids were obtained when the W-type component was replaced by a V_2-type one (R = OMe). However, in this case, a triangular metallomacrocycle was formed, as the by-product, since the self-assembly of the V_2-type component and Zn^{II} or Cd^{II} ions could not fully be suppressed, whereas the butterfly- and bowtie-like structures were synthesized *via* a stepwise method [89]. First, an M-type component was formed by linking two V-type monomers *via* Ru^{II} ions (R = COOMe). The metallo-supramolecular "butterfly" was assembled from this M-type component, a K-type derivative (R = OC_6H_{13}) and Zn^{II} ions in a 2.1:4 ratio. The "bowtie" assembly was obtained if the K-type component was replaced by the corresponding X-type one. Essentially representation constitutional isomers, the "butterfly" and "bowtie" structures, could not be distinguished by 1D ESI MS;

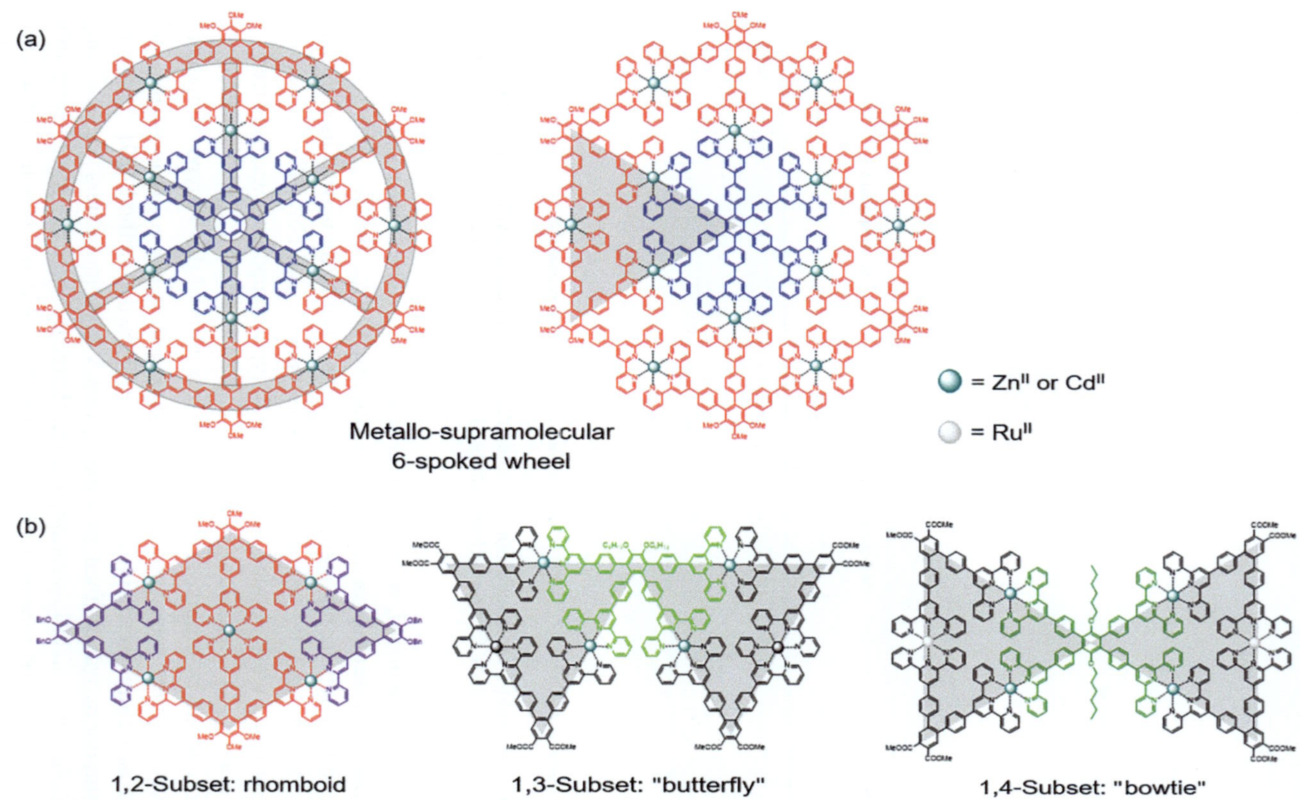

Figure 10.6 (a) Schematic representation of the metallo-supramolecular "spoked wheel" showing the analogy to a "real" 6-spoked wheel (left) and the equilateral trigonal subunit (right). (b) Schematic representation of the three possible subsets of the 6-spoked wheel containing two such equilateral triangles.

Figure 10.7 Evolution of the Sierpiński triangle [90] / with permission of IOP Publishing.

whereas, ESI-TWIM MS allowed for a distinction due to the different CCSs, which were mostly expressed for the molecules in the low-charged states.

The Sierpiński triangle (or gasket or sieve) represents one of the most well-known self-similar sets, *i.e.*, its mathematically generated pattern is reproducible at any magnification or reduction. As shown in Figure 10.7, this specific fractal has an overall shape of an equilateral triangle, which is subdivided recursively into smaller equilateral triangles [91]. Newkome *et al.* aimed for the translation of the fractal theory into metallo-supramolecular chemistry and the synthesis of "metallofractals," which were reminiscent of the Sierpiński triangle, was performed. This strategy has been referred to as "mathematical mimicry" [92]. In one elegant approach, K- and V-type components (R = OMe, for both) were co-assembled with Cd^{II} ions (1:1:3 ratio) to afford a metallofractal, which corresponded to the 2nd-generation Sierpiński triangle (Figure 10.8a). The structure of the assembly was unequivocally confirmed by ESI-TWM MS as well as microscopy imaging. Both individual triangles and dimers, which resulted from the stacking of metallofractal molecules, were observed (Figure 10.8a).

Unlike the former assembly, in which the four triangles were connected facially, a fourfold vertex connectivity was realized for the related [4]triangulane (Figure 10.9a) [93]. In this assembly, three X-type monomers were used to establish the connection points ($R = OC_{16}H_{33}$). The stepwise synthesis involved the initial formation of a Ru^{II}-containing M-type component from two V-type building blocks ($R = OC_{22}H_{45}$). Next, the same strategy, as already explained in the context of the hexagonal gasket

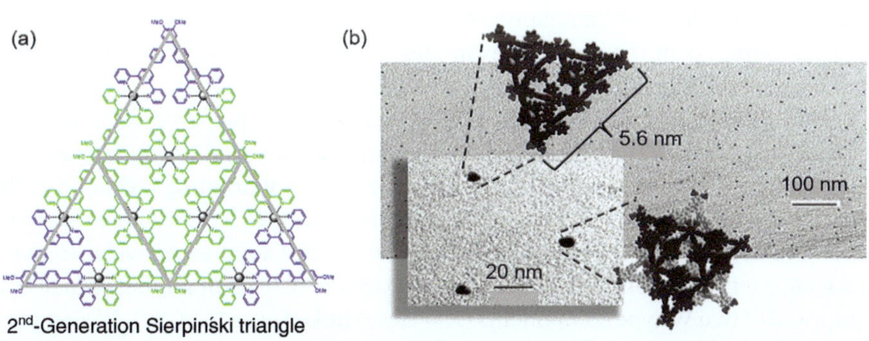

Figure 10.8 (a) Schematic representation of a metallofractal, which corresponds to the 2nd-generation Sierpiński triangle. (b) Representative TEM images of the metallofractal showing individual molecules as well as aggregated ones; the corresponding computer-generated models of the species are also shown [92] / with permission of John Wiley & Sons.

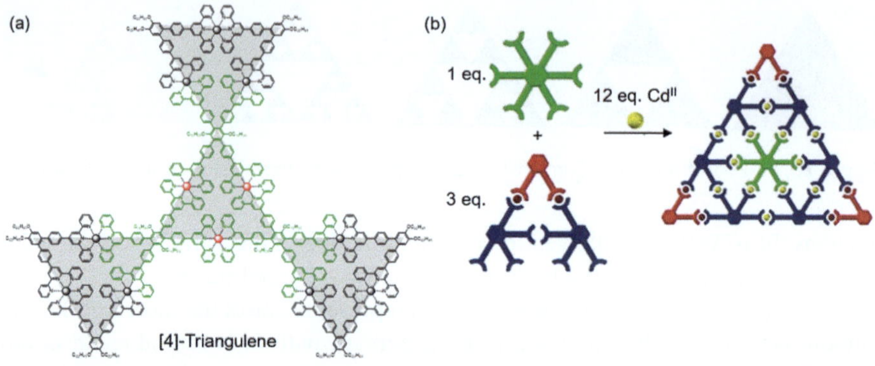

Figure 10.9 (a) Schematic representation of a metallo-supramolecular [4]traingulane [93]. (b) Schematic representation of the formation of a metallo-supramolecular triangle [94] / with permission of John Wiley & Sons.

(*vide supra*), was used to form the outer triangles, which were finally assembled with Zn^{II} ions to provide the desired [4]triangulane in quantitative yield (to be precise, the branched structure should actually represent a [3]rotane [95]). In order to retain solubility in polar-aprotic solvents, solubilizing side chains had to be introduced [93]. Similar to the relationship between the rhomboid and the 2nd-generation Sierpiński triangle, the [4]triangulane can be related to the "bowtie" motif: The former employs facial connectivity, whereas the subunits of the latter are connected *via* their vertexes.

One further triangle-based construct, strictly speaking, a non-Sierpiński-type system, was reported by Li, Wang *et al.* (Figure 10.9b) [94]. This compound, which represents an intermediate between the 2nd- and 3rd-generation Sierpiński triangles, was also prepared *via* a stepwise self-assembly route. The corner monomers (*i.e.*, 60° V-type components with R = OMe) were linked to two K-type components (R = OC_6H_{13}) *via* Ru^{II} ions. The thusly obtained tetratopic building block provided, in combination with the structure-directing C6 monomer, the metallo-supramolecular triangle upon self-assembly under stoichiometric control with Cd^{II} ions. As in the previous examples, the final self-assembly step occurred with both very high product selectivity and yield.

Within the same context, a giant hollow hexagon was prepared *via* a two-step self-assembly strategy [96]. The hexagon was formed by in-total 12 equilateral triangles, which were linked *via* the edges and vertices in an alternating fashion, *i.e.*, the combination of the rhomboid and butterfly motives spanned a ring, whose outer and inner rims represented a hexagon and a regular 6-point star, respectively. To reach this goal, a tetratopic building block was prepared first. This monomer, which was comprised of two W-type components (R = OMe) linked *via* a $[Ru(tpy)_2]^{2+}$ complex, was combined with a K-type component (R = $OC_{12}H_{25}$) in a 1:1 ratio. In the presence of 4 eq. of Fe^{II} ions, the hollow hexagon, whose diameter was *ca.* 11 nm, was formed with very high efficiency (Figure 10.10a). In subsequent work, these same authors also targeted the synthesis of a metallo-supramolecular "Star of David" (hexagram) [97]. However, the initial attempt, which utilized the pre-organized $[Ru(V)(X)]^{2+}$

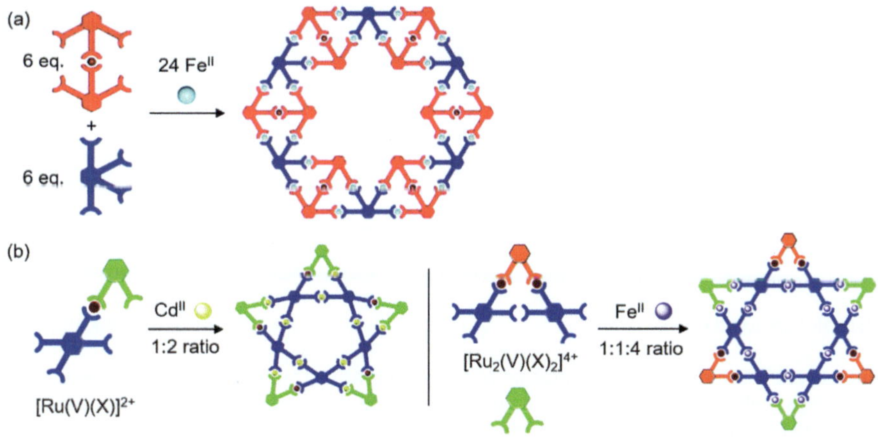

Figure 10.10 Schematic representation of the formation of (a) a giant hollow hexagon as well as (b) a pentagram and hexagram *via* the stepwise metallo-supramolecular self-assembly of molecular components [52] / with permission of Royal Society of Chemistry.

monomer, selectively afforded the pentagram upon treatment with Fe^{II} or Cd^{II} ions instead (Figure 10.10b). Thus, the synthesis strategy had to be revised: The dinuclear $[Ru_2(V)(X)_2]^{4+}$ building block provided the desired directivity and, thus, the hexagram was formed by the co-assembly of the former and the V-type component in the presence of Fe^{II} or Cd^{II} ions, as corroborated by ESI-TWIM MS and microscopy imaging (V: R = OMe, X: R = $OC_{12}H_{23}$; Figure 10.10b).

The snowflake motif, specifically the one based on rhombic subunits, has also attracted interest, as a synthesis goal, in metallo-supramolecular chemistry. For this purpose, Wang, Li *et al.* utilized three molecular components, *i.e.*, the C6 and V-type components (R = OMe), which represented the core and the corners of the supramolecular structures, respectively, as well as various Y-type components [98, 99]. Thereby, the dimensions of the "snowflakes" were determined by the arm length of the latter building block. In all cases, the V- and Y-type components were combined first with Ru^{II} ions to provide mononuclear $[Ru(V)(Y)]^{2+}$ complexes, which were then assembled with the C6 component and Zn^{II} ions in the 6:1:12 ratio to give the respective nanosized "snowflakes." The preference in forming these sophisticated structures, as the thermodynamically most favored architectures, was due to the geometry-directing properties of C6 on the one hand and the reversibility of the Zn^{II}–tpy connection on the other. Moreover, the kinetic lability of the $[Zn(tpy)_2]^{2+}$ complexes enabled the dynamic exchange of ligands, when mixtures of differently sized "snowflakes" were analyzed by ^{1}H-NMR spectroscopy and ESI-TWIM MS (Figure 10.11). Hybrid snowflakes, whose sizes were in between those of the original ones, were evidenced by these measurements.

Li *et al.* presented the synthesis of metallo-supramolecular hexagon wreaths by the one-pot self-assembly of a tritopic or tetratopic building block with Zn^{II} ions under stoichiometric control (Figure 10.12) [100]. These assemblies were of interest from a

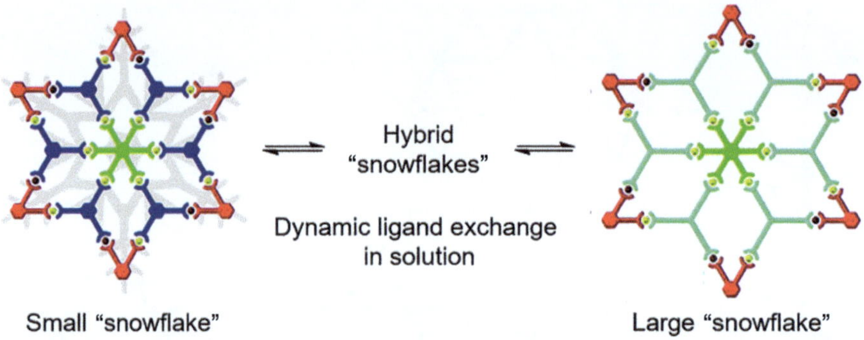

Figure 10.11 Schematic representation of two differently sized metallo-supramolecular "snowflakes," which exhibited dynamic ligand exchange if mixed in solution [98] / with permission of American Chemical Society.

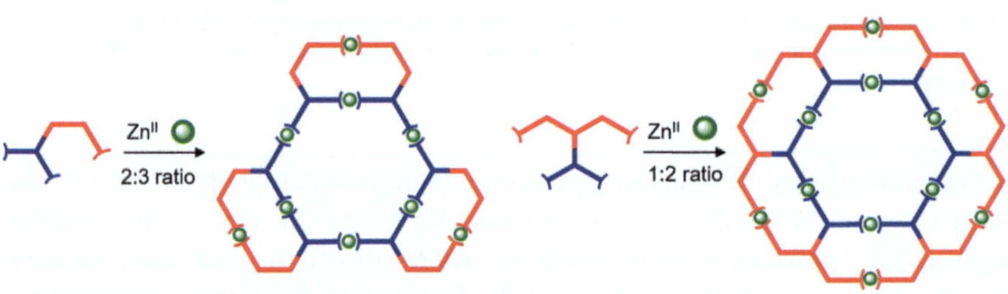

Figure 10.12 Schematic representation of the formation of metallo-supramolecular hexagon wreaths [100] / with permission of American Chemical Society.

topological point of view due to their self-similarity at different levels (*i.e.*, recursion of small hexagons around a central one). Remarkably, the formation of ill-defined assemblies and macrocycles of various sizes was suppressed due to the "density of coordination sites" (DoCS). It has been suggested that the assembly of discrete structures becomes favored if geometric constraints are high; this might be given if the high density of coordination sites is very high. Unlike simple metallomacrocycles, these wreaths featured a significantly higher rigidity, as evidenced by ESI-TWIM MS analysis and the calculated CCS values at charge states.

In subsequent work, Li *et al.* raised the question how dissymmetric ligands behave in metallo-supramolecular self-assembly, specifically in the formation of hexagon wreaths [101]. It was intended to mimic the self-assembly in nature, where often dissymmetric building blocks are assembled into advanced structures with remarkably high selectivity [102]. However, the use of asymmetric ligands has largely been avoided in metallo-supramolecular chemistry, due to the often-laborious synthesis of such systems and the limited control in self-assembly. Two tetratopic ligands were prepared and combined with Zn^{II} to afford the corresponding metallo-supramolecular wreaths [101]. Owing to the absence of steric constraints, the less rigid ligand could adopt various conformations and, thus, types of assemblies were formed, in which the four different coordination modes were

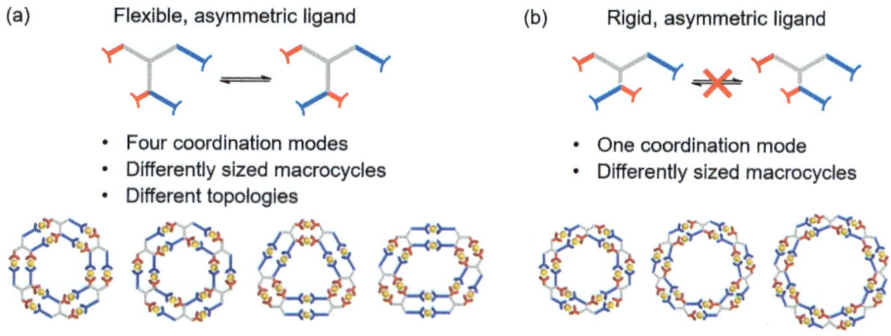

Figure 10.13 Schematic representation of the formation of metallo-supramolecular wreaths by the self-assembly of asymmetric tetratopic ligands with Zn^{II} ions [101] / with permission of American Chemical Society. (a) The flexible component enabled different conformers to be involved in coordination; (b) The rigid ligand did not allow this behavior due to steric constraints.

present (Figure 10.13a). The occurrence of these coordination modes could only be corroborated by STM imaging on the submolecule level, whereas steric constraints prevented the more compact ligand from adopting these conformations and, thus, only a single coordination mode was observed in the obtained wreaths of different sizes (STM: scanning tunneling microscopy; Figure 10.13b). These species, in turn, could readily be distinguished by ESI-TWMIN MS due to their different CCSs.

The DoCS concept was identified as a powerful tool in metallo-supramolecular chemistry due to its ability to provide excellent geometric directionality, in particular when dealing with self-assembly under thermodynamic control [55]. For example, a circle-in-circle system, *i.e.*, two concentric hexagons, was assembled from tetratopic monomers (Figure 10.14a). STM revealed that these constructs assembled further into ribbon-type nanostructures at the liquid–solid interface. In subsequent work, the ligand design was developed further and, thus, a series of nested hexagons, which were reminiscent of Kandinsky's circles, were assembled in one-step procedures (Figure 10.14b) [104]. Remarkably, in the case of the 4th-generation assembly, an octatopic building block was employed and the isolated yield was very high (*i.e.*, 82%), despite the *in-total* 24 $[Cd(tpy)_2]^{2+}$ complexes formed. The giant supramolecules afforded tubular nanostructures due to a hierarchical self-assembly process. The presence of Cd^{II} ions accounted for observed antimicrobial activity, *e.g.*, against the Gram-positive pathogen methicillin-resistant *Staphylococcus aureus*. This behavior is typical for complexes of the Cd^{II}–tpy family, for which numerous studies on their biological activity have already been provided within the last decades (see Chapter 4) [106, 107].

In order to develop the DoCS concept further and to reach the next level of structural complexity, Li *et al.* also prepared metallo-supramolecular "spiderwebs" *via* a three-component self-assembly approach (Figure 10.14c) [105]. Thereby, the aforementioned C6 component represented the assembly's core and the pentadentate building blocks, which eventually formed the meshes, were prepared by the *in situ* condensation reaction between a V_2-type and a W-type component. The 2nd-generation "spiderweb" motif was not only formed by the self-assembly of three

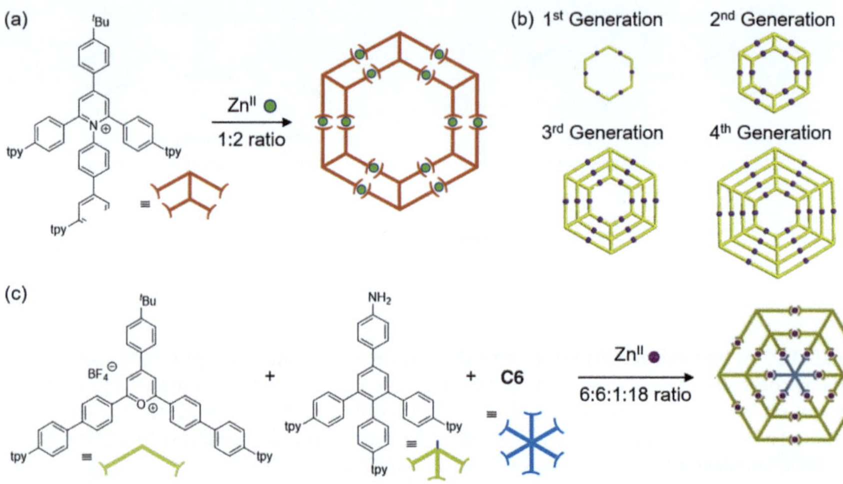

Figure 10.14 (a) Schematic representation of the one-step formation of a ring-in-ring assembly. Adapted from [103]. (b) Schematic representation of the various generation of hexagon-in-hexagon assemblies [104] / Springer Nature / CC BY 4.0. (c) Schematic representation of the self-assembly of a 2nd-generation spiderweb motif [105] / with permission of American Chemical Society.

components – the four-component self-assembly was also demonstrated. The essentially same motif was also obtained when the C6 component was replaced by two Y-type ones (R = H). In this case, two such molecules formed the core by $\pi\pi$-stacking interactions, in which the geometry-directing properties of C6 were mimicked. This example shows how the interplay of *in situ* synthesis and metallo-supramolecular self-sorting can be utilized to assemble advanced architectures in a one-pot reaction. The metallo-supramolecular "spiderwebs" were analyzed by ^{1}H NMR spectroscopy, ESI-TWIM MS and high-resolution microscopy imaging.

Very recently, Newkome, Wang, Chen *et al.* presented large metallohexagons, which possessed a coronoid-like structure [108]. Coronoids are macrocyclic conjugated hydrocarbons, which are comprised of circularly fused benzenoid rings; due to their geometrically defined cavities, they are generally considered model structures of porous graphene [109]. The edge length and the cavity size were modulated by the length of the spacers of the employed oligotopic tpy ligands (Figure 10.15a) [108]. The metallo-supramolecular assembly of matching building blocks, *i.e.*, a K-type component and a hexatopic, binuclear metalloligand, with Cd^{II} ions afforded the respective metallohexagons in very high yields (>95%; Figure 10.15b). This study clearly shows the scalability of metallo-supramolecular self-assembly since the same motif can be precisely formed in different sizes using appropriately designed ligands.

The synthesis of hexagon wreath has been mentioned before (Figure 10.12) [100]. In line with this study, the formation of symmetrical, honeycomb-like structures, in which a central metallo-hexagon was surrounded by six identical ones, was targeted. Firstly, Chan, Wang *et al.* followed a stepwise assembly strategy [110]. For this purpose, a trinuclear Ru^{II}-containing building block, which contained

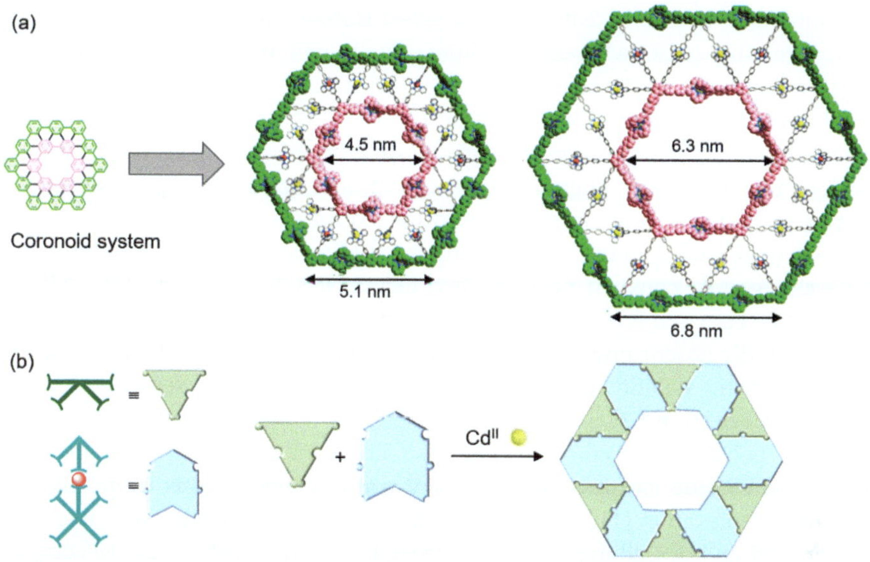

Figure 10.15 (a) Schematic representation of the coronoid-like metallo-supramolecular hexagons. (b) Schematic representation of the modular synthesis of these metallohexagons [108] / with permission of American Chemical Society.

four vacant tpy sites, was prepared; the subsequent self-assembly with Zn^{II} ions provided the target structure. The size of the honeycomb motif, as determined by TEM imaging, matched well with the calculated one (*ca.* 13.4 nm). Later, Li *et al.* employed tailor-made tetratopic tpy ligands for the same purpose [111]. These components gave the metallo-supramolecular honeycomb with a range of divalent transition-metal ions (*i.e.*, Mn^{II}, Fe^{II}, Co^{II}, Ni^{II}, Cu^{II}, Zn^{II}, Cd^{II}; Figure 10.16). From these, the assemblies containing Mn^{II}, Ni^{II}, and Cu^{II} ions represented the first examples, where these metal ions were used in the context of giant metallo-supramolecular constructs (at least in the case of tpy complexes). The gas-phase stability of the assemblies was analyzed by gMS^2, *i.e.*, the collision energies for complete dissociation of the tpy complexes were determined. The

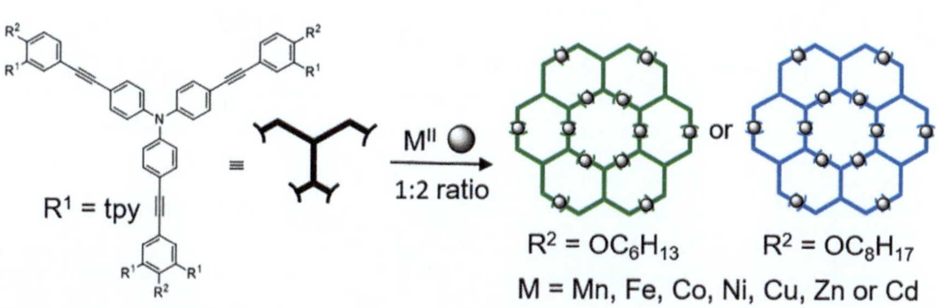

Figure 10.16 Schematic representation of the synthesis of metallo-supramolecular "honeycombs" *via* a one-pot self-assembly strategy [55] / with permission of John Wiley & Sons.

Ni^{II}-containing assembly featured the highest stability, followed by the Co^{II}- and Zn^{II}-containing ones, whereas the honeycomb, which relied on $[Mn(tpy)_2]^{2+}$ complexes, was the least stable. Moreover, the kinetics of ligand-exchange processes in solution were studied. For this purpose, two different honeycomb assemblies, which contained the same type of metal ion, were mixed and the ligand scrambling, as a function of time and temperature, was monitored by ^1H NMR spectroscopy. The slowest exchange at room temperature occurred for the Co^{II}-containing systems, whereas their Cd^{II}-containing counterparts rapidly exchanged. The difference in the exchange-rate constant for these two metals was as high as a factor of 10^{12}. Notably, the Fe^{II}- and Ni^{II}-based fractals were kinetically inert and ligand exchange was essentially absent under the experimental conditions used. Thus, the general trend in kinetic stability, as predicted by the Irving-Williams series was largely followed (see Section 1.3).

Similar polycyclic assemblies were also obtained *via* a stepwise procedure [112, 113]. For this purpose, π-extended V_2- and X-type ligands were linked by Ru^{II}-ion complexation [112]. The thusly obtained building blocks were assembled further with Zn^{II} or Cd^{II} ions to afford polycyclic constructs of increasing complexity, whereas the one-pot self-assembly of the molecular components did not afford these motifs; instead, oligomers of similar thermodynamic stability were obtained. Likewise, a series of fractals was synthesized from pre-organized Ru^{II}-containing subunits [113]. These precursors were prepared in multiple steps and enabled the subsequent formation of the respective fractals up to the 5th-generation assembly with very high selectivity (Figure 10.17a). These fractals – together with the aforementioned hexagonal wreath (Figure 10.12) – represent subsets of a regular hexagonal grid (Figure 10.17b). In the case of the larger representatives, *i.e.*, the **G4** and **G5** fractals, hierarchical self-assembly at the solid/liquid interface gave highly ordered nanofibers. These fractals also gave ordered 2D nanopatterns after slow deposition onto highly oriented pyrolytic graphite (HOPG) surfaces. These nanostructures were visualized by high-resolution TEM and STM imaging.

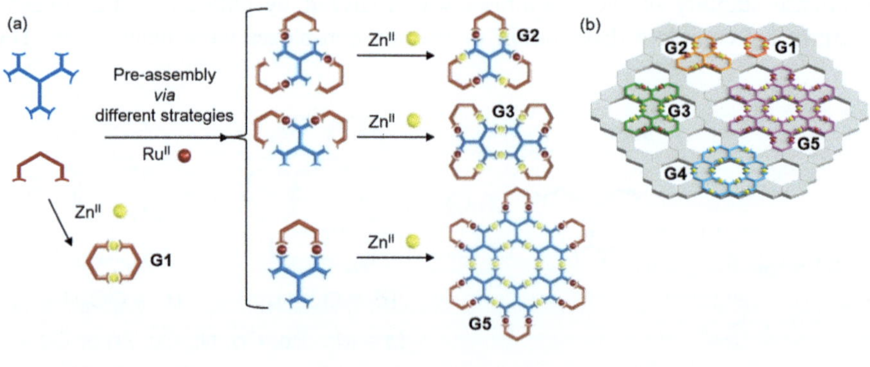

Figure 10.17 (a) Schematic representation of the self-assembly of pre-organized building blocks into fractal architectures upon self-assembly with Zn^{II} ions. (b) Schematic representation showing structural relationship between the fractals and a hexagonal grid [113] / with permission of American Chemical Society.

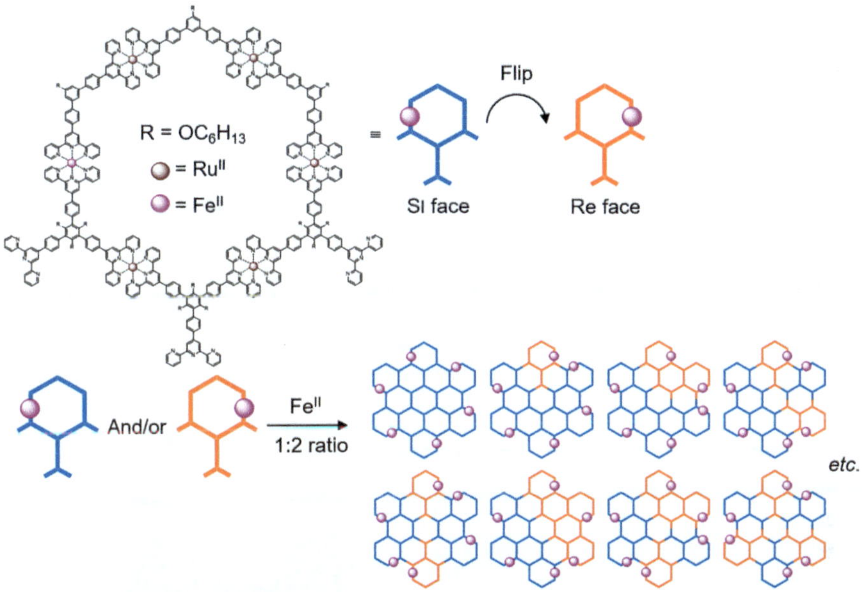

Figure 10.18 Schematic representation of the self-assembly of a hexagonal macroligand into a 2D hexagonal grid. Due to the two different possible orientations of the macroligand, different isomers were formed. Adapted from [114].

An even more extended assembly, which was based on fused hexagons, was reported by Y. Li, Zhang, Sessler, Hla, X. Li *et al.* [114]. The 20-nm-sized 2D hexagonal grid was assembled from a bimetallic macroligand, which was incorporated into the grid-type architecture *via* two different orientations (using the terminology of prochirality, these would correspond to the si and re faces, respectively; Figure 10.18). Thus, a multitude of constitutional isomers was formed in solution. Ultrahigh-vacuum, low-temperature STM imaging allowed these authors to analyze the assemblies on the submolecule level after deposition onto an Au substrate. The different isomers were distinguished due to the different HOMO–LUMO energy gaps of Fe^{II} and Ru^{II} ions (HOMO: highest occupied molecular orbital; LUMO: lowest unoccupied molecular orbital). In total, 14 different isomers were identified on the substrate by this specialized method. The thusly determined product distribution was in fair agreement with the one based on the theoretical probabilities for isomer formation.

One further approach to increase the complexity of metallo-supramolecular constructs was the evolution of rosette-like assemblies [115]. The rosettes up to the 3rd-generation representative were realized by the direct self-assembly of the respective oligotopic ligands with Cd^{II} – thereby making once more use of the DoCS concept. The **G1** species represented metallomacrocycles of various sizes (up to the hexamer), whereas the formation of the **G2** and **G3** assemblies proceeded with very high selectivity (Figure 10.19a). Besides the aspect of controlled metallo-supramolecular self-assembly into aesthetic architectures, their photophysical properties also need to be highlighted. The ligands contained a

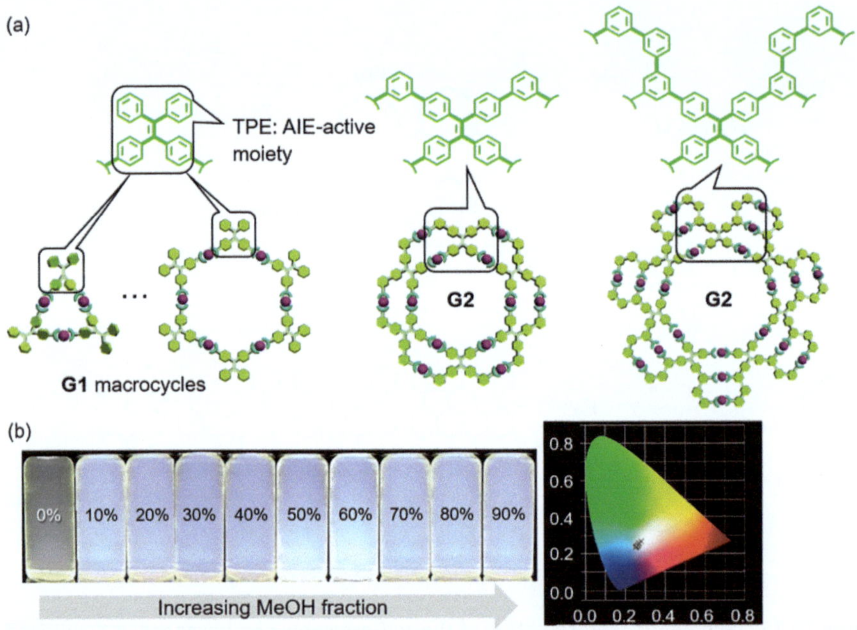

Figure 10.19 (a) Schematic representation of the TPE-containing, metallo-supramolecular rosette-like motifs. (b) Photographs of the solution of **G2** in various MeCN/MeOH solvent mixtures upon excitation at 365 nm. The corresponding CIE-1931 chromaticity diagram is also shown [115] / Springer Nature / CC BY 4.0.

tetraphenylethylene (TPE) moiety as the branching point. This chromophore is of particular interest due to its intense fluorescence, in particular, in the aggregated state or in a confined environment. Numerous covalent and supramolecular structures incorporating TPE moieties have been prepared since Tang's pioneering work on materials displaying aggregation-induced emission (AIE) [116, 117]. Owing to their extended π-conjugated structures, all three tpy-TPE ligands were highly emissive in dilute solution [115]. The luminescence properties were retained upon self-assembly with Cd^{II} ions. Thereby, the emission color depended on the rosette's size on the one hand and the solvent composition on the other. The **G2** assembly, for example, featured white-light emission over a wide range of good/poor solvent mixtures (Figure 10.19b).

Metallo-supramolecular "pinwheels," which differed in size and number of blades, were reported by Wang *et al.* (Figure 10.20) [118]. The largest "pinwheel," *i.e.*, the hexagram, contained the remarkably high count of 42 $[M(tpy)_2]^{2+}$ moieties. In order to prepare these unusual structures by metallo-supramolecular self-assembly, the respective building blocks were synthesized *via* multistep syntheses. These protocols involved the linking of molecular components by directional Ru^{II}-ion complexation on the one hand and cross-coupling reactions on the other. The latter strategy, referred to as "chemistry on the complex," is of particular virtue in the field of Ru^{II}–polypyridyl complexes, which are compatible with numerous organic reactions, due to their intrinsic kinetic and thermodynamic stability

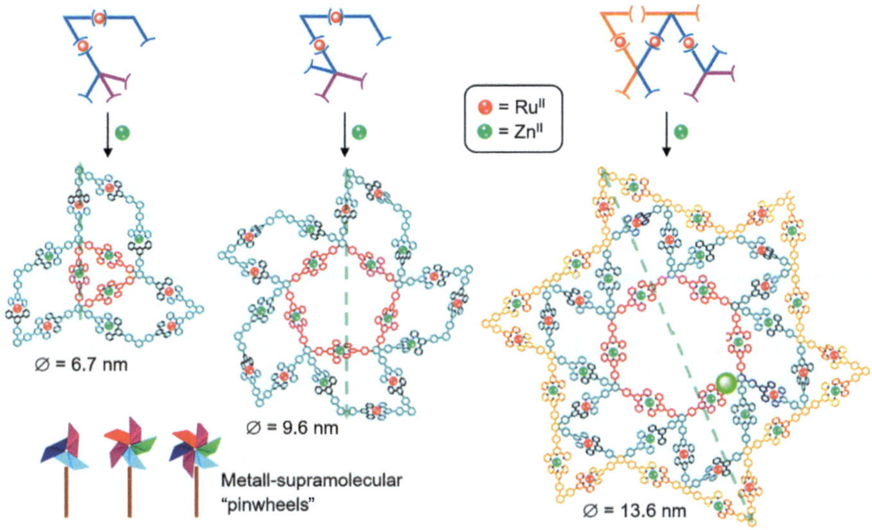

Figure 10.20 Schematic representation of the synthesis of metallo-supramolecular "pinwheels" from pre-organized metalloligands [118] / with permission of American Chemical Society.

[119]. The pinwheel-like structures were formed from these building blocks by adding Zn^{II} ions in the required stoichiometric amounts. Both the principles of geometry completion, driven by the pre-organized metalloligand structure, and DoCS accounted for the selective formation of the respective "pinwheels" in an almost quantitative fashion.

Finally, a trapezoidal architecture was reported by Wesdemiotis, Newkome *et al.* [120]. This special metallomacrocycle contained three different types of $[M(tpy)_2]^{2+}$ complexes in a predetermined pattern (M = Ru, Fe, and Zn). To reach this goal, a stepwise self-assembly approach was followed. First, the molecule's upper half was assembled from V-type and I-type components in a 2:1 ratio by directional Ru^{II}-ion complexation (for both, R = $OC_{16}H_{33}$). In a similar fashion, the lower part was assembled by the dimerization of two V_2-type molecules with Fe^{II} ions (R = OMe). Finally, the two fragments were linked by the complexation of the vacant tpy sites with Zn^{II} ions (Figure 10.21). This work demonstrates how different

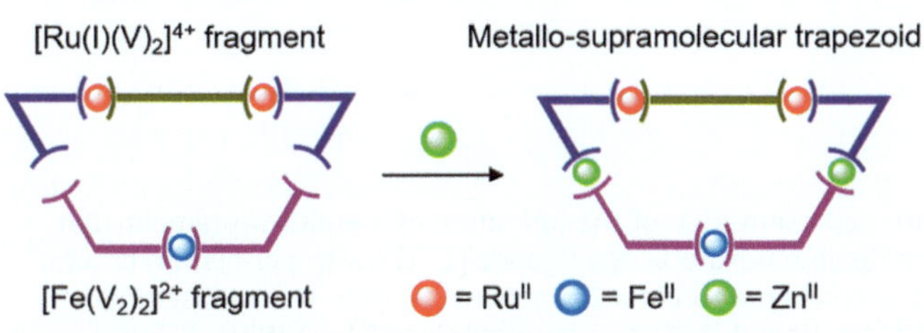

Figure 10.21 Schematic representation of the stepwise self-assembly of a trapezoidal metallomacrocycle. Adapted from [120].

transition-metal ions can be incorporated into metallomacrocyclic assemblies in a predictable fashion.

In all previous examples, the utilized tpy ligands were of rigid structure, *i.e.*, their backbones relied exclusively on sp^2 and sp-hybridized (mostly) *C*-atoms. The rigidity of these building blocks accounted for the observed selectivity and efficiency of the self-assembly in the presence of transition-metal ions, whereas flexible tpy ligands are more susceptible to form less defined architectures due to the loss of directionality of the spacers. However, Chan *et al.* demonstrated that geometrically well-defined structures, specifically ring-in-ring assemblies, can be obtained from oligotopic ligands with balanced rigidity and flexibility [121]. For this purpose, hexatopic ligands, referred to as **L1–L3**, were prepared by attaching two V_2-type components to a central V-type one *via* short alkyl chains. Depending on the length of the alkyl chain, assemblies of different topologies were obtained and eventually stabilized by them (Figure 10.22a). Independent of this, mechanistic

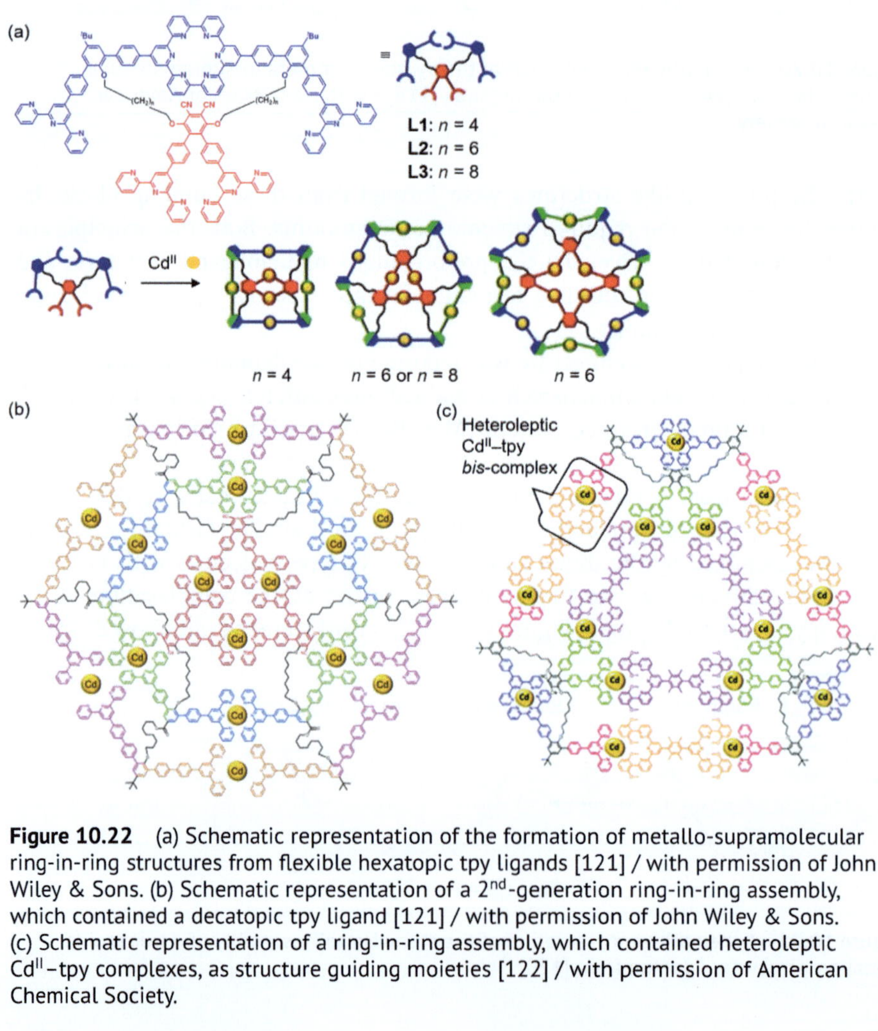

Figure 10.22 (a) Schematic representation of the formation of metallo-supramolecular ring-in-ring structures from flexible hexatopic tpy ligands [121] / with permission of John Wiley & Sons. (b) Schematic representation of a 2nd-generation ring-in-ring assembly, which contained a decatopic tpy ligand [121] / with permission of John Wiley & Sons. (c) Schematic representation of a ring-in-ring assembly, which contained heteroleptic Cd^{II}–tpy complexes, as structure guiding moieties [122] / with permission of American Chemical Society.

studies suggested that the intra-ligand complexation of a Cd^{II} ion represented the key step. The thusly rigidified ligands assembled into the ring-in-ring structures by cooperativity, whereas the alternative entropically templated pathway could be ruled out. The concept of ring-in-ring formation was also extended further to obtain the next-generation assembly. For this purpose, a decatopic tpy ligand was designed and self-assembled with Cd^{II} ions in a 1:5 ratio (Figure 10.22b). The giant assembly, which contained in total 15 Cd^{II} centers, was obtained with an isolated yield of 96%. In subsequent work, Chan *et al.* combined the hexatopic ligand **L2** with a complementary I-type component [122]. Due to steric constraints, the self-assembly of two such ligands was disabled and, thus, the formation of heteroleptic Cd^{II}–tpy *bis*-complexes, stabilized by *intra*-ligand ππ-stacking, was enabled. This concept of forming kinetically stable, heteroleptic Cd^{II}–tpy (and later Zn^{II}–tpy) complexes has been introduced by Chan *et al.* (see Section 1.4) and employed subsequently to assemble various types of metallo-supramolecular structures: diblock copolymers [123], 4- and 6-pointed stars [124, 125] as well as triangles [125].

Before turning attention to 3D metallo-supramolecular contracts, one further strategy to overcome one limitation when using benzenoid rings, as the central backbone unit, was reported recently by Wang, Zhang, Li *et al.* [126]. All the previous examples show that the benzene-based ligands facilitate the formation of triangles and hexagons, due to the provided angles of 60°, 120°, and 180°, whereas large metallomacrocycles (>6) are difficult to form, due to the absence of appropriately angled building blocks. When a metal node was integrated into the ligand's backbone, the angle between the tpy-containing arms was widened due to the complexation-induced decrease of the dihedral angle of the tpy pocket and, thus, the formation of heptagonal metallomacrocycles became thermodynamically favored (Figure 10.23). This concept was capitalized on to assemble ring-in-ring constructs, which corresponded to the aforementioned 2nd-and 3rd-generation concentric hexagons (see Figure 10.14a,b). For this purpose, the central tpy moiety of the pentatopic building block was blocked by converting it into a heteroleptic Ru^{II}–tpy *bis*-complex. The subsequent self-assembly with divalent transition-metal ions (*i.e.*, Cu^{II}, Zn^{II}, or Cd^{II}) gave the heptagon-in-heptagon structure with high selectivity and in high yield. The 3rd-generation assembly was assembled by a similar strategy. In this case, however, the central tpy moieties of the pentatopic ligand and a W-type component were linked *via* directed Ru^{II}-ion complexation (Figure 10.23). The structure, shape, and size of the metallo-supramolecular heptagons were unambiguously verified by 2D NMR spectroscopy [*i.e.*, diffusion-ordered spectroscopy (DOSY)], ESI-TWIM MS, and STM imaging.

As the elusive red thread, in all aforementioned examples, the metal–tpy complexes were part of the assemblies' edges and rims, but they were never positioned at the corners or vertices (the positions were always occupied by organic scaffolds). In order to overcome this limitation, Shi, Wang *et al.* stepped away from the usual strategy, in which the tpy moieties are connected to the (mostly) rigid backbone *via* their 4′-positions and enable only a strictly linear connectivity. These authors prepared dissymmetric *bis*-tpy ligands, which differed in the angle between the tpy sites (*i.e.*, 120° *vs.* 60°; Figure 10.24) [127]. In either case, the self-assembly of these

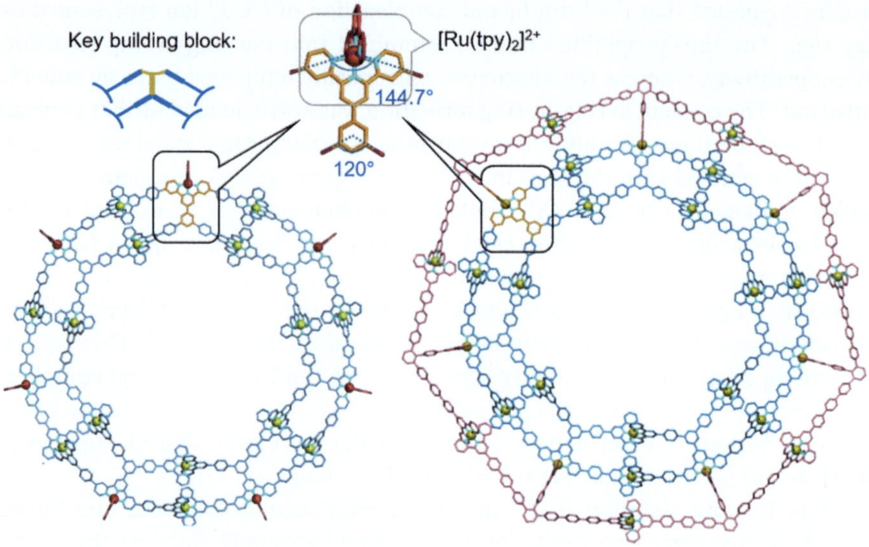

Figure 10.23 Schematic representation of heptagon-in-heptagon assemblies, which resulted from the complexation-induced widening of the angle between the ligand's tpy arms [126] / with permission of John Wiley & Sons.

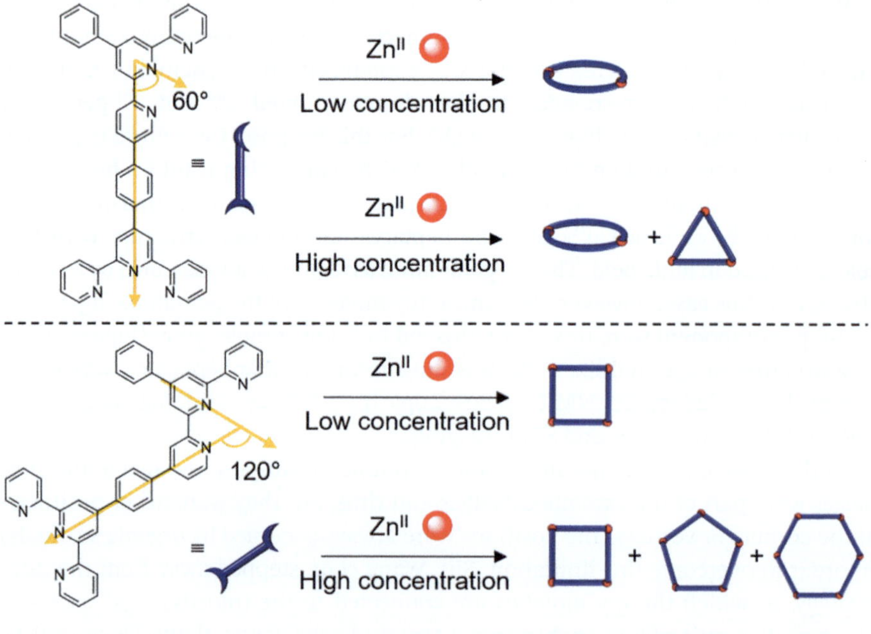

Figure 10.24 Schematic representation of the self-assembly of dissymmetric *bis*-tpy ligands with Zn^{II} ions into metallomacrocycles with $[Zn(tpy)_2]^{2+}$ complexes at the corners [127] / with permission of John Wiley & Sons.

ligands with Zn^{II} ions proceeded in a head-to-tail fashion and gave the respective metallomacrocycles with $[Zn(tpy)_2]^{2+}$ complexes at the corners. The self-assembly process was concentration dependent: In the low-concentration regime, dimers or tetramers were formed, whereas the selectivity was lost at high concentrations and mixtures of macrocycles were obtained (Figure 10.24). To the best of our knowledge, further studies that aim for placing $[M(tpy)_2]^{2+}$ complexes at the corners of metallo-supramolecular assemblies are not yet at hand.

10.3 Octahedral Tpy Complexes – 3D Fractals and Cage-Like Architectures

Self-assembled polyhedral, cage-like architectures are widely abundant in Nature. In this respect, the reader is adverted to viral capsids [128] and the protein-transport complexes [129], as two representative examples. However, when aiming for the assembly of artificial systems, one must pay particular attention to the control over the geometry of the molecular building blocks – this is even more important as when dealing "only" with 2D structures [52]. Taking this into account, researchers have accomplished the synthesis of advanced structures by the metallo-supramolecular self-assembly of tailor-made oligotopic tpy ligands. The reported 3D structures include Platonic motifs (*e.g.*, tetrahedra, cubes, octahedra, *etc.*) as well as Archimedean geometries, which both represent convex polyhedra (Figure 10.25). Such assemblies can only be obtained if one steps away from the paradigm that only building blocks featuring 60°, 120°, and 180° ligand geometries can be used. In addition to these, also different angles due to *meta*-substituted phenyl linkers, diversion from planarity due to rigid 3D vertices (*e.g.*, adamantane, triptycene, *etc.*) and the combination of directionality with flexible vertices and

(a)

(b)

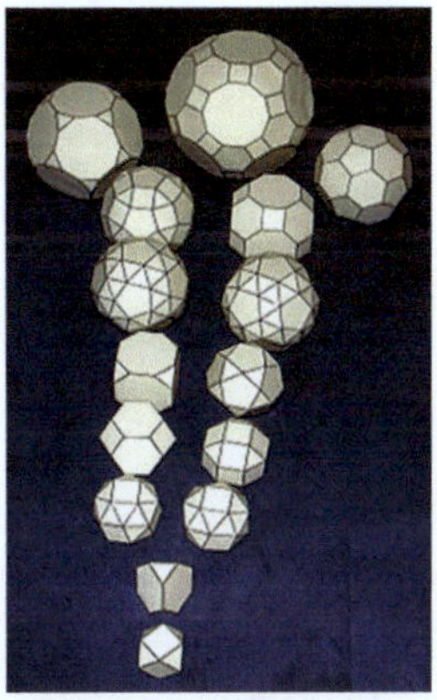

Figure 10.25 (a) Photograph of the five Platonic solids, as objects d'art in the Steinfurter Bagno, Germany [130] with permission of Zumthie / Wikimedia Commons / Public Domain. (b) Representation of the in-total 15 Archimedean solids (taking the chirality of two structures into account) [131] / Frankee 67 / Wikimedia Commens / CC BY-SA 4.0.

steric overlap represent common strategies to enforce self-assembly to proceed in a 3D-type fashion [52].

The presumably first example for a 3D metallo-supramolecular cage, which was based on tpy complexes, was published by Volmer, Mattay *et al.* [132]. This hexameric cage was assembled from a [4]-cavitand, which was equipped with tpy ligands on its upper rim, and Zn^{II} ions (Figure 10.26a). The size and shape of the assembly were deduced from SAXS measurements in solution, which matched well with the dimensions estimated by molecular-modeling studies (SAXS: small-angle X-ray scattering; Figure 10.26a). Later, Wesdemiotis, Newkome *et al.* reported a 3D spoked wheel, which was reminiscent of a bicycle wheel [135]. The structure of the original spoked wheel (Figure 10.6a) [87] was modified by replacing the geometry-directing C6 component with two Y-type ones (R = Me), which formed the hub by stacking in a staggered conformation [135]. Owing to the more globular 3D shape of this assembly, the CCS was smaller than that of the 2D spoked wheel. These same authors also synthesized a highly symmetric nanosphere by the self-assembly of a *tris*-tpy ligand with Ru^{II} ions (yield: 35%) [133]. Unlike the conventional geometry of Y-type components, the linker contained *meta*-linkages, which facilitated the formation of $[Ru(tpy)_2]^{2+}$ complexes within the 3D architecture (Figure 10.26b).

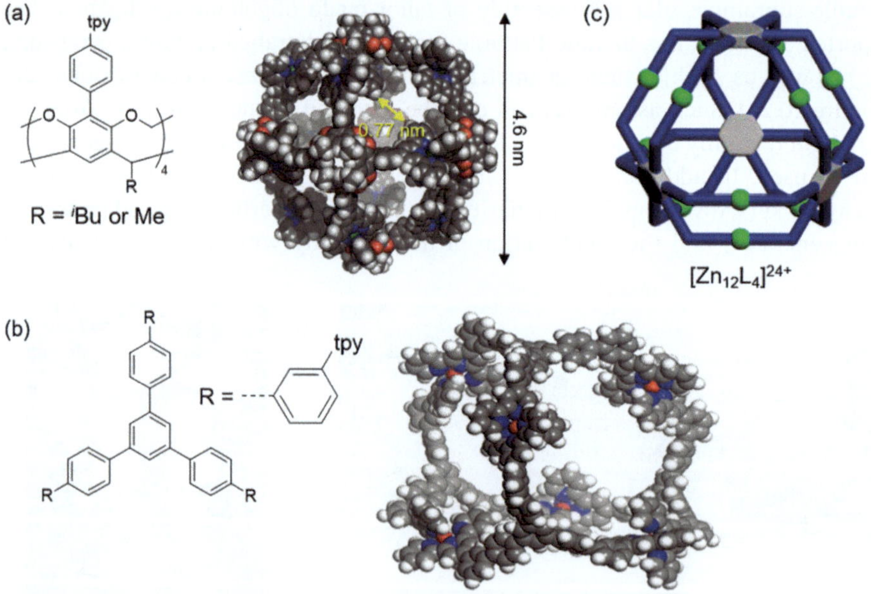

Figure 10.26 (a) Schematic representation of a tetra*kis*-tpy-equipped [4]-cavitand, which gave a hexameric cage upon self-assembly with Zn^{II} ions. The computer-generated, energy-minimized structure of the cage (R = Me) is also shown [132] / with permission of Elsevier. (b) Schematic representation of an angular *tris*-tpy ligand, which gave a metallo-supramolecular nanosphere upon self-assembly with Ru^{II} ions [133] / with permission of American Chemical Society. (c) Representation of the computer-generated, energy-minimized structure of the nanosphere, which contained a hexatopic ligand [134] / with permission of American Chemical Society.

Due to the robustness of the Ru^{II}–tpy moieties, the nanosphere retained its structure in solution without indication of disintegration, even if the pH value was varied over a broad range. The dimensions of the nanosphere, which were deduced from ESI-TWIM MS and TEM imaging, were again in good agreement with those determined by molecular-modelling studies (*i.e.*, edge-to-edge distance of *ca.* 4.1 nm; Figure 10.26b). Eventually, the nanosphere's structure was confirmed by single-crystal XRD analysis.

Unlike the angular *tris*-tpy ligand, which did not give defined metallo-supramolecular structures upon self-assembly with Fe^{II}, Zn^{II}, or Cd^{II} ions, the analogous hexatopic building block gave a nanosphere when combined with Zn^{II} ions 1:3 ratio [134]. This assembly exhibited four quasi-triangular and six rhombus-like facets (Figure 10.26c). Moreover, in MeOH/MeCN solution, the nanosphere assembled further into berry-like nanostructures, whose average hydrodynamic radius was *ca.* 30 nm. The nanosphere, whose internal volume was as high as 9,800 Å^3, acted as a nanocontainer to accommodate hydrophobic guest molecules, such as coronene.

Remarkably, if the self-assembly of the angular tritopic ligand was conducted in a stepwise fashion, different architectures were obtained [136]. The tritopic ligand was dimerized first by Ru^{II}-ion complexation. Unexpectedly, the self-assembly of this pre-organized ligand with Fe^{II} ions afforded isomeric paracyclophane-type assemblies (Figure 10.27). Thereby, the ratio of isomers depended on the reaction conditions: At ambient temperature, the isomers were formed in equal amounts, whereas at elevated temperatures, the formation of the more symmetrical, presumably thermodynamically more stable assembly was favored.

Wesdemiotis, Newkome *et al.* utilized the concept of "directed flexibility" to assemble a 3D polygon from a semiflexible hexatopic ligand and Zn^{II} ions (Figure 10.28a) [137]. The used ligand contained a tribenzo-27-crown-9 ether, to which six phenyl-tpy moieties were grafted. The metallo-supramolecular tetrahedron was comprised of four independent triangular surfaces, which were connected by the crown-ether moieties, which were located at the polygon's edges. It was emphasized that the flexibility of the crown-ether moieties enabled the directional formation of $[Zn(tpy)_2]^{2+}$ complexes, which spanned the trigonal planes. This process, which eventually gave the 3D polygon, was supported by intramolecular $\pi\pi$-stacking interactions. The host-guest chemistry of the tetrahedron was also explored: The size of the tetrahedral tetra*kis*(perfluorophenyl)borate anion nicely matched with the polygon's internal volume and, thus, was efficiently

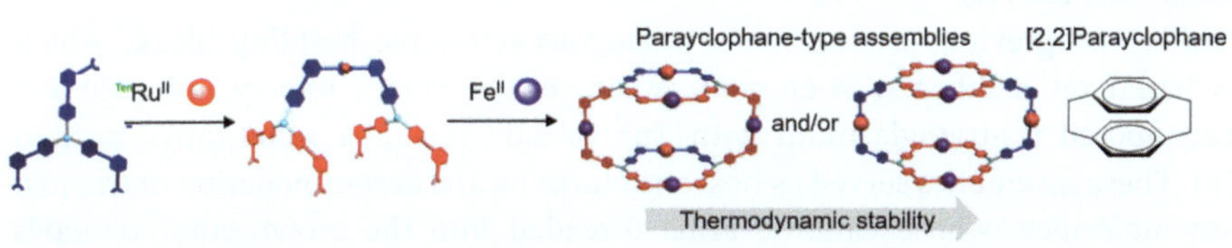

Figure 10.27 Schematic representation of the formation of metallo-supramolecular "paracyclophanes" *via* a two-step self-assembly approach [136] / with permission of American Chemical Society. For comparison, the structure of the [2.2]paracyclophane, as the most prominent paracyclophane derivative, is also shown.

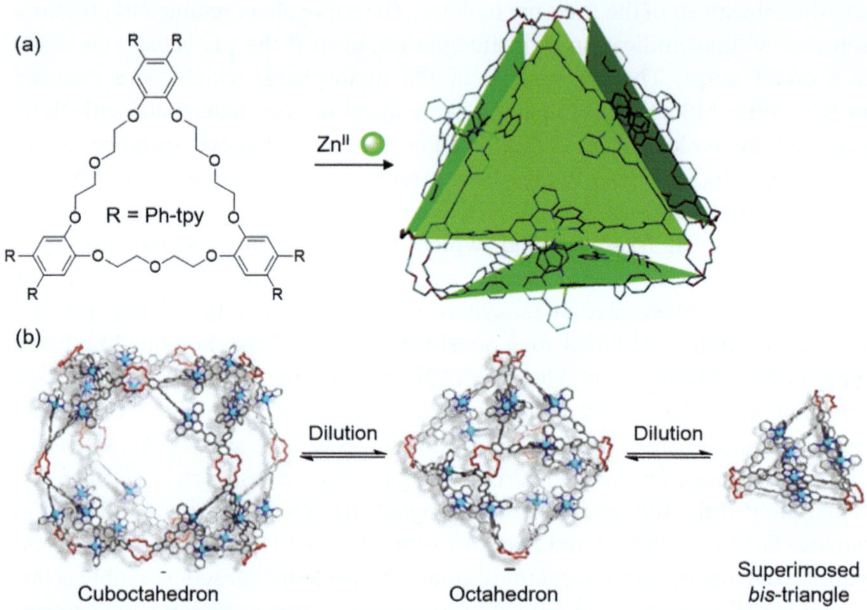

Figure 10.28 (a) Schematic representation of the formation of a metallo-supramolecular tetrahedron [137] / with permission of Royal Society of Chemistry. (b) Schematic representation of the reversible, concentration-dependent interconversion of metallo-supramolecular nanospheres [138] / with permission of American Chemical Society.

accommodated in the cavity. In related work, these same authors also opted for a dibenzo-18-crown-6 ether, as the vertex, to which four phenyl-tpy moieties were attached [138]. However, the self-assembly of this tetratopic ligand with Zn^{II} ion did not afford shape-persistent nanospheres, whereas, at high concentrations, a cuboctahedron, representing an Archimedean solid, was obtained (Figure 10.28b). Upon dilution, this assembly transformed into two identical octahedrons. These species, in turn, were each converted into superposed, *bis*-triangular complexes upon further dilution. This process, which was highly reversible, was enabled by kinetic lability of the Zn^{II}–tpy linkages on the one hand and the structural flexibility of the ligand on the other. Notably, the angle between the tpy-containing arms varied from 127° to 90° and 0° in the case of the cuboctahedron, octahedron, and triangle, respectively.

Unlike the previous example, the analogous tetratopic building block, which was based on a dibenzo-24-crown-8 ether, as the vertex moiety, afforded the superimposed *bis*-triangle motif with Zn^{II} or Cd^{II} ion in a quantitative fashion [139]. These assemblies served as host structures for the accommodation of trigonal guest molecules, whose cationic arms threaded into the crown-ether moieties located at the edges (Figure 10.29). The thusly formed host-guest complexes between a crown ether and a secondary ammonium salt are very common in supramolecular chemistry; such assemblies are referred to as rotaxanes and pseudorotaxanes (in the presence and absence of blocking groups, respectively)

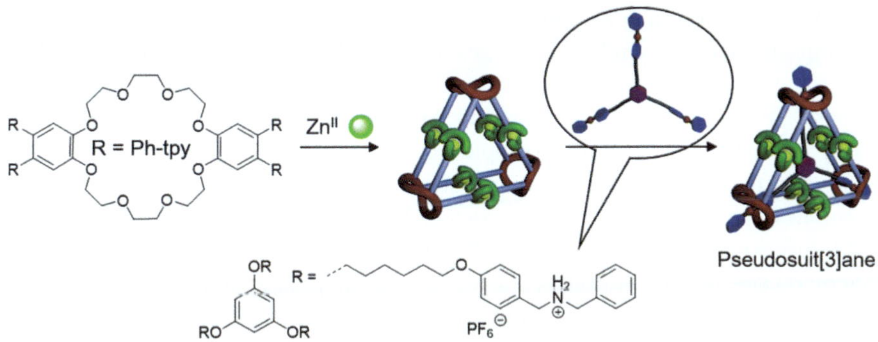

Figure 10.29 Schematic representation of the two-step assembly of a pseudosuit[3]ane [139] / with permission of Royal Society of Chemistry.

[140]. In suitanes, the formation of such host-guest complexes proceeds inside a confined environment, such as a molecule, which represents the coat or suit [141]. Since the interaction between the superimposed *bis*-triangle and the star-shaped guest afforded three crown-ether complexes, the final assembly was referred to as pseudosuit[3]ane [139]. Owing to subtle differences in kinetic stability, the formation of the Cd^{II}-containing pseudosuit[3]ane occurred more readily than that of its Zn^{II}-containing counterpart. Remarkably, the threefold threading could be reversed by adding 1,8-diazabicyclo[5.4.0]undec-7-ene (DBU), as the base, without affecting the integrity of the metallo-supramolecular assembly.

Unlike previous examples, in which flexible crown-ether scaffolds represented the vertices of the metallo-supramolecular assemblies, Li *et al.* utilized rigid-3D adamantane as the corner-directing moieties for the same purpose. Three tpy arms were grafted onto the adamantane core by Suzuki- or Sonogashira cross-coupling [142, 143]. Depending on the linker geometry, different metallo-supramolecular "Platonic solids" of the general formula $[M_{3n}L_2]$ were obtained with Zn^{II} ions in high yields (L denotes the tritopic tpy ligand; Figure 10.30): trigonal bipyramid

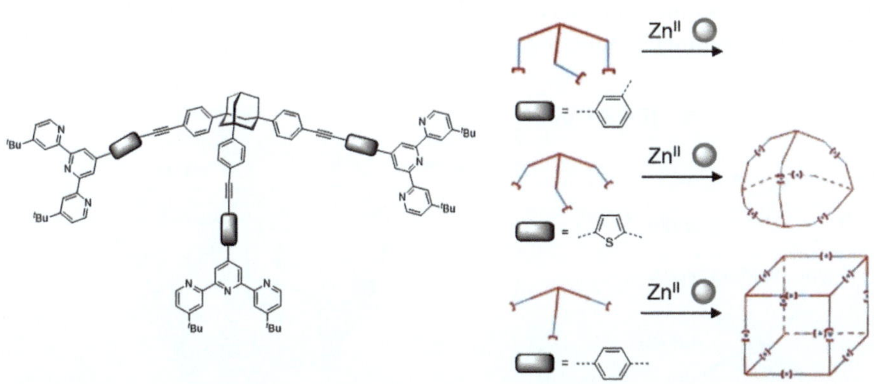

Figure 10.30 Schematic representation of the formation of "Platonic solids" by the metallo-supramolecular assembly of adamantane-centered tritopic ligands [143] / with permission of American Chemical Society.

($n = 1$), tetrahedron ($n = 2$), and cube ($n = 3$) [143]. The arms were flexible enough to adopt an adequate angle. In the case of the cube, for example, the critical 90° directionality was reached, despite the intrinsic 109° directionality, which was purported by the rigid adamantane unit. Owing to the high self-sorting capability, binary mixtures of two of the tritopic ligands afforded – upon addition of Zn^{II} ions – exclusively the corresponding polygons, whereas the occurrence of mixed assemblies, which contained both types of ligands, was ruled out on the basis of NMR spectroscopy and ESI-TWIM MS. However, the polygons represented only ephemeral species in solution: A 1:1 mixture of the preassembled tetrahedron and cube gave, within a few days, a statistical mixture of tetrahedrons, which contained both types of ligands.

The concentration-dependent interconversion of nanostructures, *i.e.*, the reversible transformation of a giant cuboctahedron into two identical octahedrons, was also witnessed by Wesdemiotis, Newkome *et al.* [144]. The assembly of the cuboctahedron motif was accomplished by the sophisticated bent X-type ligand, which – due to its 9,10-ethanoanthracene moiety – almost ideally met the geometric requirements (*i.e.*, dihedral angle: 125°, $\alpha = 60°$, $\beta = 90°$; Figure 10.31). The α and β angles enabled the formation of regular triangles and squares, respectively, whilst the proper dihedral angle of 127° facilitated the globular assembly. Thus, the cuboctahedron was formed almost free of distortion, as evidenced by synchrotron XRD analysis. Due to kinetic lability of the $[Cd(tpy)_2]^{2+}$ linkages, only this assembly exhibited the aforementioned concentration-induced switching between cuboctahedral and octahedral species, referred to as metallo-supramolecular fission and fusion. However, this transformation did not afford any stable intermediates, which could be characterized by NMR or ESI-TWIM MS, whereas, in a related system, the transformation of a cubic assembly into a tetrahedral one upon dilution proceeded *via* a prism-shaped intermediate, whose occurrence was corroborated by NMR spectroscopy [145].

One further metallo-supramolecular cuboctahedron, which could reversibly be transformed into a triangular prism, was reported by Zhang, Wang, Xie *et al.* [146]. The cuboctahedron was assembled from a tetratopic building block, which contained a geometry-directing 2,2′-(1,3-phenylene)*bis*(benzimizolium) moiety

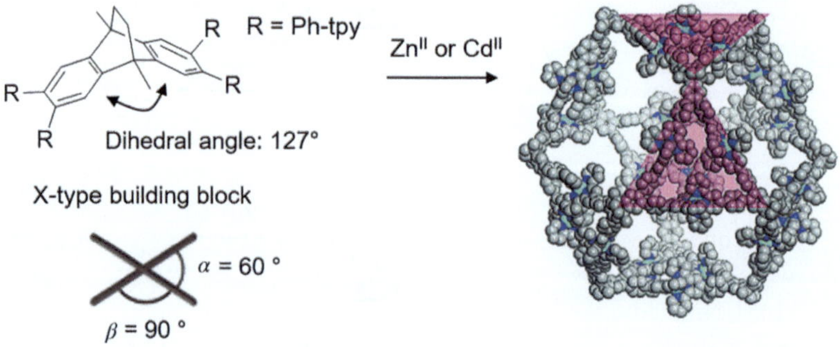

Figure 10.31 Schematic representation of the formation of metallo-supramolecular cuboctahedrons [144] / with permission of John Wiley & Sons.

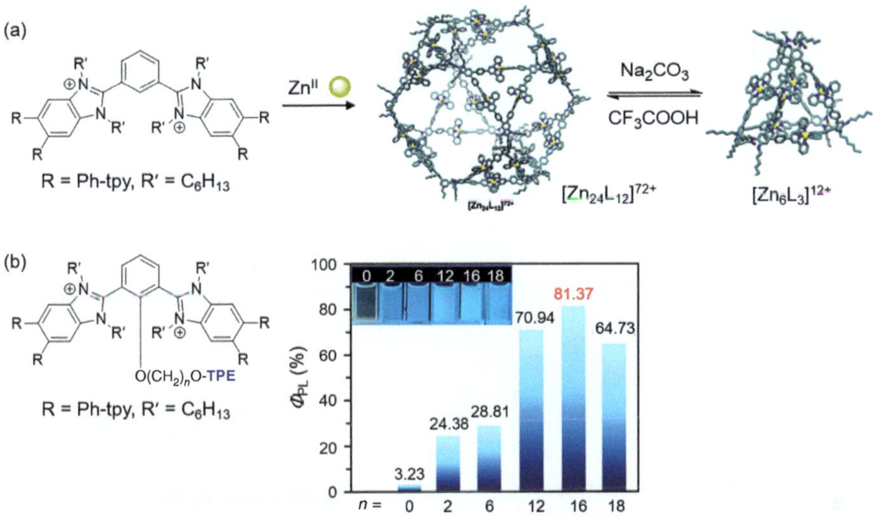

Figure 10.32 (a) Schematic representation of the formation of a metallo-supramolecular cuboctahedron, which could reversibly be converted into a triangular prism [146] / with permission of Elsevier. (b) Schematic representation of the TPE-equipped ligands, which gave emissive cuboctahedrons. Their photoluminescence quantum yield (Φ_{PL}), as a function of the alkyl-chain length, is also shown [147] / with permission of John Wiley & Sons.

(Figure 10.32a). The alkyl chains of this system induced steric hindrance and, thus, hindered the ligand from adopting conformations, which would result in different motifs upon metallo-supramolecular self-assembly. The giant cuboctahedron (diameter of 7.3 nm, total charge of +72) was stable in an acidic environment, but experienced transformation into a triangular prism upon treatment with base (Figure 10.32a). This process was attributed to the cleavage of the imidazolium rings. The resulting amide linkages were more flexible and favored the formation of the more compact assembly. This process could be reversed by the acid-catalyzed reformation of the imidazolium rings. The parent cuboctahedron was only barely emissive in DMF solution (Φ_{PL} of *ca.* 3%) [147]. However, the emission intensity was increased significantly when TPE moieties, as AIEgens, were attached to the ligand's central phenyl ring *via* short alkyl chains (Figure 10.32b). The TPE moieties aggregated within the cage's cavity and displayed AIE. The behavior was most expressed for the alkyl-chain length of 16 (Φ_{PL} of *ca.* 81%).

In extension of their work on ring-in-ring assemblies *via* the one-step self-assembly of pyridinium-based oligotopic ligands [103, 104], Li *et al.* also prepared hexagonal prisms *via* the same strategy, which relied on the DoCS concept. In these assemblies, two hexagon-in-hexagon assemblies, as already shown in Figure 10.14a, were superimposed by the complexation of their pending tpy ligands (Figure 10.33) [148]. NMR and ESI-TWIM MS studies revealed that the ring-in-ring motif was formed first and subsequently "dimerized" to afford the 3D structure. The observed antimicrobial activity against Gram-positive bacteria was presumably due to the efficient binding of the nanostructures to the bacterial membranes.

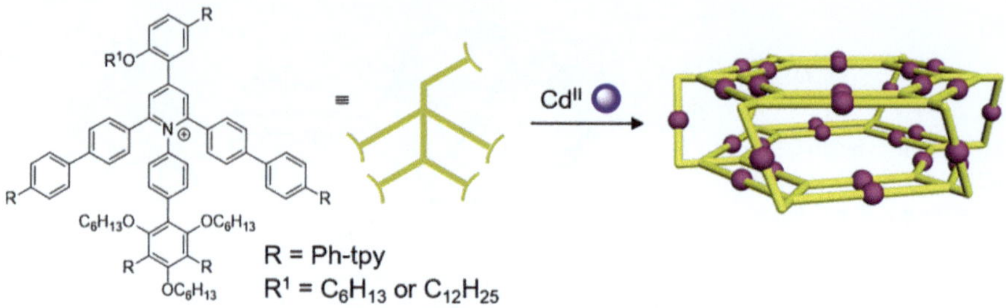

Figure 10.33 Schematic representation of the formation of a metallo-supramolecular hexagonal prism [148] / with permission of American Chemical Society.

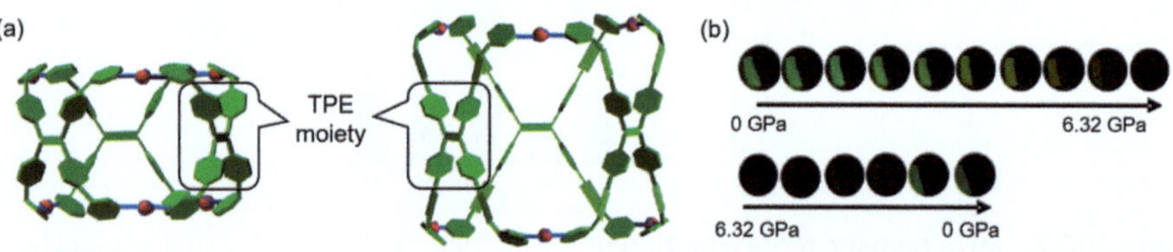

Figure 10.34 Schematic representation of the TPE-containing nanocages. (b) Photographs of the solid-state emission of the right nanocage at various pressure loadings. Reproduced with permission from [149] / with permission of Chinese Chemical Society.

Similar to the aforementioned metallo-supramolecular rosettes, which are depicted in Figure 10.19 [115], the photophysical properties of the metallo-supramolecular cages, which were reported by Xu, Wang *et al.*, were also ascribed to AIE. These constructs, which were assembled from TPE-centered tetratopic ligands and Zn^{II} ions, were highly emissive in solution (Figure 10.34a) [149]. The more extended cage, due to additional ethinyl moieties within the π-conjugated arms, featured pressure-dependent solid-state emission. With increasing pressure, the emission intensity initially increased, presumably due to rotation restriction in the compressed state; however, too dense packing resulted in emission quenching beyond 6.32 GPa (Figure 10.34b). The original emission behavior was eventually recovered when the pressure was released.

The formation of catenanes, *i.e.*, mechanically interlocked architectures consisting of (at least) two macrocycles, was observed for the rigid zigzag-shaped ditopic ligand **L1** upon its coordination with Zn^{II} ions in a 1:1 ratio [150]. In the low-concentration regime, the twisted *bis*-tpy ligand afforded a twisted metallo-macrocycle, whereas, at high concentrations, interlocked species, corresponding to [2]catenanes and [3]catenanes, represented the major products of the self-assembly process (Figure 10.35a). The same behavior in terms of catenane formation at high concentration was also partially observed for the related tritopic ligand **L2** (Figure 10.35b), whereas interlocking was completely absent for hexatopic ligand due to steric constraints. Instead, the cage-like architecture served as the host for the accommodation of calix[*n*]arenes, as guest molecules (Figure 10.35c). In particular, a high binding affinity was found for calix[6]arene: Two of these guest molecules were included more efficiently than other derivatives, *e.g.*, calix[4]arene.

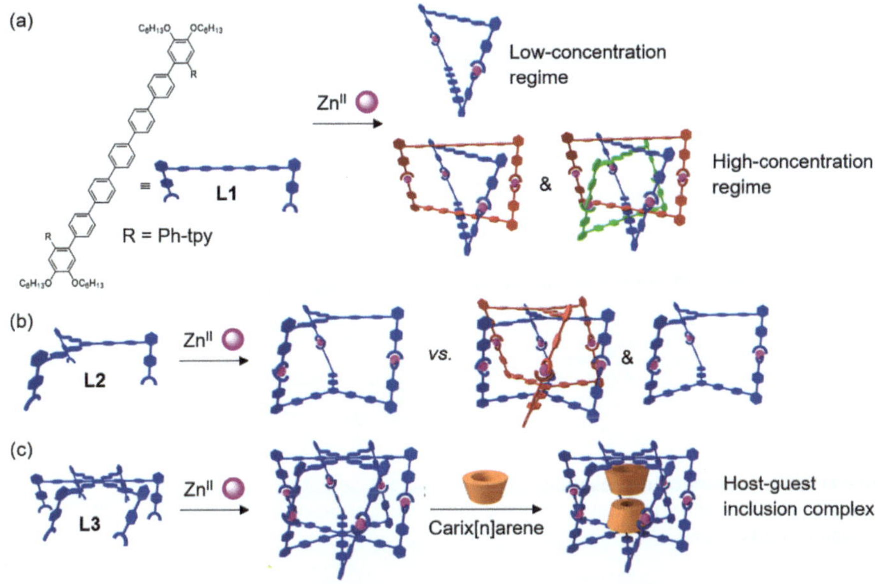

Figure 10.35 (a) Schematic representation of the concentration-dependent self-assembly of the twisted *bis*-tpy ligand **L1**. (b) Schematic representation of the concentration-dependent self-assembly of the *tris*-tpy ligand **L2**. (c) Schematic representation of the self-assembly of the hexa*kis*-tpy ligand **L3** and the subsequent formation of inclusion complexes with calix[*n*]arenes [150] / with permission of Chinese Chemical Society.

Strictly convex polyhedron, whose faces are regular polygons but do not span Platonic, Archimedean, or (anti-)prismatic solids, are referred to as Johnson solids. The complete list of such solids, which was established by the mathematician M.J. Johnson in 1966, counts in total 92 representatives [151]. One of these motifs, the D_3-symmetric triangular orthobicupola was assembled from a low-symmetry tetratopic ligand and Zn^{II} ions (1:2 ratio) in DMF solution (Figure 10.36) [152]. ESI-TWIM MS and synchrotron XRD analysis unambiguously confirmed the special structure of the assembly. However, if the self-assembly was conducted in MeOH, as the solvent, a different metallo-supramolecular construct, *i.e.*, a pentagonal prism, was obtained (Figure 10.36). Thus, the interconversion of metallo-supramolecular structures could also be induced by changing the solvent polarity. This behavior could be useful for the controlled capture and release of guest molecules in different solvent environments.

The same motif, *i.e.*, the pentagonal prism, was also obtained from the tetratopic ligand with Mn^{II} ions [153]. This assembly was identified as a powerful catalyst for the photo-oxidation of aromatic aldehydes. The assembly acted as the sensitizer, which formed singlet oxygen ($^1O_2$) with high efficiency (*ca.* 24%). Unlike for related Mn^{II}-containing macrocycles or mononuclear $[Mn(tpy)_2]^{2+}$ complexes, the pentagonal prism combined fast conversion times with recyclability. In particular, electron-poor aldehydes were quantitatively oxidized to the corresponding acids within 15 min. Remarkably, the oxidation proceeded as efficiently under

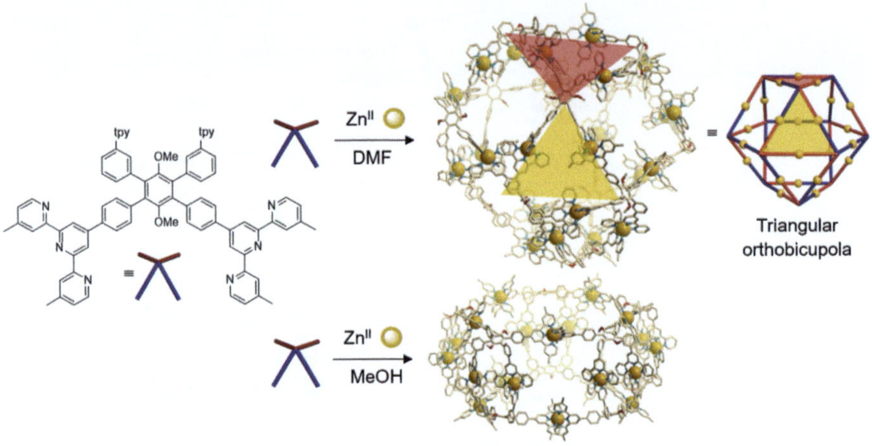

Figure 10.36 Schematic representation of the solvent-selective formation of a metallo-supramolecular triangular orthobicupola and pentagonal prism [152] / with permission of Elsevier.

solvent-free conditions under daylight in air. Thus, the Mn^{II}-containing nanocage represented a rather low-cost, eco-friendly sensitizer for the photocatalytic aldehyde oxidation.

Furthermore, an icosahedral nanocage was assembled from a regioregular corannulene-centered, pentatopic tpy ligand [154]. The structure of this construct was reminiscent of a spherical virus capsid, a soccer ball as well as a C_{60} Buckminster fullerene (Figure 10.37). Twelve ligand molecules assembled with 30 Cd^{II} ions to afford this giant assembly, whose external diameter was *ca.* 6 nm, according to microscopy imaging and SAXS measurements. The selective formation of the icosahedral motif was due to the bowl-shaped corannulene moiety on one hand and the choice of Cd^{II} ions, which provided a well-balanced ratio of thermodynamic stability and kinetic lability, on the other. The ligand's *tert*-butyl and methoxy

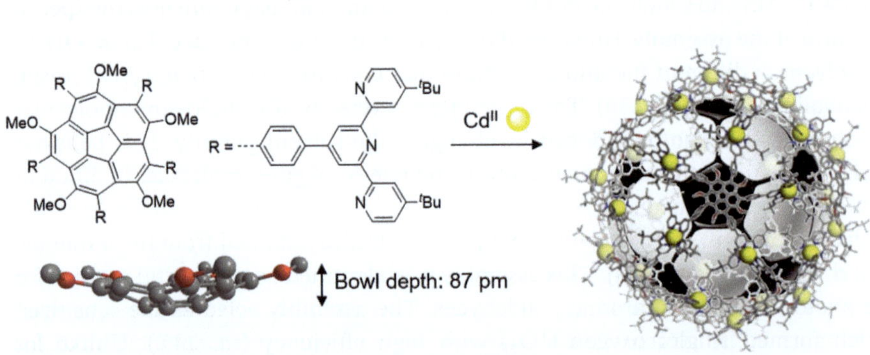

Figure 10.37 Schematic representation of the self-assembly of a giant metallo-supramolecular icosahedron [154] / Springer Nature / CC BY 4.0. The computer-generated 3D structure of pentamethoxy-substituted corannulene shows its distinct curvature, which accounts for the ligand's self-assembly behavior.

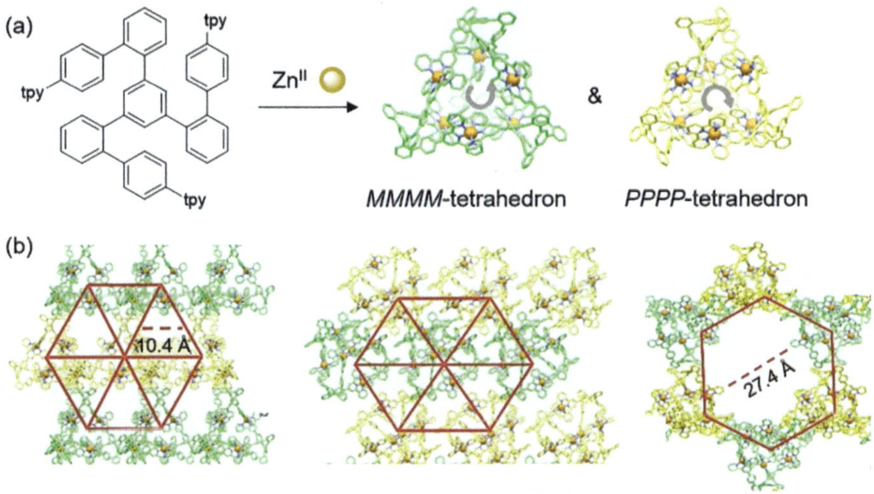

Figure 10.38 (a) Schematic representation of the formation of enantiomeric tetrahedrons from an achiral tris-tpy ligand. (b) Excerpts from the solid-state structure of the racemic tetrahedrons with PF_6^-, Cl^-/Br^-, or I^- counterions (from the left to the right) [155] / with permission of John Wiley & Sons.

substituents accounted for the nanostructure's sufficient solubility in organic solvents.

An unusual-shaped tritopic ligand provided tetrahedral nanostructures upon self-assembly with Zn^{II} ions [155]. Owing to the absence of chiral induction, the tetrahedrons were obtained as a racemic mixture (Figure 10.38a). However, in the presence of chiral camphor-10-sulfonate, chiral induction occurred and the enantiomers could be distinguished in solution by circular dichroism (CD) spectroscopy. In the presence of halide ions, the packing behavior was altered significantly (Figure 10.38b). With PF_6^-, as the counterion, the tetrahedron assembled in a closely packed hexagonal array, whereas, if Cl^- or Br^- was added, the hierarchical self-assembly afforded a highly tight hexagonal packing. In the case of the I^- ions, a graphite-like packing mode was observed. The changing in the packing mode was attributed to the high anionic charge density, thus repelling the PF_6^- counterions. This effect, which was also observed by NMR spectroscopy in solution, was absent for various other weakly coordinating anions (*e.g.*, BF_4^-, TfO^-, AcO^-, *etc.*). This study gave versatile insight into the possibilities of hierarchical self-assembly of metallo-supramolecular constructs, which goes far beyond their initial formation.

One further example for the formation of an enantiomeric pair of metallo-supramolecular nanocages was provided by Wang *et al.* [156]. Due to the steric hindrance in its structure, the triphenylene-centered ligand exhibited two distinct conformations in which the six tpy-containing arms were oriented in a clockwise or anticlockwise fashion. Synchrotron XRD analysis revealed that the self-assembled Zn^{II}-containing nanocages strictly contained only one of these conformers (Figure 10.39a). $\pi\pi$-stacking interactions, which involved the ligands'

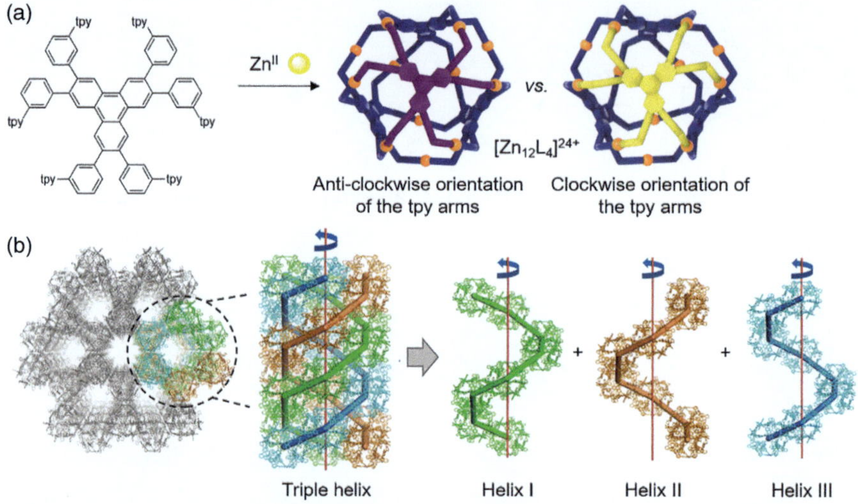

Figure 10.39 (a) Schematic representation of the formation of an enantiomeric pair of metallo-supramolecular nanocages. (b) Schematic representation of the hierarchical self-assembly of the nanocages into a triple-helix arrangement with large hexagonal channels [156] / with permission of Royal Society of Chemistry.

triphenylene cores, facilitated the hierarchical self-assembly of the nanocages into a highly ordered array. The sparse packing mode generated channels with a large diameter of *ca.* 3.6 nm. The channels were formed by the intertwined complementary triple-helix arrangement of nanocages (Figure 10.39b). Taking the descriptors for the tubular packing of spheres in biological fine structures into account, this particular motif was classified as the 3(1,1,2) mode [157, 158].

As the last notable aspect in the context of one-pot self-assembly of 3D structures, the use of dissymmetric oligotopic ligands has to be mentioned. In these ligands, the tpy moieties are not connected to the rigid backbone *via* their 4′-positions; instead, the 5-position is also utilized. Based on this concept, various sterically congested ligands have been prepared. Firstly, Wang *et al.* employed the ditopic ligand to obtain rhombic dimers with either Fe^{II}, Zn^{II}, or Cd^{II} ions (Figure 10.40a) [159]. The related tritopic ligand displayed a more complicated self-assembly behavior with these ions: In the case of Zn^{II} or Cd^{II}, a tetrameric assembly was formed exclusively, whereas a mixture of tetrameric, hexameric, and octameric macrocycles was obtained with Fe^{II} ions (Figure 10.40b). The kinetic stability of these assemblies enabled their separation by conventional column chromatography. Remarkably, only species with an even number of ligand molecules were formed. This was presumably due to the initial formation of dimers, which oligomerized into macrocycles and excluded the incorporation of single monomers, which would have afforded odd-numbered ring sizes (this assumption was based on NMR spectroscopic studies and ESI-TWIM MS). In subsequent work, K. Wang, Song, M. Wang *et al.* assembled hourglass-shaped nanocages from self-complementary ligands (Figure 10.40c) [160]. The self-assembly of these ligands with one eq. of

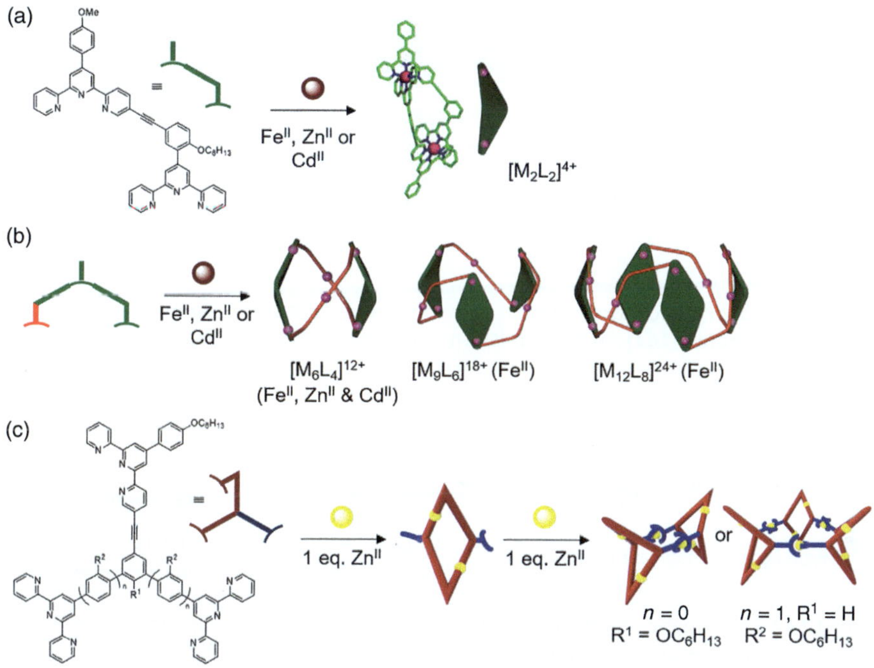

Figure 10.40 (a) Schematic representation of the metallo-supramolecular dimerization of a dissymmetric *bis*-tpy ligand [159]. (b) Schematic representation of the formation of even-numbered metallo-macrocycles from a tritopic tpy ligand [159] / with permission of John Wiley & Sons. (c) Schematic representation of hour-glass-shaped assemblies from self-complementary tritopic ligands [160] / with permission of John Wiley & Sons.

Zn^{II} ions gave dimeric species – similar to the previous example. Depending on the length of the spacer, these dimers gave hexameric or tetrameric assemblies, respectively, upon complexation of one further eq. of Zn^{II}. These assemblies, which were reminiscent of hour-glasses, went through a hierarchical self-assembly process both in solution and at the solid/liquid interface. Thereby, the nature of higher-order structures was modulated *via* the solvent polarity.

Moreover, 3D metallo-supramolecular architectures have also been assembled using template-driven methods. Preferentially, the C6 component (Figure 10.5) has been utilized as a geometry-directing building block for this purpose [55]. In one study, Wang *et al.* combined the C6 component with an antler-shaped hexatopic ligand to obtain a giant truncated tetrahedron by self-assembly with Cd^{II} ions under cautious stoichiometric control (Figure 10.41) [161]. This huge assembly featured a high density of $[Cd(tpy)_2]^{2+}$ complexes on its shell and possessed a large cavity for presumptive guest inclusion. The hierarchical self-assembly of nanocages into highly ordered aggregates resulted in emission enhancement due to AIE. In an appropriate $MeCN/CH_2Cl_2$ solvent mixture, the emission of near-white light was observed. Dynamic light scattering (DLS) measurements showed that the hydrodynamic diameter (D_h) increased from 5.6 nm (no aggregation in DMF/MeCN) to eventually 147 nm when the solvent mixture contained a 90% fraction of $CHCl_3$.

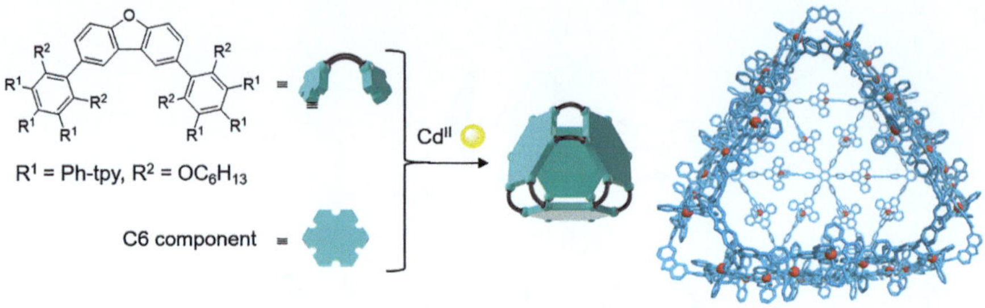

Figure 10.41 Schematic representation of the 64-component self-assembly of a giant truncated tetrahedron [161] / with permission of American Chemical Society.

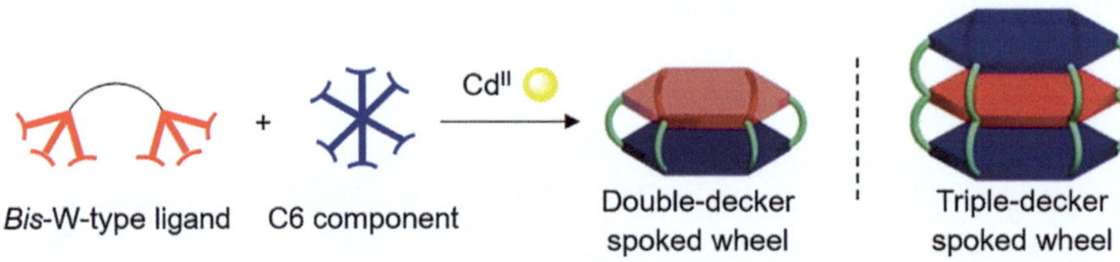

Figure 10.42 Schematic representation of the formation of metallo-supramolecular double- and triple-decker assemblies [162] / with permission of John Wiley & Sons.

However, the aggregates were only stable in solution and collapsed after deposition onto substrates for microscopy imaging.

In one further study, Newkome, Wang *et al.* prepared stacks of spoked-wheel structures, *i.e.*, double- and triple-decker motifs, by a one-pot self-assembly approach [162]. Thereby, the same strategy, which already enabled the assembly of the "normal" spoked wheel, was followed (Figure 10.6) [87]. However, in this case, the two W-type components were linked *via* flexible alkyl chains [162]. This modification allowed the formation of the 2D spoked-wheel structure, due to the geometry-directing properties of the C6 component (Figure 10.42). Two such spoked wheels were aligned vertically *via* the alkyl linkages at the vertices, thus providing double-decker structures, whose layer-to-layer distance depended on the length of the alkyl chains. The methodology was extended further toward triple-decker assemblies. For this purpose, two W-type components were attached to a third one. The self-assembly of this ligand with C6 and Cd^{II} gave the metallo-supramolecular triple-decker, which was unambiguously identified by ESI-TWIM MS and microscopy imaging. Remarkably, the C6 components facilitated the simultaneous formation of three spoked-wheel moieties per triple-decker assembly, thus highlighting once more the power of metallo-supramolecular self-sorting under thermodynamic and geometric control. The assemblies formed temperature-responsive gels in DMF. Microscopy imaging revealed that the gels were comprised of bulk networks due to intermolecular non-covalent forces (*i.e.*, $\pi\pi$-stacking, hydrophobic interactions).

As one of the first examples for the stepwise assembly of 3D metallo-supramolecular assemblies, Wesdemiotis, Newkome *et al.* reported the construction of a propeller-shaped spirane [163]. This assembly was derived from a

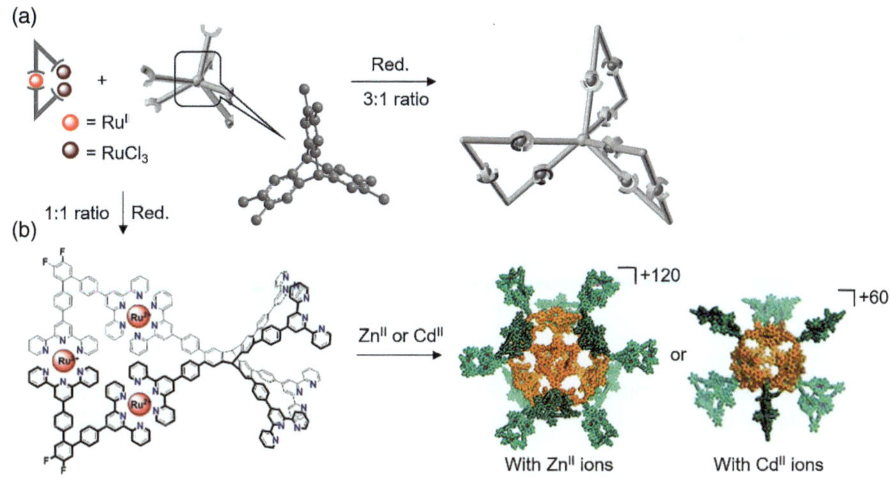

Figure 10.43 Schematic representation of the stepwise assembly of a metallo-supramolecular spirane (a) and nanocage-centered metallodendrimers (b) [163, 164] / with permission of John Wiley & Sons and American Chemical Society, respectively.

V-type component (R = OBn) and a triptycene-centered hexatopic ligand, which were combined in a 6:1 ratio *via* directional Ru^{II}-ion complexation (Figure 10.43a). First, the V-type monomers were dimerized to afford $[Ru(V)_2]^{2+}$ complexes, which represented the spirane's outer rims. Next, the vacant tpy moieties of this building block were loaded with Ru^{III} centers. The thusly obtained trinuclear $[Cl_3Ru^{III}(V)Ru^{II}(V)Ru^{III}Cl_3]^{2+}$ was reacted with the hexatopic ligand under reducing conditions to give the metallo-supramolecular spirane. Thereby, the triptycene's 3D structure enforced three different planes of directionality, from which the molecule's propeller-like shape resulted. However, if the last assembly step was performed in a 1:1 ratio (instead of the spirane-affording 3:1 ratio), a different type of architecture was obtained, *i.e.*, only one of the propeller arms was formed [164]. The self-assembly of the latter with Zn^{II} ions gave a large (Ø = 1.3 nm), highly charged (+120) metallo-supramolecular cuboctahedron, which was decorated with upright metallotriangles (Figure 10.43b). Surprisingly, a smaller, less highly charged, octahedral nanocage was formed with Cd^{II} ions. Neither of these systems exhibited the aforementioned concentration-induced transformation into another type of architecture.

Similar to the aforementioned formation of a double-decker spoked wheel [162], Wang *et al.* reported the synthesis of a stacked honeycomb-like motif (*i.e.*, a dodecagonal prism) [165]. However, unlike that example, here the linkage of the layers was accomplished by flexible alkyl chains, but *via* in-total 12 $[Cd(tpy)_2]^{2+}$ complexes. Using a preassembled pentatopic Ru^{II}-containing building block and the geometry-directing C6 component, a hexagon-fused wheel was obtained in almost quantitative yield upon self-assembly with Zn^{II} ions (Figure 10.44a). The C6 component and the *tris*-tpy moiety of the metalloligand afforded the central spoked wheel, whereas the outer rim was formed by complexation of the remote tpy arms. According to high-resolution TEM imaging, the diameter of this assembly

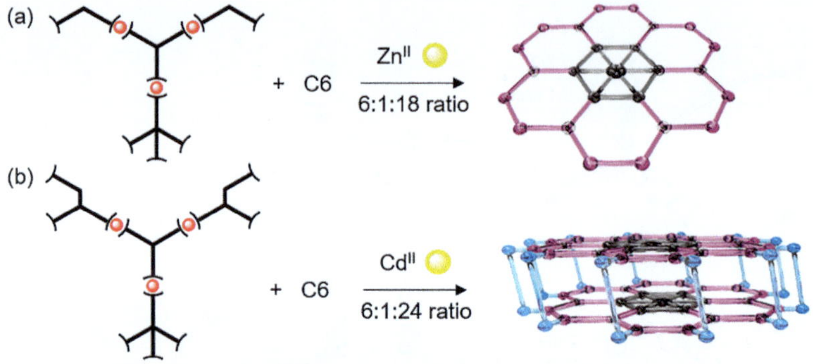

Figure 10.44 Schematic representation of the stepwise assembly of (a) a hexagon-fused wheel and (b) a dodecagonal prism (b) [165] / with permission of American Chemical Society.

was *ca.* 13.7 nm. The double-decker structure was obtained when the preassembled building block was modified by introducing two angled tpy moieties (Figure 10.44b). Remarkably, the 3D structure was formed efficiently with C6 and Cd^{II} ions, despite the enormous number of required $[Cd(tpy)_2]^{2+}$ complexes. The resulting construct had a molecular weight of *ca.* 119.5 kDa (excluding anions) and was supercharged (+168). Its structure and composition were confirmed by DOSY and ESI-TWIM MS measurements as well as high-resolution TEM imaging.

Furthermore, Wang *et al.* utilized a tetrahedral building block for the one-pot self-assembly with Zn^{II} ions [166]. This tritopic ligand was comprised of a 1,1,1-triphenylethane core, to which four tpy arms were attached. However, its self-assembly with metal ions proceeded in an ill-controlled fashion and gave a cube along with a mixture of prismatic species, as non-separable mixture of products, whereas the tailor-made hexatopic ligands – prepared *via* an intricate synthesis strategy – offered sufficient geometric strain to allow the subsequent self-assembly with Zn^{II} ions to proceed in a surprisingly efficient manner (*i.e.*, isolated yields of >90%). The cage-in-cage nanostructures, which differed in their dimensions, were each comprised of an inner cube (diameter of 5.1 nm) and outer large truncated cubes (maximum diameter of 12.0 and 13.2 nm, respectively) with six large bowl-shaped subcages distributed on six faces (Figure 10.45). It was proposed (though not shown) that such assemblies might represent versatile hosts for shape-, size-, and charge-selective accommodation of guest molecules.

More recently, the synthetic strategies to assemble Sierpiński triangles on the one hand and Platonic solids on the other *via* directional metal-ion complexation were merged to obtain Sierpiński polyhedra, *i.e.*, Platonic solids, whose faces are comprised of Sierpiński triangles [167]. For this purpose, a hexatopic metalloligand, as the geometry-directing building block, was formed. This molecule was essentially comprised of three K-type ligands, which were arranged in a triangular fashion by Ru^{II}–tpy complexation (Figure 10.46a). Making use of Chan's concept for the formation of heteroleptic $[Zn(tpy)_2]^{2+}$ or $[Cd(tpy)_2]^{2+}$ complexes [54], the former was combined with appropriately designed multitopic building blocks (Figure 10.46b).

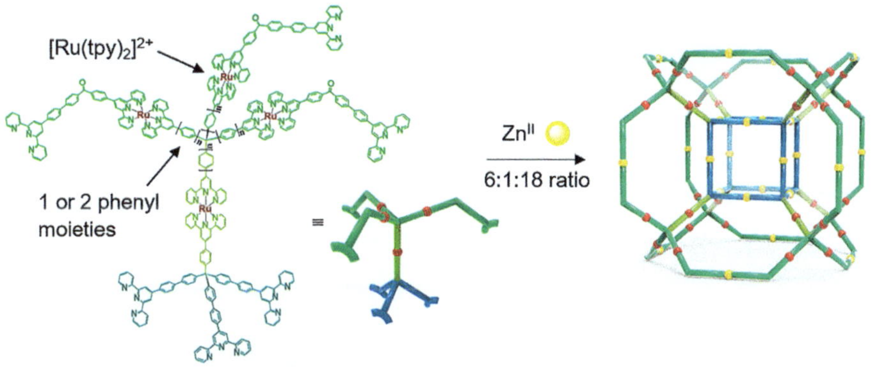

Figure 10.45 Schematic representation of the self-assembly of sophisticated hexatopic building blocks into cage-in-cage architectures [166] / with permission of American Chemical Society.

Depending on the geometry and number of tpy-containing arms, either Sierpiński triangles or Sierpiński polyhedra were obtained [167]. The triangles required a V-type component, whereas the tetrahedron (or tetrix) and octahedron were assembled with a hexatopic or tetratopic corner-directing ligand, respectively. The hexatopic ligand contained an umbrella-shaped tribenzotriquinacene core, which facilitated the self-assembly into the tetrahedral motif, due to its distinct curvature. In the case of the Sierpiński octahedron, the tetratopic ligand contained the 9,10-ethanoanthracene moiety, which provided a larger dihedral angle than the tribenzotriquinacene unit. Under cautious stoichiometric control, the respective 2D and 3D constructs were assembled in almost quantitative yield (Figure 10.46c). Thereby, the coordination of two 6,6″-substituted tpy ligands to a Zn^{II} or Cd^{II} center was excluded due to steric hindrance, whereas the heteroleptic complexes were stabilized by *intra*-ligand $\pi\pi$-stacking interactions. The structure of the Sierpiński architectures was ascertained by NMR spectroscopy, ESI-TWIM MS, microscopy imaging as well as SAXS measurements.

In subsequent work, the same tribenzotriquinacene-based building block, as in the previous example, was utilized for the formation of a huge isohedral triakis tetrahedron (height of 5.3 nm) [169]. The triakis tetrahedron is a representative of the family of Catalan solids, which are formally obtained if one adjoins a pyramid to each of the polyhedron's faces (this applies to all Archimedean solids) [170]. In terms of geometry, the triakis tetrahedron is the dual (or Kleetope) of a regular tetrahedron; its twelve faces represent isosceles triangles of identical dimensions. The assembled structure was comprised of six metallo-supramolecular rhomboids, which were connected *via* their acute vertices [169]. The rhomboid substructure itself was assembled from a pre-organized tetratopic metalloligand, a V-type component and Cd^{II} ions in a 1:2:4 ratio (Figure 10.47a). The metalloligand also served as the geometry-directing and stabilizing component for the construction of the triakis tetrahedron, which was assembled from the molecular components in the 2:3:12 ratio (Figure 10.47b). Once more, the formation of heteroleptic $[Cd(tpy)_2]^{2+}$ complexes represented one

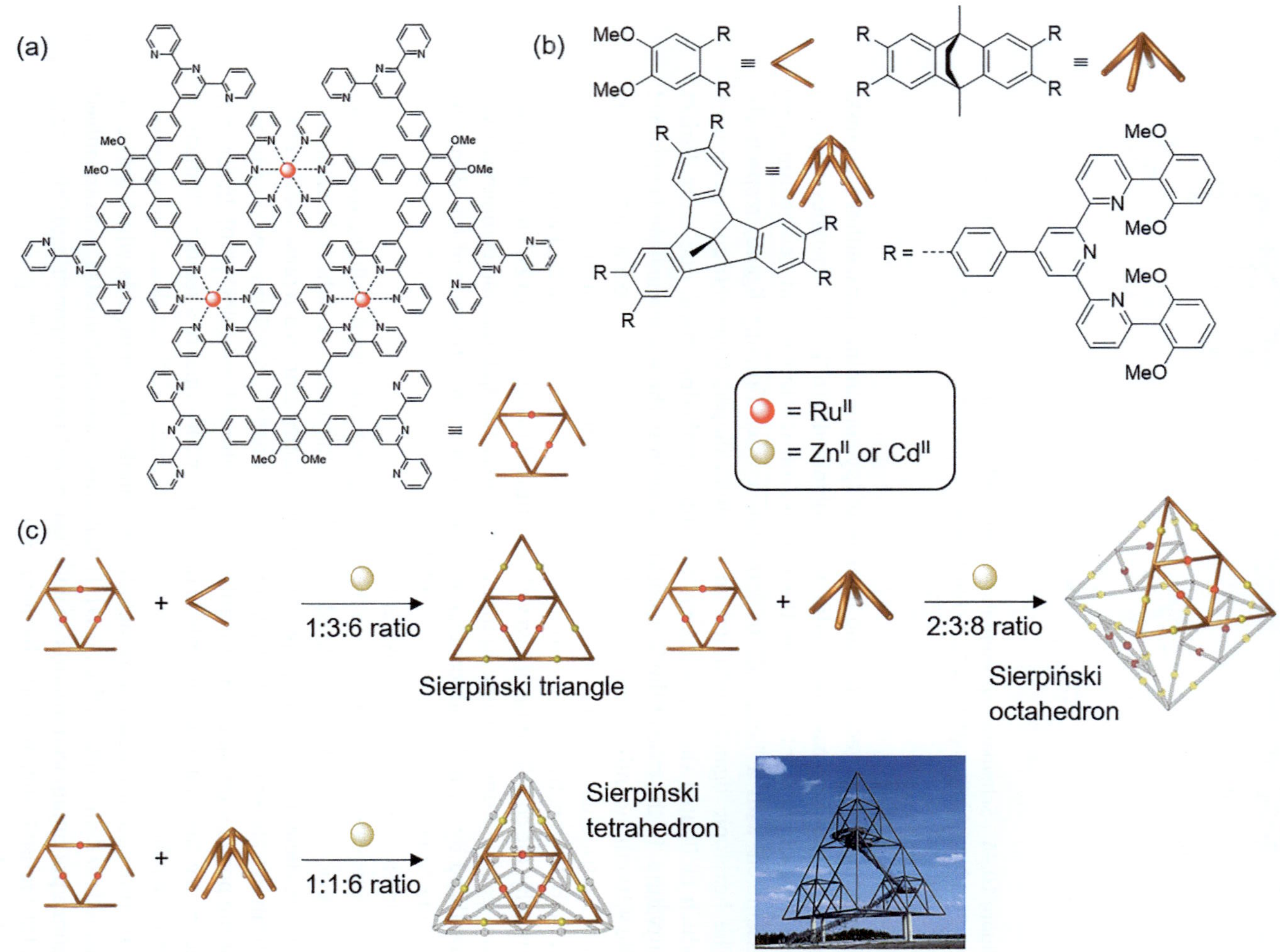

Figure 10.46 Schematic representation of the formation of Sierpiński triangles and polyhedra from a preorganized metalloligand corner-directing oligotopic ligands [167] / with permission of John Wiley & Sons. A photograph of the Tetrahedron in Bottrop, Germany, is also shown (this landmark represents a Sierpiński tetrix with a side length of 60 m). Adapted from [168] with permission of Udo / Adobe Stock Photo.

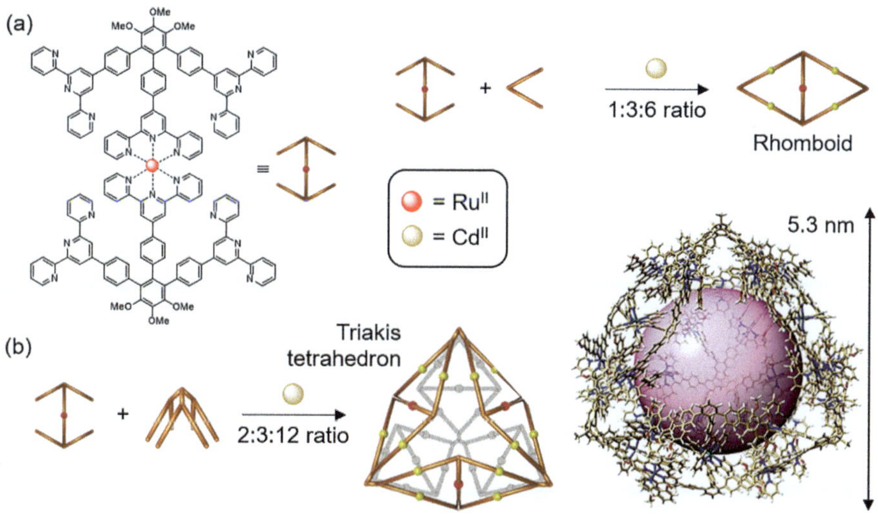

Figure 10.47 (a) Schematic representation of the formation of (a) a rhomboid and (b) a triakis tetrahedron *via* the controlled self-assembly of designer molecular components. A molecular model of the giant triakis tetrahedron is also shown (the purple sphere represents the cage's cavity volume) [169] / with permission of Elsevier.

of the key factors in controlling the self-assembly to proceed in a highly controlled fashion.

The self-assembly of molecular components into multi-cavity cages is particularly demanding and requires sophisticated building blocks as well as cautious control over the process. Even when taking this into account, the formation of metallo-supramolecular multi-cavity cages is intricate and often suffers from poor selectivity [171]. When dealing with multi-cavity cages, one has to distinguish between homomorphous and heteromorphous ones (unlike the former, the cage sizes of the latter are different). Very recently, Lin, Chen, Wang *et al.* reported the synthesis of metallo-supramolecular multi-cavity cages, whose structure was enforced by nature of the capping ligand, *i.e.*, the strain caused by its chemical structure [172]. First, a hexatopic metalloligand was prepared, which allowed the subsequent nanocage assembly to proceed on both sides (Figure 10.48a). As the complementary building blocks, a set of tritopic capping ligands was considered. These ligands differed in the nature of their core on the one hand (*i.e.*, planar *vs.* tripodal) and the length of the tpy-containing arms on the other (Figure 10.48a). Independent of the core's structure, the longer arms ($n = 1$) enabled a geometric matching with the two-sided ligand. Thus, self-assembly with Zn^{II} ions provided the corresponding homomorphous [1 + 2] double-cavity cages with high selectivity (Figure 10.48b), whereas the reduced arm length ($n = 0$) provoked a geometric mismatch and, thus, the strain-induced assembly generated heteromorphous [2 + 2] triple-cavity cages instead. Both types of multi-cavity cages accommodated planar porphyrin derivatives as guest molecules. In particular, the double-cavity cages

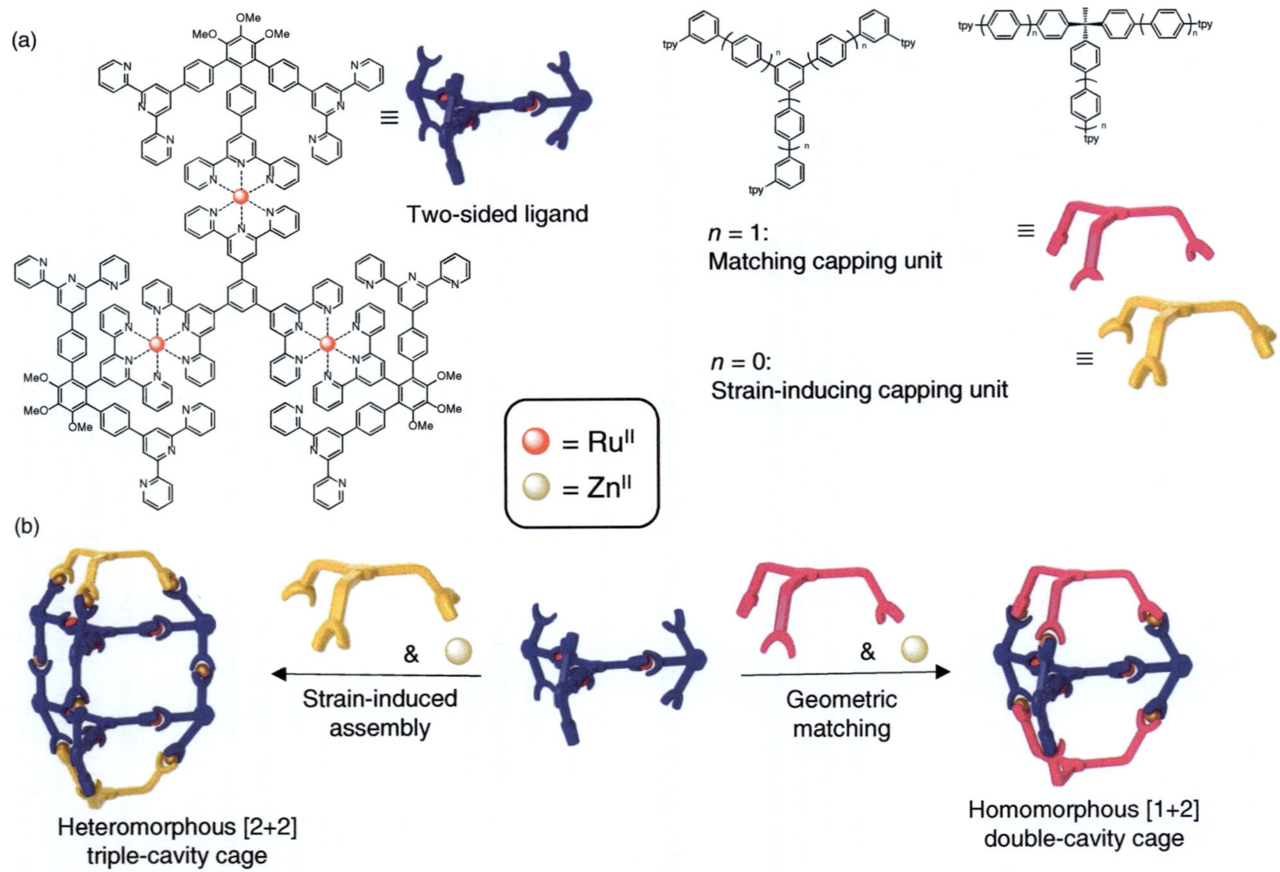

Figure 10.48 (a) Schematic representation of the two-sided and capping ligands, which were designed for the synthesis of metallo-supramolecular multi-cavity cages. (b) Schematic representation of the synthesis of such assemblies driven by geometric matching (right) or introduced strain (left) [172] / with permission of John Wiley & Sons.

were able to host up to four such guest molecules, presumably due to their larger cavities.

10.4 Metallo-Supramolecular Assemblies Based on Square-Planar Tpy Complexes

As elaborated in Sections 10.2 and 10.3, 2D and 3D metallo-supramolecular constructs can be assembled from tailor-made oligotopic tpy ligands and transition-metal ions, mainly Ru^{II}, Fe^{II}, Zn^{II}, and Cd^{II}. These divalent metal ions prefer octahedral complex geometries, thus two tpy ligands are coordinated to each metal center. Most commonly, these complexes are located at the edges or rims of the assemblies. However, nanosized architectures have also been realized by forming square-planar complexes, in which one tridentate tpy ligand and a monodentate co-ligand, mostly pyridine, saturate the coordination sphere of a transition metal ion. Mainly Pt^{II} and Pd^{II} centers have been employed in this context.

In pioneering work, Bosnich *et al.* reported various rectangular and trigonal-prismatic assemblies, which contained either Pt^{II} or Pd^{II} ions. These assemblies contained clefts in which planar guest molecules were intercalated (*e.g.*, 9-methylanthracene) [173–175]. At this stage, the preparation and host-guest chemistry of these constructs will not be discussed further and reference to various reviews [171, 176, 177] and the corresponding section in our preceding book [66] is given here for further reading. In one more recent example, Nabeshima *et al.* assembled a rectangular tetranuclear Pt^{II} complex *via* a two-step synthesis [178]. The molecular rectangle featured two clefts, which were linked by rigid butadiyne moieties (Figure 10.49); thus, the assembly was predestined for the accommodation of planar organic molecules, specifically polycyclic aromatic hydrocarbons (PAHs). In the case of small PAHs up to tetracene, binding in a 1:2 ratio was observed, *i.e.*, each cleft hosted one guest molecule, whereas the larger pentacene molecules gave

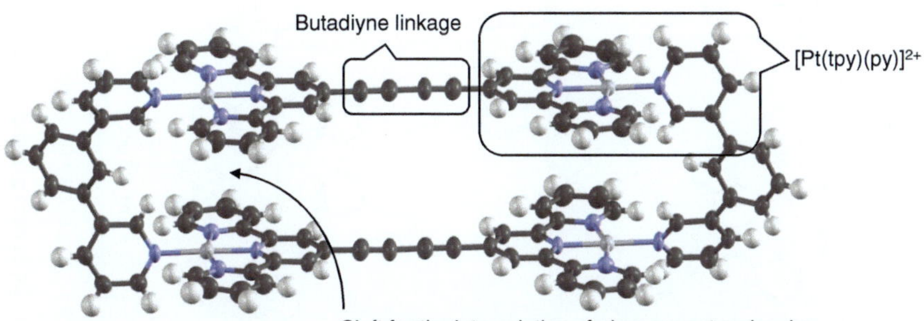

Figure 10.49 Representation of the solid-state structure of Nabeshima's rectangular tetranuclear assembly (counterions and solvent molecules were omitted for clarity). Adapted from [178].

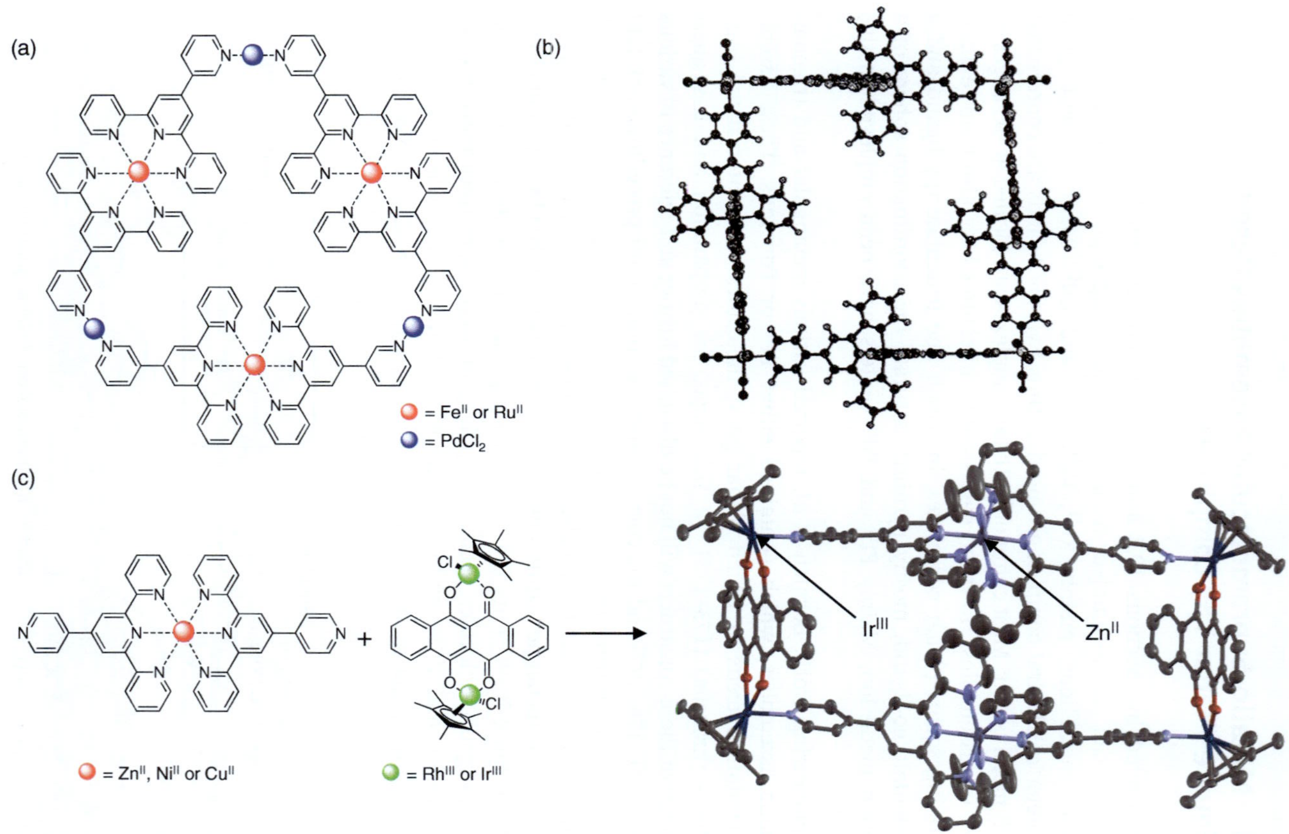

Figure 10.50 (a) Schematic representation of a heterobimetallic hexagonal macrocycle. (b) Representation of computer-generated structure of a metallo-supramolecular square, which contained $[M(tpy)_2]^{2+}$ as well as $[Re(py)_2(CO)_3Br]$ complexes [180] / with permission of American Chemical Society. (c) Schematic representation of the formation of metallo-supramolecular boxes by the self-assembly of two complexes. Representatively, the solid-state structure of the box, which contained Zn^{II} and Ir^{III} ions, is also shown (counterions and solvent molecules were omitted for clarity). Adapted from [181].

an adduct of 1:1 stoichiometry. It was assumed that the guest, whose size matched with the cavity's length, was intercalated alongside. The remarkably high binding constant (K_a) of $10^7\,M^{-1}$ resulted from the simultaneous interaction of the guest with both clefts and ππ-stacking interactions with the diyne bridges.

Various heterobimetallic macrocycles have been assembled in a stepwise fashion starting from 4′-pyridyl-tpy ligands. Firstly, 4′-pyridin-3-yl-tpy afforded the corresponding homoleptic *bis*-complexes with Fe^{II} or Ru^{II} ions [179]. The subsequent treatment with a Pd^{II} precursor afforded a hexagonal array, in which the ligand's py moieties were coordinated to $PdCl_2$ fragments (Figure 10.50a). In related work, the analogous *bis*-complexes of 4′-pyridin-4-yl-tpy were employed. Lees *et al.* assembled metallo-supramolecular squares by combining these complexes (M = Fe^{II}, Ru^{II}, or Os^{II}) with $Re(CO)_5Br$ [180, 182]. In analogy to Hanan's work, the ligands' vacant py moieties were coordinated to $[Re(CO)_3Br]$ fragments, which were located at the corners of the squares. According to molecular-modeling studies, the size of the highly regular assemblies was *ca.* 22 × 22 Å (Figure 10.50b). Related box-like assemblies were prepared from the aforementioned *bis*-complexes and dinuclear half-sandwich complexes, which contained 6,11-dihydroxy-5,12-naphthacenedione, as the bridging ligand (Figure 10.50c) [181]. These assemblies displayed a distinct host-guest chemistry, which depended on nature of metals at the corners: Only in the case of Ir^{III}, as the corner metal center, the TfO^- anions were encapsulated in the metallo-supramolecular box.

In the previous examples, preformed metal–tpy *bis*-complexes were assembled into macrocyclic architectures *via* their lateral monodentate pyridyl moieties. Wesdemiotis, Newkome *et al.* demonstrated that metallomacrocycles can also be built up elusively by $[Pd(tpy)(py)]^{2+}$ complexes [183]. A hexameric macrocycle was prepared by a two-step procedure, which involved the complexation of a Pd^{II} precursor by a V_2-type component (R = *tert*-butyl) and the subsequent metallo-macrocyclization in the presence of 4,4′-bipyridine. The high selectivity in providing the hexagonal structure, which contained in-total 12 Pd^{II} complexes, was corroborated by ESI-TWIM MS analysis. In a similar fashion, a more compact hexagonal construct was obtained from the one-post self-assembly of 3′-pyrdin-4-yl-tpy with Pd^{II} ions [184].

A different type of metallo-supramolecular architecture was reported by Stang *et al.*, who prepared twisted heterobimetallic bicyclic assemblies [185]. The hexagonal macrocycles were formed from dinuclear organoplatinum(II) precursors and angular *bis*-pyridine moieties; two of these hexagons were linked *via* a central $[Fe(tpy)_2]^{2+}$ moiety, which introduced the twist due to steric repulsion between the $[Fe(tpy)_2]^{2+}$ fragment and its 4′-aryl substituents (Figure 10.51).

As the last notable example to be highlighted within this Section, the formation of nanocages from a pre-assembled Ru^{II} complex and Pd^{II} ions was reported by Beves *et al.* [186]. These authors employed Ru^{II}–tpy building blocks, which were similar to Stang's component, but differed in the substitution pattern of the lateral py moieties. Depending on the length of the ligand's arms, different types of cages were obtained with a *cis*-fixed Pd^{II} precursor. The ligand with shorter arms (*i.e.*, $n = 0$) gave a tetrameric cage, whereas the longer arms (*i.e.*, $n = 1$) facilitated the formation of a

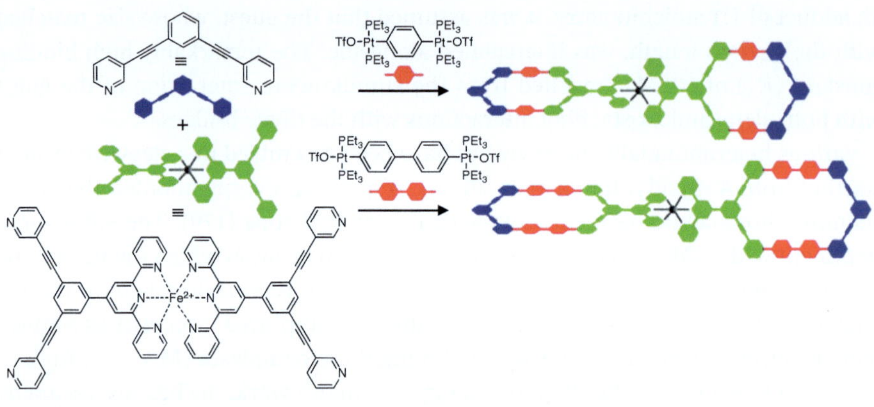

Figure 10.51 Schematic representation of the one-step formation of a twisted heterometallic bicyclic assembly [185] / with permission of American Chemical Society.

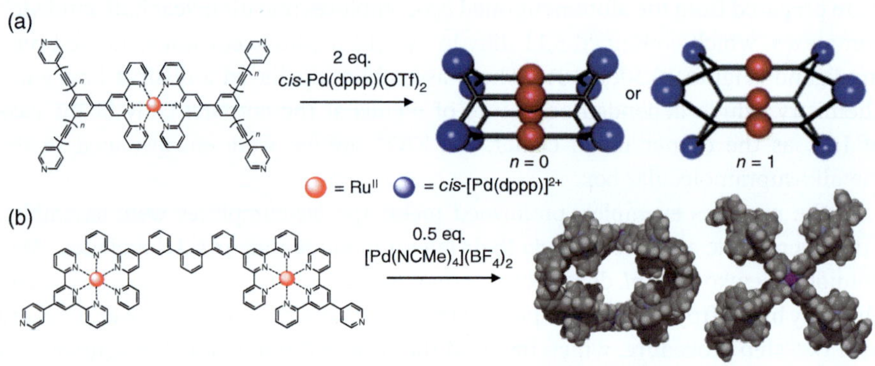

Figure 10.52 Schematic representation of the formation of nanocages by the self-assembly of preformed mononuclear or dinuclear Ru^{II} complexes with Pd^{II} ions (panels a and b, respectively) [186, 187]. The computer-generated model (two different views) of the nanocage, which contained the dinuclear metalloligand, is also shown [187] / with permission of Elsevier.

smaller, trimeric species of trigonal-prismatic geometry (Figure 10.52a). The difference in the self-assembly behavior was attributed to the slightly increased flexibility of the larger metalloligand, which allowed for different binding angles and, thus, an altered composition and geometry of the resultant cage. In one further study, Beves *et al.* prepared dinuclear complexes, in which their $[Ru(tpy)_2]^{2+}$ moieties were linked *via* 1,1′:3′,1″-triphenyl spacers. Furthermore, the complexes contained lateral pyridin-3-yl or pyridin-4-yl moieties, as designated binding sited for metal-ion complexation [187]. One of these complexes was combined with Pd^{II} ions to provide a nanocage of 4:2 stoichiometry (Figure 10.52b). Molecular-modelling studies suggested that square-planar $[Pd(py)_4]^{2+}$ complexes were formed by the coordination of each one of the metalloligands' pyridine-4-yl units to the Pd^{II} centers. These complexes represented the nanocage's caps.

10.5 Metallo-Supramolecular Assemblies Based on Lanthanide(III)–Tpy Complexes

Terpyridine complexes of lanthanide(III) ions exhibit notable magnetic or luminescent properties (see Section 7.6) [188]. The latter can, *e.g.*, be utilized for sensing or bioimaging applications. However, using Ln^{III}–tpy complexes for the precise assembly of nanostructures is demanding for two reasons. Unlike for the aforementioned octahedral or square-planar complexes of transition-metal ions, the formation of Ln^{III}–tpy complexes is less controllable due to the high and variable coordination numbers of Ln^{III} ions. Secondly, the coordination typically proceeds with insufficient stereochemical control. Thus, the formation of metallo-supramolecular constructs based on such complexes remains challenging. Although numerous Ln^{III}-containing supramolecules have been formed with oligotopic chelates, tpy ligands have still hardly been employed in this context [189].

In one example, Mazzanti *et al.* obtained a trinuclear complex by combining a tripodal *tris*-tpy ligand with Eu^{III} ions in a 2:3 ratio [190]. It was assumed that the ligand's tripodal structure provided sufficient geometric control to allow the self-assembly process to generate a discrete compound. These same authors reported the stepwise assembly of a spoked-wheel motif [191]. First, the mononuclear complex $[Eu(6ctpy)_2]^+$, in which the metal center was octa-coordinate, was formed from 6-carboxy-tpy (6ctpy) and Eu^{III} ions in 2:1 ratio. The subsequent self-assembly of this complex with Eu^{III} ions under stoichiometric control afforded the heptanuclear, pseudo-hexagonal motif, in which an octahedrally coordinated Eu^{III} center was located in the center (Figure 10.53a). It was shown further that similar assemblies could be formed if the central Eu^{III} ion was replaced by the smaller members of the Ln^{III} family. Thus, heterobimetallic spoked-wheel structures $[Lu \subset [Eu(6ctpy)_2]_6]^{9+}$ were obtained with Ho^{III}, Tm^{III}, Yb^{III}, or Lu^{III} ions (Figure 10.53b). In subsequent work, this concept was generalized by using other $[Ln(6ctpy)_2]^+$ precursors to

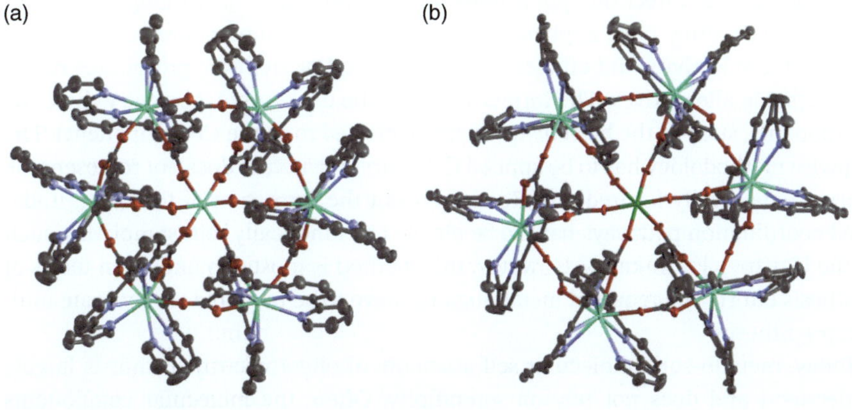

Figure 10.53 Solid state structures of the isostructural complexes (a) $[Eu \subset [Eu(6ctpy)_2]_6]^{9+}$ and (b) $[Lu \subset [Eu(6ctpy)_2]_6]^{9+}$ (hydrogen atoms, counterions and solvent molecules were omitted for clarity) Adapted from [191, 192].

obtain homometallic assemblies as well as heterobimetallic ones (Ln = Nd, Gd, or Tb) [192].

10.6 Concluding Remarks

Within the last two decades, the field of metallo-supramolecular self-assembly, *i.e.*, the formation of structurally defined, highly aesthetic nanostructures by the controlled complexation of designed oligotopic tpy ligands by transition-metal ions, has evolved tremendously. Initial research faced the challenge of overcoming the sometimes-poor control over the self-assembly process, which resulted in the formation of product mixtures containing *inter alia* oligomers and polymers as by-products. However, when using rigid ligands with pre-organized tpy sites, a significant improvement in product formation was achieved [10]. Furthermore, the choice of the transition-metal ion is of significant importance. The complexation with Ru^{II} ions generates kinetically stable *bis*-complexes and, thus, undesired linkages are trapped once formed. However, this is detrimental regarding the formation of complicated structures *via* metal-ion complexation. More labile tpy complexes, as those of Zn^{II} and Cd^{II} ions, are highly suited for this purpose. The use of such complexes allows the "correction" of structural defects and undesired linkages due to their ability to undergo metallo-supramolecular self-sorting and repair. The reversibility of complexation facilitates the formation of thermodynamically favored assemblies (corresponding to the target structure) and largely avoids the kinetic trapping of other species. However, kinetic lability might also entail decreased stability in solution and the formation of certain architectures might be concentration dependent. Interestingly, the interconversion of 2D or 3D assemblies, as a function of the concentration, has been extensively studied on various examples [52].

As summarized recently by Shi and Wang, tpy-based nanostructures can not only be formed by the direct one-pot self-assembly approach [55]. Although this strategy has successfully been applied to assembling complicated nanostructures with remarkable selectivity and efficiency, the templated or stepwise procedures represent versatile alternatives. The former relies on the use of rigid, geometry-directing components, such as the hexa*kis*-tpy ligand (referred to as the C6 component). The stepwise methodology has to be applied if the target structure does not represent the thermodynamically favored one. To circumvent the formation of the latter, undesired coordination pathways have to be blocked by kinetically robust moieties, such as the $[Ru(tpy)_2]^{2+}$ linkages. However, this method is most demanding in terms of synthesis effort: The required metalloligands have to be prepared *via* intricate multistep syntheses.

Today, metallo-supramolecular self-assembly of oligotopic tpy ligands is largely understood and does not rely on serendipity. Often, the molecular components are designed *in silico* and prepared thereafter. However, the remaining question is – what architectures can still be prepared and at which point the borderline will be reached. Of course, this is interesting from an academic point of view, but one might also ask if there are feasible applications of metallo-supramolecular constructs. Very recently, Yu, Li *et al.* published a review in which luminescent tpy-based assemblies were discussed [45]. A variety of 2D and 3D constructs have shown remarkable luminescence properties, which were due to incorporated π-conjugated systems, aggregation-induced emission, the attachment of additional luminophores, *etc.* Practical applications in the context of luminescence sensing or stimuli-responsiveness have been proposed. However, the considerable synthetic effort and the unlikelihood to prepare such assemblies on a large scale will presumably hamper their use. The same holds true regarding biomedical applications. In particular, the Cd^{II}-containing systems have displayed antibacterial properties [55]. The good antimicrobial performance was mainly due to the presence of $[Cd(tpy)_2]^{2+}$ complexes and simpler, yet as efficient complexes are more readily available [106].

Nevertheless, the esthetics of metallo-supramolecular assemblies and the demonstration of how efficient molecules can be organized into sophisticated architectures, simply by metal-ion complexation, is of significant interest. Reference is often given to Nature as the master, and researchers aim to mimic the natural self-assembly processes to obtain highly organized architectures. Future research will show if metallo-supramolecular ensembles with locally defined positions of metal centers and designed cavities/channels could function as artificial metalloenzymes, *e.g.*, for catalytic purposes.

References

1 G. M. Whitesides, J. P. Mathias, C. T. Seto, *Science* 1991, 254, 1312–1319.
2 G. M. Whitesides, B. Grzybowski, *Science* 2002, 295, 2418–2421.
3 P. J. Stang, B. Olenyuk, *Acc. Chem. Res.* 1997, 30, 502–518.
4 J.-M. Lehn, *Angew. Chem. Int. Ed.* 2013, 52, 2836–2850.
5 M. Ruben, J. Rojo, F. J. Romero-Salguero, L. H. Uppadine, J.-M. Lehn, *Angew. Chem. Int. Ed.* 2004, 43, 3644–3662.
6 J.-P. Sauvage, *Angew. Chem. Int. Ed.* 2017, 56, 11080–11093.
7 A. Coskun, J. M. Spruell, G. Barin, W. R. Dichtel, A. H. Flood, Y. Y. Botros, J. F. Stoddart, *Chem. Soc. Rev.* 2012, 41, 4827–4859.
8 J. F. Stoddart, *Chem. Soc. Rev.* 2009, 38, 1802–1820.
9 J. F. Stoddart, *Angew. Chem. Int. Ed.* 2017, 56, 11094–11125.
10 R. Chakrabarty, P. S. Mukherjee, P. J. Stang, *Chem. Rev.* 2011, 111, 6810–6918.
11 T. R. Cook, P. J. Stang, *Chem. Rev.* 2015, 115, 7001–7045.
12 K. Harris, D. Fujita, M. Fujita, *Chem. Commun.* 2013, 49, 6703–6712.
13 D. Fujita, Y. Ueda, S. Sato, N. Mizuno, T. Kumasaka, M. Fujita, *Nature* 2016, 540, 563–566.
14 G. R. Newkome, C. N. Moorefield, *Chem. Soc. Rev.* 2015, 44, 3954–3967.
15 J. M. Ludlow III, G. R. Newkome, in *Advances in Heterocyclic Chemistry*, Vol. 120, (Eds.: E. F. V. Scriven, C. A. Ramsden), Academic Press, New York, 2016, pp. 195–236.

16 S. De, K. Mahata, M. Schmittel, *Chem. Soc. Rev.* 2010, 39, 1555–1575.

17 M. L. Saha, S. Neogi, M. Schmittel, *Dalton Trans.* 2014, 43, 3815–3834.

18 C. J. Brown, F. D. Toste, R. G. Bergman, K. N. Raymond, *Chem. Rev.* 2015, 115, 3012–3035.

19 M. Frank, M. D. Johnstone, G. H. Clever, *Chem. Eur. J.* 2016, 22, 14104–14125.

20 A. J. McConnell, C. S. Wood, P. P. Neelakandan, J. R. Nitschke, *Chem. Rev.* 2015, 115, 7729–7793.

21 L.-J. Chen, H.-B. Yang, M. Shionoya, *Chem. Soc. Rev.* 2017, 46, 2555–2576.

22 S. Mukherjee, P. S. Mukherjee, *Chem. Commun.* 2014, 50, 2239–2248.

23 M. J. Wiester, P. A. Ulmann, C. A. Mirkin, *Angew. Chem. Int. Ed.* 2011, 50, 114–137.

24 D. A. Roberts, B. S. Pilgrim, J. R. Nitschke, *Chem. Soc. Rev.* 2018, 47, 626–644.

25 G. H. Clever, P. Punt, *Acc. Chem. Res.* 2017, 50, 2233–2243.

26 H. Wang, Y. Li, N. Li, A. Filosa, X. Li, *Nat. Rev. Mater.* 2021, 6, 145–167.

27 A. J. McConnell, *Chem. Soc. Rev.* 2022, 51, 2957–2971.

28 C. T. McTernan, J. A. Davies, J. R. Nitschke, *Chem. Rev.* 2022, 122, 10393–10437.

29 R. S. Forgan, J.-P. Sauvage, J. F. Stoddart, *Chem. Rev.* 2011, 111, 5434–5464.

30 W.-X. Gao, H.-J. Feng, B.-B. Guo, Y. Lu, G.-X. Jin, *Chem. Rev.* 2020, 120, 6288–6325.

31 Q.-H. Guo, Y. Jiao, Y. Feng, J. F. Stoddart, *CCS Chem.* 2021, 3, 1542–1572.

32 V. Balzani, A. Credi, F. M. Raymo, J. F. Stoddart, *Angew. Chem. Int. Ed.* 2000, 39, 3348–3391.

33 S. Erbas-Cakmak, D. A. Leigh, C. T. McTernan, A. L. Nussbaumer, *Chem. Rev.* 2015, 115, 10081–10206.

34 O. Waldmann, *Coord. Chem. Rev.* 2005, 249, 2550–2566.

35 Y. Bodenthin, G. Schwarz, Z. Tomkowicz, M. Lommel, T. M. Geue, W. Haase, H. Möhwald, U. Pietsch, D. G. Kurth, *Coord. Chem. Rev.* 2009, 253, 2414–2422.

36 Y. Liu, Y. Ma, Y. Zhao, X. Sun, F. Gándara, H. Furukawa, Z. Liu, H. Zhu, C. Zhu, K. Suenaga, P. Oleynikov, A. S. Alshammari, X. Zhang, O. Terasaki, O. M. Yaghi, *Science* 2016, 351, 365–369.

37 D. P. August, R. A. W. Dryfe, S. J. Haigh, P. R. C. Kent, D. A. Leigh, J.-F. Lemonnier, Z. Li, C. A. Muryn, L. I. Palmer, Y. Song, G. F. S. Whitehead, R. J. Young, *Nature* 2020, 588, 429–435.

38 Z.-H. Zhang, B. J. Andreassen, D. P. August, D. A. Leigh, L. Zhang, *Nat. Mater.* 2022, 21, 275–283.

39 Z. Ashbridge, S. D. P. Fielden, D. A. Leigh, L. Pirvu, F. Schaufelberger, L. Zhang, *Chem. Soc. Rev.* 2022, 51, 7779–7809.

40 F. J. Rizzuto, L. K. S. von Krbek, J. R. Nitschke, *Nat. Rev. Chem.* 2019, 3, 204–222.

41 D. Zhang, T. K. Ronson, Y.-Q. Zou, J. R. Nitschke, *Nat. Rev. Chem.* 2021, 5, 168–182.

42 L. Zhang, H. Liu, G. Yuan, Y.-F. Han, *Chin. J. Chem.* 2021, 39, 2273–2286.

43 W.-X. Gao, H.-N. Zhang, G.-X. Jin, *Coord. Chem. Rev.* 2019, 386, 69–84.

44 M. Morimoto, S. M. Bierschenk, K. T. Xia, R. G. Bergman, K. N. Raymond, F. D. Toste, *Nat. Catal.* 2020, 3, 969–984.

45 C. Guo, F. Su, P. Su, X. Yu, X. Li, *Sci. China Chem.* 2023, 66, 1940–1962.

46 D. Wu, A. C. Sedgwick, T. Gunnlaugsson, E. U. Akkaya, J. Yoon, T. D. James, *Chem. Soc. Rev.* 2017, 46, 7105–7123.

47 N. Dey, C. J. E. Haynes, *ChemPlusChem* 2021, 86, 418–433.

48 A. Brzechwa-Chodzyńska, W. Drożdż, J. Harrowfield, A. R. Stefankiewicz, *Coord. Chem. Rev.* 2021, 434, 213820.

49 N. Ahmad, H. A. Younus, A. H. Chughtai, F. Verpoort, *Chem. Soc. Rev.* 2015, 44, 9–25.

50 A. Casini, B. Woods, M. Wenzel, *Inorg. Chem.* 2017, 56, 14715–14729.

51 J. M. Ludlow III, Z. Guo, A. Schultz, R. Sarkar, C. N. Moorefield, C. Wesdemiotis, G. R. Newkome, *Eur. J. Inorg. Chem.* 2015, 2015, 5662–5668.

52 S. Chakraborty, G. R. Newkome, *Chem. Soc. Rev.* 2018, 47, 3991–4016.

53 C. Wei, Y. He, X. Shi, Z. Song, *Coord. Chem. Rev.* 2019, 385, 1–19.

54 S. Prusty, Y.-T. Chan, *Chem. Lett.* 2021, 50, 1202–1212.

55 J. Shi, M. Wang, *Chem. Asian J.* 2021, 16, 4037–4048.

56 X. Yu, C. Guo, S. Lu, Z. Chen, H. Wang, X. Li, *Macromol. Rapid Commun.* 2022, 43, 2200004.

57 M. Chiper, R. Hoogenboom, U. S. Schubert, *Macromo. Rapid Commun.* 2009, 30, 565–578.

58 A. Wild, A. Winter, F. Schlütter, U. S. Schubert, *Chem. Soc. Rev.* 2011, 40, 1459–1511.

59 A. Winter, U. S. Schubert, *Chem. Soc. Rev.* 2016, 45, 5311–5357.

60 R. Sakamoto, K.-H. Wu, R. Matsuoka, H. Maeda, H. Nishihara, *Chem. Soc. Rev.* 2015, 44, 7698–7714.

61 R. Sakamoto, K. Takada, X. Sun, T. Pal, T. Tsukamoto, E. J. H. Phua, A. Rapakousiou, K. Hoshiko, H. Nishihara, *Coord. Chem. Rev.* 2016, 320–321, 118–128.

62 R. Sakamoto, K. Takada, T. Pal, H. Maeda, T. Kambe, H. Nishihara, *Chem. Commun.* 2017, 53, 5781–5801.

63 S. M. Elahi, M. Raizada, P. K. Sahu, S. Konar, *Chem. Eur. J.* 2021, 27, 5858–5870.

64 B. Hasenknopf, J.-M. Lehn, G. Baum, D. Fenske, *Proc. Natl. Acad. Sci. U.S.A.* 1996, 93, 1397–1400.

65 V. C. M. Smith, J.-M. Lehn, *Chem. Commun.* 1996, 1996, 2733–2734.

66 U. S. Schubert, A. Winter, G. R. Newkome, *Terpyridine-Based Materials*, Wiley-VCH, Weinheim, 2011.

67 G. R. Newkome, T. J. Cho, C. N. Moorefield, R. Cush, P. S. Russo, L. A. Godínez, M. J. Saunders, P. Mohapatra, *Chem. Eur. J.* 2002, 8, 2946–2954.

68 G. R. Newkome, T. J. Cho, C. N. Moorefield, G. R. Baker, R. Cush, P. S. Russo, *Angew. Chem. Int. Ed.* 1999, 38, 3717–3721.

69 E. C. Constable, *Chem. Soc. Rev.* 2007, 36, 246–253.

70 P. R. Andres, U. S. Schubert, *Synthesis* 2004, 2004, 1229–1238.

71 M. A. R. Meier, D. Wouters, C. Ott, P. Guillet, C.-A. Fustin, J.-F. Gohy, U. S. Schubert, *Macromolecules* 2006, 39, 1569–1576.

72 Y. Li, Z. Jiang, J. Yuan, D. Liu, T. Wu, C. N. Moorefield, G. R. Newkome, P. Wang, *Chem. Commun.* 2015, 51, 5766–5769.

73 D. Liu, X. Yang, Y. Li, P. Wang, *Chem. Commun.* 2016, 52, 2513–2516.

74 J. M. Ludlow III, M. Tominaga, Y. Chujo, A. Schultz, X. Lu, T. Xie, K. Guo, C. N. Moorefield, C. Wesdemiotis, G. R. Newkome, *Dalton Trans.* 2014, 43, 9604–9611.

75 A. Schultz, X. Li, C. E. McCusker, C. N. Moorefield, F. N. Castellano, C. Wesdemiotis, G. R. Newkome, *Chem. Eur. J.* 2012, 18, 11569–11572.

76 A. Schultz, X. Li, C. N. Moorefield, C. Wesdemiotis, G. R. Newkome, *Eur. J. Inorg. Chem.* 2013, 2013, 2492–2497.

77 K. J. Endres, K. Barthelmes, A. Winter, R. Antolovich, U. S. Schubert, C. Wesdemiotis, *Rapid Commun. Mass Spectrom.* 2020, 34, e8717.

78 K. Guo, Z. Guo, J. M. Ludlow III, T. Xie, S. Liao, G. R. Newkome, C. Wesdemiotis, *Macromol. Rapid Commun.* 2015, 36, 1539–1552.

79 Y.-T. Chan, X. Li, J. Yu, G. A. Carri, C. N. Moorefield, G. R. Newkome, C. Wesdemiotis, *J. Am. Chem. Soc.* 2011, 133, 11967–11976.

80 X. Li, Y.-T. Chan, G. R. Newkome, C. Wesdemiotis, *Anal. Chem.* 2011, 83, 1284–1290.

81 H. Wang, C. Guo, X. Li, *CCS Chem.* 2022, 4, 785–808.

82 *Wacław Sierpiński*, https://en.wikipedia.org/wiki/Wac%C5%82aw_Sierpi%C5%84ski (last accessed: 2025-04-01).

83 G. R. Newkome, P. Wang, C. N. Moorefield, T. J. Cho, P. P. Mohapatra, S. Li, S.-H. Hwang, O. Lukoyanova, L. Echegoyen, J. A. Palagallo, V. Iancu, S.-W. Hla, *Science* 2006, 312, 1782–1785.

84 *Benoit Mandelbrot*, https://de.wikipedia.org/wiki/Beno%C3%AEt_Mandelbrot (last accessed: 2025-04-01).

85 K. Kuratowski, *Acta Arith.* 1972, 21, 1–5.

86 B. B. Mandelbrot, *Fractals: Form, Chance and Dimension*, W.H. Freeman Company, New York, 1977.

87 J.-L. Wang, X. Li, X. Lu, I. F. Hsieh, Y. Cao, C. N. Moorefield, C. Wesdemiotis, S. Z. D. Cheng, G. R. Newkome, *J. Am. Chem. Soc.* 2011, 133, 11450–11453.

88 X. Lu, X. Li, J.-L. Wang, C. N. Moorefield, C. Wesdemiotis, G. R. Newkome, *Chem. Commun.* 2012, 48, 9873–9875.

89 A. Schultz, X. Li, B. Barkakaty, C. N. Moorefield, C. Wesdemiotis, G. R. Newkome, *J. Am. Chem. Soc.* 2012, 134, 7672–7675.

90 R. Mehta, M. Moalic, M. Krawczyk, S. Saha, *J. Phys.: Condens. Matter* 2023, 35, 324002.

91 W. Sierpiński, *Compt. Rend. Acad. Sci. (Paris)* 1915, 160, 302–305.

92 R. Sarkar, K. Guo, C. N. Moorefield, M. J. Saunders, C. Wesdemiotis, G. R. Newkome, *Angew. Chem. Int. Ed.* 2014, 53, 12182–12185.

93 S. Chakraborty, R. Sarkar, K. Endres, T.-Z. Xie, M. Ghosh, C. N. Moorefield, M. J. Saunders, C. Wesdemiotis, G. R. Newkome, *Eur. J. Org. Chem.* 2016, 2016, 5091–5095.

94 Z. Jiang, Y. Li, M. Wang, D. Liu, J. Yuan, M. Chen, J. Wang, G. R. Newkome, W. Sun, X. Li, P. Wang, *Angew. Chem. Int. Ed.* 2017, 56, 11450–11455.

95 A. de Meijere, S. I. Kozhushkov, A. A. Fokin, I. Emme, S. Redlich, P. R. Schreiner, *Pure Appl. Chem.* 2003, 75, 549–562.

96 Y. Li, Z. Jiang, M. Wang, J. Yuan, D. Liu, X. Yang, M. Chen, J. Yan, X. Li, P. Wang, *J. Am. Chem. Soc.* 2016, 138, 10041–10046.

97 Z. Jiang, Y. Li, M. Wang, B. Song, K. Wang, M. Sun, D. Liu, X. Li, J. Yuan, M. Chen, Y. Guo, X. Yang, T. Zhang, C. N. Moorefield, G. R. Newkome, B. Xu, X. Li, P. Wang, *Nat. Commun.* 2017, 8, 15476.

98 Z. Zhang, H. Wang, X. Wang, Y. Li, B. Song, O. Bolarinwa, R. A. Reese, T. Zhang, X.-Q. Wang, J. Cai, B. Xu, M. Wang, C. Liu, H.-B. Yang, X. Li, *J. Am. Chem. Soc.* 2017, 139, 8174–8185.

99 Z. Zhang, H. Wang, J. Shi, P. Wang, C. Liu, M. Wang, X. Li, *Isr. J. Chem.* 2019, 59, 237–247.

100 M. Wang, C. Wang, X.-Q. Hao, J. Liu, X. Li, C. Xu, A. Lopez, L. Sun, M.-P. Song, H.-B. Yang, X. Li, *J. Am. Chem. Soc.* 2014, 136, 6664–6671.

101 J. Shi, Y. Li, X. Jiang, H. Yu, J. Li, H. Zhang, D. J. Trainer, S. W. Hla, H. Wang, M. Wang, X. Li, *J. Am. Chem. Soc.* 2021, 143, 1224–1234.

102 A. C. Mendes, E. T. Baran, R. L. Reis, H. S. Azevedo, *WIRES.: Nanomed. Nanobiotechnol.* 2013, 5, 582–612.

103 M. Wang, K. Wang, C. Wang, M. Huang, X.-Q. Hao, M.-Z. Shen, G.-Q. Shi, Z. Zhang, B. Song, A. Cisneros, M.-P. Song, B. Xu, X. Li, *J. Am. Chem. Soc.* 2016, 138, 9258–9268.

104 H. Wang, X. Qian, K. Wang, M. Su, W.-W. Haoyang, X. Jiang, R. Brzozowski, M. Wang, X. Gao, Y. Li, B. Xu, P. Eswara, X.-Q. Hao, W. Gong, J.-L. Hou, J. Cai, X. Li, *Nat. Commun.* 2018, 9, 1815.

105 H. Wang, Y. Li, H. Yu, B. Song, S. Lu, X.-Q. Hao, Y. Zhang, M. Wang, S.-W. Hla, X. Li, *J. Am. Chem. Soc.* 2019, 141, 13187–13195.

106 R. Abhijnakrishna, K. Magesh, A. Ayushi, S. Velmathi, *Coord. Chem. Rev.* 2023, 496, 215380.

107 A. Winter, M. Gottschaldt, G. R. Newkome, U. S. Schubert, *Curr. Top. Med. Chem.* 2012, 12, 158–175.

108 Z. Jiang, B. Chen, H. Zhao, J. Wang, Q. Dong, F. Fu, D. Liu, Y. Li, G. R. Newkome, P. Wang, M. Chen, *J. Am. Chem. Soc.* 2024, 146, 16721–16728.

109 H. Miyoshi, S. Nobusue, A. Shimizu, Y. Tobe, *Chem. Soc. Rev.* 2015, 44, 6560–6577.

110 T. Wu, J. Yuan, B. Song, Y.-S. Chen, M. Chen, X. Xue, Q. Liu, J. Wang, Y.-T. Chan, P. Wang, *Chem. Commun.* 2017, 53, 6732–6735.

111 L. Wang, B. Song, S. Khalife, Y. Li, L.-J. Ming, S. Bai, Y. Xu, H. Yu, M. Wang, H. Wang, X. Li, *J. Am. Chem. Soc.* 2020, 142, 1811–1821.

112 B. Song, S. Kandapal, J. Gu, K. Zhang, A. Reese, Y. Ying, L. Wang, H. Wang, Y. Li, M. Wang, S. Lu, X.-Q. Hao, X. Li, B. Xu, X. Li, *Nat. Commun.* 2018, 9, 4575.

113 L. Wang, R. Liu, J. Gu, B. Song, H. Wang, X. Jiang, K. Zhang, X. Han, X.-Q. Hao, S. Bai, M. Wang, X. Li, B. Xu, X. Li, *J. Am. Chem. Soc.* 2018, 140, 14087–14096.

114 Z. Zhang, Y. Li, B. Song, Y. Zhang, X. Jiang, M. Wang, R. Tumbleson, C. Liu, P. Wang, X.-Q. Hao, T. Rojas, A. T. Ngo, J. L. Sessler, G. R. Newkome, S. W. Hla, X. Li, *Nat. Chem.* 2020, 12, 468–474.

115 G.-Q. Yin, H. Wang, X.-Q. Wang, B. Song, L.-J. Chen, L. Wang, X.-Q. Hao, H.-B. Yang, X. Li, *Nat. Communi.* 2018, 9, 567.

116 Y. Hong, J. W. Y. Lam, B. Z. Tang, *Chem. Commun.* 2009, 2009, 4332–4353.

117 Y. Hong, J. W. Y. Lam, B. Z. Tang, *Chem. Soci. Rev.* 2011, 40, 5361–5388.

118 J. Wang, H. Zhao, M. Chen, Z. Jiang, F. Wang, G. Wang, K. Li, Z. Zhang, D. Liu, Z. Jiang, P. Wang, *J. Am. Chem. Soc.* 2020, 142, 21691–21701.

119 T. Mede, M. Jäger, U. S. Schubert, *Chem. Soc. Rev.* 2018, 47, 7577–7627.

120 Y. Yao, S. Chakraborty, S. Zhu, K. J. Endres, T.-Z. Xie, W. Hong, E. Manandhar, C. N. Moorefield, C. Wesdemiotis, G. R. Newkome, *Chem. Commun.* 2017, 53, 8038–8041.

121 J.-H. Fu, Y.-H. Lee, Y.-J. He, Y.-T. Chan, *Angew. Chem. Int. Ed.* 2015, 54, 6231–6235.

122 S.-Y. Wang, J.-H. Fu, Y.-P. Liang, Y.-J. He, Y.-S. Chen, Y.-T. Chan, *J. Am. Chem. Soc.* 2016, 138, 3651–3654.

123 Y.-J. He, T.-H. Tu, M.-K. Su, C.-W. Yang, K. V. Kong, Y.-T. Chan, *J. Am. Chem. Soc.* 2017, 139, 4218–4224.

124 J.-H. Fu, S.-Y. Wang, Y.-S. Chen, S. Prusty, Y.-T. Chan, *J. Am. Chem. Soc.* 2019, 141, 16217–16221.

125 S.-C. Wang, K.-Y. Cheng, J.-H. Fu, Y.-C. Cheng, Y.-T. Chan, *J. Am. Chem. Soc.* 2020, 142, 16661–16667.

126 H. Zhao, A. M. S. D. Wijerathna, Q. Dong, Q. Bai, Z. Jiang, J. Yuan, J. Wang, M. Chen, M. Zirnheld, R. Li, D. Liu, P. Wang, Y. Zhang, Y. Li, *Angew. Chem. Int. Ed.* 2024, 63, e202318029.

127 Y. Xu, H. Su, Q. Bai, F. Fang, J. Ma, Z. Zhang, X.-Q. Hao, J. Shi, P. Wang, M. Wang, *Chem. Asian J.* 2022, 17, e202200071.

128 G. Vernizzi, M. Olvera de la Cruz, *Proc. Natl. Acad. Sci. U.S.A.* 2007, 104, 18382–18386.

129 S. M. Stagg, C. Gürkan, D. M. Fowler, P. LaPointe, T. R. Foss, C. S. Potter, B. Carragher, W. E. Balch, *Nature* 2006, 439, 234–238.

130 *Platonische Körper*, https://de.wikipedia.org/wiki/Platonischer_K%C3%B6rper (last accessed: 2025-04-01).

131 *Archimedean solid*, https://en.wikipedia.org/wiki/Archimedean_solid (last accessed: 2025-04-01).

132 T. Schröder, R. Brodbeck, M. C. Letzel, A. Mix, B. Schnatwinkel, M. Tonigold, D. Volkmer, J. Mattay, *Tetrahedron Lett.* 2008, 49, 5939–5942.

133 T.-Z. Xie, S.-Y. Liao, K. Guo, X. Lu, X. Dong, M. Huang, C. N. Moorefield, S. Z. D. Cheng, X. Liu, C. Wesdemiotis, G. R. Newkome, *J. Am. Chem. Soc.* 2014, 136, 8165–8168.

134 M. Chen, J. Wang, D. Liu, Z. Jiang, Q. Liu, T. Wu, H. Liu, W. Yu, J. Yan, P. Wang, *J. Am. Chem. Soc.* 2018, 140, 2555–2561.

135 X. Lu, X. Li, Y. Cao, A. Schultz, J.-L. Wang, C. N. Moorefield, C. Wesdemiotis, S. Z. D. Cheng, G. R. Newkome, *Angew. Chem. Int. Ed.* 2013, 52, 7728–7731.

136 S. Chakraborty, W. Hong, K. J. Endres, T.-Z. Xie, L. Wojtas, C. N. Moorefield, C. Wesdemiotis, G. R. Newkome, *J. Am. Chem. Soc.* 2017, 139, 3012–3020.

137 J. M. Ludlow III, T. Xie, Z. Guo, K. Guo, M. J. Saunders, C. N. Moorefield, C. Wesdemiotis, G. R. Newkome, *Chem. Commun.* 2015, 51, 3820–3823.

138 T.-Z. Xie, K. J. Endres, Z. Guo, J. M. Ludlow III, C. N. Moorefield, M. J. Saunders, C. Wesdemiotis, G. R. Newkome, *J. Am. Chem. Soc.* 2016, 138, 12344–12347.

139 Y.-C. Wang, Y.-P. Liang, J.-Y. Cai, Y.-J. He, Y.-H. Lee, Y.-T. Chan, *Chem. Commun.* 2016, 52, 12622–12625.

140 D. Thibeault, J.-F. Morin, *Molecules* 2010, 15, 3709–3730.

141 A. R. Williams, B. H. Northrop, T. Chang, J. F. Stoddart, A. J. P. White, D. J. Williams, *Angew. Chem. Int. Ed.* 2006, 45, 6665–6669.

142 C. Wang, X.-Q. Hao, M. Wang, C. Guo, B. Xu, E. N. Tan, Y.-Y. Zhang, Y. Yu, Z.-Y. Li, H.-B. Yang, M.-P. Song, X. Li, *Chem. Sci.* 2014, 5, 1221–1226.

143 M. Wang, C. Wang, X.-Q. Hao, X. Li, T. J. Vaughn, Y.-Y. Zhang, Y. Yu, Z.-Y. Li, M.-P. Song, H.-B. Yang, X. Li, *J. Am. Chem. Soc.* 2014, 136, 10499–10507.

144 T.-Z. Xie, K. Guo, Z. Guo, W.-Y. Gao, L. Wojtas, G.-H. Ning, M. Huang, X. Lu, J.-Y. Li, S.-Y. Liao, Y.-S. Chen, C. N. Moorefield, M. J. Saunders, S. Z. D. Cheng, C. Wesdemiotis, G. R. Newkome, *Angew. Chem. Int. Ed.* 2015, 54, 9224–9229.

145 S. Chakraborty, K. J. Endres, R. Bera, L. Wojtas, C. N. Moorefield, M. J. Saunders, N. Das, C. Wesdemiotis, G. R. Newkome, *Dalton Trans.* 2018, 47, 14189–14194.

146 Z.-B. Tang, Y.-M. Guan, Z. Zhang, Q. Bai, P. Su, G. Chen, L.-Y. Yao, L.-H. Xu, T. Wu, G.-H. Ning, P. Wang, T.-Z. Xie, *Cell Rep. Phys. Sci.* 2022, 3, 100982.

147 M.-Y. Yu, Y.-M. Guan, Q. Bai, P. Su, P. Wang, T.-Z. Xie, *Adv. Opt. Mater.* 2023, 11, 2202148.

148 H. Wang, C.-H. Liu, K. Wang, M. Wang, H. Yu, S. Kandapal, R. Brzozowski, B. Xu, M. Wang, S. Lu, X.-Q. Hao, P. Eswara, M.-P. Nieh, J. Cai, X. Li, *J. Am. Chem. Soc.* 2019, 141, 16108–16116.

149 M. Li, S. Jiang, Z. Zhang, X.-Q. Hao, X. Jiang, H. Yu, P. Wang, B. Xu, M. Wang, W. Tian, *CCS Chem.* 2020, 2, 337–348.

150 X. Jiang, H. Yu, J. Shi, Q. Bai, Y. Xu, Z. Zhang, X.-Q. Hao, B. Li, P. Wang, L. Wu, M. Wang, *CCS Chem.* 2022, 4, 2127–2139.

151 N. W. Johnson, *Can. J. Math.* 1966, 18, 169–200.

152 T. Wu, Z. Jiang, Q. Bai, Y. Li, S. Mao, H. Yu, L. Wojtas, Z. Tang, M. Chen, Z. Zhang, T.-Z. Xie, M. Wang, X. Li, P. Wang, *Chem* 2021, 7, 2429–2441.

153 Q. Bai, Z. Jiang, Y. Guan, Q. Long, B. Liao, T.-Z. Xie, T. Wu, Z. Zhang, P. Wang, *Inorg. Chem.* 2023, 62, 5095–5104.

154 Y.-S. Chen, E. Solel, Y.-F. Huang, C.-L. Wang, T.-H. Tu, E. Keinan, Y.-T. Chan, *Nat. Commun.* 2019, 10, 3443.

155 Q. Bai, Y.-M. Guan, T. Wu, Y. Liu, Z. Zhai, Q. Long, Z. Jiang, P. Su, T.-Z. Xie, P. Wang, Z. Zhang, *Angew. Chem. Int. Ed.* 2023, 62, e202309027.

156 Z. Jiang, J. Wang, M. Chen, C. Tang, H. Zhao, Q. Dong, W.-D. Yu, Z. Jiang, B. Chen, X. Li, D. Liu, L.-Y. Yao, H. Liu, T.-Z. Xie, T. Wu, J. Yuan, K. Wang, Y. Li, P. Wang, *Inorg. Chem. Front.* 2023, 10, 621–629.

157 R. O. Erickson, *Science* 1973, 181, 705–716.

158 M. F. Pennybacker, P. D. Shipman, A. C. Newell, *Phys. D: Nonlinear Phenom.* 2015, 306, 48–81.

159 S. Guan, H. Yu, Z. Zhang, X. Jiang, J. Shi, T. Lu, C. Wang, P. Wang, M. Wang, *Macromol. Rapid Commun.* 2020, 41, 2000095.

160 Y. Xu, H. Zhang, H. Su, J. Ma, H. Yu, K. Li, J. Shi, X.-Q. Hao, K. Wang, B. Song, M. Wang, *Small* 2023, 19, 2300009.

161 D. Liu, M. Chen, K. Li, Z. Li, J. Huang, J. Wang, Z. Jiang, Z. Zhang, T. Xie, G. R. Newkome, P. Wang, *J. Am. Chem. Soc.* 2020, 142, 7987–7994.

162 D. Liu, M. Chen, Y. Li, Y. Shen, J. Huang, X. Yang, Z. Jiang, X. Li, G. R. Newkome, P. Wang, *Angew. Chem. Int. Ed.* 2018, 57, 14116–14120.

163 T.-Z. Xie, K. Guo, M. Huang, X. Lu, S.-Y. Liao, R. Sarkar, C. N. Moorefield, S. Z. D. Cheng, C. Wesdemiotis, G. R. Newkome, *Chem. Eur. J.* 2014, 20, 11291–11294.

164 T.-Z. Xie, X. Wu, K. J. Endres, Z. Guo, X. Lu, J. Li, E. Manandhar, J. M. Ludlow III, C. N. Moorefield, M. J. Saunders, C. Wesdemiotis, G. R. Newkome, *J. Am. Chem. Soc.* 2017, 139, 15652–15655.

165 G. Wang, M. Chen, J. Wang, Z. Jiang, D. Liu, D. Lou, H. Zhao, K. Li, S. Li, T. Wu, Z. Jiang, X. Sun, P. Wang, *J. Am. Chem. Soc.* 2020, 142, 7690–7698.

166 D. Liu, K. Li, M. Chen, T. Zhang, Z. Li, J.-F. Yin, L. He, J. Wang, P. Yin, Y.-T. Chan, P. Wang, *J. Am. Chem. Soc.* 2021, 143, 2537–2544.

167 J. Wang, Z. Jiang, W. Liu, Z. Wu, R. Miao, F. Fu, J.-F. Yin, B. Chen, Q. Dong, H. Zhao, K. Li, G. Wang, D. Liu, P. Yin, Y. Li, M. Chen, P. Wang, *Angew. Chem. Int. Ed.* 2023, 62, e202214237.

168 *Tetrahedron in Bottrop*, https://en.wikipedia.org/wiki/Tetrahedron_in_Bottrop (last accessed: 2025-04-01).

169 Z. Jiang, J. Wang, H. Zhang, W. Liu, Z. Wu, H. Zhao, J.-F. Yin, B. Chen, Y. Li, P. Yin, Y.-T. Chan, K. Wang, M. Chen, P. Wang, *Cell Rep. Phys. Sci.* 2023, 4, 101293.

170 M. V. Diudea, *Multi-shell Polyhedral Clusters*, Springer International, Cham, 2008.

171 R. A. S. Vasdev, D. Preston, J. D. Crowley, *Chem. Asian J.* 2017, 12, 2513–2523.

172 J. Wang, Z. Jiang, J.-F. Yin, H. Zhao, Q. Dong, K. Li, W. Zhong, D. Liu, J. Yuan, P. Yin, Y. Li, Y. Lin, M. Chen, P. Wang, *Angew. Chem. Int. Ed.* 2024, 63, e202317674.

173 J. D. Crowley, I. M. Steele, B. Bosnich, *Eur. J. Inorg. Chem.* 2005, 2005, 3907–3917.

174 A. J. Goshe, B. Bosnich, *Synlett* 2001, 2001, 0941–0944.

175 J. D. Crowley, A. J. Goshe, B. Bosnich, *Chem. Commun.* 2003, 2003, 2824–2825.

176 J. D. Crowley, B. Bosnich, *Eur. J. Inorg. Chem.* 2005, 2005, 2015–2025.

177 I. Eryazici, C. N. Moorefield, G. R. Newkome, *Chem. Rev.* 2008, 108, 1834–1895.

178 Y. Yamaki, T. Nakamura, S. Suzuki, M. Yamamura, M. Minoura, T. Nabeshima, *Eur. J. Org. Chem.* 2016, 2016, 1678–1683.

179 A. K. Pal, B. Laramée-Milette, G. S. Hanan, *RSC Adv.* 2014, 4, 21262–21266.

180 S.-S. Sun, A. J. Lees, *Inorg. Chem.* 2001, 40, 3154–3160.

181 J.-J. Liu, Y.-J. Lin, G.-X. Jin, *Organometallics* 2014, 33, 1283–1290.

182 S.-S. Sun, A. S. Silva, I. M. Brinn, A. J. Lees, *Inorg. Chem.* 2000, 39, 1344–1345.

183 S. Perera, X. Li, M. Soler, A. Schultz, C. Wesdemiotis, C. N. Moorefield, G. R. Newkome, *Angew. Chem. Int. Ed.* 2010, 49, 6539–6544.

184 S. Perera, X. Li, M. Guo, C. Wesdemiotis, C. N. Moorefield, G. R. Newkome, *Chem. Commun.* 2011, 47, 4658–4660.

185 H. Sepehrpour, M. L. Saha, P. J. Stang, *J. Am. Chem. Soc.* 2017, 139, 2553–2556.

186 J. Yang, M. Bhadbhade, W. A. Donald, H. Iranmanesh, E. G. Moore, H. Yan, J. E. Beves, *Chem. Commun.* 2015, 51, 4465–4468.

187 C. Shen, A. D. W. Kennedy, W. A. Donald, A. M. Torres, W. S. Price, J. E. Beves, *Inorg. Chim. Acta* 2017, 458, 122–128.

188 X. Yu, Y. Hu, C. Guo, Z. Chen, H. Wang, X. Li, *Supramol. Mater.* 2022, 1, 100017.

189 X.-Z. Li, C.-B. Tian, Q.-F. Sun, *Chem. Rev.* 2022, 122, 6374–6458.

190 Y. Bretonnière, M. Mazzanti, R. Wietzke, J. Pécaut, *Chem. Commun.* 2000, 2000, 1543–1544.

191 Y. Bretonnière, M. Mazzanti, J. Pécaut, M. M. Olmstead, *J. Am. Chem. Soc.* 2002, 124, 9012–9013.

192 X.-Y. Chen, Y. Bretonnière, J. Pécaut, D. Imbert, J.-C. Bünzli, M. Mazzanti, *Inorg. Chem.* 2007, 46, 625–637.

Appendix

The Various Constitutional Isomers from the Terpyridine Family

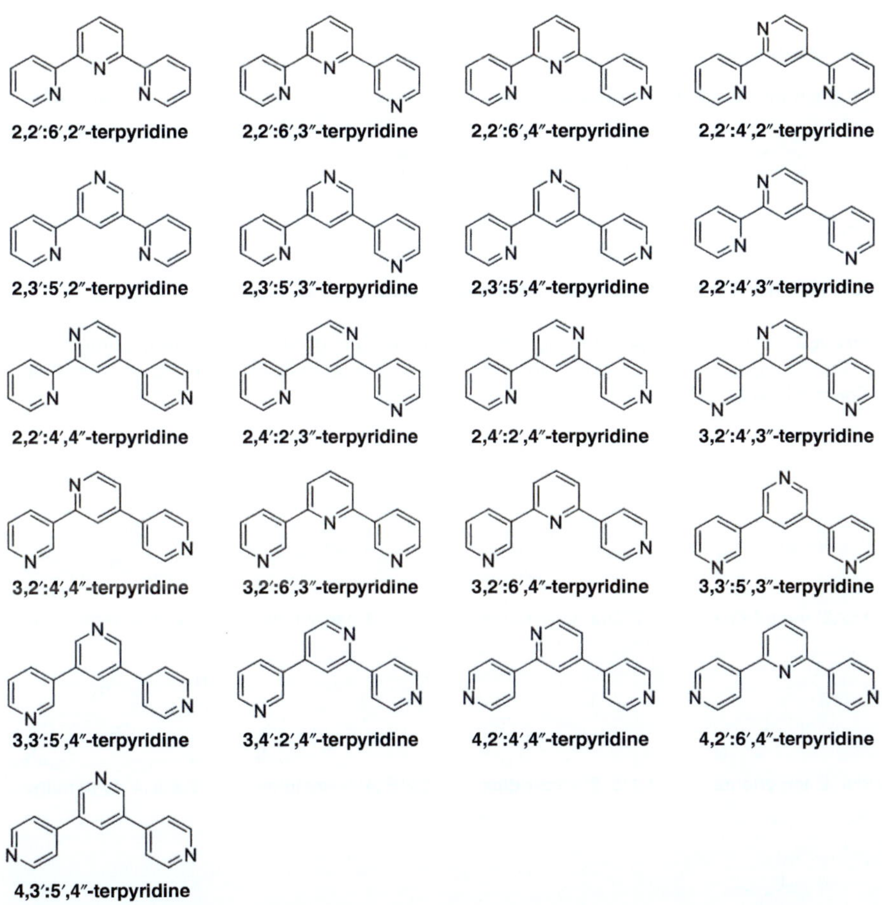

Figure A.1 Schematic representation of the various constitutional isomers from the terpyridine family, which could act as tridentate ligands. Figure redrawn according to Constable *et al.* [1] and Klein *et al.* [2].

Terpyridines: Multifaceted Ligands for Diverse Applications, First Edition.
Ulrich S. Schubert, Andreas Winter, and George R. Newkome.

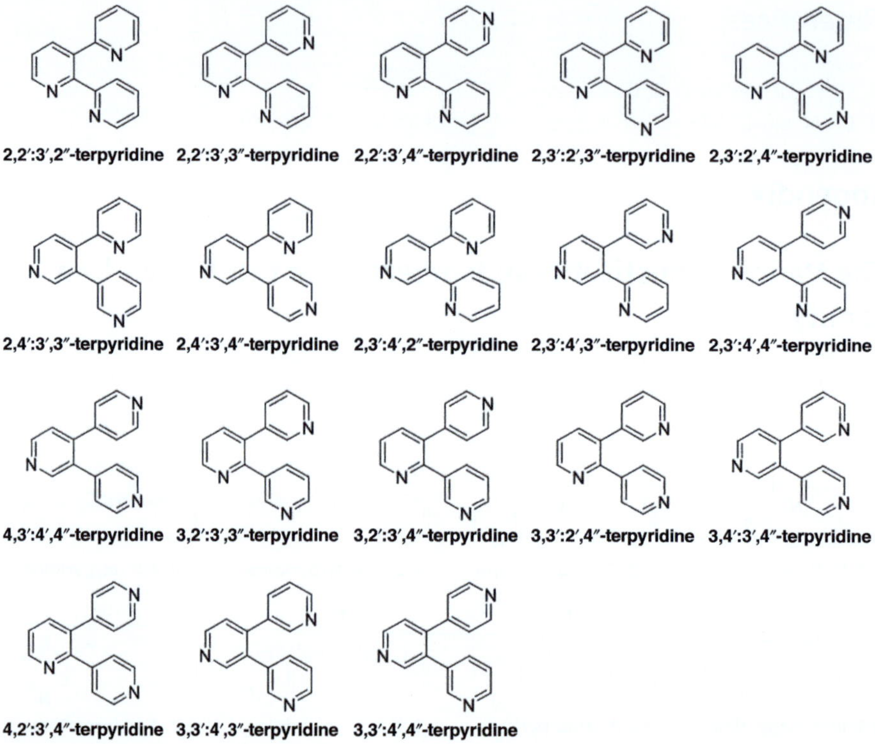

Figure A.2 Schematic representation of the remaining angled constitutional isomers from the terpyridine family, which could not act as tridentate ligands. Figure redrawn according to Constable *et al.* [1].

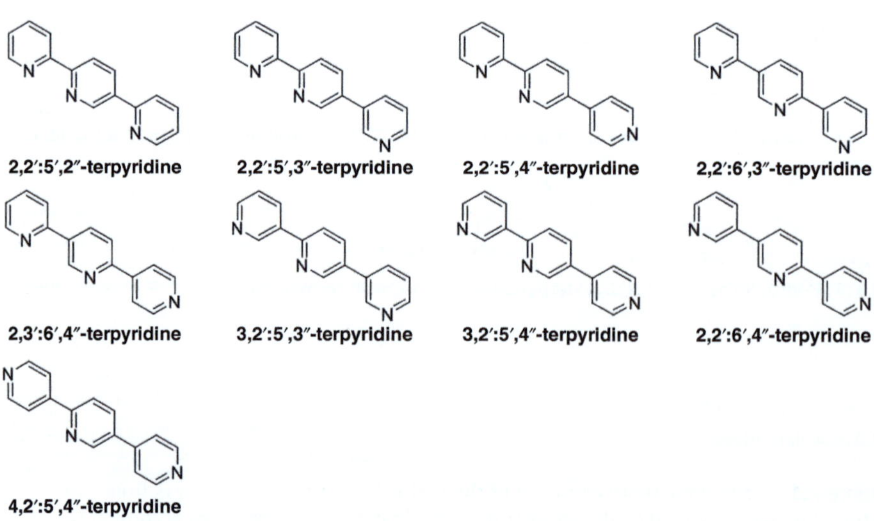

Figure A.3 Schematic representation of the various linear constitutional isomers from the terpyridine family. Figure redrawn according to Constable *et al.* [1].

References

1 C. E. Housecroft, E. C. Constable, *Molecules* 2021, 26, 3110.

2 L. Payen, L. Kletsch, T. Lapić, M. Wickleder, A. Klein, *Inorganics* 2023, 11, 174.

Index

Terpyridines: Multifaceted Ligands for Diverse Applications, First Edition.
Ulrich S. Schubert, Andreas Winter, and George R. Newkome.